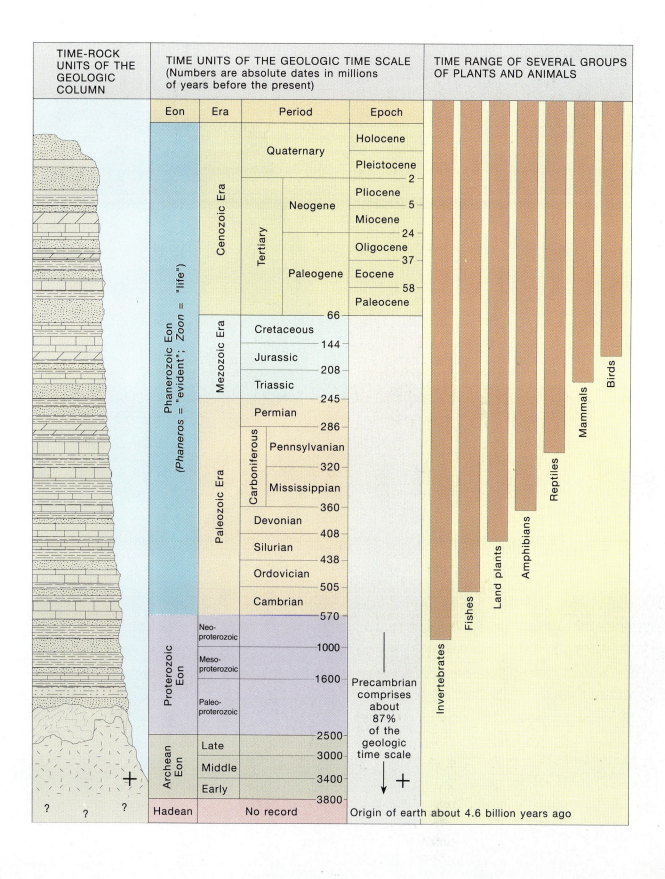

TIME-ROCK UNITS OF THE GEOLOGIC COLUMN	TIME UNITS OF THE GEOLOGIC TIME SCALE (Numbers are absolute dates in millions of years before the present)				TIME RANGE OF SEVERAL GROUPS OF PLANTS AND ANIMALS
	Eon	Era	Period	Epoch	
	Phanerozoic Eon (Phaneros = "evident"; Zoon = "life")	Cenozoic Era	Quaternary	Holocene	
				Pleistocene	
			Tertiary — Neogene	2 Pliocene	
				5 Miocene	
				24 Oligocene	
			Tertiary — Paleogene	37 Eocene	
				58 Paleocene	
		Mezozoic Era	Cretaceous	66	
			Jurassic	144	
			Triassic	208	
		Paleozoic Era	Permian	245	
			Carboniferous — Pennsylvanian	286	
			Carboniferous — Mississippian	320	
			Devonian	360	
			Silurian	408	
			Ordovician	438	
			Cambrian	505	
	Proterozoic Eon	Neo-proterozoic		570	
		Meso-proterozoic		1000	
		Paleo-proterozoic		1600	Precambrian comprises about 87% of the geologic time scale
	Archean Eon	Late		2500	
		Middle		3000	
		Early		3400	
	Hadean	No record		3800	Origin of earth about 4.6 billion years ago

Invertebrates

Fishes

Land plants

Amphibians

Reptiles

Mammals

Birds

? ? ?

EARTH

Past and Present

AN ENVIRONMENTAL APPROACH

GRAHAM R. THOMPSON, PhD
University of Montana

JONATHAN TURK, PhD

HAROLD L. LEVIN, PhD
Washington University

SAUNDERS GOLDEN SUNBURST SERIES
SAUNDERS COLLEGE PUBLISHING

Harcourt Brace College Publishers
Fort Worth   Philadelphia   San Diego   New York   Orlando
Austin   San Antonio   Toronto   Montreal   London   Sydney   Tokyo

Text Typeface: Clearface
Composition: Monotype Composition, Inc.
Publisher: John Vondeling
Developmental Editor: Lee Marcott
Managing Editor: Carol Field
Project Editor: Anne Gibby
Copy Editor: Patricia M. Daly
Manager of Art and Design: Carol Bleistine
Art Director: Caroline McGowan, Jennifer Dunn
Art and Design Coordinator: Sue Kinney
Text Designer: Nanci Kappel
Cover Designer: Larry Didona
Text Artwork: George V. Kelvin/Science Graphics
Director of EDP: Tim Frelick
Production Manager: Carol Florence
Marketing Manager: Sue Westmoreland

Cover Credit: © Tony Stone Worldwide/Nick Vedros, Vedros & Associates.
Frontispiece: Interior of a slot canyon on Colorado Plateau, Arizona. (© Tony Stone Worldwide)

Printed in the United States of America

Earth: Past and Present, An Environmental Approach

ISBN: 0-03-0982758

Library of Congress Catalog Card Number: 94-067169

4 5 6 7 8 9 0 1 2 3   0 3 2   10   9 8 7 6 5 4 3 2 1

Preface

Geologists study changes that occur on the Earth today and events that shaped its history. Traditionally, colleges and universities have offered separate courses in **physical geology**, a study of the Earth's surface and internal processes, and **historical geology**, a study of past events and the history of life. However, an increasing number of professors are teaching a course combining these two aspects of geology. We offer this book to open the widest possible window for introductory students.

Harold Levin wrote his first historical geology text, *The Earth Through Time*, in 1978. This text has been a leader in its field since then and is now in its fourth edition. Graham Thompson and Jonathan Turk published *Modern Physical Geology* in 1991, and the book has seen rapid success. The three of us now join forces to offer the present book, *Earth, Past and Present, An Environmental Approach,* an integrated look at the Earth: its history and the geological processes that affect our planet and its landscapes.

Three thousand years ago, farmers in the desert valleys of the Middle East were forced to abandon their land when salt from irrigation water poisoned the soil. As a result, the great civilizations of Mesopotamia fell. During the first century B.C., Plato wrote emotionally about soil loss and ground water depletion after the Athenian hills were deforested. Thus, since the beginning of civilization, people have understood that human activities can alter geologic systems. Today the study of interactions between humans and the environment has become a significant component of geology. Modern geologists must address environmental issues, not as advocates of any particular philosophy, but as scientists who ask questions, collect data, and seek answers based on observations and experiments. We highlight contemporary and historical environmental issues throughout the text.

We feel that it is not sufficient to teach science as a set of facts to be learned and memorized. Any study of geology would be lifeless unless students were introduced to the manner in which theories are developed. How do we know that dinosaurs dominated the Earth for 180 million years and then disappeared? What evidence indicates that continents migrate slowly around the globe, smashing into one another, crumpling the crust into mountains, and creating volcanoes and earthquakes? By describing key observations and experiments, we seek to make the course interesting and at the same time teach students how scientists function.

Geology is a uniquely visual science. City dwellers encounter sidewalks displaced by tree roots and may find fossils embedded in polished building stone. Suburbanites see exposed rock in road cuts along the highway and in outcrops in parks and woodlands. Those who live in rural environments may sit by a stream or camp in a high valley surrounded by rocky crags. Adventurous travellers may hike through parts of the American southwest among ancient sand dunes, now solidified into rock, that formed when wind blew across a sandy beach 200 million years ago. Whenever possible, we use familiar scenes as springboards to understanding events or processes that we cannot observe directly. To support this approach we have included a generous array of color photographs, many of which were taken by the authors during expeditions to remote regions of North America, Asia, the Arctic, and the Himalayas. George Kelvin, an accomplished scientific illustrator, contributed many original paintings.

While some students in introductory geology courses have backgrounds in chemistry and physics, others do not. The overall aim of this text is to portray geology accurately, but in a language and style that is readily understood by students with little or no college-level science. We have included necessary background material such as the nature and structure of atoms, so that this book needs no science prerequisites.

Teaching Options and Sequence of Topics

This book is divided into five units:

Unit 1: The Earth's Materials

Unit 2: Internal Processes

Unit 3: Surface Processes

Unit 4: The Earth Through Time

Unit 5: Special Topics

Although we believe this sequence to be logical, some instructors may prefer to alter it. As a result, we introduce the fundamental concepts of Earth materials, plate tectonics, and geological time in Chapter 1 to lay the groundwork for different sequences. The special topics in geology—mineral resources and planetary geology—can be taught separately or portions can be incorporated into specific chapters.

Special Features

Special Topics   Topics that are supplementary to the text material are presented throughout the book in the form of two types of boxes, which are set aside and highlighted in color. One box format, called *Focus On*, highlights topics that some instructors might want to include but are not essential to the sequential development of the course. Examples include "The Eruption of Mount St. Helens" and "Changing Views of Dinosaurs." Other boxes, titled *Geology and the Environment*, describe environmental topics like "Volcanoes and Climate" and "The Birthplace of Life." We feel that both types of boxes strengthen the text by providing geological anecdotes or interesting ideas that reinforce the main topic. Thus, the student is drawn into the text topics by specific examples, and at the same time the examples reinforce the basic concepts.

Chapter Review Material   Important words are highlighted in bold type in the text itself. These key words are then printed in a list at the end of each chapter for review purposes. In addition, a short summary is provided at the end of each chapter.

Questions   Two types of end-of-chapter questions are provided. The review questions can be answered in a straightforward manner from the material in the text. On the other hand, discussion questions challenge students to apply what they have learned to analyze situations not directly discussed in the text. Often there is no absolute correct answer to these questions.

Appendices and Glossary   An extensive glossary of key words is provided at the end of the book. In addition, appendices cover the following topics: physiographic provinces of the United States, periodic table and symbols for chemical elements, identifying common minerals, systems of measurement, rock symbols, and a classification of living things.

Interviews   Often students wonder, what type of life would I lead if I decide to make geology my career? With whom would I work and trade ideas? What intellectual rewards would make my life challenging? In order to answer these types of questions, we have interviewed four prominent geologists whose work is described in the text. These interviews include a brief look at both their professional and non-professional lives and are included to encourage students to look closely at a career in geology.

Ancillaries

This text is accompanied by extensive support materials.

Instructor's Manual and Test Bank

The Instructor's Manual, written by the authors with the assistance of Christine Seashore and Vicki Harder, provides teaching goals, answers to discussion questions, and a short bibliography. Sample tests are included in the manual. The Test Bank includes multiple choice, true or false, and completion questions for each chapter.

Computerized test banks are available in Macintosh® and IBM® Windows, IBM® 3.5 and 5.25-inch disks.

Saunders Geology Videodisc

The Saunders Geology Videodisc Version 2 includes over 2000 colorful, still images from 10 of Saunders' best-selling Geology, Earth Science, and Geography texts. The videodisc also includes almost one hour of live-action footage. Derived from the Encyclopedia Britannica archives, these moving images feature video clips of landscapes and geological phenomena, along with animated segments that bring geological processes to life.

LectureActive Software accompanies the videodisc. This software allows the instructors to customize lectures by giving them quick access to the video clip and still frame data on the videodisc.

A Barcode Manual is also available. The manual contains descriptions, barcodes, and text references for each still image and video clip. This allows the professor to access the images on the videodisc by either using a light pen to scan the barcodes or using the remote control to enter the frame number.

Overhead Transparencies and 35-mm Slides

The package includes 100 overhead transparencies and five hundred 35-mm slides.

Acknowledgments

We have not worked alone. The manuscript has been reviewed at several stages and the numerous careful criticisms have helped shape the book and ensure accuracy.

Larry Agenbroad, Northern Arizona University

David Anderson, San Jose State University

David Best, Northern Arizona University

Richard Bonnett, Marshall University

Rebecca Dorsey, Northern Arizona University

P. Geoffrey Feiss, the University of North Carolina at Chapel Hill

David Fitzgerald, St. Mary's University

Thomas Gillespie, Trenton State University

Bryan Gregor, Wright State University

Richard Hoare, Bowling Green State University

Stephen Mackwell, Penn State University

Jack McGregor, Western Kentucky University

J.K. Osmond, Florida State University

Charles Rich, Bowling Green State University

Monte Wilson, Boise State University

We extend our thanks to these reviewers.

We are especially grateful to Erle Kauffman, Paul Hoffman, John Horner, and Cindy Lee Van Dover for providing us with insightful interviews.

We would never be able to produce this book without professional support at home and at the offices of Saunders College Publishing. Thanks to Christine Seashore for finding photographs and contributing personal photographs, and for logistic collaboration. Special thanks to John Vondeling, Publisher. Lee Marcott, our Developmental Editor, Anne Gibby, Project Editor, Caroline McGowan and Jennifer Dunn, Art Directors, have all worked hard and efficiently to complete this project.

Graham R. Thompson
Missoula, Montana

Jonathan Turk
Darby, Montana

Harold L. Levin
St. Louis, Missouri

September 1994

Contents Overview

Contents

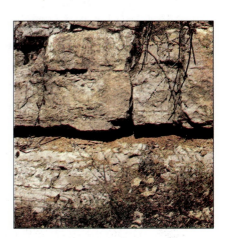

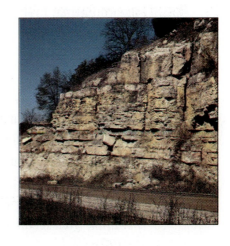

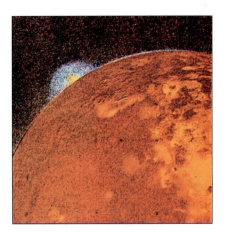

The Earth's Materials

◀ *Green tourmaline.*
(Mineralogical Museum at Harvard
University, Catalogue Number HU121761.
Photograph by Wendell Wilson.)

Geology: The Earth's Origin and History

A giant sea wave breaks as it approaches shore during a storm.
(Super Stock)

Imagine walking along a rocky coast as a storm blows in from the sea. Wind whips the ocean into whitecaps, gulls hurtle overhead, and waves crash onto shore. Blowing spray and rain soak your clothes as you scramble over the rocks. During this adventure, you have encountered the Earth's four major realms. The rocks and soil underfoot are part of the solid Earth. The rain and sea are parts of the hydrosphere, the watery part of our planet. The wind and air are portions of the atmosphere. Finally, you, the gulls, and all other forms of life in the sea, on land, and in the air are parts of the biosphere, the thin shell near the Earth's surface that is alive or supports life.

1.1 The Earth's Layers

Figure 1.1 shows that the **solid Earth** is by far the largest of the four realms. The Earth's radius is about 6400 kilometers, $1\frac{1}{2}$ times the distance from New York to Los Angeles. Nearly all of our direct contact with the solid Earth occurs at or very near its surface. The deepest wells ever drilled penetrate only about 10 kilometers, 1/640 of the total distance to the center.

The **hydrosphere** includes water in streams, lakes, and oceans; in the atmosphere; and frozen in glaciers. It also includes ground water that soaks soil and rock to a depth of 2 or 3 kilometers.

The **atmosphere** is a mixture of gases, mostly nitrogen and oxygen. It is held to the Earth by gravity and thins rapidly with altitude. Ninety-nine percent is concentrated in the first 30 kilometers, but a few traces remain even 10,000 kilometers above the Earth's surface.

The **biosphere** is the thin zone inhabited by life. It includes the uppermost solid Earth, the hydrosphere, and lower parts of the atmosphere. Although some bacteria live in rock to depths of a few kilometers, specialized organisms live in the deepest oceanic trenches, and a few windblown microorganisms are found at heights of 10 kilometers or more, the overwhelming majority of living creatures are found in a very thin layer at the Earth's surface. Land plants grow on the surface, with roots penetrating a few meters into soil. Most animals also live on the surface, fly a kilometer or two above it, or burrow a few meters underground. Sea life also concentrates near the surface of the oceans, where sunlight is available.

If you could drive a magical vehicle from the center of the solid Earth to the outer fringe of the atmosphere at 100 kilometers per hour, you would be within the Earth for 64 hours. In another 20 minutes you would pass through nearly all of the atmosphere and would enter the rarefied boundary between Earth and space.

Figure 1.1 A schematic view of the solid Earth, the hydrosphere, the atmosphere, and the biosphere.

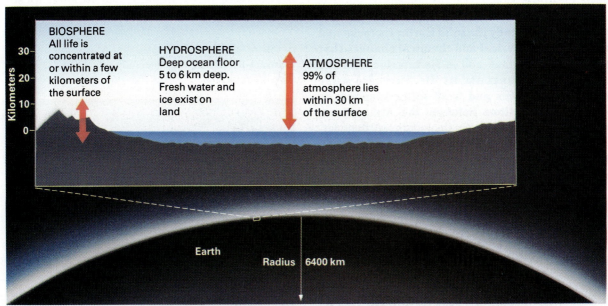

Mining salt from the Great Salt Lake, Utah.

You would pass most living organisms in a few seconds, and the entire biosphere in 6 minutes.

An understanding of the Earth is valuable simply because humans are curious creatures; we wonder about the world around us. As you drive on a highway or walk along the seacoast, you experience the world more richly if you understand how the hills and rocks formed.

Geology also has a practical side. We depend on fossil fuels—coal, oil, and natural gas—and mineral resources such as metals, sand, and gravel. We also depend on soil to support crops and other plants. The oceans provide food and sea lanes for commerce and travel. Clean, fresh water is vital to agriculture, industry, and human consumption.

Geologists study the solid Earth: its rocks and minerals, the physical and chemical changes that occur on its surface and in its interior, and the history of the planet and its life. They also study the Earth's oceans and fresh-water supplies, as well as interactions between the atmosphere, water, and the solid Earth. Geologists seek to understand both the interior and the surface of the Earth. They study our planet as it exists today and look back at its history.

1.2   The Solid Earth

The Earth's Materials: Rocks and Minerals

Below a thin layer of soil and beneath the oceans, the outer layers of the Earth are composed entirely of **rock**.

Geologists study the compositions and physical properties of rocks and the ways in which they formed. Even a casual observer sees that rocks are different from one another: Some are soft, others hard, and they come in many colors. Most rocks are composed of tiny, often differently colored grains, each of which is a **mineral** (Fig. 1.2).

The Earth's Internal Processes

The Earth is an active planet. Events and processes that occur or originate within the Earth are called **internal processes**. Earthquakes and volcanoes are familiar consequences of internal processes because they occur rapidly and dramatically. However, the same processes that cause volcanoes to erupt and earthquakes to shake the land also cause slower events such as the growth of mountain ranges and movements of continents. Engineers and city planners might consult with a geologist to ask, "What is the probability that an earthquake will occur in our city? How destructive is it likely to be? Is it safe to build skyscrapers or a nuclear power plant in the area?" Or "What is the likelihood that a volcano will erupt to threaten nearby cities?"

The Earth's Surface Processes

Most of us have seen water running over soil after a heavy rain. The flowing water dislodges tiny grains of soil and carries them downslope. After a few hours, an exposed hillside may become scarred by gullies. If you stretch your imagination over thousands or millions of years, you can envision flowing water shaping the surface of our planet, enlarging tiny gullies into great valleys and canyons (Fig. 1.3). These and other natural activities that change the Earth's surface are called **surface processes**.

Figure 1.2 Each of the differently colored specks in this rock is a mineral.

Oceans cover 71 percent of the Earth and contain 97.5 percent of its water. Thus, most of the hydrosphere is seawater. The other 2.5 percent of Earth's water is fresh. Most fresh water is frozen in glaciers. Less than 1 percent of the Earth's fresh water is found in streams, lakes, the atmosphere, and as **ground water** saturating rock and soil of the upper few kilometers of the solid Earth (Fig. 1.4). **Hydrogeologists** study the Earth's fresh water, its distribution, and its circulation among oceans, continents, and the atmosphere.

Coastlines change rapidly as a result of Earth processes. Waves erode rock, and currents carry sand from place to place. In the past 18,000 years great continental glaciers have melted, adding enough water to the oceans to raise the global sea level by about 100 meters. Enough ice remains in the Antarctic and Greenland ice sheets to raise sea level by an additional 65 meters if it were to melt completely. Some scientists now think that global temperature is rising because of changes in atmospheric composition caused by burning fossil fuels. An increase of about 5°C would melt most of the remaining ice.

Many questions arise when we think of surface processes. Why do desert landscapes differ from those in humid climates? How do river valleys change with time? Is a flood likely to destroy a housing development in this valley? How will acid rain affect a lake where people swim and fish? How will soil erosion affect our ability

Figure 1.3 Over long periods of time, running water can carve deep canyons, such as this tributary of the Grand Canyon in the American southwest.

Figure 1.4 The distribution of water in the hydrosphere. Percentages show the distribution of Earth's water among oceans, glaciers, ground water, lakes and rivers, and the atmosphere. The numbers show thousands of cubic kilometers transferred per year as precipitation, evaporation, and runoff from land.

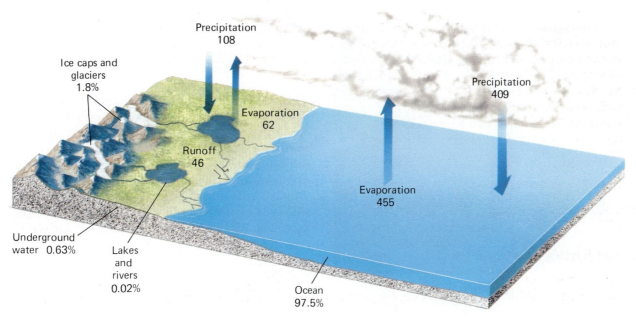

to grow crops and our food supply? What will happen to coastal cities if global temperature rises?

Environmental Geology

Human activity is too insignificant to affect most internal geological processes such as volcanic eruptions, earthquakes, and the movement of continents. In the past few decades, however, human activities have altered the surface of the planet. Today, human-caused environmental changes are found in cities, on farmlands throughout the world, on the fringes of the Sahara Desert, in the tropical rainforests, in the central oceans, and at the North Pole. Moreover, some scientists are concerned that these changes may threaten human well-being.

No simple solutions exist for these environmental challenges. In every instance, complex social and technical problems must be addressed. Consider some of the questions that a geologist might be called on to answer:

1. At many sites, toxic compounds originating from a variety of human activities have leaked into soil. It is important to know how fast these compounds will contaminate ground water and spread outward. Are drinking water resources threatened? Can the toxic materials be contained or removed?

2. If a hillside forest is cut and the land terraced and planted to grain, will the terraces hold, or will landslides destroy them?

3. An electric company wants to build a nuclear power plant. Is the proposed site safe, or is it threatened by earthquakes or floods?

4. Radioactive waste from nuclear power plants and weapons plants must be stored for a long time where

Figure 1.5   Spider preserved in amber. This spider was alive in Miocene times, about 5 to 25 million years ago.   *(Courtesy of American Museum of Natural History)*

it will not escape into the biosphere. Storage usually means burial. What types of environments are most stable and therefore safest? What is the probability that an earthquake, volcanic eruption, landslide, or flood will disturb a proposed repository and release radioactive wastes into the environment?

Experts commonly disagree on answers to these questions. In the emotional issue of radioactive waste storage, for example, some geologists believe that the major problems have been solved and that "safe" repositories have been identified. Others argue that important questions about geological hazards at "safe sites" have not yet been answered satisfactorily. When such conflict becomes public, how can you evaluate contradictory statements? No set rules exist, but surely a first step is to understand the basic science behind the human opinions.

1.3   Earth History and Geologic Time

The Earth is about 4.6 billion years old. In his book *Basin and Range,** which discusses the geology of western North America, John McPhee offers us a metaphor for the magnitude of geologic time. If the history of the Earth were represented by the old English measure of a yard (the distance from the king's nose to the end of his outstretched hand), all of human history could be erased by a single stroke of a file on his middle fingernail. Geologists study the *entire* history of the Earth, from its origins to the present.

Geologists study the history of life by studying **fossils**, remains or traces of living organisms preserved in rocks (Fig. 1.5). Studies of fossilized animals and plants

A nuclear power plant near Wiscasset, Maine.

---

* John McPhee. *Basin and Range.* New York: Farrar, Straus, and Giroux, 1981. 215 pp.

Table 1.1

The Geologic Column and Time Scale*					
Time Units of the Geologic Time Scale					Distinctive Plants and Animals
Eon	Era	Period	Epoch		
Phanerozoic Eon (Phaneros = "evident"; Zoon = "life")	Cenozoic Era	Quaternary	Recent or Holocene	"Age of Mammals"	Humans
			Pleistocene		Mammals develop and become dominant
		Tertiary — Neogene	Pliocene — 2		
			Miocene — 5		
		Tertiary — Paleogene	Oligocene — 24		
			Eocene — 37		
			Paleocene — 58		Extinction of dinosaurs and many other species
	Mesozoic Era	Cretaceous — 66		"Age of Reptiles"	First flowering plants, greatest development of dinosaurs
		Jurassic — 144			First birds and mammals, abundant dinosaurs
		Triassic — 208			First dinosaurs
	Paleozoic Era	Permian — 245		"Age of Amphibians"	Extinction of trilobites and many other marine animals
		Carboniferous — Pennsylvanian — 286			Great coal forests; abundant insects, first reptiles
		Carboniferous — Mississippian — 320			Large primitive trees
		Devonian — 360		"Age of Fishes"	First amphibians
		Silurian — 408			First land plant fossils
		Ordovician — 438		"Age of Marine Invertebrates"	First fish
		Cambrian — 505			First organisms with shells, trilobites dominant
Proterozoic		— 570			First multicelled organisms
		Sometimes collectively called Precambrian			
Archean	2500				First one-celled organisms
Hadean	3800				Approximate age of oldest rocks
	4600 ±				Origin of the Earth

* Numbers are ages in millions of years. These absolute ages are based on radiometric dating.

FOCUS ON   *Hypothesis, Theory, and Law*

On an afternoon field trip you may find several different types of rocks or watch a river flow by. But you can never see the rocks or river as they existed in the past or as they will exist in the future. Yet a geologist could tell you how the rocks formed millions or even a few billion years ago and could predict how the river valley might change in the future.

Scientists not only study events that they have never observed and never will observe, but they also study objects that can never be seen, touched, or felt. In this book we examine the core of the Earth 6400 kilometers beneath our feet, even though no one has ever visited it and we are certain that no one ever will.

Much of science is built on inferences about events and objects outside the realm of direct experience. An inference is a conclusion based on thought and reason. How certain are we that a conclusion of this type is correct?

In science, inferences are called laws, theories, or hypotheses, depending on the degree of certainty. A **law** is a formal statement of the way in which events always occur under given conditions. It is considered to be factual and always correct. A law is the most certain of scientific statements. For example, the Law of Gravity states that all objects are attracted to one another in direct proportion to their masses. We cannot conceive of any contradiction to this principle, and none has been observed. Hence, the principle is called a law.

A **theory** is less certain than a law. It is an interpretation or explanation of some aspect of the world that is supported by experimental or factual evidence but is not proved so conclusively that it is accepted as a law. For example, the theory of plate tectonics states that the outer layer of the Earth is broken into a number of plates that move horizontally relative to each other. As you will see in Chapters 4 through 8, this theory is sup-

ported by many observations and seems to have no major inconsistencies.

A **hypothesis**, or a **model**, is weaker than a theory. It is a tentative explanation of observations that can be tested by comparing it with other observations and experiments. Thus, a hypothesis or model is a rough draft of a theory that is tested against the facts. If it explains some of the facts but not all of them, it must be altered, or if it cannot be changed satisfactorily, it must be discarded and a new hypothesis developed.

Scientists develop hypotheses and theories according to a set of guidelines known as the **scientific method**, which involves three basic steps: (1) observation, (2) forming a hypothesis, and (3) testing the hypothesis and developing a theory.

Observation

All modern science is based on observation. Suppose that you observed an ocean wave carrying and depositing sand. If you watched for some time, you would see that the sand accumulates slowly, layer by layer, on the beach. You might then visit Utah or Nevada and see cliffs of layered sandstone hundreds of meters high. Observations of this kind are the starting point of science.

Forming a Hypothesis

Simple observations are only a first step along the path to a theory. A scientist tries to organize observations to recognize patterns. You might note that the sand layers deposited along the coast look just like the layers of sand in the sandstone cliffs. Perhaps you would then infer that the thick layers of sandstone had been deposited by an ancient ocean. You might further con-

---

tell us about the nature of the Earth's surface millions and even billions of years ago when the now-fossilized organisms lived.

Most of us are fascinated by dinosaurs, their size, their abundance, and their sudden disappearance 65 million years ago. If you wished to pursue this fascination, you might become a **paleontologist** and study prehistoric life, either the dinosaurs or organisms that lived before or after them.

Geologists have divided the 4.6-billion-year span of geologic time into **eons**, **eras**, **periods**, and **epochs**. These time units are arranged in a **geologic time scale** (Table 1.1). In general, the subdivisions are based on

the predominant life forms of each time. For example, during the **Proterozoic Eon** (*protero* is from a Greek root meaning "early" and *zoon* from a word meaning "life"), animals had no hard parts such as shells and bones. Most Proterozoic life forms were single celled, although a few larger organisms such as jellyfish and worms appeared toward the end of the Eon. Animals with hard shells and skeletons appeared 570 million years ago to mark the start of a new eon. These new organisms prospered, and a great many new species evolved quickly. The organisms left abundant fossils because shells and skeletons are preserved easily. The most recent 13 percent of geologic time, from 570 mil-

The sandstone layers of Arches National Park, Utah, formed and then eroded away slowly.

clude that because the ocean deposits layers of sand slowly, the thick layers of sandstone must have accumulated over a long time.

If you were then to travel, you would observe that thick layers of sandstone are abundant all over the world. Because thick layers of sand accumulate so slowly, you might infer that a long time must have been required for all that sandstone to form. From these observations and inferences you might form the hypothesis that the Earth is old.

Testing the Hypothesis and Forming a Theory

Theories differ widely in form and content, but all obey four fundamental criteria.

1. A theory must be based on a series of confirmed observations or experimental results.

2. A theory must explain all relevant observations or experimental results.

3. A theory must not contradict any relevant observations or other scientific principles.

4. A theory must be internally consistent. Thus, it must be built in a logical manner so that the conclusions do not contradict any of the original premises.

Most theories can be used to predict events that have not yet been observed, and if the theory is a good one, the predictions will be correct. When first proposed in the late 1700s, the hypothesis that the Earth is old was based only on the observation that sand layers accumulate slowly. In the past 200 years, results of many different measurements and experiments have proved consistent with this hypothesis. Today the idea that the Earth is old is a firmly grounded theory.

lion years ago to the present, is thus called the **Phanerozoic Eon** (for "evident" life). The Phanerozoic Eon is further subdivided into the **Paleozoic Era** ("ancient life"), the **Mesozoic Era** ("middle life"), and the **Cenozoic Era** ("recent life"). These terms and their divisions will be examined more fully in Chapter 15.

1.4   The Earth's Origin and the Formation of the Universe

Before we ask "How did the Earth form?" we must ask an even more fundamental question: "How did the Universe form, or did it begin at all?" One possibility is that the Universe always existed and that time has no beginning. An alternative theory is that the Universe began at a specific time and has been evolving or changing ever since. The story of the Earth's formation is very much a part of the larger story of the origin of the Universe.

In 1929 Edwin Hubble observed that all galaxies are moving away from each other. That observation implies that the galaxies must have started out at a common center. By measuring the speeds of galaxies and the distances between them, we can mentally trace their paths backward to a time when the entire Universe was compressed into a single infinitely dense point. According to modern theory, this point exploded. This cataclysmic event, called the **big bang**, marked the beginning of the Universe and the start of time. The big bang was no ordinary explosion. It is not comparable to a hydrogen bomb blast or even the catastrophic death of a massive star. This explosion created the Universe instantaneously. Matter, energy, space, and time came into existence with this single event. Most astronomers place the big bang, the start of time, and the origin of the Universe between 15 and 18 billion years ago.

Even though our estimates of the time of the big bang vary by three *billion* years, astronomers have reconstructed a picture of the first few *seconds* after the origin of the Universe. Part of this reconstruction comes from studies of how particles behave when they collide at very high velocities in modern particle accelerators. Other evidence comes from studying particles and radiant energy in space.

Immediately after the big bang, the Universe was extremely hot, about 100 billion degrees Celsius. During the first second, it cooled to about 10 billion degrees, 1000 times the temperature in the center of the modern Sun (Fig. 1.6). At such high temperatures, atoms do not exist. Most of the Universe consisted of a mixture of radiant energy, electrons, and extremely light particles called neutrinos. Protons and neutrons also began to form. After about 1.5 minutes, the temperature fell to 1 billion degrees and a few simple atomic nuclei formed, although the temperature was still too hot for atoms. During the next million years the Universe continued to cool as it expanded. When the temperature dropped

Figure 1.6   A brief pictorial summary of the evolution of the Universe.

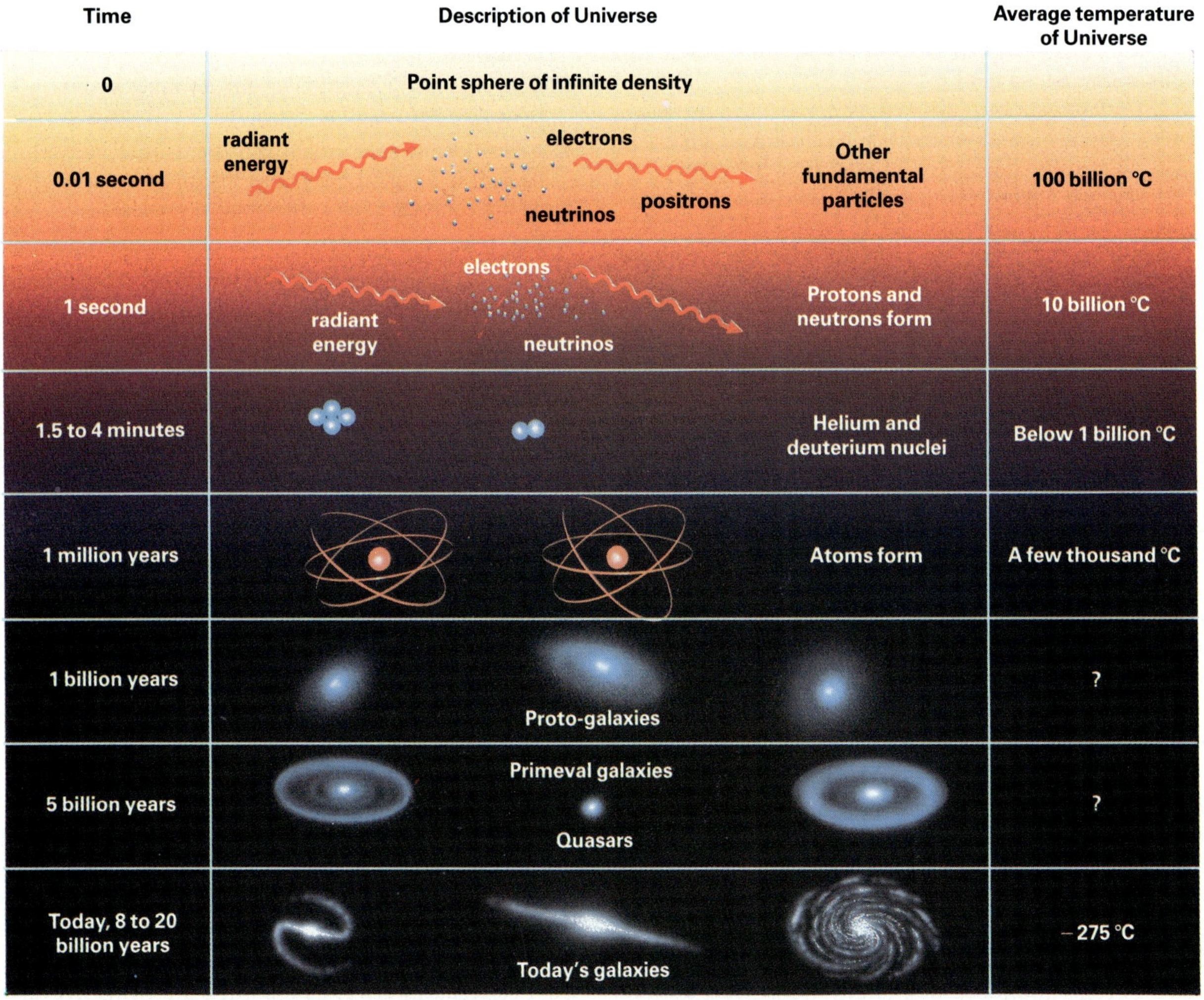

Time	Description of Universe	Average temperature of Universe
0	Point sphere of infinite density	
0.01 second	radiant energy electrons neutrinos positrons Other fundamental particles	100 billion °C
1 second	electrons radiant energy neutrinos Protons and neutrons form	10 billion °C
1.5 to 4 minutes	Helium and deuterium nuclei	Below 1 billion °C
1 million years	Atoms form	A few thousand °C
1 billion years	Proto-galaxies	?
5 billion years	Primeval galaxies Quasars	?
Today, 8 to 20 billion years	Today's galaxies	−275 °C

FOCUS ON   *Is the Big Bang Theory Correct?*

The big bang theory attempts to explain events that occurred at the start of time, long before our planet, Solar System, or Galaxy existed. How sure are we that scientists can accurately re-create a picture of the Universe during its earliest history?

For several decades, astrophysicists noted a major flaw in the big bang theory. Measurements made in the 1960s showed that the early Universe was homogeneous: Its matter and energy were distributed uniformly throughout space. Now, however, matter is obviously concentrated into galaxies, galaxies are clustered into groups, and vast expanses of empty space separate the clumps of matter. Therefore, the modern Universe is heterogeneous. There is no reasonable way to explain how an initially homogeneous Universe became heterogeneous.

This apparent contradiction sparked debate. Some scientists argued that the big bang theory must be discarded if it contains such a glaring inconsistency. Others responded that the big bang theory explains most observations and has predicted others, and therefore it should not be discarded.

When scientists disagree, they design experiments to supply new observations that will either support or contradict the theory. In 1989, the Cosmic Background Explorer (COBE) satellite was launched to make precise measurements of the early structure of the Universe. In 1992, data collected by COBE showed that matter and energy concentrated into clumps during the earliest infancy of the Universe, perhaps in the first billionth of a second. The data collected in the 1960s had been inadequate to show this initial heterogeneity. George Smoot, a researcher on the project, called these early inhomogeneities "the imprints of tiny ripples in the fabric of space-time put there by the primeval explosion."* Thus the COBE experiment showed that the heterogeneous nature of the modern Universe does *not* contradict the big bang theory.

Debate is part of science. It is important not to accept theories as fact, because sometimes they are wrong. But it is also important not to discard a theory that explains many phenomena just because some questions remain unanswered.

* M. Stroh, *Science News,* Vol. 141, p. 141, 1992.

to a few thousand degrees, atoms formed and, in a sense, the modern Universe was born. With time, matter collected into galaxies, and within the galaxies stars were born.

At present the galaxies are all flying away from one another. What will happen in the future? Think of a rocket ship taking off from a planet. If the planet's gravitational field is weak enough, the rocket escapes into space and never returns. However, if the planet has a large mass and therefore a strong gravitational field, the rocket falls back to the surface. In the same manner, if the gravitational force of the Universe is sufficient, all the galaxies will eventually slow down, reverse direction, and fall back to the center, forming another point of infinite density. This point may then explode again to form a new universe. In turn the new universe will expand and then collapse, creating a continuous chain of universes. This possibility is called the **oscillating universe cosmology**. The other possibility is that the gravitational force of the Universe is not sufficient to stop the expansion, and the galaxies will continue to fly

apart forever. Within each galaxy, stars will eventually consume all their nuclear fuel and stop producing energy. As the stars fade and cool, the galaxies will continue to separate into the cold void. This scenario is called the **forever-expanding cosmology** (Fig. 1.7). Astronomers are attempting to calculate the mass of the Universe in an effort to determine which of the two possibilities is more realistic. However, the measurements are uncertain and the final answer elusive.

1.5   Formation of the Solar System and the Earth

Five billion years ago, the matter that eventually became our Solar System existed as an immense cloud of dust and gas rotating slowly in space. Its temperature was about $-270°C$. At this extremely low temperature, particles move so slowly that even a slight force affects them. Small gravitational attractions among the dust

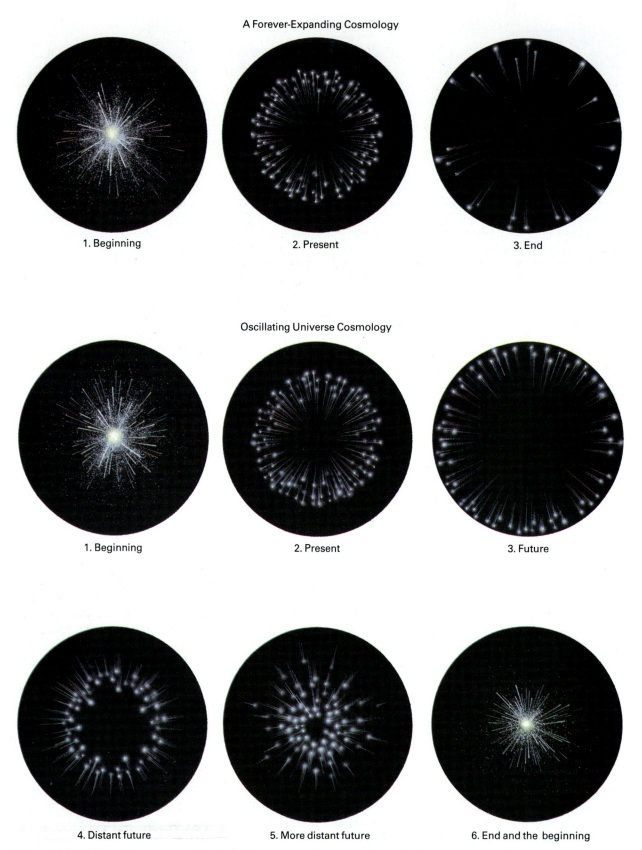

Figure 1.7 Schematic representations of the forever-expanding cosmology and the oscillating universe cosmology.

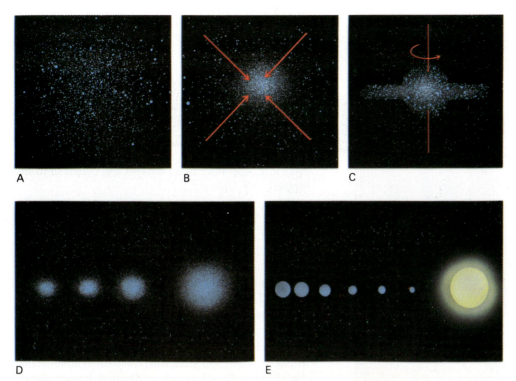

Figure 1.8 Formation of the Solar System. (A) The Solar System was originally a diffuse cloud of dust and gas. (B) The dust and gas began to coalesce as a result of gravity. (C) The shrinking mass began to rotate and formed a disk. (D) The mass broke up into a discrete protosun orbited by large protoplanets. (E) The Sun heated up until fusion temperatures were reached. The heat from the Sun drove most of the hydrogen and helium away from the closest planets, leaving small, solid cores behind. The massive outer planets remain composed mostly of hydrogen and helium.

and gas caused the cloud to condense into a sphere (Fig. 1.8). Alternatively, some astronomers suggest that a nearby star may have exploded and that the shock wave triggered the condensation (Fig. 1.9). As the condensation continued, the rotation accelerated and the sphere spread into a disk, as shown in Figure 1.8C.

More than 90 percent of the matter in the cloud gravitated toward the center of the newly formed disk. As atoms accelerated toward the center under the influence of gravity, they coalesced to form the **protosun**, the earliest form of the Sun. As the high-speed particles collided, they released heat to warm the protosun, but it was not a true star because it did not yet generate energy by nuclear fusion.

Formation of the Planets

Heat of the protosun warmed the inner region of the disk. Then as the gravitational collapse became nearly complete, the disk cooled. Gases in the outer part of the disk condensed to form small aggregates, much as

raindrops or snowflakes form when moist air cools in the Earth's atmosphere. The formation of these aggregates was the first step in the evolution of planets.

As the growing cloud rotated around the protosun, aggregates began to stick together as snowflakes sometimes do. Thus, they increased in size and developed stronger gravitational forces to attract additional particles. This growth continued until a number of small, rocky spheres formed. Their gravitational forces caused them to collide and coalesce to form mini-planets, called **planetesimals**, that ranged in size from a few kilometers to about 100 kilometers in diameter. The entire process, from the disk to the planetesimals, occurred relatively quickly in geological terms and probably required only about 10,000 to 100,000 years. As the planetesimals grew, they were attracted to one another by gravity and collided. Many small spheres coalesced to a few large ones, including the Earth. The Earth formed about 4.6 billion years ago. During the next 1 billion years, numerous additional planetesimals, comets, and small chunks of rock flying through space called **meteoroids** have continued to slam into its surface, adding to

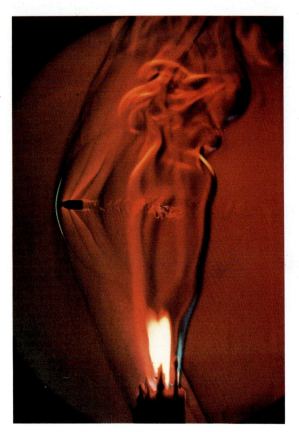

Figure 1.9 A shock wave forms as a bullet passes through hot gases generated by a burning candle. The leading edge of the shock wave is on the left. According to one theory, a shock wave created by an exploding star initiated the collapse of the cloud of dust and gas and thereby was the first step in the formation of the Solar System. *(Harold E. Edgerton, MIT; courtesy of Palm Press)*

its mass. Although meteoroids still bombard the Earth, bombardment is now so infrequent that the mass of the Earth has remained essentially constant for 3 to 4 billion years.

Formation of the Sun

At the same time that planets were evolving, changes occurred in the protosun. Gravitational attraction pulled the gases inward, creating extremely high pressure and temperature. The protosun became hottest in its center, where atoms accelerated inward most rapidly. The core of the protosun became so hot that hydrogen nuclei combined in a process called **nuclear fusion**. When hydrogen nuclei combine, they form the nucleus of the next heavier element, helium. As fusion began within the Sun, vast amounts of nuclear energy were released, in a process comparable to the continuous explosion of millions of hydrogen bombs. The onset of nuclear fusion marked the birth of the modern Sun. The

heat and light given off by our Sun are still generated by hydrogen fusion.

The Modern Solar System

In its initial form, the cloud that evolved into the Solar System must have been homogeneous; that is, it was the same throughout. However, as the Sun became hotter, many light gases, such as hydrogen and helium, boiled away from the inner Solar System and collected in the frozen outer regions.

As a result, the four planets closest to the Sun—Mercury, Venus, Earth, and Mars—are now rocky with metallic centers. Most of the gases have been lost. Thus, Mercury has virtually no atmosphere, and the atmospheres surrounding Venus, Earth, and Mars represent a tiny portion of their planetary masses. These four are called the **terrestrial planets** because they are all "Earthlike." In contrast, the four planets beyond this inner circle—Jupiter, Saturn, Uranus, and Neptune—are called the **Jovian planets** and are composed primarily of liquids and gases with small rocky and metallic cores (Fig. 1.10). Pluto, the outermost known planet, is anomalous and is more similar to Neptune's moon, Triton, than to the Jovian planets. Figure 1.11 is a schematic representation of the modern Solar System.

The Early Earth

As already explained, our Earth formed and grew by multiple collisions of smaller bodies. The bodies were drawn together by gravity, and this gravitational collapse generated heat. Additional heat was released by radioactivity. Some elements in the Earth are radioactive. When a radioactive atom breaks apart, energy is released and converts to heat. Only a tiny amount of heat is created by decay of a single radioactive atom, and radioactive atoms were dispersed throughout the early Earth. Therefore, in a given volume of rock, heat was generated very slowly. For example, ordinary granite contains small but measurable amounts of radioactive materials. If no energy were lost, it would take 500 million years to brew a cup of coffee with the heat released from 1 cubic centimeter of granite. However, the Earth is large and geologic time is long. The heat from gravitational collapse and radioactivity was retained by a thick, insulating layer of surface rocks. At the same time, the surface was heated by intense bombardment as the Earth swept up chunks of rock, comets, and other debris floating about in the early Solar System.

A few hundred million years after it formed, the Earth became so hot that it began to melt and separate into layers. Heavy molten iron and nickel seeped toward

A   B

Figure 1.10   (A) Mercury is a small planet close to the Sun. Because of this proximity, most of its lighter elements have long since been boiled off into space, and today its surface is solid and rocky. (B) Jupiter, on the other hand, is composed mainly of gases and liquids, with a small, solid core. This image is a close-up of its turbulent atmosphere.   *(NASA)*

the center and eventually collected to form a dense, hot core. Many of the Earth's light elements floated to the surface to form the relatively light rocks of the Earth's crust. The remaining material concentrated between the crust and the core to form the mantle. Shortly after it melted, the Earth cooled and most of it solidified. Today the Earth continues to cool, but it retains its layered structure.

1.6   The Structure of the Modern Earth

The modern Earth consists of three distinct layers similar to those that formed early in its history: crust, mantle, and core (Fig. 1.12).

The **crust** is a thin, rigid surface veneer. Its thickness ranges from 7 kilometers under some portions of

Figure 1.11   A schematic view of the Solar System.

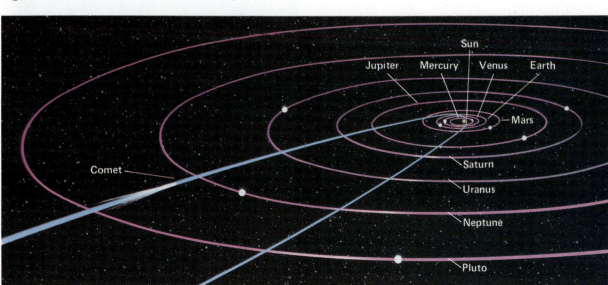

FOCUS ON

Scientific Evidence to Support the Current Theory of the Formation of the Earth

Observation	Interpretation of the Observation in Terms of the Current Theory for the Evolution of our Solar System
All the planets (except Pluto) orbit in nearly the same plane.	All the planets formed from a common planar disk.
All the planets revolve around the Sun in the same direction, which is also the direction in which the Sun rotates on its axis.	If the planets and the Sun all formed from a single rotating disk, they would all retain the direction of motion of the original disk.
The planets closest to the Sun are small and composed primarily of heavy elements. The more distant planets are larger and composed primarily of light gases.	According to our theory, all planets originally had the same composition because they formed from a common cloud. However, the Sun drove the light gases away from the closest planets, leaving small, dense, rocky spheres. This is exactly what we observe.
Flattened disks of dust and gas have been observed around several young stars (see illustration).	We cannot look backward in time to see how our Solar System evolved, but the theory of its evolution would be supported if we could see similar events occurring elsewhere today. The thin disk around the star Beta Pictoris is believed to be similar to the disk that formed the planets of our own Solar System.
Some nearby stars have been observed to wobble as if they were perturbed by the gravitational field of a nearby object, such as a planet.	In principle, a very sensitive telescope could detect a planet in orbit around a nearby star. In practice, the image of the planet is drowned out by the light of the star. However, by observing tiny aberrations in the motion of a star, we can deduce that it is affected by the gravitational field of an orbiting planet. This tells us that planetary evolution is not unique to the Solar System.

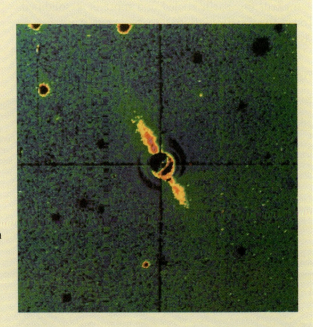

This photograph may be an image of another solar system in the process of formation. The thin disk around the central star, Beta Pictoris, is composed of bits of dust, which are believed to be similar to the material that condensed to form the planets of our own Solar System. *(University of Arizona and Jet Propulsion Laboratories)*

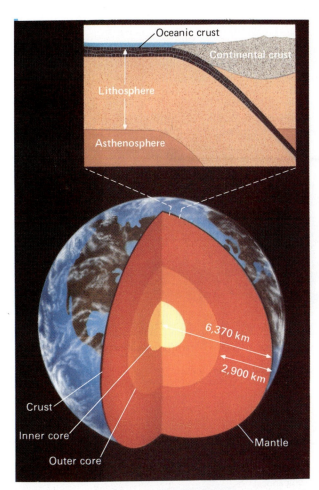

Figure 1.12   A schematic view of the interior of the Earth. The inset is a view of the outer layers.

the oceans to about 70 kilometers under the highest mountain ranges. By comparison, the radius of the entire Earth is 6400 kilometers. If you built a model of the Earth 1 meter in radius, the crust would be 1.1 to 11 millimeters thick. If the Earth were the size of an egg, the crust would be thinner than an eggshell. The rocks of the crust are rich in low-density elements that floated to the surface when the Earth was molten.

A thick, mostly solid layer called the **mantle** lies beneath the crust and surrounds the core. The mantle extends from the base of the crust to a depth of 2900 kilometers. It contains more than 80 percent of the volume of the Earth.

The uppermost portion of the mantle is relatively cool. Because the rocks are cool, they are strong and brittle. They crack when stressed, much like the rocks of the crust. The cool, strong, brittle outer portion of the Earth, including both the crust and this upper layer of mantle, is called the **lithosphere** (Greek for "rock layer"). The lithosphere extends from the Earth's surface to an average depth of 100 kilometers.

Beneath the lithosphere, but also within the mantle, lies a layer called the **asthenosphere** (Greek for "weak layer"). It extends from a depth of about 100 kilometers to about 350 kilometers below the Earth's surface. The Earth's temperature increases with depth, and the asthenosphere is hot enough that 1 or 2 percent of the rock is melted. The remaining rock, although solid, is so hot that it flows slowly without cracking—it behaves plastically. The asthenosphere has often been compared with road tar or putty, familiar materials that deform plastically. It is solid, yet when stressed it gives and flows as a fluid does.

As already mentioned, the asthenosphere extends to about 350 kilometers below the surface, whereas the base of the mantle lies at a depth of 2900 kilometers. Thus, most of the mantle lies below the asthenosphere. Below the asthenosphere, pressure is so great that even though the rock is hot, it is solid and considerably more rigid than the rock in the asthenosphere. However, the term *rigidity* is used here in a relative sense. Careful measurements show that the Earth is not perfectly spherical but slightly distorted. The rotation of the Earth causes it to bulge at the equator. Therefore, a person standing on the equator is about 50 kilometers farther from the center of the Earth than a person standing at the South Pole. If you fill a balloon with water and spin it on a tabletop, it distorts in an exaggerated model of the Earth's bulge.

The Earth's **core** is composed primarily of iron and nickel. It is divided into a solid inner core and a liquid outer core. Insulated by 2900 kilometers of overlying rock, it has retained much of the original heat acquired during the formation of the Earth. Today the temperature in the core is about 6000°C, about the same as that at the Sun's surface.

1.7   Consequences of a Hot Earth: Earthquakes, Volcanoes, and Plate Tectonics

Our planet is *not* a static, homogeneous sphere, like a bowling ball. Instead, the Earth is active and dynamic. The thin, cool, brittle lithosphere floats on the hot, plastic asthenosphere much as blocks of wood float in a tub of honey. The asthenosphere is solid (not liquid like honey), but because of its plastic nature, it can flow. We will refine our understanding of this important layer throughout this book.

Within the past 30 years the theory of **plate tectonics** has revolutionized geology. According to this theory, the lithosphere is broken into seven major plates and several smaller ones. These lithospheric plates are packed together tightly like the segments of a turtle

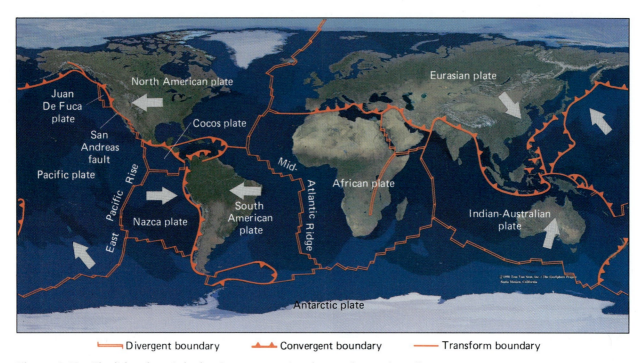

Divergent boundary   Convergent boundary   Transform boundary

Figure 1.13   The lithosphere is broken into seven major plates and several smaller ones.

shell (Fig. 1.13). In this theory, *the plates move by floating on and gliding over the plastic asthenosphere.* The boundaries between plates are zones where one lithospheric plate meets another that is moving in a different direction. As you might expect, the boundaries are geologically active. When plates rub, jerk, or jump past one another, the Earth shakes. This vibration is called an **earthquake**. In some regions (most commonly near plate boundaries), parts of the asthenosphere and lithosphere melt. The resulting liquid rock is called **magma**. Large quantities of magma rise through the lithosphere and pour out onto the Earth's surface in **volcanic eruptions**.

Three different types of plate boundaries exist (Fig. 1.14):

1. A **divergent boundary** is a zone where two or sometimes three plates separate or move apart from each other.

2. If plates on a sphere separate in some places, they must collide in others. A zone where plates collide head on is a **convergent boundary**.

3. At a **transform boundary**, plates slide horizontally past one another.

Shortly after World War II, oceanographers mapped a submarine mountain range running north–south through the whole length of the Atlantic Ocean, called the **Mid-Atlantic ridge**. The plates on opposite sides of the ridge are moving away from each other. Therefore, the Mid-Atlantic ridge is a divergent plate boundary. As

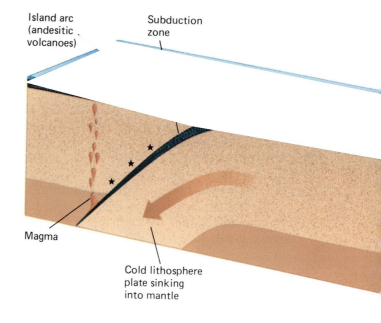

Figure 1.14 Three types of plate boundaries: (A) divergent, (B) convergent, (C) transform.

the two plates spread apart, magma from the asthenosphere oozes upward to fill the gap and form new oceanic crust (Fig. 1.15). Extensions of the Mid-Atlantic ridge continue into all other ocean basins. This global submarine mountain chain is called the **mid-oceanic ridge** (Fig. 1.16). It marks the divergent boundaries between separating plates in all ocean basins.

At this point you may ask, If the sea floor is spreading at the mid-oceanic ridge, is the Earth's crust growing and is our planet slowly expanding like a marshmallow roasting in a fire? The answer is most certainly no! Again, evidence comes from the sea floor. In several regions, notably on the west coast of South America and in the western Pacific Ocean, the sea floor sinks abruptly, forming deep **trenches**. These long, narrow oceanic trenches occur at convergent boundaries where

two tectonic plates collide. At a collision zone, one lithospheric plate dives under the other and sinks into the mantle. This downward movement of a lithospheric plate is called **subduction**, and the nearby region is called a **subduction zone** (Fig. 1.15). Volcanoes and earthquakes are common in subduction zones. The average depth of the sea floor is about 5 kilometers below sea level. At a trench, a sinking plate pulls the sea floor downward to a depth of 10 kilometers or more.

In other regions, lithospheric plates slide horizontally past each other at a transform boundary. This type of motion is occurring today along the San Andreas fault in California. The plate west of the fault is moving northwestward relative to most of North America. This horizontal movement causes most of the earthquakes for which California is famous.

Figure 1.15 New lithosphere forms at the mid-oceanic ridge and spreads outward. At the same time old lithosphere dives into the asthenosphere at a subduction zone.

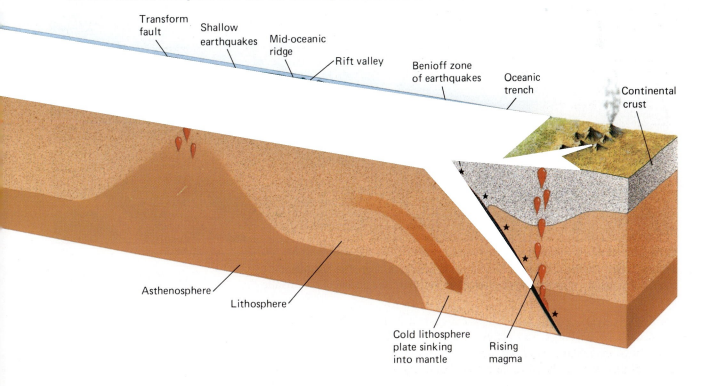

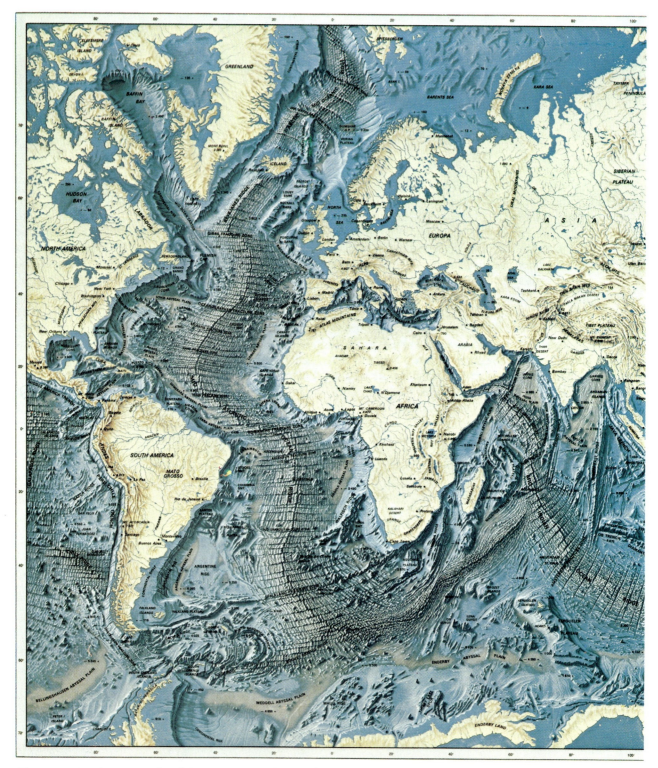

Figure 1.16   The topography of the continents and ocean floor. The Mid-Atlantic
ridge is the sinuous mountain chain snaking its way down the middle of the Atlantic
Ocean.   *(Marie Tharp)*

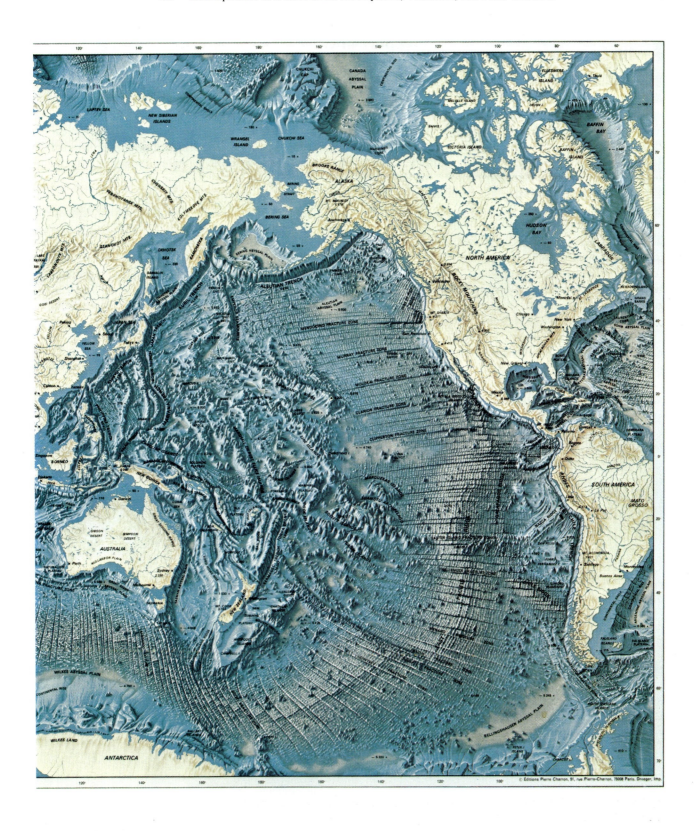

SUMMARY

Earth consists of four major realms: **the solid Earth**, the **hydrosphere**, the **atmosphere**, and the **biosphere**. The study of the solid Earth deals with rocks and minerals, **internal processes** that move continents and cause earthquakes and volcanoes, and **surface processes** that sculpt mountains and valleys. Geologists study both modern Earth processes and the Earth's history from its beginning 4.6 billion years ago to the present.

The hydrosphere consists mostly of oceans, which contain 97.5 percent of the Earth's water. Of the remaining 2.5 percent, most is frozen in the ice sheets of Antarctica and Greenland, and only 0.65 percent is fresh water in lakes, streams, and **ground water**. **Hydrogeologists** study the Earth's fresh-water supplies.

Organisms of the **biosphere**, including humans, affect Earth's surface processes and the compositions of the hydrosphere and atmosphere. Environmental geologists study the relationships among geologic processes and our environment.

Paleontologists interpret the history of life by studying fossils. The **geologic time scale** is a calendar of Earth history, based mainly on predominant life forms at different times in the past.

The **big bang theory** states that originally all matter was compressed into a single point and that this point exploded between 15 and 18 billion years ago to form the Universe and the beginning of time.

The solar system formed from a mass of dust and gases that rotated slowly in space. Within its center, gravitational attraction was so great that the gases were pulled inward with enough velocity to initiate nuclear fusion and create the Sun. In the disk, planets formed from coalescing dust and gases. In the inner planets, most of the lighter elements escaped, whereas they have remained on the outer giants.

The primordial Earth was heated by energy released from radioactive decay and by bombardment from outer space. It has since cooled so that most of it is solid, although the inner layers remain hot. The modern Earth is made up of a dense **core** of iron and nickel, a **mantle** of lower density, and a **crust** of yet lower density. The **lithosphere** contains the uppermost portion of the mantle and the crust. The **asthenosphere** is a hot, plastic layer within the upper mantle that lies just below the lithosphere.

According to the **theory of plate tectonics**, the lithosphere is broken into several plates that float on the asthenosphere and move about relative to one another. Plates separate (**diverge**) along mid-oceanic ridges and collide head on (**converge**) to cause **subduction** of one lithospheric plate beneath another. At **transform** boundaries, plates move horizontally past one another.

KEY TERMS

REVIEW QUESTIONS

1. Give a concise definition of geology.

2. Describe the relative sizes and locations of the four realms of the Earth.

3. What proportion of the Earth's water is in the seas? What proportion is in glaciers?

4. Where is the water that is not part of the oceans or the glaciers?

5. List as many of the Earth's surface processes as you can think of, and briefly describe each.

6. What are internal processes, and what are some of the effects of Earth's internal processes?

7. How old is the Earth?

8. What is ground water? Where in the hydrosphere is it located?

9. In what ways do organisms, including humans, change the Earth? What kinds of Earth processes are unaffected by humans and other organisms?

10. Briefly outline the formation of the Universe.

11. How old was the Universe when our Solar System started to evolve? How long after the start of the evolution of the Solar System did the planets take form?

12. Briefly outline the evolution of the planets.

13. How did the Sun form? How is its composition different from that of the Earth? Explain the reasons for this difference.

14. Compare and contrast the properties of the terrestrial planets with those of the Jovian planets.

15. The entire Earth was molten soon after its formation. Explain why it cooled.

16. Briefly outline the layered structure of the modern Earth.

17. What type of plate motion occurs at the mid-oceanic ridges? What type of activity leads to subduction of plates?

DISCUSSION QUESTIONS

1. What would the Earth be like if it (a) had no atmosphere? (b) had no water?

2. How might Earth be different without life?

3. Only 0.65 percent of Earth's water is fresh and liquid; the rest is salty seawater or is frozen in glaciers. What are the environmental implications of such a small proportion of fresh water?

4. Explain how the theory of the evolution of the Solar System explains the following observations: (a) All the planets in the Solar System are orbiting in the same direction. (b) All the planets in the Solar System except Pluto are orbiting in the same plane. (c) The chemical composition of Mercury is similar to that of the Earth. (d) The Sun is composed mainly of hydrogen and helium but also contains all the elements found on Earth. (e) Venus has a solid surface, whereas Jupiter is mainly a mixture of gases and liquids with a small, solid core.

5. The radioactive elements that are responsible for the heating of the Earth decompose very slowly, over a period of billions of years. How would the Earth be different if these elements decomposed much more rapidly—say, over a period of a few million years? Defend your answer.

6. Explain how the size of a terrestrial planet can affect its surface environment.

CHAPTER

2

Minerals

Gold.
(Ward's Natural Science Establishment, Inc.)

The Earth's continents are composed mostly of granite rocks. If you look closely at a piece of granite like the one in Figure 2.1, you can see many small, differently colored grains. Some grains may be pink, some black, and others white. Each grain is a separate mineral. Some rocks are made of only one mineral, but granite and most other rocks contain three or four abundant minerals plus small amounts of a few others.

2.1 What Is a Mineral?

Minerals are the substances that make up rocks. Although this statement is correct, it does not tell us much about minerals. A more informative definition is that **a mineral is a naturally occurring, inorganic solid with a characteristic chemical composition and a crystalline structure.** Thus, a mineral has five characteristics: (1) It is natural in origin, (2) it is inorganic, (3) it is solid, (4) it has a characteristic chemical composition, and (5) it has a crystalline structure.

The most important properties of a mineral are its chemical composition and its crystalline structure. They distinguish any mineral from all others. Because these two qualities are so important, they are discussed separately in the following two sections. First, however, the natural, inorganic, and solid aspects of minerals are considered briefly.

The qualification that minerals occur naturally means that mineral-like substances made in laboratories or factories are not true minerals. In one sense this distinction is artificial. A synthetic diamond can be identical to a natural one, yet natural gems are valued more highly than synthetic ones. For this reason, jewelers should always tell their customers which gems are natural and which are synthetic.

Organic substances are those produced by living organisms or are similar to ones produced by organisms. They differ chemically from inorganic substances. Although coal is a naturally occurring rock, it is not a mineral because it is derived from organisms. Similarly, oil is not a mineral because it is an organic liquid and has neither a crystalline structure nor a definite chemical composition.

2.2 The Chemical Composition of Minerals

Elements, Atoms, and Ions

In the third century B.C., the Greek philosopher Aristotle defined an **element** by stating that "everything is either an element or composed of elements." Although Aristotle's definition is still correct, a more complete modern definition is that **an element is a fundamental form of matter that cannot be broken into simpler substances by ordinary chemical processes.** Elements are the building blocks of which all other substances are composed. They are the fundamental materials of chemistry. A total of 88 elements occur naturally in the Earth's crust. **Of those 88, only eight elements—oxygen, silicon, aluminum, iron, calcium, magnesium, potassium, and sodium—make up more than 98 percent of the Earth's crust.** (All of the elements are listed in Appendix 1.) Each element is assigned a one- or two-letter symbol. The symbols for the eight most abundant elements are given in Table 2.1.

An **atom** is the basic unit of an element. It consists of a small, dense, positively charged center called a **nucleus** surrounded by a cloud of negatively charged **electrons** (Fig. 2.2). An electron is a fundamental particle; as far as we know, it is not made up of smaller components. The nucleus, however, is made up of two different kinds of particles: positively charged **protons**, and **neutrons**, which have no charge. In a neutral atom

Figure 2.1 Each of the differently colored grains in this granite is a different mineral. The pink grains are feldspar, the white ones are quartz, and the black ones are amphibole.

Table 2.1 • The Eight Most Abundant Chemical Elements in the Earth's Crust

Element	Chemical Symbol	Common Ion(s)
Oxygen	O	O^{2-}
Silicon	Si	Si^{4+}
Aluminum	Al	Al^{3+}
Iron	Fe	Fe^{2+} and Fe^{3+}
Calcium	Ca	Ca^{2+}
Magnesium	Mg	Mg^{2+}
Potassium	K	K^{1+}
Sodium	Na	Na^{1+}

the number of protons in the nucleus equals the number of electrons in the outer cloud. Therefore, the positive and negative charges balance each other so that neutral atoms have no overall electrical charge.

However, in many elements, including all eight of the most abundant ones, the neutral atoms lose or gain electrons easily. When an atom loses one or more electrons, its positive charges then outnumber its negative ones. The atom therefore becomes positively charged. If an atom gains one or more extra electrons, it becomes negatively charged. Atoms with a positive or negative charge are called **ions**.

A positively charged ion is a **cation**. All of the abundant elements except oxygen release electrons to become cations, as shown in Table 2.1. For example, each potassium atom (K) loses one electron to form a cation with a charge of $+1$. Silicon atoms lose four electrons each, forming cations with $+4$ charges. In contrast, oxygen *gains* two extra electrons to acquire a -2 charge. Atoms with negative charges are called **anions**.

Atoms and ions rarely exist as separate, isolated entities. Instead, they unite with other atoms or ions to form **compounds**. Most minerals are compounds.

Figure 2.2 An atom consists of a small, dense, positive nucleus surrounded by a much larger cloud of negative electrons.

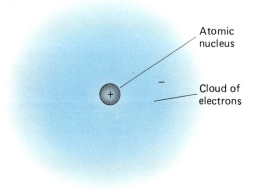

Atomic nucleus

Cloud of electrons

The atoms or ions in a compound are held together by electrical forces called **chemical bonds**.

Chemical Compositions of Minerals

Recall that a mineral has a definite characteristic composition. This means that **a mineral is made up of elements bonded together in specific proportions.** Therefore, its composition can be expressed with a chemical formula.

A few minerals consist of only a single element. For example, gold and silver are single-element minerals. Most minerals, however, are made up of two to six elements. For example, the formula of quartz is SiO_2, meaning that it consists of one atom of silicon (Si) for every two of oxygen (O). Quartz from anywhere in the Universe has that exact composition. If it had a different composition, it would be some other mineral. The compositions of some minerals, such as quartz, do not vary by even a fraction of a percent. The compositions of other minerals vary slightly, but the variations are limited, as explained in Section 2.6.

Because 88 elements occur naturally in the Earth's crust and these elements can combine in many different ways, we might expect to see an overwhelming variety of minerals on a half-day field trip. In fact, more than 2500 different minerals are known! However, only eight elements are abundant, and they commonly combine in only a few ways. Thus, only nine **rock-forming minerals** make up most rocks of the Earth's crust. This small number of common minerals makes the life of a field geologist, who must identify minerals every day, less complicated.

2.3 The Crystalline Nature of Minerals

A crystal is any substance whose atoms are arranged in a regular, orderly, periodically repeated pattern. All minerals are crystals. Halite has the composition NaCl: one sodium ion (Na^+) for every chlorine ion (Cl^-). Figure 2.3 includes a photo of halite crystals and two sketches showing halite's arrangement of sodium and chlorine ions. Figure 2.3A is an "exploded" view that allows you to see into the structure. Figure 2.3B is more realistic, showing the ions in contact. They lie in orderly rows and columns of alternating sodium and chlorine from left to right, top to bottom, and front to back. The rows and columns all intersect at right angles. This orderly arrangement is the **crystalline structure** of halite. All minerals have their atoms in orderly arrangements, although the pattern is not always as obvious as in halite. Any solid with such an orderly, repetitive arrangement of atoms is a **crystal**.

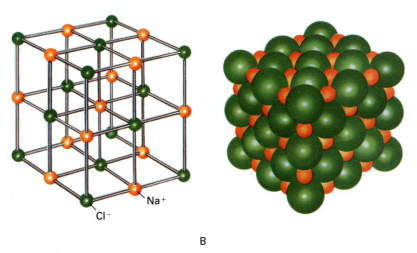

A   B   C

Figure 2.3   (A, B) The orderly arrangement of sodium and chlorine ions in halite.
(C) Halite crystals. The crystal model in (A) is exploded so that you can see into
it; the ions are actually closely packed as in (B). Note that ions in (A) and (B) form a
cube, and the crystals in (C) are also cubes.   *(C, American Museum of Natural History)*

Think of a familiar object with an orderly, repetitive pattern, such as a brick wall. The rectangular bricks repeat themselves over and over throughout the wall. Therefore, the whole wall also has the shape of a rectangle or some modification of a rectangle. In every crystal, a small group of atoms, like a single brick in a wall, repeats itself over and over.

The shape of a large, well-formed crystal such as the halite in Figure 2.3C is determined by the shape of this small group of atoms and the way in which the groups stack. For example, only certain crystal shapes can develop from a cubic group, as in halite. It is obvious from Figure 2.4A that the stacking of small cubes can produce the large cubic crystal of halite. Figure 2.4B

Figure 2.4   A and B show that different kinds of stacking of identical cubes form
different crystal shapes. Both a cubic crystal (A) and an "octahedron" (B) can form by
different kinds of stacking of identical cubes. The stacking of noncubic unit cells results
in a crystal with a different shape.

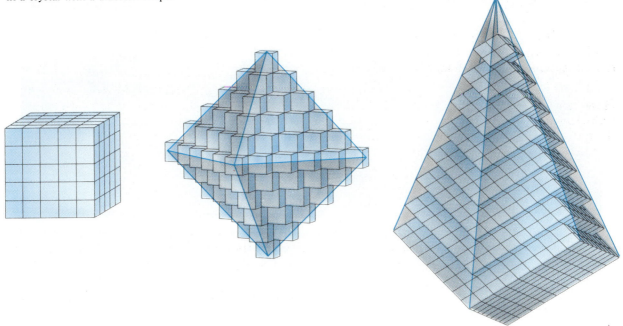

A   B   C

FOCUS ON

Chemical Bonding in Minerals

Three important types of chemical bonds hold atoms together to form minerals: ionic bonds, covalent bonds, and metallic bonds. Many physical properties of a mineral, including color, hardness, density, and the ability to conduct electricity, depend on the bond type.

The opposite electrical charges of cations and anions attract each other to form **ionic bonds**. When cations and anions bond together to form a mineral, they always combine in proportions so that the negative charges exactly equal the positive ones. Thus, minerals are always electrically neutral. For example, consider the mineral halite, which is table salt. It is composed of equal numbers of sodium cations and chlorine anions. Sodium is a soft, silvery metal that is extremely chemically reactive. If you held pure sodium in your hand, it would react with moisture in your palm and burn your skin. If you sprinkled powdered sodium into water, it would explode. Chlorine is a green poisonous gas that was used for chemical warfare during World War I. When sodium reacts with chlorine, each sodium atom loses one electron to form a cation, Na^+. The electron is

captured by a chlorine atom to form an anion, Cl^- (see illustration). When the two react, they form halite. The total charge of halite is $+1 - 1 = 0$. A thumb-sized crystal of halite contains about 10^{20} (1 followed by 20 zeros) sodium and chlorine ions, but the proportion is always 1:1.

A **covalent bond** forms when nearby atoms share their electrons. Diamond consists of a three-dimensional network of carbon atoms. Each carbon atom bonds to four neighbors by sharing electrons to form four covalent bonds. The strength of the bonds makes diamond the hardest of all minerals. In most minerals, the bonds between atoms are partly covalent and partly ionic.

In a **metallic bond**, some of the electrons are loose; that is, they are not associated with a particular atom. Thus, the metal atoms sit in a "sea" of electrons that are free to move from one atom to another. That arrangement allows the atoms to pack together as closely as possible, resulting in the characteristic high density of metals. Because the electrons are free to move, metals are excellent conductors of electricity and heat.

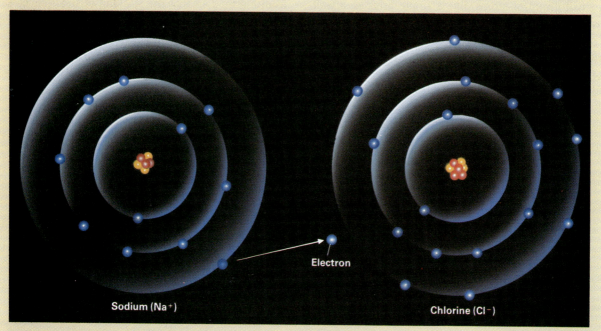

Electron

Sodium (Na⁺)

Chlorine (Cl⁻)

Chemical bonding. When sodium and chlorine atoms combine, sodium loses one electron, becoming a cation, Na^{1+}. Chlorine acquires the electron to become the anion, Cl^{1-}.

shows that a different kind of stacking of the same cubes can also produce an eight-sided crystal. Halite sometimes crystallizes with this shape. All minerals consist of small groups of atoms stacked as in halite.

Crystal faces are flat surfaces that form if a crystal grows without obstructions. The halite in Figure 2.3C has well-developed crystal faces. In nature, crystal growth is often hindered by other minerals. For this reason, minerals rarely show perfect crystal faces.

2.4   Physical Properties of Minerals

How does a geologist identify a mineral in the field? Chemical composition and crystal structure distinguish each mineral from all others. For example, halite always consists of sodium and chlorine in a one-to-one ratio, with the atoms arranged in a cubic fashion. But if you pick up a crystal of halite, you cannot see the ions. You could identify a sample of halite by measuring its chemical composition and crystal structure in the laboratory, but such methods are expensive and time-consuming. Instead, geologists commonly identify minerals by visual recognition and confirm the identification with simple tests of physical properties.

Most minerals have distinctive appearances. Once you become familiar with common minerals, you will recognize them just as you recognize any familiar object or person. For example, an apple just looks like an apple. In the same way, to a geologist quartz looks like quartz. Just as apples come in many colors and shapes, the color and shape of quartz may vary from sample to sample, but it still looks like quartz. Some minerals, however, look enough alike that their physical properties must be examined to make a correct identification.

Crystal Habit

Crystal habit is the characteristic shape of a mineral and the manner in which its crystals grow together. If a crystal grows freely, it develops a characteristic shape controlled by the arrangement of its atoms, as in the cubes of halite shown in Figure 2.3C. Figure 2.5 shows three types of crystal habits found in common minerals.

Some minerals occur in more than one habit. For example, Figure 2.6A shows quartz with a prismatic

A

B

C

Figure 2.5   (A) **Equant** garnet crystals have about the same dimensions in all directions. (B) Asbestos is **fibrous**. (C) Kyanite forms **bladed** crystals.   *(Geoffrey Sutton)*

A

B

Figure 2.6   (A) **Prismatic** quartz grows as elongated crystals. (B) **Massive** quartz shows no characteristic shape.   *(Geoffrey Sutton)*

habit, and Figure 2.6B shows massive quartz. As mentioned previously, growth of a crystal is often obstructed by other crystals. When that kind of interference occurs, the crystal cannot develop its characteristic habit. Figure 2.7 is a photomicrograph (a photo taken through a microscope) of a thin slice of granite. Notice that the crystals fit like pieces of a jigsaw puzzle. This interlocking texture developed because some crystals grew around others as the granite solidified. Because this type of interference is common, perfectly formed crystals are rare.

Cleavage

Cleavage is the tendency of some minerals to break along flat surfaces. The surfaces are planes of weak bonds in the crystal. Micas show excellent cleavage. You can peel sheet after sheet from a mica crystal as if you were peeling layers from an onion (Fig. 2.8).

Some minerals, such as mica and graphite, have one set of parallel cleavage planes. Others have two, three, or even four different sets, as shown in Figure 2.9. Some minerals, like the micas, have excellent cleavage.

Figure 2.7   A photomicrograph of a thin slice of granite. When crystals grow simultaneously, they commonly interlock and show no characteristic habit. To make this photo, a thin slice of granite was cut with a diamond saw, glued to a microscope slide, and ground to a thickness of 0.02 mm. Most minerals are transparent when such thin slices are viewed through a microscope.

Figure 2.8   Cleavage in mica. This large crystal is the variety of mica called muscovite.   *(Geoffrey Sutton)*

A   B   C

Figure 2.9   Some minerals have more than one cleavage plane. (A) Feldspar has two cleavages intersecting at right angles. (B) Calcite has three cleavage planes. (C) Fluorite has four cleavage planes.   *(Geoffrey Sutton)*

Others have poor cleavage. Many minerals have no cleavage at all because they have no planes of weak bonds to favor breakage. The number of cleavage planes, the quality of cleavage, and the angles between cleavage planes all help in mineral identification.

It is important to distinguish between a flat surface created by cleavage and a crystal face. They can appear identical because both are flat, smooth surfaces. The difference is that a cleavage surface is duplicated when a crystal is broken, whereas a crystal face is not duplicated by breakage. If you are in doubt, break the sample with a hammer (unless, of course, you want to save it).

Fracture

Fracture is the way in which a mineral breaks other than along planes of cleavage. Many minerals form characteristic shapes where they break. **Conchoi-**

Figure 2.10   Quartz shows smooth, concave, conchoidal fracture.   *(Geoffrey Sutton)*

dal fracture is breakage into smooth, curved surfaces, as shown in Figure 2.10. It is characteristic of quartz and glass. Some minerals break into **splintery** or **fibrous** fragments. Most fracture into **irregular** shapes.

Hardness

Hardness is the resistance of a mineral to scratching. Because it is controlled by the strength of bonds in the mineral, it is a fundamental property of a mineral. It is measured easily and used commonly by geologists to identify minerals. Hardness is gauged by attempting to scratch a mineral with a knife or other object of known hardness. If the blade scratches the mineral, the mineral is softer than the knife. If the knife cannot scratch the mineral, the mineral is harder.

To measure hardness more accurately, geologists use a scale based on ten fairly common minerals, numbered one through ten. Each mineral is harder than those with lower numbers on the scale, so 10 (diamond) is the hardest and 1 (talc) is the softest. The scale is known as the **Mohs hardness scale**, after F. Mohs, the Austrian mineralogist who developed it in the early nineteenth century.

Table 2.2 shows that a mineral scratched by quartz but not by orthoclase has a hardness between 6 and 7. Because the minerals of the Mohs scale are not always handy, it is useful to know the hardness values of common materials. A fingernail has a hardness of slightly more than 2, a copper penny about 3, a pocketknife blade slightly more than 5, window glass about 5.5, and a steel file about 6.5. If you practice with a knife and the minerals of the Mohs scale, you can develop a "feel" for hardnesses of minerals 5 and under by how easily the blade scratches them.

GEOLOGY AND THE ENVIRONMENT

Asbestos and Cancer

Asbestos is an industrial name for a group of minerals that crystallize as long, thin fibers. The two most common types are fibrous habits of the minerals **chrysotile** and **amphibole**. The fibers of chrysotile form tangled, curly bundles, whereas amphibole asbestos occurs as straight, sharply pointed needles.

Asbestos fibers are commercially valuable because they are flameproof, chemically inert, and extremely strong. For example, a chrysotile fiber is eight times stronger than a steel wire of equivalent diameter. Asbestos fibers have been woven into brake linings, protective clothing, insulation, shingles, tile, pipe, and gaskets but now are allowed only in brake pads, shingles, and pipe.

In the early 1900s, asbestos miners and others who worked with asbestos learned that prolonged exposure to the fibers caused **asbestosis**, an often lethal lung disease. Later, in the 1950s and 1960s, it became clear that asbestos also causes lung cancer and other forms of cancer. One reason that so much time passed before the cancer-causing properties of asbestos were recognized is that cancer commonly does not develop until decades after the first exposure to asbestos.

Experiments have shown that lung diseases are caused by the fibrous nature of asbestos, not by its chemical composition. For example, forms of amphibole of identical composition can occur with both a fibrous and a nonfibrous habit. In a laboratory study, a group of rodents was exposed to fibrous amphibole and another group to identical amounts of nonfibrous amphibole. The group exposed to the fibrous type developed cancers, but the other group did not.

Another experiment with rodents showed that amphibole asbestos is a more effective cause of lung cancer than is chrysotile. Apparently the curly chrysotile fibers are more easily expelled from the lungs, whereas the sharp amphibole needles remain in the lung. Addition-ally, the incidence of cancer among chrysotile workers is proportionally lower than among those working with amphibole asbestos. Although it is not clear how asbestos causes cancer, it is clear that the fibrous habit is important and that the sharp needles of the less common amphibole asbestos are more dangerous than chrysotile.

In response to growing awareness of its health effects, the Environmental Protection Agency (EPA) banned the use of asbestos in construction in 1978. However, the ban did not address the issue of what should be done with the asbestos already installed. In 1986 Congress passed a ruling called the Asbestos Hazard Emergency Response Act, requiring that all schools be inspected for asbestos. Public response has resulted in hasty programs to remove asbestos from schools and other buildings. The EPA estimates that removal of asbestos from schools and public and commercial buildings will cost between $50 and $150 billion. But what is the real level of hazard?

Most asbestos is the less dangerous chrysotile. More important, most asbestos in buildings is already woven tightly into cloth, and often the surface has been further stabilized by painting. Therefore, the fibers are not free to blow around. The levels of airborne asbestos in most buildings are no higher than that in outdoor air. Many scientists argue that asbestos insulation poses no health danger if left alone, but when the material is removed it is disturbed and asbestos dust escapes. Not only are workers endangered, but airborne asbestos persists in the building for months after completion of the project.

Thus, when assessing the health effects of asbestos, we must understand how it is transported and incorporated into living tissue. Asbestos is unquestionably deadly in a mine where rock is drilled and blasted and dust hangs heavy in the air. However, in a school or commercial building it may be harmless until, in the interest of public safety, workers release fibers as they disturb the insulation during removal.

Specific Gravity

Specific gravity is the weight of a substance relative to that of an equal volume of water. If a mineral weighs 2.5 times as much as an equal volume of water, its specific gravity is 2.5. You can estimate a mineral's specific gravity simply by hefting a sample in your hand. If you practice with known minerals, you can develop a feel for specific gravity. Most common minerals have specific gravities of about 2.7. Metals have much greater specific gravities; for example, gold has the highest of all minerals, 19. Silver is 10.5 and copper 8.9.

Color

Color is the most obvious property of a mineral but is commonly unreliable for identification. If all minerals were pure and had perfect crystal struc-

Table 2.2 • Mohs Hardness Scale

Minerals of Mohs Scale	Common Objects
1. Talc	
2. Gypsum	Fingernail
3. Calcite	Copper penny
4. Fluorite	
5. Apatite	
	Knife blade
	Window glass
6. Orthoclase	
	Steel file
7. Quartz	
8. Topaz	
9. Corundum	
10. Diamond	

tures, then color would be reliable. However, both small amounts of chemical impurities and imperfections in crystal structure can alter color dramatically. For example, the mineral corundum is composed of aluminum oxide, Al_2O_3. It is normally a cloudy, translucent, brownish to bluish mineral. Addition of a small amount of chromium produces a beautiful, clear, red gem known as ruby. A small quantity of iron or titanium turns corundum into the striking blue gem called sapphire.

Streak

Streak is the color of a fine powder of a mineral. Streak is more reliable than the color of the mineral itself for identification. Streak is measured by rubbing a mineral across a piece of unglazed porcelain known as a streak plate. If the mineral is softer than the porcelain, which has a hardness of about 7, the mineral leaves a streak of powder on the plate.

Luster

Luster is the manner in which a mineral reflects light. A mineral with a metallic look, irrespective of color, has a **metallic luster**. The luster of nonmetallic minerals is usually described by self-explanatory words such as **glassy**, **pearly**, **earthy**, and **resinous**.

2.5   Mineral Classification

Geologists classify minerals according to their anions (negatively charged ions). A simple anion is a single negatively charged ion such as O^{2-}. Alternatively, two or more atoms can bond together firmly and acquire a negative charge to form a complex anion. Two common examples are silicate, $(SiO_4)^{4-}$, and carbonate, $(CO_3)^{2-}$.

Each mineral group is named after the anion in the minerals of the group. For example, the oxides all contain O^{2-}, the silicates contain $(SiO_4)^{4-}$, and the carbonates contain $(CO_3)^{2-}$. Common and useful mineral groups and important minerals in each group are listed in Table 2.3.

2.6   The Rock-Forming Minerals

The nine rock-forming minerals are the most abundant minerals in rocks. Because they are so common, they are the minerals you are most likely to find and identify. Notice that seven are silicates.

Silicates

All minerals containing silicon and oxygen are called **silicates**. The silicate group makes up more than 95 percent of the Earth's crust. Silicate minerals are so abundant for two reasons. First, they are made up principally of the two most plentiful elements in the crust, silicon and oxygen. Second, silicon and oxygen bond together readily. The seven most abundant silicate minerals are feldspar, quartz, pyroxene, amphibole, mica, the clay minerals, and olivine. Except for quartz, each of these minerals is actually a group whose members have very similar chemical compositions and crystal structures.

To understand the silicate minerals, remember three principles:

Opal is one of the many varieties of silica and has a composition similar to that of quartz. *(Ward's Natural Science Establishment, Inc.)*

Table 2.3 • Important Mineral Groups

Group	Member	Formula	Economic Use
Oxides	Hematite	Fe_2O_3	Ore of iron
	Magnetite	Fe_3O_4	Ore of iron
	Corundum	Al_2O_3	Gemstone, abrasive
	Ice	H_2O	Solid form of water
	Chromite	$FeCr_2O_4$	Ore of chromium
Sulfides	Galena	PbS	Ore of lead
	Sphalerite	ZnS	Ore of zinc
	Pyrite	FeS_2	Fool's gold
	Chalcopyrite	$CuFeS_2$	Ore of copper
	Bornite	Cu_5FeS_4	Ore of copper
	Cinnabar	HgS	Ore of mercury
Sulfates	Gypsum	$CaSO_4 \cdot 2H_2O$*	Plaster
	Anhydrite	$CaSO_4$	Plaster
	Barite	$BaSO_4$	Drilling mud
Native elements	Gold	Au	Electronics, jewelry
	Copper	Cu	Electronics
	Diamond	C	Gemstone, abrasive
	Sulfur	S	Sulfa drugs, chemicals
	Graphite	C	Pencil lead, dry lubricant
	Silver	Ag	Jewelry, photography
	Platinum	Pt	Catalyst
Halides	Halite	$NaCl$	Common salt
	Fluorite	CaF_2	Used in steel making
	Sylvite	KCl	Fertilizer
Carbonates	Calcite	$CaCO_3$	Portland cement
	Dolomite	$CaMg(CO_3)_2$	Portland cement
	Aragonite	$CaCO_3$	Portland cement
Hydroxides	Limonite	$FeO(OH) \cdot nH_2O$**	Ore of iron, pigments
	Bauxite	$Al(OH)_3 \cdot nH_2O$	Ore of aluminum
Phosphates	Apatite	$Ca_5(F,Cl,OH)(PO_4)_3$	Fertilizer
	Turquoise	$CuAl_6(PO_4)_4(OH)_8 \cdot 4H_2O$	Gemstone
Silicates	(See Fig. 2.12 for silicate minerals.)		

* The $2H_2O$ in the chemical formula of gypsum means that water is incorporated into the mineral structure.
** nH_2O means that varying amounts of water are incorporated into the structure.

1. Every silicon atom surrounds itself with four oxygens. The bonds between the silicon and its four oxygens are very strong.

2. The silicon atom and its four oxygens form a pyramid called the **silicate tetrahedron**, with silicon in the center and oxygens at the four corners (Fig. 2.11). **The silicate tetrahedron is the fundamental building block of all silicate minerals.** The silicate tetrahedron has a negative charge, forming the $(SiO_4)^{4-}$ complex anion.

3. Silicate tetrahedra link together by sharing oxygens.

Thus, two tetrahedra share a single oxygen, bonding the two tetrahedra together.

Silicate minerals fall into five classes based on five different ways in which tetrahedra share oxygens. Each class contains at least one of the rock-forming minerals.

1. In **independent tetrahedra** silicates (Fig. 2.12A), adjacent tetrahedra do not share oxygens.

2. In the **single-chain** silicates (Fig. 2.12B), each tetrahedron links to two others by sharing oxygens. This forms a continuous chain of tetrahedra.

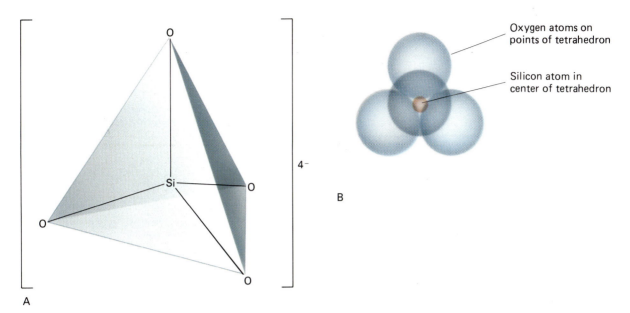

Figure 2.11 The silica tetrahedron consists of one silicon atom surrounded by four oxygens. It is the fundamental building block of all silicate minerals. (A) A ball-and-stick representation. (B) A proportionally accurate model.

3. The **double-chain** silicates consist of two single chains cross-linked by the sharing of more oxygens (Fig. 2.12C). Minerals using both the single- and double-chain structures grow crystals that are elongate parallel to the chains.

4. In the **sheet** silicates, each tetrahedron links to three others in the same plane, forming a continuous sheet of tetrahedra (Fig. 2.12D). Mica is a sheet silicate. All of the atoms in each sheet are strongly bonded, but each sheet is only weakly bonded to those above and below. Therefore, it is easy to peel sheet after sheet from a mica crystal.

5. In the **framework** silicates, each tetrahedron shares all four of its oxygens with adjacent tetrahedra (Fig. 2.12E). Because tetrahedra share oxygens in all directions, minerals using the framework structure tend to grow blocky crystals that have the same dimensions in all directions.

Each silicate tetrahedron is negatively charged. However, all minerals are electrically neutral. Therefore, cations must enter the structures of most silicate minerals to balance the negative charges. The lone exception is quartz, SiO_2, in which the positive charges on the silicons exactly balance the negative ones on the oxygens.

Rock-Forming Silicate Minerals

Feldspar (Fig. 2.13A) makes up more than 50 percent of the Earth's crust and is the most abundant mineral.

It is a major component of nearly all common rocks. Feldspar is a group of minerals with similar crystal structures and compositions. Individual minerals within the group are named according to whether they contain potassium, sodium, or calcium. **Orthoclase** is the most common type of potassium feldspar. Feldspar containing calcium and sodium is called **plagioclase**. Plagioclase and orthoclase often look alike and can be difficult to tell apart.

Quartz (Fig. 2.13B) is pure SiO_2. It is the only silicate mineral that contains no cations other than silicon. It is widespread and abundant in continental rocks but rare in oceanic crust and the mantle.

Pyroxene (Fig. 2.13C), like feldspar, is a group of similar minerals. It is a major component of oceanic crust and the mantle and is abundant in some rocks of the continents. **Amphibole** (Fig. 2.13D) is also a group of minerals with similar properties. It is common in many rocks of the continents. Pyroxene and amphibole can resemble each other so closely that they are difficult to tell apart.

Mica (Fig. 2.13E) has a platy habit and perfect cleavage. Both result from the sheet linkages of silicate tetrahedra. Mica is common in continental rocks. The **clay minerals** (Fig. 2.13F) are similar to mica in structure, composition, and platy habit. Individual clay crystals are so small that they can barely be seen with a good optical microscope. Most clay forms when other minerals weather at the Earth's surface. Thus, clay is

Text continued on page 38

Class	Arrangement of SiO₄ tetrahedron	Unit composition	Mineral Examples
A Independent tetrahedra		$(SiO_4)^{4-}$	Olivine: The composition varies between Mg_2SiO_4 and Fe_2SiO_4
B Single chains		$(SiO_3)^{2-}$	Pyroxene: The most common pyroxene is augite, $Ca(Mg, Fe, Al)(Al, Si)_2O_6$
C Double chains		$(Si_4O_{11})^{6-}$	Amphibole: The most common amphibole is hornblende, $NaCa_2(Mg, Fe, Al)_5(Si, Al)_8O_{22}(OH)_2$.
D Sheet silicates		$(Si_2O_5)^{2-}$	Mica, clay minerals, chlorite eg: muscovite $KAl_2(Si_3Al)O_{10}(OH)_2$
E Framework silicates		SiO_2	Quartz: SiO_2 Feldspar: As an example, potassium feldspar is $KAlSi_3O_8$

A   B   C

D   E   F

Figure 2.13 The seven rock-forming silicate minerals.
(A) Feldspar, represented here by orthoclase feldspar.
(B) Quartz. (C) Pyroxene. (D) Amphibole. (E) Black
biotite is one common type of mica. White muscovite
(Fig. 2.8) is the other. (F) Clay. (G) Olivine. *(Geoffrey Sutton)*

G

Figure 2.12 The five silicate structures are based on sharing of oxygens among silica
tetrahedra. (A) Independent tetrahedra share no oxygens. (B) In single chains, each
tetrahedron shares two oxygens with adjacent tetrahedra, forming a chain. (C) A double
chain is a pair of single chains linked together by additional oxygen sharing. (D) In the
sheet silicates, each tetrahedron shares three oxygens with adjacent tetrahedra. (E) A
three-dimensional framework silicate forms when all four oxygens of each tetrahedron
are shared.

A

B

Figure 2.14 Calcite (A) and dolomite (B) are the only two rock-forming minerals that are not silicates. *(Ward's Natural Science Establishment, Inc.)*

abundant at and near the Earth's surface and is an important component of soil.

Olivine (Fig. 2.13G) occurs in small quantities in both continental and oceanic rocks. However, olivine and pyroxene make up most of the mantle.

Rock-Forming Nonsilicate Minerals

Two minerals that are not silicates—**calcite**, $CaCO_3$, and **dolomite**, $CaMg(CO_3)_2$—are abundant enough to qualify as rock-forming minerals (Figs. 2.14A and B). Both are carbonates, and both are common in near-surface rocks of the continents. Calcite and dolomite make up the rocks called "carbonate rocks," or sometimes simply "limestones."

Most carbonate rocks start out as shell fragments and other hard parts of marine organisms. When you see limestone cliffs in the Canadian Rockies, in New York State, or anywhere else, you know that the region once lay beneath the sea.

2.7   Other Important Minerals

We study the rock-forming minerals because they make up most of the Earth's crust. A small number of other minerals are important for economic reasons or because they are commonly found in small quantities. The most notable of these less abundant minerals are listed and described in Appendix 2.

Ore minerals are minerals from which metals or other elements can be recovered profitably. Thus, they are minerals that contain commercially valuable elements or compounds. Native gold and native silver are ore minerals composed of pure metals. Most other metals exist in nature as compounds. The industrially important metals copper, lead, and zinc are obtained from chalcopyrite, galena, and sphalerite, respectively. Halite is mined for table salt, and gypsum is mined for the manufacture of plaster and sheetrock.

A **gem** is a mineral that is prized for its beauty rather than for industrial use. Depending on its value, a gem can be either precious or semiprecious. Precious gems include diamond, emerald, ruby, and sapphire. Several varieties of quartz, including amethyst, agate, jasper, and tiger's eye, are semiprecious gems. Garnet, olivine, topaz, turquoise, and many other minerals sometimes occur as aesthetically pleasing semiprecious gems.

Accessory minerals are minerals that are seen often, but usually only in small amounts. Although common, they are not abundant enough to classify as rock-forming minerals. Chlorite, garnet, limonite, magnetite, and pyrite are among the most common accessory minerals.

Galena is a common ore of lead. Here it occurs as nearly perfect cubic crystals.   *(Ward's Natural Science Establishment, Inc.)*

SUMMARY

Minerals are the substances that make up rocks. A mineral is a naturally occurring inorganic solid with a definite chemical composition and a crystalline structure. Each mineral consists of specific chemical elements bonded together in certain proportions, so that its chemical composition can be given as a chemical formula. The **crystalline structure** of a mineral is the orderly, periodically repeated arrangement of its atoms. The shape of a crystal is determined by the arrangement of its atoms. Every mineral is distinguished from others by its chemical composition and crystal structure.

Most common minerals are easily recognized and identified visually, and identification is aided by observing a few physical properties, including **crystal habit**, **cleavage**, **fracture**, **hardness**, **specific gravity**, **color**, **streak**, and **luster**.

Although more than 2500 minerals are known in the Earth's crust, only the nine **rock-forming minerals** are abundant in most rocks. They are **feldspar**, **quartz**, **pyroxene**, **amphibole**, **mica**, **the clay minerals**, **olivine**, **calcite**, and **dolomite**. The first seven on this list are **silicates**; their structures and compositions are based on the **silicate tetrahedron**, in which a silicon atom is surrounded by four oxygens. Silicate tetrahedra link together by sharing oxygens to form the basic structures of the silicate minerals. The silicates are the most abundant minerals because silicon and oxygen are the two most abundant elements in the Earth's crust and bond together readily to form the silicate tetrahedron.

Ore minerals and **gems** are important for economic reasons. **Accessory minerals** are found commonly, but in small amounts.

KEY TERMS

Mineral 25	Compound 26	Cleavage 30	Independent tetrahedra 34
Element 25	Chemical bond 26	Fracture 31	Single-chain silicates 34
Atom 25	Rock-forming mineral 26	Hardness 31	Double-chain silicates 35
Nucleus 25	Crystalline structure 26	Mohs hardness scale 31	Sheet silicates 35
Electron 25	Crystal 26	Specific gravity 32	Framework silicates 35
Proton 25	Ionic bond 28	Color 32	Ore mineral 38
Neutron 25	Covalent bond 28	Streak 33	Gem 38
Ion 26	Metallic bond 28	Luster 33	Accessory mineral 38
Cation 26	Crystal faces 29	Silicate 33	
Anion 26	Habit 29	Silicate tetrahedron 34	

ROCK-FORMING MINERALS

Feldspar	Amphibole	Clay minerals	Calcite
Quartz	Mica	Olivine	Dolomite
Pyroxene			

REVIEW QUESTIONS

1. What properties distinguish minerals from other substances?

2. Explain why oil and coal are not minerals.

3. What does the chemical formula for quartz, SiO_2, tell you about its chemical composition? What does $KAlSi_3O_8$ tell you about orthoclase feldspar?

4. What is an atom? An ion? A cation? An anion? What roles do they play in minerals?

5. What is a chemical bond? What role do chemical bonds play in minerals?

6. Every mineral has a crystalline structure. What does this mean?

7. What factors control the shape of a well-formed crystal?

8. What is a crystal face?

9. What conditions allow minerals to grow well-formed crystals? What conditions prevent their growth?

10. List and explain the physical properties of minerals most useful for identification.

11. Why do some minerals have cleavage and others do not? Why do some minerals have more than one set of cleavage planes?

12. Why is color often an unreliable property for mineral identification?

13. List the rock-forming minerals. Why are they called "rock forming"? Which are silicates? Why are so many of them silicates?

14. Draw a three-dimensional view of a single silicate tetrahedron. Draw the five different arrangements of tetrahedra found in the rock-forming silicate minerals. How many oxygen ions are shared between adjacent tetrahedra in each of the five configurations?

15. Make a table with two columns. In the left column list the basic silicate structures. In the right column list one or more rock-forming minerals with each structure.

DISCUSSION QUESTIONS

1. Diamond and graphite are two minerals with identical chemical compositions, pure carbon (C). Diamond is the hardest of all minerals, and graphite is one of the softest. If their compositions are identical, why do they have such profound differences in physical properties?

2. List the eight most abundant chemical elements in the Earth's crust. Are any unfamiliar to you? List familiar elements that are not among the eight. Why are they familiar?

3. Table 2.1 shows that silicon and oxygen together make up nearly 75 percent by weight of the Earth's crust. But silicate minerals make up more than 95 percent of the crust. Explain the apparent discrepancy.

4. Quartz is SiO_2. Why does no mineral exist with the composition SiO_3?

5. If you were given a crystal of diamond and another of quartz, how would you tell which is diamond?

6. Would you expect minerals found on the Moon, Mars, or Venus to be different from those of the Earth's crust? Explain your answer.

CHAPTER

3

Rocks

Mount Asgaard on Baffin Island is composed of granite,
one of the most abundant rocks in continental crust.

The Earth is almost entirely rock. A few meters of soil conceal bedrock in most places on land, but beneath the soil the Earth is hard, solid rock. If you were to dive to the sea floor and dig through a layer of mud, again you would find solid rock. If you could tunnel 2900 kilometers to the boundary between the Earth's mantle and core, you would have to excavate hard, solid rock all the way. The outer core is molten, but the inner core is solid.

Even casual observation reveals that rocks are not all alike. The peaks of the Sierra Nevada in California are hard granite. The red cliffs of the Utah desert are sandstone so soft that if you scrape the rock with a knife, tiny sand grains pile up at your feet. If you were to climb to the top of Mount Everest, you would find rock called limestone, made up of clamshells and the remains of other small marine animals.

In this chapter we will study rocks: how they form and what they are made of. In later chapters we will use our understanding of rocks to interpret the geological history of the Earth. As an example of how rocks tell us about Earth history, consider the limestone of Mount Everest. Because it contains marine fossils, it must have formed in the sea. Some force pushed it up to its present high elevation after it formed. But such deductions lead to other questions. What forces lift rock to create mountains? Where did the vast amounts of sand in the Utah sandstone come from? Why are the sand grains so easily released from the rock? How did the granite of the Sierra Nevada form, and why is it so hard and strong? Although these questions ask about the nature of rocks, the answers involve the processes that formed the rocks and the geological history of each region.

3.1 Types of Rocks and the Rock Cycle

Geologists separate rocks into three groups according to how they form: igneous rocks, sedimentary rocks, and metamorphic rocks. Under certain conditions, rocks of the upper mantle and lower crust melt, forming a hot liquid called **magma. Igneous rock forms when this magma cools and solidifies.** Igneous rock is the most abundant kind of rock. It makes up much more than half of the Earth's crust. **Granite** and **basalt** are the two most common igneous rocks.

Rocks of all kinds decompose, or **weather**, at the Earth's surface. Weathering breaks large rocks into smaller fragments such as gravel, sand, and clay. Streams, wind, glaciers, and gravity carry this weathered material, called **sediment**, downhill and deposit it at lower elevations. The sand on a beach and mud on a mud flat form in this way. **With time, sand, mud, and other kinds of sediment become cemented together to form sedimentary rock.** When beach sand is cemented, it turns to sandstone; mud becomes shale. Sedimentary rock makes up less than 5 percent of the Earth's crust. However, because sediment accumulates on the Earth's surface, sedimentary rocks form a thin veneer covering about 80 percent of the continents. Therefore, it is easy to get the impression that sedimentary rocks are more abundant than they really are. The most common sedimentary rocks are **shale**, **sandstone**, and **limestone**.

Tectonic activity can force a portion of the Earth's crust downward. As the crust sinks, rocks that once were at the surface become buried by a thick pile of sediment accumulating in the depression. When sediment or rock is buried in this way, both temperature and pressure increase. The higher temperature and pressure change both the minerals and the texture of the rock. These changes are called **metamorphism**, and the rock is termed a **metamorphic rock**. Metamorphic changes can also occur when a rock is crushed or otherwise deformed during mountain building. **Schist**, **gneiss**, and **marble** are common metamorphic rocks.

Igneous, sedimentary, and metamorphic rocks seem to be permanent features of the Earth over a human life span and even over the range of human history. Archeologists have used biblical descriptions of rocky peaks to locate ancient ruins. But historical records go back only a few thousand years, whereas geologic time extends back 4.6 billion years. Over this much greater length of time, rocks change. In geologic time it is common for Earth processes to convert an igneous rock to a sedimentary rock or a sedimentary rock to a metamorphic rock. In turn, when a metamorphic rock becomes hot enough, it melts to form magma; the magma then cools to form a new igneous rock. Thus, no rock is permanent over geologic time; instead, all

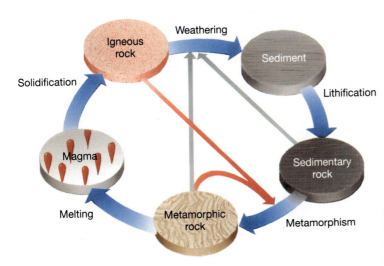

Figure 3.1   The rock cycle shows that rocks of the crust change continuously over geologic time. The arrows show paths that rocks can follow as they change.

rocks change slowly from one of the three rock types to another. This continuous transformation is called the **rock cycle** (Fig. 3.1).

Although the term *rock cycle* implies an orderly progression from one type of rock to another, such a regular sequence does not necessarily occur. Shortcuts are common, as shown by the arrows cutting across the circle of Figure 3.1. For example, a sedimentary or metamorphic rock may weather to form sediment. An igneous rock may be metamorphosed. The rock cycle simply expresses the concept that rocks are not permanent, but change continuously over geologic time.

IGNEOUS ROCKS

3.2   Magma: The Source of Igneous Rocks

If you drilled a well deep into the crust, you would find that the Earth's temperature rises about 30°C for every kilometer of depth. Below the crust, in the upper mantle, temperature continues to rise, but not as rapidly. In the upper mantle, between depths of 100 and 350 kilometers, the temperature is so high that in certain places large amounts of rock melt to form magma. Recall from Chapter 1 that this layer between 100 and 350 kilometers in depth is called the asthenosphere. It is weak, soft, and plastic because it is so hot.

The temperature of magma varies from about 600° to 1400°C, depending on its chemical composition and the depth at which it forms. In comparison, an iron bar turns red hot at about 600°C and melts at slightly over 1500°C. Blacksmiths easily heat iron to redness on the glowing embers of a coal forge.

When rock melts to form magma, it expands by about 10 percent. Therefore, magma is of lower density than the solid rock around it. Because of its lower density, magma starts to rise as soon as it forms, just as a hot balloon rises in the atmosphere. As the magma rises, it enters the cooler, lower-pressure environment near the Earth's surface. When temperature and pressure drop sufficiently, the liquid solidifies to form solid igneous rock.

Because the Earth melted shortly after it formed, the crust initially consisted entirely of igneous rock. Later geological activity has modified the original igneous crust to form sedimentary, metamorphic, and younger igneous rocks. However, about 95 percent of the Earth's crust is still igneous rock or metamorphosed igneous rock. Even though today much of this igneous

Gneiss is a common metamorphic rock. Baffin Island, Northwest Territories, Canada.

foundation is buried by a relatively thin layer of sedimentary rock, igneous rocks are easy to find because they make up some of the world's most spectacular mountains (Fig. 3.2).

3.3 Classifications of Igneous Rocks

Igneous rocks are divided into two groups on the basis of how they form. **Intrusive igneous rocks** form when magma solidifies *within* the Earth, before it can rise all the way to the surface. Intrusive rocks are sometimes called **plutonic rocks** after Pluto, the ancient Roman god of the underworld. An intrusive rock mass is called a **pluton**.

A stream of molten lava and a fire fountain pour from Pu'u 'O'o vent, Hawaii, 1986. The lava solidifies to form an extrusive igneous rock. *(J. D. Griggs, USGS)*

Figure 3.2 Granite cliffs rise above Lost Twin Lake, Big Horn Mountains, Wyoming. Granite is an intrusive igneous rock.

Extrusive igneous rocks form when magma erupts and solidifies on the Earth's surface. Because extrusive rocks are so commonly associated with volcanoes, they are also called **volcanic rocks**. Vulcan was the Roman god of fire.

Textures of Igneous Rocks

The **texture** of a rock refers to the size, shape, and arrangement of its mineral grains, or crystals. Some igneous rocks consist of mineral grains that are too small to be seen with the naked eye; others are made up of thumb-size or even larger crystals. The common names used for igneous rock textures are summarized in Table 3.1.

Table 3.1 • Igneous Rock Textures Based on Grain Size

Grain Size	Name of Texture
No mineral grains (obsidian)	Glassy
Too fine to see with naked eye	Very fine grained
Up to 1 millimeter	Fine grained
1–5 millimeters	Medium grained
More than 5 millimeters	Coarse grained
Relatively large grains in a finer-grained matrix	Porphyritic

Figure 3.3   Obsidian is natural volcanic glass. It contains no crystals. *(Geoffrey Sutton)*

Figure 3.4   Basalt is a dark, fine-grained volcanic rock. Southeastern Idaho.

One of the most striking differences between volcanic and plutonic rocks is the contrast in their textures. Volcanic rocks are usually fine grained, whereas plutonic rocks are medium or coarse grained. The difference exists because crystals grow slowly as magma solidifies. Volcanic magma cools rapidly on the Earth's surface and solidifies before crystals have time to grow to a large size. In contrast, plutonic magma cools slowly within the crust, and crystals have a long time to grow to larger sizes.

Extrusive (Volcanic) Rocks

Some volcanic magma may solidify within a few hours of erupting, forming volcanic glass called **obsidian** (Fig. 3.3). Glass has no crystalline structure; the atoms or ions have no orderly arrangement because the magma solidified before they could align themselves to form crystals. If magma solidifies somewhat more slowly, over a period of days to a few years, crystals begin to form, but they do not have time to grow to large sizes. The result is a very fine-grained rock, one in which the crystals are too fine to be seen with the naked eye. Basalt is the most abundant example of a very fine-grained volcanic rock (Fig. 3.4).

Intrusive (Plutonic) Rocks

Plutonic igneous rocks form when magma solidifies deep within the crust. Overlying rock insulates the magma like a thick blanket, keeping the magma hot so that it solidifies slowly, over hundreds of thousands or even millions of years. As a result, crystals have a long time to grow, and they form large grains. Therefore, most plutonic rocks are medium to coarse grained. Granite, the most abundant rock in the continental crust, is a medium- or coarse-grained, plutonic igneous rock. When you look at granite, you see individual grains of different colors. Each grain is a separate mineral.

If magma rises slowly through the crust, some mineral grains may solidify and grow while most of the magma remains molten. If this mixture of magma and crystals suddenly erupts onto the surface, the magma cools quickly to form a **porphyry**, a fine-grained rock with large crystals embedded within it (Fig. 3.5).

Classification Based on Minerals and Texture

Geologists use both minerals and texture to name igneous rocks. For example, **granite** is mostly feldspar and quartz, and it is medium or coarse grained. Any igneous rock with these minerals and texture is granite. **Rhyolite** contains the same minerals but is very fine grained

Figure 3.5   Porphyry is igneous rock containing large crystals embedded in a fine-grained matrix. This is a rhyolite porphyry with large, pink feldspar crystals. *(Geoffrey Sutton)*

A   B

Figure 3.6   Although granite (A) and rhyolite (B) contain the same minerals, they
have very different textures because granite cools slowly and rhyolite cools
rapidly.   *(Geoffrey Sutton)*

(Fig. 3.6). The same magma that erupts onto the Earth's
surface to form rhyolite also forms granite if it solidifies
slowly within the crust.

Thus, igneous rocks are classified in pairs. The
members of each pair contain the same minerals but
have different textures. The texture depends mainly on
whether the rock is volcanic or plutonic in origin. Figure
3.7 shows the mineralogy and textures of common igneous rocks.

The same rocks and their chemical compositions
are shown in Figure 3.8. Granite and rhyolite contain
large amounts of *si*licon and *al*uminum, and so are

Figure 3.7   The minerals and textures of common igneous rocks. A mineral's
abundance in a rock is proportional to the thickness of its colored band beneath the
rock name. Use the numbers on the left side of the figure to estimate the relative
abundance of each mineral in the rock listed at the top of the figure.

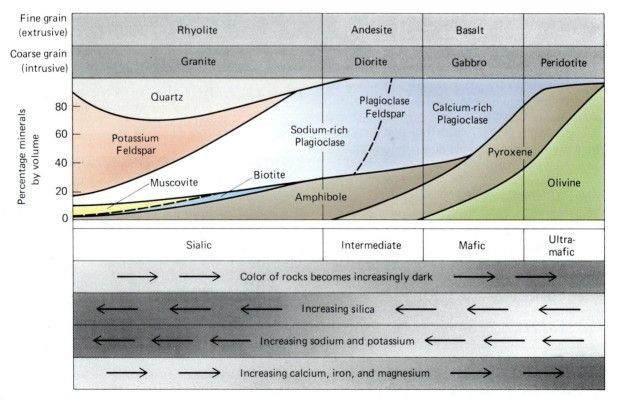

Common Igneous Rocks

Descriptive Terms	Sialic (granitic)	Intermediate (andesitic)	Mafic (basaltic)	Ultramafic
Intrusive	Granite	Diorite	Gabbro	Peridotite
Extrusive	Rhyolite	Andesite	Basalt	
Composition	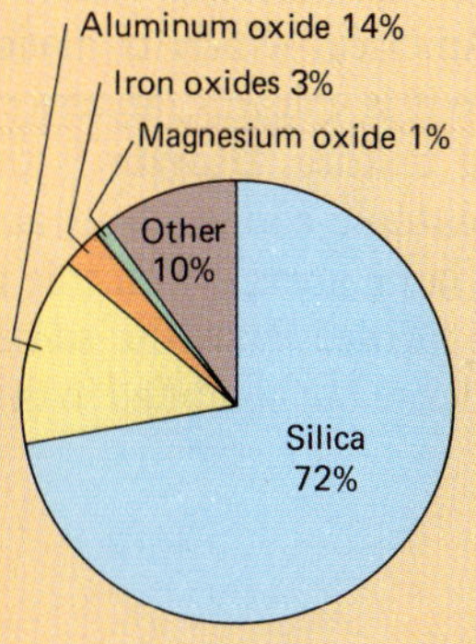 Aluminum oxide 14% Iron oxides 3% Magnesium oxide 1% Other 10% Silica 72%	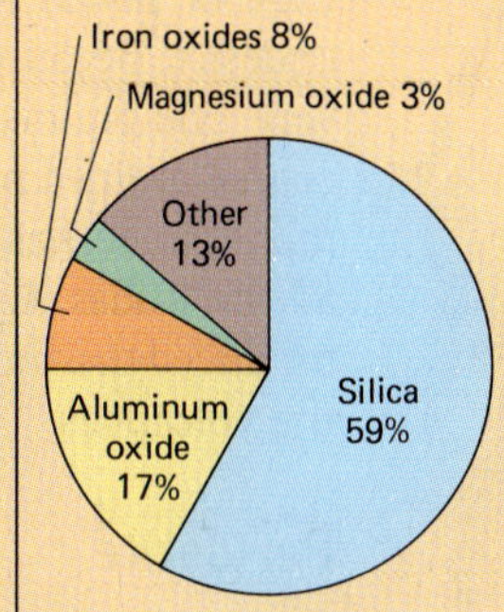 Iron oxides 8% Magnesium oxide 3% Other 13% Aluminum oxide 17% Silica 59%	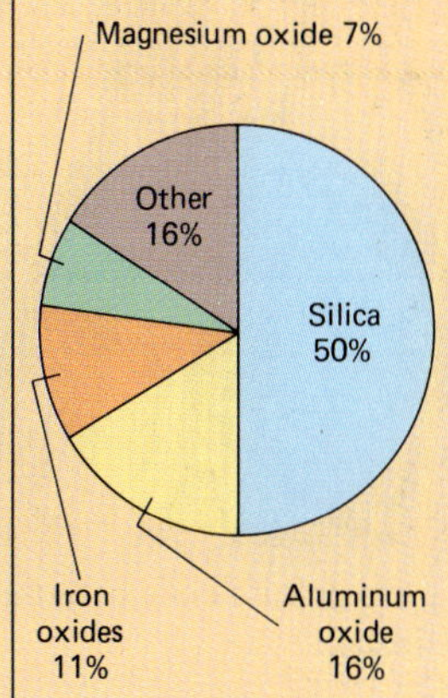 Magnesium oxide 7% Other 16% Silica 50% Iron oxides 11% Aluminum oxide 16%	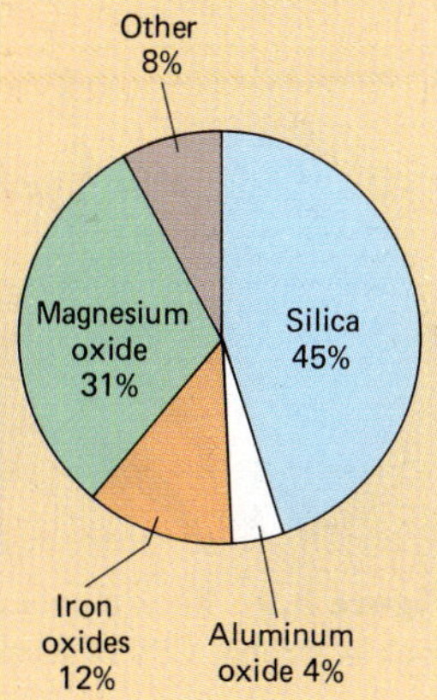 Other 8% Magnesium oxide 31% Silica 45% Iron oxides 12% Aluminum oxide 4%
Major minerals	Quartz Potassium feldspar Sodium feldspar (plagioclase)	Amphibole Intermediate plagioclase feldspar	Calcium feldspar (plagioclase) Pyroxene	Olivine Pyroxene
Minor minerals	Muscovite Biotite Amphibole	Pyroxene	Olivine Amphibole	Calcium feldspar (plagioclase)
Most common color	Light colored	Medium gray or medium green	Dark gray to black	Very dark green to black

Figure 3.8 Chemical compositions, minerals, and typical colors of common igneous rocks.

called **sialic**, or **felsic**, rocks. Basalt and gabbro are called **mafic** rocks because of their high magnesium and iron content. The word *mafic* is derived from *mag*nesium and *ferrum*, the Latin word for iron. Rocks with especially high magnesium and iron concentrations are called **ultramafic**. Rocks with compositions intermediate between those of granite and basalt are called **intermediate rocks.**

3.4   The Most Common Igneous Rocks

Granite and Rhyolite

Granitic rocks are the most common rocks of the Earth's continents. They are found nearly everywhere beneath the relatively thin veneer of sedimentary rocks that covers most of the continents. Geologists often call granitic rocks **basement rocks** because they make up the foun-

dation of a continent. Granite is hard and resistant to weathering; it forms steep, sheer cliffs in many of the world's great mountain ranges. Such cliffs are sought out by mountaineers for the steepness and strength of the rock (Fig. 3.9).

Rhyolite contains the same minerals as granite but has a fine-grained texture because it is volcanic. When granitic magma rises toward the Earth's surface, some may erupt from a volcano to form rhyolite while the remainder solidifies beneath the volcano, forming granite. Most obsidian forms from magma with a granitic (rhyolitic) composition.

Basalt and Gabbro

Basalt is a dark, very fine-grained volcanic rock. It is about half plagioclase and half pyroxene. Most oceanic crust is basalt. As we will learn in Chapter 4, the basalt of oceanic crust erupts beneath the sea along a great submarine mountain chain known as the mid-oceanic

Figure 3.9 Rock climbers prize granite cliffs because the rock is strong. City of Rocks, Idaho.

Andesite and Diorite

Andesite is a volcanic rock intermediate in composition between basalt and granite. It is commonly gray or green and consists of plagioclase and dark minerals (usually biotite, hornblende, or pyroxene). It is named for the Andes Mountains, the volcanic chain on the western edge of South America, which is made up mostly of andesite. Because it is volcanic, andesite typically has a very fine-grained texture. **Diorite** is the medium- to coarse-grained plutonic equivalent of andesite. It commonly underlies large areas of volcanic andesites, such as the Andes. It formed from the same magmas that produced the andesite but solidified in the crust beneath the volcanoes.

Peridotite

Peridotite is an ultramafic igneous rock that is rare in the Earth's crust. However, most of the upper mantle is peridotite. It is coarse grained and composed of olivine and usually contains pyroxene, amphibole, or mica, but no feldspar. Figures 3.7 and 3.8 show that peridotite has the lowest silica content of all the important igneous rocks.

Recognizing and Naming Igneous Rocks

Once you learn to identify the nine rock-forming minerals, it is easy to name a plutonic rock using Figure 3.7 because the minerals are large enough to be seen. It is

ridge. Basalt magma also erupts in great volumes on continents. In some cases the fluid magma may flood thousands of square kilometers of land, forming a large **basalt plateau** (Fig. 3.10). **Gabbro** is mineralogically identical to basalt but has larger mineral grains because it is a plutonic rock. Gabbro is uncommon at the Earth's surface, although it is abundant in deeper parts of oceanic crust.

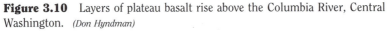

Figure 3.10 Layers of plateau basalt rise above the Columbia River, Central Washington. *(Don Hyndman)*

Table 3.2 • Sizes and Names of Sedimentary Particles and Clastic Rocks

Diameter (mm)	Sediment		Clastic Sedimentary Rock
256 ⎯ 64 ⎯ 2 ⎯	Boulders Cobbles Pebbles	Gravel (rubble)	Conglomerate (rounded particles) or breccia (angular particles)
⎯	Sand		Sandstone
$\frac{1}{16}$ ⎯	Silt	Mud	Siltstone ⎫ Claystone ⎬ Mudstone or shale ⎭
$\frac{1}{256}$ ⎯	Clay		

harder to name volcanic rocks because the minerals are usually too small to identify. In these cases, a field geologist often uses color to make a tentative identification. Figure 3.8 shows that rhyolite is usually light in color: white, tan, red, and pink are common. Andesite is gray or green, and basalts are black.

SEDIMENTARY ROCKS

3.5   Sediment

All rocks at the Earth's surface disintegrate slowly by weathering. Solid rock decomposes to sediment: gravel, sand, and clay. Rain, streams, wind, glaciers, and gravity erode this sediment, carry it off, and deposit it elsewhere.

Figure 3.11   Streams and glaciers carry sediment to lower elevations. Eventually most sediment reaches the ocean.

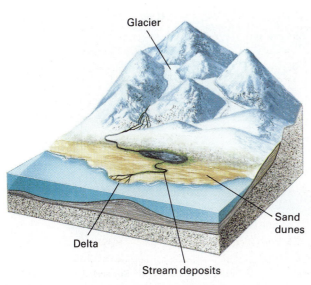

Eventually the loose particles become cemented to form sedimentary rock.

Particles that have been physically carried and deposited by water, wind, glaciers, and gravity are called **clastic sediment**. Clastic sediment includes beach sand, pebbles and cobbles in a river bed, dust in the air, mud in a puddle, boulders carried by a glacier, and shell fragments on a beach. Clastic sediment and clastic sedimentary rocks are named according to the sizes of the clastic particles (Table 3.2).

Clastic sediment is the most common, but not the only, kind of sediment. As rocks weather to gravel, sand, or clay, rainwater also dissolves some of the rock. The water then carries the dissolved material away and into streams. In certain environments the dissolved ions then precipitate directly to form **chemical sediment**. Large deposits of halite (table salt) form in this way. In addition, plants and animals may form an **organic sedimentary rock**, such as coal, directly.

Erosion and Transport of Sediment

After rocks weather, rain, streams, wind, glaciers, and gravity erode the sediment and carry it downhill. Streams and rivers carry the greatest amount of sediment. Because nearly all streams eventually empty into the oceans, most sediment accumulates along coastlines, near the edge of a continent (Fig. 3.11).

As clastic particles tumble downstream, their sharp edges are worn off and they become **rounded** (Fig. 3.12). Particles ranging in size from coarse silt to boulders are rounded during transport. Clay and fine silt do not become as well rounded because they are so small and light that water, and even air, cushion them as they bounce along.

Streams and wind also separate sediment according to size, a process called **sorting**. Figure 3.13 is a profile of a stream flowing from the mountains to the plains. Near its source, the stream is steep, and the water flows

Figure 3.12   Cobbles become rounded as they are carried by a stream. These cobbles are of approximately equal size as a result of sorting. Chaba River, Alberta, Canada.

Figure 3.13   A stream deposits large particles in its steeper headwaters, and smaller ones on the nearly level plain.

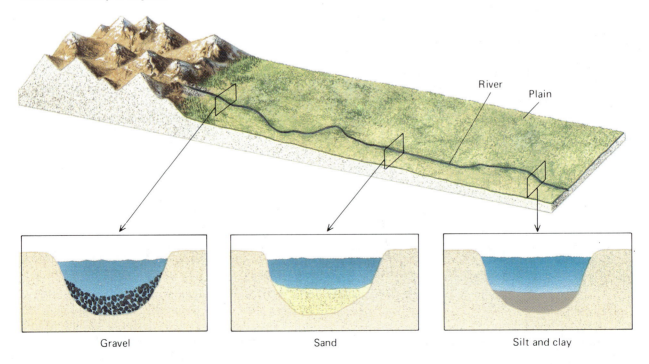

Gravel   Sand   Silt and clay

Figure 3.14 Glaciers deposit poorly sorted sediment containing particles of all sizes, from clay to boulders. Western Montana.

rapidly and with much energy. Therefore, the stream carries and deposits large boulders in its upper portion. But the stream levels out where it flows into the valley below. As steepness decreases, stream energy diminishes. Here, only smaller particles are carried and deposited. Thus, the largest particles are usually found near the headwaters of a stream, and the sediment becomes progressively smaller downstream.

Wind transports only small particles: sand, silt, and clay. Therefore, sand dunes and other wind-deposited sediment are well sorted. A glacier, in contrast, is a solid and able to carry particles of all sizes, from boulders to clay. As a consequence, glaciers deposit poorly sorted sediment (Fig. 3.14). Thus, a geologist can determine how sediment was transported and deposited by observing how well sorted it is.

Deposition of Sediment

Deposition of clastic sediment occurs when transport stops, usually because the wind or water slows down and loses energy or, in the case of glaciers, when the ice melts. Deposition of dissolved ions occurs when they precipitate directly from solution or are extracted from solution by an organism to form a shell or skeleton.

Sediment is deposited in many environments. A stream deposits sediment in its stream bed and flood plain and on its delta where it enters a lake or the ocean. Currents redistribute the sediment on beaches and the

sea floor. Wind may deposit sand and silt on land, forming dunes. Glaciers deposit large volumes of sediment wherever they melt. Calcium dissolved in a stream may be carried to the sea, where a clam or coral uses it to form its shell. When the organism dies, the shell may contribute to a growing pile of sediment on a beach or near a reef.

Lithification of Sediment

Lithification refers collectively to all the processes that convert loose sediment to hard rock. If you fill a measuring cup with sand, you can still add a substantial amount of water. The water fills empty spaces, or **pore space**, among the sand grains (Fig. 3.15A). Most sediment contains pore space. When sediment is deposited in water, the pores fill with water. Commonly, sand and

Figure 3.15 (A) Pore space is the space between sediment grains. (B) Compaction squashes the grains together, reducing the pore space and lithifying the sediment by interlocking the grains. (C) Cement fills the remaining pore space, lithifying the sediment by gluing the grains together.

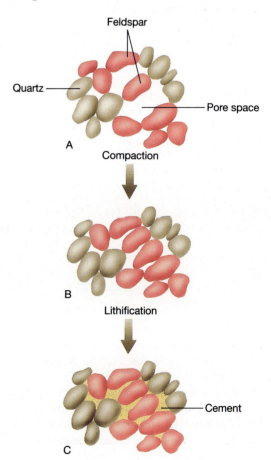

FOCUS ON

Bowen's Reaction Series

N. L. Bowen was a geologist who worked during the first part of this century. He studied the order in which minerals crystallize from a cooling magma. Because magma forms deep within the crust, where it cannot be studied, Bowen made artificial magma by melting powdered rock samples in a container called a **bomb**, a strong, hollow, steel cylinder that can be sealed with a threaded cap.

The bomb is heated until the powder melts and is then cooled to the temperature and pressure chosen for the experiment. It is left at that temperature and pressure long enough for minerals to crystallize from the melt. Commonly a few months are required for the minerals to form.

Bowen allowed a sample of artificial magma to cool slowly in his bomb until it had partly solidified. At that point the sample consisted of a mixture of crystals plus the melt that had not yet crystallized. He then plunged the bomb into cold water or oil, causing the sample to cool so rapidly that the crystals were preserved and the uncrystallized melt solidified as glass. He removed the sample from the bomb and identified the minerals with a microscope.

Bowen repeated the experiment many times, allowing his artificial magmas to form crystals at different temperatures before quick-cooling them. By identifying the minerals that had formed at each temperature, he was able to determine the order in which minerals crystallize from a cooling magma.

For example, Bowen found that as basalt magma cools, crystals of olivine and calcium-rich plagioclase form first at high temperatures. After the first crystals form, the crystallization of additional minerals upon further cooling can follow one of two paths, depending on what happens to the first crystals.

Crystals Separating from the Magma In a natural magma chamber the olivine and the calcium-rich plagioclase may settle to the bottom of the chamber because they are denser than the liquid. As they collect at the bottom of the magma chamber, they become isolated from the magma and cannot react with it.

In this case, because olivine and plagioclase are poor in silica, the remaining melt becomes enriched in silica. Therefore, as the magma cools further, minerals richer in silica form. The final bit of melt to solidify has the composition of granite: potassium feldspar, sodium plagioclase, and quartz. In this way, a cooling basalt magma produces a large amount of basalt, but it also can produce a small amount of granite.

Crystals Mixing with the Magma Alternatively, in a natural magma chamber, currents in the magma may prevent the crystals from settling to the bottom, thereby keeping them in contact with the liquid. Bowen found that if the crystals remained mixed with the magma, they continued to react with the liquid as it cooled.

Although it may seem contrary to intuition, Bowen discovered that crystals of olivine that had formed at high temperature dissolved back into the melt as it cooled. At the same time, crystals of pyroxene formed. Bowen reasoned that, as the magma cooled, the olivine crystals reacted with the melt to form the pyroxene crystals. At the same time, the early formed calcium-rich plagioclase crystals reacted with the melt to form plagioclase with less calcium and aluminum and more sodium and silica.

When he started with an artificial basalt magma, Bowen found that all of the magma had solidified by the time most of the olivine had reacted to form pyroxene, and the calcium-rich plagioclase had reacted to form plagioclase of intermediate composition. He had produced an artificial basalt consisting mainly of pyroxene and intermediate plagioclase with a small amount of olivine, despite the fact that initially only olivine and calcium-rich plagioclase crystallized.

Bowen also experimented with artificial magmas of intermediate and granitic compositions. He discovered that, as those melts cooled, pyroxene reacted with the melt to form hornblende. Hornblende then reacted to form biotite. Plagioclase continued to react with the melt to become progressively enriched in sodium and silica and depleted in calcium and aluminum with falling temperature. At the lowest temperatures, potassium

---

similar sediments have about 20 to 40 percent pore space.

As more sediment accumulates, the weight of the overlying layers compresses the buried sediment. Some of the water is forced out, and the pore space shrinks (Fig. 3.15B). This process is called **compaction**. If the grains have platy shapes, as in clay and silt, compaction alone may lithify the sediment because the platy grains interlock like pieces of a puzzle.

As sediment is buried and compacted, water circulates through the remaining pore space. This water commonly contains dissolved ions. The dissolved materials

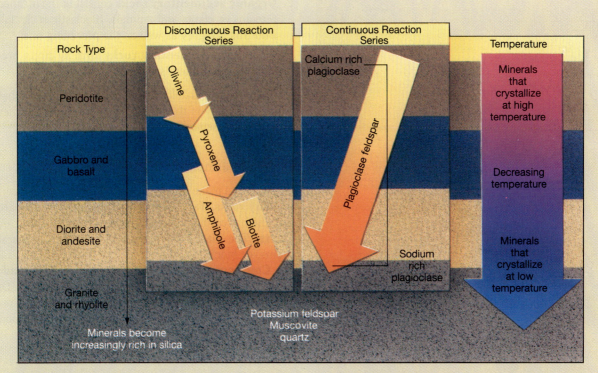

Bowen's reaction series shows the order in which minerals crystallize from a cooling magma and then react with the magma as it cools to form minerals lower on the diagram.

feldspar, muscovite, and quartz formed as the last of the magma crystallized.

Bowen summarized his work in a Y-shaped figure, now known as **Bowen's reaction series** (see illustration). It is called a reaction series to emphasize that minerals formed at high temperatures react with cooling magma to form other minerals.

The right side of Bowen's reaction series describes the reactions that occur between plagioclase and magma. As the temperature decreases, plagioclase reacts continuously with the melt to become progressively enriched in sodium and silica and depleted in calcium and aluminum. Thus, this arm of the Y is known as a **continuous reaction series**.

The left side of Bowen's reaction series shows that olivine, the first mineral to form, reacts with the melt to

form pyroxene as the temperature decreases. Then pyroxene dissolves back into the magma as hornblende (amphibole) forms. The reactions continue until biotite, the last mineral in the series, forms as hornblende dissolves. This arm of the Y is called a **discontinuous reaction series** because each of the minerals has a different crystal structure.

The minerals of both arms of the Y become progressively higher in silica content with decreasing temperature.

The rock names in the illustration are placed on the same level as the minerals that make up each of those rocks. The placement of each rock type also shows that basalt forms at relatively high temperatures, andesite at intermediate temperatures, and granite at relatively low temperatures.

precipitate in the pore spaces, bonding the grains together to form a hard rock (Fig. 3.15C). This process is **cementation**. Calcite, quartz, and iron oxides are the most common cements in sedimentary rocks. The type of cement affects the nature of the rock. The red sandstone in Figure 3.16 gets its color from red iron

oxide cement. Quartz cement forms the toughest sedimentary rocks.

The time required for lithification of loose sediment varies greatly, depending on the availability of cement and water to carry the cement. In some heavily irrigated areas of southern California, calcite precipitated from

Figure 3.16   Red iron oxide cement colors the red sandstone of Indian Creek, Utah.

irrigation water has cemented soils within a few decades. In the Rocky Mountains, some glacial deposits less than 20,000 years old are cemented by calcite. In contrast, sand and gravel deposited in southwestern Montana between 30 and 40 million years ago can still be dug with a hand shovel.

3.6   Types of Sedimentary Rocks

Sedimentary rocks are broadly divided into three categories based on the type of sediment of which they are made.

1. **Clastic sedimentary rocks** are composed of lithified clay, silt, sand, and gravel. The clastic particles are rock and mineral fragments that have been transported physically and deposited. This category includes rocks made up of broken shells and other organic fragments, called **bioclastic rocks**, indicating their biological origin.

2. **Organic sedimentary rocks** consist of the lithified remains of plants and animals. Bioclastic rocks fall into this category as well as the preceding one.

3. **Chemical sedimentary rocks** form by direct precipitation of minerals from solution.

Limestone and dolomite are sedimentary rocks that can form by any of the foregoing three processes; they are discussed separately.

Clastic Sedimentary Rocks

Clastic rocks make up more than 80 percent of all sedimentary rocks. Table 3.2 shows that clastic rocks are classified according to the size of their particles, or **clasts**. The clasts may be rock fragments, mineral grains, or bioclastic fragments.

Conglomerate (Fig. 3.17) is lithified gravel. **Sandstone** consists of lithified sand grains (Fig. 3.18). Of the nine rock-forming minerals, quartz is the most resistant to weathering. Feldspar and the other common

Figure 3.17   Conglomerate is lithified gravel. Near Soldier Summit, Utah.

A

B

Figure 3.18 Sandstone is lithified sand. (A) A sandstone cliff above the Colorado River, Canyonlands, Utah. (B) A close-up of sandstone. Notice the well-rounded sand grains.

minerals succumb to chemical attack and physical abrasion during weathering and transport. In contrast, about all that happens to quartz grains during weathering and transport is that they become rounded. Consequently, most sandstones consist of rounded quartz grains.

Shale is a fine-grained clastic sedimentary rock (Fig. 3.19A). It consists of clay minerals and small amounts of quartz and feldspar. Shale has thin bedding called **fissility**, along which the rock splits easily (Fig. 3.19B). Clay minerals, like micas, have platy shapes. The plates stack like dishes or sheets of paper. The fissility of shales results from parallel alignment of the clay plates.

Shale makes up about 70 percent of all sedimentary rocks (Fig. 3.20). Its abundance reflects the vast quantities of clay produced by weathering. Shale is usually gray to black due to the presence of decayed remains of plants and animals commonly deposited with clay. This organic material in shale is the source of most oil and natural gas.

A

B

Figure 3.19 Shale is made up mostly of platy clays. Therefore, it shows very thin layering called fissility. (A) An outcrop of shale near Drummond, Montana. (B) A close-up of shale.

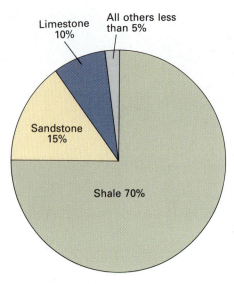

Figure 3.20   The relative abundances of sedimentary rocks.

A

B

Figure 3.22   (A) A coal bed near Price, Utah. (B) Organic-rich shale interbedded with coal contains abundant plant fossils.

Organic Sedimentary Rocks

Organic sedimentary rocks such as chert and coal form by lithification of organic sediment. Most limestone and dolomite are organic in origin as well, but they are described separately in a following section.

Chert is pure quartz (Fig. 3.21). Microscopic examination of most chert shows that it derives its silica from the siliceous skeletons of marine organisms. Silica from volcanic ash is also present in many chert formations.

Coal is lithified plant remains. When plants die, they usually decompose by reaction with oxygen. How-

ever, in warm swamps and other environments where plants grow rapidly, dead plants can accumulate so quickly that the available oxygen is used up before decay is complete. The partly decayed plant remains form **peat**. As peat is buried and compacted by overlying sediment, it converts to coal, a hard, black, combustible rock commonly associated with rocks containing abundant plant fossils (Fig. 3.22).

Chemical Sedimentary Rocks

When rocks weather, some elements such as calcium, sodium, potassium, and magnesium dissolve. Those dissolved ions are transported by ground water and streams to the oceans or to saline lakes such as Great Salt Lake of Utah. **Evaporites** are sedimentary rocks formed when evaporation of water concentrates dissolved ions to the point that they precipitate from solution. The most common minerals found in evaporite deposits are **gypsum**

Figure 3.21   Red nodules of chert in light-colored limestone.

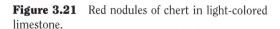

GEOLOGY AND THE ENVIRONMENT

Radon

Radon is a radioactive gas. Invisible, odorless, and tasteless, it occurs naturally in bedrock and soil. It seeps from the ground into homes and other buildings, where it concentrates and causes an estimated 5000 to 20,000 cancer deaths per year among Americans. The risk of dying of radon-caused lung cancer in the United States is about 0.4 percent over a lifetime, much greater than the risk of dying from cancer caused by asbestos, pesticides, or other air pollutants and nearly as high as the risk of dying in an auto accident or from a fall or fire at home.

However, Americans are not all exposed to equal amounts of radon. Some homes contain very low concentrations of the gas; others have high concentrations. The variations in concentration are due to two factors: geology and home ventilation.

Radon is one of a series of radioactive elements formed by the radioactive decay of uranium. Thus, radon forms wherever uranium occurs. Uranium occurs naturally in tiny amounts in all types of rock, but it concentrates in some types and occurs in only miniscule amounts in others. Because uranium concentrates in granite and shale, radon levels are highest in poorly ventilated homes built on granite or shale bedrock or on soil derived from these rocks. In other cases, building materials contain rocks with high uranium contents. The highest home radon concentrations ever measured have occurred in houses built on the Reading Prong, a uranium-rich granite pluton that extends from Reading, Pennsylvania, through northern New Jersey and into New York. The air in one home in this area contained 700 times as much radon as the EPA "action level"—the concentration at which the Environmental Protection Agency recommends that corrective measures be taken to reduce the amount of radon in indoor air.

A homeowner may ask two questions: "What is the radon concentration in my house?" and "If it is high, what can be done about it?" Because radon is radioactive, it can be measured with a simple detector available at most hardware stores and from local government agencies for about $25.

If the detector indicates excessive radon in a home, measures can be taken to remove it. Radon gas forms by slow radioactive decay of uranium in soil or bedrock beneath a house. After radon seeps from the ground or foundation into the basement, it circulates throughout the house. Thus, with time, it accumulates in indoor air. An unventilated house seals out the weather and therefore accumulates radon. Three types of solutions have proved effective. The first is simply to extend a ventilation duct from the basement directly outside the house. In this way, air from the basement does not circulate throughout the entire house, and the basement air is changed frequently so that radon does not accumulate at high levels. The second solution is to ventilate the entire house so that indoor air is refreshed continually. An open window suffices. However, an open window allows hot air to escape and thereby increases fuel consumption. A third solution is to pump outside air into the house to keep indoor air at a slightly higher pressure than the outside air. This positive pressure prevents gas from seeping from soil or bedrock into the basement.

It is impossible to avoid exposure to radon completely because it is everywhere, in outdoor air as well as in homes and other buildings. But it is relatively easy and inexpensive to minimize exposure and thus avoid a major cause of lung cancer.

(CaSO$_4 \cdot$ 2H$_2$O) and **halite** (NaCl). Evaporites constitute only a small proportion of all sedimentary rocks but can be important sources of salt and other materials (Fig. 3.23).

Carbonate Rocks: Limestone and Dolomite

Carbonate rocks are made up primarily of the minerals **calcite** and **dolomite**. They are called carbonates because both minerals contain the carbonate ion, $(CO_3)^{-2}$. Calcite-rich carbonate rocks are called **limestone**, whereas carbonate rocks rich in the mineral dolomite are also called **dolomite**. Some geologists use the term *dolostone* for the rock name to distinguish it from the mineral.

Seawater contains much dissolved calcium carbonate. Clams, corals, some types of algae, and many other marine organisms make their shells and other hard body parts of calcium carbonate. After the organisms die, waves or ocean currents break up and transport the shell fragments to form bioclastic sediment. Limestone formed by lithification of such sediment is called **bioclastic limestone**. The name indicates that both biological and clastic processes contributed to its origin.

A

B

Figure 3.23 (A) Bonneville Salt Flats, Utah. (B) Close-up of a salt crust, Death Valley, California.

Most limestones are bioclastic. The bits and pieces of shells appear as fossils in the rock (Fig. 3.24). Bioclastic rocks fit equally well into both the clastic and organic classes of sedimentary rocks. **Coquina** is bioclastic limestone consisting wholly of coarse shell fragments cemented together. **Chalk** is a very fine grained, soft, earthy, white to gray bioclastic limestone made of the shells and skeletons of microorganisms that spend their lives floating near the surface of the oceans. When they die, their remains sink to the bottom and accumulate to form chalk. The famous White Cliffs of Dover in England are made of chalk. In some cases, limestone can precipitate directly from solution as a chemical sediment, but such rocks are uncommon.

Dolomite is a common sedimentary rock. Because it is so abundant, we would expect to see dolomite forming commonly today; yet we do not. There is no place in the world where dolomite is forming today in large amounts. This dilemma is known among geologists as **the dolomite problem**, and it has been the cause of a tremendous amount of research.

The general consensus among geologists is that most dolomite does not originate as a primary sediment or rock. Instead, it forms when magnesium-rich solutions derived from seawater percolate through older beds of limestone. Magnesium ions replace half of the calcium in the calcite, converting the limestone beds to dolomite.

Figure 3.24 Most limestone is lithified shell fragments and other remains of marine organisms. (A) A limestone mountain in British Columbia, Canada. (B) A close-up of shell fragments in limestone.

A

B

A   B

Figure 3.25   (A) Horizontal sedimentary bedding, Mexican Hat, Utah. (B) Thin sedimentary clay beds and fine sand that accumulated in a glacial lake in Montana.

3.7   Sedimentary Structures

Nearly all sedimentary rocks contain **sedimentary structures**, features that developed during or shortly after deposition of the sediment. These structures often contain important clues that help us understand how the sediment was transported and deposited.

The most obvious and widespread sedimentary structure is **bedding**, or **stratification**: layering that develops as sediment is deposited (Fig. 3.25). Bedding forms because sediment is almost always deposited in a layer-by-layer process. It may result from differences between layers in texture, mineral composition, color, or cementation. Most sedimentary beds were originally horizontal because most sediment is deposited on nearly level surfaces.

Flowing water or wind piles sand in small, parallel ridges called **ripples** (Fig. 3.26A). **Ripple marks** are

Figure 3.26   (A) Modern ripples on a mud flat along the Oregon coast. (B) Ripple marks in billion-year-old mud rocks in eastern Utah.

A   B

A

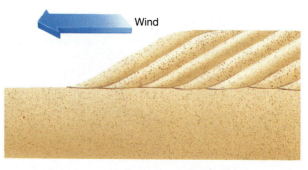

Wind

Wind

B

Figure 3.27   (A) Cross-bedding preserved in lithified ancient sand dunes in Arches National Park, Utah. (B) The development of cross-bedding in sand as a dune migrates.

often preserved in sandy sedimentary rocks (Fig. 3.26B). A **sand dune** is a large-scale version of a ripple.

Cross-bedding is an arrangement of small beds lying at an angle to the main sedimentary layering (Fig. 3.27A). Figure 3.27B shows that cross-beds form as sand grains tumble down the steep, downstream face of a sand dune or ripple. Cross-bedding forms in both wind-blown and water-transported sediments.

Graded bedding is a type of bedding in which the largest grains collect at the bottom of a layer and the grain size decreases toward the top (Fig. 3.28). Graded beds commonly form when sediment containing a mix-

ture of different particle sizes settles to the bottom of a body of water. The larger grains settle rapidly and concentrate at the base of the bed. Finer particles settle more slowly and accumulate in the upper parts of the bed.

Mud cracks are polygonal cracks that form when mud shrinks as it dries (Fig. 3.29). They indicate alternate wetting and drying—for example, on an intertidal mud flat, where the sediment is flooded by water at high tide and then exposed at low tide.

Figure 3.28   A graded bed in Tonga, southwest Pacific. Larger grains collected near the bottom, and smaller particles settled near the top of the bed.   *(Peter Ballance)*

Figure 3.29   Mud cracks form when mud shrinks as it dries. They are often preserved in ancient rocks.

Occasionally, very delicate sedimentary structures are preserved in rocks. Geologists have found the imprint of a single raindrop that fell on mud a billion years ago, and the imprint of a cubic salt crystal that formed as a puddle evaporated. Mud cracks, raindrop imprints, and salt crystal imprints all show that the mud must have been deposited in shallow water and was intermittently exposed to air.

Fossils are any remains or traces of a plant or animal preserved in rock—any evidence of past life. They, too, are sedimentary structures; they are discussed in detail in Chapter 15.

METAMORPHIC ROCKS

3.8 Metamorphism

A potter forms a delicate vase from moist clay. She places the new piece in a kiln and slowly heats it to 1000°C. As the temperature rises, the clay minerals decompose. The atoms from the disintegrating clay minerals recombine to form new minerals that make the vase strong and hard. The breakdown of the clay minerals, growth of new minerals, and hardening of the vase all occur without melting. The reactions in a potter's kiln are called **solid-state reactions** because they occur in solid materials.

Metamorphism (from the Greek words for "changing form") is the process by which rocks and minerals change because of changes in temperature, pressure, or other environmental conditions. Like the changes in the vase as it is fired in the kiln, metamorphism occurs in solid rocks. Both the texture and the minerals can change as a rock is metamorphosed.

A potter firing a kiln. *(Christine Seashore)*

Textural Changes

A rock is an aggregate of individual mineral grains. As a rock is metamorphosed, the grains grow and their shapes change. For example, fossils give fossiliferous limestone its texture (Fig. 3.30A). Both the fossils and the cement between them are made of small calcite crystals. If the limestone is buried and heated, the calcite grains grow larger. In the process, the fossils are usually destroyed.

The resulting metamorphic rock, called **marble** (Fig. 3.30B), is still made of calcite, but its texture is now one of large, interlocking grains. Although its minerals are the same as those of limestone, marble is a coarse-grained rock that can be polished to create a

Figure 3.30 Metamorphism converts fossiliferous limestone (A) to marble (B), which has a very different texture, although both are made of the mineral calcite. *(Geoffrey Sutton)*

A

B

Metamorphism formed these white marble beds in Vermont.

smooth, lustrous surface and therefore is prized for sculpture.

Tectonic forces commonly deform rocks at the same time that the rocks are metamorphosed by rising temperature. When deformation and metamorphism occur together, the metamorphic rocks develop a layered texture called **foliation**. The metamorphic layers range from a fraction of a millimeter to a meter or more in thickness. This metamorphic layering results from alignment of platy minerals such as mica that grow during metamorphism. Metamorphic foliation superficially resembles sedimentary bedding, but it is quite different.

Mineralogical Changes

In the example of limestone, metamorphism changed the rock's texture but not its minerals. In many other cases, metamorphism forms new and different minerals from the original ones. For example, a typical shale

contains clay, quartz, and feldspar (Fig. 3.31A). When heated, the clay minerals are altered, as they are when the potter's vase is fired in the kiln. The atoms of the clay recombine to form new minerals. Figure 3.31B shows a rock called hornfels that forms when both new textures and new minerals develop during metamorphism of shale.

Thus, two types of metamorphic reactions occur. In one, the original mineral grains simply grow larger. In the other, the original minerals are transformed into new minerals that grow in their place.

As a general rule, when a parent rock (the original rock) contains only one mineral, metamorphism forms a rock composed of the same mineral but with a coarser texture. The metamorphism of limestone to marble is one example. In a similar manner, metamorphism of a quartz sandstone forms **quartzite**, a rock composed of recrystallized quartz grains. In contrast, metamorphism of a parent rock containing several minerals usually forms new and different minerals. Shales commonly contain clay minerals as well as quartz and feldspar. During metamorphism, shales always grow entirely new minerals as well as new textures.

The Causes of Metamorphism

The outer layers of the Earth move constantly. If you could watch the Earth over hundreds of millions of years, you would see continents moving around the globe, colliding, and then splitting apart. Huge mountain ranges would rise in the collision zones, only to erode to level plains with the passage of time. You would see new ocean basins form and old ones disappear. Some of these processes force rocks downward into deeper regions of the crust, burying them under 5, 10, or even 20 kilometers of sediment.

When rocks are buried, they become hotter and the pressure on them increases. These new conditions cause chemical and physical changes in the same way that the heat of a potter's kiln alters moist clay. Let us look

Figure 3.31 When shale (A) is metamorphosed to hornfels (B), both a new texture and new minerals form. The white spots in (B) are metamorphic minerals. *(Geoffrey Sutton)*

A

B

briefly at each of the factors that cause metamorphism, to see how they work and what kinds of changes they cause.

Temperature

Recall that the Earth's crust gets hotter by an average of 30°C for each kilometer of depth. Heat causes metamorphism. Think of a layer of clay deposited in a sedimentary basin. If the basin sinks as a result of tectonic processes, it continues to fill with more sediment. As more sediment accumulates, the clay layer is buried. If several kilometers of sediment pile up on top of the layer of clay, the temperature rises enough to decompose the clay minerals. The atoms from the clays then recombine to grow new minerals that are stable at the higher temperature.

Pressure

Minerals are also sensitive to pressure. If atoms in a crystal are squeezed together very tightly, the bonds between the atoms can break. The atoms then reorganize to form a new mineral that is stable under the higher pressure. Most minerals are more sensitive to temperature changes than to pressure variations. Nevertheless, pressure also plays an important role in metamorphism.

Migrating Fluids

Recall that sediment commonly contains water in the pore space between the grains. Water is also present in most rocks of the Earth's crust and in magma. This water usually contains dissolved ions and flows slowly through rock. The water and ions can react with rock, decomposing original minerals and replacing them with new ones. In this way, migrating fluids can also cause metamorphism.

Deformation

Tectonic plates move and smash together, creating tremendous forces. Rocks bend and break in response. **Deformation** is the change in shape of rocks in response to tectonic forces. It occurs in tectonically active regions—areas where rocks move.

When metamorphism occurs without deformation, mineral grains grow with random orientation (Fig. 3.32A). In contrast, Figure 3.32B shows originally

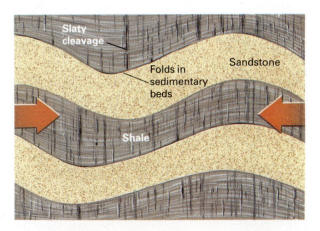

Figure 3.32 (A) Minerals grow randomly when metamorphism occurs without deformation. (B) If metamorphism occurs while tectonic forces deform shale into folds, mica flakes grow perpendicular to the force (*arrows*). The rock then breaks easily parallel to the mica flakes. This breakage is called slaty cleavage. (C) Slaty cleavage cuts across folded bedding. (*Carl Mueller*)

flat-lying shale beds being squeezed (deformed) into **folds** during metamorphism. The clays decompose and platy minerals such as mica grow in their place. The mica grows with its flat surfaces perpendicular to the force squeezing the rock. This parallel arrangement of minerals forms layering called **slaty cleavage** (Fig. 3.32C). Notice that the slaty cleavage often cuts across the original sedimentary bedding. Rocks with slaty cleavage break neatly along the newly formed planes. Any kind of metamorphic layering, such as slaty cleavage, is a form of **foliation**.

Metamorphic Grade

The **metamorphic grade** of a rock is the intensity of metamorphism that formed the rock. Temperature is the most important factor in metamorphism, and therefore grade mostly reflects the temperature of metamorphism. Because different minerals form as a rock becomes hotter, the minerals reflect the metamorphic grade of a rock. Recall that temperature increases with depth in the Earth; therefore, a general relationship exists between depth and metamorphic grade (Fig. 3.33). Low-grade metamorphism occurs at shallow depths, less than 10 kilometers beneath the surface, where tempera-

Figure 3.33 Different grades of metamorphism commonly occur at different temperatures, pressures, and depths below the Earth's surface. The blue arrow traces the path of increasing temperature and pressure with depth in a normal part of the crust.

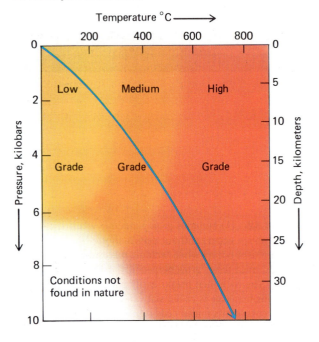

ture is no higher than 300° to 400°C. High-grade conditions are found deep within the continental crust and in the upper mantle, 40 to 55 kilometers below the Earth's surface. The temperature there is 600° to 800°C, close to the melting point of rock. High-grade conditions can develop at shallower depths, however, in areas adjacent to rising magma or hot, intrusive rocks. For example, metamorphic rocks are forming today beneath Yellowstone Park, where hot magma lies close to the Earth's surface.

3.9   Types of Metamorphism and Metamorphic Rocks

Metamorphism is divided into three general categories on the basis of the cause of metamorphism.

Contact Metamorphism

Contact metamorphism results from the intrusion of hot magma into cooler rocks. The country rock (the rock intruded by the magma) may be of any type—sedimentary, metamorphic, or igneous. The intrusion heats the adjacent rock, causing old minerals to decompose and new ones to form. The highest-grade metamorphic rocks form at the contact, where the temperature is greatest. Lower-grade rocks develop farther out. The zone of metamorphism forms a metamorphic halo around the intrusion (Fig. 3.34). Contact metamorphic halos can range in width from less than a meter to hundreds of meters, depending on the size and temperature of the intrusion and the effects of water or other fluids.

Because contact metamorphism commonly occurs without deformation, the rocks are **nonfoliated**; that is, they have no metamorphic layering.

Hornfels (Fig. 3.31B) is a hard, dark, fine-grained rock usually formed by contact metamorphism of shale. **Tactite**, also called **skarn** (Fig. 3.35), forms by contact metamorphism of limestone.

Regional Metamorphism

Regional metamorphism affects broad regions of the Earth's crust, usually in areas of tectonic activity where mountains are being built. Because it is commonly accompanied by deformation, the rocks are foliated. It is the most common and widespread type of metamorphism.

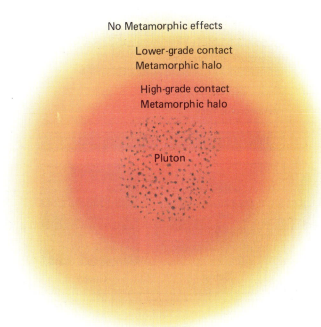

Figure 3.34   A contact metamorphic halo surrounding a pluton.

Recall that shale is the most abundant type of sedimentary rock. It consists mostly of platy clay minerals lying parallel to bedding planes. The clay grains are too small to be seen with the naked eye. Shale changes in a regular sequence as temperature rises and rocks are deformed during regional metamorphism (Fig. 3.36).

As regional metamorphism begins, clay minerals decompose and new minerals grow perpendicular to the squeezing direction, as already described. Thus, slaty

Figure 3.35   Tactite containing garnet (brown) and calcite (white).   *(Geoffrey Sutton)*

cleavage develops. Rock formed in this manner is called **slate** (Fig. 3.36).

With increasing metamorphic grade and continuing deformation, the crystals grow larger and foliation becomes very well developed. Rock of this type is called **schist** (Fig. 3.36). Schist first forms between low and intermediate metamorphic grades.

At high metamorphic grades, light- and dark-colored minerals often separate into bands 1 centimeter or more thick, to form a rock called **gneiss** (pronounced "nice") (Fig. 3.36).

Foliated metamorphic rocks such as schist and gneiss form when tectonic forces deform rocks at the same time that rising temperature is causing new minerals to grow. Thus, any kind of rock—sedimentary, igneous, or older metamorphic—can become schist or gneiss.

Marble and quartzite form during both regional and contact metamorphism. Marble forms from limestone, and quartzite from quartz sandstone.

Hydrothermal Metamorphism

Hydrothermal metamorphism, also called **hydrothermal alteration**, is the changes in rock caused by migrating hot water and by ions dissolved in the hot water.

Most hydrothermal alteration is caused by circulating **ground water**, water contained in soil and bedrock. Cold ground water sinks through fractures in bedrock to depths of a few kilometers, where it is heated by the hotter rocks. Shallow magma or a hot, shallow pluton can enhance the heating of shallow ground water. Water is a chemically active fluid that attacks and dissolves many minerals. If the water is hot, it decomposes minerals even more rapidly. In some hydrothermal environments, water reacts with sulfur minerals in the rock to form sulfuric acid, making the solution even more corrosive. Upon heating, the water expands and rises back toward the surface through other fractures, altering the rocks adjacent to the fractures as it goes (Fig. 3.37).

Hydrothermal solutions are corrosive enough to dissolve metals such as copper, gold, lead, zinc, and silver from hot rocks or magma. Most rocks contain low concentrations of many metals. For example, gold makes up 0.0000002 percent of average crustal rock; copper makes up 0.0058 percent, and lead 0.0001 percent. Although the metals are present in the rock in low concentrations, hydrothermal solutions percolate slowly through vast volumes of rock, dissolving and accumulating large amounts of the metals. The metals are then deposited when the solutions encounter changes in temperature, pressure, or chemical

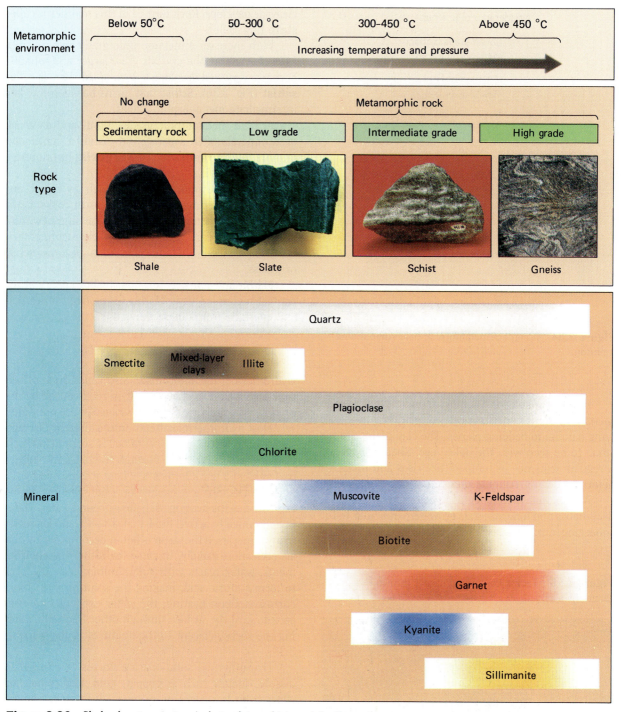

Figure 3.36 Shale changes progressively to slate, schist, and finally gneiss as metamorphic grade increases. The lower part of the figure shows when old minerals decompose and new ones grow as metamorphic grade increases. *(Geoffrey Sutton, Don Hyndman, and Hubbard Scientific Co.)*

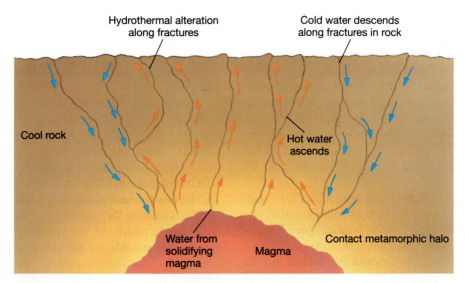

Figure 3.37   Ground water descending through fractured rock is heated by magma and ascends through other fractures, causing hydrothermal metamorphism in nearby rock.

environment (Fig. 3.38). In this way, hydrothermal solutions scavenge metals from average crustal rocks and then deposit them locally to form ore.

If the metals precipitate in fractures in rock, a hydrothermal vein deposit forms. Ore veins range from less than a millimeter to several meters wide. They can be incredibly rich. Single gold or silver veins have yielded several million dollars' worth of ore.

The same hydrothermal solutions that flow rapidly through open fractures to form rich ore veins may also soak into large volumes of country rock around the fractures. If metals precipitate in the rock, they may create a very large but much less concentrated **disseminated ore deposit.** Because they may form from the same solutions, vein deposits and disseminated deposits are often found together. In many early mining districts, the miners dug shafts and tunnels to follow the rich veins. After the veins were exhausted, later miners used huge power shovels to extract low-grade ore from disseminated deposits surrounding the veins.

Figure 3.38   Hydrothermal ore deposits form when hot ground water deposits ions in concentrated zones.

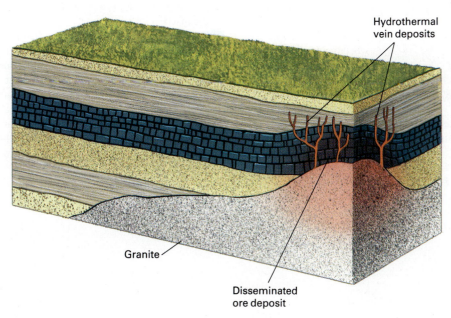

SUMMARY

Geologists divide rocks into three groups depending on how they formed. **Igneous rocks** solidify from **magma**. **Sedimentary rocks** form by **lithification** of **sediment** that collects at the Earth's surface. **Metamorphic rocks** form when any rock is changed by temperature, pressure, or deformation. The **rock cycle** summarizes processes by which rocks continuously recycle in the outer layers of the Earth, forming new rocks from old ones.

Intrusive, or **plutonic**, igneous rocks are medium- to coarse-grained rocks that solidify within the Earth's crust. **Extrusive**, or **volcanic**, igneous rocks are fine-grained rocks that solidify from magma erupted onto the Earth's surface. **Granite** and **basalt** are the two most common igneous rocks.

Sediment forms by weathering of rocks and minerals. It includes all solid particles such as rock and mineral fragments, organic remains, and precipitated minerals. It is **transported** by streams, glaciers, wind, and gravity; **deposited** in layers; and eventually **lithified** to form sedimentary rock. **Shale**, **sandstone**, and **limestone** are the most common kinds of sedimentary rock.

When a rock is heated, when pressure increases, or when the rock is deformed, both its minerals and its textures change in a process called **metamorphism**. **Contact metamorphism** affects rocks heated by a nearby igneous intrusion. **Regional metamorphism** affects large regions of the crust during mountain building. **Hydrothermal metamorphism** is caused by hot solutions soaking through rocks and is often associated with emplacement of ore deposits. **Slate**, **schist**, **gneiss**, and **marble** are common metamorphic rocks.

Common Sedimentary Rocks	Metamorphic Rocks that Form from the Common Sedimentary Rocks		
Shale	Slate	Schist	Gneiss
Sandstone	Quartzite		
Limestone	Marble		

Common Igneous Rocks

	Felsic	Intermediate	Mafic	Ultramafic
Volcanic	Obsidian Rhyolite	Andesite	Basalt	
Plutonic	Granite	Diorite	Gabbro	Peridotite

KEY TERMS

Magma *42*
Igneous rock *42*
Granite *42*
Basalt *42*
Sedimentary rock *42*
Metamorphic rock *42*
Rock cycle *43*
Intrusive igneous rock *44*
Extrusive igneous rock *44*
Plutonic rock *44*
Volcanic rock *44*
Porphyry *45*

Sialic (felsic) *47*
Mafic *47*
Sediment *49*
Clastic sediment *49*
Rounding *49*
Chemical sediment *49*
Organic sedimentary rock *49*
Sorting *49*
Deposition *51*
Lithification *51*
Pore space *51*

Compaction *52*
Cementation *53*
Clasts *54*
Fissility *55*
Sedimentary structure *59*
Bedding *59*
Stratification *59*
Ripples *59*
Ripple marks *59*
Sand dune *60*
Cross-bedding *60*
Graded bedding *60*
Mud cracks *60*

Foliation *62*
Deformation *63*
Slaty cleavage *64*
Metamorphic grade *64*
Contact metamorphism *64*
Regional metamorphism *64*
Hydrothermal alteration *65*

REVIEW QUESTIONS

1. What are the three main kinds of rock in the Earth's crust?

2. How do the three main types of rock differ from each other?

3. What is magma, and where does it originate?

4. Describe how igneous rocks are classified and named.

5. What are the two most common kinds of igneous rock?

6. Describe and explain the differences between a plutonic and a volcanic rock.

7. What is sediment? How does it form?

8. Where in your own area would you look for rounded and sorted sediment?

9. Describe how sediment becomes lithified.

10. What are the differences among shale, sandstone, and limestone?

11. Explain why almost all sedimentary rocks are layered, or bedded.

12. How does cross-bedding form?

13. What do mud cracks and raindrop imprints in shale tell you about the water depth in which the mud accumulated?

14. What is metamorphism? What factors cause metamorphism?

15. What kinds of changes occur in a rock as it is metamorphosed?

16. Explain the concept of metamorphic grade.

17. How do contact metamorphism and regional metamorphism differ, and how are they similar?

18. Explain what the rock cycle tells us about Earth processes.

DISCUSSION QUESTIONS

1. Magma usually begins to rise toward the Earth's surface as soon as it forms. It rarely accumulates as large pools in the upper mantle or lower crust, where it originates. Why?

2. In the San Juan Mountains of Colorado, parts of the range are made up of granite, and other parts are volcanic rocks. Explain why these two types of igneous rock are likely to be found together.

3. Suppose you were given a fist-size sample of igneous rock. How would you tell whether it is volcanic or plutonic in origin?

4. How would you tell whether another rock sample is igneous, sedimentary, or metamorphic in origin?

5. Why is shale the most common sedimentary rock?

6. How does sediment become rounded?

7. How does sediment become sorted?

8. One sedimentary rock is composed of rounded grains, and the grains in another are angular. What can you tell about the history of the two rocks from these observations?

9. How would you distinguish between contact metamorphic rocks and regional metamorphic rocks in the field?

10. What is metamorphic foliation? How does it differ from sedimentary bedding?

11. What happens to bedding when sedimentary rocks undergo regional metamorphism?

12. How can granite form during metamorphism?

Jack Horner, curator of paleontology at Montana State University's Museum of the Rockies, is the model for the hero in Michael Crichton's best seller, Jurassic Park. *Horner also served as a consultant to the Steven Spielberg film.*

John R. Horner was born in 1946 in Shelby, Montana. His father operated a large gravel quarry and his mother was a homemaker. He developed an early interest in fossils and geology during youthful forays on the High Plains around Shelby, where the remains of marine animals, dinosaurs, and ice age mammals are plentiful. His study of geology at the University of Montana was interrupted by service in Vietnam with the U.S. Marine Corps from 1966 to 1968. He returned to the University of Montana in 1968. He was Research Assistant in the Department of Geological Sciences at Princeton University from 1975 to 1982 and also was Museum Scientist at the American Museum of Natural History in New York City from 1978 to 1982. While at Princeton and the American Museum, he returned occasionally to the area near his home in Montana during summer field seasons to explore and recover the fossils of large animals that he had noticed as a child. Many of his discoveries are now prized specimens in the Princeton and American Museum of Natural History collections. During this time, he began to concentrate on dinosaurs, and he realized that the rocks near Shelby contain abundant and well-preserved dinosaur remains. In 1982 he was

appointed Curator of Paleontology at the Museum of the Rockies in Bozeman, Montana, where he has continued his research on the nature and living habits of dinosaurs. He was awarded an Honorary Doctorate of Science from the University of Montana in 1986.

He is author of many publications, most dealing with dinosaurs and their social habits, and of two popular books about dinosaurs. He has received several large National Science Foundation grants in support of his research. In 1986 he was awarded the MacArthur Foundation Fellowship, commonly known as the "Genius Award." It is a large cash grant given to unusually creative and productive individuals. He has been featured in many national news and science magazines, including *Reader's Digest, Natural History, Omni, Life, National Geographic, People,* and *U.S. News and World Report.*

What initiated your interest in geology and paleontology?

I really don't know how my interest in geology and paleontology started. My father owned a gravel quarry, and he says that when I was very young, I was always sorting the rocks out of the big gravel pile into what I thought were groups of different kinds of rocks.

My father would take me to places that he had ridden horseback when he was a rancher, and he would show me areas where he had found what he thought were dinosaur bones. I was eight years old when he first took me to one of these places, and I collected a couple of bone fragments. Then we went back to the same area when I was in high school, and I collected two partial dinosaur skeletons.

Through high school, I was really interested in science. I spent most of my time working on science projects. In my senior year I did a project on fossils. I was trying to figure out why the dinosaur remains found in the Judith River Formation in Montana were different from the ones found in the Judith River Formation in Canada. On the Canadian side, there are articulated dinosaurs and lots of duck-billed bones. But even at that time I knew that most of my dinosaur bones from Montana were flatheaded kinds and were strewn all over. I never could find an articulated one. I saw this difference as a problem, but I was unable to resolve it then. In fact, I didn't resolve it until I published a paper just two years ago on it. Now we know that the dinosaur bone beds in Alberta are of a slightly different age from those in Montana, so they are stratigraphically not the same. They also had different environments of deposition.

> I have a learning disability called dyslexia. It's a problem that didn't stop me from wanting to learn, but it did stop me from being able to.

You are now one of the most highly visible and best known paleontologists/geologists in the country. Yet in high school and as a university student you got low grades. How do you reconcile your success as a scientist with your grades as a student?

My academic record at the university was even worse than my high school grades. After my first year and a half at the University of Montana, my cumulative grade point average was a 0.06. If I went to college now, I would still have the same problem. I have a learning disability called dyslexia. It's a problem that didn't stop me from wanting to learn, but it did stop me from being able to. It made it almost impossible for me to absorb information that was assigned to me. I just couldn't assimilate the material fast enough. It got to the point where I didn't care what the grade was. I would learn as much as I could, and if at the end of the course that was D work for the professor, then I got a D. If I felt that the course seemed really interesting and if it was something that I thought I could get more out of, I'd just take it again. If I thought I knew enough for what I wanted to do, then I wouldn't take it again regardless of my grade. I didn't know what the problem was at the time, but I did know that I really wanted to learn.

I spent a year and a half at the university, and then in 1966 I got drafted into the Marine Corps. I served with the Special Forces in Vietnam. I got out of the Marine Corps in February 1968 and went back to the University of Montana. My grades were a lot better when I came back, and I started taking zoology and geology courses. But my grades were still lousy by university standards, and that's when they started throwing me out of school every quarter. Each time I could always demonstrate that I was bringing my grades up so they had to let me back in. I took every geology course that was offered and all the zoology courses that looked appropriate, as well as botany, physical geography, and a few anthropology courses. I never finished my undergraduate degree, but I did eventually get a doctorate from the University of Montana.

What did you do after college? What was your first job?

After I took all the courses I was interested in, I went back to Shelby where my brother and I bought the gravel company from my father, who was retiring. I had been there for about a year and a half when I began sending letters to all the museums in the English-speaking world to apply for jobs from janitor up to curator. I didn't really care what it was, I would have taken anything. I got three responses: from the Los Angeles County Museum for the position of Chief Preparator, from the Royal Ontario Museum in Toronto for Assistant Curator, and from Princeton for a preparator and research assistant. I took the job at Princeton and worked there for seven years.

In 1979, I was talking to the Director of the Museum of the Rockies and he told me that he had heard about all the dinosaur eggs we were finding in Montana around Choteau and Shelby. He asked if we would donate some to the Museum of the Rockies so I donated a clutch of eggs. In 1981, I saw the Director again, and I told him I was from Montana and that I really wanted to work there. He offered me the Curator of Paleontology position at the Museum of the Rockies and I came home.

Tell us about your work at the Museum of the Rockies.

The paleontology crew at the museum includes one curator (me), four full-time preparers, a thin-section histology technician, an illustrator, a collection manager, a computer illustrator who does all of the mathematical simulation, six graduate students, and a staff of about 15 part-time people.

My research is primarily on dinosaur behavior, ecology, and evolution. My graduate students are all geologically oriented to do field studies in stratigraphy and sedimentology. One of the students is doing comparative studies of different kinds of bone beds we have found. A bone bed is one geologic horizon on which lots of specimens occur together. For example, one at Choteau appears to be a volcanic ash kill, and our evidence shows that over 10,000 animals are buried there. All of the bone beds we work on cover at least one square mile.

Much of your work has changed the traditional views of dinosaurs, how they behaved, and even what they were. Tell us what the traditional view is and how your work has affected this view.

Originally, scientists decided, on the basis of certain cranial features, that the dinosaurs should be classified as reptiles. Once they had been placed in this group, there were certain characteristics, certain little labels, that were automatically assigned to dinosaurs simply because they were called reptiles. For example, modern reptiles are cold-blooded, therefore dinosaurs were considered cold-blooded. Modern reptiles are slow moving, so dinosaurs were probably slow. Modern reptiles drag their tails, so dinosaurs must have dragged their tails.

You have to realize that in the early days of dinosaur discoveries, people didn't really study dinosaurs; they collected their remains for museums. It wasn't until the last 15 years or so that people started to consider how dinosaurs actually lived. In those 15 years we've come to find out that the original classification is probably wrong. Dinosaurs don't belong to the reptile group. They are much more like birds, and it is likely that modern-day birds evolved from dinosaurs.

Most reptiles dig a hole in the ground, lay their eggs in it, cover it up, and then leave. So it was assumed dinosaurs did the same thing. But we now know that dinosaurs, like modern birds, put a lot of time and care into building a nest, laying the eggs, and guarding the eggs. Our evidence shows that dinosaurs even guarded their young after they hatched by herding or flocking in large groups. These large groups of dinosaurs were very similar to what we observe in modern herding animals or flocking birds, which are not just aggregations of animals, but actually structured groups with certain individuals in charge.

The animals that we find in large groups, such as the duck-billed dinosaurs and the horned dinosaurs, all have some type of cranial display features such as horns on their heads. We know that the horns of modern mammals are a primary adaptation for determining hierarchies within a society. Modern horned animals, such as elk, live in big groups, and generally the males use their horns for male-to-male combat to determine the individual hierarchy. We see a similar thing with the horned dinosaurs.

What would large herds of dinosaurs eat? How could so many of them live in a square mile? What was their environment like?

Well, one of the interesting things is that the dinosaurs that lived in large herds, hadrosaurs and ceratopsians, did not evolve until Late Cretaceous time. That coincides with the evolution of the angiosperms, deciduous plants that can be stripped one season and still grow back the next. So, I don't think the large groups or herds existed before angiosperms appeared. But once these plants existed as a food source, then all that's required of these big groups is for them to migrate with the seasons. About 75 to 80 million years ago, the western part of North America was a linear continent extending north to south. The Rocky Mountains were young and were actively building up. So there was a mountain barrier on the west and an ocean to the east with a coastal plain in between. The dinosaurs would migrate north to south on that coastal plain with the seasons. For their size and stride lengths, it appears that they could have easily walked 1500 to 2000 miles each year following seasonal shifts in temperature and food supply.

The only constraint on the migrating dinosaurs was the nesting period. The dinosaurs had to wait at least a month somewhere for the incubation of their eggs. Most of the incubation, or growth of the fetuses, may have occurred in the mothers' bodies prior to egg laying. After the eggs were laid, it was probably a relatively short time (possibly only two or three weeks) before the eggs hatched. Another three

cent suggestions that they were furry, hairy, or covered with feathers?

I think the babies had some kind of downy cover. All our evidence suggests that dinosaurs were warm-blooded just like we are. In fact, they were probably warmer blooded than we are. Their fast growth rate suggests a very high metabolism, suggesting a relatively warm internal body temperature. So, the babies had to have some kind of insulatory mechanism to keep the heat in. But our bone histology shows that the rapid growth ended at about 20 feet in length. Then we see rest lines in the bones suggesting that the metabolism slowed way down. At this point, the dinosaurs were still creating heat, but much less than before.

So this means that their rate of food consumption decreased when they became large. This would address those who argue that dinosaurs could not be warm-blooded because there wouldn't be enough food for many animals of that size. If there are hundreds of dinosaurs or possibly more nesting in one area, there would have to be a large food supply for the young, and the adults would be eating too. Well, I don't think the adults were eating at all. It's like when birds are feeding their young—they don't eat. They just haul in food for the babies. So all that was needed was a supply large enough for the young, and it really makes little difference how far the adults had to walk to get food. For example, penguins often go a hundred miles to get food for their babies.

What good is a study of geology to someone who doesn't plan to become a geologist?

I think our environment is going to be the next century's biggest topic. And you cannot have a good handle on the environment unless you understand both biology and geology. So as the environment becomes more and more of an issue, there are going to be more and more jobs in the field of environmental geology. I think we are going to see the pendulum swing toward fields that figure out how to save our world. We're going to have to understand animals as animals, how rivers work, and similar concepts. A strong understanding of geology and biology is what most people are going to need to address environmental concerns.

or four weeks at the most would have been needed for the nesting period. During this short time the young would grow large enough to walk with the adults. We're guessing from preliminary information that duck-billed dinosaurs hatched out of their eggs at about 18 to 20 inches long and grew to about 45 to 50 inches long, possibly more, by the time they left the nest. But at that same growth rate, they would have grown to 9 to 12 feet the first year. The maximum size of an adult is about 35 feet. This suggests that the growth rate was a little faster than an ostrich, but a lot slower than most birds.

What do you think about the traditional view of dinosaurs being green and scaly as opposed to more re-

Internal Processes

◄ *Lava pouring from a lava tube in Hawaii*
(J.D. Griggs/ USGS)

Plate Tectonics

*Tectonic forces folded these limestone beds and raised the
Canadian Rockies to form a great mountain range.*

Science usually creeps forward by innumerable little discoveries, each won by months or years of hard work in the field or laboratory. Occasionally, however, scientists gather all the little advances into a new idea or a new way of looking at old ideas to initiate a major scientific revolution. Such a revolution has recently occurred in geology. Its effects are as exciting and important to geologists as Einstein's theory of relativity was to physicists earlier in this century.

The theory that has evolved from this revolution is called the **plate tectonics theory**. Briefly, it states that the Earth's outer layer, a 100-kilometer-thick shell of rigid rock called the lithosphere, is broken into independent segments called **plates** (Fig. 4.1). The plates float on a layer of hot, soft, plastic rock called the **asthenosphere**. Plates move horizontally across the Earth's surface by gliding over the asthenosphere, like sheets of ice floating back and forth on a pond.

A **fault** is any fracture in rock along which movement has occurred. A **plate boundary** is a fault separating one plate from another. Because the plates move relative to one another as they float over the asthenosphere, they collide at some of the plate boundaries. At other boundaries, two plates move apart from one another; and at yet other boundaries, two plates move horizontally past one another. California's famous San Andreas fault is an example of this type. Because 100-kilometer-thick plates bump and grind together at a plate boundary, you can easily imagine that a boundary is a geologically active place. Earthquakes, volcanoes, mountain building, and similar geological activities and features concentrate near plate boundaries.

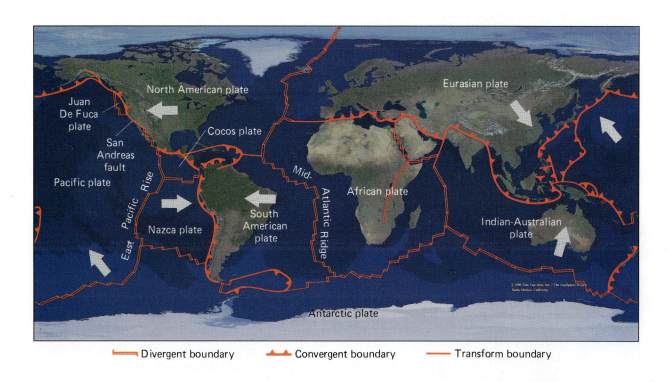

Divergent boundary   Convergent boundary   Transform boundary

Figure 4.1   Lithospheric plate boundaries are shown in red. The major plates are the African, Eurasian, Indian-Australian, Antarctic, Pacific, North American, and South American. A few of the smaller plates are also shown. Gray arrows indicate directions of plate movement.   *(Tom Van Sant/Geosphere Project)*

4.1 Alfred Wegener and the Origin of an Idea

In the early twentieth century, a young German scientist named Alfred Wegener noticed that the African and South American coastlines on opposite sides of the Atlantic Ocean seemed to fit as if they were adjacent pieces of a jigsaw puzzle (Fig. 4.2). He then launched an idea whose pursuit became one of the most fascinating chronicles in the history of science.

Born in 1880, Wegener began his career studying meteorology. While a university student, he took part in an exploratory expedition to Greenland in 1906. He developed a fascination for that ice-covered island in the North Atlantic that continued throughout his life and was eventually responsible for his premature death.

Although his formal profession was meteorology, Wegener was intrigued by the matching of continents across the Atlantic Ocean. He realized that the jigsaw-like fit suggested that the continents had once been joined together and had later split and drifted apart. But, as a good scientist, he recognized that this similarity alone did not prove that a single large continent had split apart to form the modern continents. Therefore, he began seeking additional evidence in 1910 and continued work on the project until his death in 1930. The idea that Wegener developed and pursued for 20 years is known as the theory of **continental drift**.[1]

Wegener's Evidence of Continental Drift

Wegener accumulated evidence of many different kinds indicating that all of the Earth's continents once had been joined and formed a single, large landmass.

Fit of Continental Coastlines

Studying world maps, Wegener noticed that not only did the continents on both sides of the Atlantic fit together well, but that other continents, when moved properly, also fit like additional pieces of the same jigsaw puzzle. He constructed a map of the Earth based on the fit of continents. On his map, all the continents were joined together, forming one supercontinent, which he called **Pangaea** (Fig. 4.3), from the Greek root words for "all lands." The northern part of Pangaea is commonly called **Laurasia** and the southern part **Gondwanaland**.

Evidence from Fossils

Mesosaurus was a small, aquatic, sharp-toothed reptile that looked like an alligator. *Mesosaurus* fossils have been found in South America and Africa, but nowhere else in the world. It is difficult to imagine how a slow, wallowing, fresh-water swamp dweller could swim thousands of kilometers across the Atlantic Ocean to populate two different continents. However, Wegener noted that on his reconstructed map of Pangaea, the fossil sites were immediately adjacent to one another.

Wegener also mapped the location of the fossil remains of a fern called *Glossopteris*. Fossils of this plant were found in Antarctica, Africa, Australia, South America, and India. Why would the same species be found on continents separated by thousands of kilometers of ocean? To solve this riddle, Wegener plotted localities where those fossils had been found on his Pangaea map. Again, all of the fossil localities are in the same region of Pangaea (Fig. 4.4).

The fossil evidence led Wegener to an additional interpretation of Pangaea's history. He observed that rocks of Ordovician through Triassic age (505 through 208 million years ago) from both Africa and South America contained identical fossils. Rocks younger than

Figure 4.2 The African and South American coastlines appear to fit together like adjacent pieces of a jigsaw puzzle.

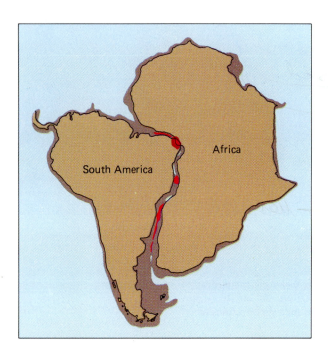

South America

Africa

---

[1] The theory of continental drift is similar to the theory of plate tectonics in that both ideas involve the movement of continents. But Wegener never postulated the existence of lithospheric plates or many of the ramifications of the plate tectonics theory.

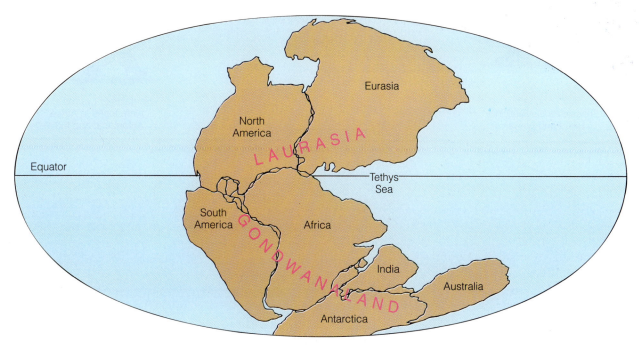

Figure 4.3 Wegener's Pangaea, a supercontinent that existed about 200 million years ago. The northern part of Pangaea is called Laurasia and the southern part is called Gondwanaland.

Triassic age, however, showed development of different species on the two continents. Wegener reasoned that the Atlantic Ocean basin had begun to open and tear Pangaea apart at the end of the Triassic period. Rocks that had formed before this separation contained identical fossils because the plants and animals had evolved and spread throughout Pangaea. Rocks formed after Pangaea split contained different fossils because evolution had followed different paths on the separated continents. Wegener was unable to determine the absolute time when the Atlantic Ocean began opening because radiometric dating had not yet been developed.

Fossil sand dunes are preserved in these desert sandstones of Arches National Park in Utah.

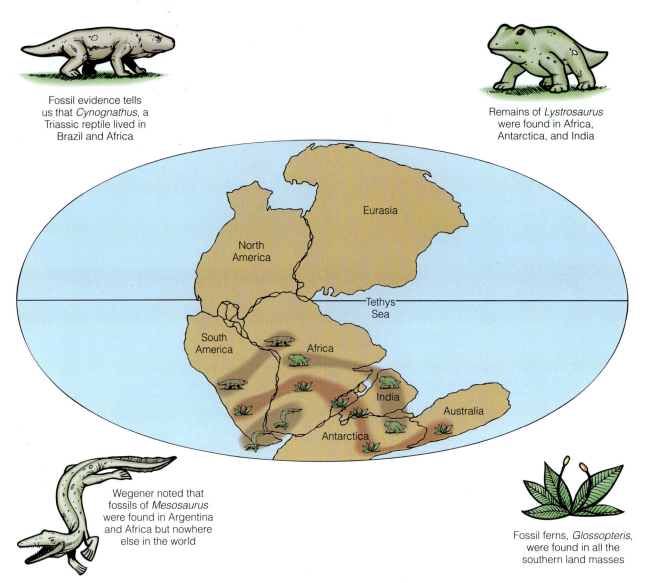

Fossil evidence tells us that *Cynognathus*, a Triassic reptile lived in Brazil and Africa

Remains of *Lystrosaurus* were found in Africa, Antarctica, and India

Wegener noted that fossils of *Mesosaurus* were found in Argentina and Africa but nowhere else in the world

Fossil ferns, *Glossopteris*, were found in all the southern land masses

Figure 4.4   Geographic distributions of plant and animal fossils indicate that a single supercontinent, called Pangaea, existed about 200 million years ago.

However, we now know that Pangaea started to split at the end of the Triassic period, slightly more than 200 million years ago.

Evidence from Paleoclimatology

Paleoclimatology is the study of ancient climates. As both a meteorologist and a geologist, Wegener knew that certain types of sedimentary rocks form in certain climatic zones of the Earth. Glaciers and glacial sediment, for example, concentrate in high latitudes. Deserts and the rocks that form in deserts cluster around latitudes 30° north and south. Coral reefs and coal swamps thrive in near-equatorial tropical climates.

Thus, sedimentary rocks reflect the latitudes at which they formed (Fig. 4.5).

Wegener plotted pre-Triassic sedimentary rocks that indicated climate and latitude on maps showing the modern distribution of continents. Figure 4.6A shows his map of 300-million-year-old glacial deposits. The light blue area shows how large the ice mass would have been if the continents had been in their present positions. Notice that the glacier would have crossed the equator, and glacial deposits would have formed in tropical and subtropical zones. Figure 4.6B shows the same glacial deposits plotted on Wegener's Pangaea map. Here they are neatly clustered about the South Pole. Other types of climate-sensitive sedimentary rocks

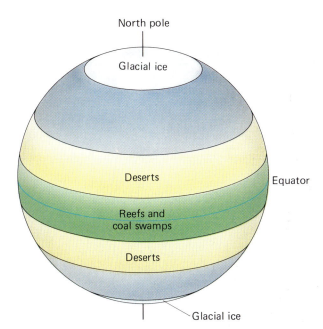

Figure 4.5   Climatic zones and many rock-forming environments are closely related to latitude.

are plotted on a Pangaea map in Figure 4.7. As in the case of glacial deposits, the rock distribution is more sensible on the Pangaea map than on a map showing the modern distribution of continents.

Geologic Evidence

Wegener noticed several instances in which an uncommon rock type or distinctive sequence of rocks on one side of the Atlantic Ocean was identical to rocks on the other side. When he plotted the rocks on a Pangaea map, those on the east side of the Atlantic were continuous with their counterparts on the west side. For example, the Cape Fold Belt of South Africa consists of a sequence of deformed rocks similar to rocks found in the Buenos Aires Province of Argentina. Plotted on a Pangaea map, the two sequences of rocks appear as a single, continuous belt of folded rocks. Figure 4.8 shows this relationship along with two other localities of geologic continuity between South America and Africa.

Wegener's Mechanism for Continental Drift: The Tragic Flaw

Wegener's theory of continental drift was so revolutionary that skeptical scientists demanded an explanation of *how* continents could move. They wanted an explanation of the *mechanism* of continental drift. Wegener had concentrated on developing evidence that continents had *drifted*, not on *how they moved*. Finally, perhaps out of exasperation and largely as an afterthought to what he considered the important part of his theory, Wegener suggested two alternative possibilities: first, that continents plow their way through oceanic crust, shoving it aside as a ship plows through water; and second, that continental crust slides over oceanic crust. These suggestions turned out to be an ill-considered and fatal step for his theory. Physicists immediately proved that both of Wegener's mechanisms were impossible. Oceanic crust is too strong for continents to plow through it. The attempt would be like trying to push a matchstick boat through heavy tar. The boat, analogous

Reefs produce fossil-rich limestone beds in tropical climates. *(Larry Davis)*

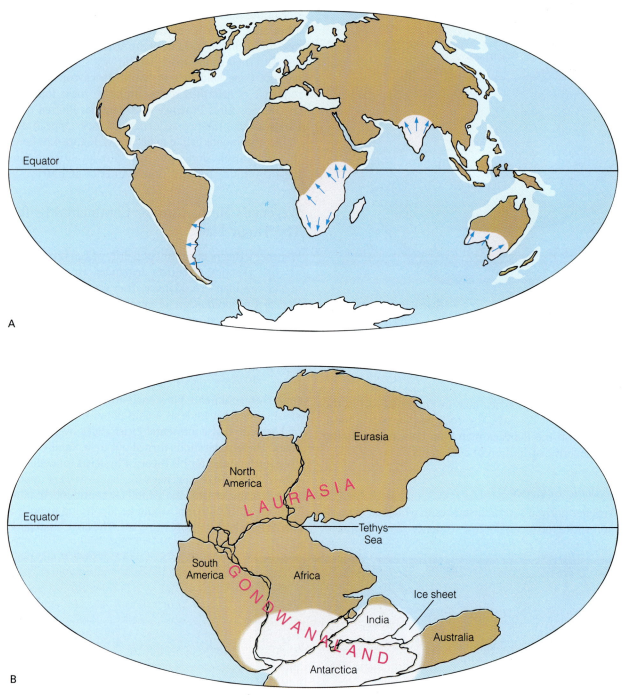

Figure 4.6 (A) Three-hundred-million-year-old glacial deposits plotted on a map showing the modern distribution of continents. Arrows show directions of ice movement. (B) The same glacial deposits plotted on a map of Pangaea.

to the continents, would break apart. Furthermore, frictional resistance is too great for continents to slide over oceanic crust.

The calculations of the physicists were adopted quickly by most scientists as proof that Wegener's theory of continental drift was wrong. Notice, however, that those calculations only proved that the *mechanism* proposed by Wegener was incorrect. They did not disprove, or even consider, the huge mass of evidence indicating that the continents were once joined together. This obvious distinction appears to have been missed by many. During the 30-year period from about 1930 to

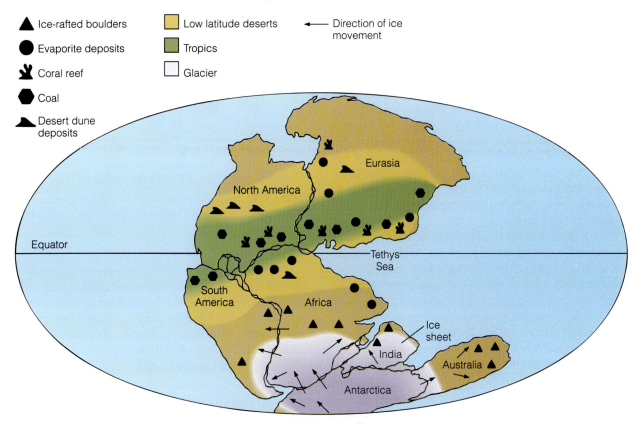

Figure 4.7   Climate-sensitive sedimentary rocks plotted on a map of Pangaea.

Figure 4.8   Locations of distinctive rock types in South America and Africa, plotted on a portion of a Pangaea reconstruction.

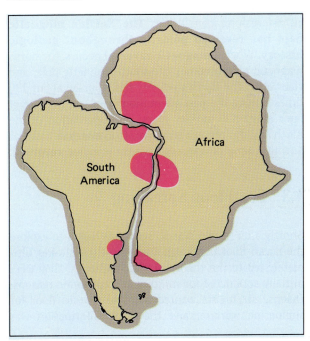

1960, a few geologists debated the continental drift theory, but most ignored it.

Wegener's fascination for Greenland led him to undertake a third expedition to the ice sheet in 1930. One of the shortest routes of the newly available air travel from northern Europe to North America involved flying over Greenland. To make the flights safer, it was necessary to have weather information from Greenland. Part of Wegener's mission was to establish a weather station near the center of the ice cap. In the late summer and autumn of 1930, Wegener and his companions freighted supplies and equipment by dogsled to establish the station, which was to be manned by a single meteorologist through the Arctic winter. Upon looking at what seemed a meager pile of supplies and considering the long winter ahead, the man who was to stay announced that the supplies were insufficient and that he would not stay unless he had more food and gear. The days were becoming short, and winter was coming rapidly. To save the expedition, Wegener and his assistant, an Eskimo named Rasmus Willimsen, set off on Wegener's fiftieth birthday to obtain additional supplies. Wegener was never seen again. Willimsen's body was found, frozen, the following summer by a search party.

An Assessment of Wegener's Theory

As you will see in the following sections, much of the theory of continental drift is similar to plate tectonics theory. Modern evidence indicates that the continents *were* together, just as Wegener's Pangaea map showed. Furthermore, recent data validate Wegener's interpretation that Pangaea started to split at the end of Triassic time, about 200 million years ago. He accumulated a vast amount of accurate data that firmly supported his theory that the continents were once joined together to form Pangaea and later drifted apart. The results of his work are a contribution to science of which any geologist would be proud.

Why, then, was the theory of continental drift nearly universally rejected? Most scientists of the day were apparently stuck in their intuitive notion that rocks are too hard and solid to permit the movement of continents. When physicists proved Wegener's mechanism impossible, the scientific community was only too willing to reject his data as well. Therefore, his careful work and well-thought-out conclusions were ignored.

As we shall see in the following sections, the plate tectonics theory resurrected Wegener's theory of continental drift and offers a reasonable mechanism for the movement of continents. The resurrection occurred unexpectedly in the 1960s as a result of new information about the Earth's magnetic field and the way in which rocks preserve a record of the Earth's magnetic history. Today, most geologists recognize the excellence of Wegener's contributions.

4.2   Rock and Earth Magnetism

Early navigators learned that no matter where they sailed, a needle-shaped magnet aligned itself in a north–south orientation. Thus, they learned that the Earth exhibits magnetic behavior and has a magnetic north pole and a magnetic south pole.

All iron-bearing minerals are permanent magnets. Although their magnetism is much weaker than that of a magnet used to stick cartoons on your refrigerator door, it is strong enough to be measured with laboratory instruments.

When magma solidifies, iron-bearing minerals crystallize and become permanent magnets. When an iron-bearing mineral cools within the Earth's magnetic field, its magnetic field aligns parallel to the Earth's field (Fig. 4.9). That is, the mineral's magnetic field points toward the North Pole just as a compass needle does. Thus, minerals in an igneous rock record the orientation of the Earth's magnetic field at the time the rock cooled.

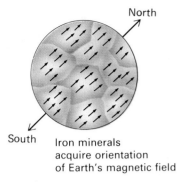

Figure 4.9   Iron-bearing minerals in cooling igneous rock acquire a permanent magnetic orientation parallel to the Earth's magnetic field.

Many sedimentary rocks also preserve a record of the orientation of the Earth's magnetic field at the time the sediment was deposited. As sedimentary grains settle through water in lakes or oceans, iron-bearing grains tend to settle with their magnetic axes parallel to the Earth's field. Even silt settling through air orients parallel to the external magnetic field.

Reversals of the Earth's Magnetic Field

The **polarity** of a magnetic field is the orientation of its positive, or north, end and of its negative, or south, end. Because many rocks record the orientation of the Earth's magnetic field at the time the rocks formed, we can construct a history of the Earth's field by studying magnetic orientations in rocks from many different ages and places. When geologists constructed a history of the Earth's magnetic field in this way, they arrived at an astonishing conclusion: **The Earth's magnetic field has reversed polarity throughout geologic history.** When a **magnetic reversal** occurs, the north magnetic pole becomes the south magnetic pole, and vice versa. The Earth's polarity has reversed about 130 times during the past 65 million years, an average of once every half-million years. The orientation of the Earth's field at present is referred to as **normal**, and that during a time of opposite polarity is called **reversed**.

4.3   Sea-Floor Spreading

Shortly after World War II, scientists began to explore the ocean floor. Although these sea-floor studies ultimately led to the theory of plate tectonics, they were initially conducted for military and economic reasons. Defense strategists wanted maps of the sea floor for submarine warfare, and the same information was needed to lay undersea telephone cables. As they mapped the sea floor, oceanographers discovered a long submarine mountain range in the middle of the Atlantic Ocean. This range is called the **Mid-Atlantic ridge** (Fig. 4.10). Further studies showed that the Mid-Atlantic ridge is part of a continuous submarine mountain chain called the **mid-oceanic ridge**, which girdles the entire globe and is by far the Earth's largest and longest mountain chain.

One remarkable feature of the Mid-Atlantic ridge is that it lies right in the middle of the Atlantic Ocean basin, halfway between Europe and Africa to the east and North and South America to the west. As you learned in Chapter 3, most oceanic crust is basalt. Basalt is an iron-rich igneous rock and therefore becomes magnetic as it cools. Thus, sea-floor basalt records the orientation of the Earth's magnetic field at the time the basalt cooled.

Figure 4.11 shows magnetic orientations of sea-floor rocks along a portion of the Mid-Atlantic ridge known as the Reykjanes ridge. The black stripes represent rocks with **normal polarity**; i.e., their magnetic orientation parallels that of the Earth today. The intervening stripes represent rocks with **reversed polarity**, in which the magnetic orientation is exactly oppo-

The granite of Mount Asgaard, Baffin Island, contains iron-bearing minerals that record the orientation of the Earth's magnetic field at the time the rock cooled.

Figure 4.10 The Mid-Atlantic ridge is a submarine mountain chain in the middle of the Atlantic Ocean. It is a segment of the mid-oceanic ridge, which circles the globe like the seam on a baseball. *(Marie Tharp)*

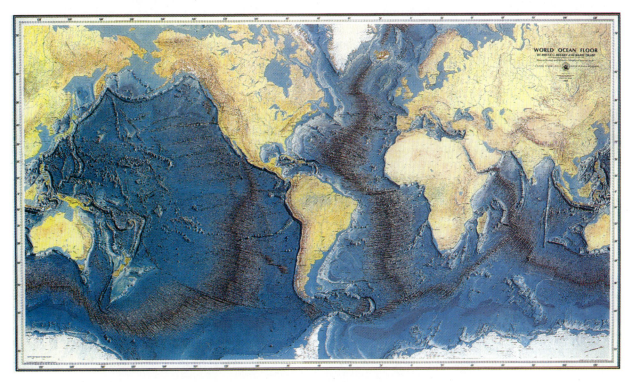

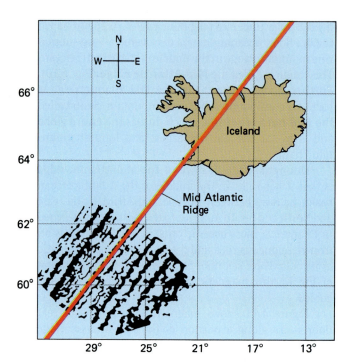

Figure 4.11 The Mid-Atlantic ridge, shown in red, runs through Iceland. Magnetic orientations of sea-floor rocks near the ridge are shown in the lower left portion of the map. The black stripes represent sea-floor rocks with normal magnetic polarity, and the intervening stripes represent rocks with reversed polarity. The stripes form a symmetrical pattern of alternating normal and reversed polarity on each side of the ridge. *(After Heirtzler et al., 1966, Deep-Sea Research, Vol. 13.)*

site that of the Earth today. Notice that the stripes form a pattern of alternating normal and reversed polarity. Note also that this pattern is symmetrical about the axis of the ridge. The central stripe is black, indicating that the rocks of the ridge axis have a magnetic orientation parallel to that of the Earth today.

What caused this pattern of alternating stripes of normal and reversed polarity in the rocks of the sea floor, and why are they symmetrically distributed about the Mid-Atlantic ridge? In the mid-1960s, geologists suggested that a sequence of events creates the symmetrical pattern of magnetic stripes on the sea floor:

1. New sea floor forms continuously from basaltic magma rising beneath the ridge axis. The new oceanic crust then spreads outward from the ridge. This movement is analogous to two broad conveyor belts moving away from one another. As a result of this process, the sea floor at the Mid-Atlantic ridge is very young and becomes progressively older with greater distance from the ridge.

2. As the new sea floor cools, it acquires the orientation of the Earth's magnetic field.

3. The Earth's magnetic field reverses orientation on an average of every half-million years.

4. Thus, the alternating magnetic stripes on the sea floor simply record the succession of reversals in the Earth's magnetic field that take place as the sea floor spreads away from the ridge (Fig. 4.12).

To return to our analogy of the conveyor belt, imagine that a can of white spray paint is mounted above a black conveyor belt. Alternating white and black stripes across the width of the belt are produced if paint is sprayed intermittently as the conveyor belt moves beneath it. When the paint sprayer is off, a black stripe forms, analogous to a time of normal polarity. When the paint sprayer is on, a white stripe forms, symbolizing a time of reversed polarity.

Figure 4.12 As new oceanic crust cools at the mid-oceanic ridge, it acquires the magnetic orientation of the Earth's field. Alternating stripes of normal (blue) and reversed (green) polarity record reversals in the Earth's magnetic field that occur as the crust spreads outward from the ridge.

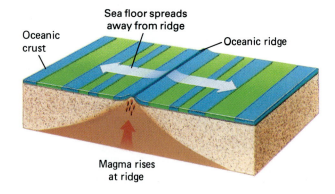

At about the same time that oceanographers discovered magnetic stripes on the sea floor, they also began to sample the mud covering the deep-sea floor. Mud is thinnest at the Mid-Atlantic ridge and becomes progressively thicker at greater distances from the ridge. If mud falls to the sea floor at about the same rate everywhere, the fact that it is thinnest at the ridge confirms the idea that the sea floor is youngest there and becomes progressively older away from the ridge.

Oceanographers soon recognized similar magnetic stripes and sediment trends along other portions of the mid-oceanic ridge. As a result, they proposed the theory of **sea-floor spreading** as a model for the origin of all oceanic crust. However, the sea-floor spreading theory offered no further enlightenment regarding Wegener's evidence for continental drift, nor did it explain other geological phenomena such as earthquakes, volcanic activity, and mountain building. In a short time, however, geologists combined the sea-floor spreading theory with Wegener's evidence and other data to develop the plate tectonics theory. The plate tectonics theory offers us a single model to explain how and why continents move, sea floor spreads, mountains rise, earthquakes shake our planet, and volcanoes erupt.

4.4   The Earth's Layers

The Earth is a layered planet, and the theory of plate tectonics depends on an understanding of the Earth's layers.

The Crust

Figure 4.13 shows a cross-sectional view of the Earth, and Table 4.1 summarizes the properties of its layers. The thinnest layer is the outer shell, called the **crust**. Both the thickness and the composition of the crust vary. Oceanic crust ranges from 7 to 10 kilometers in thickness and is composed mostly of basalt. Continental crust is much thicker than oceanic crust. The continents not only rise above the ocean floor, but also extend below it. Thus, continents have roots that protrude into the mantle. In mid-continent regions, the crust is about 20 to 40 kilometers thick, whereas under major mountain ranges its thickness increases to as much as 70 kilometers. Most continental crust is granitic in composition.

The granitic rock of continental crust is rigid. It is important to understand that rigidity is relative. Imagine that you could cut out the state of Kansas from the crust to a depth of 35 kilometers and place it on top of a neighboring state. It could neither support its own weight nor hold its shape. The edges would fracture and crumble and flow out over the Great Plains like honey.

The Mantle

The **mantle** is almost 2900 kilometers thick and makes up about 80 percent of the Earth's total volume. Its chemical composition is nearly constant throughout. However, the Earth's temperature and pressure increase with depth, and these changes cause the structure of the minerals and the physical properties of the mantle to vary with depth. The upper part of the mantle consists of two layers.

Figure 4.13   The Earth is a layered planet. The inset is drawn on an expanded scale to show near-surface layering.

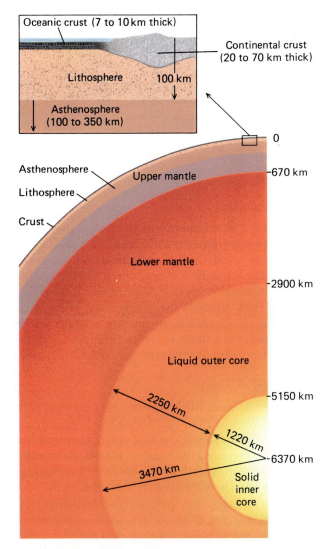

Table 4.1 • The Layers of the Earth

	Layer	Composition	Depth	Properties
Crust	Oceanic crust Continental crust	Basalt Granite	7–10 km 20–70 km	Cool, rigid, and brittle Cool, rigid, and brittle
Lithosphere	Lithosphere includes the crust and the uppermost portion of the mantle	Varies; the crust and the mantle have different compositions	About 100 km	Cool, rigid, and brittle
Mantle	Uppermost portion of the mantle included as part of the lithosphere Asthenosphere Remainder of upper mantle Lower mantle	Entire mantle is ultramafic rock. Its mineralogy varies with depth	Extends from 100 to 350 km Extends from 350 to 670 km Extends from 670 to 2900 km	Hot and plastic, 1% or 2% melted Hot, under great pressure, rigid, and brittle High pressure forms minerals different from those of the upper mantle
Core	Outer core	Iron and nickel	Extends from 2900 to 5150 km	Liquid
	Inner core	Iron and nickel	Extends from 5150 km to the center of the Earth	Solid

The Lithosphere

The outer 100 kilometers of the Earth, including both the uppermost mantle and the crust, make up a layer called the **lithosphere** (Greek for "rock layer"). The lithosphere is close enough to the surface that it is relatively cool. As a result, the rock in this layer is rigid. Although the compositions of the crust and upper mantle are different, this 100-kilometer-thick zone behaves as a single layer. Most earthquakes originate in the rigid rock of the upper part of the lithosphere.

The Asthenosphere

At a depth of about 100 kilometers, the cool, rigid rock of the lithosphere gives way to the hot, soft, plastic rock of the **asthenosphere**. The rock of this layer is so hot that it flows readily, even though it is solid. To visualize a solid that can flow, think of Silly Putty™ or road tar on a hot day. The asthenosphere extends from the base of the lithosphere to a depth of about 350 kilometers. At the base of the asthenosphere, increasing pressure causes the mantle to become more rigid and less plastic, and it remains in this state all the way down to the core.

Isostasy

The lithosphere is of lower density than the asthenosphere. Therefore, it floats on the asthenosphere, much as an iceberg or a block of wood floats on water. This concept of a floating lithosphere is called **isostasy**.

To illustrate isostasy, imagine three icebergs of different sizes floating in the ocean. When ice floats in seawater, approximately 10 percent of the berg rises above water level, whereas the remaining 90 percent is underwater. Of the three bergs, the largest has the most material underwater, but it also has the highest peak (Fig. 4.14). The lithosphere behaves in a similar manner. Continents have "roots" that extend into the mantle. The roots beneath major mountain ranges are deeper than the roots beneath continental plains, just as the bottom of a large iceberg is deeper than the base of a smaller one.

If our theory of isostasy is a good one, it should explain why the ocean floor is lower than the continents. Even if both the ocean floor and the continents were made of the same material, the thinner oceanic lithosphere would be expected to float lower in the astheno-

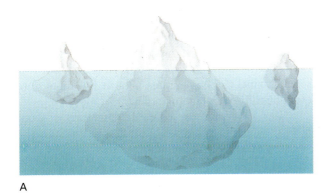

A

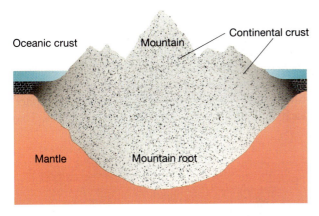

Oceanic crust

Mountain

Continental crust

Mantle

Mountain root

B

Figure 4.14 The principle of isostasy. (A) The largest of the three icebergs has the most material underwater and also the most above. (B) In an analogous manner, continental crust extends more deeply into the mantle beneath high mountains than it does under lower areas of a continent. Thinner and denser oceanic crust floats lower than continental crust.

sphere, just as a thin iceberg sits lower in the water than a thick one does. But, in addition, oceanic crust is made up of basalt, which is denser than granite. The denser oceanic lithosphere settles down even farther in the asthenosphere, giving rise to the large difference in elevation between ocean basins and continents.

If you have ever loaded equipment onto a small boat, you may have noticed that it settles lower in the water as cargo is added, whereas when the boat is unloaded, it rises. The lithosphere behaves in a similar manner. But how is "cargo" added or subtracted from the Earth's lithosphere? One example is provided by the growth and melting of large glaciers. When glaciers grow, the added weight of ice forces the continents to sink. The central portion of Greenland, which lies under a 3000-meter-thick glacier, has been depressed so far that it actually lies below sea level. On the other hand, when glaciers melt, the continents rebound, or rise up. Geologists have discovered Ice Age beaches along many coastlines in the Northern Hemisphere that now lie

high above sea level because the land rose as the glaciers melted. This vertical movement in response to changing burdens is called **isostatic adjustment** (Fig. 4.15). The largest human-made structures alter the distribution of weight on the lithosphere sufficiently to trigger a detectable isostatic adjustment. When dams are built, the newly formed lakes are heavy enough to cause the lithosphere to sink a small but measurable amount. This sinking often causes swarms of small earthquakes. Mountains also rise isostatically as rain, glaciers, and landslides erode their tops. This topic is discussed further in Chapter 8.

The Core

Recall from Chapter 1 that during the early history of the Solar System, the Earth was a homogeneous mixture of dust and gas. As the planet evolved, it became hotter until it began to melt. Once the Earth melted, most of the heavy elements gravitated toward the core and most of the lighter ones floated toward the surface. As a result, the **core** consists mostly of the heavy metals iron and

Figure 4.15 The weight of an ice sheet causes continental crust to sink isostatically.

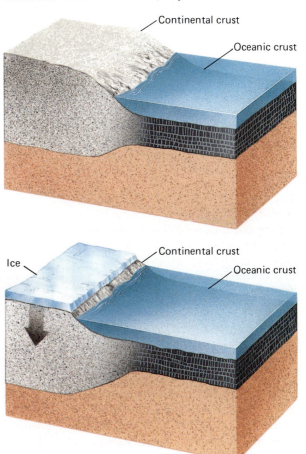

Continental crust

Oceanic crust

Ice

Continental crust

Oceanic crust

The Himalayas rise in the collision zone between two tectonic plates. Peaks above Askole, Pakistan.

nickel. It is a sphere with a radius of about 3470 kilometers, larger than the planet Mars. The outer core is molten because of the high temperatures near the Earth's center. The inner core is even hotter, but it is solid because the higher pressures in the inner core overwhelm the temperature effect.

Near its center the core's temperature is about 6000°C, which is as hot as the surface of the Sun. The pressure exceeds 1 million atmospheres.

4.5   Plates and Plate Tectonics

The plate tectonics theory is a model of the Earth in which the lithosphere "floats" on the hot, plastic asthenosphere. The lithosphere is broken into seven large plates and several small ones, resembling segments of a turtle's shell. The plates move across the Earth's surface, each in a different direction from its neighbors. They

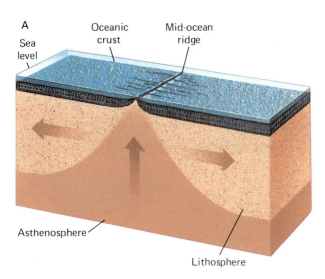

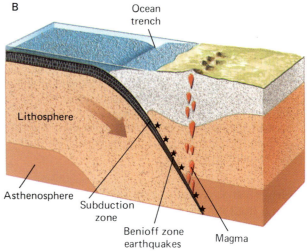

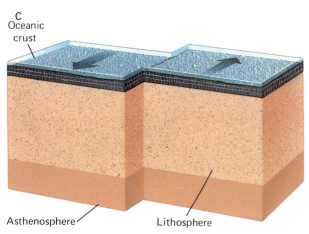

Figure 4.16   Three types of plate boundaries exist: (A) Two plates separate at a divergent plate boundary. New lithosphere forms as hot asthenosphere rises to fill the gap where the two plates spread apart. The lithosphere is relatively thin at this type of boundary. (B) Two plates collide at a convergent plate boundary. If one or both plates carry oceanic crust, the dense oceanic plate sinks into the mantle in a subduction zone. Here an oceanic plate is sinking beneath a less-dense continental plate. Magma rises from the subduction zone, and a trench forms where the subducting plate sinks. The stars mark Benioff zone earthquakes. (C) At a transform plate boundary, rocks on opposite sides of the fracture slide horizontally past each other.

glide slowly over the weak, plastic asthenosphere at rates ranging from less than 1 to about 18 centimeters per year. **Plate tectonics** is the study of the movement of lithospheric plates and the consequences of those motions.

Because the plates move in different directions, they bump and grind together at their boundaries. Just imagine two 100-kilometer-thick slabs of rock colliding along a boundary a few thousand kilometers in length. No convenient analogy exists to describe such a collision. If a billion large bulldozers were to drive into a giant city made of all the buildings on Earth, the force would be tiny compared with that resulting from a collision of two tectonic plates. Because of the great forces generated at plate boundaries, mountain building, volcanic eruptions, and earthquakes occur where two plates meet. These events are called **tectonic** activity, from the ancient Greek word for "construction." Tectonism constructs and modifies the Earth's crust. In contrast to plate boundaries, interior portions of lithospheric plates are usually tectonically quiet because they are far from the zones where two plates interact.

Simple logic tells us that one plate can move relative to an adjacent plate in three different ways (Fig. 4.16). At a **divergent plate boundary**, two plates move apart, or separate. At a **convergent plate boundary**, two plates move toward each other and collide. At a **transform plate boundary**, two plates slide horizontally past each other. Each of these three types of boundaries creates different tectonic features. Table 4.2 summarizes characteristics of each type of plate boundary and lists modern examples of each.

4.6   Divergent Plate Boundaries

A divergent boundary, also called a **spreading center** and a **rift**, occurs where two plates move apart horizontally (Fig. 4.17). As the two plates separate, hot, plastic asthenosphere rock flows upward to cool and form new lithosphere in the gap left by the diverging plates. The rising asthenosphere melts partly, forming basalt magma that oozes to the surface. Rifts occur in both oceanic and continental crust.

Table 4.2 • Plate Boundaries

Type of Boundary	Types of Plates Involved	Topography	Geologic Events	Modern Examples
Divergent	Ocean-ocean	Mid-oceanic ridge	Sea-floor spreading, shallow earthquakes, rising magma, volcanoes	Mid-Atlantic ridge
	Continent-continent	Rift valley	Continents torn apart, earthquakes, rising magma, volcanoes	East African rift
Convergent	Ocean-ocean	Island arcs and ocean trenches	Subduction, deep earthquakes, rising magma, volcanoes, deformation of rocks	Western Aleutians
	Ocean-continent	Mountains and ocean trenches	Subduction, deep earthquakes, rising magma, volcanoes, deformation of rocks	Andes
	Continent-continent	Mountains	Deep earthquakes, deformation of rocks	Himalayas
Transform	Ocean-ocean	Major offset of mid-oceanic ridge axis	Earthquakes	Offset of East Pacific rise in South Pacific
	Continent-continent	Small deformed mountain ranges, deformations along fault	Earthquakes, deformation of rocks	San Andreas fault

Divergent Boundaries in Oceanic Crust: The Mid-Oceanic Ridge

As shown in Figure 4.17, the mid-oceanic ridge is a divergent boundary in oceanic crust. The basalt magma that oozes out onto the sea floor at the ridge forms oceanic crust on top of the new lithosphere. Approximately 6.5×10^{18} tons of new oceanic crust form in this way each year! Note that it is not merely oceanic crust that spreads outward from the mid-oceanic ridge, as the sea-floor spreading theory proposed. Rather, the entire lithosphere spreads, carrying the sea floor on top of it in piggyback fashion.

At the spreading center, the new lithosphere is hot and therefore of low density. It then cools and becomes more dense as it spreads outward from the ridge axis. The new, hot, low-density rock at the spreading center "floats" at a higher elevation than the older and cooler rock on both sides. Thus, the mid-oceanic ridge is elevated above surrounding sea floor because it is made of the newest, hottest, and lowest-density lithosphere.

Although the mid-oceanic ridge is the Earth's longest mountain chain, we do not normally see it because it lies below sea level. It winds through all of the Earth's ocean basins, much like the seam on a baseball. Occasionally it rises above sea level to form islands such as Iceland.

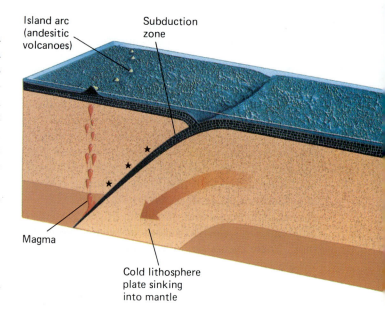

Tectonic forces are pulling the African continent apart along the East African rift, Kenya. *(Amos Turk)*

Divergent Boundaries in Continental Crust: Continental Rifting

A continent can also be pulled apart at a divergent boundary; the process is called **continental rifting**. Continental rifting is occurring today along a north–south fault zone in eastern Africa called the East African rift. If the rifting continues, eastern Africa will eventually separate from the main portion of the continent. Basalt magma will rise to fill the growing gap, forming a new ocean basin between the separating portions of Africa. Continental rifting may also be occurring in North America along the Snake River plain, which extends westward from Yellowstone National Park to southeastern Oregon. Elongate depressions called **rift valleys** develop along continental rifts because continental crust becomes stretched, fractured, and thereby thinned as it is pulled apart. Both the East African rift and the Snake River plain form great valleys in continental crust.

4.7 Convergent Plate Boundaries

A convergent boundary develops where two plates are moving horizontally toward each other and therefore are colliding (Fig. 4.16B). Collisions can occur (1) between a plate carrying oceanic crust and another carrying a continent, (2) between two plates carrying oceanic crust, and (3) between two continental plates.

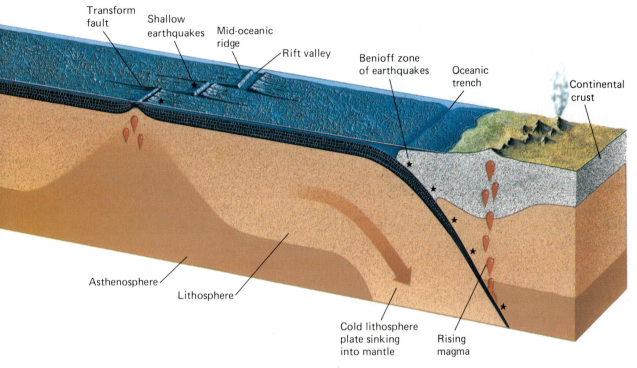

Figure 4.17   The outer few hundred kilometers of the Earth. In the center of the drawing, new lithosphere forms at a spreading center. At the sides of the drawing, old lithosphere sinks into the mantle at subduction zones. The lithospheric plates move away from the spreading center by gliding over the weak, plastic asthenosphere.

Convergence of Oceanic Crust with Continental Crust

Recall that oceanic crust is denser than continental crust. The difference in density determines what happens in a collision. Think of a boat colliding with a floating log. The log is denser than the boat, so it sinks beneath the boat. When a continental plate collides with a denser oceanic plate, the oceanic plate sinks beneath the continental plate and dives into the mantle. This process is called **subduction**.

A **subduction zone** is a long, narrow belt where a lithospheric plate dives into the mantle. On a worldwide scale, the rate at which old lithosphere sinks into the mantle at subduction zones is equal to the rate at which new lithosphere forms at spreading centers. In this way, a perfect global balance is maintained between the creation of new lithosphere and the destruction of old lithosphere.

Only lithosphere covered with oceanic crust can sink into the mantle at a subduction zone. Continental crust is of lower density than oceanic crust and cannot sink. Attempting to stuff a continent down a subduction zone would be like trying to flush a marshmallow down a toilet: It just would not go because it is too light. The oldest sea-floor rocks on Earth are only about 200 million years old because oceanic crust is continuously recycled back into the mantle at subduction zones. In contrast, geologists have found 3.8-billion-year-old rocks on continents. These old rocks exist because continental crust is not consumed by subduction.

Today, oceanic plates are subducting beneath continental crust along the western edge of South America and the coasts of Oregon, Washington, and British Columbia (see Fig. 4.1). Earthquakes, active volcanoes, and rising mountains all characterize these regions. Each of these tectonic activities is discussed in a following chapter. We describe them briefly here.

The Andes Mountains are rising today as a result of subduction of an oceanic plate beneath the western edge of South America. Condoriri, Bolivia. *(Karl Mueller)*

Subduction and Earthquakes

As a lithospheric plate sinks into the mantle, it scrapes against the opposite plate, causing numerous earthquakes. These quakes trace the path of the subducting plate as it sinks into the mantle (Fig. 4.16B). This zone of earthquakes is called a **Benioff zone**, after one of the geologists who first recognized it. The deepest earthquakes known occur in Benioff zones at a depth of about 700 kilometers. Below 700 kilometers, subducting plates must become so hot that they flow in a plastic manner rather than fracture. Earthquakes are common in western South America and along the coasts of Oregon, Washington, and British Columbia (see Chapter 5).

Subduction and Volcanoes

Because oceanic crust is covered by the sea, the upper part of a subducting plate consists of water-soaked sea-floor mud and basalt. As a subducting plate sinks, this water escapes into the asthenosphere. As you learned earlier in this chapter, the asthenosphere is so hot that it is soft and plastic—in fact, it is very close to melting. *Addition of water to very hot rock can cause the rock to melt.* Thus, water from the subducting plate causes melting in the asthenosphere (Fig. 4.16B). In this way, huge quantities of magma form in subduction zones. The magma rises through the overlying lithosphere. Some solidifies within the crust to form coarse-grained igneous rocks such as granite, whereas some erupts

onto the Earth's surface to form volcanic rocks. Igneous activity is common in the Cascade Range of Oregon, Washington, and British Columbia and in western South America. The relationship between subduction and igneous activity is discussed further in Chapter 6.

Subduction and Mountain Building

Many of the world's great mountain chains, including the Andes and parts of the mountains of western North America, formed at subduction zones. Several factors contribute to growth of mountain chains at subduction zones. The great volume of magma rising through the Earth's crust thickens the crust, causing mountains to rise. Volcanic eruptions pour huge amounts of lava onto the surface, constructing chains of volcanoes. Additionally, the Earth's crust crumples and buckles into mountain ranges where two lithospheric plates collide. The growth of mountains at subduction zones is described in greater detail in Chapter 8.

Oceanic Trenches

An **oceanic trench** is a long, narrow trough in the sea floor formed where a subducting plate bends downward to sink into the mantle (Fig. 4.16B). A trench can form wherever subduction occurs—where oceanic crust sinks beneath either continental or oceanic crust. Trenches are the deepest parts of the ocean basins. The deepest point on Earth is in the Mariana trench, in the southwestern Pacific Ocean north of New Guinea, where the sea floor is as much as 10.9 kilometers below sea level, compared with the average sea-floor depth of about 5 kilometers.

Convergence of Two Plates Carrying Oceanic Crust

Subduction also occurs where two oceanic plates collide. Recall that new oceanic lithosphere is hot and therefore of low density when it first forms at the mid-oceanic ridge. It cools and becomes denser as it ages and spreads away from the ridge. When two oceanic plates collide, the older, cooler, and denser plate subducts into the mantle. Magma forms and rises toward the Earth's surface, as it does at a subduction zone adjacent to a continent. The rising magma erupts onto the sea floor to form submarine volcanoes. With time, the volcanoes grow above sea level to form a chain of volcanic islands, called an **island arc**, along the subduction zone. The western Aleutian Islands and many of the island chains in the southwestern Pacific Ocean are island arcs. Both the trenches and island arcs of the southwestern Pacific show up clearly in Figure 4.10.

Mount Rainier, Washington, is one of several active and recently active volcanoes erupting above a subduction zone in the Pacific Northwest.   *(Don Hyndman)*

Convergence of Two Plates Carrying Continental Crust

If two colliding plates are both covered with continental crust, subduction cannot occur because continental crust is too light to sink into the mantle. In this case, the two continents collide and crumple against each other, forming huge mountain chains in the collision zone. The Himalayas, the Alps, and the Appalachians all formed as results of continental collisions. These processes are discussed further in Chapter 8.

4.8   Transform Plate Boundaries

A **transform plate boundary** forms where two plates slide horizontally past one another (Fig. 4.16C). California's San Andreas fault is a transform boundary between two major lithospheric plates, the North American plate and the Pacific plate. Although earthquakes are common at this type of boundary, igneous activity is not. Transform plate boundaries are discussed in Chapter 5.

4.9   Anatomy of a Plate

Plate tectonics theory provides a mechanism for the movement of continents that was lacking in Wegener's theory of continental drift. Recall that Wegener proposed that continents plowed through or slid over ocean crust and that these proposed mechanisms were discredited. Now geologists understand that continents are just large masses of granite sitting on top of some litho-

spheric plates. When lithospheric plates move, they carry the continents along with them. This is the mechanism that was not understood in Wegener's day. Continents do not plow through or slide over oceanic crust. They simply ride as passengers on the thick lithospheric plates.

The nature of a tectonic plate can be summarized as follows:

1. A plate is a segment of the lithosphere; thus, it includes the uppermost mantle and all of the overlying crust.

2. A portion of a plate with continental crust composing its uppermost layer is thicker than one bearing oceanic crust (Fig. 4.18). The average thickness of the part of a plate carrying oceanic crust is about 75 kilometers, whereas that of a plate bearing continental crust is about 125 kilometers. Lithosphere may be as little as 20 kilometers thick at an oceanic spreading center.

3. A plate is hard, rigid, or nearly rigid rock.

4. A plate "floats" on the underlying hot, plastic asthenosphere and glides horizontally over it.

5. A plate behaves like a large slab of ice floating on a pond. It may flex slightly, as thin ice does when a skater goes by, allowing minor vertical movement such as isostatic adjustment. In general, however, each plate moves as a single, large, intact sheet of rock.

6. A plate margin is tectonically active. Earthquakes and faulting are common at all plate boundaries. Igneous activity—volcanic eruptions and intrusion of magma—is frequent at subduction zones and spreading centers. Intense deformation of the Earth's crust occurs at transform boundaries and subduction

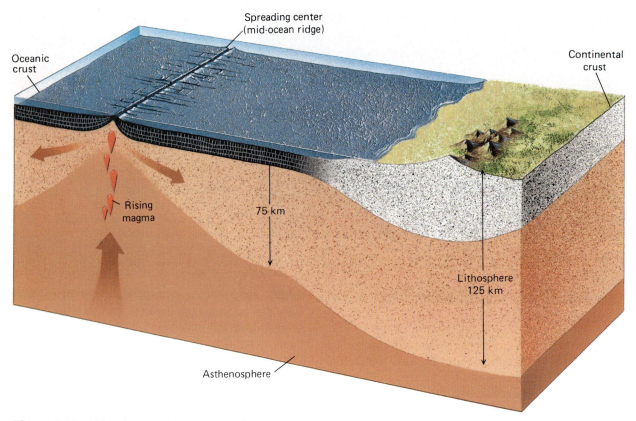

Figure 4.18 Lithosphere carrying continental crust is typically thicker than lithosphere bearing oceanic crust.

zones. In contrast, the interior of a lithospheric plate is generally tectonically stable.

Plate Velocity

How rapidly does a lithospheric plate move? Calculations based on several methods show that plates move away from spreading centers at rates that vary from 1 to 18 centimeters per year. In recent years, plate motion has been measured directly by surveying techniques that bounce laser beams off the Moon and satellites and by other methods that use radio waves originating outside our Galaxy (Fig. 4.19).

4.10   The Search for a Mechanism

Geologists have accumulated ample evidence that lithospheric plates move, and can even measure how fast they move. But *why* do plates move? At present, one of the most active and exciting areas of research in geology is the search for the cause of plate motions.

Mantle Convection

Although the mantle is solid rock (except for small, partially melted zones in the asthenosphere), it is so hot that over geologic time it flows slowly as a stiff fluid. Hot rock from deep in the mantle rises to the base of the lithosphere. At the same time, cooler upper-mantle rock sinks. This circular flow of solid rock is called **mantle convection**.

Mantle convection is thought to be closely related to movement of lithospheric plates. However, it is not clear whether convection of the mantle causes the plates to move or, conversely, movement of the plates is the cause of mantle convection.

Mantle Convection as the Cause of Plate Movement

Convection occurs when a fluid is heated from below, as in a pot of soup on a stove. The soup at the bottom of the pot expands as it is heated and becomes less dense than the soup at the top. Because it is less dense, it rises. When the hot soup reaches the top of the pot, it

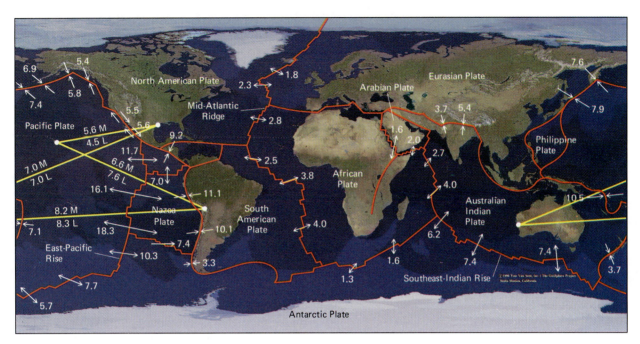

Figure 4.19   Plate velocities in centimeters per year. Numbers along the mid-oceanic ridges indicate the rates at which the two plates are separating, based on magnetic reversal patterns on the sea floor. The arrows indicate the directions of plate motions. The yellow lines connect stations used to measure present-day rates of plate motions with satellite laser ranging methods. The numbers followed by L are the present-day rates measured by laser. The numbers followed by M are the rates measured by magnetic reversal patterns.   *(Modified from NASA report, Geodynamics Branch, 1986, Tom Van Sant, Geosphere Project)*

flows along the surface as cooler soup sinks to take its place (Fig. 4.20). If the heat source persists, this cool, sinking soup is then warmed. It rises, and the convection continues.

Mantle convection might occur in a manner similar to that in the soup pot. In this model, the base of the mantle is heated from below, perhaps by the hot core. In turn, the heating causes mantle convection. Imagine a block of wood floating on a tub of honey. If you heated the honey from below so that it started to convect, the horizontal flow of honey along the surface would drag the block of wood along with it. Some geologists suggest that lithospheric plates are dragged along in a similar manner by a convecting mantle (Fig. 4.21).

Plate Movement as a Cause of Mantle Convection

Other geologists have suggested that movement of lithospheric plates might be the cause, rather than the effect, of mantle convection. Return to our analogy of the block of wood and the tub of honey. If you dragged the block of wood across the honey, friction between

Figure 4.20   Soup convects because it is heated from the bottom of the pot.

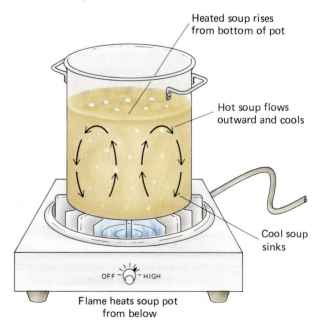

Heated soup rises from bottom of pot

Hot soup flows outward and cools

Cool soup sinks

Flame heats soup pot from below

OFF — HIGH

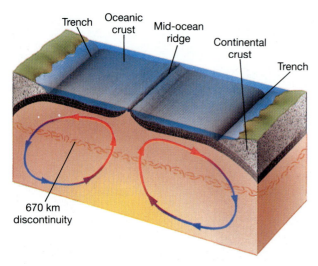

Figure 4.21 According to one explanation, lithospheric plates are dragged along by mantle convection.

the block and the honey would make the honey flow. Similarly, if some force caused the plates to move, their motion might cause the mantle to flow.

But what force would move the plates? A plate may simply glide downhill, away from a spreading center. Because newly formed lithosphere is hot and of low density, it is thin and sits at relatively high elevation at the spreading center. That is why the mid-oceanic ridge is topographically high. As the lithospheric plate spreads away from the ridge, it cools and thickens. Therefore, both the surface and the base of the lithosphere slope downward from the spreading center.

The average slope of the surface of the mid-oceanic ridge is about 0.6 percent. However, because the lithosphere thickens as it spreads away from the ridge, the slope of the base of the lithosphere beneath the ridge is much steeper (Fig. 4.22). The average slope of the base of the lithosphere beneath the ridge is about 8 percent, steeper than almost any paved road in North America. Calculations show that if the slope is as slight as 0.03 percent, a plate could glide away from a spreading center at a rate of a few to several centimeters per year.

Mantle Plumes, Meteorite Impacts, and Plate Movement

A recent suggestion for the cause of lithospheric plate motion is that a **mantle plume** initiates plate movement. A mantle plume is a vertical column of plastic rock rising through the mantle like hot smoke from an industrial smokestack (Fig. 4.23). Plumes originate deep within the mantle, perhaps even at the core–mantle

boundary. When a plume reaches the base of the lithosphere, it spreads outward, dragging the lithosphere apart and initiating a spreading center.

The suggestion that mantle plumes initiate plate movement raises another question: What causes mantle plumes? Geologists have recently suggested that a mantle plume (as well as several other major geologic features) forms when a large meteorite strikes the Earth with enough energy to blast a huge crater in the crust. Mantle rock directly below the crater begins to flow upward in isostatic adjustment to fill the crater, and the upwelling of a mantle plume begins. Once started, such upwelling may continue for millions of years.

Synthesis of a Single Model for the Cause of Plate Movement

At present, a combination of these models is the most appealing explanation for plate movement. In this synthesis, lithospheric spreading is initiated by a mantle plume. The mantle plume may have begun rising because of an unusually hot region near the core–mantle boundary or, alternatively, in response to a meteorite impact.

Once lithospheric spreading begins, the lithosphere–asthenosphere boundary steepens and the plates continue to slide downhill away from the spreading center (Fig. 4.22). At this point, subduction must also begin; if the plate is growing at one end, it (or some other lithospheric plate) must be consumed back into the mantle elsewhere. As the old, cold lithosphere sinks into the mantle at a subduction zone, it pulls on the rest of the plate, like a weight on the edge of a tablecloth.

In this synthesis, plate motion is initiated by a mantle plume, which may be the result of deep heating or a meteorite impact. However, once the plate begins to move over the asthenosphere, the motion becomes self-perpetuating and may continue for tens or even hundreds of millions of years.

4.11   Supercontinent Cycles

Recall that Wegener's Pangaea broke apart at the end of the Triassic period, approximately 200 million years ago. Although 200 million years is a long time span in terms of human life, it is a small fraction of the geologic history of our planet. What happened before the Pangaea supercontinent split apart?

Recently, Paul Hoffman of the Geologic Survey of Canada has suggested that at least three times during the Earth's history all continents were joined together

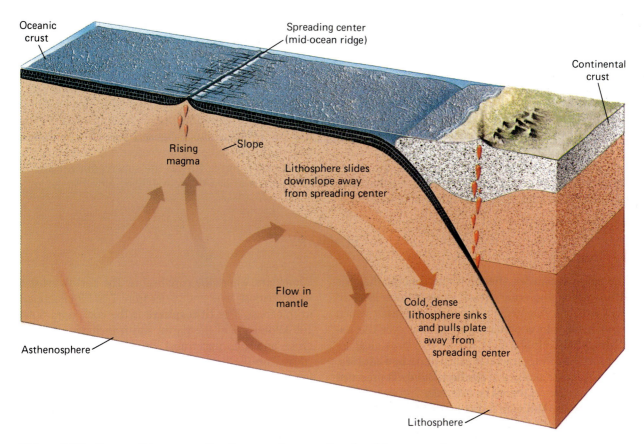

Figure 4.22 Two possible causes, other than mantle convection, for plate movement. (1) A plate glides down an inclined surface on the asthenosphere. (2) A cold, dense plate sinks at a subduction zone, pulling the rest of the plate along with it. In this drawing, both mechanisms are shown operating simultaneously.

Figure 4.23 A mantle plume rises as a vertical column and spreads at the base of the lithosphere.

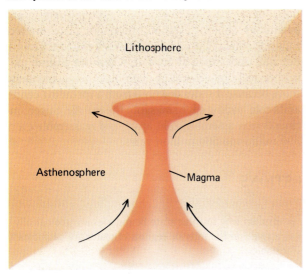

in a single supercontinent like Pangaea and then split apart. Hoffman has estimated that it takes about 300 million to 500 million years for a supercontinent to assemble, split apart, and then reassemble.

In this model, rifting breaks up a supercontinent and the fragments begin to separate. But because the Earth is a sphere, the continental fragments migrate halfway around the globe and then collide on the far side to reassemble as a new supercontinent. Thus, the breakup of one supercontinent leads to the assembly of a new one.

Prior to about 2 billion years ago, the land masses of the Earth consisted mainly of small island chains and microcontinents scattered about the globe and separated by ocean basins. Then movements of tectonic plates swept all of the islands and microcontinents together into a single great landmass. This accretion took about 200 million years to complete. Thus, Hoffman suggests, by 1.8 billion years ago, all of the Earth's continental crust was joined together in a supercontinent we called Pangaea I (after Alfred Wegener's Pangaea).

In Hoffman's model, the mantle beneath the new supercontinent soon began to warm up because the vast layer of continental crust acted as a giant blanket and kept heat from escaping. The warming mantle then started to rise as a mantle plume beneath the supercontinent. Another possibility is that a major meteorite impact initiated a mantle plume. In any case, the plume spread out at the base of the supercontinent, tearing it into several fragments, each riding on its own lithospheric plate. In this way Pangaea I broke apart about 1.3 billion years ago.

After Pangaea I split, the fragments of continental crust migrated halfway around the Earth and then reassembled, forming a second supercontinent called Pangaea II, about 1 billion years ago. In turn this continent fractured, and the continents migrated around the globe and reassembled into Pangaea III about 300 million years ago, 70 million years before the appearance of dinosaurs. Pangaea III was the supercontinent discovered and described so accurately by Alfred Wegener.

SUMMARY

The **plate tectonics theory** is the concept that the **lithosphere**, the outer, 100-kilometer-thick layer of the Earth, is segmented into seven major **plates**, which move relative to one another by gliding over the **asthenosphere**. Most of the Earth's major geological activity occurs at huge fractures called **plate boundaries**. Alfred Wegener's theory of **continental drift** preceded the plate tectonics theory by 4 decades and was similar in many ways. Atlhough it was based on accurate data, Wegener's theory was rejected because of a faulty explanation of how continents move.

The Earth's magnetic field reverses its orientation about every half-million years. **Rock magnetism** records the orientation of the Earth's field at the time rocks form. A pattern of alternating **normal** and **reverse magnetic polarity** in sea-floor rocks is arranged symmetrically about the mid-oceanic ridge. Comparison of this pattern with prehistoric reversals of the Earth's magnetic field led to the theory of **sea-floor spreading**. This theory states that new oceanic crust forms continuously at mid-oceanic ridges and spreads outward. The sea-floor spreading theory was expanded into the plate tectonics theory, which states that the entire lithosphere, not merely oceanic crust, forms and spreads outward from the mid-oceanic ridge.

The **core** is mostly iron and nickel and consists of a liquid outer layer and a solid inner sphere. The **mantle** extends from the base of the crust to a depth of 2900 kilometers, where the core begins. The Earth's crust is its outermost layer and varies from 7 to 70 kilometers in thickness. The **lithosphere** is the cool, rigid outer 100 kilometers of the Earth, and it includes all of the crust and the uppermost mantle. The lithosphere "floats" on the hot, plastic **asthenosphere**, which extends from 100 to 350 kilometers in depth.

The concept that the lithosphere floats on the asthenosphere is called **isostasy**. When weight is added to or subtracted from portions of the crust, it rises or falls. This vertical movement in response to changing burdens is called **isostatic adjustment**.

Tectonic plates move at rates that vary from 1 to 18 centimeters per year. Three types of plate boundaries exist: (1) New lithosphere forms and spreads outward at a **divergent boundary**, or **spreading center**; (2) two lithospheric plates collide at a **convergent boundary**, which develops into a **subduction zone** if at least one plate carries oceanic crust; and (3) two plates slide horizontally past each other at a **transform plate boundary**. Interior parts of lithospheric plates are tectonically stable. The cause or causes of plate motion are not well understood at present. A **mantle plume** may initiate plate movement. A plate may then continue to move because it slides downhill from a spreading center as its cold leading edge sinks into the mantle. Supercontinents may assemble, split apart, and reassemble every 300 million to 500 million years.

REVIEW QUESTIONS

1. Summarize the important aspects of the plate tectonics theory.

2. Describe the lithosphere and the asthenosphere.

3. How many major tectonic plates exist? List them.

4. Describe the three types of tectonic plate boundaries.

5. Explain why tectonic plate boundaries are geologically active and the interior regions of plates are geologically stable.

6. Describe the similarities and differences among the theories of continental drift, sea-floor spreading, and plate tectonics.

7. Why did Wegener consider that fossils of *Mesosaurus* found on both sides of the Atlantic Ocean constituted good evidence that the two fossil localities were once joined as a single continental mass?

8. Explain how Wegener's fossil evidence indicated when Pangaea began to rift apart.

9. How does a sedimentary rock record the orientation of the Earth's magnetic field at the time the sediment was deposited?

10. Describe the Mid-Atlantic ridge and the mid-oceanic ridge.

11. What is the magnetic orientation of a rock with normal polarity? What is reverse polarity?

12. What do rocks with normal and reverse polarity tell us about the history of the Earth's magnetic field?

13. Explain how a magnetic pattern in sea-floor rocks led to the theory of sea-floor spreading.

14. Why are the oldest sea-floor rocks only about 200 million years old, whereas some continental rocks are more than 3 billion years old?

15. Draw a cross-sectional view of the Earth. Label all the major layers and the thickness of each.

16. Describe the physical properties of each of the Earth's layers.

17. What properties of the asthenosphere allow the lithospheric plates to glide over it?

18. How is it possible for the solid rock of the mantle to flow and convect?

19. How might a mantle plume fracture the lithosphere?

20. Explain how convection of the mantle could cause movement of lithospheric plates. Explain how movement of lithospheric plates could cause mantle convection.

21. In Paul Hoffman's model, why do supercontinents break up within a few hundred million years after they form?

22. How many supercontinents have formed in Hoffman's model?

DISCUSSION QUESTIONS

1. Describe the similarities between Wegener's continental drift theory and modern plate tectonics theory. What are the major differences between the two theories?

2. Discuss the various mechanisms that have been suggested for the movement of tectonic plates. Attempt to decide which mechanism, if any, is preferable.

3. Although most earthquakes occur at plate margins, occasionally very large erthquakes occur within lithospheric plates. How might this happen?

4. The interior of the Moon is much cooler than that of the Earth, and the lunar crust is thicker than that of Earth. Compare the probable geologic effects of the impact of a large meteorite on the Moon with the effects of a similar impact on the Earth.

5. Describe and discuss Paul Hoffman's model of supercontinent cycles. Does the model seem plausible in light of the data presented in this chapter?

6. Discuss how microcontinents might have formed.

7. Discuss how and why supercontinents form.

8. The breakup of the first supercontinent was accompanied by intrusion of many granite plutons into continental crust. Most granites form in continental crust above subduction zones. Develop and discuss a model in which granite magma forms during rifting of a supercontinent.

KEY TERMS

Plate tectonics theory 77
Plate 77
Fault 77
Plate boundary 77
Continental drift 78
Pangaea 78
Laurasia 78
Gondwanaland 78
Paleoclimatology 80
Polarity 84

Mid-Atlantic ridge 85
Mid-oceanic ridge 85
Normal polarity 85
Reversed polarity 85
Sea-floor spreading 87
Crust 87
Mantle 87
Lithosphere 88
Asthenosphere 88
Isostasy 88

Isostatic adjustment 89
Core 89
Tectonic 91
Divergent plate boundary 91
Convergent plate boundary 91
Transform plate boundary 91
Spreading center 91

Rift 91
Continental rifting 92
Rift valley 92
Subduction 93
Subduction zone 93
Benioff zone 94
Oceanic trench 94
Island arc 94
Mantle convection 96
Mantle plume 98

Earthquakes and the Earth's Structure

California National Guardsmen patrol a Hollywood street and apartment damaged by the January 1994 earthquake.
(Wide World Photos)

San Francisco area residents were startled when the ground began to shake on October 17, 1989. A woman driving home from work noticed that the road suddenly started to roll like sea waves. Mesmerized, she continued driving until the pavement fractured and rose in front of her. She slammed on the brakes to avoid crashing into it. Others were not so lucky. In Oakland the upper tier of a double-deck freeway collapsed, crushing motorists on the lower tier. Another resident reading quietly in his room heard the dishes on the table begin to rattle. Pictures on the wall swung back and forth, and the floor began to shake. The motion was so foreign to his sense of normal earthly stability that he did not recognize what was happening. He wondered, "Who is shaking my house?" Within a few moments he realized, "The Earth is shaking; it's an earthquake!" Structural damage varied greatly throughout the Bay Area. Most buildings survived, but some were destroyed. Sixty-five people died.

About a million earthquakes occur worldwide every year. Most are too mild to be felt and are detected only with sensitive instruments. Many shake houses and rattle windows but cause little damage. A few destroy buildings or even entire cities. An average of 10,000 earthquake fatalities and billions of dollars in damage occur every year. Major historical earthquakes are listed in Table 5.1. Although earthquakes are among the most dramatic of all geological events, geologists cannot reliably predict when or where a quake will occur.

5.1 What Is an Earthquake?

An **earthquake** is a sudden motion or trembling of the Earth. The motion is caused by the abrupt release of slowly accumulated energy in rocks. What is the source of this energy, how does it accumulate in rocks, and why does it suddenly cause the Earth to shake?

Recall from Chapter 4 that the lithosphere is a layer of rock, about 100 kilometers thick, that floats on the hot, plastic rock of the asthenosphere. The entire lithosphere is broken into seven large tectonic plates and several smaller ones. As the plates glide over the hot, plastic asthenosphere, they slip past one another along immense fractures called plate boundaries. California's famous San Andreas fault is this kind of fracture. Although tectonic plates move continuously at rates between 1 and 18 centimeters per year, slippage along the fractures is not smooth and continuous.

Commonly, as two lithospheric plates move past one another, rocks along the fracture are held tightly by friction and cannot move. Thus, although the two plates continue to move past each other, no slipping occurs at the fault. *The interior of a plate can move whereas its edges do not because rock can stretch like an elastic band.* We do not commonly think of rock as elastic, but if you drop a rock onto a cement floor, it bounces. Although the magnitude of the bounce is small, the rock behaves like other elastic objects, such as tennis balls.

Imagine that you are trying to break a stick by gradually bending it over your knee (Fig. 5.1). At first

Figure 5.1 A stick stores elastic energy as it is bent and then releases the energy when it snaps.

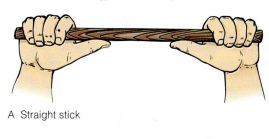

A. Straight stick

B. Stress applied

C. Fracture

Table 5.1 • Major Historical Earthquakes

Year	Date	Region	Deaths	Richter Magnitude*	Comments
1556		China, Shensi	830,000		More deaths than in any other natural disaster in history
1663	Feb. 5	Canada, St. Lawrence River	?		Chimneys broken as far away as Massachusetts
1811	Dec. 16	Missouri, New Madrid	Several		Three shocks; second largest historical earthquake in contiguous 48 states
1857	Jan. 9	California, Fort Tejon	?		San Andreas fault rupture
1868	Aug. 16	Ecuador and Colombia	Ecuador, 40,000 Colombia, 30,000		
1886	Aug. 31	South Carolina, Charleston-Summerville	About 60		Last major quake in eastern United States
1896	June 15	Japan, Riku-Ugo	22,000		Giant tsunami
1906	April 18	California, San Francisco	650	8.25†	San Francisco fire
1923	Sept. 1	Japan, Kwanto	143,000	8.2†	Great Tokyo fire
1960	May 21–30	Southern Chile	5700	8.5–8.7	Largest sequence of earthquakes ever recorded
1964	March 28	Alaska	131	8.6	Damaging tsunami
1970	May 31	Peru	66,000	7.8	Great rockslide destroyed the town of Yungay
1971	Feb. 9	California, San Fernando	65	6.5	$550 million damage
1976	July 27	China, Tangshan	250,000	7.6	Great economic damage; not predicted
1985	Sept. 19	Mexico, Michoacan	9500	7.9	More than $3 billion damage; 30,000 injured; small tsunami
1988	Dec. 7	Armenia	28,000	6.9	Death toll large due to poor construction
1989	Oct. 17	California, Loma Prieta	65	6.9	May be harbinger of additional Bay Area quakes

* Richter magnitudes were not recorded prior to 1935.

† Estimated magnitudes based on reported damage.

the two ends of the stick move, but the section bent over your knee is stationary. If you continue to pull, the stick suddenly snaps in two. The energy that you put into the stick by bending it was stored as **elastic energy**. When the stick breaks, that energy converts to sound and motion as the two halves of the stick spring back to become straight again.

In a similar manner, when two tectonic plates move past one another, rocks near the plate boundary stretch and store elastic energy. When the energy overcomes the friction that is keeping the rocks from moving, the rocks jump along the fault (Fig. 5.2). The rocks may move from a few centimeters to a few meters, depending on the amount of stored energy. The sudden movement releases the elastic energy, and the rocks spring back to their original shape. This rapid motion sets up vibrations that travel through the Earth like vibrations in a bell struck by a hammer. The vibrations are felt as an earthquake.

5.2   Earthquakes and Tectonic Plate Boundaries

Recall from Chapter 4 that a fault is a fracture in rock along which movement has occurred. In most cases, an earthquake starts on an old fault that has moved many times in the past and will move again in the future, simply because it is easier for rocks to move along an old fracture than for tectonic forces to create a new fault. Therefore, earthquakes occur over and over again in the same places, along the same faults. Although many faults occur within tectonic plates, the largest and most active faults are the boundaries between tectonic plates. Therefore, as Figure 5.3 shows, most earthquakes occur along plate boundaries. Long, quiet intervals separate the sudden movements.

Earthquakes and Transform Plate Boundaries: The San Andreas Fault

The San Andreas fault runs parallel to the California coast (Fig. 5.3). It is the boundary between two plates that are sliding horizontally past one another. The Pacific plate is moving northwest relative to the North American plate at a rate of about 3.5 centimeters per year. Along some portions of the fault, rocks on opposite sides slip past one another at a continuous, snail-like pace. This type of movement is called **fault creep**. In Hollister, California, old houses were inadvertently built directly over the fault. Slowly, millimeter by millimeter, fault creep has torn the houses in two. The movement occurs without violent and destructive earthquakes be-

cause the rocks move only a little bit at a time, continuously and slowly.

Along other portions of the San Adreas fault, friction binds rocks on opposite sides of the fault, keeping those parts of the fault motionless for decades. Nevertheless, the plates continue to move past one another. Elastic energy accumulates as the movement deforms rock near the fault. Because the plates move past one another at 3.5 centimeters per year, 3.5 meters of elastic deformation accumulate over a period of 100 years. When the elastic energy exceeds the friction binding the fault, the rock slips suddenly along the fault and snaps back to its original shape. During the San Francisco earthquake of 1906, rocks along the fault suddenly moved 4.5 to 6 meters.

Subduction Zone Earthquakes

In a subduction zone, two tectonic plates collide. At the collision boundary, one cold, lithospheric plate dives beneath the opposite plate and sinks slowly into the mantle. In many places the diving plate does not sink smoothly by fault creep, but with intermittent slips and

The San Andreas fault slices the Earth's surface in San Luis Obispo County, California.   *(R. E. Wallace, USGS)*

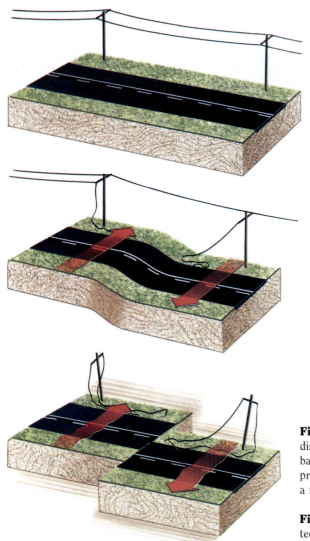

B

Figure 5.2 (A) A rock stores elastic energy when it is distorted by a tectonic force. When the rock fractures, it snaps back to its original shape, creating an earthquake. In the process, the rock has moved along the fracture. (B) Fractures in a roadway in Santa Cruz following the 1989 quake.

Figure 5.3 The Earth's major earthquake zones coincide with tectonic plate boundaries. Each yellow dot represents an earthquake that occurred between 1961 and 1967. *(Tom Van Sant, Geosphere Project)*

A

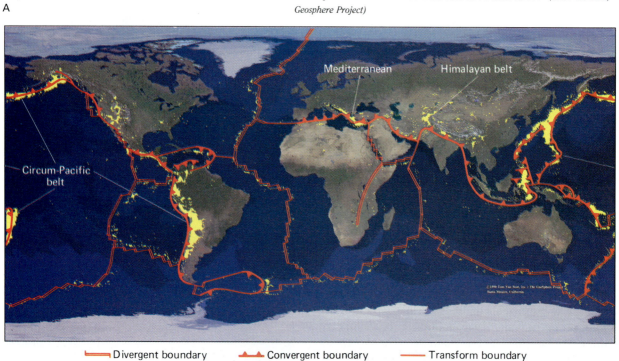

The 1964 Alaska earthquake destroyed much of Anchorage.   *(Ward's Natural Science Establishment, Inc.)*

jerks, giving rise to numerous earthquakes. Most of the earthquakes occur in the Benioff zone near the top of the sinking plate, where it scrapes past the opposing plate. Thus, earthquake distribution enables geologists to locate the top of the subducting plate. Subduction zone earthquakes of this type occur frequently along the west coast of South America, and in Japan, Alaska, the Aleutian Islands, and other places where subduction is active today. Many of the world's most destructive earthquakes occur in subduction zones.

A long time is required for a cold subducting plate to become hot and plastic as it sinks into the mantle. Therefore, a sinking plate remains brittle and generates earthquakes even when it has sunk to a depth of a few hundred kilometers. For this reason, the deepest earthquakes occur in subduction zones. Although very few occur below about 350 kilometers, the deepest earthquake ever recorded occurred at a depth of 700 kilometers. The fact that no deeper earthquakes occur indicates that at greater depths rocks become too hot and plastic to fracture. Instead, they flow as a viscous fluid.

Earthquakes at Spreading Centers

Many earthquakes occur along the mid-oceanic ridge as a result of fractures and faults that form as the two plates separate. Blocks of oceanic crust drop downward along the faults, forming the rift valley in the center of the ridge (Fig. 5.4). Only shallow earthquakes occur along the mid-oceanic ridge because here the asthenosphere rises to within 20 to 30 kilometers of the Earth's surface and is too hot and plastic to fracture.

Figure 5.4   Blocks of oceanic crust drop downward, forming a rift valley in the center of the mid-oceanic ridge, as two lithospheric plates spread apart. Earthquakes are common along the faults.

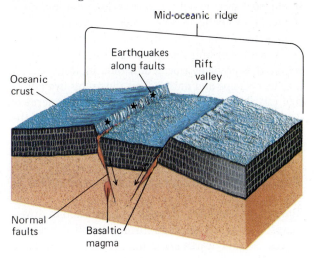

GEOLOGY AND THE ENVIRONMENT

Earthquake Danger in the Central and Eastern United States

No major earthquakes have ocurred in the central or eastern United States in the past 100 years,[1] and no lithospheric plate boundaries are known in these regions. Therefore, one might conclude that earthquake danger is insignificant. Indeed, building codes in most major central and eastern cities do not require earthquake-resistant construction.

Today this complacency is being questioned. According to one seismologist,

> We were told by theory that plate interiors should be quiet areas. Then people started building nuclear power plants and started to look at the details of data and discovered, in fact, that there is a lot of seismicity in interior plates and that we know very little about it.[2]

The largest historical earthquake sequence in the contiguous 48 states occurred not in California, but in Missouri. In 1811 and 1812, three shocks (with estimated Richter magnitudes between 7.3 and 7.8) altered the course of the Mississippi River and made church bells ring 1500 kilometers away in Washington, D.C. Another large quake (with an estimated Richter magnitude of 7.0) occurred in Charleston, South Carolina, in 1886.

Geologists have remeasured distances between old survey pins near New York City and found that the pins have moved significantly during the past 50 to 100 years. This motion indicates that the crust in this region is being deformed. If a major quake were to occur near New York City today, the consequences could be disastrous.

Causes of earthquakes in plate interiors are not as well understood as those at plate boundaries, and interior quakes are nearly impossible to predict. Some occur where thick piles of sediment were deposited to form great river deltas. The weight of sediment cannot be supported by the underlying lithosphere, and the lithosphere fractures. Other interior quakes may be caused by plate movement. As the plates glide horizontally over the asthenosphere, they may undergo some vertical motion above irregularities, or "bumps," in that plastic zone. Alternatively, under special circumstances, stress from a plate boundary may be transmitted hundreds of kilometers into the interior of a plate to produce an earthquake.

---

[1] A major earthquake is defined as one with a magnitude of 7.0 or higher on the Richter scale.

[2] Leonardo Seeber of Columbia University, as quoted in the *New York Times*, March 1, 1988, p. 87.

5.3   Earthquake Waves

If you have ever bought a watermelon, you know the challenge of trying to pick a ripe, juicy one without being able to look inside. One trick is to tap the melon gently with your knuckle. If you can hear a sharp, clean sound, it is probably ripe; a dull thud indicates that it may be overripe and mushy. This illustrates two points that can also be applied to the Earth: (1) The energy of your tap is transmitted through the melon, and (2) the quality of the sound is affected by the nature of the interior of the melon.

Waves transmit energy from one place to another. Thus, a drumbeat travels through air to your ear as waves, and the Sun's heat travels to Earth as waves. Similarly, a tap travels through a watermelon in waves. Waves that travel through rock are called **seismic waves**. Seismic waves are initiated naturally by earth-

quakes. They can also be produced artificially by explosions. **Seismology** is the study of earthquakes and of the nature of the Earth's interior based on evidence provided by seismic waves.

An earthquake begins at the **focus**, where rocks move suddenly along a fault. The point on the Earth's surface directly above the focus is the **epicenter**.

An earthquake creates two basic types of waves. **Body waves** radiate outward from the focus in concentric spheres and travel through the Earth's interior (Fig. 5.5). **Surface waves** radiate away from the epicenter and travel along the Earth's surface like the waves that form when you throw a rock into a calm lake.

Body Waves

Two main types of body waves travel through the Earth's interior. A **primary wave**, or **P wave**, forms by alter-

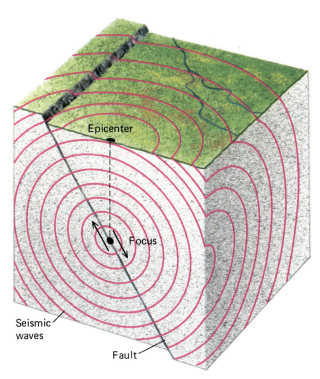

Figure 5.5 Body waves radiate outward from the focus of an earthquake.

nate compression and expansion of the rock (Fig. 5.6A). Consider a long spring such as the popular Slinky™ toy. If you stretch a Slinky and strike one end, a compressional wave travels along its length. Sound also travels as a compressional wave. A ringing bell produces a sound wave in air, which is a type of P wave. Both liquids and solids transmit P waves. Next time you take a bath, immerse your head until your ears are underwater and listen as you tap the sides of the tub with your knuckles. In a similar manner the music from a neighbor's radio in the apartment next door travels easily through the walls when you are trying to study geology.

P waves travel at speeds between 4 and 7 kilometers per second in the Earth's crust and at about 8 kilometers per second in the upper mantle. In comparison, the speed of sound in air is only 0.34 kilometer per second, and the fastest jet fighters fly at about 0.85 kilometer per second. Therefore, even the slowest P waves in the Earth travel more than ten times faster than the speed of sound in air, and more than five times faster than a jet fighter. P waves are called primary waves because they are so fast that they are the first, or primary, waves to reach an observer.

A second type of body wave, called a **secondary wave**, or **S wave**, can be illustrated by tying a rope to

Figure 5.6 Two different types of body waves travel through the Earth. Their characteristics are shown by a spring and a rope. (A) A compressional, or P, wave travels along the spring. The particles in the spring move parallel to the direction in which the wave itself travels. (B) An S wave travels along a rope, but the particles in the rope move perpendicular to the direction in which the wave travels.

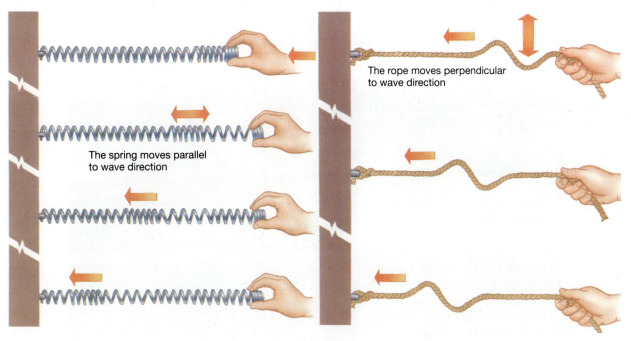

A   B

a wall, holding the end, and giving it a sharp up-and-down jerk (Fig. 5.6B). An S-shaped wave moves from your hand to the wall.

S waves are slower than P waves, traveling at speeds between 3 and 4 kilometers per second in the crust. As a result, they arrive after P waves and are the secondary waves to reach an observer on Earth.

Unlike P waves, S waves move *only through solids*. Because molecules in liquids and gases are only weakly bound together, they slip past each other and thus cannot transmit S waves. As we shall see in later sections, this difference is important for geologists studying the interior of the Earth. For example, we know that the Earth's outer core is liquid because P waves travel through it but S waves do not.

Surface Waves

Surface waves are called **L waves** (think of long waves) and travel more slowly than either type of body wave. Two types of surface waves occur simultaneously (Fig. 5.7). One is an up-and-down motion. Recall from the introduction to this chapter that a woman driving home during the 1989 San Francisco earthquake noticed that the road was rolling like an ocean wave. This motion was one form of L wave. The second type of L wave is

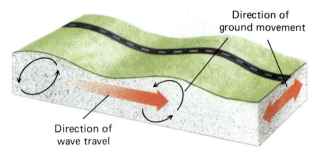

Figure 5.7   Surface waves move up and down, like ocean waves, and also from side to side.

a side-to-side vibration. Think of a snake writhing from side to side as it crawls along the ground.

Measurement of Seismic Waves

Earthquakes are detected and measured with a device called a **seismograph**. To understand how a seismograph works, imagine writing a letter while riding in an airplane. If the plane hits turbulence, your handwriting becomes wiggly. Because the paper is on a tray that in turn is connected to the frame of the aircraft, the paper moves with the plane when the plane bounces. But your

Earthquakes cause structural damage.   *(Ward's Natural Science Establishment, Inc.)*

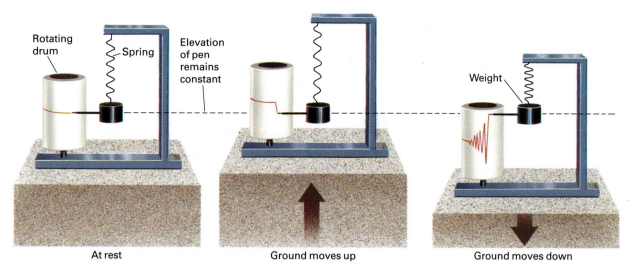

Rotating drum / Spring	Elevation of pen remains constant	Weight
At rest	Ground moves up	Ground moves down

Figure 5.8   A seismograph records ground motion during an earthquake. When the ground is stationary, the pen draws a straight line across the rotating drum. When the ground rises abruptly during an earthquake, it carries the drum up with it. But the spring stretches so the weight and pen hardly move. Therefore, the pen marks a line lower on the drum. Conversely, when the ground sinks, the pen marks a line higher on the drum. During an earthquake, the pen traces a jagged line as the drum rises and falls.

hand is connected to your body by a series of movable joints that flex when the plane lurches. Inertia keeps your hand stationary as the plane moves back and forth beneath it, so your hand does not move as erratically as the plane. The paper jiggles back and forth beneath a relatively motionless hand, and your handwriting becomes erratic.

An early type of seismograph works on the same principle. Imagine a weight suspended from a spring. A pen attached to the weight is aimed at the zero mark on a scale (Fig. 5.8). The scale is attached firmly to bedrock by solid metal braces, but the weight and pen hang from the flexible spring. During an earthquake, the scale jiggles up and down, but inertia keeps the suspended pen stationary. As a result, the scale moves up and down beneath the pen. If the scale is replaced by a rotating drum then the pen records earthquake motion on the drum. This record of Earth vibration is called a **seismogram**. Modern seismographs detect motion electronically.

An earthquake measuring station generally contains at least three seismographs. One instrument records vertical motions as just described. Two other seismographs record east–west and north–south horizontal movements. In this manner, the seismograph station records waves of different types and with different directions of motion.

5.4   Locating the Source of an Earthquake

If you have ever watched an electrical storm, you may have used a simple technique for estimating the distance between you and the place where the lightning is striking. You watch for the flash of the lightning bolt and then count the seconds until you hear the thunder. When an electrical discharge occurs, the thunder and lightning are produced simultaneously. But light travels so rapidly that it reaches you virtually instantaneously, whereas sound travels much more slowly, at 340 meters per second. Therefore, if the time interval between the flash and the thunder is 1 second, then the lightning struck about 340 meters away.

The same principle is used to determine the distance from a recording station to an earthquake. Recall that P waves travel faster than S waves and that L waves are slower yet. If a seismograph is located close to an earthquake epicenter, the different waves are separated by only short times for the same reason that the thunder and lightning come close together when a storm is close. On the other hand, if a seismograph is located farther away from the earthquake, the S waves arrive at correspondingly later times after the P waves arrive, and the L waves are even farther behind, as shown in Figure 5.9.

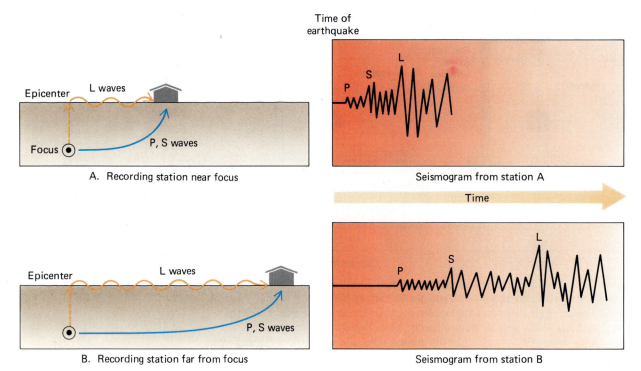

Figure 5.9 The time intervals between arrivals of P, S, and L waves at a recording station increase with distance from an earthquake.

Figure 5.10 A time-travel curve. With this graph you can calculate the distance from a seismic station to the epicenter of an earthquake. In the example shown, a 3-minute delay between the first arrivals of P waves and S waves corresponds to an earthquake with an epicenter 1900 kilometers from the seismic station.

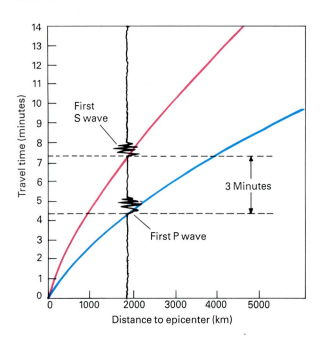

Geologists use **time-travel curves** to quantify this general relationship between distance from an earthquake epicenter and arrival times of the different types of waves. To make a time-travel curve you must know both when and where an earthquake started. A number of seismic stations at different locations on the Earth then record the time of arrival of earthquake waves, and a graph such as the one shown in Figure 5.10 is drawn. This graph can then be used to measure the distance between a recording station and an earthquake whose epicenter is unknown. Time-travel curves were first drawn from data obtained from natural earthquakes. However, a problem with using a natural earthquake is that its location and time are not always known precisely. In the 1950s and 1960s, geologists studied seismic waves from atomic bomb tests to refine their data. During these tests, both the location and timing of the events were known.

Figure 5.10 shows us that if the first P wave arrives 3 minutes before the first S wave, the recording station is about 1900 kilometers from the epicenter. But this distance by itself gives no information about direction; it does not indicate whether the earthquake originated to the north, south, east, or west. To solve this problem and to pinpoint the location of an earthquake, geologists compare data from three separate recording stations.

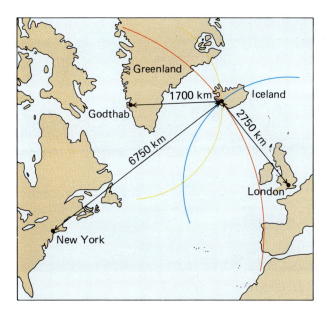

Figure 5.11 Locating an earthquake's epicenter. The distance from each of three seismic stations to the epicenter is determined from time-travel curves. The three distance circles intersect at the epicenter.

Imagine that a seismic station in New York records an earthquake with an epicenter 6750 kilometers away. From this information, geologists know that the epicenter must lie on a circle surrounding New York at a distance of 6750 kilometers. Part of this circle is shown by the red arc in Figure 5.11. But the same epicenter is also reported to be 2750 kilometers from a seismic station in London and 1700 kilometers from one in Godthab, Greenland. If three such circles are drawn, one for each recording station, the arcs intersect at only one point, and that is the epicenter of the quake.

5.5 Earthquakes and Humans

Measurements of Earthquake Strength

Earthquakes vary from gentle tremors that cannot be detected without a seismograph to destructive giants that create large-scale movements of the Earth's surface and catastrophic losses of life and property. Before seismographs were in common use, earthquake strength was evaluated on scales based on human experience and damage to buildings. One such scale, devised in 1902, is called the **Mercalli scale** (Table 5.2). This scale of earthquake **intensity** measures the effects of an earthquake at a particular place. Although useful, the Mercalli scale is qualitative and subjective. For example, some

Table 5.2 • Modified Mercalli Intensity Scale of 1931 (Abridged)

I. Not felt except by a very few under especially favorable circumstances.

II. Felt only by a few persons at rest, especially on upper floors of buildings.

III. Felt quite noticeably indoors, especially on upper floors, but many people do not recognize it as an earthquake. Vibration like passing truck.

IV. During the day felt indoors by many, outdoors by few. At night some awakened. Dishes, windows, doors disturbed; walls make cracking sound. Sensation like heavy truck striking building.

V. Felt by nearly everyone; many awakened. Some dishes, windows, and so on, broken; a few instances of cracked plaster; unstable objects overturned. Disturbance of trees, poles, and other tall objects sometimes noticed. Pendulum clocks may stop.

VI. Felt by all; many frightened and run outdoors. Some furniture moved; a few instances of damaged chimneys. Damage slight.

VII. Everybody runs outdoors. Damage negligible in buildings of good construction, slight to moderate in well-built ordinary structures, considerable in poorly built or badly designed structures.

VIII. Damage slight in specially designed structures, considerable in ordinary substantial buildings, great in poorly built structures. Fall of chimneys, factory stacks, columns, monuments, walls.

IX. Damage considerable in specially designed structures; well-designed frame structures thrown out of plumb; damage great in substantial buildings, with partial collapse. Ground cracked conspicuously. Underground pipes broken.

X. Some well-built wooden structures destroyed; most masonry and frame structures destroyed; ground badly cracked. Considerable landslides from river banks and steep slopes.

XI. Few if any (masonry) structures remain standing; bridges destroyed. Broad fissures in ground; underground pipelines completely out of service. Earth slumps and land slips in soft ground.

XII. Damage total. Waves seen on ground surface. Lines of sight and level distorted. Objects thrown upward into the air.

people exaggerate their experiences and others underestimate them. Therefore, it may be difficult to decide whether a given quake was a level V ("felt by nearly everyone") or a level VI ("many frightened and ran outdoors"). In addition, human experience and structural damage depend on a variety of factors including distance from the focus, the rock and soil where the quake occurs, and the quality of construction in the area.

Destruction caused by the Manjil, Iran, earthquake of June 21 and 22, 1990. The earthquake had a magnitude of 7.7 on the Richter scale and killed more than 50,000 people. *(M. Shandiz/Sygma)*

A quantitative scale based on seismograph measurements was first refined by Charles Richter in 1935, and today the **Richter scale** is used almost universally. It measures the **magnitude** of an earthquake from the amplitude (height)[1] of the largest wave recorded on a seismograph.

The Richter scale is logarithmic; an increase of one unit on the scale represents a tenfold increase in the amplitude of a seismic wave. Thus, the amplitude of the largest seismic wave produced by a magnitude 7 quake is 10 times greater than that from a magnitude 6 quake and 100 times greater than that from a magnitude 5 quake. An increase of one unit—for example, from 7 to 8—on the Richter scale corresponds approximately to a 30-fold increase in energy released during the quake. Thus, a magnitude 8 quake releases 30 times as much energy as a magnitude 7 quake and 900 times as much as a magnitude 6 quake. An earthquake with a magnitude of 6.5 has an energy of about 10^{21} (10 followed by 21 zeros) ergs.[2] The atomic bomb dropped on the Japanese city of Hiroshima at the end of World War II released about this much. The upper limit of the magnitude of an earthquake is determined by the strength of rocks. A strong rock can store more elastic energy before it fractures than a weak rock. The largest earthquakes

ever observed had magnitudes of 8.5 to 8.7, about 900 times greater than the energy released by the Hiroshima bomb.

Earthquake Damage

Ground Motion

During an earthquake, waves travel both along the surface of the Earth and through subterranean rock. The ground undulates; buildings and bridges may topple and roadways fracture. Most of the people killed during an earthquake are crushed by falling debris.

As mentioned previously, earthquake damage depends not only on the magnitude of the quake and its proximity to population centers, but also on rock and soil types and the quality of the construction in the affected area. The quake in Armenia in 1988 and the San Francisco quake in 1989 both measured 6.9 on the Richter scale. More than 28,000 people died in the Armenian quake, whereas the death toll in the Bay area was only 65. The tremendous mortality in Armenia occurred because buildings were not engineered to withstand earthquakes. Proper engineering and construction in an earthquake-prone area demand both a foundation on stable rock or soil and a structure that can withstand movement.

Permanent Alteration of Landforms

As explained previously, the Earth's surface moves both horizontally and vertically during an earthquake. As

---

[1] Amplitude is the height of a wave, as measured by the distance from the flat zero point to the top of the crest.

[2] An erg is the standard unit of energy in scientific usage. One erg is a small amount of energy. Approximately 3×10^{12} ergs are needed to light a 100-watt light bulb for 1 hour. However, 10^{21} ergs is a very large number and represents a considerable amount of energy.

Figure 5.12   Horizontal displacement along a fault offset this fence during the 1906 San Francisco earthquake. *(G. K. Gilbert/USGS)*

much as 6 meters of horizontal displacement occurred during the 1906 San Francisco quake (Fig. 5.12). The New Madrid, Missouri, earthquake of 1811 changed the course of the Mississippi River. Permanent displacement obviously destroys any structures built across a fault. Although relatively few buildings are likely to be situated right on a fault, many roadways and pipelines cross faults and are destroyed by such displacements.

Fire

When rock moves on a fault, buried gas mains and electrical wires crossing the fault may be torn apart, leading to fire (Fig. 5.13). In addition, water mains may rupture, so fire fighters may not have water. Much of the damage from the 1906 San Francisco earthquake resulted from fires.

Figure 5.13   A fire caused by the 1989 San Francisco earthquake. *(Michael Williamson/Sygma)*

Landslides

When the Earth trembles, hillsides may shake loose, and large quantities of rock and soil can slide downslope (Fig. 5.14). An earthquake in Peru in 1970 triggered a landslide that buried the town of Yungay and killed 17,000 people.

Tsunami

Imagine an earthquake epicenter beneath the sea. When part of the sea floor drops, as shown in Figure 5.15, the water drops with it. Almost immediately, water from the surrounding area rushes in to fill the depression, forming a wave. Sea waves produced by an earthquake are often called tidal waves, but because they have nothing to do with tides, geologists prefer to call them by their Japanese name, **tsunami**.

In the open sea, a tsunami is so flat and spread out that it is barely detectable. Typically, the crest may be only 1 to 3 meters high, and successive crests may be

Figure 5.14   The landslide that buried the town of Yungay, Peru. *(USGS)*

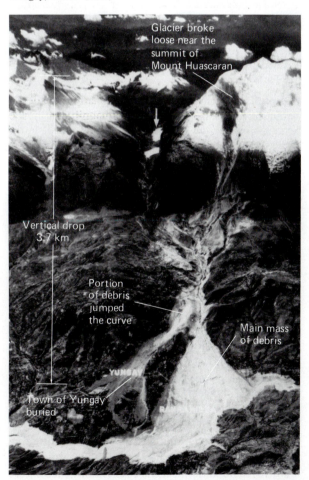

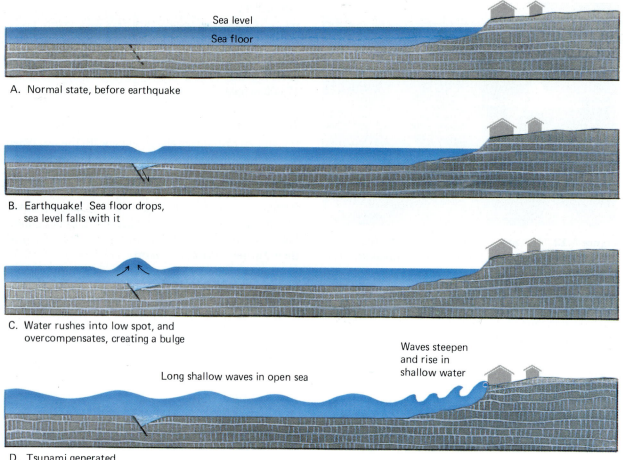

A. Normal state, before earthquake

B. Earthquake! Sea floor drops,
   sea level falls with it

C. Water rushes into low spot, and
   overcompensates, creating a bulge

D. Tsunami generated

Figure 5.15   A tsunami develops when part of the sea floor drops during an earthquake. Water rushes in to fill the low spot, but the inertia of the rushing water forces too much water into the area, creating a bulge in the water surface. The long, shallow waves can build up into destructive giants when they reach shore.

more than 100 to 150 kilometers apart. However, a tsunami travels rapidly and is very destructive when it reaches land. A tsunami may attain speeds of 750 kilometers per hour (450 miles per hour). When the wave approaches shore, its base drags against the ocean floor and slows down. This compresses the wave, and the distance between successive crests decreases as the wave height increases rapidly. The rising wall of water then flows inland. A tsunami can flood the land for as long as 5 to 10 minutes before it withdraws.

One of the worst tsunamis in history struck the eastern coast of Japan in 1896. The wave was probably formed by a submarine earthquake in the western Pacific Ocean. As it approached shore, the water rose about 30 to 35 meters (about 100 feet) above high-tide level, flooding villages and killing 26,000 people. The same wave was 3 meters high when it reached Hawaii, where it destroyed buildings near the coast; minor effects were

felt on the west coast of North America. The wave then bounced off the coast and sped westward, back across the Pacific Ocean, until it reached New Zealand and Australia. By then it had lost enough energy that it was no longer destructive.

5.6   Earthquake Prediction

Geologists attempt to predict earthquakes and identify earthquake hazard areas to advise the public of danger and to reduce loss of life and property damage. **Long-term predictions** distinguish areas of high earthquake potential from those of lower potential but do not specify exactly where or when an earthquake will occur. **Short-term predictions** attempt to pinpoint the time and location of future quakes but are much more difficult to make accurately.

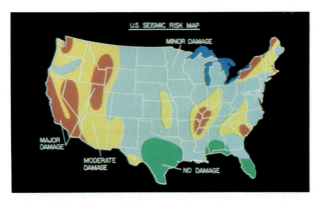

This map shows potential earthquake damage in the United States. The predictions are based on records of frequency and magnitude of historical earthquakes. *(Ward's Natural Science Establishment, Inc.)*

Let us first look at long-term earthquake predictions in two earthquake-prone regions of North America.

Case History I: California and the San Andreas Fault Zone

About 10 percent of the U.S. population and industrial resources are located in California. About 85 percent of the people and industry in California are concentrated along the west coast from San Francisco to San Diego. This area straddles the great San Andreas fault zone.

The fault zone is the boundary between the North American lithospheric plate to the east and the Pacific Ocean plate to the west. At present, the Pacific plate is moving northwest relative to the North American plate at a rate of about 3.5 centimeters per year. This motion has produced many earthquakes in recent times. For example, geologists measured 10,000 earthquakes in 1984 alone, although most could be detected only with seismographs. More severe quakes occur periodically. One shook Los Angeles in 1857, and another destroyed San Francisco in 1906. A large quake in 1989 was centered south of San Francisco and another rocked Los Angeles in January 1994. The fact that the San Andreas fault zone is part of a major boundary between lithospheric plates tells us that more great earthquakes are inevitable in the future. Can the next large one be forecast with enough reliability to evacuate threatened areas and reduce loss of life?

Motion along a fault is not uniform. Three types of motion can occur:

1. On segments of the fault where creep occurs, the plates slip past one another smoothly and without major earthquakes.

A police officer stands on a car near a section of freeway that collapsed during the January 1994 earthquake in the San Fernando Valley. *(Fred Prouser/REUTERS)*

2. In other segments of the fault, the plates pass one another in a series of small "hops," causing numerous small, nondamaging earthquakes.

3. In still other segments of the fault, plates become locked together for tens to hundreds of years and then produce catastrophic earthquakes when they break free.

In 1979, William McCann and his coworkers at Columbia University outlined the **seismic gap theory** to forecast earthquakes. This theory starts with the simple premise that all segments of a fault such as the San Andreas eventually must move by the same amount because the fault is the boundary between two tectonic plates that are moving past each other. The theory continues with the assumption that when an earthquake occurs, the accumulated elastic energy in that region is released. However, strain continues to build in nearby regions where earthquakes have not occurred. Thus, after a quake, the probability of a second quake in the same place in the near future is low. Alternatively,

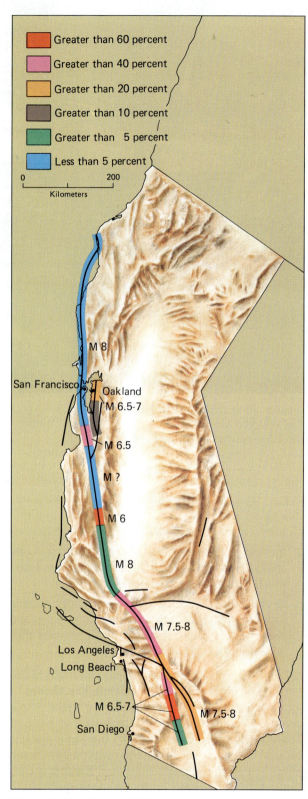

Greater than 60 percent
Greater than 40 percent
Greater than 20 percent
Greater than 10 percent
Greater than 5 percent
Less than 5 percent

0   200
Kilometers

San Francisco

Oakland
M 6.5-7

M 8

M 6.5

M ?

M 6

M 8

M 7.5-8

Los Angeles
Long Beach

M 6.5-7

M 7.5-8

San Diego

Figure 5.16 The probability of an earthquake along segments of the San Andreas fault.

(*R. L. Wesson and R. E. Wallace, "Predicting the Next Great California Earthquake,"* Scientific American, *Feb. 1985.*)

regions along an active fault that have not experienced a recent major quake are at high risk of a large earthquake in the near future (Fig. 5.16).

In 1991, David Jackson and Yan Kagan of UCLA challenged the seismic gap theory. Using a catalogue of all the recorded earthquakes of magnitude 7 or higher along both coasts of the Pacific Ocean, they showed no statistical correlation between earthquake activity and seismic gaps. On one hand, many earthquakes occurred in the same region in quick succession while on the other hand regions that had been quake-free for a long time didn't show a statistically higher probability of major earthquakes.

To understand some of the problems inherent in the argument over the seismic gap theory, we must appreciate that the San Andreas fault is part of a complex system of parallel and crisscrossed faults that transfer forces in complex and only partially understood ways.

In April 1992, a magnitude 6.1 earthquake occurred along a new fault in the Mojave desert, northeast of Palm Springs, California. Then, another earthquake with a magnitude of 7.3 struck in late June near Landers. Geophysicists Steven Jaume and Lynn Sykes of Columbia University calculated that the earth motion twisted rock to the west and thus *increased* the probability of an earthquake on the San Andreas fault.

Then, in January 1994, a magnitude 6.6 earthquake struck the San Fernando valley just north of Los Angeles. Fifty-five people died and property damage was estimated at $8 billion (total damage, which included lost work and business revenues, was much higher.) This quake occurred on a buried fault west of the San Andreas fault. Geologists are now measuring earth motion in an effort to calculate forces and stresses in rock near the epicenter. Ultimately, they are trying to forecast whether a great quake is imminent along the San Andreas fault itself. Although the 1994 quake was deadly and expensive, it was *not* a great earthquake. An 8.0 magnitude quake would release 125 times the energy of the San Fernando disaster. If such a shock were to occur during rush hour traffic, the death toll could be in the thousands.

Case History II: The Pacific Northwest

Recall from Chapter 4 that subduction is occurring today along the coasts of Oregon, Washington, and southern British Columbia in the Pacific Northwest. There, an oceanic plate is diving beneath the continent. Magma generated in the subduction zone rises to erupt from Mount St. Helens, Mount Rainier, and other Cascade Range volcanoes. One would also expect many earthquakes at such an active plate boundary. Small earthquakes occur occasionally in the region, but no large ones have occurred in the past 150 to 200 years.

Why are earthquakes relatively uncommon in this tectonically active region? The answer to this question is important because the Pacific Northwest is densely populated and heavily industrialized. Two possible answers exist. Subduction may be occurring slowly and continuously by fault creep. Alternatively, rocks along the fault may be locked together by friction, accumulating a huge amount of elastic energy that will be released in a giant, destructive quake sometime in the future.

Recently geologists have discovered evidence of great prehistoric earthquakes in the Pacific Northwest. A major coastal earthquake commonly creates violent sea waves that deposit a layer of mud and sand along the coast. Six such layers have formed in this region within the past 7000 years. Other evidence indicates that the coastline dropped suddenly by 0.5 to 2 meters as each of the sediment layers formed. This information suggests that continuous movement of plates along that subduction zone is not accommodated by creep or frequent small earthquakes. Instead, friction has locked the plates together for an average of 1150 years between large, devastating earthquakes. Thus, many geologists anticipate another major, destructive earthquake in the Pacific Northwest sometime in the future.

On April 25, 1992, a magnitude 6.9 earthquake rocked northern California and southern Oregon. Centered near Eureka, California, it caused damage locally and was felt as far away as Reno, Nevada. Two aftershocks of magnitudes 6.5 and 6.0 followed on the next day. These earthquakes occurred in the area where the Pacific Northwest subduction zone joins the San Andreas fault system. Although they caused some damage, these quakes were not the destructive "major" events predicted for the region.

Short-Term Prediction

Long-term forecasts give enough information to identify hazard areas and to establish building codes, but they do not provide a warning for evacuation of an area just prior to an earthquake. Short-term forecasts require a reliable early warning system—a signal or group of signals that immediately precede an earthquake.

Foreshocks are small earthquakes that precede a large quake by an interval ranging from a few seconds to a few weeks. The cause of foreshocks can be explained by a simple analogy. If you try to break a stick by bending it slowly, you may hear a few small cracking sounds just before the final snap. If foreshocks consistently preceded major earthquakes, they would be a reliable tool for short-term forecasts. However, only about half

The 1971 San Fernando, California, quake damaged this road. *(USGS)*

of the major earthquakes in recent years were preceded by a significant number of foreshocks. At other times, swarms of small shocks that could have been foreshocks were recorded, but a large quake did not follow.

Another approach to earthquake prediction is to measure changes in the shape of the land surface. California seismologists monitor rising bulges and other unusual Earth movements with tiltmeters and laser surveying instruments. The concept behind these studies is that distortions of the crust precede major earthquakes. Some earthquakes have been predicted successfully with this method, but in other instances predicted quakes did not occur, or quakes occurred that had not been predicted.

Chinese scientists reported that just prior to the 1975 quake in the city of Haicheng, snakes crawled out of their holes, chickens refused to enter their coops, cows broke their halters and ran off, and even well-trained police dogs became restless and refused to obey commands. Some researchers in the United States have attempted to quantify the relationship between animal behavior and earthquakes. In one study near San Francisco, scientists asked animal trainers and zoo keepers to report unusual behavior. The scientists received a flurry of calls—*after* a magnitude 5.7 earthquake—from people who reported that they had neglected to call earlier, but now that they thought about it, they remembered that some animals had behaved strangely. However, there was no significant increase in calls *before* the quake, so no predictions were made.

Figure 5.17   If you place a pencil in water, the pencil appears bent. It actually remains straight, but our eyes are fooled because light rays bend, or refract, as they cross the boundary between air and water.

5.7   Studying the Earth's Interior

Let us return to our image of tapping a watermelon to decide if it is ripe. Just as sound waves tell us about the interior of a watermelon, geologists use seismic waves to study the Earth's interior. In fact, it was by studying the behavior of seismic waves passing through the Earth that geologists discovered the Earth's layers. To under-

stand how seismic waves tell us about the Earth's interior, we must consider several properties of waves:

1. The speed with which a seismic wave travels depends on the kinds of rock that it travels through. In addition, temperature, density, rigidity, and other properties of rock affect wave speed.

2. A wave **refracts** (bends) and sometimes **reflects** (bounces back) as it passes from one transmitting medium into another. If you place a pencil in a glass half filled with water, the pencil appears bent. Of course, the pencil does not really bend; it is the light rays that bend. Light rays slow down when they pass from air to water, and as their speed changes they refract (Fig. 5.17). A mirror reflects light from your face when you look into it. Seismic waves both refract and reflect as they pass from one medium into another.

3. P waves are compressional waves and travel through all media—gases, liquids, and solids—whereas S waves can travel only through solids.

During the late 1800s and the eartly 1900s, geologists discovered that earthquake waves both refract and reflect at certain depths within the Earth. These changes show that the Earth is composed of three major layers: the core, the mantle, and the crust.

Figure 5.18   Seismic waves curve gently as they pass through the Earth. They also bend sharply where they cross major layer boundaries in the Earth's interior. The blue S waves do not travel through the liquid outer core, and therefore direct S waves are observed only within 105° of the epicenter. The yellow P waves bend sharply at the core–mantle boundary to create a shadow zone of no direct P waves from 105° to 140°.

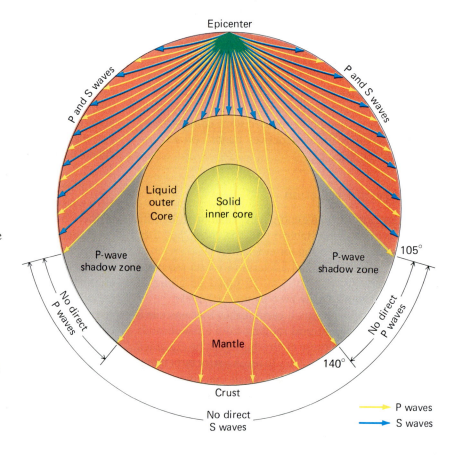

Discovery of the Core

If the entire Earth were composed of one type of rock, then P and S waves would travel everywhere and would be detected anywhere on the planet. However, Figure 5.18 shows that S waves do not pass through the Earth's outer core. Because S waves travel through solids but not through liquids, the Earth's outer core must be liquid.

Neither S nor P waves arrive in a "shadow zone" between 105° and 140° from an epicenter. Beyond 140°, direct P waves arrive, but direct S waves do not (Fig. 5.18). The shadow zone results from the change in physical properties at the core–mantle boundary. Earthquake waves curve gently as they pass through the Earth. But P waves refract sharply as they pass from the mantle into the core, as shown in Figure 5.18. As a result, no P waves arrive in the shadow zone.

P waves refract sharply again when they pass from the outer core to the inner core, indicating another radical change in physical properties of the Earth's interior. In this case, the change in direction results from an abrupt transition from the molten outer core to the solid inner core. Thus, seismic data tell us that the core is composed of an inner solid sphere surrounded by an outer liquid shell. The entire core is composed mostly of iron and nickel.

Discovery of the Crust–Mantle Boundary

In 1909 a seismologist, Andrija Mohorovičić, discovered that seismic wave velocities increase sharply at a depth varying from 7 to 70 kilometers beneath the Earth's surface. This sudden velocity change indicates an abrupt transition in rock type at that depth. The upper layer is the Earth's crust, and the lower layer is the mantle. The boundary between the two is called the **Mohorovičić discontinuity** or the **Moho**, in honor of its discoverer.

The Lithosphere–Asthenosphere Boundary

Recall that the lithosphere is the 100-kilometer-thick rigid outer shell of the Earth. It includes both the crust and the uppermost part of the mantle and floats on the hot, plastic asthenosphere. Geologists discovered the boundary between the two layers when they realized that both the speed and strength of seismic waves decrease abruptly at a depth of 100 kilometers. The sudden changes occur as the waves pass through the liquid magma pools and the hot, soft rock of the asthenosphere.

Thus, the Earth's layers and the nature of the material in each are known from the behavior of seismic waves as they travel through our planet.

SUMMARY

An **earthquake** is a sudden motion or trembling of the Earth caused by the release of **elastic energy** stored in rocks. About a million earthquakes occur each year, but only a small fraction are strong enough to be felt. Most earthquakes occur along the boundaries between moving lithospheric plates. Earthquake energy is released at the **focus** of the earthquake, where rocks move abruptly. The **epicenter** of a quake is the point on the Earth's surface directly over the focus. Earthquake energy travels through rock as **seismic waves**, which occur as **body waves** and **surface waves**. Two main types of body waves transmit earthquake energy. A **P wave** travels through solids, liquids, and gases, but an **S wave** is transmitted only through solids and travels more slowly than a P wave. Surface waves travel along the Earth's surface.

An earthquake is detected and measured by a **seismograph**. The epicenter and focus of an earthquake can be identified using a **time-travel curve** to determine the distances from the earthquake source to three or more seismograph stations.

The **Mercalli scale** measures earthquake **intensity** on the basis of damage and human response to a quake. The **Richter scale** measures earthquake **magnitude** on the basis of the amplitude of ground movement caused by a quake. Earthquake damage is caused by ground motion, alteration of the Earth's surface, fire, landslides, and **tsunamis**.

Long-term forecast of earthquakes is based on studies of the earthquake history of an area. The **seismic gap** theory, proposed in 1979, is now being questioned. Short-term forecast is based on occurrences of **foreshocks**, bulges and other changes in the land surface, and animal behavior.

The Earth's internal structure and properties are known by studies of earthquake wave velocities and **refraction** and **reflection** of seismic waves as they pass through the Earth.

REVIEW QUESTIONS

1. Explain how energy is stored in rocks and then released during an earthquake.

2. Why do most earthquakes occur at the boundaries between tectonic plates?

3. Explain why fault creep along a segment of a fault provides evidence that an earthquake may soon occur in a nearby segment where creep is *not* occurring.

4. Describe the differences among P waves, S waves, and L waves.

5. Explain how a seismograph works. Sketch what a seismogram would look like before and during an earthquake.

6. Describe how the epicenter of an earthquake is located.

7. Describe the similarities and differences between the Mercalli and Richter scales. What does each actually measure, and what information does each provide?

8. List five different mechanisms for earthquake damage. Discuss each briefly.

9. Discuss the scientific reasoning behind long-term and short-term earthquake forecast.

10. What is the Moho? How was it discovered?

11. Describe how the behavior of seismic waves led to the discovery of the Earth's core.

12. Describe how the behavior of seismic waves led to the discovery of the boundary between the lithosphere and the asthenosphere.

DISCUSSION QUESTIONS

1. If rock is caught between two lithospheric plates moving in different directions, it may bend in a plastic manner or, alternatively, it may fracture. Which type of movement is more likely to cause an earthquake? Explain.

2. Using the graph in Figure 5.10, determine how far away from an earthquake you would be if the first P wave arrived 5 minutes before the first S wave.

3. Using a map of the United States, locate an earthquake that is 1000 kilometers from Seattle, 1300 kilometers from San Francisco, and 700 kilometers from Denver.

4. It has been suggested that engineers should inject large quantities of liquids into locked portions of the San Andreas fault. Proponents of the plan believe that these liquids would reduce friction by lubricating the fault, and consequently the fault would creep slowly and a major earthquake would be averted. If you were the mayor of San Francisco, would you encourage or discourage the injection of fluids into the fault? Defend your stance.

5. Significant earthquakes have occurred in Parkfield, California, in 1857, 1881, 1901, 1922, 1934, and 1966. Draw a graph with the dates on the vertical Y-axis and the numbers of the events (simply 1, 2, 3, and so on) spaced evenly on the X-axis. Use your graph to predict when the next earthquake might occur in Parkfield.

6. Imagine that geologists forecast a major earthquake in a densely populated region. The forecast may be right or it may be wrong. City planners may heed it and evacuate the city, or they may ignore it. The possibilities lead to four combinations of forecasts and responses, which can be set out in a grid as follows:

Does the forecasted earthquake really occur?

		Yes	No
Is the city evacuated?	Yes		
	No		

For example, the space in the upper left corner of the grid represents the situation in which the predicted earthquake occurs and the city is evacuated. For each space in the square, outline the consequences of that sequence of events.

KEY TERMS

Volcanoes and Plutons

The September 1972 eruption from the East Rift of Kilauea Volcano, Hawaii.
(D.W. Peterson/USGS)

In the spring of 1980, 1 cubic kilometer of rock, lava, and ash exploded from Mount St. Helens in western Washington. The blast flattened and burned surrounding forests, and the sky grew so dark with pulverized rock that motorists 150 miles from the mountain had to use their headlights at noon. But this eruption was minor compared with others in history. In 1815 Mount Tambora in the southwestern Pacific Ocean exploded, throwing out 100 cubic kilometers of rock and ash and killing 12,000 people; Mount Pelée in the Caribbean exploded in 1902, killing 29,000. Yet even the most violent historical eruptions have been small compared with some prehistoric ones. A volcanic explosion in Yellowstone Park, Wyoming, 1.9 million years ago ejected approximately 2500 cubic kilometers of rock, lava, and ash. In contrast to these violent explosions, Hawaiian volcanoes erupt gently enough that tourists approach closely to photograph flowing lava and to watch fire fountains erupting into the sky.

Volcanoes are common in some regions and unheard of in others. Eighteen recently active volcanoes form high peaks in the Cascade Range in northern California, Oregon, and Washington, but no volcanoes exist in the central or eastern United States. Why are some parts of the Earth volcanically active, whereas other parts are not? And why do some volcanoes explode violently but others erupt gently? To answer these questions, consider how magma forms and behaves.

6.1   Formation of Magma

As explained in Chapter 3, igneous rock forms in a two-step process. First, part of the upper mantle or crust melts to form magma. The magma then solidifies to form new igneous rock. But magma does not form just anywhere. It forms in three geologic environments: subduction zones, spreading centers, and mantle plumes. Let us consider each of these environments to see why and how rock melts to make magma.

Magma Production in a Subduction Zone

At a subduction zone, two 100-kilometer-thick plates of lithosphere collide. One of the plates dives beneath the other, sinking hundreds of kilometers into the mantle. Three processes combine to form great amounts of magma in a subduction zone (Fig. 6.1).

Heating

Everyone knows that a solid melts when it becomes hot enough. Butter melts in a frying pan, and snow melts under the spring sun. Recall that the asthenosphere is very hot—so hot that it is plastic. When a tectonic plate sinks into the mantle at a subduction zone, it creates friction as it scrapes past the opposite plate. The friction raises the temperature of the hot asthenosphere, bringing it closer to its melting point. Oddly, however, increasing temperature is the least important of the three causes of melting in a subduction zone.

Addition of Water to the Asthenosphere

In general, a wet rock melts at a lower temperature than an identical dry rock. Therefore, addition of water can melt a rock that is already close to its melting point. In this way, asthenosphere rock melts if water is added to it.

A subducting plate is covered by oceanic crust, and the mud and fractured basalt of oceanic crust are soaked with seawater. As wet oceanic crust dives into the mantle, it becomes hotter because the Earth's temperature increases with depth. Eventually the water boils to form steam. The steam rises, adding water to the hot asthenosphere directly above the sinking plate. This addition of hot water melts portions of the asthenosphere, forming huge quantities of magma (Fig. 6.1). This process is the most important cause of magma formation in a subduction zone (Fig. 6.2).

Pressure-Relief Melting

In Chapter 3 you learned that a rock expands by about 10 percent when it melts to form magma. If the rock is near the Earth's surface it can expand and melt easily because there is little pressure preventing it from doing so. However, in the asthenosphere, pressure is so great from the weight of the overlying 100 kilometers of lithosphere that expansion and therefore melting are more difficult. However, if the pressure should somehow decrease, the hot rock in the asthenosphere can expand and melt. In other words, a drop in pressure can melt a hot rock. Melting caused by decreasing pressure is called **pressure-relief melting** (Fig. 6.2).

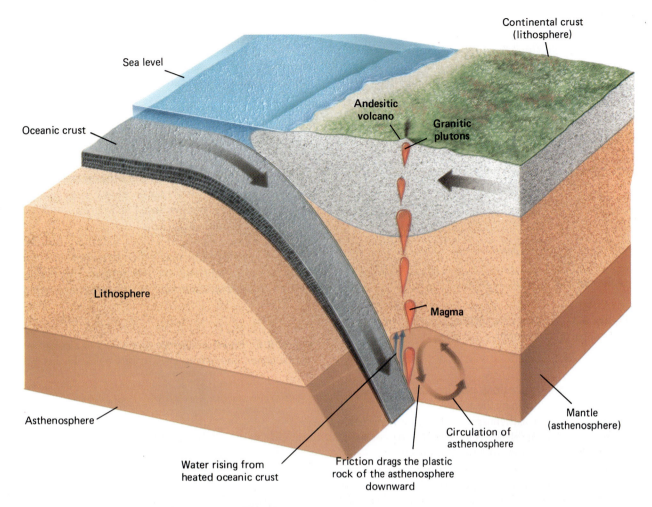

Figure 6.1   Three factors contribute to melting of the asthenosphere and production of magma at a subduction zone: (1) Friction heats rocks in the subduction zone; (2) water rises from oceanic crust on top of the subducting plate; and (3) circulation in the asthenosphere decreases pressure on hot rock.

Figure 6.2   Increasing temperature, addition of water, and decreasing pressure all melt rock to form magma.

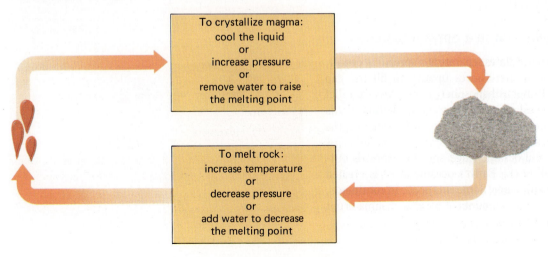

To crystallize magma:
cool the liquid
or
increase pressure
or
remove water to raise
the melting point

To melt rock:
increase temperature
or
decrease pressure
or
add water to decrease
the melting point

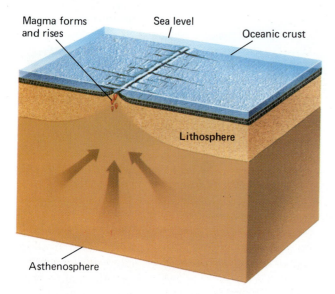

Magma forms and rises

Sea level

Oceanic crust

Lithosphere

Asthenosphere

Figure 6.3   Pressure-relief melting occurs where hot asthenosphere rises beneath a spreading center.

A subducting plate drags plastic asthenosphere rock down with it, as shown by the arrows in Figure 6.1. Rock from deeper in the asthenosphere then flows upward to replace the sinking rock. As this rock rises, pressure decreases and pressure-relief melting occurs.

To summarize, parts of the asthenosphere above a subducting plate melt by three processes: (1) addition of water from subducting oceanic crust, (2) pressure-relief melting, and (3) frictional heating. These three factors combine to form huge amounts of magma in subduction zones at depths of about 100 kilometers, where the subducting plate passes from the lithosphere into the asthenosphere. Therefore, igneous rocks are common features of a subduction zone. The volcanoes of the Pacific Northwest and the granite cliffs of Yosemite are examples of igneous rocks formed at subduction zones.

Magma Production in a Spreading Center

As two lithospheric plates separate at a spreading center, soft, hot asthenosphere oozes upward to fill the gap (Fig. 6.3). As the hot asthenosphere rises, pressure-relief melting forms vast quantities of magma. Melting of this type below the mid-oceanic ridge system forms magma that rises to become new oceanic crust. The basalt crust forms at the mid-oceanic ridge and then spreads outward. Thus, all of the Earth's oceanic crust is created at the mid-oceanic ridge. Some rifts occur in continents, and there, too, great amounts of basaltic magma erupt onto the Earth's surface.

Magma Production at a Hot Spot

The third environment in which magma commonly forms, known as a **hot spot**, is a volcanic region at the Earth's surface directly above a rising plume of hot, plastic mantle rock. As a mantle plume rises, pressure-relief melting forms magma that flows upward to erupt at the Earth's surface (Fig. 6.4).

A mantle plume originates deep in the mantle, and its magma then rises through the lithosphere to erupt and form a volcano. If a mantle plume rises beneath the sea, volcanic eruptions build submarine volcanoes and volcanic islands. Because a lithospheric plate continues to move over the relatively stationary asthenosphere, a mantle plume may generate magma continuously as a plate migrates over the plume. This process forms a chain of volcanic islands that becomes progres-

Figure 6.4   Pressure-relief melting occurs in a rising mantle plume, and magma rises to form a volcanic hot spot.

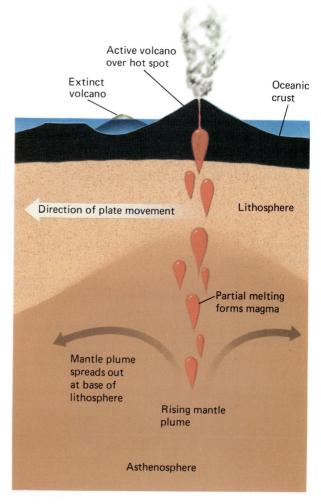

Active volcano over hot spot

Extinct volcano

Oceanic crust

Direction of plate movement

Lithosphere

Partial melting forms magma

Mantle plume spreads out at base of lithosphere

Rising mantle plume

Asthenosphere

A lava fountain erupts from Pu'u O'o Vent on Hawaii, June 1986. *(USGS, J. D. Griggs)*

sively younger toward one end, such as the Hawaiian Islands.

Granitic and Basaltic Magmas

Melting of the asthenosphere normally produces basaltic magma. Many volcanoes erupt basaltic magma directly from this source. But some volcanoes erupt granitic magma, and most of the world's plutons are of granitic composition. How does granitic magma form?

Granitic magma forms when basaltic magma rises into continental crust. This occurs where a subduction zone, rifting, or a mantle plume forms basaltic magma beneath a continent. Recall that continental crust is mostly granite. Granite melts at a temperature 200° to 300°C lower than the temperature of basaltic magma. Thus, when basaltic magma rises into granitic continental crust, it melts vast amounts of the granite, forming granitic magma. Most of the granite plutons of the world, including those of the Appalachians, the Rocky Mountains, Yosemite Valley, and the Alps, formed by this two-step process.

6.2 Why Some Magma Erupts from a Volcano and Other Magma Solidifies Below the Surface

Once magma forms, it rises toward the Earth's surface because it is less dense than surrounding rock. As it rises, two changes occur: (1) It cools as it enters shallower and cooler levels of the Earth, and (2) pressure drops because the weight of overlying rock decreases. Cooling and decreasing pressure have opposite effects on rising magma: Cooling tends to solidify it, but dropping pressure tends to keep it liquid.

So does magma solidify or remain liquid as it rises toward the Earth's surface? The answer depends on the type of magma. Basaltic magma commonly rises all the way to the surface to erupt from a volcano. In contrast,

granitic magma usually solidifies within the crust, although under special conditions it, too, can rise to erupt from a volcano. These special conditions, described in Section 6.6, cause extremely violent volcanic explosions.

The contrasting behavior of granitic and basaltic magmas is a result of their different compositions. Granitic magma contains about 70 percent silica, whereas the silica content of basaltic magma is only about 50 percent. In addition, granitic magma generally contains

The granite of Yosemite National Park is part of the Sierra Nevada batholith. *(Don Hyndman)*

127

10 to 15 percent water, whereas basaltic magma contains only 1 to 2 percent water. These differences are summarized in the following table.

Typical Granitic Magma	Typical Basaltic Magma
70% silica 10 to 15% water	50% silica 1 to 2% water

Effects of Silica on Magma Behavior

In the silicate minerals, silicate tetrahedra link together to form the chains, sheets, and framework structures described in Chapter 2. In magma, silicate tetrahedra link together in a similar manner. They form long chains and similar structures if silica is abundant in the magma, but shorter chains if less silica is present. Therefore, granitic magma contains longer chains of silicate tetrahedra than does basaltic magma. When granitic magma flows, the long chains become tangled, making the magma stiff, or viscous. **Viscosity** is resistance to flow. Basaltic magma, with its shorter silicate chains, is less viscous and flows easily. Because of its fluidity, it rises rapidly to erupt at the Earth's surface. Granitic magma, in contrast, rises slowly because of its stiffness. Therefore, it commonly solidifies within the crust before reaching the surface.

Effects of Pressure and Water on Magma Behavior

Another difference is that granitic magma contains more water than basaltic magma. Recall that water lowers the temperature at which magma solidifies. Thus, if dry granitic magma solidifies at 700°C, the same magma with 10 percent water may remain liquid at 600°C.

At high temperatures water tends to escape from magma as steam. Deep in the crust, where granitic magma forms, high pressure prevents the water from escaping. But when the magma rises, pressure decreases and water escapes. Because the water escapes, the solidification temperature of the magma rises. Thus, loss of water from rising granitic magma causes it to solidify within the crust. For this reason, most granitic magmas solidify and stop rising at depths of 5 to 20 kilometers beneath the Earth's surface.

Because basaltic magmas have only 1 to 2 percent water to begin with, any loss is relatively unimportant. As basaltic magma rises, it remains liquid all the way to the surface. Thus, basalt volcanoes are common.

6.3   Plutons

For reasons already described, granitic magma usually solidifies within the Earth's crust to form a mass of intrusive rock called a **pluton** (Fig. 6.5). Thus, a pluton

Figure 6.5 Forms of igneous rocks.

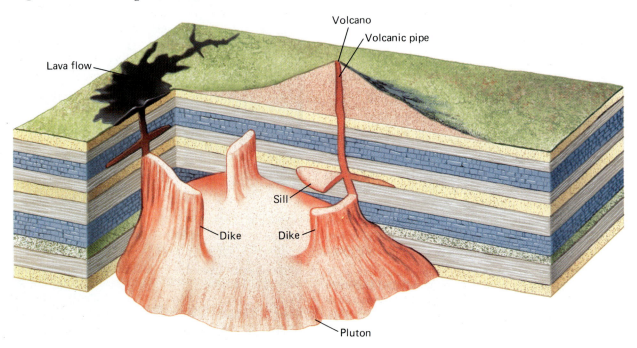

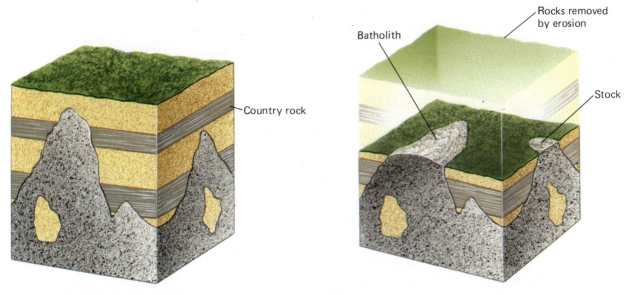

Figure 6.6 A batholith is a pluton with more than 100 square kilometers exposed at the Earth's surface. A stock is similar to a batholith but has a smaller surface area.

is emplaced in **country rock** (any previously existing rock). Occasionally basaltic magma also solidifies as intrusive rock. Basaltic plutons tend to form thin, sheetlike masses because basaltic magma is fluid.

To form a pluton a huge mass of granitic magma must rise through solid continental crust. How can such a large bubble of magma rise through solid rock? If you place oil and water in a jar, screw the lid on tightly, and shake it, oil droplets disperse throughout the water. Set the jar down and the droplets coalesce slowly to form larger bubbles, which rise to the surface. The water is displaced easily as the bubbles rise. Although the rock of continental crust is solid, at the high temperatures in the lower crust it also behaves in a plastic manner. As magma rises, it shoulders aside the hot, plastic crustal rock, which then flows back slowly to fill in behind the rising bubble.

After a pluton forms, it may be pushed upward by tectonic forces and exposed by erosion. A **batholith** is a pluton exposed over more than 100 square kilometers of the Earth's surface (Fig. 6.6). A large batholith may be as much as 20 kilometers thick, but an average one is about 10 kilometers thick. Most batholiths are granite. A **stock** is similar to a batholith but is exposed over a smaller area.

Many mountain ranges contain large granite batholiths that are the remains of plutons once emplaced deep within the crust and later uplifted and eroded. Most of the great granite batholiths of the world formed in the immense regions of magma production at subduction zones.

Whereas a pluton pushes country rock aside as it rises, magma may also flow into a fracture or between layers in country rock. A **dike** is a tabular, or sheetlike, intrusive igneous rock that forms when magma oozes into a fracture in country rock (Fig. 6.7). Dikes cut across sedimentary layers or other features in country rock and range from less than a centimeter to more than a kilometer thick (Fig. 6.8A). Dikes are commonly more resistant to weathering than surrounding rock. Where this is the case, country rock may be eroded away, leaving the dike standing alone on the surface.

If magma oozes between layers of country rock, a **sill** forms. It is a tabular intrusive rock that lies parallel to the grain, or layering, of country rock rather than cutting across the layers (Fig. 6.8B). Sills also range in thickness from less than a centimeter to more than a kilometer and may extend for tens of kilometers in length and width.

6.4   Volcanoes and Other Volcanic Landforms

Volcanic eruptions create a wide variety of landforms, including several different kinds of volcanoes and lava plateaus. Many islands, including all of the Hawaiian Islands, Iceland, and most of the islands of the southwestern Pacific Ocean, were constructed entirely by volcanic eruptions. Occasionally a violent eruption destroys a volcanic peak, as happened in the 1980 eruption of

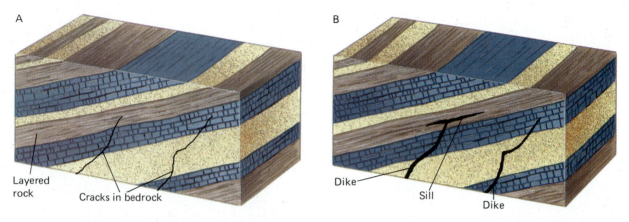

Figure 6.7 A dike cuts across the grain of country rock. A sill is parallel to the grain, or layering, of country rock.

Mount St. Helens, when most of the mountaintop was blown away.

Structures and Textures of Lava Flows

Hot magma shrinks as it cools and solidifies. The shrinkage pulls the rock apart, forming cracks that grow as the rock continues to cool.

In Hawaii geologists have watched fresh lava cool and solidify. When a solid crust only 0.5 centimeter thick forms on the surface of the glowing liquid, five- or six-sided cracks begin to develop. As the lava cools and solidifies from the surface down, the cracks grow downward to hotter zones where the last bit of magma is solidifying. Such cracks, called **columnar joints**, are common in lava flows and shallow sills. They are regularly spaced and intersect to form five- or six-sided columns (Fig. 6.9).

Lava can develop different textures depending on its viscosity and rate of cooling. Lava with low viscosity may continue to flow as it cools and stiffens, forming basalt with smooth, glassy-surfaced, wrinkled, or "ropy" ridges. This type of lava is called **pahoehoe** (Fig. 6.10). If the viscosity of the lava is higher, it may solidify partially as it flows to form a slow-moving mixture of solid rock and liquid lava. As a result, **aa** lava has a

Figure 6.8 (A) A pink pegmatite dike cuts across the grain of country rock on Baffin Island. (B) A black basalt sill lies parallel to sedimentary layering in Grand Canyon.

A

B

A   B

Figure 6.9 (A) Columnar joints in basalt of the Columbia River basalt plateau, Washington. (B) A top view of columnar joints in basalt. *(Don Hyndman)*

Figure 6.10 A car buried in pahoehoe lava, Hawaii. *(Kenneth Neuhauser)*

A lava stream pours from the mouth of a lava tube on Hawaii, December 1986. *(USGS, J. D. Griggs)*

Figure 6.11   Aa lava near Shoshone, Idaho, showing numerous gas bubbles frozen into the solid rock.

jagged, rubbly, broken surface (Fig. 6.11). Aa lava commonly contains numerous holes and resembles Swiss cheese. The holes are gas bubbles that formed in the magma and became frozen in the rock as it solidified. Pahoehoe and aa are Hawaiian names for the lava types.

When basaltic magma erupts under water, the rapid cooling causes it to contract and form spheroidal **pillow lava** (Fig. 6.12). Pillow lavas are abundant in the upper layers of oceanic crust, where they form when basaltic magma oozes onto the sea floor from cracks in the mid-oceanic ridge.

When a volcano erupts explosively, it may eject both liquid magma and solid rock fragments. A rock formed from explosively erupted rock particles or magma is called a **pyroclastic rock**. The smallest particles are glassy pieces of **volcanic ash**, ranging up to 2 millimeters in diameter. Mid-sized particles called **cinders** range from about 2 to 64 millimeters. Still larger fragments called **volcanic bombs** form when blobs of molten lava spin through the air as they solidify and therefore take the form of spindles or spheroids (Fig. 6.13).

Although the words *ash* and *cinder* are used to describe these volcanic particles, they are *not* the same as the ashes and cinders produced by a conventional fire. In December 1989, Mount Redoubt volcano, south-

Figure 6.12   Lava pillows in Oregon.

Figure 6.13   A volcanic bomb. The streaky surface formed as the blob of magma whirled through the air.

west of Anchorage, Alaska, erupted. The pilot of a Boeing 747, with 231 passengers aboard, ignored warnings about the rising ash cloud and flew into it. The abrasive particles were sucked into the jet engines and quickly ground the sharp blades of the compressor into small stubs. Exposed to the intense heat in the combustion chamber, the volcanic ash particles melted. Moments later, the liquid solidified to form glass on the cooler turbine blades. As a result, all four engines stalled and the jet fell 4000 meters in 8 minutes. Finally the pilot was able to restart the engines and fly the disabled craft back to Anchorage, where he landed safely.

Flood Basalts

The gentlest, least catastrophic type of volcanic eruption occurs when magma is so fluid that it oozes from cracks and flows over the land like water. When a large flow of this type solidifies, it forms a broad plain or plateau. Basaltic magma commonly pours out as a **flood basalt**, which is so named because it covers the landscape like a flood. The plateaus created by such lava flows are called **lava plateaus**. The Columbia River plateau in eastern Washington, northern Oregon, and western Idaho is an extensive lava plateau (Fig. 6.14). This

Figure 6.14   (A) The Columbia River basalt plateau covers much of Washington, Oregon, and Idaho. (B) Columbia River basalt in eastern Washington. Each layer is a separate lava flow.   *(Larry Davis)*

A

B

Table 6.1 • Characteristics of Different Types of Volcanoes

Type of Volcano	Form of Volcano	Size	Type of Magma	Style of Activity	Examples
Basalt plateau	Flat to gentle slope	100,000 to 1,000,000 km² in area; 1 to 3 km thick	Basalt	Gentle eruption from long fissures	Columbia River Plateau
Shield volcano	Slightly sloped, 6° to 12°	Up to 9000 m high	Basalt	Gentle, some fire fountains	Hawaii
Cinder cone	Moderate slope	100 to 400 m high	Basalt or andesite	Ejections of pyroclastic material	Paricutín, Mexico
Composite volcano	Alternate layers of flows and pyroclastics	100 to 3500 m high	Variety of types of magmas and ash	Often violent	Vesuvius, Mount St. Helens, Aconcagua
Caldera	Cataclysmic explosion leaving a circular depression called a caldera	Less than 40 km in diameter	Granite	Very violent	Yellowstone, San Juan Mountains

sequence of basalt flows contains 350,000 cubic kilometers of rock, is 3000 meters thick in places, and covers 200,000 square kilometers. It formed over a period of 3 million years, from 17 to 14 million years ago, as layer upon layer of basaltic magma oozed from fissures in eastern Washington and Oregon. Each flow formed a layer between 15 and 100 meters thick. Similar large basalt plateaus are found in western India, northern Australia, Iceland, Brazil, Argentina, and Antarctica.

Volcanoes

If lava is too viscous to spread out as a lava plateau, it builds up into a hill or mountain called a **volcano**. Volcanoes differ widely in shape, structure, and size (Table 6.1). Lava and rock fragments erupt from an opening called a **vent**. In many volcanoes the vent is located in a **crater**, a bowl-like depression at the summit of the volcano.

Shield Volcanoes

If basaltic magma is too viscous to form a lava plateau, but still quite fluid, it will heap up slightly to form a gently sloping volcanic mountain called a **shield volcano** (Fig. 6.15). The sides of a shield volcano generally slope away from the vent at angles between 6° and 12° from horizontal. As a reference, a ski slope for beginning to intermediate skiers is about 10°, and an expert slope may be as steep as 30°. Unless erosion has formed deep gullies, you find no challenging ski runs on a shield volcano; it is simply not steep enough.

When a shield volcano erupts, the fluid lava usually flows gently over the lip of the crater or from **fissures**, linear cracks in the sides of the volcano. Although the Hawaiian shield volcanoes erupt regularly, the eruptions are normally gentle and rarely life threatening. Lava flows occasionally overrun homes and villages, but the flows advance slowly enough to give people time to evacuate.

Cinder Cones

A **cinder cone** is a small volcano, as high as 400 meters, made up of pyroclastic fragments blasted out of a central vent at high velocity. A cinder cone forms when large amounts of gas accumulate within rising magma. When the gas pressure builds up sufficiently, the entire mass erupts explosively, hurling cinders, ash, and molten magma into the air. The particles then fall back around the vent to accumulate as a small mountain of volcanic debris.

Figure 6.15   Mount Skjoldbreidier in Iceland bears the typical profile and low-angle slopes of a shield volcano.   *(Science Graphics, Inc./Ward's Natural Science Establishment, Inc.)*

As the name implies, a cinder cone is symmetrical. It also can be quite steep (about 30°), especially near the vent where ash and cinders pile up (Fig. 6.16). A cinder cone is usually active for only a short period of time because once the gas escapes, the driving force behind the eruption is removed. Because the pyroclastic fragments are not cemented together, a cinder cone erodes easily and quickly. Therefore, it is a geologically transient feature of the landscape.

About 350 kilometers west of Mexico City a broad plain is dotted with numerous extinct cinder cones. Prior to 1943, a small hole existed in the ground in one of the level portions of the plain. The hole had been there for as long as anyone could remember, and people grew corn just a few meters away. In February 1943, as two farmers were preparing their field for planting, smoke and sulfurous gases rose from the hole. As night fell, hot, glowing rocks flew skyward, creating spectacular arcing flares like a giant fireworks display. By morning, a 40-meter-high cinder cone had grown in the middle of the cornfield. For the next 5 days, pyroclastic material erupted 1000 meters into the sky and the cone grew to 100 meters in height. Within a few months, a fissure opened at the base of the cone, extruding lava that buried the town of San Juan Parangaricutiro. Two years later the cone had grown to a height of 400 meters. After 9 years the eruptions ended, and today the volcano, called El Paricutin, is dormant.

Composite Cones

Some of the most beautiful and spectacular volcanoes in the world are **composite cones**, sometimes called **stratovolcanoes**. They form by a series of alternating lava flows and pyroclastic eruptions. An explained previously, pyroclastic eruptions form steep but unconsolidated slopes. When these loose cinders are covered by lava flows, the hard lava rock protects them from erosion (Fig. 6.17).

A composite cone grows as a result of many eruptions occurring over a long time. Many of the highest mountains of the Andes are composite cones, as are many of the most spectacular mountains of the west

Figure 6.16   Cinder cones of Wupatki Sunset Crater National Monument, Arizona.   *(Wupatki National Monument)*

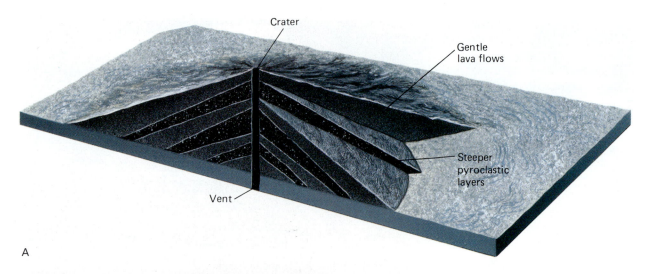

A

B

Figure 6.17   (A) A schematic cross section of a composite cone showing alternating layers of lava and pyroclastic material. (B) Steam and ash pouring from Mount Ngauruhoe, a composite cone in New Zealand.   *(Don Hyndman)*

coast of North America. Two examples are Mount St. Helens and Mount Adams (Fig. 6.18). Mount St. Helens erupted in 1980. Mount Adams has been dormant in recent times but could become active at any moment; repeated eruption is a trademark of a composite volcano.

Volcanic Necks and Pipes

After an eruption, the vent of a volcano may fill with magma that later cools and solidifies. Commonly this **volcanic neck** is harder than surrounding rock. Given enough time, the slopes of the volcano may erode, leaving only the tower-like neck exposed (Fig. 6.19).

In some locations, cylindrical dikes called **pipes** extend from the asthenosphere to the Earth's surface. They are conduits that once carried magma on its way to erupt at a volcano, but they are now filled with the last bit of magma that solidified within the conduit. They are both fascinating and economically important. For unknown reasons, most known pipes formed between 70 and 140 million years ago, and most intruded into continental crust older than 2.5 billion years. Pipes are interesting because the rock found in them, called **kimberlite**, originated in the asthenosphere. It is among the few direct samples of the upper mantle available to geologists.

The best evidence indicates that pressure in the asthenosphere was so great that the kimberlite magma shot upward through the Earth's crust at very high, perhaps even supersonic, speed. Under such intense pressures, the small amount of carbon in some pipes crystallized as diamond, making those pipes commercially important. The most famous diamond-rich kimberlite pipes are in South Africa. The diamonds crystallized more than 200 kilometers beneath the surface and were carried upward with the rising magma to depths of a few kilometers or less, where they can be mined.

6.5   Igneous Activity and Plate Tectonics

Igneous Activity at Subduction Zones

We learned in Section 6.1 that magma forms abundantly in subduction zones. It then rises to form both volcanoes and plutons directly over the subducting tectonic plate. Figure 6.20 shows that most of the world's volcanoes are located near subduction zones. The "ring of fire" is a zone of concentrated volcanic activity encircling the Pacific Ocean basin. About 75 percent of the Earth's active volcanoes lie in the ring of fire.

Figure 6.18   Mount Adams in Washington is a composite cone.   *(Don Hyndman)*

All 18 volcanoes of the Cascade Range, from Mount Baker near the Canadian border to Mount Lassen in northern California, have been active in the past 2 million years, fired by subduction of oceanic plates beneath the continent. Mount St. Helens erupted in 1980 and Mount Lassen in 1915. About 7000 years ago, a Cascade volcano, posthumously named Mount Mazama, exploded, blasting 10 cubic kilometers of rock and ash

Figure 6.19   Shiprock, New Mexico, is a volcanic neck. The great rock was once the core of a volcano. The softer flanks of the cone have now eroded away. A dike several kilometers long extends to the left.   *(Dougal McCarty)*

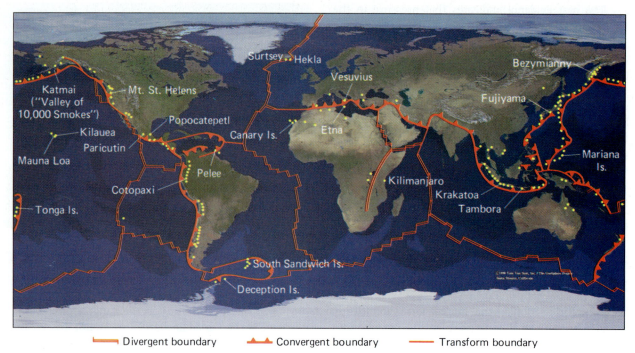

Divergent boundary   Convergent boundary   Transform boundary

Figure 6.20   This map shows the close relationship between subduction zones (convergent boundaries) and major volcanoes, indicated by the yellow dots. *(Tom Van Sant/Geosphere Project)*

into the air. The explosion blew some of the mountain away, and the remainder collapsed into the hole formed by the blast, leaving what is now Crater Lake, Oregon (Fig. 6.21). The ash is found in soils over much of western and central North America.

Some volcanoes that have not erupted recently show signs of activity today. Mount Baker has recently experienced earthquake swarms, and gas regularly escapes from its crater. Mount Rainier has steam caves under its summit glaciers. It is reasonable to predict that the Cascades will see additional eruptions, although it is difficult to predict where, when, or how large. It is important to remember that, despite the loss of life and tremendous damage, the eruption of Mount St. Helens was very small compared with other known eruptions from similar volcanoes.

Explosive eruptions are also common elsewhere in the ring of fire. On June 9, 1991, Mount Pinatubo in the Philippines began ejecting a gray-green cloud of ash, rock, and smoke 15 kilometers into the sky. By the end of June, 338 people had been killed and several towns evacuated because of continued explosions. As many as 200 earthquakes a day were felt. Both the U.S. Clark Air Base, 15 kilometers from the volcano, and Subic Bay Naval Station sustained heavy damage. Clark was evacuated because of the threat to personnel and aircraft. Mount Mayon, also in the Philippines, erupted violently, killing several people in February 1993.

Volcanic eruptions are the most obvious form of igneous activity at a subduction zone. If subduction occurs beneath a continent, the basaltic magma formed by melting of the asthenosphere rises into the granitic continental crust. Because the basaltic magma is much hotter than the melting temperature of the granite, large amounts of granitic magma form and rise further into the continental crust.

This magma cools slowly and solidifies to form plutons. Eventually mountain building elevates many of these regions, and erosion strips away overlying rock, exposing the plutons as batholiths. The granite batholiths of the Sierra Nevada of California, the Coast Range of western Canada, and the Appalachian Mountains of eastern North America formed beneath great chains of subduction zone volcanoes. In some places, the overlying rocks have now eroded away, exposing the granite at the Earth's surface.

Igneous Activity at Rift Zones

The island of Iceland in the North Atlantic Ocean is about the size of Virginia and supports a quarter of a million people. It lies directly over the Mid-Atlantic ridge and thus is on a spreading boundary between two tectonic plates. The island formed by repeated volcanic eruptions at the ridge. Iceland has experienced numerous eruptions since its settlement by Vikings before A.D.

Figure 6.21 Crater Lake, Oregon. *(Crater Lake National Park Administration)*

1000. Some have covered villages with ash and cinders, although human injury and death have been rare. Hot springs and hot rock resulting from recent volcanism are used to generate electrical power for the island, and the spectacular volcanic scenery is a major tourist attraction.

Although eruptions on Iceland are readily visible, most of the volcanic activity at the mid-oceanic ridge occurs unseen as submarine lava oozes from cracks in the ridge axis onto the sea floor. The rocks of the sea floor are described further in Chapter 7.

Rifting can also occur in continental crust. When it does, pressure-relief melting in the mantle forms basaltic magma below the continent. The basaltic magma rises into lower granitic continental crust and melts it to form granitic magma, as described in Section 6.1. Thus, continental rifting commonly forms both basalt volcanoes and granite plutons. Continental rifting is occurring today along a north–south trend through eastern Africa in a zone called the East African rift. It may also be occurring in Idaho's Snake River plain and in central Colorado and New Mexico along the Rio Grande River, where the Rio Grande rift is marked by numerous young volcanoes.

Igneous Activity at Mantle Plumes and Hot Spots

The island of Hawaii is composed of several overlapping volcanoes above a mantle plume. The youngest, Kilauea, frequently erupts basaltic lava, commonly for periods of weeks or months. As in Iceland, lava flows occasionally destroy homes and agricultural land, but they rarely cause injury or death because eruptions are compara-

tively gentle and the flows advance slowly enough that people can evacuate threatened areas. Because the eruptions are relatively safe, at least from a distance, and because they continue for long periods of time, tourists flock to the island to see the eruptions.

Mantle plumes also rise beneath continents. As in the case of continental rifting, basaltic magma forms in a rising mantle plume beneath the continent. The magma melts lower continental crust as it ascends, forming granitic magma, which in turn rises to create a continental hot spot, such as Yellowstone National Park.

6.6 Violent Magma: Ash-Flow Tuffs and Calderas

As described in Section 6.2, granitic magma usually solidifies within the Earth's crust. Under certain conditions, however, granitic magma can rise all the way to the Earth's surface, where it characteristically erupts with great violence. This violent behavior contrasts sharply with that of basaltic magma, which typically erupts with relative calm. Why does some granitic magma rise all the way to the Earth's surface, and why does it erupt with such violence?

The granitic magmas that do rise to the surface and explode probably start out with less water than "normal" granitic magma, perhaps only a few percent, like basaltic magma. With such a low water content, granitic magma rises to the Earth's surface for the same reasons that basaltic magma does.

"Dry" granitic magma rises more slowly than basaltic magma, however, because of its higher viscosity.

FOCUS ON

The May 18, 1980 Eruption of Mount St. Helens

The 1980 eruption of Mount St. Helens was recorded by photographers on the ground and in the air, and it dominated media reports for weeks after it occurred. About 2 months before the eruption, the volcano, which had last erupted in 1857, began to experience small to moderate-sized earthquakes. Puffs of steam and volcanic ash were emitted from a newly formed crater on the summit. The activity was great enough to convince geologists to install seismographs and tiltmeters around and on the volcano. The seismographs were emplaced to detect earthquakes caused by moving bodies of magma that might be associated with an impending eruption. Tiltmeters detect small, local tilting of the Earth's surface that might be caused by magma moving upward in the volcano and swelling the flanks.

In early April the emissions of steam and ash ceased, although earthquakes continued. The tiltmeters detected swelling on the north side of the volcano, which by early May had grown to an easily visible, ominous bulge rising as fast as 1.5 meters per day. Eventually it measured nearly 1 by 2 kilometers and had swelled outward by more than 100 meters. Warned by the geologists, officials barred the public from the area around the volcano and evacuated most of the local inhabitants.

Two strong earthquakes rocked Mount St. Helens at 8:27 and 8:31 on Sunday morning, May 18. The bulge

Burned and uprooted trees approximately 13 kilometers from Mount St. Helens.   *(Larry Davis)*

As it rises, the pressure decreases, allowing the small amount of water to separate and form gas bubbles in the liquid magma (Fig. 6.22A). The water forms gas rather than liquid because the temperature is high enough to turn the water to steam.

As an analogy, think of a bottle of beer or soda pop. When the cap is on and the contents are under pressure, carbon dioxide gas is dissolved in the liquid. Because the gas is dissolved, no bubbles are visible. If you remove the cap, the pressure is reduced. As a result, the gas escapes from solution and bubbles rise to the surface. If the conditions are favorable, the frothy mixture of gas and liquid then erupts through the opening. In a similar manner, gas bubbles form in rising magma. The gas rises to the uppermost layer of magma to create a highly pressurized, frothy, expanding mixture of gas and liquid magma. The temperature of this mixture may be as high as 900°C.

The eruption of Mount St. Helens.   *(USGS, R. P. Hoblitt)*

had grown so steep that the second earthquake caused it to break away from the mountain, forming an immense landslide of rock, soil, and glacial ice. The landslide roared down the mountain and, in doing so, relieved the pressure on the gas-charged magma that had been causing the bulge. The magma then exploded out through the side of the mountain where the bulge had been. This horizontal blast flattened trees in a 400-square-kilometer area on the north side of the mountain (Fig. 1). Large trees as far as 25 kilometers away were knocked over. The landslide, combined with large volumes of new volcanic ash from the eruption, poured down the Toutle River and into the Columbia River, filling their channels.

A vertical plume of ash, rocks, and gas was blown upward to a height of 18 kilometers (Fig. 2). Static electricity generated by the billowing plume caused light-

ning strikes in the surrounding forest. Mountaineers on Mount Adams, 50 kilometers to the east, saw sparks jump from their ice axes, were pelted with falling debris, and felt the intense heat from the blast. The eruption blew out the entire north side of the volcano, and the top 410 meters of the mountain were removed.

An airborne ash cloud spread directly eastward on the prevailing winds at an average speed of about 60 kilometers per hour. It darkened the sky and turned the air gritty and unpleasant for several days as far away as western Montana. Some ash fall was noted in Minnesota and Oklahoma. The fine, abrasive ash made breathing difficult, destroyed crops, and ruined automobile engines. Damage resulting from the eruption was estimated to be hundreds of millions of dollars. A total of 63 people were known or presumed to have been killed.

As the magma body rises to within a few kilometers of the Earth's surface, it creates a blister, or dome, by uplifting and fracturing overlying rocks. The highly pressurized mixture of gas and magma then explodes through the fractures, blowing great streams of liquid magma, gas, and rocks several kilometers into the sky (Fig. 6.22B). In other cases, the gas-charged magma may simply ooze through the fractures and flow rapidly over the land, like rootbeer foam overflowing the edge

of a mug. During an eruption of this type, some of the frothy magma may solidify to form **pumice**, a rock so full of gas bubbles that it floats on water.

The amount of material erupted depends on the size of the magma body and the amount of gas available. In a large and extremely violent eruption, the column of rising pyroclastic material may reach a height of 12 kilometers above the Earth's surface and be several kilometers in diameter. A cloud of fine ash may be blown

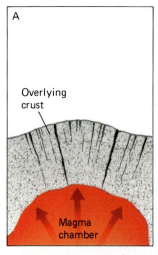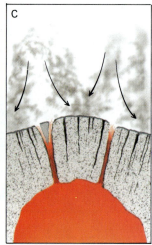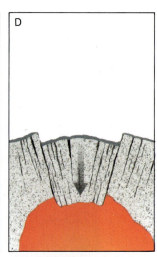

Figure 6.22   (A) When granitic magma rises to within a few kilometers of the surface, it stretches and fractures overlying rock. Gas separates from the magma and rises to the upper part of the magma body. (B) The gas-rich magma explodes through fractures, rising in a vertical column of hot ash, rock fragments, and gas. (C) When the gas is used up, the column collapses and spreads outward as a high-speed ash flow. (D) Because so much material has erupted from the top of the magma chamber, the roof collapses to form a caldera.

much higher, into the upper atmosphere. The column may be held up for hours or even days by the force of additional material streaming out of the magma chamber. Several recent eruptions of Mount Pinatubo blasted ash columns high into the atmosphere and held them up for hours. In contrast, the smaller 1980 Mount St. Helens eruption blasted out almost all of its pyroclastic material nearly instantaneously.

Ash Flows

When the gas in the upper part of the magma is used up, the eruption ceases abruptly. Because no more material is streaming upward to support it, the frothy column of ash, rock, and gas falls back to the Earth's surface (Fig. 6.22C). Although the collapsing column contains some solid particles, it behaves as a fluid falling from a great height.

The falling column spreads outward over the Earth's surface from its point of impact. Because the fluid is denser than air, it flows down stream valleys. Such a flow is called an **ash flow**. Small ash flows travel at speeds up to 200 kilometers per hour. Large flows have traveled distances exceeding 100 kilometers. One large flow leaped over a 700-meter-high ridge as it crossed from one valley into another. Ash flows racing across the land at night glow brightly because of their high temperature. For this reason, an ash flow is also called **nuée ardente,** from the French term for "glowing cloud."

When an ash flow comes to a stop, most of the gas escapes into the atmosphere, leaving behind a chaotic mixture of volcanic ash and rock fragments. The rock formed by such a process is an **ash-flow tuff** (Fig. 6.23). **Tuff** includes all pyroclastic rocks—that is, rocks composed of volcanic ash or other material formed in a volcanic explosion. Some ash flows are hot enough to

Figure 6.23   Ash-flow tuff forms when an ash flow comes to a stop. The fragments in the tuff are pieces of rock that were carried along with the volcanic ash and gas. *(Geoffrey Sutton)*

GEOLOGY AND THE ENVIRONMENT

Volcanoes and Climate

When Mount St. Helens erupted, the ash darkened the sky over a wide area. In Yakima, Washington, which is 140 kilometers from the mountain, people had to turn their car headlights on at noon. In Missoula, Montana, 620 kilometers from Mount St. Helens, people observed an eerie darkness, or dry fog. Clouds of volcanic ash and gas reflect light and heat from the Sun out into space, which leads to a cooling and darkening of the Earth's surface. The immediate effects are easy to document. However, it is more difficult to assess the long-term climatological effects of major volcanic eruptions.

The largest volcanic eruption in recent history occurred in 1815 when Mount Tambora in the southwestern Pacific Ocean exploded, ejecting approximately 100 times as much magma and ash into the atmosphere as did Mount St. Helens. The following year, 1816, was one of the coldest years in recent history and has been recorded as the "year without a summer" and the year "eighteen hundred and froze to death." Crop failures (compounded by the devastation caused by the Napoleonic Wars) led to widespread famine in Europe, a period that has been called "the last great subsistence crisis in the western world." The question remains: Was this cold period caused by the Mount Tambora eruption, or was

it merely a coincidence that the two events occurred together?

The figure shows a plot of global temperatures before and after eight major volcanic events that have occurred in recent times. This graph clearly shows a statistical correlation between global cooling and volcanic eruptions. Meteorological models also demonstrate that high-altitude dust and gas act as an umbrella to prevent sunlight from reaching the Earth. Some scientists think that the recent eruptions of Mount Pinatubo in the Philippines may have caused global cooling.

As explained in the text, historical eruptions have been small compared with some in the more distant past. What would happen if a huge eruption like those that created the Yellowstone calderas were to occur? In a worst-case scenario, dust and sulfur aerosols would block out so much sunlight that daytime would be just a little brighter than a full-moon night for nearly a year after the eruption. However, other calculations indicate that a bright, sunny day a year later would be about as bright as a normal overcast day. In either case, the effects on climate and agriculture would be significant, perhaps catastrophic. But on the positive side, only a few years would pass before normal climatological conditions returned.

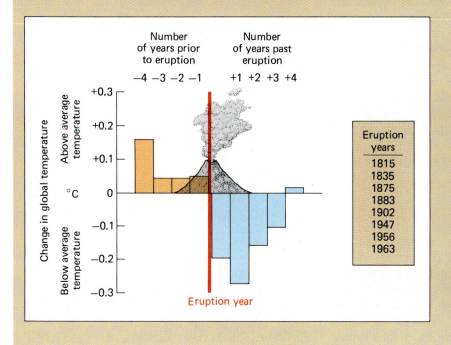

Temperature changes in the Northern Hemisphere in the 4 years immediately before and after eight large recent eruptions. *(Michael Rampino, Annual Review of Earth and Planetary Science 16 [1988]:73–99.)*

Figure 6.24   This welded tuff formed when an ash flow became hot enough to melt and flow as a plastic mass. The streaky texture was created when rock fragments similar to those in Figure 6.23 melted and smeared out as the rock flowed. *(Geoffrey Sutton)*

melt partially after they stop moving. A tough, hard rock called **welded tuff** forms when this mixture solidifies. Welded tuff often shows spectacular structures and textures formed by plastic flow of the partly melted ash (Fig. 6.24). Other ash-flow tuffs are unwelded.

A single large eruption may eject a few hundred to a few thousand cubic kilometers of ash. To visualize this quantity of material, think of a cube of rock with a volume of 1000 cubic kilometers. The perimeter of its base would be 40 kilometers, or slightly less than the length of a marathon race. A world-class distance runner could circle it in a little over 2 hours. Its height would be 10,000 meters, 2000 meters higher than Mount Everest. The largest known ash-flow tuff from a single eruption is located in the San Juan Mountains of southwestern Colorado. It has a volume greater than 3000 cubic kilometers—three times the size of the imaginary cube. Another of comparable size has been mapped in southern Nevada.

Calderas

Think of what must happen in the magma chamber when such an immense volume of material suddenly explodes skyward: Nothing remains to hold up the overlying rock. Therefore, the roof of the magma chamber collapses (Fig. 6.22D). Most large magma bodies are roughly circular when viewed from above. Consequently, roof collapse usually forms a circular depres-

sion called a **caldera**. A large caldera may be 40 kilometers in diameter and have walls as much as a kilometer high. Some calderas fill up with ash-flow tuff as the ash column collapses; others show remarkable preservation of the circular depression and steep walls. We usually think of volcanic landforms in terms of gracefully symmetrical peaks. The topographic depression of a caldera is an interesting exception to this notion. Figure 6.25 shows the abundance of ash-flow tuffs and related rocks in western North America.

Magma that explodes to form ash-flow tuffs and calderas commonly erupts more than once, with substantial time intervals between eruptions. Following an initial eruption, the remaining upper part of a magma body has been depleted in gas and has lost its explosive potential. However, lower portions of the magma continue to release gas, which rises and builds pressure anew to begin another cycle of eruption. Time intervals

Figure 6.25   Calderas (red dots) and ash-flow tuffs (orange areas) in western North America.

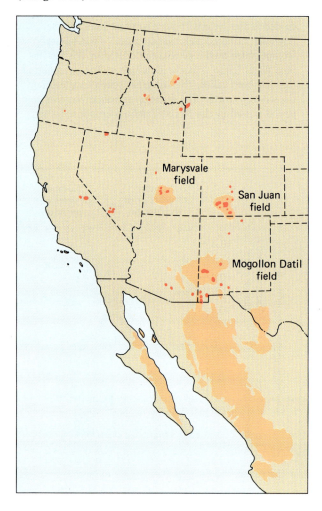

between successive eruptions vary from a few thousand to about half a million years.

Yellowstone National Park

Yellowstone National Park in Wyoming and Montana is the oldest national park in the United States. Its geology consists of three large, overlapping calderas and the ash-flow tuffs that erupted from them (Fig. 6.26). The oldest eruption occurred 1.9 million years ago and ejected 2500 cubic kilometers of pyroclastic material. The next major eruption occurred 1.3 million years ago. The most recent, 0.6 million years ago, produced the Yellowstone caldera in the very center of the park by ejecting 1000 cubic kilometers of ash and other debris.

Consider the periodicity of the three Yellowstone eruptions. They occurred at 1.9, 1.3, and 0.6 million years ago. Intervals of 0.6 to 0.7 million years separated the eruptions. It has been 0.6 million years since the last eruption. The numerous geysers and hot springs indicate that hot magma still lies beneath the park, and seismographs frequently detect movement of liquid magma near the surface. The periodicity of Yellowstone eruptions, the presence of shallow magma, and the well-known tendency of magma of this type to erupt multiple times all suggest that a fourth eruption may be due. Geologists would not be surprised if an eruption occurred at any time. However, arguments based on periodicity are only approximate. Even if the periodicity were exactly 0.6 million (600,000) years, a 1 percent

Figure 6.26   Calderas and ash-flow tuffs of Yellowstone Park. The Big Bend Ridge caldera formed during the eruption of the Huckleberry Ridge Tuff 1.9 million years ago. The Henry's Fork caldera formed during the eruption of the Mesa Falls Tuff 1.3 million years ago. The Yellowstone caldera formed during the eruption of the 0.6-million-year-old Lava Creek Tuff. Portions of older calderas are obliterated by younger calderas. Dashed boundaries of the Huckleberry Ridge and Mesa Falls Tuffs are covered by younger rocks. *(Figure modified from Hildreth and others,* Journal of Geophysical Research *89 [1984].)*

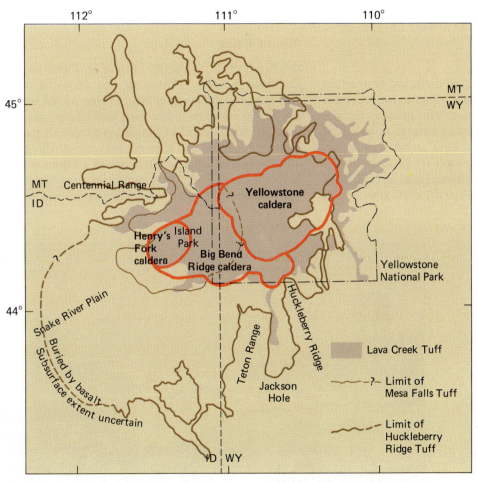

Table 6.2 • Some Notable Volcanic Disasters Since the Year A.D. 1000 Involving 5000 or More Fatalities

Volcano	Country	Year	Primary Cause of Death				
			Pyroclastic Flow	Debris Flow	Lava Flow	Posteruption Starvation	Tsunami
Kelut	Indonesia	1586		10,000			
Vesuvius	Italy	1631			18,000		
Etna	Italy	1669			10,000		
Lakagigar	Iceland	1783				9,340	
Unzen	Japan	1792					15,190
Tambora	Indonesia	1815	12,000			80,000	
Krakatoa	Indonesia	1883					36,420
Pelée	Martinique	1902	29,000				
Santa Maria	Guatemala	1902	6,000				
Kelut	Indonesia	1919		5,110			
Nevada del Ruiz	Colombia	1985		>22,000			

error is 6000 years. Thus, it is conceivable that the next eruption will not occur for several thousand years. It is also possible that Yellowstone has seen its last eruption.

6.7 Volcanoes and Human Settlements

Volcanic eruptions have caused death and destruction throughout history. The most destructive eruptions have been of the type described in the previous section, in which gas-charged magma explodes from a shallow granitic magma chamber to form an ash flow. When Mount St. Helens erupted in 1980, it exploded with the force of 500 atomic bombs of the size used on Hiroshima in World War II. The blast ejected 1 cubic kilometer of rock and ash, leveled forests, and, despite geologists' warnings that an eruption was imminent, killed 63 people. Yet compared with other historical eruptions, Mount St. Helens was small. Table 6.2 summarizes the major known volcanic disasters since A.D. 1000.

Mount Pelée

On May 2, 1902, the coastal town of St. Pierre on the Caribbean Island of Martinique was destroyed completely by an ash flow from Mount Pelée. All but two of the 29,000 residents of the town died nearly instantly when an 800°C cloud of gas and volcanic ash roared down the volcano and through town at speeds up to 100 kilometers per hour. Only one of 18 ships in the harbor escaped, and it lost many crew members. The magma responsible for the eruption was viscous, silica

The smoldering ruins of St. Pierre, May 14, 1902, following the May 8 eruption of Mount Pelée. *(Institute of Geological Sciences, London)*

rich, and charged with gas. Ironically, one of the survivors was a convicted murderer imprisoned in a dungeon when the eruption occurred.

Krakatoa

Before August of 1883, Krakatoa was an 800-meter-high island lying between Java and Sumatra in the southwestern Pacific Ocean. In the days before August 27, several small volcanic explosions occurred on the island. On August 27, the entire island exploded with an amount of energy equivalent to 100 million tons of TNT. The sound of the explosion was heard in Australia, more than 4000 kilometers away. Although Krakatoa itself was unpopulated, the volcanic explosion formed an immense tsunami between 35 and 40 meters high that radiated outward, inundating coastal villages on nearby islands and killing about 35,000 people. Most of the island disappeared in the explosion, leaving in its place a basin 100 meters deep.

It is thought that much of Krakatoa's explosive energy developed as seawater flowed into cracks formed by the rising magma body. The seawater heated rapidly and then exploded as it approached the shallow magma beneath the island.

Mount Vesuvius

In A.D. 79 Mount Vesuvius erupted, destroying the Roman cities of Pompeii and Herculaneum and several neighboring villages near what is now Naples, Italy. Prior to that eruption the volcano had been inactive for more than 2000 years—so long that vineyards had been cultivated on the sides of the mountain all the way to the summit. During the eruption, a pyroclastic flow streamed down the flanks of the volcanic cone, burying the cities and towns under 5 to 8 meters of hot ash. Pompeii was found and excavated 17 centuries later. The excavations revealed molds of inhabitants trapped by the ash flow as they attempted to flee or find shelter. Some of the molds even appear to have preserved facial expressions of terror. Mount Vesuvius returned to relative quiescence, only to become active again in 1631. It was active frequently from 1631 to 1944, and in this century it has experienced eruptions in 1906, 1929, and 1944.

SUMMARY

Magma forms in three geologic environments: subduction zones, spreading centers, and mantle plumes. Heat, addition of water, and **pressure-relief melting** cause rocks to melt and form magma. Granitic magma typically solidifies within the Earth's crust, whereas basaltic magma usually erupts from a **volcano** at the surface. This contrast in behavior of the two types of magma is due to differences in silica and water content.

Any intrusive mass of igneous rock is a **pluton**. A **batholith** is a pluton with more than 100 square kilometers of exposure at the Earth's surface. A **dike** and a **sill** are both sheetlike plutons. Dikes cut across layering in country rock, and sills run parallel to layering.

Magma may flow onto the Earth's surface as **lava flows** or may erupt explosively as **pyroclastic** material. Fluid lava forms **lava plateaus** and **shield volcanoes**. A pyroclastic eruption may form a **cinder cone**. Alternating eruptions of fluid lava and pyroclastic material from the same vent form a **composite cone**. About 70 percent of the Earth's volcanic activity occurs along a circle of subduction zones in the Pacific Ocean called the **ring of fire**. When granitic magma rises to the Earth's surface, it usually erupts explosively, forming **ash-flow tuffs** and **calderas**. Eruptions of this type have caused widespread death and destruction throughout history.

KEY TERMS

Pressure-relief melting *124*
Hot spot *126*
Viscosity *128*
Pluton *128*
Country rock *129*
Batholith *129*
Stock *129*
Dike *129*
Sill *129*

Columnar joint *130*
Pahoehoe *130*
Aa *130*
Pillow lava *132*
Pyroclastic rock *132*
Volcanic ash *132*
Cinder *132*
Volcanic bomb *132*
Flood basalt *133*
Lava plateau *133*

Volcano *134*
Vent *134*
Crater *134*
Shield volcano *134*
Fissure *134*
Cinder cone *134*
Composite cone *135*
Stratovolcano *135*
Volcanic neck *136*
Pipe *136*

Kimberlite *136*
Pumice *141*
Ash flow *142*
Nuée ardente *142*
Tuff *142*
Welded tuff *144*
Caldera *144*

REVIEW QUESTIONS

1. List and describe the major tectonic environments in which magma forms.

2. List and explain the most important factors in the melting of rock to form magma.

3. What happens to most basaltic magma after it forms?

4. What happens to most granitic magma after it forms?

5. Explain why basaltic magma and granitic magma behave differently as they rise toward the Earth's surface.

6. How large is a batholith?

7. Explain the difference between a dike and a sill.

8. How do columnar joints form in a basalt flow?

9. How do a shield volcano, a cinder cone, and a composite cone differ from one another? How are they similar?

10. Which type of volcanic mountain has the shortest life span? Why is this structure a transient feature of the landscape?

11. How does a composite cone form?

12. What is a volcanic neck? How is it formed?

13. Why do diamonds occur naturally in kimberlite pipes?

14. What is the ring of fire? Why do most of the Earth's volcanoes occur in this zone?

15. Explain why and how granitic magma forms ash-flow tuffs and calderas.

16. What is pumice, and how does it form?

17. How does welded tuff form?

18. How does a caldera form?

19. How much pyroclastic material can erupt from a large caldera?

20. Explain why additional eruptions in Yellowstone Park seem likely. Describe what such an eruption might be like.

21. What is a tsunami?

DISCUSSION QUESTIONS

1. If you were handed a sample of rock but given no information about the geologic environment from which it came, would it be possible to determine whether it originated in a batholith, a dike, or a sill? Defend your answer.

2. How could you distinguish between a sill exposed by erosion and a lava flow?

3. Many mountains are composed of granite that is exposed at the Earth's surface. Does this observation prove that granite forms at the Earth's surface?

4. Are basalt plateaus made of extrusive or intrusive rocks? How could you distinguish between a basalt plateau and an uplifted batholith?

5. How does the huge mass of magma that eventually solidifies to form a batholith make its way upward through the Earth's crust?

6. Much of the surface of the Moon is scoured by meteor craters. However, some regions, called seas or maria, are flat expanses of rock. Outline a plausible geologic explanation for these maria. Be sure to include a chronology of events in your sequence.

7. Sometimes gases dissolve in a liquid, and at other times they form bubbles within it. Discuss the difference between these two conditions and its relevance to the geology of volcanoes.

8. Imagine that you detect a volcanic eruption on a distant planet but have no other data. What conclusions could you draw from this single bit of information? What types of information would you search for to expand your knowledge of the geology of the planet?

9. Explain why some volcanoes have steep, precipitous faces, but many do not.

10. Parts of the San Juan Mountains of Colorado are composed of granite plutons, and other parts are volcanic rock. Explain why these two types of rock are likely to occur in proximity.

11. In what natural environments in addition to kimberlite pipes might diamonds be found?

12. Compare and contrast the danger of living 5 kilometers from Yellowstone National Park with the danger of living an equal distance from Mount St. Helens. Would your answer differ for people who live 50 kilometers or those who live 500 kilometers from the two regions?

13. Study a geologic map of the place where you grew up. Evaluate the possible threat of a catastrophic volcanic eruption in that area.

The Geology of the Ocean Floor

Corals grow abundantly in the warm, clean water of shallow seas.
(Superstock)

If you were to ask most people to describe the difference between continents and oceans, they might show surprise and reply, "Why, obviously oceans are water and continents are land!" This is true, of course, but to a geologist a more important distinction exists. He or she would explain that the rocks beneath the oceans are very different from those of the continent's crust. The accumulation of seawater in the world's ocean basins is a result of that difference.

Recall that the lithosphere floats on the asthenosphere. The concept of isostasy tells us that thin, dense lithosphere should sink to a low elevation, whereas thicker, lighter lithosphere should float to a higher level. The upper part of the lithosphere is made of either oceanic or continental crust. Oceanic crust is dense basalt and is only about 10 kilometers thick. In contrast, continental crust is granite, which is of lower density and averages about 30 kilometers thick. As a result of these differences, continents float at high elevations, and the ocean basins sink to low elevations. The Earth's oceans collect in the huge depressions formed by oceanic crust.

The sizes of all ocean basins change over geologic time because new oceanic crust forms at spreading centers such as the Mid-Atlantic ridge, and old sea floor is consumed at subduction zones. At present, the Atlantic Ocean is growing while the Pacific is shrinking. These changes affect ocean currents and the transport of heat across the globe. Therefore, such tectonic changes affect climate and life on Earth.

7.1 Studying the Ocean Floor

For years geologists and oceanographers wondered, "What does the sea floor look like?" and "What types of sediment or rock make up oceanic crust?" Seventy-five

Figure 7.1 A sediment core is retrieved from the sea floor for study. *(Ocean Drilling Program, Texas A&M University)*

years ago, scientists had better maps of the Moon than of the sea floor. The Moon is clearly visible in the night sky, and we can view its surface with a telescope. The sea floor, on the other hand, lies at an average depth of nearly 5 kilometers and in places is more than 10 kilometers deep. Today oceanographers use a variety of techniques to study the sea floor, including sampling, remote sensing, and direct observation from deep-diving submarines.

Sampling Techniques

Several devices collect samples of sediment and rock from the ocean floor. A **rock dredge** is an open-mouthed steel net dragged along the sea floor behind a research ship. The dredge breaks rocks from submarine outcrops and hauls them to the surface. Sediment near the surface of the sea floor can be sampled by a **coring device**, a weighted, hollow steel pipe lowered on a cable from a research vessel. The weight drives the pipe into the soft sediment, which is forced into the pipe. The sediment **core** is retrieved from the pipe after it is winched back to the surface. If the core is taken and removed from the pipe carefully, even the most delicate sedimentary layering is preserved (Fig. 7.1).

Both sediment and rock samples are retrieved by **sea-floor drilling** methods that were developed for oil exploration and recovery. Large drill rigs are mounted

Figure 7.2 The *Joides Resolution,* a deep-sea drilling ship. *(Ocean Drilling Program, Texas A&M University)*

on offshore platforms and on research vessels (Fig. 7.2). The drill cuts cylindrical cores from both sediment and rock, which are then brought to the surface for study.

Remote Sensing

Remote sensing methods do not require direct physical contact with the ocean floor, and for some studies this approach is both effective and economical. The **echo sounder** is the principal tool for mapping sea-floor topography. It emits a sound signal from a research ship and then records the signal after it bounces off the sea floor and travels back up to the ship (Fig. 7.3A). The water depth is calculated from the time required for the sound to make the round trip. The ship steers a carefully navigated course with the echo sounder operating continuously, and in this way a topographic profile of the ocean floor is constructed. Many such profiles are then combined to produce a topographic map of the sea floor. The **seismic profiler** works in the same way but uses a higher-energy signal. In addition to bouncing off the sea floor, some of the sound waves penetrate its surface and reflect off layers within the underlying sediment and rock. This gives a picture of the layering

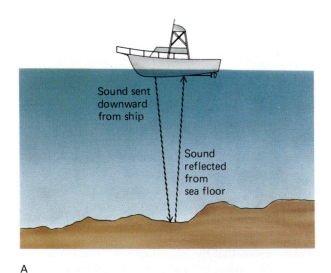

A

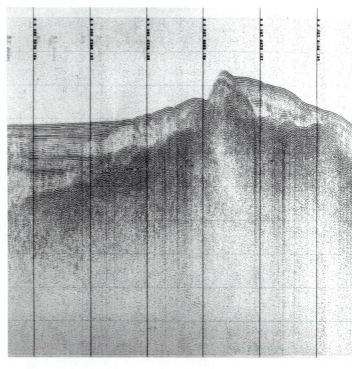

B

Figure 7.3 (A) Measuring the topography of the sea floor with an echo sounder. A sound signal generated by the echo sounder bounces off the sea floor and back up to the ship, where its travel time is recorded. (B) A seismic profiler record of sediment layers and basaltic ocean crust in the Sea of Japan. *(Ocean Drilling Program, Texas A&M University)*

Figure 7.4 *Alvin,* a research submarine capable of diving to portions of the ocean floor. Scientists on board control robot arms to collect sea-floor rocks and sediment. *(Red Catanach, Woods Hole Oceanographic Institution)*

and structure of sediment and rocks of oceanic crust as well as sea-floor topography (Fig. 7.3B).

In the early 1970s, small, specially designed research submarines were developed jointly by French and American scientists. The submarines carry researchers to the ocean floor to view, photograph, and sample sea-floor rocks and sediment directly (Fig. 7.4). In the 1980s and 1990s, scientists began using deep-diving robots and laser imagers to sample and photograph the sea floor. Robots are cheaper and safer than submarines, and laser images penetrate up to eight times farther through water than conventional cameras.

7.2 The Mid-Oceanic Ridge

The extensive use of submarines during World War II made it essential to have topographic maps of the sea floor. Those maps, made with early versions of the echo sounder, were kept secret by the military. When they became available to the public after peace was restored, scientists were surprised to learn that the ocean floor has at least as much topographic diversity and relief as the continents (Fig. 7.5). Great undersea mountain ranges, broad plains, isolated high peaks, and deep valleys make up a varied and fascinating landscape beneath the oceans. Plate tectonics theory now enables us to explain the origins of these features.

The mid-oceanic ridge is a continuous submarine mountain chain that encircles the globe (Fig. 7.5B). It is the largest mountain system on Earth, with a total length of more than 80,000 kilometers and a width of more than 1500 kilometers in most places. The ridge rises an average of 3 kilometers above the surrounding deep-sea floor. It covers more than 20 percent of the Earth's surface, nearly as much as all continents combined.

This submarine mountain chain is completely unlike mountain ranges found on continents. Continental ranges are usually composed of granite and folded and faulted sedimentary and metamorphic rocks. In contrast, the mid-oceanic ridge is made up entirely of undeformed basalt.

A **rift valley** 1 to 2 kilometers deep and several kilometers wide splits the ridge crest. In 1974, French and American scientists used a small research submarine to dive into the rift valley in the Atlantic Ocean. They saw gaping vertical cracks up to 3 meters wide on the floor of the rift. Nearby were basalt flows so young that they were not covered by any sediment. The cracks form when oceanic crust separates at the ridge axis. Basalt magma then rises through the cracks and flows onto the floor of the rift valley. This basalt becomes new oceanic crust as two lithospheric plates spread outward from the ridge axis. Thus, the rift valley is the boundary between two diverging plates. Scientists also saw cliffs hundreds of meters high formed by faulting as the two lithospheric plates separated.

The mid-oceanic ridge rises high above the surrounding sea floor because new lithosphere forming at the ridge axis is hot and therefore of relatively low density. Its low density causes it to float higher than older and cooler lithosphere on both sides of the ridge. Because the new lithosphere cools as it spreads away from the ridge, it becomes denser. Therefore its sinks to lower elevations, forming the deeper sea floor (Fig. 7.6).

Shallow earthquakes commonly occur at the mid-oceanic ridge as a result of fracturing and faulting of oceanic crust as the two plates separate (Fig. 7.7). Blocks of new oceanic crust drop downward along the faults, forming the rift valley.

We have portrayed the mid-oceanic ridge as a mountain range bisected by a deep rift valley, snaking its way beneath the world's major oceans. Closer examination shows that the rift valley and the ridge are not really continuous, but rather are cut and offset by numerous fractures called **transform faults** (Fig. 7.8).

Transform faults extend through the entire thickness of the lithosphere. They develop because the mid-oceanic ridge is not a single, continuous spreading center, but rather consists of many short segments. Each segment is sightly offset from adjacent segments by the cross-cutting transform faults. Thus, transform faults are original features of a mid-oceanic ridge; they begin to form at the same time sea-floor spreading begins.

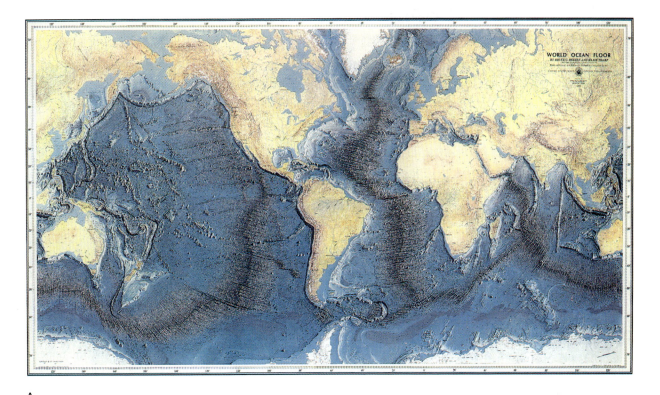

A

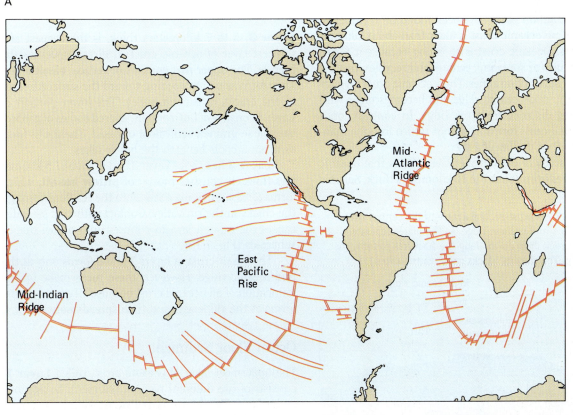

B

Figure 7.5 (A) Topography of the ocean floors. *(Copyright © Marie Tharp)* (B) A map of the ocean floor showing the mid-oceanic ridge in red. Double lines indicate the ridge axis; single lines indicate transform faults. Note that the deep-sea mountain ranges shown in (A) correspond to the mid-oceanic ridge system shown in (B).

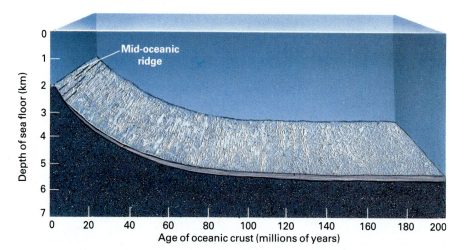

Figure 7.6 The sea floor becomes deeper as it grows older. At the mid-oceanic ridge, new lithosphere is buoyant because it is hot and of low density. It ages, cools, and becomes denser as it moves away from the ridge and consequently sinks. The central portion of the sea floor lies at a depth of about 5 kilometers.

7.3 Sediment and Rocks of the Deep-Sea Floor

The Earth is 4.6 billion years old, and rocks as old as 4.1 to 4.2 billion years have been found on continents. However, no oceanic crust is older than about 200 million years. Oceanic crust is so young because it forms continuously at spreading centers and recycles into the mantle at subduction zones. Thus, oceanic crust is youngest at the mid-oceanic ridge and becomes older with increasing distance from the ridge. In contrast, once continental crust forms, it cannot return into the mantle because of its low density.

The study of earthquake waves shows that oceanic crust varies from about 7 to 10 kilometers thick. Seismic

Figure 7.7 A cross-sectional view of the central rift valley of the mid-oceanic ridge. As the plates separate, blocks of rock drop down along the fractures to form the rift valley. The moving blocks cause earthquakes.

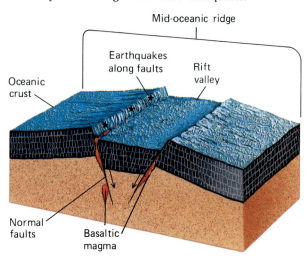

profiling and sea-floor drilling show that it consists of three layers. The lower two are basalt and the upper is sediment (Fig. 7.9).

Basaltic Oceanic Crust

Layer 3, 5 to 7 kilometers thick, is the deepest and thickest layer of oceanic crust. It directly overlies the mantle. The upper part consists of vertical basalt dikes. They form as magma oozing toward the surface freezes in the cracks of the rift valley. The lower portion of Layer 3 consists of horizontally layered bodies of gabbro, the coarse-grained equivalent of basalt. These sills form as magma cools beneath the basalt dikes.

Layer 2 lies above Layer 3 and is about 1.5 kilometers thick. It consists mostly of **pillow basalt**, which forms as basalt magma oozes onto the sea floor through the cracks in the rift valley. Contact with cold seawater causes the molten lava to contract into pillow-shaped spheroids (Fig. 7.10).

The basalt crust of layers 2 and 3 forms only at the mid-oceanic ridge. However, these rocks make up the foundation of all oceanic crust because all oceanic crust forms at the ridge axis and then spreads outward.

Ocean Floor Sediment

The uppermost layer of oceanic crust, called **Layer 1**, consists of two different types of sediment: terrigenous and pelagic. **Terrigenous sediment** is derived directly from land. It is composed of sand, silt, and clay eroded from the continents and deposited on the ocean floor near the continents by submarine currents. **Pelagic sediment**, on the other hand, is found even in deep basins far from continents. It is a gray and red-brown mixture of clay and the remains of tiny plants and animals that live in the surface waters of the oceans. When

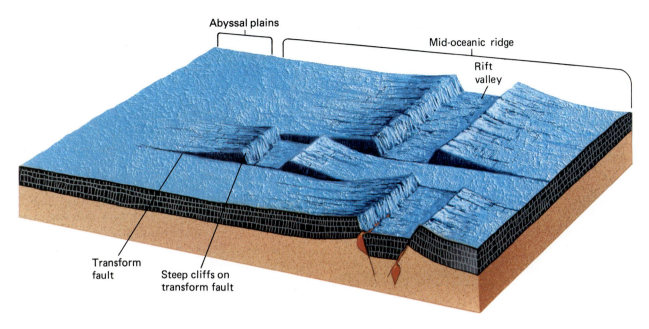

Figure 7.8   Transform faults offset segments of the mid-oceanic ridge. Adjacent segments of the ridge may be separated by steep cliffs 3 or 4 kilometers high. Note the flat abyssal plain far from the ridge.

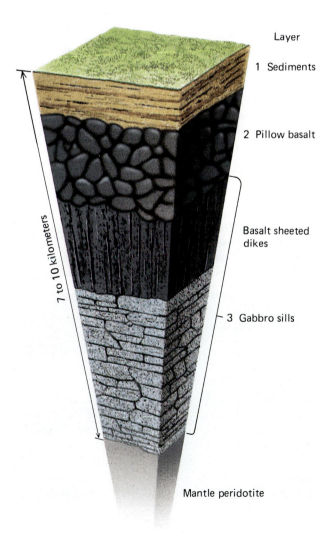

Figure 7.9   The three layers of oceanic crust. Layer 1 consists of mud. Layer 2 is pillow basalt. Layer 3 consists of vertical dikes overlying gabbro sills. Below Layer 3 is the upper mantle.

Figure 7.10   Sea-floor pillow basalt in the Cayman trough.   *(Woods Hole Oceanographic Institution)*

these organisms die, their remains settle slowly to the ocean floor. Much of the clay in the mixture was transported from continents by wind.

Pelagic sediment accumulates at a rate of about 2.5 millimeters every 1000 years. At the axis of the mid-oceanic ridge where oceanic crust is newly formed, there is no sediment. The sediment layer becomes thicker with increasing distance from the ridge because the sea floor becomes older with increasing distance from the ridge axis (Fig. 7.11). Close to shore, pelagic sediment gradually merges with the much thicker layers of terrigenous sediment, which can be 3 or 4 kilometers thick.

Parts of the ocean floor beyond the mid-oceanic ridge are flat, level, featureless submarine surfaces called the **abyssal plains** (Fig. 7.8). They are the flattest surfaces on Earth. Seismic profiling shows that the surface of the basaltic crust is rough and jagged throughout the ocean. On the abyssal plains, however, this rugged profile is buried by sediment. Thus, the extraordinarily level surfaces of the abyssal plains result from accumulation of mud in the deep ocean. The mud covers the hills

Figure 7.11   Deep-sea mud becomes thicker with distance away from the mid-oceanic ridge.

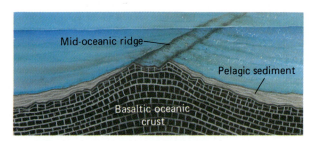

and fills in the valleys of the basaltic crust to create a smooth surface. If you were to remove all of the sediment, you would see the same rugged topography that you see near the mid-oceanic ridge.

7.4 Geological Resources on the Ocean Floor

Manganese Nodules

About 25 to 50 percent of the Pacific Ocean floor is covered with golf-ball- to bowling-ball-size **manganese nodules** (Fig. 7.12). The average nodule contains 20 to 30 percent manganese, 6 percent iron, about 1 percent each of copper and nickel, and lesser amounts of other metals. (Much of the remaining 60 to 70 percent consists of anions chemically bonded to the metal ions.) The metal ions may be introduced into seawater from volcanic activity at the mid-oceanic ridge. The dissolved ions then precipitate to form the nodules by chemical reaction between seawater and pelagic sediment.

The nodules grow by about ten layers of atoms per year, which amounts to 3 millimeters per million years. Curiously, they are found only on the surface of, but never within, the sediment on the ocean floor. Because sediment accumulates much faster than the nodules grow, why doesn't it bury the nodules as they form? Photographs show that animals churn up sea-floor sediment. Worms burrow into it, and other animals pile sediment against the nodules to build protective shelters. Some geologists suggest that these activities constantly lift the nodules onto the surface.

Figure 7.12   Manganese nodules covering the sea floor on the Blake Plateau east of Florida. The nodules range from a few centimeters to about 30 centimeters in diameter, so this photograph covers a few hundred square meters.   *(Frank Manheim, Woods Hole Oceanographic Institution)*

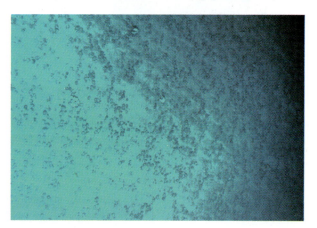

A trillion or more tons of manganese nodules lie on the sea floor. They contain several valuable industrial metals that could be harvested without drilling or blasting. One can imagine a scenario in which exploration is performed using undersea television cameras and the nodules are collected by a vacuum diver, a scoop, or a robot. But despite its tremendous wealth, the ocean floor is not the easiest environment in which to operate complex machinery, so at present, exploitation of manganese nodules is not profitable.

Ore Deposits Formed by Submarine Volcanic Activity

In volcanically active regions of the sea floor—near the mid-oceanic ridge and submarine volcanoes—warm rock heats seawater as it seeps into cracks in oceanic crust. The warm water then circulates through the fractured basalt. In this way, hot, salty seawater migrates through vast volumes of rock. The rate at which seawater circulates through oceanic crust is equivalent to the entire combined volume of the oceans passing through once every 8 million years! Recall from Chapter 3 that hot brine dissolves metals from rock. The hot seawater picks up metals as it circulates through oceanic crust. Then, as it rises through the upper layers of oceanic crust, it cools, and chemical conditions change. As a result, huge deposits of iron, copper, lead, and zinc precipitate within sea-floor sediment and upper layers of basalt. The metal-bearing solutions can be seen as jets of black water, called **black smokers**, spouting from fractures in the mid-oceanic ridge (Fig. 7.13). The black color is caused by precipitation of fine-grained metal sulfide minerals as the solutions cool upon contact with seawater.

Submarine metal deposits formed in this manner can be highly concentrated. For example, whereas on land an ore deposit containing 2.5 percent zinc is rich enough to mine commercially, some undersea zinc deposits contain as much as 55 percent zinc. The cost of operating machinery beneath the sea is so great that such undersea deposits are currently not profitable to mine. However, in some places ore deposits formed underwater have been uplifted by tectonic forces. The ancient Romans mined copper, lead, and zinc ores of this type in the Apennine Mountains of Italy. This geological wealth contributed to their political and military ascendancy. In modern times, rich deposits of lead, zinc, and silver formed in underwater volcanic environments are mined in Australia, North America, and other continents.

In addition to these metal deposits, geologists have discovered vast undersea petroleum reserves along continental margins.

Figure 7.13   A black smoker spouting from the East Pacific rise. Seawater becomes hot as it circulates through the hot basalt near a spreading center. The hot water dissolves sulfur, iron, zinc, copper, and other ions from the basalt. The ions then precipitate as "smoke," consisting of tiny mineral grains, when the hot solution meets cold ocean water. The hot, mineral-rich water sustains thriving plant and animal communities.   *(Dudley Foster, Woods Hole Oceanographic Institution)*

7.5   Continental Margins

Continental margins are regions where continental crust meets oceanic crust. Two principal types of continental margins exist. A **passive continental margin** is characterized by a firm connection between continental and oceanic crust. Little tectonic activity occurs at this type of boundary. Continental margins on both sides of the Atlantic Ocean are passive margins.

In contrast, an **active continental margin** is characterized by subduction of an oceanic lithospheric plate beneath a continental plate. At an active continental margin, subduction typically occurs at or very close to the edge of the continent, and the subducting plate descends at an angle for hundreds of kilometers beneath the continent. The west coast of South America is an example of an active margin. Each type of continental margin has its own characteristic features.

Passive Continental Margins

Consider the passive margin of eastern North America. Recall from Chapter 4 that, about 200 million years ago, all of the Earth's continents were joined, forming the supercontinent Pangaea. Shortly thereafter, Pangaea began to break up into the continents as we know them

GEOLOGY AND THE ENVIRONMENT

Life in the Sea

On land, most photosynthesis is conducted by multicellular plants such as mosses, ferns, grasses, and trees. In turn, large animals such as cows, deer, elephants, and bison consume the plants. In contrast, most of the photosynthesis and consumption in the ocean are carried out by small organisms called **plankton** (drifters). Many plankton are microscopic; others are up to a few centimeters long. **Phytoplankton** are plankton that conduct photosynthesis like land-based plants. Therefore, they are the ultimate source of food for aquatic animals. When you look across the surface of the ocean, you do not notice the phytoplankton, but they are so abundant that they supply about 50 percent of the oxygen in our atmosphere. **Zooplankton** are tiny animals that feed on the phytoplankton. The larger and more familiar marine plants such as seaweed and animals such as fish, sharks, whales, and dolphins play a relatively small role in oceanic photosynthesis and consumption.

Coastal waters above the continental shelf are generally 200 meters or less deep. In some regions this shallow zone extends up to 500 kilometers from the shore; in others the zone is much narrower. In contrast, the central oceans average about 5 kilometers deep.

Coastal areas support large populations of marine organisms. In addition, many deep-sea fish spawn in shallow water within a kilometer or two of shore. These shallow zones are hospitable to life because they have (1) easy access to the deep sea, (2) lower salinity than the open ocean, (3) a high concentration of nutrients originating from land and sea, (4) shelter, and (5) abundant plant life rooted to the sea floor in addition to the phytoplankton floating on the surface. As a result, about 99 percent of the marine fish caught every year are harvested from the shallow waters of the continental shelves. In many areas these fisheries are threatened by pollution.

The productivity of the central oceans is limited because light is available only at the surface, whereas gravity pulls nutrients downward toward the sea floor. As a result, key nutrients, especially nitrates and phosphates, are scarce in upper layers of the ocean where photosynthesis can occur. Productivity is so low that the central oceans have been likened to a great desert.

For many years, scientists assumed that little life exists in the depths of the open oceans because they are too dark for photosynthesis. However, improved diving and sampling techniques have revealed fascinating communities on the deep-ocean floor. It is now clear that the ocean floor, at all depths, supports populations of both large animals and smaller organisms such as bacteria. These organisms eat the carcasses and wastes of plants and animals that sink from the surface layers.

The Maine coast is part of the passive continental margin of eastern North America.

today. As Pangaea was heated and pulled apart, its crust stretched and thinned (Fig. 7.14A). Eventually the lithosphere fractured and began to separate where the crust was thinnest. Basalt magma rose through the cracks and flowed out onto the splitting continent. Continued eruption of basalt formed new oceanic crust between the separating fragments of Pangaea as they drifted apart (Fig. 7.14B). However, no further tectonic activity occurred along the ocean–continent boundary (hence the term *passive continental margin).*

Continental Shelf

Streams and rivers carry sediment from land to the sea and deposit it on coastal deltas. Then ocean currents redistribute the sediment along the coast. As sediment accumulated on the passive margin of the east coast of North America, it built a shallow, gently sloping submarine surface on the submerged edge of the continent.

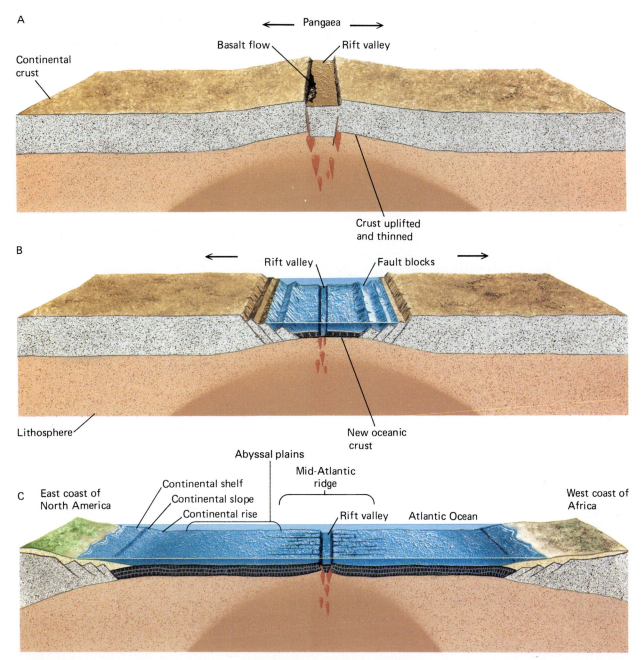

Figure 7.14 Development of the passive continental margin of eastern North America. The Atlantic Ocean basin formed as Pangaea rifted apart. (A) Pangaea is elevated over a rising mantle plume. (B) The crust thins as a result of faulting and erosion as a rift valley forms. Pangaea tears apart, and rising basalt magma forms new oceanic crust between the two halves of continental crust. (C) As the new Atlantic Ocean basin widens, sediment from the continents accumulates to form a broad continental shelf–slope–rise complex and buries the faults.

The Andes have risen along the active continental margin of western South America.

This surface is a **continental shelf** (Fig. 7.15). Its depth increases gradually from the shoreline to about 200 meters at the outer shelf edge. A continental shelf on a passive margin is often a large feature. The continental shelf off the coast of southeastern Canada is about 500 kilometers wide. Parts of the shelves of Siberia and northwestern Europe are even wider.

Most continental shelves are covered by young sediment carried to the continental margin by modern rivers. Thick layers of shale, limestone, and sandstone lie beneath the younger sediment. Many of these older sedimentary rocks contain oil. Some of the world's richest offshore petroleum reserves occur in the North Sea between England and Scandinavia, in the Gulf of Mexico, and in the Beaufort Sea on the northern coast of Alaska and western Canada. In recent years, extensive exploration and development of offshore petroleum reserves have taken place on continental shelves. Deep drilling for oil has revealed that beneath these sedimentary rocks of the continental shelves lie granitic continental crustal

Figure 7.15 A passive continental margin is characterized by a broad continental shelf, slope, and rise formed by accumulation of sediment eroded from the continent. In some areas, salt deposits form when the land is uplifted or sea level falls. In tropical areas, reefs may also grow on the continental shelf.

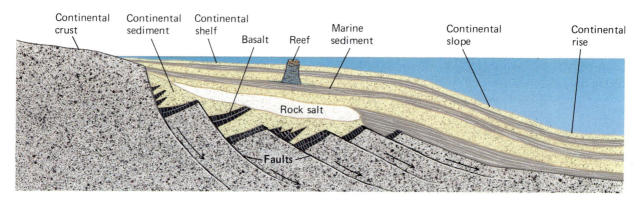

rocks, confirming that the continental shelves are truly parts of the continents even though they are covered by seawater.

Continental Slope and Rise

At the outer edge of a continental shelf, the sea floor steepens suddenly as its depth increases from 200 meters to about 5 kilometers. This steep region of the sea floor averages about 50 kilometers wide and is called the **continental slope** (Fig. 7.15). It is formed by sediment much like that of the shelf. Its steeper angle is due primarily to rapid thinning of underlying continental crust as it approaches the junction with oceanic crust.

The steepness of a continental slope usually decreases as it merges gradually with the deep-ocean floor. This region, called the **continental rise**, consists of an apron of sediment that was transported across the continental shelf and down the slope. It came to rest on the deep-ocean floor at the foot of the slope, and it is the terrigenous sediment described in Section 7.3. The continental rise averages a few hundred kilometers wide. Typically, it joins the deep-sea floor at a depth of about 5 kilometers.

In essence, then, the continental shelf–slope–rise complex on a passive continental margin is a topographic surface formed by accumulation of sediment near the continental margin. The sediment is derived by

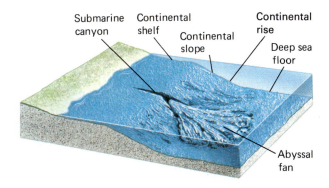

Figure 7.16 A submarine canyon and an abyssal fan.

erosion of the continent and creates a smooth, sloping surface at the junction of continental and oceanic crust.

Submarine Canyons and Abyssal Fans

Deep, V-shaped, steep-walled valleys called **submarine canyons** are eroded into continental shelves and slopes. They look like submarine stream valleys. These canyons typically start on the outer edge of a continental shelf and continue across the slope to the rise (Fig. 7.16). At their lower ends submarine canyons commonly lead into **abyssal fans** (sometimes called **submarine fans**), large, fan-shaped accumulations of sediment deposited on the continental rise.

Oil has formed beneath shallow continental shelves in many parts of the world. Offshore oil rigs provide stable platforms to extract the oil. *(Schlumberger, Inc.)*

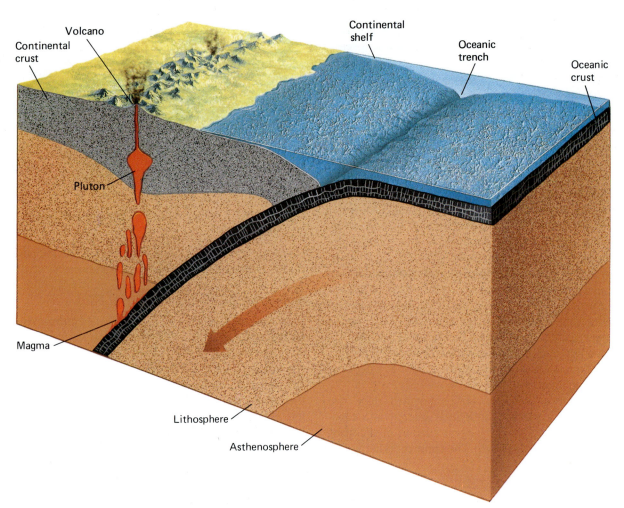

Figure 7.17   At an active continental margin an oceanic plate sinks beneath a continental plate, forming an oceanic trench.

Most geologists now agree that submarine canyons are cut by **turbidity currents**. A turbidity current develops when loose, wet sediment resting on the continental shelf or slope starts to slip downslope, drawn by gravity. The movement may be triggered by an earthquake or simply by oversteepening of the slope as sediment accumulates. When the sediment starts to move, it mixes with water. Because the mixture of sediment and water is fluid and denser than water alone, it flows downslope as a turbulent, chaotic avalanche over the surface of the shelf and slope. A turbidity current can travel at speeds greater than 100 kilometers per hour and for distances up to 700 kilometers. Sediment-laden water traveling at such speeds has tremendous erosive power. Once a turbidity current erodes a small channel into the continental shelf and slope, subsequent currents tend to use the same channel, just as intermittent

surface streams use the same channel year after year. Over time, the currents erode a deep submarine canyon into the continental margin. Turbidity currents slow down when they reach the level, deep-sea floor beyond the continental slope. The sediment accumulates there to form an abyssal fan.

Active Continental Margins

An active continental margin forms where an oceanic plate subducts beneath a continental plate. A long, narrow, steep-sided depression called a **trench** forms on the sea floor where the oceanic plate bends downward and dives into the mantle (Fig. 7.17). The deepest place on Earth is in the Mariana trench, north of New Guinea in the southwestern Pacific, with a maximum depth of nearly 11 kilometers below sea level. Depths of 8 to 10 kilometers are common in other trenches. Trenches form wherever subduction occurs, at an active continental margin or in the middle of an ocean basin where two oceanic plates converge.

An active continental margin commonly has a much narrower continental shelf and a considerably steeper continental slope than does a passive margin. The continental rise is absent because sediment flows into the trench instead of accumulating on the ocean floor. The landward wall of the trench forms the continental slope of an active margin.

Subduction at an active continental margin causes earthquakes, mountain building, and volcanic eruptions. The west coast of South America is an example of such an environment.

7.6 Island Arcs

In many parts of the Pacific Ocean and elsewhere, two oceanic plates are colliding. When oceanic plates converge, one subducts beneath the other, diving into the mantle and forming a mid-oceanic trench (Fig. 7.18). As we learned in Chapter 4, when a plate subducts, the overlying asthenosphere melts, forming magma. In an ocean–ocean collision, the magma rises and erupts on the sea floor to form submarine volcanoes next to the trench. The volcanoes eventually grow to become a chain of islands, called an **island arc**. The western Aleutian Islands are an example of an island arc. Many others are found in the southwestern Pacific.

7.7 Seamounts, Oceanic Islands, and Aseismic Ridges

A **seamount** is a submarine mountain that rises 1 kilometer or more above the surrounding sea floor. An **oceanic island** is a seamount that protrudes above sea level. Both seamounts and oceanic islands are common in all ocean basins, but they are particularly abundant in the southwestern Pacific Ocean. Seamounts and oceanic islands occur most commonly as isolated peaks on the ocean floor. However, some occur in chains of mountains called **aseismic ridges**. *Aseismic* means "no earthquake activity," as opposed to the mid-oceanic ridge and island arcs, where earthquakes are common. The Hawaiian Island–Emperor Seamount Chain is an example of an aseismic ridge. Dredge samples show that seamounts, like oceanic islands and the ocean floor itself, are made of basalt.

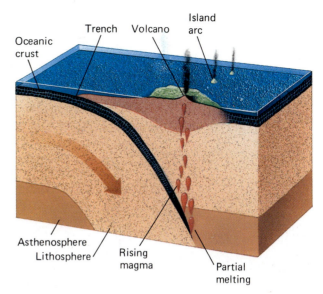

Figure 7.18 An island arc forms at a convergent boundary between two oceanic plates. One of the plates sinks, generating magma that rises to form the islands.

Seamounts and oceanic islands are submarine volcanoes probably formed at hot spots above mantle plumes. Isolated seamounts and islands must have formed over plumes that persisted for only a short time. In contrast, aseismic ridges, such as the Hawaiian Island–Emperor Seamount Chain, formed over long-lasting plumes. In this case the lithospheric plate on which the volcanoes formed migrated over the plume as the magma continued to rise. Each volcano formed directly over the plume and then became extinct as the moving plate carried it away from the plume. As a result, the seamounts and islands become progressively younger toward one end of the chain (Fig. 7.19).

After a volcanic island or seamount forms, it begins to sink. Three factors contribute to the sinking.

1. If the hot spot feeding the volcanic eruptions cools and stops supplying magma, the lithosphere beneath the island cools, becomes denser, and contracts. Alternatively, the island migrates away from the hot spot if the lithospheric plate moves. This also causes cooling, contraction, and sinking of the island.

2. The weight of the newly formed volcano causes isostatic sinking.

3. Erosion lowers the top of the volcano.

These three factors commonly result in gradual transformation of a volcanic island to a seamount (Fig. 7.20). If the Pacific Ocean plate continues to move at its present rate, the island of Hawaii may sink beneath the sea within 10 to 15 million years.

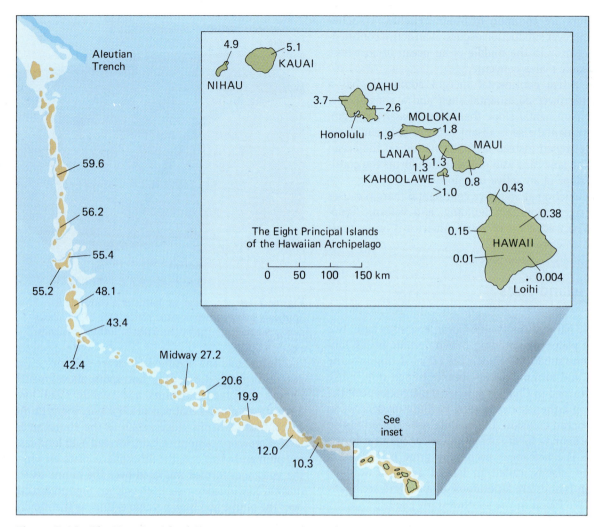

Figure 7.19 The Hawaiian Island–Emperor Seamount Chain. The ages, in millions of years, are for the oldest volcanic rocks found on each island or seamount. The islands become progressively older as they migrate away from the mantle plume.

Figure 7.20 The Hawaiian Islands and Emperor Seamounts sink as they move away from the mantle plume.

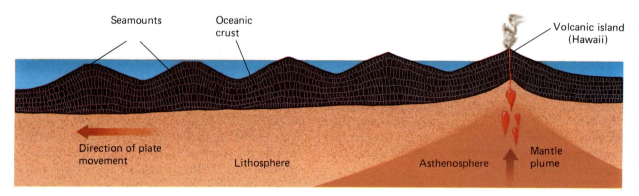

SUMMARY

The major geological difference between a continent and an ocean basin is that the continent is composed of relatively thick, low-density granite, whereas oceanic crust is mostly thin, dense basalt. Thin, dense oceanic crust lies at low topographic levels and forms ocean basins. Because of the great depth and remoteness of the ocean floor and oceanic crust, our knowledge of them comes mainly from **sampling** and **remote sensing**.

The **mid-oceanic ridge** is a submarine mountain chain that extends through all of the Earth's major ocean basins. A **rift valley** runs down the center of the ridge, and the ridge and rift valley are both offset by numerous **transform faults**. The mid-oceanic ridge forms at the center of lithospheric spreading, where new oceanic crust is added to the sea floor.

Abyssal plains are flat areas of the deep-sea floor where the rugged topography of the basaltic oceanic crust is covered by deep-sea sediment. Oceanic crust varies from about 7 to 10 kilometers thick and consists of three layers. The top layer is sediment, which varies from zero to a few kilometers thick. Beneath this lies about 1.5 kilometers of **pillow basalt**. The deepest layer of oceanic crust is from 5 to 7 kilometers thick and consists of basalt dikes on top of gabbro sills. The base of this layer is the boundary between oceanic crust and mantle. The age of sea-floor rocks increases regularly away from the mid-oceanic ridge. No oceanic crust is older than about 200 million years because it recycles into the mantle at subduction zones.

A **passive continental margin** includes a **continental shelf**, a **slope**, and a **rise** formed by accumulation of sediment in the region where continental crust joins oceanic crust. **Submarine canyons**, eroded by **turbidity currents**, notch continental margins and commonly lead into **abyssal fans**, where the turbidity currents deposit sediments on the continental rise. An **active continental margin**, where oceanic crust subducts beneath the margin of a continent, usually includes a narrow continental shelf and a continental slope that steepens rapidly into a **trench**. A trench is an elongate trough in the ocean floor formed where oceanic crust dives downward at a subduction zone. Trenches are the deepest parts of ocean basins.

Island arcs are common features of some ocean basins, particularly the southwestern Pacific. They are chains of volcanoes formed at subduction zones where two oceanic plates collide. **Seamounts, oceanic islands**, and **aseismic ridges** also form in oceanic crust as a result of volcanic activity over mantle plumes.

KEY TERMS

Rock dredge *150*	Pelagic sediment *154*	Continental shelf *160*	Island arc *163*
Coring device *150*	Abyssal plains *156*	Continental slope *161*	Seamount *163*
Echo sounder *151*	Manganese nodules *156*	Continental rise *161*	Oceanic island *163*
Seismic profiler *151*	Black smoker *157*	Submarine canyon *161*	Aseismic ridge *163*
Mid-oceanic ridge *152*	Passive continental	Abyssal fan (submarine	
Rift valley *152*	margin *157*	fan) *161*	
Pillow basalt *154*	Active continental	Turbidity current *162*	
Terrigenous sediment *154*	margin *157*	Trench *162*	

REVIEW QUESTIONS

1. Describe the main differences between oceans and continents.

2. Sketch a cross section of the mid-oceanic ridge, including the rift valley.

3. Describe the dimensions of the mid-oceanic ridge.

4. Explain why the mid-oceanic ridge is elevated topographically above the surrounding ocean floor. Why does its elevation gradually decrease away from the ridge axis?

5. Explain the origin of the rift valley in the center of the mid-oceanic ridge.

6. Why is heat flow unusually high at the mid-oceanic ridge?

7. Why are the abyssal plains characterized by such low relief?

8. Sketch a cross section of oceanic crust from a deep-sea basin. Label, describe, and indicate the approximate thickness of each layer.

9. Describe the two main types of sea-floor sediment. What is the origin of each type?

10. Compare the ages of oceanic crust with the ages of continental rocks. Why are they so different?

11. Sketch a cross section of both an active continental margin and a passive continental margin. Label the features of each. Give approximate depths.

12. Explain why a continental shelf is made up of a foundation of granitic crust, whereas the deep-ocean floor is composed of basalt.

13. Why does an active continental margin typically have a steeper continental slope than a passive margin? Why does an active margin typically have no continental rise?

14. Explain the relationships among submarine canyons, abyssal fans, and turbidity currents.

15. Why are turbidity currents often associated with earthquakes or with large floods in major rivers?

16. Explain the origins of and differences between seamounts and aseismic ridges.

17. Why do oceanic islands sink after they form?

DISCUSSION QUESTIONS

1. How and why does an oceanic trench form?

2. The east coast of South America has a wide continental shelf, whereas the west coast has a very narrow shelf. Discuss and explain this contrast.

3. Seismic data indicate that continental crust thins where it joins oceanic crust at a passive continental margin, such as on the east coast of North America. Other than that, we know relatively little about the nature of the junction between the two types of crust. Speculate on the nature of that junction. Consider rock types, geologic structures, ages of rocks, and other features of the junction.

4. Discuss the topography of the Earth in an imaginary scenario in which all conditions are identical to present ones except that there is no water. In contrast, what would be the effect if there were enough water to cover all of the Earth's surface?

Mountains and Geologic Structures

The Flatirons of Boulder, Colorado are tilted layers of sandstone at the boundary between the High Plains and the Colorado Rockies.

Among all the Earth's topographic features, mountains stand out for sheer beauty and their projection of nature's power. Green forests and alpine meadows blanket the lower slopes of the great ranges. High on the mountainsides the last stunted trees stand fast against the wind and cold amid rubble deposited by alpine glaciers only a few thousand years ago. Higher yet, rock walls may rise vertically for 1000 meters or more. The highest peaks sparkle with glacial ice and snow, remnants of the last alpine ice age. Because of this beauty and the feeling of closeness to nature, mountain ranges have become popular vacation areas. The European Alps, the American Rockies, the Sierra Nevada of California, the Tetons of Wyoming, and the Appalachians draw millions of visitors each year. Even the less accessible Himalayas and South American Andes are standard on the itineraries of more adventurous tourists.

8.1   Mountains and Mountain Ranges

A **mountain** is any part of the Earth's surface sufficiently elevated above its surroundings to have a distinct summit. Normally, to qualify as a mountain, a peak must rise at least 300 meters above the surrounding terrain. A **mountain range** is a series of mountains or mountain ridges that are grouped closely and that formed at about the same time and under the same geologic conditions. Commonly, several mountain ranges cluster together in an elongate zone called a **mountain chain**. Mountains usually occur in ranges and chains because the forces that create them operate over large regions of the Earth's crust rather than in small, isolated localities.

8.2   Plate Tectonics and Mountain Building

Most of the Earth's tectonic activity occurs at lithospheric plate boundaries, where two moving plates interact with each other. Tectonic activity builds mountains. Therefore, most of the Earth's mountains and mountain ranges have formed at plate boundaries (Fig. 8.1). Mountain building is commonly accompanied by folding and

Figure 8.1   Lithospheric plate boundaries and major mountain ranges. Arrows indicate directions of plate movements.   *(Tom Van Sant/Geosphere Project)*

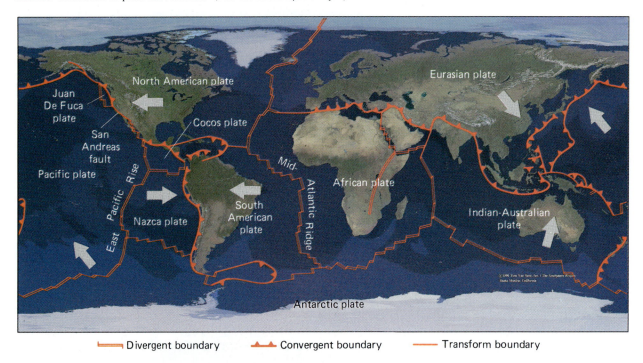

Divergent boundary   Convergent boundary   Transform boundary

A   B

Figure 8.2   (A) Folds in limestone, Canadian Rockies, Alberta. (B) Small-scale folds, McCarty's Mountain, Montana.

faulting of rocks, earthquakes, volcanic eruptions, intrusion of plutons, and metamorphism.

Recall from Chapter 4 that three different types of boundaries separate tectonic plates: divergent boundaries, transform boundaries, and convergent boundaries.

1. At a divergent boundary, or rift, two plates spread apart as new lithosphere forms between them. Such a boundary can form in either oceanic or continental crust. The world's longest mountain chain, the mid-oceanic ridge, formed in this tectonic environment. The highest mountains on the African continent, Mounts Kilimanjaro and Kenya, are volcanoes formed along the East African rift.

2. At a transform boundary, two plates slide horizontally past each other. The San Andreas fault is a transform boundary in continental crust in western California. Movement along this fault has uplifted the San Gabriel Mountains. However, few major mountain ranges develop along transform boundaries.

3. At a convergent boundary, two plates meet in a head-on collision. Most of the great continental mountain ranges grew at convergent plate boundaries.

Before we continue our discussion of mountain building, consider what happens to rocks caught between two moving tectonic plates.

8.3   Folds, Faults, and Joints

If you push or pull on any solid object with enough force, it either bends or breaks. If you step on a supple, green stick, it bends; if you step on a dry twig, it snaps. Similarly, tectonic forces bend and fracture rocks. Whether a rock bends or breaks depends on the rock type, temperature, and force applied to it.

Enormous tectonic forces develop at a convergent plate boundary where two 100-kilometer-thick slabs of lithosphere collide. Rocks trapped between colliding plates bend and fracture. In some cases the collision deforms rocks tens or even hundreds of kilometers away from the actual boundary. In addition, huge masses of magma rise through the lithosphere at a convergent plate boundary. They rise by shouldering aside hot, plastic country rock, deforming it as they force their way upward through the Earth's crust. Because great mountain chains grow at convergent plate boundaries, rocks in mountainous regions are commonly broken and bent by these processes. Less intense forces also deform rocks at divergent and transform plate boundaries.

A **geologic structure** is any feature produced by deformation of a rock. Folds, faults, and joints are the most common types of structures. A **fold** is a bend in rock (Fig. 8.2). A **fault** is a fracture along which rock on one side has moved relative to rock on the other

Figure 8.3 A small fault in sedimentary rocks near Kingman, Arizona. *(Ward's Natural Science Establishment, Inc.)*

Figure 8.4 Vertical joints in sandstone, Indian Creek, Utah.

side (Fig. 8.3). A **joint** is a fracture without movement of rock (Fig. 8.4). Joints and faults often occur as sets of many parallel fractures.

Folds

If you hold a sheet of clay between your hands and push it together, the clay deforms into a sequence of folds (Fig. 8.5). This simple demonstration illustrates three important characteristics of folds in rock:

1. Folding usually results from tectonic compression.
2. Folding always shortens distances in rock. Notice in Figure 8.6 that the distance between two points, A and A', is shorter in the folded rock than it is before folding.

3. Folds often occur as a set of several or many folds (Fig. 8.7). But exceptions do exist. Figure 8.8 shows a single fold, called a **monocline**, that developed where sedimentary rocks sagged over an underlying fault.

Depending on the type of rock, temperature, and the forces involved, folds may be large or small. Often,

Figure 8.6 (A) Horizontally layered sedimentary rocks. (B) A fold in the same rocks. The forces that folded the rocks are shown by the arrows. Notice that points A and A' are closer after folding.

Figure 8.5 If you squeeze layers of clay, the clay deforms into folds.

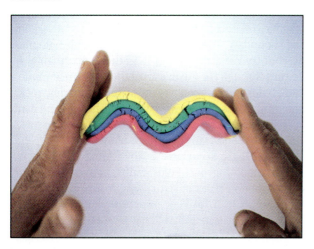

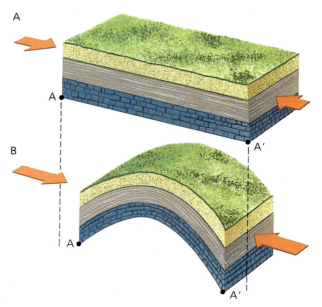

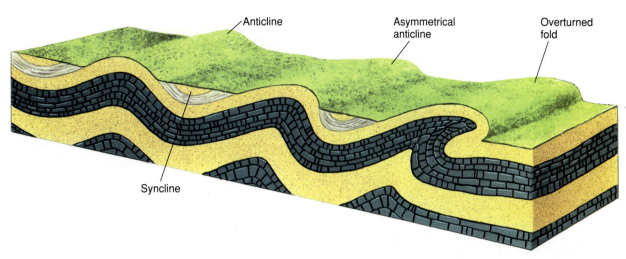

Figure 8.7 Folds in rock, showing both anticlines and synclines. Folds can be symmetrical, as shown on the left, or asymmetrical, as shown in the center. If a fold has tilted beyond the perpendicular, it is said to be overturned.

sets of folds involve rock layers thousands of meters thick and cover hundreds of square kilometers of land. In other instances, folds may occur in a thin layer and be only a few centimeters or less apart. Figure 8.7A shows that a fold arching upward is called an **anticline**, and one arching downward is a **syncline**.

Domes and basins are types of folds that form primarily by rising or sinking of the lithosphere rather than by lateral compression. A **dome** is a circular or elliptical landform resembling an inverted bowl. If a dome occurs in sedimentary rock, the layering dips away from the center of the dome in all directions (Fig. 8.9). A similarly shaped syncline is called a **basin**. Domes and basins can be small structures a few kilometers in diameter or less, but commonly they are much larger. Large domes and basins involve regional warping of the entire continental crust. The Black Hills of South Dakota are a large structural dome. The Michigan basin covers much of the state of Michigan.

Faults

Tectonic forces may cause rocks on opposite sides of a fracture to slip past each other to create a fault. As described in Chapter 5, this movement may be gradual. Alternatively, if the two sides of the fault become locked until elastic energy accumulates, the rock may move suddenly, generating an earthquake. Some faults are a single fracture in rock. On large faults that move hundreds of meters or many kilometers, movement may occur along numerous closely spaced fractures, collectively called a **fault zone** (Fig. 8.10).

Many faults and fault zones move repeatedly for two reasons: (1) Tectonic stress commonly continues

Figure 8.8 A monocline in southeastern Utah.

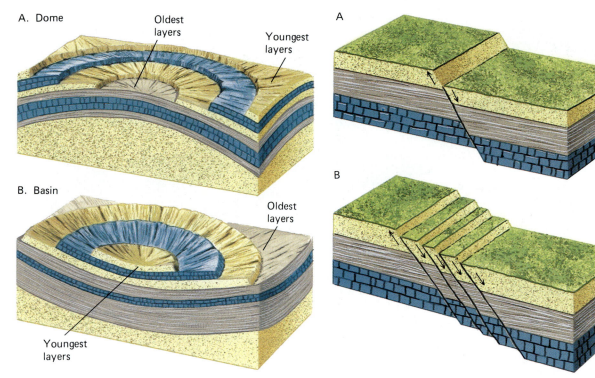

Figure 8.9 (A) Sedimentary layering dips away from a dome in all directions, and the outcrop pattern is circular or elliptical. (B) Layering dips toward the center of a basin.

Figure 8.10 (A) Movement along a single fracture characterizes faults with relatively small displacement. (B) Movement along numerous fractures in a fault zone is typical of faults with large displacement.

to be active in the same place over long periods of time; and (2) once a fault forms, it is easier for movement to occur again along the same fracture than for a new fracture to develop nearby.

Ore deposits often concentrate in faults. In the early days of mining, miners dug shafts and tunnels along faults to get at the ore. Most faults are not vertical but

dip into the Earth at an angle. Therefore, most faults have an upper and a lower side. Miners referred to the side of a fault that hung over their heads as the **hanging wall** and the side they walked on as the **footwall**. These names are still commonly used (Fig. 8.11).

A fault in which the hanging wall has moved down relative to the footwall is called a **normal fault**. Notice

Figure 8.11 A normal fault forms where tectonic forces stretch the Earth's crust. The overhanging side of the fault is called the hanging wall, and the side beneath the fault is called the footwall.

Figure 8.12 A normal fault in sedimentary rocks. The offset of bedding shows the direction and amount of rock movement along the fault. The large arrows show the forces that caused the fault; the small arrows show the direction of rock movement.

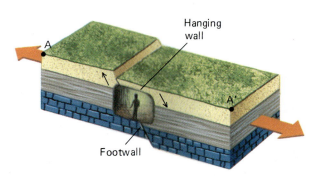

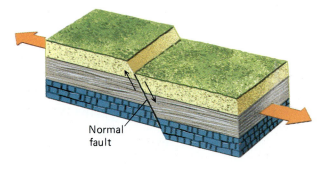

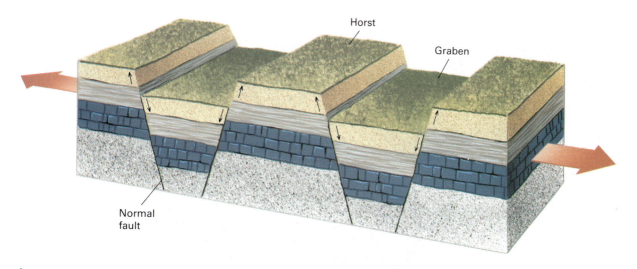

Horst

Graben

Normal
fault

A

B

Figure 8.13 (A) Horsts and grabens commonly form where tectonic forces stretch the Earth's crust. (B) In this photo of the Inyo Mountains of California, the valley is a graben and the mountain ranges in the foreground and background are horsts.

that the horizontal distance between points on opposite sides of the fault, such as A and A′ in Figure 8.12, is greater after normal faulting occurs. Hence, normal faults form where tectonic forces are stretching the Earth's crust, pulling it apart.

Figure 8.13 shows a wedge-shaped block of rock called a **graben** dropped downward between a pair of normal faults. The word *graben* comes from the German word for "grave" (think of a large block of rock settling downward into a grave). If tectonic forces stretch the crust over a large area, many normal faults may develop, allowing numerous grabens to settle downward between the faults. The blocks of rock between the downdropped grabens then appear to have moved upward relative to the grabens; they are called **horsts**.

Normal faults, grabens, and horsts are common where the crust is being pulled apart at spreading centers, such as the mid-oceanic ridge and the East African rift zone. They are also common where tectonic force stretches a single plate, as in the Basin and Range of Nevada and nearby parts of western North America.

In regions where tectonic force squeezes the crust, geologic structures must accommodate crustal shortening. A fold is one structure that shortens crustal rocks. A **reverse fault** is another (Fig. 8.14). In a reverse fault the hanging wall has moved up relative to the footwall. The distance between points A and A′ is clearly shortened by the faulting.

A **thrust fault** is a special type of reverse fault that is nearly horizontal (Fig. 8.15). In some thrust faults,

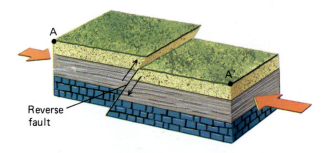

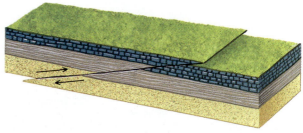

A

Figure 8.14 (A) A reverse fault accommodates crustal shortening and reflects squeezing of the crust, shown by large arrows. (B) A small reverse fault in Zion National Park, Utah.

B

Figure 8.15 (A) A thrust fault is a low-angle reverse fault. (B) A small thrust fault near Flagstaff, Arizona. *(Ward's Natural Science Establishment, Inc.)*

the rocks of the hanging wall have moved many kilometers over the footwall rocks. For example, all of the rocks of Glacier National Park in northwestern Montana slid eastward along a thrust fault to their present location. Their original position, before thrusting, was 50 to 100 kilometers to the west. This thrust is one of many that formed from about 180 to 45 million years ago as the mountains of western North America were being built. Most of those thrusts moved large slabs of rock, some even larger than that of Glacier Park, from west to east in a zone reaching from Alaska to Mexico.

A **strike-slip fault** is one in which the fracture is vertical, or nearly so, and rocks on opposite sides of the

fracture have moved horizontally past each other (Fig. 8.16). A transform plate boundary is a strike-slip fault. The famous San Andreas fault is actually a zone of many parallel strike-slip faults in western California. As explained previously, it is a boundary between two lithospheric plates: The Pacific plate is moving northwestward relative to the North American plate. The true San Andreas fault is just one of many faults in the fault zone.

Joints

A **joint** is a fracture similar to a fault except that rocks on either side of the fracture have not moved. Parallel joints form when tectonic force is sufficient to fracture the rock but not to move it and cause faulting.

Faults and joints are important in engineering, mining, and quarrying because they are planes of weakness in otherwise strong rock. Dams constructed in faulted or jointed rock often leak, not because the dams themselves have holes but because water flows around

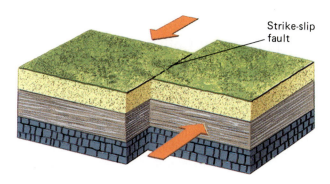

A

Figure 8.16 (A) A strike-slip fault is nearly vertical, but movement along the fault is horizontal. The large arrows show direction of movement. (B) Recent movement along the San Andreas strike-slip fault in southern California has produced distinctive topographic features. *(USGS, R. E. Wallace)*

B

the dam through the fractures. You can commonly see seepage caused by such leaks in canyon walls downstream from a dam.

Geologic Structures and Plate Boundaries

Tectonic forces stretch the crust and pull it apart at a divergent boundary but squeeze and shorten it at a convergent boundary. At a transform boundary, rock is sheared as one plate slips horizontally past another. Because the forces are different, different kinds of geologic structures commonly develop at each type of plate boundary.

Normal faults and grabens are typical of divergent boundaries at mid-oceanic ridges and continental rifts, where the crust is pulled apart. A valley runs down the middle of the mid-oceanic ridge. It is a graben bounded by normal faults on both sides. Normal faults and grabens are the most common structures in the East African rift zone.

At most convergent plate boundaries, where two plates collide, compressional forces squeeze rocks, forming folds and reverse and thrust faults. Broad regions of folds and thrust faults formed in the mountains of western North America from about 180 to 45 million years ago, while collision and subduction were occurring along the west coast. Folds and thrust faults are common in the Appalachian Mountains, the Alps, and the Himalayas (Fig. 8.17).

At other convergent plate boundaries, however, subduction is accompanied by crustal stretching and normal faulting, for reasons that geologists do not un-

Figure 8.17 Wildly folded sedimentary rocks on the Nuptse-Lhotse wall, seen from an elevation of 7600 meters on Mount Everest. *(Galen Rowell)*

derstand well. The Andes Mountains of western South America are one of the Earth's highest mountain ranges. They formed as a result of subduction of the Pacific plate beneath the western edge of the South American plate. The two plates are converging, yet large grabens—structures that reflect crustal stretching—occur just west of the mountains. Thus, although convergent plate boundaries normally squeeze rocks, forming folds and reverse and thrust faults, this is not always the case.

A transform plate boundary is an immense strike-slip fault that cuts through the entire lithosphere. Frictional drag along the fault may cause folding, faulting, and uplift of nearby rocks. Forces of this type have formed the San Gabriel Mountains along the San Andreas fault zone. However, strike-slip faulting does not create lofty mountains. Thus, the San Gabriels and other mountains formed by similar processes are not major ranges.

With the exception of the mid-oceanic ridge, which formed at divergent plate boundaries, most of the world's major mountain ranges formed at convergent plate boundaries. Recall that there are three types of convergent plate boundaries: ocean–ocean, ocean–continent, and continent–continent. Let us examine each in turn.

8.4   Island Arcs: Mountain Chains Formed by Collision Between Two Oceanic Plates

Where two oceanic plates converge, one subducts beneath the other and dives into the mantle, forming an oceanic trench. Magma forms in the subduction zone and erupts onto the sea floor. Eventually submarine volcanoes grow into an **island arc**, a chain of volcanic islands next to the trench. However, the geology of an island arc is not simply that of a chain of volcanoes rising from the sea floor.

A layer of mud 0.5 kilometer or more thick covers the sea floor. This sediment is saturated with water and consequently is soft and not very dense. Because of its low density, most of it cannot descend into the mantle with the subducting slab. Instead, it is scraped from the subducting slab and jammed against the growing island arc. The process is like a bulldozer blade scraping soil from bedrock. As the bulldozer advances, more and more soil piles up on the blade.

Occasionally, slices of basalt from the oceanic crust, and even pieces of the upper mantle, are scraped off and mixed in with the sea-floor sediment. The ocean-

Figure 8.18   An island arc forms during a collision between two plates, each carrying oceanic crust. A subduction complex contains slices of oceanic crust and upper mantle scraped from upper layers of a subducting plate.

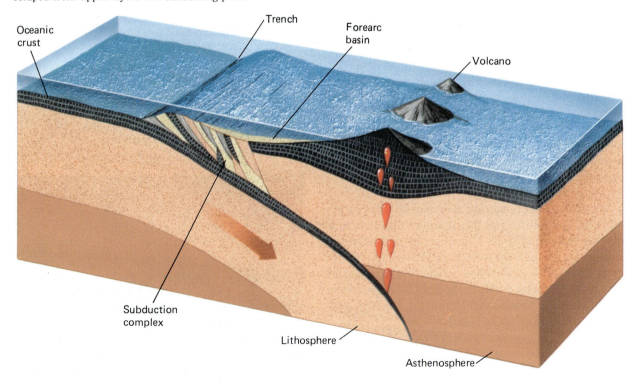

Modern subduction beneath the west coast of South America fires volcanic eruptions throughout the Andes. Here, Chile's Villarrica volcano erupts into a moonlit sky in November 1980. *(Noel Cramer)*

floor sediment and slices of rock are highly deformed, sheared, faulted, and metamorphosed in the bulldozer process. This material, termed a **subduction complex**, becomes a part of the island arc, as shown in Figure 8.18. Thus, an island arc is a volcanic mountain chain plus associated sedimentary and metamorphic rocks. Island arcs are abundant in the Pacific Ocean, where convergence of oceanic plates is common. The western Aleutian Islands and most of the island chains of the southwestern Pacific basin are island arcs.

8.5 The Andes: A Mountain Chain Formed by Collision Between an Ocean Plate and a Continental Plate

Imagine that you have the opportunity to spend a season mountaineering and trekking through the Andes in western South America and, shortly thereafter, another season traveling through the Himalayas between Asia and India. Initially you might be struck by similarities between the two mountain chains. The immensity and great height of the peaks are the most striking features of both; the high summits are covered with snow and ice. The Andes have 49 peaks with elevations above 6000 meters (nearly 20,000 feet). The highest is Aconcagua, at 6962 meters. The Himalayas have 14 peaks above 8000 meters (26,000 feet), including Mount Everest, the highest peak on Earth, at 8848 meters.

Another similarity between the two chains is that both rise abruptly from adjacent, low-lying regions. The Andes rise almost immediately from the Pacific coast of South America and thus start nearly at sea level. The Himalayas rise from the low plains of India to the south. Furthermore, both mountain chains are deeply eroded by glaciers that were once much more extensive than they are today. Both are still capped by immense alpine glaciers.

Thus, the two mountain chains might seem nearly identical except for local variations in terrain and culture. However, as a student of geology, you can distinguish among igneous, sedimentary, and metamorphic rocks. After a month in the Andes, you would have seen all three, with igneous rocks, particularly volcanic rocks, being the most abundant. In fact, if you hiked the entire length of the Andes and were then asked to summarize the geology in a single phrase, you might reply, "It's a pile of volcanic rocks!"

After a month or two of trekking through the Himalayan chain, again you would also have seen igneous, sedimentary, and metamorphic rocks. But folded and thrust-faulted sedimentary rocks dominate Himalayan geology.

The fact that the Andes is predominantly a chain of volcanic rocks, whereas the Himalayas contain mostly sedimentary rocks, suggests that the two mountain chains must have been built by different geologic processes. Both chains formed at convergent plate boundaries. However, the Andes rose where a tectonic plate

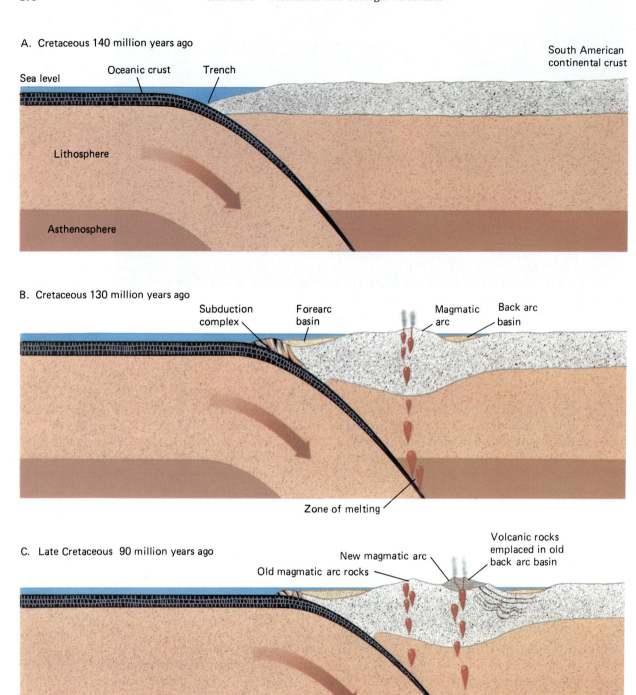

Figure 8.19   Development of the Andes, seen in cross section looking northward. (A) As the South American lithospheric plate began to move westward in early Cretaceous time, about 140 million years ago, subduction began and a trench formed at the west coast of the continent. (B) By 130 million years ago, the subducting plate crossed the lithosphere–asthenosphere boundary. Igneous activity began and a subduction complex formed. (C) In late Cretaceous time, the trench and region of igneous activity both migrated eastward.

carrying oceanic crust collided with another carrying continental crust, whereas the Himalayas developed where two continents collided.

During Triassic time all of the Earth's continents were gathered in the supercontinent called Pangaea. When Pangaea broke apart about 200 million years ago, it first split into two large pieces, a southern portion called Gondwanaland and a northern piece called Laurasia. During the next 60 million years, each of those two fragments also broke apart as the Atlantic Ocean began to open. At that time, the lithospheric plate that included South America started moving westward. To accommodate the westward motion, oceanic lithosphere began to subduct and dive into the mantle beneath the west coast of South America (Fig. 8.19A).

By about 130 million years ago, the diving plate had reached the lithosphere–asthenosphere boundary. Melting then began in the asthenosphere above the descending plate, forming vast amounts of basaltic magma (Fig. 8.19B). The magma rose to the base of the South American continent where it, in turn, melted the granitic crust. The basaltic and granitic magmas then rose into the continent. Some solidified within the crust to form plutons. However, vast quantities of magma continued upward to erupt at the surface, forming volcanoes. This intrusive and volcanic activity occurred along the entire length of western South America, but in a band only a few tens of kilometers wide, directly over the zone of melting.

As oceanic crust dived beneath the continent, slices of sea-floor mud and rock were scraped from the subducting plate and jammed onto the edge of the continent. This process formed a subduction complex similar to that of an island arc.

As the crust beneath the Andes thickened, it was heated by rising magma and grew weak and soft. Eventually it became so thick and soft that rocks began to spread outward under their own weight, as a mound of cool honey would. The spreading formed a great belt of thrust faults and folds along the east side of the Andes (Fig. 8.19C).

The Andes, then, are a relatively narrow mountain chain consisting predominantly of volcanic and plutonic rocks produced by subduction at a continental margin. The chain also contains extensive sedimentary rocks on both sides of the mountains. These rocks formed from sediment eroded from the rising peaks. Some of those rocks were folded, faulted, and metamorphosed by tectonic forces and rising magma. The Andes are a good general example of subduction at a continental margin, and this type of plate margin is called an **Andean margin**.

8.6   The Himalayan Mountain Chain: A Collision Between Continents

The **Himalayan mountain chain** separates the Earth's two most populous nations, China and India (Fig. 8.20). The world's highest mountains, including Mount Everest and K2, are in the Himalayas. Today, if you stood in southern Tibet and looked southward, you would see the high peaks of the Himalayas. Beyond this great mountain chain are the rainforests and hot, dry plains of India. If you had been able to look south from the same place 100 million years ago, you would have seen only ocean. At that time India lay south of the equator, separated from Tibet by thousands of kilometers of open ocean. The Himalayas had not yet begun to rise.

Formation of an Andean-Type Margin

When Pangaea split into Laurasia and Gondwanaland, the two were separated by open ocean. About 120 million years ago a large, triangle-shaped piece of lithosphere split off from Gondwanaland. It began drifting northward toward Asia at high speed (geologically speaking), perhaps as fast as 20 centimeters per year (Fig. 8.21A). The northern part of this lithospheric plate carried oceanic crust, but the Indian subcontinent lay on its southern corner (Fig. 8.21B).

Figure 8.22A shows that India and southern Asia were connected by oceanic crust before India began moving northward. As the Indian plate started to move, oceanic crust began to subduct at Asia's southern margin (Fig. 8.22B). As a result, magma formed, volcanoes erupted, and granite plutons were emplaced in southern Tibet. At this point southern Tibet was an Andean-type continental margin, and it continued to be so from about 120 to 50 million years ago, while India drew closer to Asia.

Continent–Continent Collision

By about 50 million years ago, all of the oceanic lithosphere between India and Asia had been consumed by subduction (Fig. 8.22C). Then the two continents collided. Because both are continental crust, neither could subduct deeply into the mantle. The collision did not stop the northward movement of India, but it did slow it down. India had been speeding northward at 20 centimeters per year and suddenly slowed to about 5 centimeters per year.

Figure 8.20 The Himalayas separate the Indian subcontinent from southern Asia. A view into the Annapurna Sanctuary, Nepal.

Continued northward movement of India was accommodated in two ways. The leading edge of India began to slide under Tibet in a process called **underthrusting**. As a result, the thickness of continental lithosphere in the region doubled (Fig. 8.22C). When underthrusting began, igneous activity in Tibet stopped because the subducting plate was no longer diving into the mantle. As India slid beneath Tibet, the edge of Tibet scraped the soft Indian sedimentary rocks from harder basement rock and pushed them into folds and thrust faults. These deformed sedimentary rocks make up the greatest part of the Himalayas (Fig. 8.22D).

The second way in which India continued moving northward after colliding with Tibet was by crushing Tibet and wedging China out of the way along huge strike-slip faults. India has pushed southern Tibet 1500 to 2000 kilometers northward since the beginning of the collision. The forces have created major mountain ranges and basins north of the Himalayas.

The Himalayas Today

The underthrusting of India beneath Tibet and the squashing of Tibet have produced thick continental crust under the Himalayas and the Tibetan Plateau to the north. Consequently, the region floats isostatically at a high elevation. Even the valleys lie at elevations of 3000 to 4000 meters, and the Tibetan Plateau has an average elevation of 4000 to 5000 meters. One reason the Himalayas contain all of the Earth's highest peaks is simply that the bases of the peaks are at such a high elevation. From its base to the summit, Mount Everest is actually smaller than Alaska's Denali (Mount McKinley), North America's highest peak. Mount Everest rises about 3300 meters from base to summit, whereas Denali rises about 4200 meters. The difference in elevation of the respective summits lies in the fact that the base of Mount Everest is at about 5500 meters, but Denali's base is at 2000 meters.

Comparisons of older surveys with newer ones show that the tops of some Himalayan peaks are now rising rapidly—perhaps as fast as 1 centimeter per year. If this rate were to continue, Mount Everest would double its height in about 1 million years, a short time compared with many other geologic events. Is this scenario likely? How high will Mount Everest be a million years from now? Mountain height is determined by a tug of war between different forces and processes; some tend to make a mountain peak rise, whereas others reduce its height.

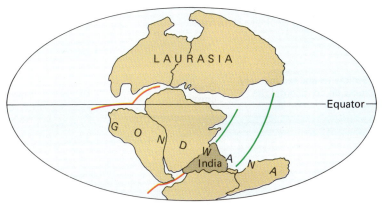

A 200 Million years ago

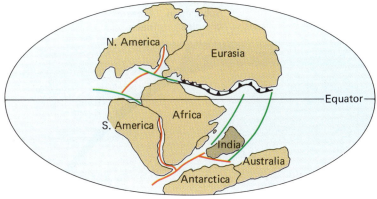

B 120 Million years ago

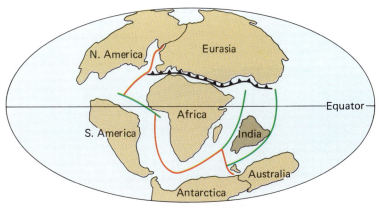

C 80 Million years ago

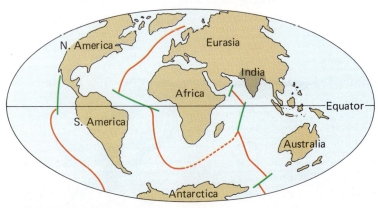

D 40 Million years ago

Figure 8.21 (A) Gondwanaland and Laurasia formed shortly after 200 million years ago as a result of the breakup of Pangaea. Notice that India was initially part of Gondwanaland. (B) About 120 million years ago, India broke off from Gondwanaland and began drifting northward. (C) By 80 million years ago, India was an island drifting toward the equator. (D) By 40 million years ago, it had moved 4000 to 5000 kilometers northward and collided with Asia.

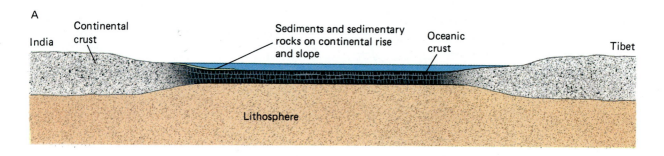

A

India
Continental crust
Sediments and sedimentary rocks on continental rise and slope
Oceanic crust
Tibet
Lithosphere

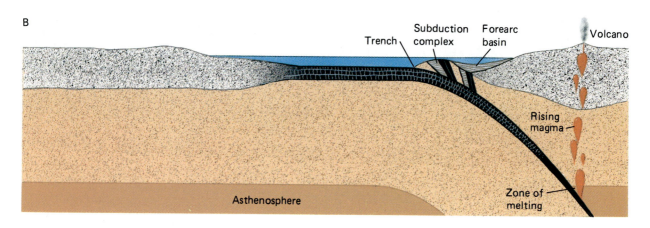

B

Trench
Subduction complex
Forearc basin
Volcano
Rising magma
Zone of melting
Asthenosphere

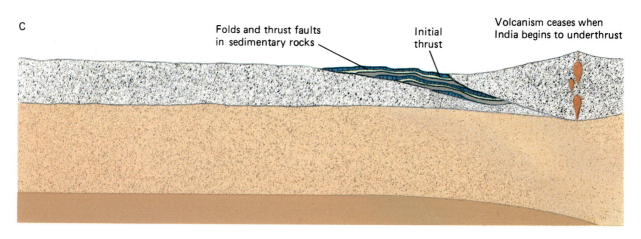

C

Folds and thrust faults in sedimentary rocks
Initial thrust
Volcanism ceases when India begins to underthrust

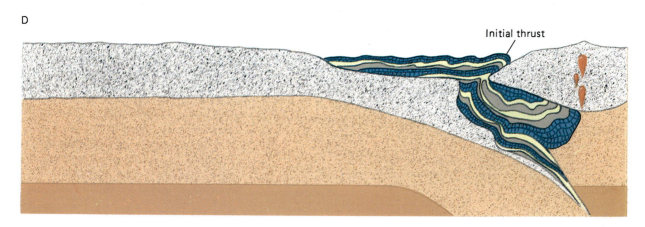

D

Initial thrust

What Processes Lower Himalayan Peaks?

Streams, glaciers, and landslides erode the peaks as they grow, carrying the sediment into adjacent valleys. John Shroder, a University of Nebraska geologist who studies Himalayan geology, stated, "If I am studying landslides, I might have to wait in the U.S. for years to see one. But in the Himalayas I'll get a good one a couple of times a summer. It may be hazardous to life and limb, but at least you'll get what you're looking for."

The strength of rock, and its ability to hold up high mountains, is limited. As a mountain range grows higher and heavier, it exerts greater force on the underlying lithosphere. At a certain point, the deeper part of the lithosphere becomes hot and plastic and begins to flow outward from beneath the mountains. The situation is similar to pouring cold honey onto a tabletop. At first, the honey piles up into a high, steep mound. But soon it begins to flow outward under its own weight, lowering the top of the mound. If the newly formed, steep mound of honey were covered with a layer of brittle chocolate frosting, the frosting would crack and slip apart in normal faults as the honey spread outward. The upper few kilometers of rocks of the Himalayas are like the frosting, and normal faulting is common in parts of the Himalayas. The faults form as the mountains sink under their own weight (Fig. 8.23).

What Processes Heighten the Himalayas?

Frequent moderate earthquakes and occasional large and destructive ones show that India and Asia continue to collide today. This collision and the resultant underthrusting lift the Himalayas.

Recall that the lithosphere floats on the underlying asthenosphere. As a result, mountains rise when their tops are removed by erosion. Thus, paradoxically, removal of material from the tops of the mountains causes them to rise upward. This isostatic adjustment may be

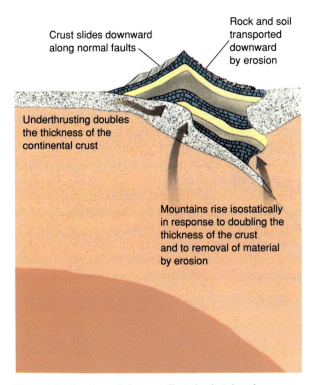

Figure 8.23 Several factors affect the height of a mountain range. Today the Himalayas are being uplifted by continued underthrusting. At the same time, erosion removes the tops of the peaks, and the sides of the range slip downward along normal faults. As these processes remove weight from the range, the mountains rise by isostatic adjustment.

the primary cause of the rapid rise observed today in parts of the Himalayas.

Summary

Notice that the development of the Himalayan chain occurred in two steps. First, an Andean-type continental

---

Figure 8.22 These cross-sectional views show the Indian and Asian plates before and during the collision between India and Asia. (A) Shortly before 120 million years ago, India, southern Asia, and the intervening ocean basin were parts of the same lithospheric plate. (In this figure, the amount of oceanic crust between Indian and Asian continental crust is abbreviated to fit the diagram on the page.) (B) When India began moving northward, the plate broke and subduction began at the southern margin of Asia. By 80 million years ago, an oceanic trench and subduction complex had formed. Volcanoes erupted, and granite plutons formed in the region now called Tibet. (C) By 40 million years ago, India had collided with Tibet. The leading edge of India was underthrust beneath southern Tibet. (D) Continued underthrusting and collision between the two continents has crushed Tibet and created the high Himalayas. India continues to underthrust and crush Tibet today, and the mountains continue to rise.

margin formed as oceanic crust subducted beneath southern Asia. At that time, the geology of southern Asia was similar to the present geology of the Andes. Only later, after all the oceanic crust between the two continents was consumed by subduction, did continent–continent collision begin. The two-step nature of the process is common to all continent–continent collisions because ocean basins separating continents must first be subducted before the continents can collide.

When two continents collide, they weld into a single mass of continental crust. The junction is called a **suture zone**, or **continental suture**. Continental sutures are often recognized by a sudden difference between rock types, ages, and structures on opposite sides of the suture zone. Such a geologic discontinuity occurs because the rocks on one side of the suture originated on one continent, whereas the rocks on the other side

formed on the other continent, perhaps at different times and by different processes. Another criterion for recognizing a suture is that rocks of the suture zone are commonly deformed and sheared by the collision. In addition, bits and scraps of oceanic crust and mantle occur in suture zones, preserving traces of the oceanic lithosphere, which was mostly consumed by subduction.

The Himalayan chain is only one example of mountain building by continent–continent collision. The Appalachian Mountains formed when eastern North America collided with Europe, Africa, and South America between 400 and 250 million years ago. The European Alps formed during repeated collisions between northern Africa and southern Europe beginning about 30 million years ago. The Urals of northwestern Asia formed by a similar process about 250 million years ago.

SUMMARY

Mountain chains and **ranges** form by tectonic activity at boundaries between lithospheric plates. Volcanic eruptions, intrusion of granite, metamorphism, earthquakes, and folding and faulting of rocks commonly accompany growth of a mountain chain.

Folds, **faults**, and **joints** are **geologic structures** that develop as mountains are built. Folds, **reverse faults**, and **thrust faults** usually form where tectonic forces squeeze rocks together. **Normal faults** usually form where rocks are pulled apart. **Strike-slip faults** form where blocks of crust slip horizontally past each other along vertical fractures.

Mountains form at all three types of plate boundaries. The mid-oceanic ridge and East African rift ranges formed at divergent boundaries. The San Gabriel Mountains of California rose along the San Andreas transform

plate boundary. But most of the great continental ranges, including the Andes, the Appalachians, the Alps, and the Himalayas, grew at convergent plate boundaries. If two converging plates both carry oceanic crust, a volcanic **island arc** forms. If one plate carries oceanic crust and the other carries continental crust, an **Andean margin** develops, with numerous plutons and volcanoes.

When two plates carrying continental crust collide, an Andean margin first develops as oceanic crust separating the two continents subducts. Later, when all oceanic crust is consumed, the continents themselves collide, forming vast regions of folded and thrust-faulted sedimentary and metamorphic rocks and of older plutonic and volcanic rocks.

KEY TERMS

REVIEW QUESTIONS

1. Explain the differences among a mountain, a mountain range, and a mountain chain.

2. Explain why the interiors of lithospheric plates are normally tectonically inactive, whereas plate boundaries are tectonically active. List the types of tectonic activity that normally occur at plate boundaries.

3. Describe types of tectonic activity that are specific to each of the three different types of plate boundaries. What kinds of tectonic activity are common to all three types of plate boundaries?

4. Describe the tectonic environments of the three different types of convergent plate boundaries.

5. What is a geologic structure? What are the three main types of structures?

6. What is the difference between a fault and a joint?

7. What is the difference between an anticline and a syncline?

8. At what type of tectonic plate boundary would you expect to find normal faults?

9. How are faults related to earthquakes?

10. Why does movement often occur repeatedly on a single fault or along a fault zone?

11. Why are thrust faults, reverse faults, and folds commonly found together?

12. Draw a cross-sectional sketch of an anticline–syncline pair. Use your sketch to explain how folds accommodate crustal shortening and tectonic squeezing of rocks.

13. Draw a cross-sectional sketch of a normal fault. Label the hanging wall and the footwall. Use your sketch to explain how a normal fault accommodates crustal extension. Sketch a reverse fault and show how it accommodates crustal shortening.

14. In what sort of a tectonic environment would you expect to find a strike-slip fault?

15. Give examples of mountain chains formed at each of the three types of convergent plate boundaries.

16. What mountain chain has formed at a divergent plate boundary? Explain the main differences between this chain and those developed at convergent boundaries.

17. What mountain ranges have formed at transform plate boundaries?

18. Where are island arcs forming today? In what tectonic environment do they form?

19. Describe the main geologic features of an island arc.

20. Describe the similarities and differences between the Andes and the Himalayan chain. Why do the differences exist?

21. Sketch a cross section of an Andean-type plate boundary to a depth of several hundred kilometers. Show the positions of the subducting plate, trench, subduction complex, volcanoes and plutons, and earthquakes.

22. Draw a series of cross-sectional sketches showing the evolution of a Himalayan-type plate boundary. Why does this boundary start out as an Andean-type margin?

DISCUSSION QUESTIONS

1. Why do most major continental mountain chains form at convergent plate boundaries? What topographic and geologic features characterize divergent and transform plate boundaries in continental crust? Where do these types of boundaries exist in continental crust today?

2. Discuss the relationships among types of lithospheric plate boundaries, the kinds of tectonic forces that occur at each type of plate boundary, and the main types of geologic structures you might expect to find in each environment.

3. Discuss the rock types, tectonic forces, and geologic structures that you might expect to find in rocks of a subduction complex.

4. Compare and explain the similarities and differences between the Andes Mountains and the Himalayan chain.

How would the Himalayas 60 million years ago have compared with the modern Andes?

5. Where would you be more likely to find large quantities of igneous rocks in the Himalayan chain—in the northern parts of the chain, near Tibet, or southward, near India? Discuss why.

6. The largest mountain in the Solar System is Olympus Mons, a volcano on Mars. It is 25,000 meters high, nearly three times the elevation of Mount Everest. Speculate on the factors that might permit such a large mountain on Mars.

7. If you were studying photographs of another planet, what features would you look for to determine whether the planet is or has been tectonically active?

Cindy Lee Van Dover received a B.S. in environmental science from Rutgers University (1977), followed by an M.S. in ecology (1985) from the University of California, Los Angeles. In 1980 she published the first of three articles on decapod crustaceans from the Indian River region of Florida. Then, in 1982, she participated in the Oasis Expedition to hydrothermal vents in the East Pacific rise. This was only the beginning of her relationship with the deep ocean and specifically hydrothermal vent communities. She took part in several other expeditions—aboard the research vessel *Atlantis II* and, as a scientist, aboard the submersible *ALVIN*—to places such as the Rose Garden and Gorda ridge.

In 1989 Van Dover completed a Ph.D. in biological oceanography in the Massachusetts Institute of Technology and Woods Hole Oceanographic Institution Joint Program. Since then she has held the position of postdoctoral investigator with the Biology Department and is currently a Visiting Investigator in the Department of Marine Chemistry and Geochemistry at Woods Hole Oceanographic Institution in Woods Hole, Massachusetts. Also in 1989, Van Dover began her pilot training, and by 1990 she had piloted *ALVIN* to a variety of underwater locations throughout the Pacific Ocean. Later that year, she participated in a joint United States–Soviet Union dive aboard the *MIR-2* submarine in Monterey Bay, off Baja California, Mexico, and in the nearby Guaymas basin. At the controls of *ALVIN,* Van Dover explores hydrothermal vents with dexterity and thoroughness, combining her expert knowledge of sea life with her Navy-certified piloting skills. Most recently, she has employed data gathered from *ALVIN* expeditions to study the effects of sludge from deep-sea waste dumping and the ecology of hydrothermal vent communities.

Van Dover is a member of the Deep Submersible Pilots' Association and the American Geophysical Union. In addition to her considerable scholarly contributions, she has published several nontechnical articles on the undersea world as seen from her unusual perspective. In 1988 she was honored as one of *Ms.* magazine's women of the year. Her commitment to science and her pioneering spirit have melded to help unlock the mysteries of the deep ocean, one of the few remaining earthly frontiers.

Where were you born and where did you attend school?

I was born and grew up in New Jersey, about 5 miles from the coast, in Eatontown; I went to a regional high school there, which happened to have a summer marine biology program and excellent teachers. That's how I got into marine science to begin with. As an undergraduate, I went to Rutgers University. After that, I spent some years working up and down the East Coast at various marine labs, as a technician. Eventually, I traveled west to earn my master's degree at UCLA. Then, back once more on the East Coast, I worked as a technician at the Marine Biological Lab in Woods Hole before starting work on my Ph.D. in the Woods Hole Oceanographic Institution–Massachusetts Institute of Technology Joint Program in biological oceanography.

So, from an early time in your career, you've been involved in marine biology or biological oceanography.

I always liked animals—especially invertebrates, because they come in so many different forms. Bugs and spiders I still think are creepy, an inheritance from my mother, but the animals that live in water are wonderful; some look more like plants than animals, others have fantastic features—feather-like appendages or multiple pairs of tiny eyes. Marine animals can be the stuff of science fiction, and I wanted to know more about them.

Then, too, I always wanted to be on a boat when I was little, even though I would get terribly seasick. I don't know where the sense of the romance of the sea comes from, but I certainly had it. Partly the animals fascinated me. But I admit that the adventure, the sense of independence or strength, the chance to prove myself against the sea lies somewhere behind my wanting to sail.

> As an ecologist, it seemed important to get a broad perspective of what the sea floor and hydrothermal vents are like.

What did you do your Ph.D. on?

My Ph.D. was on the ecology of deep-sea hydrothermal vents. I started off my dissertation work studying food webs at vents—who eats whom. Geologists had just discovered hydrothermal vents on the Mid-Atlantic ridge, back in 1985. Thousands of shrimp swarm over the sulfide chimneys there. One scientist described them as looking like maggots swarming on a hunk of rotten meat—a not very poetic, but very accurate, description. My advisor was given some specimens; he passed them on to me: "Find out what they are doing, how they are making a living."

As part of my dissertation work, I joined an *ALVIN* expedition to vent sites off the coast of Washington. I had learned that the chief scientist was planning to use an electronic camera developed here in Woods Hole. At the local bar on a Friday afternoon, I talked with the engineers who built the camera and learned that it ought to be able to detect the sort of light we supposed might be generated at a high-temperature vent. The chief scientist was willing to configure the camera appropriately, position it in front of a black smoker, turn out the submersible lights, and collect a time exposure of the ambient light. He and one of his colleagues were able to do this successfully, capturing a spectacular image of a glow emitted by the hot water.

Can you tell us some more about what these black smokers are and something about the animal or plant life associated with them?

Black smokers occur along mid-ocean ridge spreading centers, which are amazing places. Along the spreading centers, the Earth's plates are pulling apart and new ocean crust is forming. It is a very black, high-contrast environment with incredible terrain, basically a linear volcano; pillowed and ponded lavas cover the sea floor. It is a very fluid-looking, dynamic landscape, with lava in frozen pools and swirls, buckles, and fractures. Fissures there are often big enough for the submersible to drive into; exposed along the walls are the histories of eruptions, flows on top of flows.

Where the sea floor is moving apart, there is a lot of earthquake activity, a lot of cracking of the basalt. Seawater fills those cracks and actually reaches, if not the magma chamber, then very, very hot rock. It is heated, reacts with the rock, and becomes modified. The heated water is buoyant and rises up through the crust to exit at the sea floor with a very interesting chemistry—enriched in metal sulfides and depleted in magnesium and oxygen. It is the chemical characteristics of the vent water that support microorganisms, the bacteria at the base of the food web. Life at vents is based *not* on photosynthesis, like we are used to in shallow-water and terrestrial systems, but on chemosynthesis. Instead of using light energy, the microorganisms use energy in reduced compounds to fix carbon. Instead of plants, the bacteria are the "grasses" of the hydrothermal vent community. The bacterias live both as free-living organisms suspended in the water column and on surfaces of rocks, sulfides, and animals; they also live as endosymbionts in tube worms and bivalves, supplying nutrients to their invertebrate hosts.

Can you describe the macrofauna around a black smoker a little bit more? What, in addition to tube worms, lives there?

In some places tube worms are the dominant organism. They are beautiful; they have bright red plumes and very white tubes. Living with them are a variety of crabs and other crustaceans. There can also be large golden-brown mussels and giant white clams with blood-red bodies occurring in large numbers at the base of black smokers, in the cracks and crevices where warm, sulfide-laden vent fluids are flowing.

How did you originally become interested in learning to pilot the ALVIN?

I think I published something like a dozen papers on the biology or ecology of vent fauna, but I'd only dived once to one vent. I was writing about animals I never saw alive, in their natural setting. I would go on lots of cruises, out at sea for a month or more each time, yet felt very lucky to get one dive. A simple look around me showed me who dove the most: the pilots. They could dive every other day, every three days, depending on the number of pilots. If you are a scientist, you dive only at the specific site you wrote the proposal for—you can't go all over the globe, diving anywhere. But the pilots get to go to all the different sites. As an ecologist, it seemed important to get a broad perspective of what the sea floor and hydrothermal vents are like. So that was the scientific motivation. I also wanted to do it just because of the adventure of it.

How did you first get involved with the ALVIN group?

ALVIN is Navy certified, which means they have to get recertified at intervals and prove that they are operating a safe sub. Traditionally, repairs on *ALVIN* were supervised by a shop manager, and he knew how to maintain the thing. There was little in the

way of procedures written down. But the Navy insisted they put together a maintenance manual. I came along at the right time with the talent to write and organize a document—I'd just finished my Ph.D. dissertation as proof of that ability. I began as a volunteer and was eventually hired.

What sort of training do you need to drive, or pilot, the submarine?

The training is all on the job, and when I trained, it was all done independently; the schematics and instrument manuals were in files and I was encouraged to work my way through them all. In order to qualify, you have to pass a qualifying dive with the chief pilot, proving you can competently operate the sub. You also have to pass a series of qualifying boards, with the two most important being an Engineering Board and a Navy Board. During the Engineering Board, you stand in front of the engineers who designed and built the submarine. They have you draw and explain the schematics of every major system in the submarine. It was, at my exam, an unfriendly board. I passed. The Navy Board was mostly related to safety procedures and was a thoroughly professional exchange.

Have you ever had to troubleshoot beneath the surface?

All the time. We constantly work with untested scientific gear that needs troubleshooting. Some of the failures are more scary than others. You learn there are certain failures you cannot continue a dive with; when one of those happens, you really pay attention.

As an example, once I had a 10K leak indication, which meant that I had water in one of the outside electrical boxes. These boxes are oil-filled; water in a box could short out the system and cause serious damage. When I saw that 10K indication, my pulse shot up rapidly. You see, at first you don't know where the 10K leak is. It could be in a battery box— bad news; these are 120-volt batteries and they don't like salt water. There are a series of toggle switches to go through that will locate where the problem is. You go through them one at a time, systematically, fast. It is tense. You know once you find out where the problem is, you'll know what to do. Still, it's tense. In this particular case, the problem was in the variable ballast system. I just shut that system down and continued the dive. On deck we discovered that a single drop of water had closed the circuit in the leak detector. But there was no way of knowing the prob-

Tube worms and other organisms living on sulfide chimneys at the Juan de Fuca ridge. *(Cindy Lee Van Dover)*

lem was minor under water. The moment of a failure is always a challenge; you have to be on top of things, know your submarine; you have to know those schematics to know what to do next.

What do you see as your professional direction in the next decade?

I hope to continue to be a research scientist and to continue to study the ecology of the sea floor. It is an important field if we are to appreciate the world in which we live—not the small proportion that we ourselves occupy and modify, but the vast ocean environment that buffers our world and makes it livable. Continuing work on hydrothermal vents is an exciting prospect, pure scientific adventure, where the rules have not even begun to be defined. I'm also involved in some of the sewage sludge disposal issues off New Jersey. I'm interested in trying to trace the entry of sewage sludge into the benthic community of the deep sea. Dumping in deep water was initially started with the belief that none of the discharged materials would be detectable on the sea floor—the particles were so fine, they would be diluted and dispersed in the surface waters. My advisor and some of his colleagues thought maybe that wasn't the case and obtained funding to investigate. My contribution to the project is to use stable isotopes as tracers of the sewage sludge. The technique requires a hypothesis that the organisms have two potential, isotopically distinct

food sources. In this case, the sewage sludge has a very distinctive sulfur isotopic composition compared to normal sources of deep-sea organic sulfur. We collected benthic invertebrates—sea stars, sea urchins, etc.—from within the dump site and from a reference area upstream along the same depth contour. From the isotopic data for one species, an urchin, there was a strong sewage-sludge signal in the dump-site specimens. In fact, as much as 25 to 30 percent of the organic sulfur in dump-site urchins may be derived from sewage sludge.

What do you think that means? Is it good or bad?

In a way, it is good in that the sewage sludge provides a source of nutrition for the urchins in a food-limited environment. But the urchins may pay a price in accumulating contaminants. There is the chance for magnification in the food web. It might also select for opportunistic species and change the structure of the benthic community. What strikes me as significant is the fact that the original thought was "Oh, we don't need to worry about this; it can't possibly get there." But sewage sludge is reaching the sea floor. If we can still be so naive about what we are doing to the environment, there is a lesson: We need to assess what we are doing a little more carefully, we need to know more about the deep sea.

What would you tell a student who might be interested in further investigating or even professionally pursuing a field of biological oceanography or marine biology, your field?

You will have a life that is exciting and adventurous. There truly are unimaginable discoveries waiting for us in the oceans. But you only get there through hard work and creative thinking. You don't do this for money or prestige or glory. Each scientist has a philosophy, a way of viewing the world and tackling questions. My approach is probably more unorthodox than most, and I pay my dues for that. I don't hesitate to cut across boundaries, to venture into fields that I know nothing about in order to get at some aspect of my work. Becoming an *ALVIN* pilot is a good example of that; I had no technical skills appropriate for that job. My first cruise with the *ALVIN* group, I sailed as an electrician, of all things! But I learned, and learned fast. Have the confidence in yourself to go after what you want; have the consciousness of your abilities to know how far you can go.

Surface Processes

◀ *Trail Creek, Idaho. Autumn colors.* *(© 1993, David R. Stoecklein)*

Weathering, Soil, and Erosion

Rain and surface runoff have eroded gullies into the Death Valley landscape.

Tectonic processes create mountain ranges and other large-scale landforms. Surface processes—forces generated by wind, flowing water, gravity, and ice—then reshape the Earth's surface into familiar landscapes. If you live along a rocky coastline or near sandy beaches and low-lying barrier islands, you can watch waves crash against the rocks or carry sand along the coastline. If you live in a broad river valley or in mountainous country with narrow, V-shaped valleys, you may occasionally observe swollen flood waters erode the stream banks or deposit mud in the valley floor. Throughout the remainder of this unit, you will study the forces that sculpt the Earth around you.

9.1   Surface Processes

Landforms are the features that make up the shapes of the Earth's land surface. Some landforms, such as a mountain chain or a plain, can be very large; others, such as a hill or stream valley, may be small.

Surface processes are all of the processes that work on the Earth's surface to create and modify landforms. They include weathering, erosion, transport, and deposition. **Weathering is the decomposition and disintegration of rocks and minerals at the Earth's surface by both mechanical and chemical processes.** It converts solid rock to sediment—gravel, sand, clay, and soil. Weathering involves little or no movement of the decomposed rocks and minerals.

Erosion is the removal of weathered rocks and minerals from the place where they formed. Flowing water, wind, glaciers, and gravity erode weathered material from its place of origin. Then these same agents—water, wind, ice, and gravity—may transport the sediment great distances and deposit it in layers at the Earth's surface.

Weathering, erosion, transport, and deposition typically occur in an orderly sequence. For example, water freezes in a crack in granite, loosening a grain of quartz. A hard rain erodes the grain and washes it into a stream. The stream then transports the quartz to the seashore and deposits it as a grain of sand on a beach.

Surface processes create landforms by wearing away mountain ranges, sculpting coastlines, and carving valleys (Fig. 9.1A). They also build landforms by depositing sediment (Fig. 9.1B). Although the processes work slowly from the perspective of human life, in geologic time they can erode an entire mountain range to a flat plain, excavate a deep canyon, or build a large river delta.

9.2   Weathering

The physical and chemical environment at the Earth's surface is corrosive to most materials. A pocketknife rusts when it is left out in the rain. For similar reasons rocks decompose naturally. Thus, over the centuries, stone cities have fallen into ruin.

If you visit the remains of ancient Greece or Rome, you can see two types of changes. First, large building stones have broken into smaller fragments (Fig. 9.2). **Mechanical weathering** is the physical disintegration of rock into smaller pieces. For example, plant roots may grow in cracks in building stones, enlarging the cracks and eventually toppling the walls. In cold climates, water expands as it freezes in cracks, fracturing rocks and reducing stone buildings to rubble. Such mechanical processes break rocks into smaller pieces, but they do not alter the chemical compositions of the rocks and minerals.

If you look closely at a building stone in an ancient city, you may see a second type of disintegration. Its face may be pitted and discolored, and once-sharp edges are rounded. In addition, the rock may be soft and earthy rather than hard and solid as it was when originally quarried. **Chemical weathering** occurs when air and water attack rocks chemically. The changes are similar to rusting in that a chemically weathered rock contains different minerals and has a different chemical composition from the original rock. Certain kinds of air pollution accelerate chemical weathering. Thus, decomposed building stones can be seen in most modern industrial cities as well as in ancient cities. Just as building stones in ancient cities decompose both mechanically and chemically, rocks in natural settings weather by both processes.

Mechanical and chemical weathering reinforce one another. For example, chemical processes generally act on the surface of a solid object. Therefore, a chemical process speeds up if the surface area increases. Think of a burning log; the fire starts on the outside and works its way inward. If you want the log to burn faster, split it in half to increase its surface area. Mechanical weathering cracks rocks, exposing more surface area for chemical agents to work on (Fig. 9.3).

Mechanical Weathering

Recall that mechanical weathering does not alter the chemical nature of rocks and minerals; it simply breaks them into smaller pieces. Think of breaking a rock with a

A

B

Figure 9.1 Surface processes develop landforms by both erosion and deposition. (A) Spring Creek, a tributary of the Colorado River in Grand Canyon, Arizona, carved this narrow canyon in bedrock. (B) A small desert stream deposited this fan delta in Laguna Salada, Baja California, Mexico. *(Karl Mueller)*

Figure 9.2 In these Roman ruins mechanical weathering has broken and toppled large building stones, and chemical weathering has rounded and pitted the stones. *(Italian State Tourist Office)*

194

This boulder weathered in place.

Both chemical and mechanical processes have weathered this boulder.

hammer: The fragments are smaller than, but otherwise identical to, the original rock. Six processes weather rocks mechanically: frost wedging, salt cracking, abrasion, biological activity, pressure release fracturing, and thermal expansion and contraction. In addition, chemical weathering causes minerals to expand, and this expansion also breaks rocks mechanically into smaller pieces.

Frost Wedging

Water collects in cracks and crevices in rocks. If the outside temperature drops below 0°C, the water freezes. Water expands when it freezes. Thus, water freezing in a crack pushes the rock apart in a process called **frost wedging**. The ice may cement the rock fragments together, but when it melts, rocks may tumble from a steep outcrop. If you walk through the mountains during a season when water freezes at night and thaws during the day, be careful. Rocks tumble from cliffs when the rising sun melts the ice. Mountaineers commonly travel in the early morning, before rocks begin to fall.

Anyone who has spent time in the mountains has noticed large piles of broken, angular rocks at the bases of cliffs. These piles are called **talus slopes** (Fig. 9.4). The rocks in talus slopes have fallen from the cliffs, mainly as a result of frost wedging.

Salt Cracking

Growing salt crystals can also crack rocks. In areas where ground water is salty, salt water may seep into

Figure 9.3 When mechanical weathering breaks rocks, more surface area is available for chemical weathering.

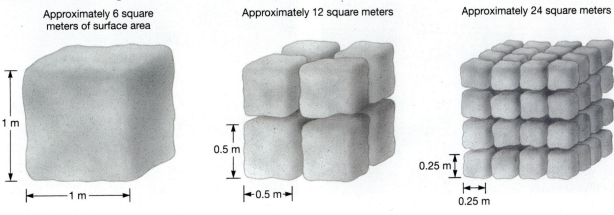

Approximately 6 square meters of surface area

1 m

1 m

Approximately 12 square meters

0.5 m

0.5 m

Approximately 24 square meters

0.25 m

0.25 m

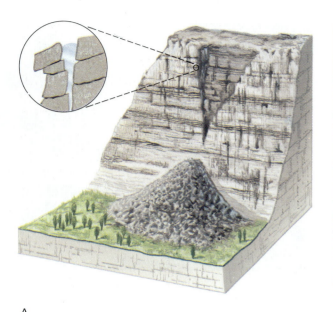

A

B

Figure 9.4   (A) Frost wedging dislodges rocks from cliffs to create talus slopes.
(B) Talus slopes along the San Juan River, Utah.

cracks or between mineral grains in rock. When the water evaporates, the dissolved salts crystallize (Fig. 9.5). The growing crystals exert forces great enough to widen cracks and push grains apart. Many sea cliffs are pitted by **salt cracking**. These features exist only within the reach of salt spray from the breaking waves. Salt cracking is also common in deserts, where ground water is often salty.

Figure 9.5   White salt crystallized when salty ground water evaporated from this rock face in Grand Canyon, Arizona.

Abrasion

Have you ever looked at rocks in a stream or on a beach? They have been smoothed and rounded by collisions with other rocks, silt, and sand carried by the moving water. As the rocks and smaller grains collide, their sharp edges and corners wear away. The mechanical wearing and grinding of rock surfaces by friction and impact is called **abrasion**. Abrasion produces rocks with a characteristic smooth, rounded appearance (Fig. 9.6).

Like water, wind by itself is not abrasive, but it hurls dust and sand against rocks to carve unusual and beautiful landforms (Fig. 9.7). Glaciers also abrade

Figure 9.6   Abrasion rounded these rocks in a stream bed in Yellowstone National Park, Wyoming.

rocks. A glacier picks up particles ranging in size from clay to boulders and drags them across bedrock, abrading both the rock fragments embedded in the ice and the bedrock beneath (Fig. 9.8).

Biological Activity

If soil collects in a crack, a seed may fall there and grow. Roots work their way down into the crack. As the roots grow, they push the rock apart (Fig. 9.9). City dwellers can observe this effect in sidewalks, which are often cracked by tree roots.

Pressure Release Fracturing

Kilometers beneath the Earth's surface, rock is under immense pressure from the weight of overlying rocks. Imagine granite 15 kilometers deep. At that depth, pressure is about 5000 times that at the Earth's surface. Imagine further that tectonic forces push this granite up to form a mountain range. As overlying rock erodes gradually, pressure on the granite decreases and the

Figure 9.7 Windblown sand sculpts exotic figures in bedrock. Grand Canyon, Arizona.

Figure 9.8 Rocks embedded in the base of a glacier can wear grooves in bedrock.

rock expands. But because the granite is brittle, it cracks as it expands. This form of mechanical weathering is called **pressure release fracturing**.

Granite commonly fractures by **exfoliation** (Fig. 9.10), a process in which large concentric plates split off the rock like the layers of an onion. The plates may be only 10 or 20 centimeters thick at the surface, but they thicken with depth. Exfoliation fractures are usually absent below a depth of 50 to 100 meters. The fractures appear to result from expansion of the granite near the Earth's surface. Some geologists think that exfoliation results from expansion caused by pressure release. Others attribute it to expansion caused by chemical weathering near the Earth's surface.

Chemical Weathering as a Cause of Mechanical Weathering

Feldspar makes up about half of most granitic rocks. Feldspar weathers chemically by combining with water to form clay. The clay has greater volume than the original feldspar as a result of the addition of water. As the clay expands, it pushes against nearby mineral grains and breaks them free. Thus, chemical weathering causes mechanical weathering.

Thermal Expansion and Contraction

At the Earth's surface, rocks warm and cool as air temperature changes. A rock expands when it is heated and contracts when it cools. If you heat an object rapidly,

197

Figure 9.9   As this tree grew from a crack in bedrock, its roots widened the crack.

its surface expands faster than its interior and the object may fracture. For example, if you line a campfire with certain types of rocks, such as granite, they commonly break as you cook your dinner. Similarly, a hot object may crack when it cools rapidly.

In mid-latitude mountains and deserts, daily temperature may fluctuate between −5°C and +25°C. Is this 30° difference sufficient to fracture rock? Geologists disagree on the answer. In one laboratory experiment, granite was subjected to repeated, rapid temperature changes of more than 100°C and did not fracture. This result suggests that thermal change may not be important in mechanical weathering. On the other hand, daily heating and cooling over hundreds of thousands

of years may fracture rock. Alternatively, thermal effects may be important only during occasional catastrophic events. If you walk over rock-covered soil in an area recently burned by a hot forest fire, you can see that the rocks have fractured as a result of the much greater temperature changes resulting from the fire.

Chemical Weathering

Rock seems durable when it is observed over a human lifespan. Return to your childhood haunts and you will see that rock outcrops in woodlands or parks have not changed. Yet over a longer time, rocks decompose chemically at the Earth's surface. You need only visit a ceme-

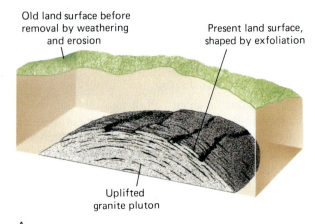

Old land surface before
removal by weathering
and erosion

Present land surface,
shaped by exfoliation

Uplifted
granite pluton

A

Figure 9.10   (A) Development of an exfoliation dome. The exfoliated slabs are only a few centimeters to a few meters thick. (B) Exfoliated granite in Pinkham Notch, New Hampshire.

B

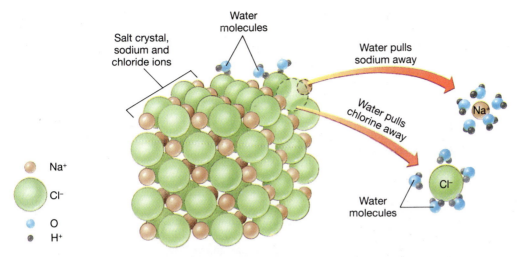

Figure 9.11   Halite dissolves in water because the attraction between water molecules and the sodium and chloride ions are stronger than the chemical bonds in the crystal.

tery to see the effects of chemical weathering. Carvings on old headstones are commonly faint and poorly preserved, whereas younger inscriptions are sharp and clear.

Many natural chemicals attack rock. Water, acids, bases, and oxygen are the most important ones.

Weathering by Solution

If you put salt (halite) in water, it dissolves and the ions disperse into the water to form a **solution** (Fig. 9.11). Halite dissolves so easily that it exists only in deserts. Common minerals also dissolve in water, but more slowly. Acids and bases increase the rates at which most minerals dissolve.

Acids and Bases

In pure water, a small proportion of water molecules break apart to form an equal number of hydrogen ions (H^+) and hydroxyl ions (OH^-).[1] If a chemical compound is added to increase the proportion of hydrogen ions, the resultant solution is an **acid**; if the hydroxyl ion concentration is increased, the solution is a **base**.

Acids and bases are much more corrosive than pure water because of the nature of hydrogen and hydroxyl ions. Think of an atom on the surface of a crystal. It is held in place because it is attracted to other atoms in the crystal by electrical forces. But at the same time,

electrical forces in the outside environment are pulling the atom away from the crystal. The result is a tug-of-war. If the atom is bonded to the crystal more strongly than it is attracted to the outside environment, then the crystal remains intact. If outside attractions are stronger, then the crystal dissolves as its atoms are pulled away (Fig. 9.11). Acids and bases dissolve minerals faster than pure water because the electrically charged hydrogen and hydroxyl ions pull atoms from crystals more effectively than water molecules do.

Natural water is never pure. Atmospheric carbon dioxide reacts with rainwater to form a weak acid called carbonic acid. The solution chemically attacks rocks and minerals. In addition, when water flows across or beneath the surface of the Earth, it dissolves ions from minerals. In some instances, these ions make the water acidic; in other cases, the water becomes basic. In either case, the resulting solution has a greater ability to dissolve rocks and minerals.

Reactions with Oxygen

Many elements react rapidly with molecular oxygen, O_2. In our everyday experience, an iron bar reacts with water and oxygen in the atmosphere to rust. Rusting is one manifestation of a more general process called **oxidation**.[2]

About 21 percent of the Earth's atmosphere is oxygen, and as a result oxidation is common. It usually

---

[1] Hydrogen ions react instantaneously and completely with water, H_2O, to form the hydronium ion, H_3O^+, but for the sake of simplicity we will consider the hydrogen ion, H^+, an independent entity.

[2] Oxidation is defined properly as the loss of electrons from a compound or element during a chemical reaction. In the weathering of common minerals, this usually occurs when the mineral reacts with molecular oxygen.

GEOLOGY AND THE ENVIRONMENT

Acid Precipitation

As discussed in the text, natural rainfall is slightly acidic. In recent years, however, industrial pollution has dramatically increased the acidity of rainfall. During most of the year, rainfall in the northeastern United States and in many industrial regions in Europe is from 10 to 100 times more acidic than in unpolluted environments. In a few isolated instances, acidity of rain has been even higher. For example, a rainstorm in Baltimore in 1981 was 1000 times as acidic as natural rainfall and was about as acidic as vinegar. A fog in southern California in 1986 was 10,000 times as acidic as natural rainfall, approaching the acidity of some solutions of hydrochloric acid used as toilet bowl cleaners.

Sulfur dioxide (SO_2) and oxides of nitrogen (NO and NO_2) are released when coal and petroleum are burned. Sulfur dioxide reacts in moist air to produce sulfuric acid. Similarly, oxides of nitrogen react in air to produce nitric acid. Both sulfuric and nitric acids are strong acids. They dissolve in water droplets in the atmosphere and fall to Earth as **acid precipitation**.

Acid precipitation kills trees and fish; reduces the growth of certain agricultural crops; corrodes paint, steel bridges, textiles, and other materials; and can be directly responsible for injury to the health of humans, livestock, and wildlife. The damage to masonry is most severe for limestone and marble, which are soluble in acidic solutions. In Athens, Greece, acid rain has caused widespread weathering of priceless ancient carvings and sculpture. Government officials have removed many statues from their outdoor settings and protected them in sealed glass cases in museums. In the United States, the environmental cost of acid rain is estimated at several billion dollars per year.

Emissions from factories, power plants, and automobiles can be reduced if proper air pollution control devices are installed. The problem is mainly one of economics. Generally, limited pollution control is relatively inexpensive, but an essentially pollution-free environment is more costly. People object to the added expense. However, the damage caused by pollution is often more expensive to society than the cost of pollution control.

---

turns useful material to waste. Wood burns (oxidizes) to form ashes, iron oxidizes to rust, and so on. Rocks and minerals oxidize when their iron reacts with oxygen.

Chemical Weathering of Common Rocks and Minerals

Granite

Recall that continents are made up mostly of granite and similar rocks. Granite consists mostly of feldspar and quartz. When water and atmospheric gases attack granite, the feldspar crystals dissolve. Most of the dissolved ions recombine quickly to form soft, earthy clay. Commonly, the clay replaces feldspar on such a fine scale that weathered granite may look as if its feldspar crystals are unaltered. When you attempt to scratch them, however, they are soft and crumble to dust. The feldspar has converted completely to clay but preserves the shape and appearance of feldspar crystals.

As feldspar weathers to clay, the clay crystals absorb water in a process called **hydration**. As explained previously, the clay then expands against nearby mineral grains and pushes the crystals apart. Thus, hydration, which is a form of chemical weathering, causes mechanical weathering. Some granites are so severely weathered by hydration that their mineral grains are loose to depths of several meters and can be pried out with a fingernail (Fig. 9.12). Many geologists think that the expansion of granite caused by this process is responsible for exfoliation.

Feldspar and the other rock-forming minerals, except for quartz, all contain soluble cations such as potassium, sodium, or calcium, in addition to silicon and oxygen. Water and acids readily dissolve these ions and therefore decompose the minerals. However, quartz is pure silica, SiO_2. Silica is relatively insoluble, and thus quartz resists chemical weathering.

When granite weathers, the feldspar and other minerals, except for quartz, decompose to clay. As they decompose, they release the unaltered quartz grains from the rock. Therefore, quartz accumulates as sand grains in soil. Quartz is also resistant to mechanical weathering. Therefore, when streams carry eroded soil to the seacoast, the quartz grains survive the journey and accumulate on beaches. For this reason, the sand on most beaches is quartz (Fig. 9.13).

Limestone

Limestone is a common sedimentary rock composed of the mineral calcite, which dissolves even in mild acid. Thus, limestone dissolves in rainwater. When acidic rainwater seeps into cracks in limestone, it dissolves the rock, enlarging the cracks to form subterranean caves and caverns. For this reason, most caves form in limestone. The origins of caves and related features are explored in Chapter 11.

9.3 Soil

Bedrock breaks into smaller fragments as it weathers, and much of it decomposes to clay and sand. Thus, a layer of loose rock fragments mixed with clay and sand overlies bedrock. This material is called **regolith**. Scientists in different disciplines use slightly different definitions for **soil**. In engineering and construction, the terms *soil* and *regolith* are interchangeable. However, soil scientists define *soil* as upper layers of regolith that support plant growth. That is the definition we use here.

Components of Soil

Soil is a mixture of mineral grains, organic material, water, and gas. The mineral grains include clay, silt, sand, and rock fragments. Clay grains are so small and closely packed that water and gas do not flow through them readily. Pure clay is so impermeable that plants growing in clay soils often become waterlogged and

Figure 9.13 Most beach sand is quartz. Lake Tahoe, California.

suffer from lack of oxygen. In contrast, water flows easily through sandy soil. The most fertile soil is **loam**, a mixture of sand, clay, silt, and generous amounts of organic matter.

When plant or animal matter dies and falls, it retains its original shape until it begins to decay. Thus, if you walk through a forest or prairie, you can find bits of leaves, stems, and flowers on the surface. This material is called **litter** (Fig. 9.14). When litter decomposes so that you can no longer determine the origins of the individual pieces, it becomes **humus**. Humus is an essential component of most fertile soils. Scoop up some forest or rich garden soil with your hand. Soil rich in humus is light and spongy and absorbs water readily. It soaks up so much moisture that it swells after a rain and shrinks during a dry spell. This alternate shrinking and swelling keeps the soil loose, allowing roots to grow easily. A rich layer of humus also insulates deeper soil from heat and cold and reduces water loss by evaporation.

Soil nutrients are chemical elements necessary for plants to grow. Some examples are phosphorus, nitrogen, and potassium. Humus holds nutrients in soil and makes them available to plants.

Intensive agriculture commonly destroys humus by exposing it to erosion and oxidation. Farmers then replace the lost nutrients with chemical fertilizers to maintain crop growth. But loss of humus reduces the natural ability of soil to conserve water and nutrients. Water can then flow over the surface, eroding the soil. In addition, the muddy runoff carries fertilizer and pesticide residues, contaminating streams and ground water.

Figure 9.12 Coarse grains of quartz and feldspar accumulate directly over weathering granite. Lens cap for scale.

GEOLOGY AND THE ENVIRONMENT

Chemical Fertilizers and Soil Deterioration

Chemical fertilizers have increased global food production and have thus helped feed the expanding human population. For this reason, modern farmers all over the world use fertilizers in ever-increasing amounts. But the long-term effects of chemical fertilizers also cause serious concern. In types of intensive agriculture where chemical fertilizers are used, crop residues such as straw are removed and soil humus is thus depleted continuously. When humus is lost, soil retains less water, nutri-

ents leach from the soil, and the agricultural system becomes dependent on irrigation and chemical fertilizers.

One solution to humus depletion is to use agricultural wastes or other organic matter as fertilizers. Natural fertilizers such as manure are usually more expensive than chemicals. However, many soil scientists calculate that if one were to view profit and loss over a period of decades, not just one or two growing seasons, it would be economically advantageous to use natural fertilizers.

Soil Profiles

If you dig down through undisturbed soil, you can see several layers, or **soil horizons** (Fig. 9.15). The uppermost layer of mature soil is the **O horizon**, named for its organic component. This layer is mostly litter and humus with a small proportion of minerals. The next layer down, called the **A horizon**, is a mixture of humus, sand, silt, and clay. The thicker layer including both O and A horizons is often called **topsoil**. A kilogram of average fertile topsoil contains about 30 percent by weight organic matter, including approximately 2 trillion bacteria, 400 million fungi, 50 million algae, 30 million protozoa, and thousands of larger organisms such as insects, worms, and mites.

The third layer, the **B horizon**, or subsoil, is a transitional zone between topsoil and weathered parent rock below. Roots and other organic material occur in the B horizon, but the amount of organic matter is low. The lowest layer, called the **C horizon**, consists of partially weathered bedrock. It lies directly on unweathered parent rock. This zone contains very little organic matter.

When rainwater falls on soil, it sinks into the O and A horizons. As it travels through the topsoil, it partially dissolves minerals and carries the dissolved ions to lower levels. This downward movement of dissolved material is called **leaching**. The A horizon is sandy because water also carries clay downward but leaves the sand grains behind. Because materials are

Figure 9.14 Litter is organic matter that has fallen to the ground and started to decompose but still retains its original form.

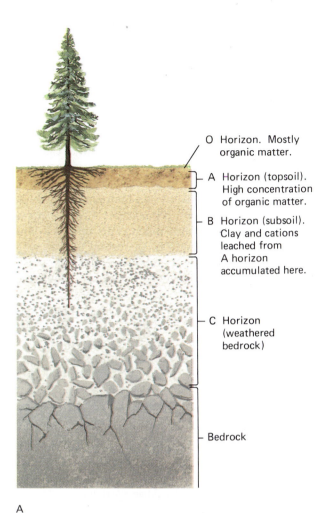

O Horizon. Mostly
   organic matter.

A Horizon (topsoil).
   High concentration
   of organic matter.

B Horizon (subsoil).
   Clay and cations
   leached from
   A horizon
   accumulated here.

C Horizon
   (weathered
   bedrock)

Bedrock

A

B

Figure 9.15   (A) A well-developed soil commonly shows
several distinct horizons. (B) Soil horizons are often
distinguished by color and texture. The dark upper layer is
the A horizon; the lighter lower layer is the B
horizon.   *(Soil Conservation Service)*

removed from the A horizon, it is called the **zone of
leaching**.

Dissolved ions and clay carried downward from the
A horizon accumulate in the B horizon, which is therefore called the **zone of accumulation**. This layer retains moisture because of its high clay content. Although moisture retention may be beneficial, if too
much clay accumulates, a dense, waterlogged soil can
develop.

Soil-Forming Factors

Why is one soil rich and another poor, or one sandy
and another loamy? What factors contribute to the character of a specific soil? Five **soil-forming factors** control how soil develops as parent material weathers.

Parent Rock

The type of parent rock exerts a strong influence on
soil. As explained earlier, granite contains mostly feldspar and quartz. When it weathers, its feldspar converts
to clay, but the quartz grains resist chemical weathering
and become sand. Thus, sandy soil commonly forms on

weathering granite. In contrast, basalt contains much
feldspar but no quartz, and soil formed from basalt is
likely to be clay-rich but not sandy. In addition to texture, the parent material provides nutrients to soil, so
nutrient abundance or deficiency depends in part on
the chemical composition of the parent rock.

Time

In a geologically young soil, weathering of feldspar and
other minerals may be incomplete, and so the soil is
likely to be sandy. As a soil matures and more feldspar
decomposes, the soil's clay content increases. Thus
many young soils consist mostly of slightly weathered
minerals inherited from bedrock.

Addition of new material also changes the character
of soil. A stream may deposit layers of sand or mud, or
wind may blow in dust. The foreign minerals mix with
the residual soil, changing its composition and texture.

Climate

Rain seeps downward through soil, but other factors
pull the water back upward. Roots suck soil water toward

Figure 9.16 Capillary action causes colored water to soak upward through the pores in a paper towel.

the surface, and water near the surface evaporates. In addition, water is electrically attracted to soil particles. If the pore size is small enough, water can be drawn upward by **capillary action**. Capillary action can be demonstrated by placing the corner of a paper towel in water and watching the water rise (Fig. 9.16).

During a rainstorm, water percolates down through the A horizon, dissolving soluble ions such as calcium, magnesium, potassium, and sodium. In arid and semiarid regions, when the water reaches the B horizon, capillary action and plant roots then draw it back up toward the surface, where it evaporates. As it evaporates or is taken up by plants, many of its dissolved ions precipitate in the B horizon, encrusting the soil with salts. A soil of this type is a **pedocal** (Fig. 9.17). Such a process often deposits enough calcium carbonate to form a hard cement called **caliche** in the soil.

In a wet climate, ground water leaches soluble ions from both the A and B horizons. The less soluble elements such as aluminum, iron, and some silicon remain behind, accumulating in the B horizon to form a soil type called a **pedalfer** (Fig. 9.17). The subsoil in a pedalfer commonly is rich in clay, which is mostly aluminum and silicon, and has the reddish color of iron oxide.

Figure 9.17 Pedocals, pedalfers, and laterites form in different climates.

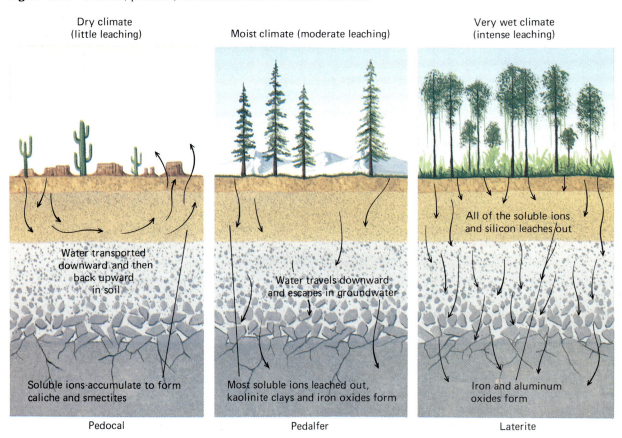

Dry climate (little leaching)

Moist climate (moderate leaching)

Very wet climate (intense leaching)

Water transported downward and then back upward in soil

Water travels downward and escapes in groundwater

All of the soluble ions and silicon leaches out

Soluble ions accumulate to form caliche and smectites

Most soluble ions leached out, kaolinite clays and iron oxides form

Iron and aluminum oxides form

Pedocal

Pedalfer

Laterite

In regions of very high rainfall, such as a tropical rainforest, so much water seeps through the soil that nearly all the cations are leached away. Only very insoluble aluminum and iron minerals remain (Fig. 9.17). Soil of this type is called a **laterite**. Laterites are often colored rust-red by iron oxide. A highly aluminous laterite, called **bauxite**, is the world's main source of aluminum ore.

Rates of Growth and Decay of Organic Material

As explained previously, humus is decomposed plant material, and a thick layer of humus makes rich soil. The most fertile soils are those of prairies and forests in temperate latitudes. There, plant growth and decay are balanced so that thick layers of humus form. In the tropics, organic material decays rapidly, so little humus accumulates. The Arctic is so cold that plant growth and decay are slow. Therefore, litter and humus form slowly and Arctic soils contain little organic matter.

Slope Angle and Aspect

Soil generally migrates downslope in response to gravity. Therefore, if all other factors are equal, soil is thinner and poorer on a hillside, and the valley floor has the deepest and richest soil.

Aspect is the orientation of a slope with respect to compass direction. Exposure of a slope to the Sun affects soil formation. For example, in the semiarid West, thick soils and dense forests cover the north exposure slopes of hills, but thin soils and grass dominate southern exposures. The reason for this difference is that in the Northern Hemisphere more water evaporates from the sunny southern slopes. Therefore, fewer plants grow, weathering occurs more slowly, and soil development is retarded. The moister northern slopes weather more deeply to form thicker soils.

9.4 Erosion

Natural Soil Erosion

Weathering decomposes bedrock, and plants add organic material to the regolith to create soil at the Earth's surface. However, soil does not continue to accumulate and thicken throughout geologic time. If it did, the Earth would be covered by a mantle of soil hundreds or thousands of meters thick, and rocks would be unknown at the Earth's surface. Instead, flowing water, wind, and glaciers erode soil as it forms (Fig. 9.18). In addition, some weathered material simply slides downhill under the influence of gravity. In fact, all forms of

Figure 9.18 Flowing water has eroded gullies in this hillside. *(Don Hyndman)*

erosion combine to remove soil about as fast as it forms. For this reason, soil is usually only a few meters thick or less in most parts of the world.

Once soil erodes, the sediment begins a long journey as it is carried downhill by the same agents that eroded it: streams, glaciers, wind, and gravity. During the journey the sediment may come to rest in a stream bed, a sand dune, or a lake bed, but those environments are usually temporary stops. Sooner or later most sediment erodes again and is carried further downhill, until finally it is deposited where the land meets the sea. There it remains and is buried by younger sediment until it lithifies to form sedimentary rocks. Table 9.1 shows the sediment load of some major rivers.

Erosion and transport of sediment by streams, ground water, glaciers, wind, and ocean currents are

Table 9.1 • Sediment Load of Selected Major Rivers

River	Countries	Annual Sediment Load (million metric tons)
Yellow	China	1600
Ganges	India	1455
Amazon	Brazil	363
Mississippi	United States	300
Irrawaddy	Burma	299
Kosi	India	172
Mekong	Southeast Asia	170
Nile	Egypt	111

Source: S. A. El-Swaify and E. H. Dangler, "Rainfall Erosion in the Tropics: A State of the Art," in American Society of Agronomy, *Soil Erosion and Conservation in the Tropics* (Madison, Wisc.: 1982).

GEOLOGY AND THE ENVIRONMENT

Tropical Rainforests

Most laterite soils form in the tropics, where the climate is hot and rainfall is high. Tropical rainforests support a rich and varied community of plants and animals, but paradoxically their laterite soils are not suitable for agriculture. When litter falls to the forest floor, it decays quickly in the hot, moist environment, and growing plants absorb most of the nutrients. Any excess nutrients are leached out by the great amount of water seeping into the ground. As a result, the soil contains little humus and nutrients.

When the rainforest is cut, heavy tropical rains rapidly leach any remaining nutrients from the soil. Farmers can grow crops for a few years, but soon the small supply of nutrients is consumed and the soil becomes useless for agriculture. Because the nutrients are lost, it is then impossible for the natural rainforest vegetation to regenerate, and the abandoned soil lies bare and unprotected from erosion.

Between 1945 and 1990, more than half of the Earth's tropical rainforests were cut, partly for timber and partly for agriculture (see figure). Today, deforestation of the tropics continues at a rate of about 7 million hectares per year, or about 13 hectares per minute, day and night, throughout the year. Only about one third of this land is capable of supporting sustained agriculture. In some regions the exposed clayey soil bakes in the hot sun to a bricklike texture, and the area becomes a desolate wasteland. This loss is especially tragic because the rainforests are a valuable renewable resource. If the nuts and fruits were harvested and trees selectively cut for fuel and lumber, the forests could provide local people

Destruction of a tropical rainforest in Costa Rica.

with continuous economic rewards. Beyond direct economics, many scientists fear that if vast regions of the forest are cut, the carbon dioxide–oxygen balance in our atmosphere may be affected, resulting in global climate change. In addition, millions of species of plants and animals in the rainforest are rapidly becoming extinct, reducing the biological diversity of our planet.

the subjects of the following five chapters. In the remainder of this chapter, we will describe mass wasting: erosion by gravity.

9.5 Mass Wasting

Mass wasting is the movement of Earth material downslope, primarily under the influence of gravity. Look up toward a hill or mountain and think about the bedrock and soil near the top. Gravity constantly pulls them downward, but on any given day the rock and soil are not likely to slide or tumble down the slope. They are held up by their own strength and by friction. Even so, a slope can become unstable and mass wasting can occur. For example, stream erosion can undermine a rock cliff so much that it collapses. Water can add weight and lubricate soil on a slope and cause it to slide. Mass wasting occurs naturally in any hilly or mountainous terrain. Steep slopes are especially vulnerable, and scars from recent movement of rock and soil are common in mountainous country.

In recent years, the human population has increased dramatically. As land has become overpopulated, the character of human settlements has changed. In poor countries, more and more people try to scratch

GEOLOGY AND THE ENVIRONMENT

Soil Erosion and Agriculture

In nature, soil erodes approximately as rapidly as it forms. However, improper farming, livestock grazing, and logging can accelerate erosion. Plowing removes plant cover that protects soil. Logging often removes forest cover, and the machinery breaks up the protective litter layer. Similarly, intensive grazing can strip away protective plants. Rain, wind, and gravity then erode the exposed soil. Meanwhile, soil continues to form by weathering at its usual slow, natural pace. Thus, increased rates of erosion caused by agricultural practices can lead to net soil loss.

When farmers use proper conservation measures, soil can be preserved indefinitely or even improved. Some regions of Europe and China have supported continuous agriculture for centuries without soil damage. However, in recent years, marginal lands on hillsides, in tropical rainforests, and along the edges of deserts have been cultivated. These regions are particularly vulnerable to soil deterioration, and today soil is being lost at an alarming rate throughout the world.

Soil is eroding more rapidly than it is forming on about 35 percent of the world's croplands. About 23 trillion kilograms of soil (about 25 billion tons) are lost every year. The soil lost annually would fill a train of freight cars long enough to encircle our planet 150 times. Table 9.1 shows the sediment loads of several major rivers. Although much of the sediment results from natural erosion, the sediment loads of major rivers have been rising, and this increase stems from excessive erosion caused by logging and improper agriculture.

In the United States, approximately one third of the topsoil that existed when the first European settlers arrived has been lost. Erosion is continuing in the United States at an average rate of about 10.5 tons per hectare[1] per year. However, in some regions the rate is considerably higher, and yearly losses of more than 25 tons per hectare are common. To put this number in perspective, a loss of 25 tons per hectare per year would lead to complete loss of the topsoil in about 150 years.

---

[1] One hectare equals 2.47 acres.

out a living in mountains once considered too harsh for agriculture. In rich nations, people have moved into the hills to escape congested cities. As a result, permanent settlements have grown in previously uninhabited steep terrain. Many slopes are naturally unstable. Construction or agriculture destabilize others.

Every year, small landslides destroy homes and farmland. Occasionally, an enormous landslide buries towns, killing thousands or even tens of thousands of people. Mass wasting causes billions of dollars' worth of damage every year. The total global property damage from mass wasting in a single year approximately equals the damage caused by earthquakes in 20 years. In many instances, losses occur because people do not recognize danger zones. In other cases, damage is caused, or at least aggravated, by improper planning and construction.

Consider three real examples of mass wasting that have affected humans:

1. A movie star builds a mansion on the edge of a picturesque California cliff. After a few years, the

An expensive landslide in Hong Kong. *(Hong Kong Government Information Services)*

cliff collapses and the house slides into the valley (Fig. 9.19A).

2. A ditch carrying irrigation water across a hillside in Montana leaks water into the ground. After years of seepage, the muddy soil slides downslope and piles against a house at the bottom of the hill (Fig. 9.19B).

3. In the springtime, when snow is melting, water saturates soil and bedrock on a steep mountain slope. Suddenly, an earthquake shakes the overburdened slope, triggering a landslide that buries a campground and kills several people (Fig. 9.19C).

Why does mass wasting occur? Is it possible to avoid or predict such catastrophes and thus to reduce property damage and loss of life?

Factors that Control Mass Wasting

Imagine that you are a geological consultant on a construction project. The developers want to construct a road at the base of a hill, and they wonder whether landslides or other forms of mass wasting are likely. What factors should you consider in your evaluation?

Steepness of the Slope

Obviously, the steepness of a slope is a factor in mass wasting. If frost wedging dislodges a rock from a vertical cliff, the rock tumbles to the valley below. However, a similar rock is less likely to roll down a gentle hillside. The relationship between slope steepness and mass wasting can be illustrated by placing a block of wood on a

A

Figure 9.19 Examples of mass wasting. (A) A few days after this photo was taken, the corner of the house hanging over the gully fell in. *(J. T. Gill, USGS)* (B) A landslide, triggered by a leaking irrigation ditch, threatens a house in Darby, Montana. (C) This landslide near Yellowstone Park buried a campground, killing 26 people. *(Don Hyndman)*

B

C

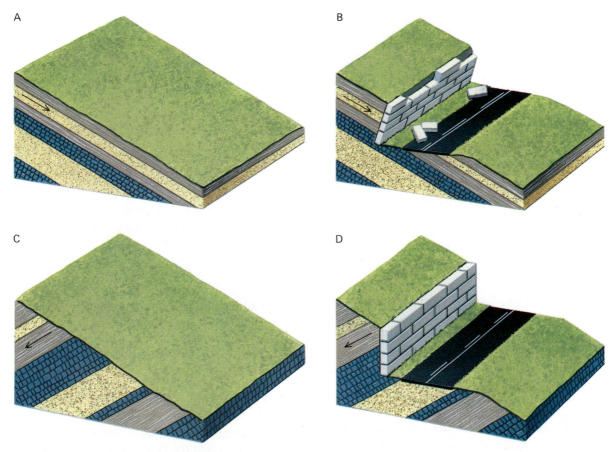

Figure 9.20 (A) Sedimentary rock layers dip parallel to this slope. (B) If a road cut undermines the slope, the dipping rock provides a good sliding surface, and the slope may fail. (C) Sedimentary rock layers dip at an angle to this slope. (D) The slope may remain stable even if it is undermined.

board and slowly tilting one end of the board. When the board is nearly level, friction holds the block in place. However, if you tilt the board beyond a certain critical angle, the block slides or tumbles.

Type of Rock and Orientation of Rock Layers

The block of wood is a coherent mass; either the entire block slides, or none of it moves. A hillside does not behave in the same way. Any portion of a slope can move. For example, if sedimentary rock layers dip in the same direction as a slope, the upper layers may slide over the lower ones. Imagine a hill underlain by shale, sandstone, and limestone layered parallel to the slope, as shown in Figure 9.20A. If the base of the hill is undermined (Fig. 9.20B), the upper portion is left hanging and may slide along the layer of weak shale. In contrast, if the rock layers dip at an angle to the hillside,

the slope may be stable even if it is undercut (Figs. 9.20C and D).

Several processes can undermine a slope. A stream or ocean waves can erode its base. Road cuts and other types of excavation can create instability in the same manner. Therefore, geologists and engineers must consider not only a slope's stability before construction, but how the project might alter its stability.

The Nature of Unconsolidated Materials

The **angle of repose** is the maximum slope or steepness at which loose material remains stable. Recall that when chunks of rock break from a cliff, they fall to the bottom, where they collect to form a talus slope. Because rocks in talus are angular and irregular, they interlock and jam together, allowing talus to maintain a high angle of repose, up to 45°. The slope cannot be any steeper

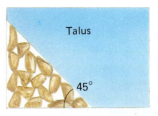 

Figure 9.21 The angle of repose is the maximum slope that can be maintained by a specific material.

because if it were, the rock would slide. In contrast, rounded sand grains do not interlock as well as angular talus and therefore have a lower angle of repose (Fig. 9.21).

Water and Vegetation

To understand how water affects slope stability, think of a sand castle. Even a novice sand castle builder knows that it is impossible to build steep-sided towers and wall with dry sand; the sand must be moistened first (Fig. 9.22). But if you add too much water, the castle collapses. Small amounts of water bind sand grains together because electrical charges on water molecules attract sand grains. However, excess water adds too much weight to a slope and lubricates the sand. When some soils become saturated with water, they flow downslope, just as the sand castle collapses. In addition, if water collects on impermeable clay or shale, it may

Figure 9.22 The angle of repose depends on both the type of material and its water content. Dry sand forms low mounds, but if you moisten the sand, you can build steep, delicate towers with it.

provide a weak, slippery layer so that overlying rock or soil can move easily.

Roots hold soil together and plants absorb water; therefore, a highly vegetated slope is more stable than a similar bare one. Many forested slopes that were stable for centuries have slid when the trees were removed during logging, agriculture, or construction.

Mass wasting is common in deserts and in regions with intermittent rainfall. For example, southern California has dry summers and occasional heavy winter rainfalls. Vegetation is sparse because of summer drought and wildfires. When winter rains fall, bare hillsides often become saturated and slide. Mass wasting occurs for similar reasons during rare but intense storms in deserts.

Earthquakes and Volcanoes

Earthquakes and volcanic eruptions have triggered many devastating landslides. An earthquake can cause mass wasting by shaking the ground, causing an unstable slope to slide. A volcanic eruption may melt snow and ice near the top of a volcanic mountain. The water then soaks into the slope to release a landslide.

9.6 Types of Mass Wasting

Mass wasting can occur slowly or rapidly. Sometimes rocks fall freely down the face of a steep mountain. In other instances, material moves downslope so slowly that the movement is unnoticed by casual observers. Mass wasting falls into three categories: flow, slide, and fall (Fig. 9.23). To understand these categories, think again of building a sand castle. If you add too much water to the sand, it flows like molasses down the face of the structure. During **flow**, loose, unconsolidated regolith moves as a fluid, analogous to wet concrete. Flow can be slow; some slopes flow at a speed of about 1 centimeter per year or less. On the other hand, mud with a high water content can flow almost as rapidly as water.

If you undermine the base of a sand castle, a block of sand may break away from the rest of the castle and move downslope as a coherent unit. This type of movement is called **slide**. Slide is usually faster than flow, but it still may take several seconds for the block to slide down the face of the castle.

If you took a huge handful of sand out of the bottom of a sand castle, the whole tower would topple. This rapid, free-falling motion is called **fall**. Fall is the most rapid type of mass wasting. In extreme cases, rock can fall at a speed dictated solely by the force of gravity and air resistance.

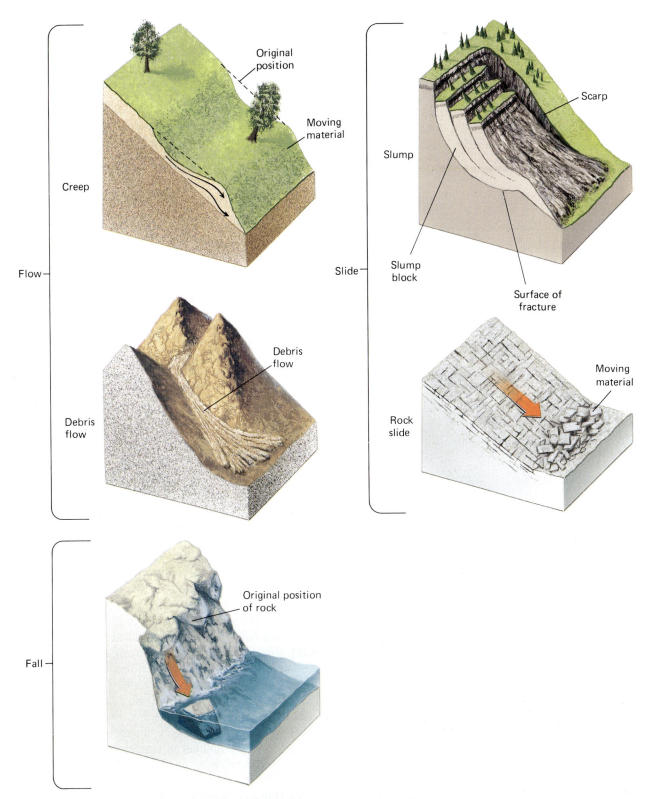

Figure 9.23   Flow, slide, and fall are the three categories of mass wasting.

Table 9.2 • Some Categories of Mass Wasting

Type of Movement	Description	Subcategory	Description	Comments
Flow	Individual particles move downslope independently of one another, not as a consolidated mass. Typically occurs in loose, unconsolidated regolith.	Creep	Slow, visually imperceptible movement	
		Debris flow	More than half the particles larger than sand size; rate of movement varies from less than 1 m/year to 100 km/hr or more	Often occurs in conjunction with slump. Common in arid regions with intermittent heavy rainfall, or can be triggered by volcanic eruption.
		Earthflow and mudflow	Movement of fine-grained particles with large amounts of water	
		Solifluction	Movement of waterlogged soil situated over permafrost	Can occur on very gradual slopes
Slide	Material moves as discrete blocks, can occur in regolith or bedrock.	Slump	Downward slipping of a block of earth material, usually with a backward rotation on a concave surface	Often triggers flow; trees located on slump blocks remain rooted
		Rockslide	Usually rapid movement of a newly detached segment of bedrock	
Fall	Material falls freely in air; typically occurs in bedrock.	—	—	Occurs only on steep cliffs

Table 9.2 outlines the characteristics of flow, slide, and fall. Details of these three types of mass wasting are explained in the following sections.

Flow

Types of flow include creep, debris flow, earthflow, mudflow, and solifluction.

Creep

As the name implies, **creep** is a slow, downhill movement of rock or soil. Individual particles move independently of one another, and the slope does not move as a consolidated mass. Typically, movement occurs at a rate of about 1 centimeter per year—so slowly that the motion cannot be detected without careful observation. When a slope creeps, the surface moves more rapidly than deeper layers (Fig. 9.24). As a result, anything with roots or a foundation tilts downhill. For example, if you look at a hillside cemetery, you may note that older headstones are tilted, whereas newer ones are vertical

(Fig. 9.25). Over the years, soil creep has tipped the older monuments, but the newer ones have not yet had time to tilt.

Creep also tilts fences and telephones poles, and in some instances it may tear entire buildings apart. When soil creep tilts trees, they develop curved trunks as a result of their natural tendency to grow upward. The result is a J-shaped appearance called pistol butt (Fig. 9.26). If you ever contemplate buying hillside land for a home site, examine the trees. If they have pistol-butt bases, the slope is probably creeping.

Creep often accelerates as heavy rain or snowmelt adds weight and lubrication to soil. In addition, movement can result from freeze–thaw cycles that occur mainly in spring and fall in temperate regions. Recall that water expands when it freezes. When wet soil freezes, it expands outward away from the slope (Fig. 9.27). However, when the soil thaws, the particles fall downward vertically. The expansion and falling result in a net displacement downslope. The displacement in any one cycle is small, but the soil may freeze and thaw

Figure 9.24   Creep has bent this layering in sedimentary rocks in a downslope direction.   *(Ward's Natural Science Establishment, Inc.)*

Figure 9.25   During creep, the soil surface moves more rapidly than deeper layers, so objects embedded in the soil tilt downhill.

once a day for a few months each year, leading to a total movement of a centimeter or more per year.

Debris Flows, Mudflows, and Earthflows

Debris flows, **mudflows**, and **earthflows** all involve downslope flow of wet soil or regolith as a plastic or semifluid mass. Think of what can happen when heavy rain falls on an unvegetated slope, as in a desert. The rain mixes rapidly with soil to form a slurry of mud and rocks. A slurry consists of water and solid particles, but it flows as a liquid. Wet concrete is a familiar example

of a slurry. It flows easily and is routinely poured or pumped from a truck.

The advancing front of a flow often forms a tongue-shaped lobe (Fig. 9.28). Slow-moving flows travel at a rate of about 1 meter per year, but others can move as fast as a car speeding along an interstate highway. The destructive potential of a flow depends on its speed and the consistency of the slurry. Flows can pick up boulders and automobiles and smash houses, filling them with mud or even dislodging them from their foundations.

Different types of flows are characterized by the sizes of the solid particles. A **debris flow** consists of a mixture of clay, silt, sand, and rock fragments in which

Figure 9.26   If a hillside creeps as a tree grows, the tree develops pistol butt.

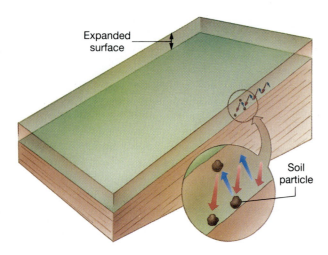

Figure 9.27 When soil freezes, it expands outward, away from the slope. But when it thaws again, the soil particles move vertically downward, resulting in a net downslope migration of the soil.

more than half of the particles are larger than sand. In contrast, mudflows and earthflows are predominantly sand and mud. Some **mudflows** have the consistency of wet concrete, and others are more fluid. Because of their high water content, mudflows may race down stream channels at speeds up to 100 kilometers per hour. **Earthflows** have less water than mudflows and are therefore less fluid.

Solifluction

In temperate regions, soil moisture freezes in winter and thaws in summer. However, in very cold regions such as the Arctic, the Antarctic coast, and some mountain ranges, a layer of permanently frozen soil or subsoil, called **permafrost**, lies about a half meter to a few meters beneath the surface. Because ice is impermeable, summer meltwater cannot percolate downward as it does in temperate and tropical regions, and it therefore collects on the ice layer below it. This leads to two unique characteristics in these soils:

1. Water cannot penetrate the ice layer, so it collects near the surface. As a result, even though many Arctic regions receive little annual precipitation, bogs and marshes are common.

2. Ice, especially ice with a thin film of water on top, is extremely slippery. Therefore, Arctic regions are particularly susceptible to mass wasting of soil overlying the permafrost.

Solifluction is a type of mass wasting that occurs when water-saturated soil moves over permafrost (Fig. 9.29). Solifluction can occur even on very gentle slopes.

Figure 9.28 A debris flow in the Cascade Range of Washington has a characteristic lobe-shaped form.

Slide

In many cases, a large block of rock or soil, sometimes an entire hillside, fractures and moves. The material does not flow as a fluid, but rather slides downslope as a coherent mass or as several blocks that remain intact. There are two types of slide: slump and rockslide.

A **slump** occurs when a gently curved fracture forms in rock or regolith. Overlying blocks of material slide downhill along the fracture, as shown in Figure 9.30. Trees remain rooted in the moving blocks. However, because the blocks rotate on the concave fracture, trees on the slumping blocks are tilted backward. Thus, you can distinguish slump from creep by the orientations of the trees. Slump tilts trees uphill, whereas creep tilts them downhill. When the blocks reach the bottom of the slope, they pile up to form a broken, jumbled, hummocky topography.

It is useful to be able to identify slump because it often recurs in the same place or on adjacent slopes. Thus a slope that shows evidence of past slump is not a good place to build a house. Figure 9.19B shows a slump that almost destroyed a home in Darby, Montana.

In the western United States, irrigation water commonly flows through unlined ditches. If the ground is porous, water seeps from the ditch into the soil and bedrock. The slump in Figure 9.19B resulted from water

Figure 9.29 Arctic solifluction is characterized by lobes and a hummocky surface. *(R. B. Colton, USGS)*

seeping through an irrigation ditch cut across an unstable slope. Alternatively, excessive irrigation can trigger slump. A Los Angeles man left his sprinkler on when he went away for a vacation. His property was on a hillside, and when he returned, he found that not only was his lawn gone, but his house had slid downslope as well.

During a **rockslide**, or **rock avalanche**, a fracture occurs in bedrock and the overlying rock slides downslope. Characteristically, the rock breaks into small pieces of rubble, and a turbulent mass of broken rock tumbles down the hillside.

A classic example of a rock avalanche occurred in the hills above the Gros Ventre River near the town of Kelly, Wyoming. Before the slide, a layer of sandstone rested on shale, which in turn was supported by a thick bed of limestone (Fig. 9.31). The sedimentary layers dipped 15° to 20° toward the river and parallel to the slope. Over time, the Gros Ventre River had undermined

the sandstone, leaving the slope unsupported above the river. Only a trigger was needed to release the hillside. Snowmelt and heavy rains provided the trigger in the spring of 1925. Water seeped into the ground, saturating the soil and bedrock and increasing their weight. The water collected on the shale, forming a slippery surface. Finally the sandstone layer broke loose and began to slide over the shale. In a few moments, approximately 38 million cubic meters of rock tumbled into the valley. The sandstone crumbled into small blocks, forming a 70-meter-high natural dam across the Gros Ventre River. (For comparison, a 20-story building is about 70 meters high.) But a dam made of rockslide debris is generally unstable. Two years later the lake overflowed the dam, washing it out and creating a flood downstream that killed several people.

Fall

If a rock dislodges from a steep cliff, it falls rapidly under the influence of gravity. Recall from our discussion of weathering that when water freezes and thaws, the alternate expansion and contraction can dislodge rocks from cliffs and thus cause rockfall. Rockfall also occurs when a cliff is undermined. For example, if ocean waves undercut a steep shoreline (Fig. 9.32), rock above the waterline may tumble.

9.7 Mass Wasting Triggered by Earthquakes and Volcanoes

Mass wasting often occurs when a trigger mechanism releases an unstable slope. Water commonly triggers landslides because it adds weight, lubricates, and

Figure 9.30 In slump, blocks of soil or rock remain intact as they move downslope.

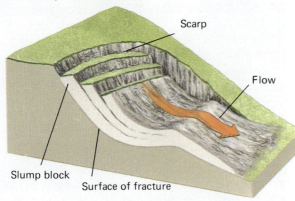

Scarp

Flow

Slump block

Surface of fracture

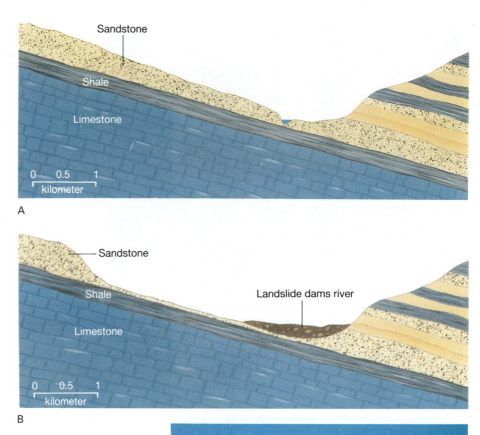

A

B

Figure 9.31 Profiles of the Gros Ventre hillside (A) before and (B) after the slide. (C) About 38 million cubic meters of rock and soil broke loose and slid downhill during the Gros Ventre slide.

C

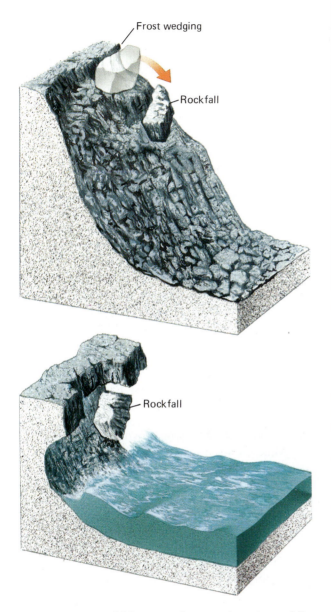

Frost wedging

Rockfall

Rockfall

Figure 9.32 Rockfall commonly occurs in spring or fall when freezing water dislodges rocks from cliffs. Undercutting of cliffs by waves, streams, or construction can also cause rockfall.

Figure 9.33 The eruption of Nevado del Ruiz in Colombia triggered a mudflow that buried the town of Armero, killing 20,000 people. *(Wide World Photos/Associated Press)*

reduces shear strength. Earthquakes and volcanic eruptions also initiate mass wasting. In fact, in many cases, an earthquake or eruption itself causes comparatively little damage, but the resulting mass wasting is devastating. Consider the following two case histories.

Case History I: The Madison River Slide, Montana

In August 1959, an earthquake occurred just west of Yellowstone National Park. This region is sparsely populated, and most of the buildings in the area are wood-frame structures that can withstand quakes. Thus, the earthquake itself caused little property damage and no loss of life. However, the ground motion triggered a massive rockslide from the top of Red Mountain above the Madison River. About 30 million cubic meters of rock broke loose and slid into the valley below, burying a campground and killing 26 people. The slide's momentum carried it more than 100 meters up the mountain on the opposite side of the valley. The mass of falling rock compressed large quantities of air, creating intense winds as the air escaped from beneath the falling debris. The winds lifted a car off the ground and carried it more than 10 meters into nearby trees. The debris dammed the Madison River, forming a lake behind the dam that was later named Quake Lake. Figure 9.19C shows the pile of debris and some of the damage caused by this slide.

Case History II: Nevado del Ruiz, Colombia

In November 1985, the volcanic mountain Nevado del Ruiz erupted in central Colombia. The eruption caused only minor damage, but heat from the ash and lava melted large quantities of ice and snow that lay on the mountain. The rushing water mixed with ash and debris on the mountainside, forming a mudflow that raced down gullies and stream valleys to the town of Armero, 48 kilometers from the mountain. Twenty thousand people were buried and killed in Armero, and additional loss of life and property damage occurred in a dozen or so other villages in nearby valleys (Fig. 9.33).

SUMMARY

Weathering is the decomposition and disintegration of rocks and minerals by mechanical and chemical processes. **Erosion** is the physical removal of weathered rock or soil by moving water, wind, glaciers, or gravity. After rock or soil has been eroded from the immediate environment, it may be **transported** great distances and is eventually **deposited**.

Mechanical weathering can occur by **frost wedging**, **salt cracking**, **abrasion**, **biological activity**, **pressure release**, **thermal expansion** and contraction, or expansion caused by **chemical weathering**.

Chemical weathering occurs when chemical reactions decompose minerals. Some minerals react with oxygen to form new minerals. This process is known as **oxidation**. A few minerals are **soluble** in water. **Acids and bases** enhance solubilities of minerals. Rainwater is slightly acidic as a result of reactions between water and atmospheric carbon dioxide. Weathering of feldspar and other common minerals, except quartz, produces clay. Weathering releases quartz grains from rock to form sand. Limestone dissolves to form caverns as rainwater seeps through cracks in the rock.

Soil is the layer of weathered material overlying bedrock. **Sand**, **silt**, **clay**, and **humus** are common components of soil. Water **leaches** soluble ions downward through soil. Clay is also transported downward by water. The uppermost layer of soil, called the **O horizon**, consists mainly of **litter** and humus. The **A horizon** is the **zone of leaching**, and the **B horizon** is the **zone of accumulation**.

Five major factors control the character of soil: **parent rock**, **time**, **climate**, **vegetation**, and **slope angle and aspect**. In dry climates, **pedocals** form. In pedocals, leached ions precipitate in the B horizon, where they accumulate and may form **caliche**. In moist climates, soluble ions are removed from the soil, and **pedalfer** soils develop with high concentrations of less soluble aluminum and iron. **Laterite** soils form in very moist climates where leaching removes all of the soluble ions.

Mass wasting is the downhill movement of rock and soil under the influence of gravity. The stability of a slope and the severity of mass wasting depend on (1) steepness of the slope, (2) orientation and type of rock layers, (3) nature of unconsolidated materials, (4) climate and vegetation, and (5) earthquakes or volcanic eruptions.

Mass wasting falls into three categories: flow, slide, and fall. During **flow**, a mixture of rock, soil, and water moves as a viscous fluid. **Creep** is a slow type of flow that occurs at a rate of about 1 centimeter per year. A **debris flow** consists of a mixture in which more than half the particles are larger than sand. **Earthflows and mudflows** are mass movements of predominantly fine-grained particles mixed with water. Earthflows have less water than mudflows and are therefore less fluid. **Solifluction** is a type of flow that occurs when water-saturated soil moves over permafrost.

Slide is the movement of a coherent mass of material. **Slump** is a type of slide in which the moving mass travels on a concave surface. In a **rockslide**, a newly detached segment of bedrock slides along a tilted bedding plane or fracture. **Fall** occurs when particles fall or tumble down a steep cliff. Earthquakes and volcanic eruptions trigger devastating mass wasting. Damage to human habitation can be averted by proper planning and construction.

KEY TERMS

REVIEW QUESTIONS

1. Explain the differences among weathering, erosion, transportation, and deposition.

2. Explain the differences between mechanical weathering and chemical weathering. Give examples of each.

3. Explain how chemical weathering can cause mechanical weathering.

4. Explain how mechanical weathering can speed up chemical weathering.

5. Explain frost wedging. What landforms are created by frost wedging?

6. Explain how thermal expansion can fracture a rock.

7. What are the components of healthy soil? What is the function of each component?

8. Characterize the four major horizons of a mature soil.

9. List the five soil-forming factors and briefly discuss each one.

10. Imagine that soil forms on granite in two regions, one wet and the other dry. Will the soil in the two regions be the same or different? Explain.

11. Explain how soils formed from granite change with time.

12. What is a laterite soil? How does it form? Why is it unsuitable for agriculture?

13. Explain how a small amount of water might increase slope stability, whereas mass wasting might occur on the same slope during heavy rainfall or rapid snowmelt.

14. Discuss the differences among flow, slide, and fall. Give examples of each.

15. Compare and contrast creep, debris flow, and mudflow.

16. Why is solifluction more likely to occur in the Arctic than in temperate or tropical regions?

17. Compare and contrast slump and rockslide.

18. Explain how trees are bent but not killed by slump. How are trees affected by rockslide?

DISCUSSION QUESTIONS

1. What process is responsible for each of the following observations and phenomena? Is the process mechanical or chemical? (a) A board is sawn in half. (b) A board is burned. (c) A cave forms when water seeps through limestone. (d) Calcite precipitates from a hot underground spring. (e) Meter-thick sheets of granite peel off a newly exposed pluton. (f) In mountains of the temperate region, rockfall is more common in the spring than in mid-summer.

2. Arctic regions are cold most of the year, and summers are short. Therefore, decomposition of organic matter is slow. In the temperate regions, decay is much more rapid. How does this difference affect the fertility of the soils?

3. As a class, debate the relative merits of chemical and organic fertilization. If you live in a rural region, call your local county agricultural service (such as the Agricultural Stabilization and Conservation Service) and the Soil Conservation Service and ask their opinions. If possible, interview local farmers and record their opinions.

4. In some regions subsoil lies on top of a layer of topsoil. Suggest a plausible process to explain this inverted soil profile. Where would you look to find such a soil?

5. Show how time interacts with the other soil-forming factors listed in the text.

6. Explain how exfoliation might lead to mass wasting.

7. The Moon is considerably smaller than the Earth, and therefore its gravitation is less. It has no atmosphere and therefore no rainfall. The interior of the Moon is cool, and thus it is geologically inactive. Would you expect mass wasting to be a common or an uncommon event in mountainous areas of the Moon? Defend your answer.

8. Explain how wildfires might contribute to mass wasting.

9. What types of mass wasting (if any) would be likely to occur in each of the following environments? (a) A very gradual (2 percent) slope in a heavily vegetated tropical rainforest. (b) A steep hillside composed of alternating layers of conglomerate, shale, and sandstone, in a region that experiences distinct dry and rainy seasons. The dip of the rock layers is parallel to the slope. (c) A steep hillside composed of clay in a rainy environment in an active earthquake zone.

10. Identify a hillside in your city or town that might be unstable. Using as much data as you can collect, discuss the magnitude of the potential danger. Would the earth or rock movement be likely to affect human habitation?

11. Explain how the mass wasting triggered by earthquakes and volcanoes can have more serious effects than the earthquake or volcano itself. Is this always the case?

CHAPTER

10

Streams

*Deer Creek, a tributary of the Colorado River
in Grand Canyon, has eroded its channel through solid bedrock.*

About 1.3 billion cubic kilometers of water exist at the Earth's surface. Of this huge quantity, however, 97.5 percent is salty seawater, and just under 1.8 percent is frozen into the great ice caps of Antarctica and Greenland. Only about 0.65 percent is fresh water in streams, lakes, and underground reservoirs.

All organisms that live on land need fresh water to survive. Rivers and streams serve as arteries for transportation; they can be dammed to produce energy and are used for irrigation and industry. Fresh-water fisheries yield protein for human consumption. In addition, although the Earth's landforms are initially created by tectonic processes, flowing water is the most important agent in modifying them.

10.1   The Water Cycle

The Earth's water moves constantly. It evaporates from the sea, falls as rain, and flows over land as it returns to the ocean. The constant circulation of water among sea, land, and the atmosphere is called the **hydrologic cycle**, or the water cycle (Fig. 10.1).

Water **evaporates** from sea and land to become water vapor in the atmosphere. Water also evaporates directly from plants as they breathe in a process called **transpiration**. Atmospheric moisture then returns to the Earth's surface as **precipitation**: rain, snow, hail, and sleet. **Runoff** is the water that flows back to the oceans over the surface of the land.

10.2   Characteristics of a Stream

When rain falls on land, most of it soaks into the ground, although during a hard rain some may run over the soil. Surface runoff and some of the ground water seeps into channels to flow over the Earth's surface. Water flowing in a channel has a variety of names, such as creek, brook, rivulet, stream, and river. To avoid confusion, the term **stream** is used for all water flowing in a channel, regardless of size. The term **river** is used commonly for any large stream fed by smaller streams called **tributaries**. Most streams flow year round, even when it is not raining, because they are fed primarily by ground water.

Normally a stream flows in its **channel**. The floor of the channel is called the **bed**, and the sloping edges of the channel are the **banks**. When rainfall is heavy or when snow melts rapidly, a flood may occur. During a flood, a stream overflows its banks and water covers the adjacent land, called a **flood plain**.

Stream Flow

A slow stream flows at 0.25 to 0.5 meter per second (1 to 2 kilometers per hour), whereas a steep, flooding stream may race along at about 7 meters per second (25 kilometers per hour). Three factors control stream velocity: (1) the gradient of the stream bed; (2) the flow, or discharge, of the stream; and (3) the channel characteristics.

Gradient

Obviously, if all other factors are equal, water flows more rapidly down a steep slope than a gradual one. **Gradient** is the vertical drop of a stream within a certain horizontal distance. The lower Mississippi River has a shallow gradient and drops only 10 centimeters per kilometer. In contrast, a tumbling mountain stream may drop 40 meters or more per kilometer.

Discharge

Discharge is the volume of water flowing downstream per unit time. It is measured in cubic meters per second (m^3/sec). The velocity of a stream increases when its discharge increases. Thus, a steam flows faster during flood, even though its gradient is unchanged.

The largest river in the world is the Amazon, with a discharge of 150,000 m^3/sec. In contrast, the Mississippi River, the largest in North America, has a discharge of about 17,500 m^3/sec, approximately one ninth that of the Amazon.

The discharge of a stream can change dramatically from month to month or even during a single day. For example, the Selway River, a mountain stream in Idaho, has a discharge of 100 to 130 m^3/sec during early summer, when the mountain snowpack is melting rapidly. During the dry season in late summer, the discharge drops to about 10 to 15 m^3/sec (Fig. 10.2). In extreme cases, discharge can vary almost instantaneously. A desert stream bed may be completely dry in mid-summer. But a sudden thunderstorm can send a wall of water rushing violently down the stream bed. After a few hours, the flow may die off to a gentle trickle as the stream dries up again.

Channel Characteristics

Flowing water is slowed by friction between the water and the stream bed and banks. Consequently, water

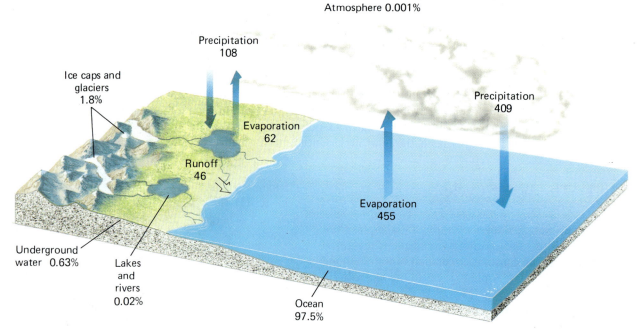

Figure 10.1 The hydrologic cycle shows that water is recycled constantly among the sea, the atmosphere, and the land. Numbers are thousands of cubic kilometers of water transferred each year. Percentages are proportions of total global water in different portions of the Earth's surface.

flows more slowly near the banks than near the center of a stream. If you paddle a canoe down a straight stream channel, you move faster when you stay away from the banks. The amount of friction depends on the roughness and shape of the channel. Boulders on the banks or in the stream bed increase friction and slow down a stream, whereas the water flows more rapidly if the bed and banks are smooth.

10.3   Stream Erosion

A stream erodes soil and bedrock, transports sediment, and eventually deposits the sediment. These processes shape the Earth's surface by cutting stream valleys, creating flood plains, and building deltas in oceans and lakes. Ultimately, stream erosion can level an entire mountain range.

Processes of Stream Erosion

A stream weathers and erodes its bed and banks by three different processes: hydraulic action, abrasion, and solution (Fig. 10.3).

Hydraulic action is the ability of flowing water to dislodge fragments of rock and grains of sediment. To demonstrate hydraulic action, aim a garden hose into bare dirt. In a short time the water erodes a small hole, moving soil and small pebbles. Similarly, a stream dislodges solid particles in its bed and banks, especially when it is flowing swiftly.

Abrasion is the mechanical wearing away of rock by a stream. Water itself is not abrasive, but sand and other sediment carried by a stream wear away boulders

Rapids on Idaho's Selway River during spring runoff.

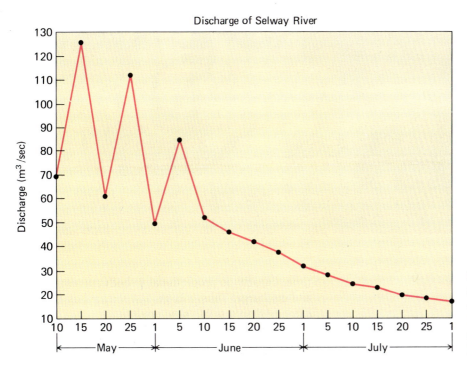

Discharge of Selway River

Figure 10.2 Discharge of the Selway River in northern Idaho during spring and early summer, 1988. The discharge drops to about 10 to 15 m³/sec during the late summer. *(U.S. Forest Service)*

and bedrock in the stream bed. Thus, a stream loaded with sediment can be thought of as flowing sandpaper. As the moving sediment abrades rock in the stream bed, the sediment itself becomes worn and rounded. For this reason, most stream-transported sand and gravel are well rounded. Abrasion can produce **potholes** and other rounded depressions in a stream bed. If cobbles are

caught in a small hollow in a rocky stream bed, the current swirls them around and around, enlarging the hollow (Fig. 10.4).

As explained in Chapter 9, many minerals dissolve in water, a process called **solution**. Streams dissolve rocks and minerals in the stream bed and then carry away the dissolved ions.

Figure 10.3   A stream weathers and erodes its channel by hydraulic action, abrasion, and solution.

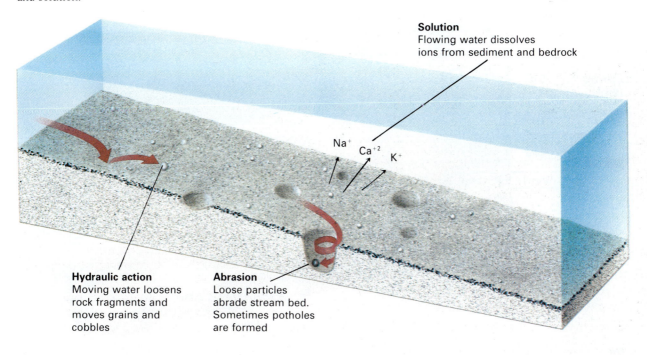

Solution
Flowing water dissolves ions from sediment and bedrock

Na^+ Ca^{+2} K^+

Hydraulic action
Moving water loosens rock fragments and moves grains and cobbles

Abrasion
Loose particles abrade stream bed. Sometimes potholes are formed

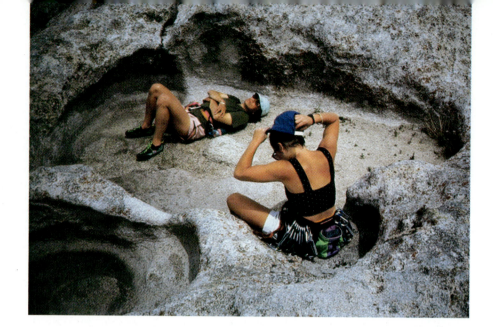

Figure 10.4 Circulating pebbles wear potholes in a stream bed.

Ability of a Stream to Erode and Transport Sediment

A rapidly flowing stream has more energy to pick up and transport sediment than a slow stream. The **competence** of a stream is a measure of the largest particle it can carry. A fast-flowing stream can carry cobbles and boulders in addition to small particles. A slow stream carries only silt and clay.

The **capacity** of a stream is the total amount of sediment it can carry past a point in a given amount of time. Capacity is proportional to both current velocity and discharge. Thus, a large, fast stream has a greater capacity than a small, slow one.

Because the ability of a stream to erode and carry sediment is proportional to its velocity and discharge, most erosion and sediment transport occur during the few days each year when the stream is in flood. Relatively little erosion and sediment transport occur during the remainder of the year. To see this effect for yourself, look at any stream during low water. It will most likely be clear, indicating little erosion or sediment transport.

Figure 10.5 A stream transports sediment in three ways. It carries dissolved ions in solution, silt and clay in suspension, and larger particles as bed load.

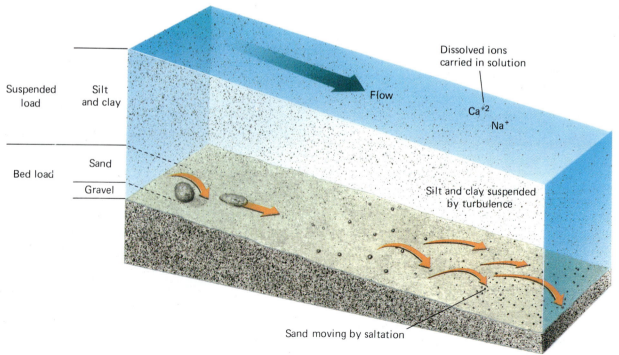

GEOLOGY AND THE ENVIRONMENT

Salinization

Most of the world's large rivers contain about 110 to 120 parts per million (ppm) dissolved ions (100 ppm is equal to one hundredth of one percent). Thus, 1 liter of average river water contains 1/10 of a gram of dissolved ions. Although most rivers do not taste salty, they carry tremendous amounts of material to the oceans in this manner. When river water is used to irrigate crops, much of it evaporates, and the dissolved ions precipitate as salts in the soil. Over a period of years, the salts accumulate in a process called **salinization**. If desert or semidesert soils are irrigated for long periods of time, they can become too salty for crops.

The great civilizations of Mesopotamia, where much of Western art, science, and literature originated, were built in a desert between the Tigris and Euphrates rivers of western Asia. The agriculture of this region depended on an extensive irrigation system, one of the great achievements of early civilization. Today much of this once fertile region is barren, eroded, and desolate.

Archaeologists dig up ancient irrigation canals, farming tools, and grinding stones in the desert. This ecological catastrophe was the outcome of salinization due to irrigation.

In the United States, salinization is severe in regions such as the San Joaquin Valley in California, where intensive irrigation is used to convert desert and semidesert to rich farmland. In some soils, enough calcium carbonate has precipitated from irrigation water to form a rock-hard layer called **caliche**. If the land is to be used, the caliche must be broken apart and removed at great expense with heavy machinery. Caliche development can be slowed by removing excess salty water with a drainage system, but it is expensive to install tiles and pipes to collect and dispose of salty water. One technique for preventing salinization may be to use less irrigation water by moistening the roots of individual plants with perforated pipes rather than by sprinkling an entire field. However, drip irrigation systems are expensive to maintain and therefore applicable only to cash-intensive vegetables.

Look at the same stream later, when it is flooding. It will probably be muddy and dark, indicating that much sediment is eroded from the bed and banks and carried by the current.

10.4   Sediment Transport

After a stream erodes grains of rock or soil, the flowing water may carry the sediment downstream. A stream transports sediment in three forms: dissolved load, suspended load, and bed load (Fig. 10.5).

Dissolved Load

A stream's **dissolved load** consists of ions dissolved in the water. Dissolved ions travel wherever the water goes. A stream's ability to carry dissolved ions depends on its discharge and its chemistry, not its velocity. Thus, dissolved ions exist even in lake and ocean waters; that is why the seas and some lakes are salty.

Suspended Load

Even a slow stream can carry fine sediment. If you shake up loamy soil in a jar of water and then let the jar sit,

the sand grains settle quickly. But the smaller silt and clay particles remain suspended in the water as **suspended load**, giving it a cloudy appearance. Suspension is a mode of transport in which water turbulence keeps fine particles mixed with the water and prevents them from settling to the bottom. Clay and silt are small

Suspended sediment colors the Colorado River in Grand Canyon.

Meanders, mid-channels bars, and point bars in the Bitterroot River, western Montana.

enough that even the slight turbulence of a slow stream keeps them in suspension. A rapidly flowing stream can carry sand in suspension.

Bed Load

Cobbles and boulders are too heavy to be carried in suspension by a stream. However, a fast current can roll or drag these large particles along the stream bed as **bed load**. During a flood, when stream velocity is highest, even large boulders can be transported in this way. Sand also moves as bed load. If the stream velocity is sufficient, however, sand grains bounce along in a series of short leaps or hops. This type of movement is called **saltation** (Fig. 10.5).

The world's muddiest rivers, the Yellow River in China and the Ganges River in India, each carry more than 1.5 billion tons of sediment to the ocean every year. The sediment load of the Mississippi River is about 450 million tons per year. Most rivers carry the largest proportion of their sediment in suspension, a smaller proportion in solution, and the smallest proportion as bed load.

10.5   Sediment Deposition

A rapidly flowing stream transports all sizes of particles, from clay to boulders. When the current slows down, it deposits them in the stream bed. However, most stream deposits are only temporary. Eventually the stream erodes the sediment again and transports it downslope toward the sea.

Many processes contribute to the stepwise movement of sediment. Consider a boulder perched on a rock ledge above a mountain stream. An avalanche or landslide may carry the boulder downslope into the stream channel. Spring floods may carry it downstream, or alternatively it may weather slowly into smaller cobbles and grains. The sediment weathered from the boulder may remain in place for years or centuries, or seasonal snowmelt may turn the quiet creek into a raging torrent that carries the sediment onward. The sediment is deposited once again when the flood waters recede or the stream flattens out on a shelf of resistant bedrock. It may rest for another year until the next spring, or another century until the next catastrophic flood. Or it may be trapped in a natural lake for millennia until the landscape evolves.

Stream deposits fall into three categories: (1) **Channel deposits** form in the stream channel itself; (2) **alluvial fan and delta deposits** form where the stream slows abruptly; and (3) **flood plain deposits** accumulate on the flood plain adjacent to a stream channel.

Channel Deposits

A **bar** is an elongate mound of sediment in a stream channel. Bars form when the stream is no longer able to carry the amount of sediment it has been carrying.

Figure 10.6   Current velocity varies in a winding stream. In a curved section the current is fastest on the outside of a bend. In a straight section the current is fastest in the center of the channel. The shaded zone in each cross section shows the area with fastest flow.

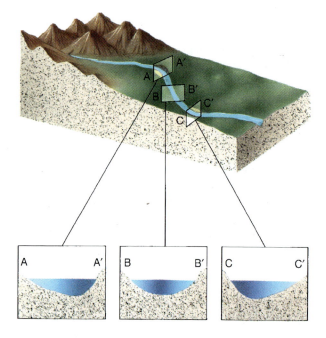

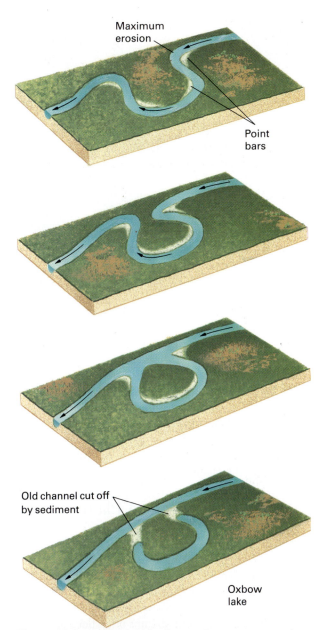

Figure 10.7   An oxbow lake forms when a stream erodes through a meander neck.

of the stream bends. The water on the outside of the curve moves faster than the water on the inside. The faster water on the outside of the bend has a greater erosive ability, and therefore a stream erodes its bank on the outside of a curve. At the same time, the slower water on the inside of the bend deposits sediment. Because such deposits are located on the inside point of a bend, they are called **point bars**.

As a result of this uneven erosion and deposition, the initial bend becomes accentuated. When the current bunces off one bend, it is deflected to the other side of the channel, where a second curve develops downstream from the first. Eventually, a series of bends called **meanders** forms (Fig. 10.7). As erosion continues, the curves may become so pronounced that the outside of one meander approaches that of another. Given enough time, the stream cuts across the gap and creates a new channel, abandoning a portion of its old bed. The old meander becomes isolated from the flowing water by sediment to become an **oxbow lake**.

A **braided stream** is one that flows in many shallow, interconnecting channels (Fig. 10.8). It develops where more sediment is supplied to a stream than it can carry. The excess sediment is deposited in the channel. The accumulation of sediment gradually forces the stream to overflow its banks and erode new channels. As a result, a braided stream flows simultaneously in several channels and shifts back and forth across its flood plain as old channels fill and new ones form.

Braided streams are common in both deserts and glacial environments because both provide abundant sediment. In a desert, large amounts of sediment are

Figure 10.8   The braided channel of the Chaba River in the Canadian Rockies.

Mid-channel bars are the sandy and gravelly deposits that form within stream channels. Bars of all types are transient features of a stream. They may form in one year and erode the next. If a river is used extensively for shipping, tugboat captains must be careful because bars form, disappear, and move. Rivers used for commercial navigation are charted frequently, and detailed maps of the bottoms are redrawn regularly.

Imagine a winding stream such as that in Figure 10.6. Now consider the cross section A-A′ through one eroded and transported during occasional heavy rains because little or no vegetation exists to prevent erosion. Glacial erosion supplies huge amounts of sediment to streams flowing from a glacier.

Alluvial Fans and Deltas

A stream slows down suddenly and deposits much of its sediment in two environments. If a stream flows from a steep mountain front onto a flat plain, its velocity decreases abruptly. It deposits particles of all sizes, from cobbles to fine silt, in a fan-shaped landform called an **alluvial fan**. Alluvial fans are common in many arid and semiarid mountainous regions (Fig. 10.9).

A stream also slows abruptly where it enters the still water of a lake or ocean. Here its sediment settles out to form a nearly flat plain called a **delta**. Most gravel has already been deposited farther upstream, so deltas consist mostly of sand, silt, and clay. Part of the delta lies above water level, and the remainder consists of a shallow underwater plain. Deltas are sometimes fan shaped, resembling the Greek letter Δ.

Deltas are shifting, changing environments. As more sediment accumulates, old channels fill and are abandoned while new channels develop, as in a braided stream. As a result, the stream splits into many channels called **distributaries** (Fig. 10.10). A large delta may spread out in this manner until it covers thousands of square kilometers.

Figure 10.11 shows changes in the Mississippi delta over the past 5000 to 6000 years. During this period the main channel has changed position seven times.

Figure 10.9 An alluvial fan in the Canadian Rockies.

A satellite photo of the Mississippi River delta. *(NASA)*

What will happen in the future? A river is dynamic; in its natural state it continues changing. If the Mississippi River were left alone, it would probably abandon the lower 500 kilometers of its present path and cut into the channel of the Atchafalaya River to the west. However, the mouth of the Mississippi is heavily industrialized, and it is impractical to allow the river to change its course, flooding towns in some areas and leaving shipping lanes and wharves high and dry in others. Therefore, engineers have built great systems of levees to stabilize the channels.

10.6   Floods and Flood Plains

Large rivers flood as a result of continuous, heavy rains, often augmented by spring snowmelt in the mountains. For example, in 1937, heavy rain fell almost continuously for 25 days in the Ohio River drainage. In response, the river rose 12 meters above its normal level.

Mountain and desert streams often flood as a result of rapid snowmelt or intense thunderstorms. In 1976 a series of summer storms saturated soil and bedrock near Rocky Mountain National Park, northwest of Denver. Then a large thunderhead dropped 19 centimeters of rain in 1 hour in the headwaters of Big Thompson Canyon. The Big Thompson River flooded, filling its narrow valley with a deadly, turbulent wall of water. Some people tried to escape by driving toward the mouth of the canyon, but traffic clogged the two-lane roadway and trapped motorists in their cars, where they drowned. A few residents tried to escape by scaling the steep canyon walls; some of them were caught by the rising waters. Within a few hours, 139 people had died and five were missing. By the next day, the flood was over.

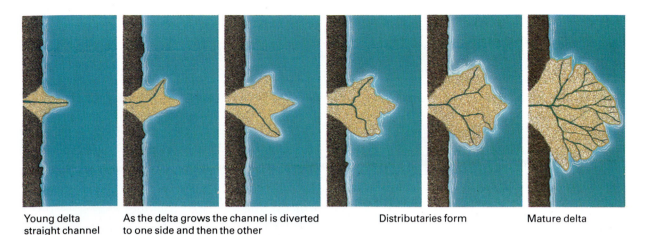

Young delta | As the delta grows the channel is diverted | Distributaries form | Mature delta
straight channel | to one side and then the other

Figure 10.10   A delta grows larger with time.

Many streams flood regularly, some every year. A given stream floods to a higher level in some years than in other years. A 100-year flood is the largest flood that occurs in a given stream an average of once every 100 years. A 10-year flood is the largest that occurs on the average of once every 10 years. In any stream, small floods are more common than large ones. For example, a stream may rise 7 meters above its banks during a 100-year flood, but only 2 meters during a 10-year flood. Thus, a 100-year flood is higher and larger, but less frequent, than a 10-year flood.

Does a 100-year flood mean that if a large flood occurs in 1994, one of equal size will occur again in 2094? No, not at all. The 100-year cycle is a measure of probability: The chance of a 100-year flood occurring in any given year is 1 in 100. Think of a roulette wheel with 100 slots. Imagine that one is marked F for a 100-year flood, and 99 are marked NF for no flood (or just a small flood). Now you spin the wheel. The chance on any spin is 1 out of 100 that you will land on F. If you land on F on one spin, you might land on F again on

Text continued on page 234

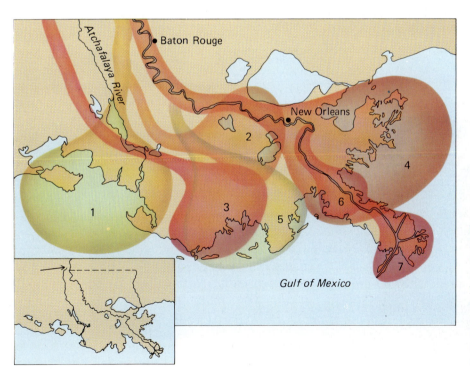

Figure 10.11   The Mississippi River has flowed into the sea by seven different channels during the past 6000 years. As a result, the modern delta is composed of seven smaller deltas formed at different times. The oldest delta is numbered 1, and the current delta is 7.

GEOLOGY AND THE ENVIRONMENT

The 1993 Mississippi River Floods and Flood Control

A flood plain is a desirable place for human habitation. Floods deposit rich, fertile soil on the flood plain, creating excellent agricultural land. In addition, rivers are important transportation corridors, and as a result cities and towns have grown on the banks of many rivers. But floods are part of the natural cycle of a stream. Therefore, a conflict arises. Poeple want to live and work along streams to take advantage of their benefits, but they don't want the inconvenience and economic loss associated with floods.

In addition to this conflict, many kinds of human development increase the severity of floods. For example, compare a forest and a parking lot during a heavy rain. In the forest, vegetation and soil absorb large amounts of moisture. In a parking lot, however, all of the rain runs off the pavement directly into nearby streams. As a result, floods intensify when a portion of a drainage basin is urbanized.

Figure 1 Residents of Hartsburg, Missouri, boating along Main Street flooded by the Missouri River. *(Stephen Levin)*

Figure 2 Floodwaters pouring from the channel of the Missouri River through a broken levee onto the flood plain, Boone County, Missouri. *(Stephen Levin)*

During the late spring and summer of 1993, unusually heavy rains soaked the upper Midwest, raising the
levels of the Mississippi River and its tributaries. In mid-
July, 2.5 centimeters of rain fell in 6 minutes in Papil-
lion, Nebraska. Thirteen centimeters fell in already saturated central Iowa in a day. In Fargo, North Dakota, the
Red River, fed by a day-long downpour, rose 1.2 meters
in 6 hours, flooding the town and backing up sewage

Figure 3   Artificial levees retain sediment, thus raising
the stream bed.

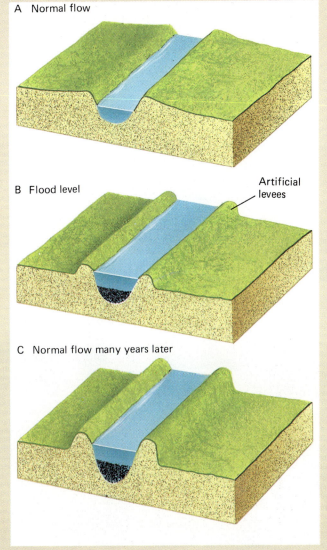

A   Normal flow

B   Flood level

Artificial
levees

C   Normal flow many years later

into homes and the Dakota Hospital. In St. Louis, Missouri, the Mississippi crested 14 meters above normal
and 1 meter above the highest previously recorded flood
level. At its peak, the flood inundated nearly 44,000
square kilometers in at least a dozen states.

The summer floods of 1993 wreaked most of their
damage (totaling close to $10 billion and 45 deaths) on
cities, towns, and farms built on the flood plain of the
Mississippi River and its tributaries (Fig. 1). Human-
made structures increased the damage in some places,
but in others artificial levees and other flood control
structures are credited with saving entire towns and preventing millions of dollars in damage.

An **artificial levee** is a wall built of earth, rocks,
or concrete along the banks of a stream to prevent rising water from spilling out of the stream channel onto
the flood plain. In the past 70 years, the U.S. Army
Corps of Engineers has spent billions of dollars building
flood control structures, including 11,000 kilometers of
levees along the banks of the Mississippi and its tributaries (Fig. 2).

As the Mississippi River crested in July 1993, flood
waters surged through low areas of Davenport, Iowa,
built on the flood plain. However, the business district
of nearby Rock Island, Illinois, on the same flood plain,
remained mostly dry. In 1971, Rock Island had built levees to protect low-lying areas of the town, whereas Davenport had not built levees. Hannibal, Missouri, had just
completed levee construction when the flood struck. The
$8 million project is credited with saving the Mark
Twain home and museum as well as much surrounding
land from flooding.

Unfortunately, two major problems plague flood
control projects that rely on artificial levees: Levees are
temporary solutions to flooding, and in some cases they
cause higher floods along nearby reaches of the river.

In the absence of levees, when a stream floods, it
deposits mud and sand on the flood plain. When artificial levees are built, the stream cannot overflow, so it deposits the sediment in its channel, raising the level of
the stream bed. After several floods the entire stream
may rise above its flood plain, contained only by the levees (Fig. 3). This configuration creates the potential for
a truly disastrous flood because if the levee is breached,
the entire stream then flows out of its channel and onto
the flood plain. As a result of levee building and channel
sedimentation, portions of the Yellow River in China
now lie 10 meters above the flood plain. Thus, levees

(continued)

may solve flooding problems in the short term, but in a longer time frame they may cause even larger and more destructive floods.

Engineers have tried to solve the problem of channel sedimentation by building artificial channels across meanders. When the stream is thus straightened, its velocity increases, and therefore it scours more sediment from its channel. This solution, however, has its drawbacks. When a stream bed is straightened, the total volume of the channel is reduced. Therefore, the channel cannot contain excess water, and flooding is likely to increase downstream.

Levees protecting one portion of a flood plain may also cause higher flood levels in unprotected portions of

the flood plain just upstream. A flooding river spreads horizontally over its flood plain, temporarily forming a wide path of flowing water. But when levees constrict the river into its narrow channel, they form a partial dam, causing the waters to rise to even higher flood levels upstream from the levees (Fig. 4).

Streams are complex systems. If they are dammed, if the channels are dredged and straightened, or if levees are built, the natural balance of the stream is disrupted. Sometimes the end result is that a "solution" to one problem may cause new and perhaps greater problems.

Figure 4   Levees constrict a river, raising flood levels upstream.

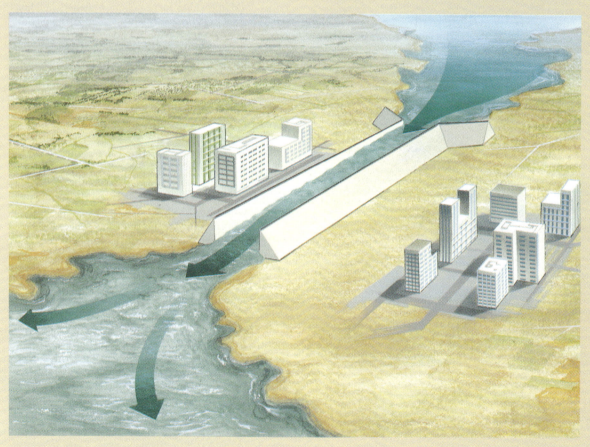

GEOLOGY AND THE ENVIRONMENT

The Impact of Dams

Since antiquity, people have understood that they can create a reservoir by building a dam across a stream. In a region where water is abundant in the spring but scarce in the summer, a reservoir provides flood control and stores water for irrigation and domestic uses. In addition, when a lake forms behind a dam, the potential energy of the water can be harnessed to generate electricity. Energy produced in this manner is called **hydroelectric energy**. Today, about 5 percent of the total world consumption of energy is supplied by hydroelectric generators. Although dams are beneficial to humans, they also create undesired side-effects.

Loss of Water

The reservoir formed by a dam provides more surface area for evaporation and more bottom area for seepage into bedrock than did the stream that preceded it. For example, about 270,000 cubic meters of water per year evaporate from Lake Powell, the lake behind the Glen Canyon Dam. Because salt does not evaporate, the remaining water becomes more saline, and its use for irrigation hastens the rate of salinization of the soil.

Silting

Sediment is deposited when a stream slows down as it enters a reservoir. Rates of sediment accumulation in reservoirs vary. In a few instances, horribly expensive mistakes have been made. The reservoir behind the Sanmenxia Dam on the Yellow River in China filled in just 4 years, rendering the dam and reservoir useless. The Tarbela Dam in Pakistan took 9 years to build, and the reservoir is expected to fill with sediment in 20 years. Other reservoirs have longer life expectancies. Lake Mead behind Hoover Dam in Arizona and Nevada lost 6 percent of its capacity in its first 35 years. At rates such as this, the lakes behind high dams can last hundreds of years, but that is not forever.

Erosion

Flowing water has a certain amount of energy that is used for transporting sediment and overcoming resistance to flow. Stream sediment settles out in a reservoir, and as a result the water that flows through or over a dam is clear. Downstream from the dam, the stream has more energy than it needs to flow because it is not carrying any sediment. Therefore, it flows faster and erodes the stream bed to increase its sediment load again. This increased erosion may cause both recreational and economic losses. The beautiful beaches on the Colorado River in Grand Canyon are eroding rapidly from water management practices of the Glen Canyon Dam. If erosion leads to a deeper main valley, erosion of the side valleys increases because their temporary base level is lowered. This increased erosion causes loss of agricultural soil.

Risk of Disaster

Flood plains are attractive for farming, industry, and commerce, especially when dams promise flood control, irrigation water, and electric power. But unusually heavy rainfall can fill a reservoir and overflow a dam. Furthermore, dams have been known to break. Under such circumstances, the population in the flood plain is vulnerable to disaster.

Recreational and Aesthetic Losses

Dams are often built across narrow canyons to minimize engineering and construction costs. But when the canyons are flooded, rare and sometimes unique ecosystems are destroyed. In addition, whitewater recreation areas are destroyed. The flooding of canyons, however, provides lakes that can be used for fishing and other types of water sports.

Ecological Disruptions

A river is an integral part of the ecosystem that it flows through, and when the river is altered, the ecosystem is altered as well. Many of the changes are detrimental. For example, before the Aswan Dam was built, the Nile River flooded every spring, depositing nutrient-rich sediments over the Nile flood plain. When the dam was built and the flood waters were controlled, this source of free fertilizer was eliminated. Although the energy produced by the dam is used to manufacture commercial fertilizers, many of the poorer farmers in the valley cannot afford to purchase what they once received free from the environment. In addition, increased erosion below the dam has destroyed many farms.

The aftermath of the Big Thompson Canyon flood.

A steep mountain stream cutting a V-shaped valley, Yellowstone National Park, Wyoming.

the very next try, or you might spin 200 times before it happens again. It is a game of chance.

As a stream rises to flood stage, both its discharge and velocity increase. Therefore, it erodes its bed and banks. When a stream floods, water flows out of its channel and onto its flood plain. The current slows abruptly where water leaves the channel to flow onto the flood plain. Because it slows down so quickly, the current deposits sand and silt on the banks of the stream. This sediment forms ridges called **natural levees** at the margins of the channel (Fig. 10.12).

A meandering stream in Maine occupies only a small part of its flood plain.

Farther out on the flood plain, the floodwater carries mostly clay and silt. This finer sediment accumulates on flood plains to form fertile soils.

10.7   Downcutting and Base Level

A stream creates a valley by eroding bedrock and carrying off the sediment. The flowing water erodes both downward into the stream bed and laterally against the banks. Downward erosion is called **downcutting**. How deeply can a stream erode its channel? How deep can a valley become? The **base level** of a stream is the deepest level to which it can erode its bed. For most streams,[1] the lowest possible level of downcutting is sea level, which is called the **ultimate base level**. This concept is straightforward. Water can only flow downhill. If a stream were to cut its way down to sea level, it would stop flowing and hence would no longer erode its bed or banks.

In addition to ultimate base level, a stream may have a number of **local**, or **temporary**, **base levels**. For example, where a stream flows into a lake, its current stops and erosion ceases because the stream has reached a temporary base level. A layer of rock that resists ero-

---

[1] We say "for most streams" because a few empty into valleys that lie below sea level.

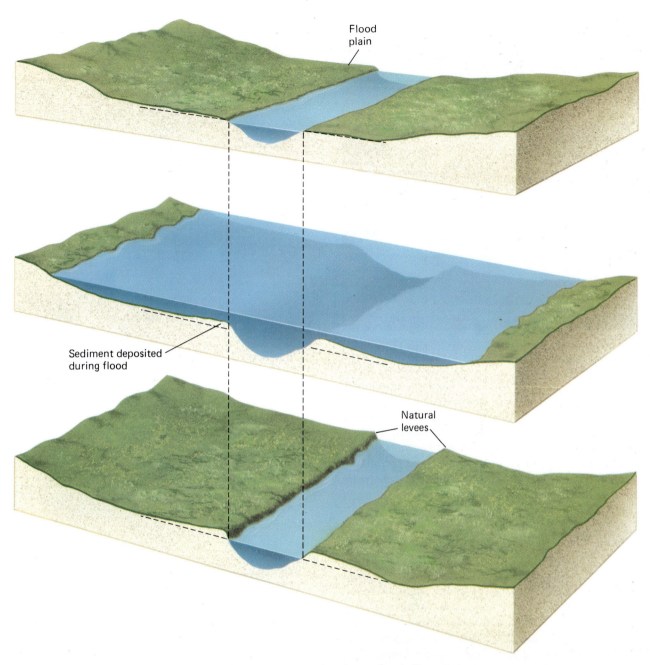

Figure 10.12   A natural levee forms as a flooding stream deposits sand and silt on its banks. Silt and clay accumulate on the flood plain.

sion may also establish a temporary base level because it resists downcutting and flattens the stream gradient. Thus, the stream slows down and erosion decreases. The top of a waterfall is an example of a temporary base level established by resistant rock. Beneath the waterfall, less resistant rock is eroded more easily. Niagara Falls is formed by a resistant layer of dolomite over softer shale. As the shale erodes, the harder dolomite cap is undermined and collapses. Thus both the shale and dolomite erode and the cliff face retreats upstream. Niagara Falls has retreated 11 kilometers since its formation.

If a stream has numerous temporary base levels, it erodes its bed in the steep places where flow is rapid and deposits sediment in the low-gradient stretches where it flows more slowly (Fig. 10.13B). Over time, erosion and

Mass wasting of the steep banks of the Kicking Horse River, British Columbia.

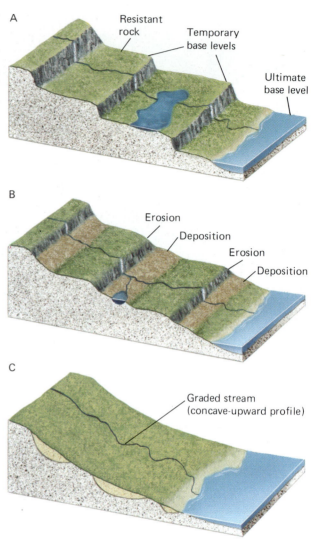

Figure 10.13 An ungraded stream (A) has many temporary base levels. With time, the stream smooths out the irregularities (B) to develop a graded profile (C).

deposition smooth out the irregularities in the gradient. The resultant **graded stream** has a smooth, concave profile (Fig. 10.13C). Once a stream becomes graded, there is no net erosion or deposition and the stream profile remains unchanged. An idealized graded stream such as this does not actually exist in nature, but many streams come close.

A steep mountain stream usually downcuts rapidly compared with the rate of lateral erosion. As a result, it cuts a relatively straight channel with a steep-sided, V-shaped valley. The steep valley walls erode by mass wasting, and the stream removes the fallen sediment, enlarging the valley. In contrast, a low-gradient stream is closer to base level, and its downcutting is therefore slower. It mainly erodes laterally, wandering back and forth across the flood plain. Such a stream forms a wide valley with a flat bottom and broad flood plain. Meanders and oxbow lakes are common.

10.8   Drainage Basins

Only a dozen or so major rivers flow into the sea along the coastlines of the United States (Fig. 10.14). Each is fed by a number of tributaries, and the tributaries are in turn fed by smaller streams. Adjacent river systems are separated by mountain ranges or other raised areas. These high places are called **drainage divides**. The region that is ultimately drained by a single river is called a **drainage basin**. It can be large or small and is bounded by drainage divides. For example, the Rocky Mountains separate the Colorado and Columbia drainage basins to the west from the Mississippi and Rio Grande basins to the east.

In the most common type of drainage basin, the pattern of tributaries resembles veins in a leaf. This arrangement is called a **dendritic drainage pattern** (Fig. 10.15A). Tributaries join the main stream at V-shaped junctions in dendritic drainage systems. You can determine the direction of flow because the V usually

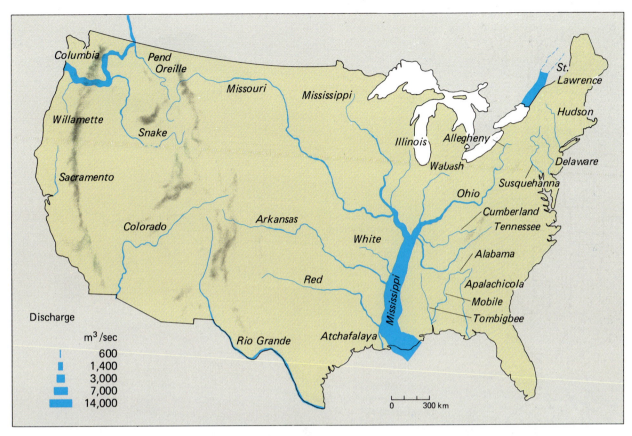

Figure 10.14   Major drainage basins of the United States.

points downstream. The next time you fly in an airplane, look out the window and try to determine the directions of stream flow in the drainage systems below. Dendritic patterns occur in regions where bedrock is relatively uniform, and streams therefore take the shortest route downslope.

Bedrock is not uniform in some regions, however. For example, Figure 10.16 shows a layer of sandstone lying over granite. The rocks have been faulted and tilted. As a result, outcropping bands of easily eroded sandstone alternate with bands of resistant granite. Streams follow the softer sandstone rather than cutting across the hard granite and form straight, parallel channels. A drainage pattern characterized by parallel channels intersected at right angles by tributaries is called a **trellis** or **rectangular pattern** (Fig. 10.15B). A trellis pattern can also develop where bedrock is criss-crossed by a series of parallel faults or joints that intersect at right angles. A **radial** drainage pattern forms where a number of streams flow from a mountaintop (Fig. 10.15C).

A satellite view of a dendritic drainage, South Yemen. *(NASA)*

Map view

Perspective view

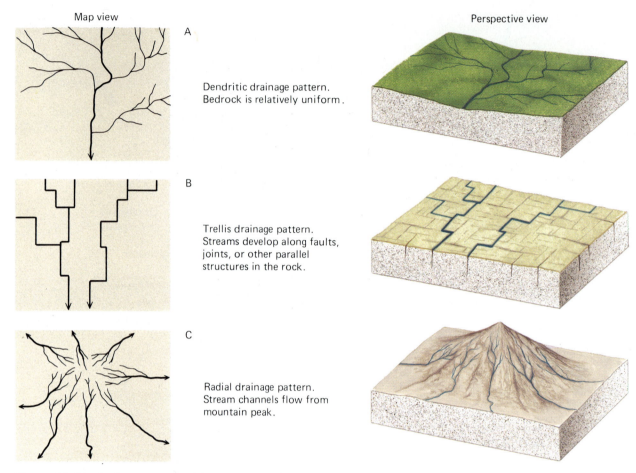

A

Dendritic drainage pattern.
Bedrock is relatively uniform.

B

Trellis drainage pattern.
Streams develop along faults,
joints, or other parallel
structures in the rock.

C

Radial drainage pattern.
Stream channels flow from
mountain peak.

Figure 10.15   Drainage patterns. (A) Dendritic. (B) Trellis or rectangular. (C) Radial.

10.9   The Evolution of Valleys and Rejuvenation

According to a model popular in the first half of this century, streams erode mountain ranges and create landforms in a particular sequence. At first, the streams

Figure 10.16   Parallel faults and contrasting rock types may cause a trellis or rectangular drainage pattern to develop.

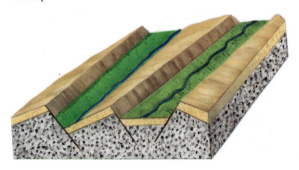

cut steep, V-shaped valleys. Over time, mass wasting and lateral erosion widen the valleys into broad flood plains. Eventually, the entire landscape flattens, forming a large, featureless plain called a **peneplain** (Fig. 10.17).

This model seems to make good sense based on what we have learned of stream erosion, but it is invalid because it tells only half the story. Streams do continuously erode the landscape, flattening mountains and widening flood plains. But at the same time, tectonic activity may uplift the land and interrupt the simple, idealized sequence.

Consider the Himalayas. Today, streams, glaciers, and mass wasting are eroding deep valleys. The huge amounts of sediment carried from the Himalayas to the sea by the Indus and Ganges Rivers are evidence of this erosion. However, at the same time, the lithospheric plate carrying India continues to ram into Asia, raising the peaks. In addition, the mountains rise isostatically as they erode. The net result of all these processes is that the mountain peaks are rising more rapidly than erosion is wearing them down.

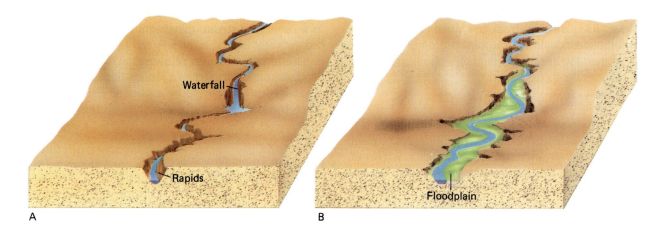

A

B

Waterfall

Rapids

Floodplain

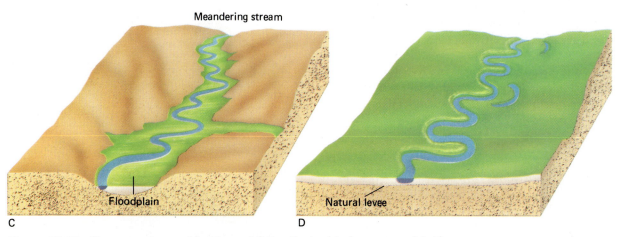

Meandering stream

C

D

Floodplain

Natural levee

Figure 10.17   Stream erosion would widen and flatten the land to form a peneplain if no tectonic changes occurred.

Stream terraces above the
Fraser River, British Columbia.

A stream is **rejuvenated** when any change increases its erosive ability. Rejuvenation can result from tectonic forces uplifting the land, a drop in base level, or an increase in regional rainfall.

Incised Meanders and Stream Terraces

Imagine a stream meandering across a broad flood plain. Now suppose that tectonic forces uplift the entire plain, increasing the stream's gradient and velocity. As a result, the stream begins to cut downward into its bed. Eventually the stream cuts downward below its original flood plain, and if the stream keeps its meandering course as it erodes into its bed, **incised meanders** develop (Fig. 10.18). The incised stream abandons its old flood plain well above its new channel, and the abandoned flood plain is called a **stream terrace** (Fig. 10.19). Both incised meanders and stream terraces reflect rejuvenation.

In the preceding example, rejuvenation resulted from tectonic uplift, but it could equally well have resulted from a climate change. For example, if the climate

becomes wetter, the stream's discharge may increase. As a result, the stream may begin to downcut into its bed, forming incised meanders and terraces. Lowering of base level also causes rejuvenation. For example, a flood plain may form when a stream flows on a temporary base level supported by a layer of resistant rock. When the stream finally erodes through this resistant

Figure 10.19 Formation of terraces. (A) A stream has formed a broad flood plain. (B) Tectonic uplift or climatic change causes the stream to erode its bed. As the stream cuts downward, the old flood plain becomes a terrace above the new stream level. (C) A new flood plain forms at the lower level.

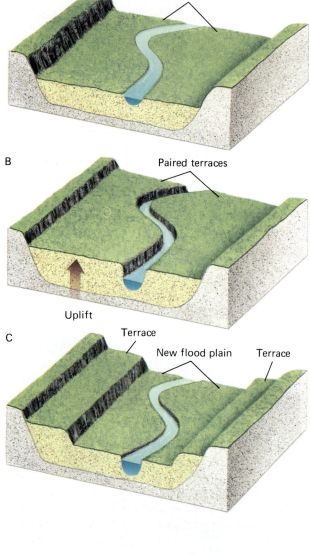

Figure 10.18 If a valley with a meandering stream is raised by tectonic uplift, the meanders may become incised into the rising flood plain.

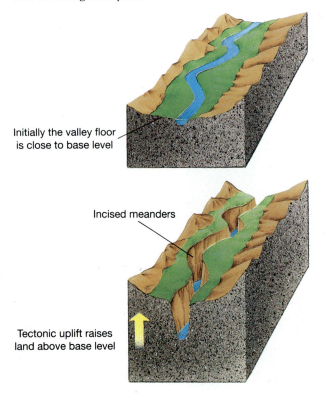

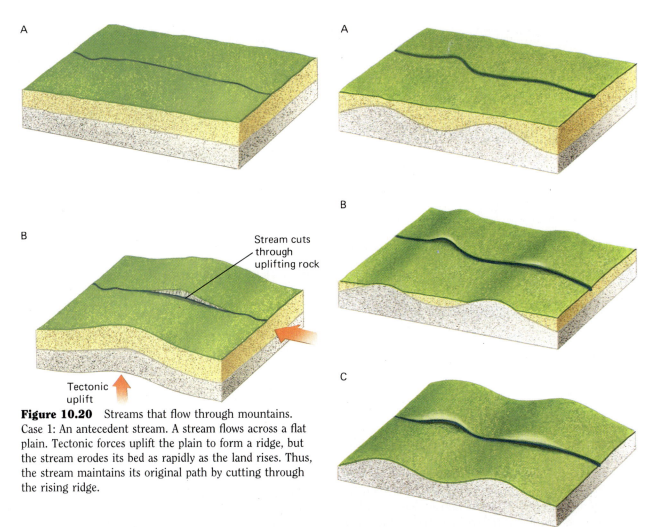

Figure 10.20   Streams that flow through mountains. Case 1: An antecedent stream. A stream flows across a flat plain. Tectonic forces uplift the plain to form a ridge, but the stream erodes its bed as rapidly as the land rises. Thus, the stream maintains its original path by cutting through the rising ridge.

layer, its gradient may increase and it may start downcutting rapidly.

Streams that Flow Through Mountain Ranges

In some regions, a stream may flow right through a mountain range, ridge, or high plateau. Why doesn't it flow around the mountains rather than cutting directly through them? Again, tectonic activity may be responsible. Imagine a stream flowing across a plain. If tectonic forces build a mountain range or high plateau in the middle of the plain, but the uplift is slow enough that the stream can cut through the rising bedrock, the stream retains its original course through the newly formed mountains or plateau (Fig. 10.20). In this case the stream is said to be **antecedent** because it existed before the mountains or plateau rose. The Colorado River cut its way through more than 1600 meters of sedimentary rock, forming the Grand Canyon, as the

Figure 10.21   Streams that flow through mountains. Case 2: A superposed stream. A stream flows over young sedimentary rock that has buried an ancient mountain range. As the stream erodes downward, it cuts into the old mountains, maintaining its course.

Colorado Plateau rose. Most of this erosion may have occurred within the past 2 to 3 million years, a remarkably rapid rate.

A stream also may cut through an uplift in resistant bedrock by the process of **superposition**. If a ridge of hard bedrock is covered with younger sedimentary rocks, a stream downcutting into the sedimentary rocks is unaffected by the buried feature. Eventually, however, the stream may downcut to the ridge. At this point the channel may be well established, and the stream may cut through the hard ridge rather than be diverted to one side or the other (Fig. 10.21).

A

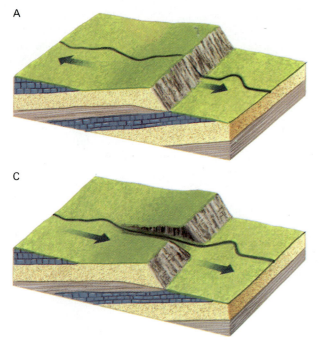

C

B

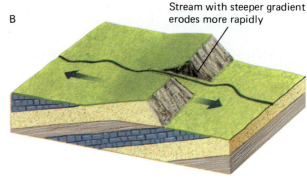

Stream with steeper gradient erodes more rapidly

Figure 10.22   Stream piracy. The stream on the steeper side of the ridge erodes downward more rapidly than the stream on the opposite, gentler slope. Eventually, the steeper stream cuts through the ridge to intersect the higher, gentler stream. The higher stream then reverses direction to flow into the lower one.

Stream Piracy

Consider what may happen when two streams flow in opposite directions from a drainage divide. One of the steams may cut downward faster than the other. This can occur if one side of the range is steeper than the other, if one side receives more rainfall than the other, or if rock on one side is softer than rock on the other. In any of these situations, the stream that is eroding more rapidly may cut its way backward into the divide. This process is called **headward erosion**. If headward erosion continues, the more deeply eroded stream may cut through the drainage divide. The higher stream then reverses direction and flows into the more deeply cut stream that has just penetrated the divide. Hence, one stream eventually captures the drainage of the other. This sequence of events is called **stream piracy** (Fig. 10.22).

SUMMARY

Only about 0.65 percent of Earth's water is fresh water in streams, lakes, and underground reservoirs. The rest is salty seawater and glacial ice. **Evaporation**, **precipitation**, and **runoff** continuously recycle water among land, sea, and the atmosphere in the **hydrologic cycle**.

A **stream** is any body of water flowing in a channel. The floor of the channel is called the **bed**, and the sides of the channel are the **banks**. The velocity of a stream is determined by its **gradient**, **discharge**, and channel shape.

A stream **erodes** rock and soil, **transports** sediment downslope, and **deposits** it. The relative rates of these three processes are determined by the velocity and discharge of the stream. A stream erodes its bed and banks by **hydraulic action**, **abrasion**, and **solution** and carries sediment as **dissolved load**, **suspended load**, and **bed load**.

A stream deposits sediment in three categories. (1) **Channel deposits**, called **bars**, form within the river channel itself. Curves in the channel are accentuated by uneven erosion and deposition, leading to the formation of **point bars**, **meanders**, and **oxbow lakes**. **Braided streams** develop when more sediment is supplied to a stream than it can transport. (2) A **delta** forms at the mouth of a stream, and an **alluvial fan** forms where a fast mountain stream slows as it enters a larger valley. (3) **Flood plain deposits** occur on the valley floor surrounding the stream. Floods create **natural levees** along the stream bank.

Ultimate base level is the lowest elevation to which a stream can erode its bed. This is usually sea level. A lake or resistant rock can form a **local**, or **temporary**, **base level**. A **graded stream** has a smooth, concave profile.

Downcutting, lateral erosion, and mass wasting combine to form a stream valley. Mountain streams downcut rapidly and form V-shaped valleys, whereas lower-gradient streams form wider valleys by lateral erosion and mass wasting.

A drainage basin can have a **dendritic**, **trellis**, or **radial** pattern, depending on the bedrock geology.

Streams erode mountains and deposit sediment in lowlands, to flatten rugged topography. Tectonic activity usually interrupts this process, **rejuvenating** a landscape. **Incised meanders** and **stream terraces** often form where a stream downcuts into its bed as tectonic uplift occurs. In some instances, tectonic uplift can cause a stream to cut through a mountain range or resistant bedrock. **Headward erosion** can lead to **stream piracy**.

KEY TERMS

Hydrologic cycle *221*
Evaporation *221*
Transpiration *221*
Precipitation *221*
Runoff *221*
Stream *221*
River *221*
Tributary *221*
Channel *221*
Bed *221*
Bank *221*
Flood plain *221*
Gradient *221*

Discharge *221*
Hydraulic action *222*
Abrasion *222*
Pothole *223*
Competence *224*
Capacity *224*
Dissolved load *225*
Suspended load *225*
Bed load *226*
Saltation *226*
Bar *226*
Point bar *227*
Meander *227*

Oxbow lake *227*
Braided stream *227*
Alluvial fan *228*
Delta *228*
Distributary *228*
Natural levee *234*
Downcutting *234*
Base level *234*
Ultimate base level *234*
Local base level *234*
Graded stream *236*
Drainage divide *236*
Drainage basin *236*

Dendritic drainage
  pattern *236*
Trellis pattern *237*
Radial pattern *237*
Peneplain *238*
Rejuvenation *240*
Incised meander *240*
Stream terrace *240*
Antecedent *241*
Superposition *241*
Headward erosion *242*
Stream piracy *242*

REVIEW QUESTIONS

1. In which physical state (solid, liquid, or vapor) does most of the Earth's free water exist? Which physical state accounts for the least?

2. Assume that there are two streams with equal gradients but different velocities. Why can the rapidly flowing stream carry sand and cobbles, whereas the slow-moving stream can carry only silt and clay?

3. Why does the Mississippi River carry more sediment than a small mountain stream?

4. Distinguish among the three types of stream erosion: hydraulic action, solution, and abrasion.

5. List and explain three ways in which sediment can be transported by a stream. Which type of transport is independent of stream velocity? Explain.

6. Why do braided streams often develop in glacial and desert environments?

7. How are alluvial fans and deltas similar? How do they differ?

8. Give two examples of natural features that create temporary base levels. Why are they temporary?

9. Draw a profile of a graded stream and an ungraded stream.

10. Explain how a stream forms and shapes a valley.

11. In what type of terrain would you be likely to find a V-shaped valley? Where would you be likely to find a meandering stream with a broad flood plain?

12. How can a stream become rejuvenated? Give an example of a landform created by a rejuvenated stream.

13. Explain the difference between an antecedent and a superposed stream.

DISCUSSION QUESTIONS

1. Describe the ways in which (a) a rise and (b) a fall in the average global temperature could affect the Earth's hydrologic cycle.

2. A stream is 50 meters wide at a certain point. A bridge is built across the stream, and the abutments extend into the channel, narrowing it to 40 meters. Discuss the changes that might occur as a result of this constriction.

3. Obtain a copy of a hydrograph of a local river from the county extension agent, the Coast Guard, or the Forest Service. (A hydrograph is a continuous record of discharge over time.) How does it compare with the hydrograph of the Selway River shown in Figure 10.2?

The Selway is a small stream in the mountains of Idaho. Explain any differences between the two hydrographs.

4. Defend the statement that most stream erosion occurs when the stream is flooding.

5. Describe the difference between a natural levee and an artificial one. How is each formed? How effective is each in reducing floods?

6. What type of drainage pattern would you expect in the following geologic environments? (a) Platform sedimentary rocks. (b) A batholith fractured by numerous faults. (c) A flat plain with a composite volcano in the center. Explain your answers.

CHAPTER

11

Ground Water

*Old Faithful Geyser in Yellowstone National Park
spouts hot ground water about once every hour.*
(Superstock)

If you drill a hole into the ground in most places, after a few days the bottom fills with water. The water appears even if no rain falls and no streams flow nearby. The water that seeps into the hole is part of the vast reservoir of subterranean **ground water** that saturates the Earth's crust in a zone between a few meters and a few kilometers below the surface.

Globally, 30 times more water is stored as ground water than in all streams and lakes combined. This resource is exploited by digging wells and pumping the water to the surface. Ground water provides drinking water for more than half of the population of North America and is a major source of water for irrigation and industry. Before the development of advanced drilling and pumping technologies, human impact on ground water was minimal. Today, however, deep wells and high-speed pumps can extract ground water more rapidly than natural processes replace it. This situation is common in the central and western United States. In addition, surface contaminants seep into ground water. Such pollution is often hard to detect and remedy.

11.1   Characteristics of Ground Water

Porosity, Permeability, and Ground Water

Ground water fills small cracks and voids in soil and bedrock. **Porosity** is the proportional volume of rock or soil that consists of open spaces. Igneous and metamorphic rocks, such as granite, gneiss, and schist, have low porosities unless they are fractured. However, many sedimentary rocks can be quite porous. Sandstone can have 5 to 20 percent porosity (Fig. 11.1). The porosity of loose sediment and soil can be even greater, reaching upper limits of 50 percent in sand and even 90 percent in clay.

Although porosity tells us how much water rock or soil can retain, it tells us nothing about the rate at which water can flow through the pores. **Permeability** is a measure of the speed at which water can travel through porous soil or bedrock.

Soil and loose sediment such as sand and gravel are both porous and permeable. Thus, they can hold a lot of water, and it flows easily through them. Sandstone and conglomerate can also have high permeabilities. Although clay and shale are porous and can hold a large amount of water, the pores in these fine-grained materials are so small that water flows very slowly through them. Thus, clay and shale have low permeabilities.

The Water Table

When rain falls on dry soil, it usually soaks into the ground. Water does not descend into the crust indefinitely, however. Below a depth of a few kilometers, rock is at high enough temperature and pressure to be plastic. The weight of overlying rock closes the pores, and rock

at this depth is both nonporous and impermeable. Water accumulates on this impermeable barrier, filling all the pores in the rock and soil above it. This completely wet layer of soil and bedrock above the barrier is called the **zone of saturation**. The **water table** is the top of the zone of saturation (Fig. 11.2). Above the water table lies the unsaturated zone, or **zone of aeration**. In this layer, the rock or soil may be moist but not saturated. The **soil moisture belt** is the soil layer. It holds more water than the unsaturated zone below and supplies much of the water needed by plants.

If you dig into the unsaturated zone, the hole does not fill with water. However, if you dig below the water table into the zone of saturation, you have dug a well, and the water level in a well is at the level of the water table.

An **aquifer** is any body of rock or soil that can yield economically significant quantities of water. An aquifer must be both porous and saturated; that is, it must contain water. It must also be permeable so that water flows into a well to replenish water that is pumped out. High-quality aquifers commonly occur in sand and gravel, sandstone, limestone, and highly fractured igneous and metamorphic rock. Think of an aquifer as a sponge that water seeps through, not as an underground pool or stream.

Movement of Ground Water

In a few regions, underground rivers flow through caverns, but they are the exception, not the rule. Nearly all ground water seeps slowly through interconnected pores in bedrock and soil. Typically, ground water flows at about 4 centimeters per day (about 15 meters per year), although flow rates may be much faster or slower. The rate depends on the permeability of soil or rock or

Well-sorted sediment

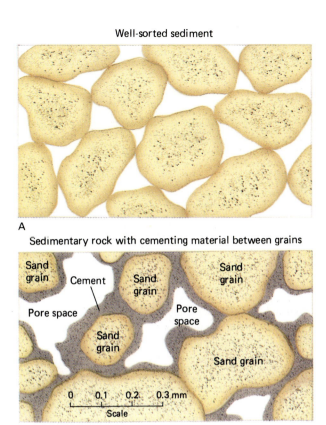

A

Poorly sorted sediment

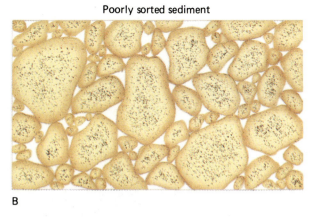

B

Sedimentary rock with cementing material between grains

Sand grain
Cement
Pore space
Sand grain
Sand grain
Pore space
Sand grain
Sand grain

0   0.1   0.2   0.3 mm
Scale

C

Figure 11.1 Different materials have different amounts of open pore space between grains. (A) Well-sorted sediment consists of equal-size grains and has a high porosity, about 30 percent in this sample. (B) In poorly sorted sediment, small grains fill the spaces among the large ones, and porosity is lower. In this drawing it is about 15 percent. (C) Cement partly fills pore space in sedimentary rock, lowering the porosity.

Figure 11.2 The water table is the top of the zone of saturation near the Earth's surface. It intersects the land surface at lakes and streams and is the level of standing water in a well.

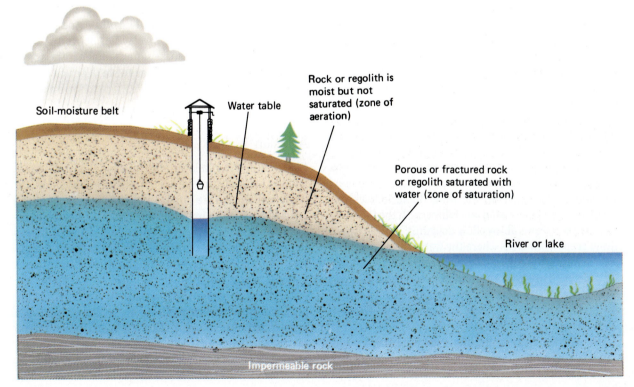

Soil-moisture belt

Water table

Rock or regolith is moist but not saturated (zone of aeration)

Porous or fractured rock or regolith saturated with water (zone of saturation)

River or lake

Impermeable rock

A desert stream lies above the water table. Courthouse Wash, Arches National Park, Utah. (See page 288 for a photo of the same stream during the dry season.)

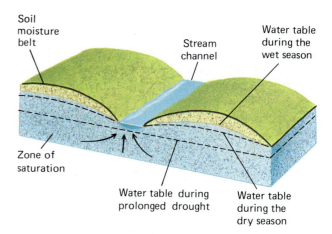

Figure 11.3 The water table follows topography, rising beneath hills and sinking beneath valleys. It also rises during a rainy season and falls during dry times.

on the nature of fractures in bedrock. Water can flow rapidly through large interconnected fractures.

The water table rises and falls with the seasons. During a wet season, such as spring in a temperate climate, rain seeps into the ground and the water table rises. During a dry season, the water table falls. That is why it is common for the water level in wells to rise and fall through the year and in wet years and years of drought.

In a temperate climate, ground water seeps into streams. Figure 11.3 shows a stream and water table in hilly country. In general, the water table follows the contours of the land, rising and falling with the topography. Just as streams always flow downhill, ground water always seeps from areas where the water table is highest toward areas where it is lowest. Streams follow gulleys and valleys. Therefore, ground water normally flows toward streams and seeps into stream beds, which is why streams continue to flow even when rain has not fallen for weeks or months.

In contrast, in an arid climate the water table commonly lies below stream beds and water seeps downward from a stream to the water table. Most of the time, desert stream channels are dry. When streams do flow, the water often originates in wetter environments in nearby mountains, although sometimes desert storms fill the channels for short periods of time. Thus a desert stream feeds the ground-water reservoir, but in temperate climates, ground water feeds the stream.

Springs

A **spring** is a place where ground water flows or seeps onto the surface. A spring forms wherever the water table intersects the land (Fig. 11.4). In some places, a layer of impermeable rock or clay lies above the main water table, creating a local saturated zone, the top of which is called a **perched water table**. A perched water table can also intersect the land surface to form a spring (Fig. 11.4B).

Artesian Wells

Figure 11.5 shows a layer of permeable sandstone sandwiched between two layers of impermeable shale. On the left, the strata are tilted. An inclined aquifer bounded top and bottom by impermeable rock is called an **artesian aquifer**. Water in the lower part of the aquifer is under pressure from the weight of water above. Therefore, if a well is drilled into the sandstone, water rises without being pumped. If pressure is sufficient, the water spurts out onto the surface. A well of this kind is called an **artesian well**. As an analogy, think of a water-filled hose. If one end is held high, the lower end sealed, and you puncture the hose below the high point, water squirts out.

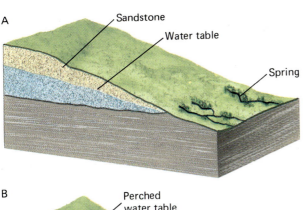

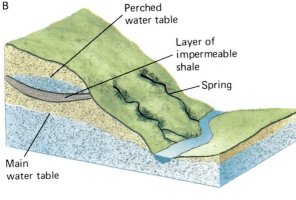

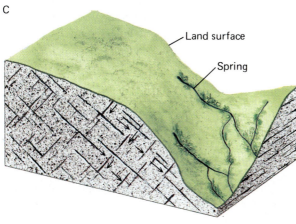

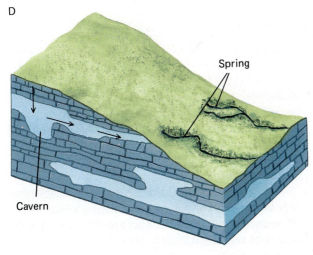

11.2   Use of Ground Water

Ground water is a valuable resource for a variety of reasons:

1. It is abundant. As mentioned at the beginning of this chapter, 30 times more fresh water exists underground than in surface reservoirs.

2. Because ground water moves so slowly, it is stored in the saturated zone and remains available during dry periods.

3. In some regions, ground water flows from wet environments to arid ones, making water available in deserts.

Most ground water used by humans is extracted from wells. Most successful wells are dug or drilled in valleys, close to the level of streams or lakes. If pumps extract ground water continuously and rapidly, the water table can fall significantly. The first disturbance occurs near the well. If water is withdrawn faster than it flows into the well from the aquifer, a **cone of depression** forms (Fig. 11.6). When the pump is turned off, ground water flows back toward the well in a matter of days or weeks if the aquifer has good permeability, and the cone of depression disappears. On the other hand, if water is continuously removed more rapidly than it flows to the well through the aquifer, then the cone of depression grows and the water table drops over a broad area (Fig. 11.6C). In an extreme case, the water table can drop over an entire aquifer.

Subsidence

Excessive removal of ground water can cause **subsidence**, the sinking or settling of the Earth's surface. When water is withdrawn from an aquifer, rock or soil particles may shift closer to each other, filling some of the space left by the lost water. As a result, the volume of the aquifer decreases and the overlying ground subsides. Removal of oil from petroleum reservoirs has the same effect.

Figure 11.4   Springs form where the water table intersects the land surface. This situation can occur where (A) the land surface intersects a contact between permeable and impermeable rock layers; (B) a layer of impermeable rock or clay lies "perched" above the main water table; (C) water flows from fractures in otherwise impermeable bedrock; and (D) water flows from caverns onto the surface.

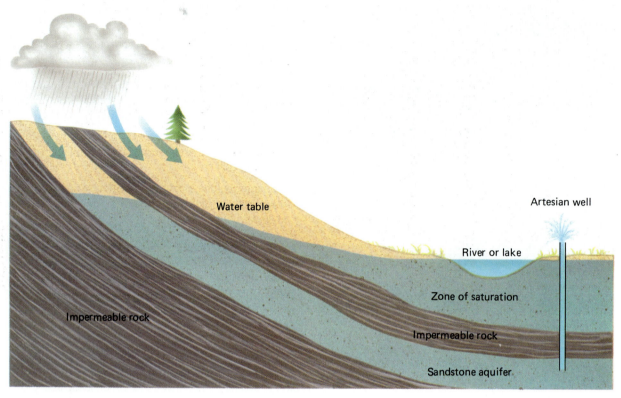

B

Figure 11.5 (A) An artesian aquifer forms where a tilted layer of permeable rock, such as sandstone, lies sandwiched between layers of impermeable rock, such as shale. Water rises in an artesian well without being pumped. (B) A hose with a hole shows why an artesian well flows spontaneously.

Subsidence rates can reach 5 to 10 centimeters per year, depending on the rate of ground-water removal and the nature of the aquifer. Some areas in the San Joaquin Valley of California have sunk several meters. The problem is particularly severe when it occurs beneath a city. For example, Mexico City is built on an old marsh. Over the years, as the weight of buildings and roadways has increased and much of the ground water has been removed, parts of the city have settled 1 to 2 meters. Many millions of dollars have been spent to maintain this complex city on its unstable base. Similar problems are occurring in cities in the United States, including Phoenix, Arizona, and Houston, Texas.

Unfortunately, subsidence is not a reversible process. When rock and soil contract, their porosity is usually reduced permanently so that ground-water reserves cannot be recharged completely even if water becomes abundant again.

An artesian well spouts into the air. *(D. R. Crandell, USGS)*

Western agriculture depends heavily on irrigation.

11.3 Depletion of Ground Water: Irrigation on the High Plains

Many parts of the high plains in western and mid-western North America receive scant rainfall. Early settlers planted crops and suffered in times of drought. In the 1930s, two sequences of events combined to change agriculture in this region. One was a great drought that destroyed crops and exposed the soil to erosion. Dry winds blew across the land, eroding the parched soil and carrying it for hundreds and even thousands of kilometers. Thousands of families lost their farms, and the region was dubbed the Dust Bowl. The second event was the arrival of inexpensive technology. Electricity came to rural regions, and relatively cheap pumps and irrigation systems were developed. With the specter of drought fresh in people's memories and the tools to avert future calamities available, the age of modern irrigation began.

Figure 11.7 shows a map and cross section of the Ogallala aquifer in the central high plains. The aquifer extends almost 900 kilometers from the Rocky Mountains eastward across the prairie, and from Texas into South Dakota. It consists of porous sandstone and

Figure 11.6 (A) A well is drilled into an aquifer. (B) A pump draws water from the well faster than it can flow into the well through the aquifer, and a cone of depression forms. (C) If the pump continues to extract water more rapidly than it flows to the well, the water table falls.

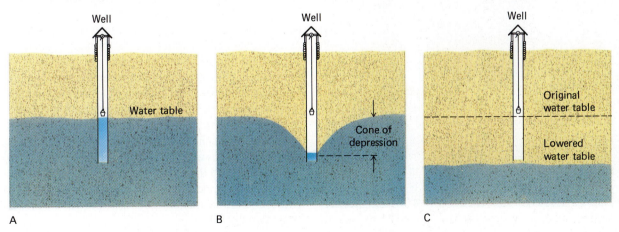

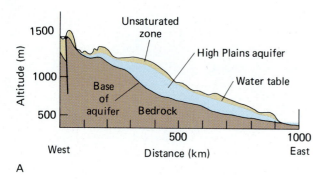

Figure 11.7 The Ogallala aquifer supplies water to much of the high plains. (A) A cross-sectional view of the aquifer shows that much of its water originates in the Rocky Mountains and flows slowly as ground water beneath the high plains. (B) A map showing the extent of the aquifer.

New Jersey suffers some of the worst ground-water pollution in North America as a result of heavy industry. An oil refinery, New Jersey.

conglomerate within 350 meters of the surface. The aquifer averages about 65 meters thick and contains a vast amount of water. Between 1930 and 1980, about 170,000 wells were drilled into the Ogallala aquifer, and extensive irrigation systems were installed throughout Kansas, Nebraska, Oklahoma, the Texas panhandle, and parts of neighboring states.

Today, farmers and hydrologists are concerned that the Ogallala aquifer is being depleted. They estimate that half of the water has already been removed from parts of the aquifer, and pumping rates are increasing. As explained earlier, water moves through aquifers at an average rate of about 15 meters per year. Most of the water in the Ogallala aquifer accumulated when the last Pleistocene ice sheet melted. But because the high plains receive little rain, the aquifer is mostly recharged by rain and snowmelt in the Rocky Mountains, hundreds of kilometers to the west. At a flow rate of 15 meters per year, ground water takes 60,000 years to travel from the mountains to the eastern edge of the aquifer. Under such conditions, deep ground water is, for all practical purposes, nonrenewable. Just as coal and petroleum are called fossil fuels, deep ground water is sometimes called fossil water. The removal of deep ground water is therefore analogous to mining.

If the present pattern of water use continues, wells in the Ogallala aquifer beneath the high plains will dry up early in the next century. In the mid-1980s, about 5 million hectares of land were irrigated from this aquifer. (This is an area about the size of the states of Massachusetts, Vermont, and Connecticut combined.) About 40 percent of the cattle in the United States are fed with corn and sorghum raised in this region, and large quantities of grain and cotton are grown there as well. If the aquifer is depleted and another source of irrigation water is not found, productivity in the central high plains is expected to decline by 80 percent. Farmers will go bankrupt, and food prices throughout the nation will rise.

Mill wastes in Anaconda, Montana, have been polluting ground water for decades.

11.4 Ground-Water Pollution

Hooker Chemical Company's main plant was located in Niagara Falls, New York. Early in the 1940s, Hooker purchased an abandoned canal called Love Canal. During the following years, the company disposed of approximately 19,000 tons of chemical wastes from manufacturing processes by loading them into 55-gallon steel drums and storing them in the canal. In 1953, the company covered one of the dump sites with dirt, sold the land to the Board of Education of Niagara Falls for $1, and a school was built on the site.

The steel drums eventually began to leak, and the chemicals seeped into ground water. In the spring of 1977, heavy rains raised the water table and turned the area around Love Canal into a muddy swamp. But it was no ordinary swamp; poisonous compounds from the leaking drums mingled with the water and soil. The toxic fluid soaked the playground, seeped into basements, and saturated gardens and lawns. Children who attended the school and adults who lived nearby developed epilepsy, liver malfunctions, miscarriages, skin sores, rectal bleeding, severe headaches, and birth defects.

The awareness generated by Love Canal and the fear that other such episodes could occur were so great that in December 1979, the U.S. Congress passed the Comprehensive Environmental Response, Compensation, and Liability Act, commonly known as Superfund. This law provides an emergency fund to clean up hazardous waste sites. The Environmental Protection Agency (EPA) has estimated that between 1950 and 1975, 5.5 billion metric tons of hazardous waste were spilled onto the land or buried in dump sites throughout the United States. A total of 20,766 sites were identified in the initial tally. By 1989 the General Accounting Office estimated that there may have been as many as 400,000 hazardous waste sites. The massive scale of the problem has overwhelmed the EPA and other government agencies. By 1991, 1245 sites had been targeted for cleanup, but only 63 of these projects had been completed.

Sources of Ground-Water Pollution

Sewage from septic tanks and cesspools may contaminate ground water. These sources contribute mainly bacterial and viral contamination, although chemical contamination may also occur if people flush paints, pesticide residues, or other household chemicals down the toilet.

Improperly sealed landfills, dumps, and hazardous waste disposal sites contribute chemical pollutants. Some industrial chemicals are poisonous even in trace amounts; others are suspected of causing cancer or birth defects. Landfills and dumps can also release bacterial and viral contamination.

Wastewater from factories, farms, and municipal sewage treatment plants is often stored in basins, pits, ponds, or lagoons. About 100,000 to 150,000 such sites exist in the United States. Many are unlined, and the soils beneath them are permeable. Therefore, the polluted water seeps downward to contaminate ground water.

Mine and mill wastes commonly contain high concentrations of sulfur and metals. Some of these metals

Agricultural chemicals are a major source of ground-water pollution.

and compounds leach into the ground water, and many, such as arsenic, cobalt, lead, and other heavy metals, are toxic.

Approximately 2 million underground storage tanks exist in the United States. Half hold gasoline at service stations, and the remainder contain chemicals. Many of the tanks are untreated steel, and, like the drums at Love Canal, they eventually rust and leak.

Liquid chemical wastes are sometimes injected into deep wells below an aquifer to dispose of them without contaminating the aquifer. If the wastes are corrosive or if the pressure is high, the injection pipe may leak. The wastes may then enter ground water.

Agricultural practices also create troubling ground-water pollution. Pesticides and herbicides are sometimes heavily applied (or misapplied) to deal with infestations of insects or weeds, and they often percolate down to the ground water. Fertilizers also leach into aquifers, contributing inorganic contaminants such as nitrates and phosphates.

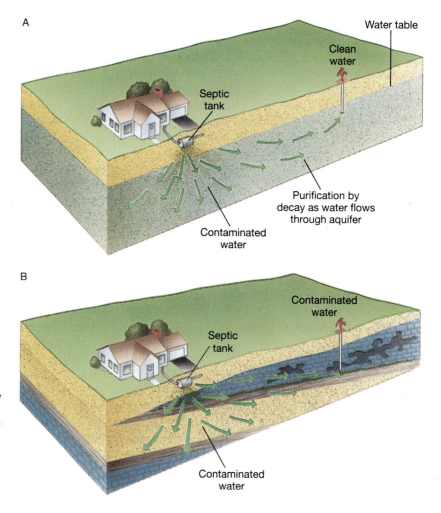

Figure 11.8 (A) Water moves slowly through permeable sandstone, giving natural processes enough time to purify it before it reaches a well. (B) But water flows through cavernous limestone too rapidly to be purified and pollutes a nearby well.

GEOLOGY AND THE ENVIRONMENT

The Lipari Landfill: Pitman, New Jersey

The Lipari Landfill was opened in 1958, before modern environmental laws were enacted. It was situated in an abandoned sand and gravel pit near Pitman, New Jersey. Three million gallons of chemical wastes were dumped legally into the site before it was closed and covered in 1971. But engineers only sealed the surface, and wastes seeped into the ground water. By the mid-1970s, nearby Alcyon Lake turned orange and purple. Local residents said that fumes rising from the soil brought tears to their eyes and sometimes even a bittersweet taste on their tongues. In 1983, 4 years after the Superfund Legislation was enacted, the EPA called the Lipari the worst hazardous waste site in the United States. New Jersey Senator Frank Lautenberg proclaimed that "Lipari is a symbol" and promised rapid action.

EPA engineers reasoned that although the sand and gravel in the dump site is porous, a clay layer that lies beneath the porous material is not. Therefore they could isolate the landfill by digging a trench around its perimeter to the clay and then filling the trench with concrete. However, the year after the wall was finished, approximately 2600 gallons of polluted water leaked outside the perimeter. Perhaps the wall had cracked, the bond between the wall and the clay was permeable, or the clay layer itself was fractured.

Next, the EPA sunk numerous wells into the landfill and into the ground outside the wall. They pumped clean water into some of the wells and removed polluted water from others. The engineers then pumped the polluted water to a purification plant. This process, begun in 1989, is expected to continue for 7 to 10 years.

Local citizens are not happy. They contend that the plan is analogous to pouring water into a leaky bathtub and that the process will increase ground-water contamination. In addition, some of the wastes in the landfill were sealed in metal drums. What will happen if the drums rust through after the flushing project is completed? These critics argue that the only solution is to dig up the landfill, remove the polluted material, and recover the sealed drums. In addition, they recommended dredging Alcyon Lake. But where do you put thousands of cubic yards of contaminated sediment from the landfill and the lake? In another landfill?

As explained in the text, the Lipari Landfill is only one of tens or hundreds of thousands of hazardous waste sites in the United States. There are no easy solutions. Even partial clean-up will cost tens of millions of dollars. People cannot agree on how much they are willing to pay and how much pollution they are willing to live with.

Solving the Problem of Ground-Water Pollution

Natural processes decompose some ground-water pollutants. In a septic system, domestic sewage is flushed into a large tank. The solids settle to the bottom of the tank and are pumped out every 5 years or so. The wastewater flows out of the tank and into a series of underground trenches filled with gravel. The wastewater trickles through the gravel into soil, where the organic matter decomposes into harmless byproducts.

For purification to be effective, wastewater must move slowly through well-aerated soil or rock. Sandy soils and sandstone have the proper permeability for effective purification. If movement is slow and oxygen is available, decay organisms can feed on the sewage, decomposing it to harmless byproducts before the polluted water travels very far (Fig. 11.8A). On the other hand, imagine installing a septic tank in coarse gravel, fractured granite, or cavernous limestone. In any of these situations, the polluted water would flow rapidly through the large openings. Not enough time would elapse for the decay organisms to decompose the sewage, so wells or streams quite a distance from the source could become polluted (Fig. 11.8B).

Whereas sewage decomposes naturally, many chemicals such as pesticides, paints, and solvents are nonbiodegradable; that is, microorganisms do not decompose them. Once a nonbiodegradable compound contaminates ground water, it may persist for a long time. Recall that ground water flows slowly, so pollutants are not readily diluted or washed away. In addition, ground water does not have access to as much air as surface water does. Therefore, oxidation of persistent chemicals is less effective and less complete underground than it is in surface water.

Because natural processes are not effective in removing chemical contaminants from ground water,

Figure 11.9   Stalactites and stalagmites in a limestone cavern.   *(Hubbard Scientific Co.)*

clean-up can be expensive (see "Geology and the Environment: The Lipari Landfill"). To avoid creating a new problem for future generations, it is important to prevent contamination in existing sites. Thus, underground storage containers should be lined and leak detectors installed. Landfills and waste treatment ponds should be built over a layer of impermeable clay or plastic so the contaminants are isolated from aquifers. Factories should be monitored closely to ensure that hazardous wastes are disposed of responsibly.

Approximately 50 percent of the U.S. population depends on ground water as its primary source of drinking water. The EPA has established maximum tolerance levels for a variety of chemicals that may be present in water drawn from wells. According to the Association of Ground Water Scientists and Engineers, approximately 10 million Americans drink water that does not meet EPA standards.

11.5   Caverns and Karst Topography

Just as streams erode valleys and form flood plains, ground water also creates landforms. Recall from Chapter 9 that rainwater reacts with atmospheric carbon dioxide to produce a slightly acidic solution that is capable of dissolving the calcite that forms limestone. This

reaction is reversible: The dissolved ions can precipitate to form calcite again.

Caverns

A **cavern** forms when slightly acidic water seeps into a crack in limestone bedrock, dissolving the rock and enlarging the crack. Mammoth Cave in Kentucky and Carlsbad Caverns in New Mexico are two famous caverns formed in this way. The largest chamber in Carlsbad Caverns is taller than the U.S. Capitol building and is broad enough to accommodate 14 football fields. Most caverns form at or below the water table. If the water table drops, the chambers are opened to air.

If you entered a cavern, you would notice features obviously formed by deposition, not dissolution. Long, pointed structures hang from the ceilings and rise from the floors. Collectively, all mineral deposits formed in caves by the action of water are called **speleothems**.

When a solution of water, dissolved calcite, and carbon dioxide percolates through the ground, it is under the pressure of water in the cracks above it. If a drop of this solution seeps into the ceiling of a cavern, the pressure decreases suddenly because the drop comes in contact with the air. The high humidity of the cave prevents the water from evaporating rapidly, but the lowered pressure allows some of the carbon dioxide to escape as a gas. When the carbon dioxide escapes, the drop becomes less acidic. This decrease in acidity causes some of the dissolved calcite to precipitate as the water drips from the ceiling.

Figure 11.10   Columns in Carlsbad Caverns.   *(Tom Till)*

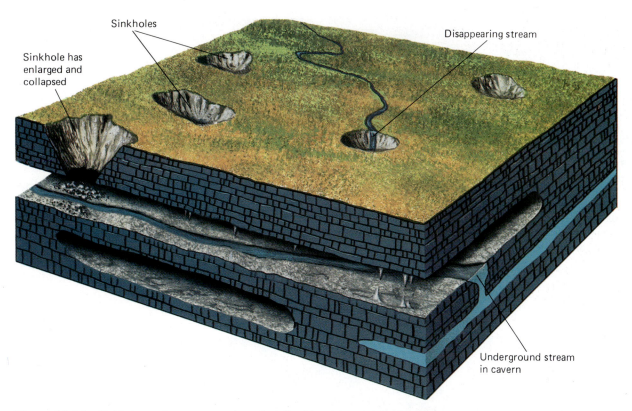

Figure 11.11   Sinkholes and caverns are characteristic of karst topography. Streams commonly disappear into sinkholes and flow through the caverns to emerge elsewhere.

Over time, beautiful and intricate speleothems form. **Stalactites** hang icicle-like from the ceiling of a cavern (Fig. 11.9). Only a portion of the dissolved calcite precipitates as the drop seeps out of the ceiling. When the drop falls to the floor, it spatters, and the impact releases more carbon dioxide. The acidity of the drop decreases further, and another minute amount of calcite precipitates. Thus, **stalagmites** build from the floor upward to complement the stalactites. Because stalagmites are formed by splashing water, they tend to be broader than stalactites. As the two features continue to grow, they may eventually join to form a **column** (Fig. 11.10).

Sinkholes

If the roof of a cavern collapses, a **sinkhole** forms on the Earth's surface. A sinkhole can also form as limestone dissolves from the surface downward (Fig. 11.11). A well-documented sinkhole formed in May 1981 in Winter Park, Florida. During the initial collapse, a three-bedroom house, half a swimming pool, and six Porsches in a dealer's lot all fell into the underground cavern. Within a few days, the sinkhole was about 200 meters wide and 50 meters deep and had devoured additional buildings and roads (Fig. 11.12).

Although sinkhole formation is a natural event, the problem can be intensified by human activities. The Winter Park sinkhole formed when the water table dropped, removing support for the ceiling of the cavern. The water table fell as a result of a severe drought augmented by excessive removal of ground water by humans.

Karst Topography

Karst topography forms in broad regions underlain by limestone and other readily soluble rocks. Caverns and sinkholes are common features of karst topography. Surface streams often pour into sinkholes and disappear into caverns. In the area around Mammoth Caves in Kentucky, streams are given names such as Sinking Creek, an indication of their fate.

The word *karst* is derived from a region in Croatia where this type of topography is well developed. Karst landscapes are found in Alabama, Tennessee, Kentucky, southern Indiana, and northern and central Florida. Extensive karst landscapes also occur in China.

Figure 11.12 This sinkhole in Winter Park, Florida, collapsed suddenly in May 1981, swallowing several houses and a Porsche agency. *(Wide World Photos/Associated Press)*

11.6 Hot Springs and Geysers

At numerous locations throughout the world, hot water naturally flows to the surface to produce **hot springs**. Ground water can be heated in three different ways:

1. The Earth's temperature increases by about 25°C per kilometer in the upper portion of the crust. Therefore, if ground water descends through cracks to depths of 2 to 3 kilometers, it is heated by 50° to 75°. The hot water then rises because it is less dense than cold water. The springs at Warm Springs, Georgia, are probably heated in this manner. However, it is unusual for fissures to descend so deep into the Earth, and this type of hot spring is uncommon.

An aerial view of karst topography, Winter Park, Florida. The lakes are sinkholes.

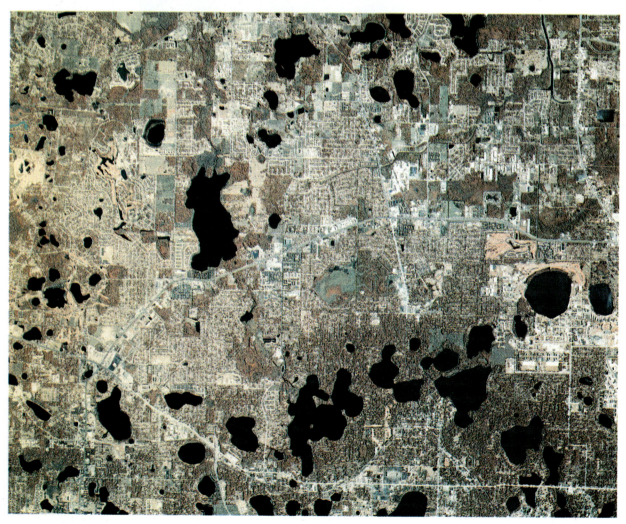

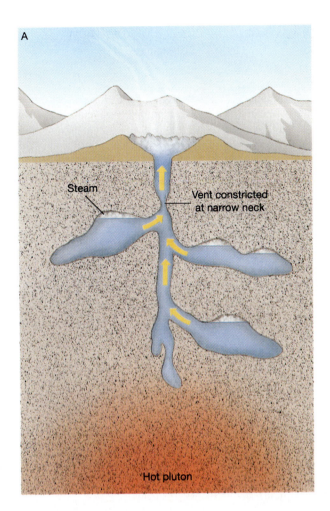
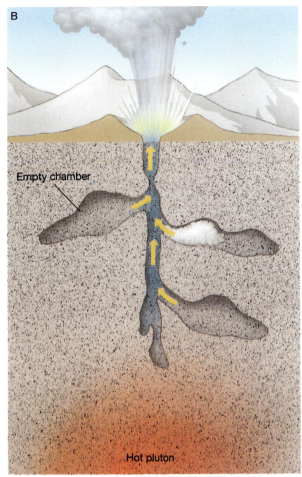

Figure 11.13 For a geyser to erupt, (A) ground water seeps into underground chambers and is heated by hot igneous rock. Foam constricts the geyser's neck, trapping steam and raising pressure. (B) When the pressure exceeds the strength of the blockage, the constriction blows out. Then the hot ground water flashes into vapor and the geyser erupts.

2. In regions of recent volcanism, magma or hot igneous rock may remain near the surface and can heat ground water at relatively shallow depths. Hot springs heated in this way are common throughout western North and South America, because these regions have been tectonically active in the recent past and remain active today. Hot springs are reminders of recent volcanic activity. Shallow magma heats the hot springs and geysers of Yellowstone National Park.

3. Many hot springs have the odor of rotten eggs from small amounts of hydrogen sulfide (H_2S) dissolved in the hot water. These springs are heated by chemical reactions. Sulfide minerals, such as pyrite (FeS_2), react chemically with water. The reactions release heat. Hydrogen sulfide, produced by the reaction, rises with the heated ground water and gives it the strong odor.

Most hot springs bubble gently to the surface or flow from cracks in bedrock. However, a **geyser** violently erupts hot water and steam. Geysers generally form in open cracks and channels in hot underground rock. In the first step, ground water seeps into the empty channels (Fig. 11.13). The hot rock then heats the water. Gradually, bubbles of water vapor form and start to rise, just as they do in a heated teakettle. If part of the channel is constricted, the bubbles may accumulate in the narrow neck and form a temporary barrier, increasing pressure. When the pressure rises, some of the bubbles are forced upward past the constriction. On the surface,

Steamboat Geyser, Yellowstone National Park. *(National Park Service)*

Figure 11.14 A schematic view of the Fenton Hill, New Mexico, dry geothermal energy plant.

this movement appears as short bursts of steam and spurts of water. Below ground, when the bubbles rise, the pressure at the constriction is reduced suddenly. The water, which was already hot, flashes into vapor, which explodes, blowing steam and hot water skyward. In Iceland, tourists used to make geysers blow by throwing soap into them to encourage the formation of bubbles.

Steam from a geothermal field in California generates electricity in this Pacific Gas and Electric plant. *(PG&E)*

The most famous geyser in North America is Old Faithful in Yellowstone Park, which erupts on the average of once every 65 minutes. Old Faithful is not as regular as people like to believe; the frequency of eruptions varies from about 30 to 95 minutes apart.

11.7 Geothermal Energy

Hot ground water can be used to drive turbines and generate electricity, or it can be used directly to heat homes and other buildings. Energy extracted from the Earth's heat is called **geothermal energy**. The United States is the largest producer of geothermal electricity in the world, with a capacity of 2200 megawatts. This amount is equivalent to the power output of two large nuclear reactors and provides about 1 million people with all of their electrical needs. However, this amount of energy is minuscule compared with the potential of geothermal energy. The United States Geological Survey estimates that the upper 4 kilometers of rock beneath the United States contain heat energy equivalent to 3000 trillion barrels of oil. This is enough energy to run the country for the next 200,000 years at current energy consumption rates.

The major problem with current methods of extracting energy from the Earth is that they work only where deep ground water is naturally heated. Only a limited number of such "wet" sites exist, where abundant ground water and hot rock or magma occur at the same place. However, many "dry" sites exist where rising magma has heated rocks close to the surface, but little ground water is available. Wells can be drilled in these regions, and water pumped into the wells can be circulated through the hot rock and then extracted.

Scientists and engineers are developing methods for extracting energy from dry Earth heat at a pilot project at Fenton Hill, New Mexico. They drilled two separate wells side by side (Fig. 11.14). Water is pumped down one, called the injection well, to a depth of about 4 kilometers. The pump forces the water into hot, fractured granite at the bottom of the well, and then into the extraction well, where it returns to the surface. The scientists have succeeded in pumping water into the well at 20°C and extracting it 12 hours later at 190°C.

As long as petroleum prices are low, there is little incentive to develop expensive geothermal projects. The economic picture is expected to change within the next few decades as petroleum resources become limited and prices rise.

SUMMARY

Of the rain that falls on land, most evaporates and runs off in streams, but some seeps into soil and bedrock to become **ground water**. Ground water saturates the upper few kilometers of soil and bedrock to a level called the **water table**. **Porosity** is the proportion of rock or soil that consists of open space. **Permeability** reflects the speed with which fluid can move through pores in soil or bedrock. An **aquifer** is a body of rock that can yield economically significant quantities of water. An aquifer is both porous and permeable.

Most ground water moves slowly, about 4 centimeters per day. **Springs** occur where the water table intersects the surface of the land. Dipping layers of permeable and impermeable rock can produce an **artesian aquifer**. If a well is drilled into an artesian aquifer, the water rises and may even spout up above the ground surface.

If water is withdrawn from a well faster than it can be replaced by the aquifer, a **cone of depression** forms. If rapid withdrawal continues, the water table falls.

Polluted ground water may be purified slowly by natural processes.

Caverns form where ground water dissolves limestone. **Speleothems** form by precipitation of calcite from ground water in caves. A **sinkhole** forms when the roof of a limestone cavern collapses. **Karst topography**, with numerous caves, sinkholes, and subterranean streams, is characteristic of limestone regions.

Hot springs develop when hot ground water rises to the surface. Ground water can be heated by (1) the geothermal gradient, (2) shallow magma or a cooling pluton, or (3) chemical reactions between ground water and sulfide minerals. Hot springs have been tapped to produce **geothermal energy**, and "dry" sites are now being explored.

KEY TERMS

REVIEW QUESTIONS

1. (a) Draw a cross section of soil and shallow bedrock, showing the zone of saturation, water table, and zone of aeration. (b) Explain each of the preceding terms.

2. What is an aquifer, and how does water reach it?

3. Explain why bedrock or regolith must be both porous and permeable to be an aquifer.

4. Compare the movement of ground water in an aquifer with that of water in a stream.

5. How does an artesian aquifer differ from a normal one? Why does water from an artesian well rise without being pumped?

6. Describe problems that can arise from excessive use of ground water.

7. Why is depletion of ground water likely to have longer-lasting effects than depletion of a surface reservoir?

8. Explain how land can subside when ground water is depleted. If the removal of ground water is stopped, will the land rise to its original level again? Defend your answer.

9. Explain how water from a septic tank is purified by natural processes. Explain why purification is more effective in some types of bedrock or regolith than in others.

10. What is karst topography? How can it be recognized? How does it form?

11. Explain three mechanisms for the formation of hot springs.

DISCUSSION QUESTIONS

1. Which of the wells in Figure 11.15 would you expect to contain water? Which would you expect to be polluted? Explain your answers.

2. Imagine that you live on a hill 25 meters above a nearby stream. You drill a well 40 meters deep and do not reach water. Explain.

3. Explain why many desert streams dry up in the summer, whereas streams in humid environments do not go dry even during periods of drought.

4. The ancient civilization of Mesopotamia fell after its agricultural system collapsed because of problems resulting from failure of irrigation techniques. In his book *Cadillac Desert,* Marc Reisner describes the transformation of the western United States from its natural semidesert condition to its modern agricultural wealth. He maintains that this system, like others that preceded it, cannot be sustained indefinitely. Argue for or against Reisner's hypothesis.

5. Would you expect to find a cavern in granite? Would you expect to discover a cavern in shale? Defend your answers.

6. Why can't stalactites or stalagmites form when a cavern is filled with water?

7. Contrast problems of ground-water pollution in a region of karst topography with those of a sandstone-aquifer region.

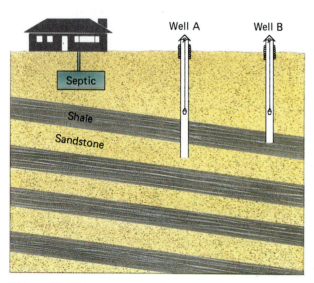

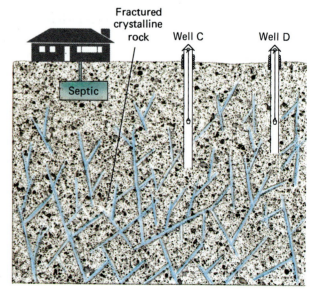

Figure 11.15

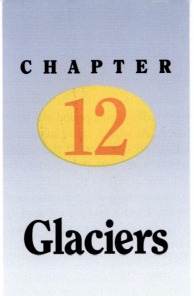
Glaciers

*Alpine glaciers have eroded mountain landscapes
and deposited moraines on Baffin Island, Canada.*

One hundred thousand years ago, the world was free of ice except for the polar ice caps of Antarctica and Greenland. Then, in a period of a few thousand years, the Earth's climate cooled by a few degrees. As winter snow failed to melt in summer, the polar ice caps grew and spread into lower latitudes. At the same time, glaciers formed near the summits of high mountains, even near the equator. They flowed down mountain valleys and out onto nearby lowlands. When the glaciers reached their maximum size, 18,000 years ago, they covered one third of the Earth's continental surface.

About 15,000 years ago, the Earth's climate warmed again by a few degrees. As a result, the glaciers melted rapidly. A flowing glacier picks up huge amounts of rock and soil. When it melts, it deposits that sediment. The smooth, low, rounded hills of parts of upper New York State, Wisconsin, and Minnesota are piles of gravel deposited by great ice sheets as they melted. The barren sediment deposited by the melting ice was unprotected by plants. Strong winds blowing from the dying glaciers eroded the sand and dust and then deposited it again. As a result, great fields of sand dunes and windblown silt formed in the wake of the glaciers.

Geologists often speak of events that occurred millions or billions of years ago. In contrast, the most recent continental glaciers reached their maximum sizes only 18,000 years ago. To put this time in perspective, Cro-Magnon artists were painting on cave walls in central Europe 35,000 years ago, and agricultural societies developed about 10,000 years ago.

12.1 Types of Glaciers

In most temperate regions, winter snow melts entirely during the summer. However, in certain cold, wet environments, winter snow does not melt completely but accumulates year after year. It becomes denser as new snow buries it to ever greater depths. If snow survives through one summer, it converts to loosely packed, rounded ice grains called **firn** (Fig. 12.1). Mountaineers like firn because the sharp points of their ice axes and crampons sink into it easily and hold firmly. If firn is buried deeper in the snowpack and is subjected to thawing and freezing, it converts to **glacial ice**, a mass of interlocking ice crystals. The crystals pack together so tightly that water cannot percolate through the ice.

A **glacier** is a massive, long-lasting accumulation of compacted snow and ice. Glaciers form only on land, wherever the amount of snow that falls in winter exceeds the amount that melts in summer. Glaciers in mountain regions flow slowly downhill. Glaciers on level land flow outward under their own weight, just as cold honey poured onto a tabletop spreads outward slowly.

Alpine Glaciers

Mountains are generally colder and wetter than adjacent lowlands. Near the mountain summits, winter snowfall is deep and summers are short and cool. These conditions create **alpine glaciers** (Fig. 12.2). Alpine glaciers exist on every continent—in the Arctic and Antarctic, in temperate regions, and in the tropics. Glaciers cover the summits of Mount Kenya in Africa and Mount Cayambe in South America, even though both peaks are near the equator. Some alpine glaciers in high latitudes flow great distances from the peaks onto adjacent low-

Figure 12.1 Newly fallen snow changes through several stages into glacial ice.

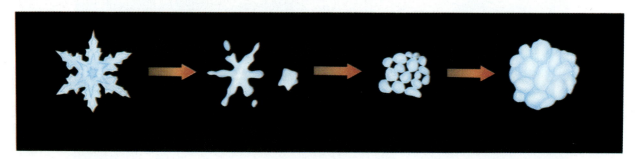

Figure 12.2 This alpine glacier flows around granite peaks in British Columbia, Canada.

lands. For example, the Kahiltna Glacier flows down the southwest side of Denali (Mount McKinley) in Alaska. It is about 65 kilometers long, 12 kilometers wide at its widest point, and about 700 meters thick. Most alpine glaciers are smaller; some are larger.

The development of an alpine glacier depends on both temperature and precipitation. The average annual temperature in the state of Washington is warmer than that in Montana. But alpine glaciers in Washington are larger and flow to lower elevations than those in Montana because more snow falls in Washington. Washington's mountains receive such heavy winter snowfall that even though summer melting is rapid, large quantities of snow accumulate every year. In Montana, snowfall is light enough that most of it melts in the summer, and thus most of Montana's mountains have no glaciers.

Continental Glaciers

In Greenland and Antarctica, winters are so long and cold and summers so short and cool that glaciers are not confined to the mountains, but cover most of the land regardless of elevation. An **ice sheet**, or **continental glacier**, covers an area of 50,000 square kilometers or more (Fig. 12.3). The ice spreads outward in all directions under its own weight. Together, the ice sheets of

Greenland and Antarctica make up 99 percent of the world's ice. The Greenland sheet is more than 2.7 kilometers thick in places and covers 1.8 million square kilometers. Yet it is small compared with the Antarctic ice sheet, which covers about 13 million square kilometers, almost 1.5 times the size of the United States. If the Antarctic ice sheet melted, the meltwater would create a river the size of the Mississippi that would flow for 50,000 years. In contrast, much of the Arctic is ocean, and because glaciers cannot form on oceans, no ice sheet exists at the North Pole. Instead, much of the Arctic Ocean is commonly covered by sea ice.

At certain times in the past, the average global temperature was lower than it is at present. Polar and temperate climates were colder, and possibly wetter, than now. Consequently, vast continental glaciers covered much of North America, Europe, Asia, and parts of the southern continents. These times, called the ice ages, are discussed later in this chapter.

12.2   Glacial Movement

Imagine that you set two poles in dry ground on opposite sides of a glacier, and a third pole in the glacier in a straight line with the other two. After a few months, the center pole would have moved downslope, forming a triangle with the other two (Fig. 12.4). This simple experiment tells us that the glacier has moved downhill. Rates of glacial movement vary with steepness, precipitation, and temperature. In the coastal ranges of Alaska,

Figure 12.3 The Antarctic ice sheet, Victoria Land. *(Mugs Stump)*

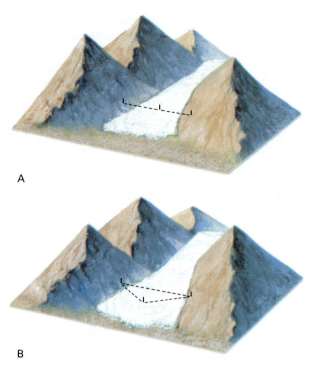

Figure 12.4 Glaciers flow. If three stakes are set in a straight line with two on land and one on a glacier (A), the stake on the ice will move (B).

where annual precipitation is high and average temperature is relatively high (for glaciers), some glaciers move several meters a day. In contrast, in the interior of Alaska, where conditions are generally cold and dry, glaciers move only a few centimeters a day. At these rates, ice flows the length of an alpine glacier in hun-

dreds to a few thousand years. In some instances, a glacier may surge at a speed of 10 to 50 meters per day.

Imagine a mass of ice as large as the Kahiltna glacier flowing downslope. How does it move? A glacier moves by two mechanisms: basal slip and plastic flow.

Basal Slip

In **basal slip**, the entire glacier slides over bedrock in the same way a bar of soap slides down a tilted board. Just as wet soap slides more easily than dry soap, an accumulation of water between bedrock and the base of a glacier accelerates basal slip.

Several factors melt ice near the base of a glacier. Earth heat rises to warm the ice near bedrock. Friction from glacial movement also generates heat. In addition, pressure at the base of a glacier can melt ice. Recall from Chapter 6 that solids melt in response to changing pressure as well as to rising temperature. When ice melts and converts to water, it shrinks. Pressure near the base of a glacier attempts to squeeze the ice into a smaller volume; therefore, ice at the base of a glacier can respond to pressure by melting. Thus, water may be present even when the temperature at the base of a glacier is slightly below 0°C. Additionally, during summer, ice on the surface of a glacier may melt. Some of that meltwater seeps downward to the glacier's base.

Plastic Flow

Near the surface of a glacier, pressure is low and the ice is brittle, like an ice cube or ice on a frozen lake. If you hit it with a hammer, it shatters. However, at depths greater than 40 to 50 meters, pressure is high enough

Figure 12.5 (A) Crevasses form in the upper, brittle zone of a glacier where the ice flows over uneven bedrock. (B) Crevasses in the Bugaboo Mountains of British Columbia.

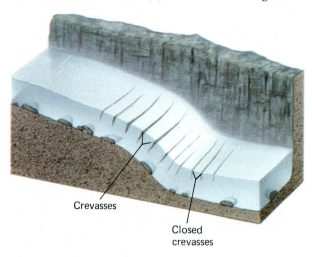

A

B

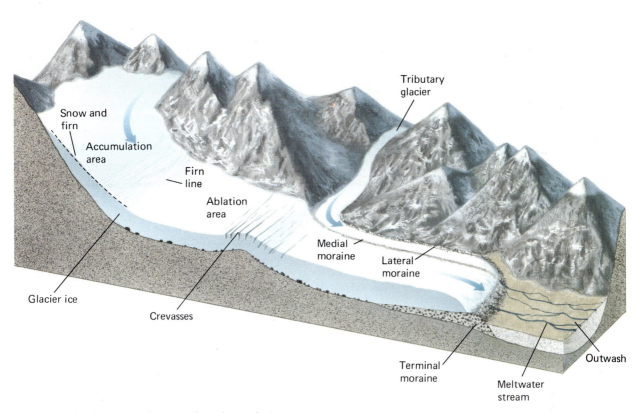

Figure 12.6   Prominent features of an alpine glacier.

that the ice flows plastically. As a result, this portion of the glacier moves by **plastic flow** in addition to basal slip. In plastic flow, glacial ice deforms and flows as a very viscous fluid rather than fracturing.

When a glacier flows over uneven bedrock, the deeper, plastic part of the glacier bends and flows over bedrock bumps. But the brittle upper zone cracks, forming **crevasses** (Fig. 12.5). Crevasses form only in the brittle upper 40 to 50 meters but do not continue into the plastic ice beneath.

An ice fall is a section of a glacier where many crevasses alternate with towering ice pinnacles. If you were a skilled mountaineer, you might climb down into a crevasse. The surrounding walls are a pastel blue, and sunlight filters through the narrow opening above. Almost certainly, you would hear and feel the ice shift and crack. With crampons and ice axes, you could scale one of the pinnacles, but it would be dangerous to do so because the glacier moves continuously and such pinnacles often topple without warning. Crevasses open and close as a glacier moves. Many mountaineers have been crushed by falling ice while traveling through ice falls.

The Mass Balance of a Glacier

Consider an alpine glacier flowing from the high mountains into a lowland valley (Fig. 12.6). At the top of the glacier, snowfall is heavy, temperatures are low for much of the year, and avalanches carry large quantities of snow from the surrounding peaks onto the ice. In the summer some snow melts, but more snow falls in winter than melts in summer. Therefore, snow piles up from year to year. The higher-elevation part of the glacier is called the **accumulation area**. There the glacier's surface is covered by snow year round, which is often powdery in winter and slushy in summer.

Lower in the valley, the temperature is higher throughout the year and less snow falls. In this lower part of a glacier, called the **ablation area** or **zone of wastage**, more snow melts in summer than accumulates in winter. When the snow melts, a surface of old, hard glacial ice is left behind. The **firn line**, or **snowline**, is the boundary between permanent snow and seasonal snow. The firn line shifts up and down the glacier from year to year. Why is there any ice at all in the ablation area? Glacial ice flows downward from the

Figure 12.7 The terminus of an alpine glacier on Baffin Island, Canada, in midsummer. Dirty, old ice can be seen in the lower part of the glacier below the firn line and clean snow higher up on the ice above the firn line. *(Steve Sheriff)*

accumulation area to the ablation area and continuously replenishes it.

Even farther down valley, the rate of glacial flow cannot keep pace with melting, so the glacier ends at its **terminus** (Fig. 12.7). The terminus is usually located on land. Streams or even large rivers flow out from the melting ice. In some regions, however, **tidewater**

Figure 12.8 A tidewater glacier flows directly into the sea at Le Conte Bay, Alaska. Icebergs form when chunks of ice break away from the ice cliffs.

glaciers extend directly into the sea. There, the terminus may be a steep ice cliff that drops off abruptly into the sea (Fig. 12.8). Giant chunks of ice break off, or **calve**, forming **icebergs**. The largest icebergs in the world calve from the Antarctic ice shelf and may be hundreds or, in rare instances, thousands of square kilometers in area. One giant covered 30,000 square kilometers, about as large as Vancouver Island in western Canada. The tallest icebergs in the world come from tidewater glaciers in Greenland. Some rise 150 meters above sea level. Because the visible portion of an iceberg is only about 10 to 15 percent of its total mass, these bergs may be 1500 meters thick.

Glaciers grow and shrink. If average annual snowfall increases or average annual temperature drops, the firn line of an alpine glacier descends to a lower elevation, and the glacier grows. At first, the terminus may remain stable. However, as the accumulation area expands and thickens, more ice flows downslope. Eventually the terminus advances farther down the valley. The lag time between a change in climate and the advance of a glacier may range from a few years to several decades, depending on the size of the glacier, its rate of motion, and the magnitude of the climate change. On the other hand, if annual snowfall decreases or the climate warms, the accumulation area shrinks and the glacier retreats.

When a glacier retreats, its ice continues to flow downhill, but the terminus melts back faster than the glacier flows downslope. In Glacier Bay, Alaska, glaciers have retreated 60 kilometers in the past 125 years, leaving a barren landscape of rock and rubble. No plants grow there, and only sea birds perch on the otherwise lifeless rock. Over the centuries, their droppings will mix with windblown silt and weathered rock to form thin soil. First lichens will grow on the bare rock, and then mosses will take hold in sheltered niches that hold soil. The mosses will be followed by grasses and then bushes as vegetation reclaims the landscape.

12.3 Glacial Erosion

A flowing glacier scours bedrock and erodes landscapes. It does not simply scrape the Earth's surface as a bulldozer does, however. In addition, water seeps into cracks in the bedrock beneath a glacier and then freezes, loosening rocks. The flowing glacier then pries the rocks loose and incorporates them in the ice. This process is called **plucking**. The rock fragments range from silt-size grains to house-size boulders (Fig. 12.9).

Once a glacier picks up rocks and other sediment, it carries them along. Ice itself is too soft to wear away

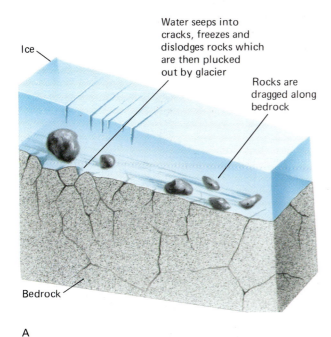

Water seeps into cracks, freezes and dislodges rocks which are then plucked out by glacier

Rocks are dragged along bedrock

Ice

Bedrock

A

B

Figure 12.9   (A) A glacier plucks rocks from bedrock and then drags them over the bedrock, abrading both the loose rocks and the bedrock. (B) Plucking formed these crescent-shaped depressions in granite at Le Conte Bay, Alaska.

bedrock, but rocks embedded in the base of the ice scrape across bedrock like a sheet of coarse sandpaper pushed by a giant's hand. As the ice flows, the embedded rocks gouge parallel grooves and scratches, called **glacial striations**, into bedrock (Fig. 12.10). Striations show the direction of ice movement. They are used to map movements of continental ice sheets that flowed over the land during the most recent ice age.

If silt or sand, rather than rock, is embedded in the base of a glacier, it abrades a smooth, shiny finish called

Figure 12.10   Rocks embedded in the base of a glacier gouge glacial striations into bedrock.

glacial polish on the bedrock surface. Abrasion grinds rocks and other coarse particles into silt-size sediment called **rock flour**. Much of this fine sediment is carried away by streams flowing from the terminus of a glacier. Glacial streams are often so muddy that they are brown or gray.

Erosional Landforms Created by Alpine Glaciers

Let us take an imaginary journey through a mountain range that was glaciated in the past but is now mostly ice free. We start with a helicopter ride to the summit of a high, rocky peak. Our first view from the helicopter is of sharp, jagged mountains rising steeply above smooth, U-shaped valleys (Fig. 12.11A).

A mountain stream commonly erodes downward into its bed, cutting a steep-sided, V-shaped valley. A glacier, however, is not confined to a narrow stream bed but instead fills its entire valley. As a result, it erodes outward against the valley walls as well as downward into its bed, scouring the sides as well as the bottom of its valley. This outward erosion forms broad, rounded, U-shaped valleys (Fig. 12.12).

We land on the summit and step out of the helicopter. Beneath us, a steep cliff drops off into a horseshoe-shaped depression gouged out of the mountainside. This depression is called a **cirque**. A small glacier at the head of the cirque reminds us of the larger mass of ice that existed there in a colder, wetter time.

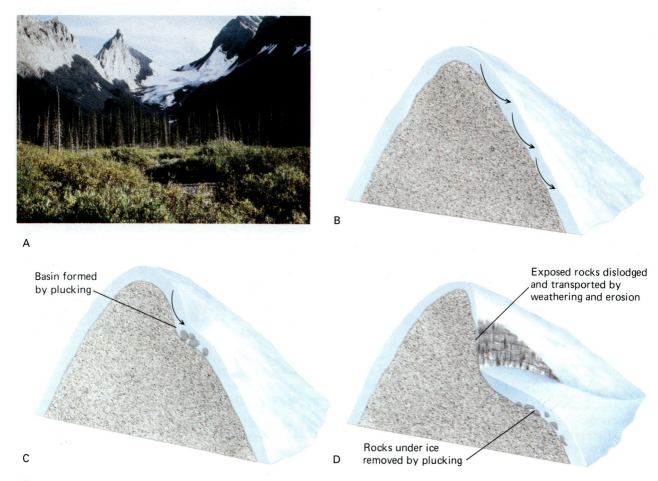

Figure 12.11 (A) A cirque rises above a mountain meadow in the Canadian Rockies. (B) Snow accumulates, and a glacier begins to flow downslope from the summit of a peak. (C) Glacial plucking erodes a small depression in the mountainside. (D) Continued glacial erosion and weathering enlarge the depression. When the glacier melts, it leaves a cirque carved in the side of the peak, as in the photograph.

To understand how a glacier forms a cirque, imagine a gently rounded mountain. As snow accumulates and a glacier forms, the ice begins to flow down the mountainside (Fig. 12.11B). Abrasion and plucking erode a small depression that grows slowly as the glacier continues to flow (Fig. 12.11C). As erosion enlarges the cirque, its walls become steeper and higher. Frost wedging releases rocks from the cirque walls. The rocks fall onto the surface of the glacier, and the flowing ice carries this debris from the cirque to lower parts of the valley (Fig. 12.11D). When the glacier finally melts, it leaves a steep-walled, rounded cirque.

As a cirque forms, the glacier often plucks a depression at its bottom. After the glacier melts, this depression fills with water, forming a small lake called a **tarn** nestled in the cirque (Fig. 12.13).

If glaciers erode three or more cirques into different sides of a peak, they may create a steep, pyramid-shaped rock summit called a **horn**. The Matterhorn in the Swiss Alps is a famous example of a horn (Fig. 12.14). If two glaciers flow along opposite sides of a mountain ridge, they erode both sides of the ridge, forming a sharp, narrow **arête** between adjacent valleys (Fig. 12.15).

Looking downward from the glaciated peak, you may see a waterfall where a small, high valley empties into a larger, deeper one. A small glacial valley lying high above the floor of the main valley is called a **hanging valley** (Fig. 12.16). To understand how a hanging valley forms, imagine these mountain valleys filled with glaciers, as they were several millennia ago (Fig. 12.17). The main glacier, flowing through the lower valley, gouged a deep trough. In contrast, the smaller tributary

Figure 12.12 A U-shaped glacial valley, Purcell Mountains, British Columbia.

Figure 12.13 A tarn lying in the bottom of a cirque, Rocky Mountain National Park, Colorado.

Figure 12.14 The Matterhorn formed as three glaciers eroded cirques into the peak from three different sides. *(Swiss Tourist Board)*

Figure 12.15 An arête in the Bugaboo Mountains in British Columbia.

271

Figure 12.16   Hanging valleys in Yosemite National Park.   *(Science Graphics/Ward's Natural Science Establishment, Inc.)*

glacier did not scour the rock as deeply. As a result, when the ice melted, the floor of the tributary was considerably higher than that of the main valley, forming an abrupt drop where it enters the main valley. Typically, many hanging valleys empty into a main valley.

In many coastal regions, deep, narrow inlets called **fjords** extend far inland. Most fjords are glacially carved valleys that were later flooded by a rising sea (Fig. 12.18).

Continental glaciers erode the landscape just as alpine glaciers do. The main difference is that continental glaciers are considerably thicker and not confined to valleys. Therefore, they scour the entire landscape and sometimes cover whole mountain ranges. Alpine glaciers are like an engraver's knife cutting thin, deep lines into the landscape, whereas continental glaciers are more closely analogous to a bulldozer. The most spectacular features of continental glaciers are those formed by deposition.

12.4   Glacial Deposits

In the 1800s, before geologists understood that continental glaciers covered vast parts of the land only 10,000 to 20,000 years ago, they recognized that large deposits of sand and gravel found in some places had been transported from somewhere else. A popular theory suggested that this material had been carried by icebergs during catastrophic floods. The deposits were called **drift** after this inferred mode of transport.

Today we know that glaciers carried and deposited drift. Although the word *drift* is a misnomer, it remains in common use. Now geologists define *drift* as all rock or sediment transported and deposited by a glacier. Glacial drift averages 6 meters (20 feet) thick over the rocky hills and pastures of New England, and 30 meters (100 feet) deep over the plains of Illinois.

Drift is subdivided into two categories. **Till** is deposited directly by glacial ice. **Stratified drift** is sediment that was first carried by a glacier and then transported and deposited by a stream.

Landforms Composed of Till

Ice is so much more viscous than water that it carries all sizes of sediment together. As a result, when a glacier melts, it deposits clay and boulders together in an unsorted, unstratified, jumbled mass.

If you travel in country that was once glaciated, you occasionally find large boulders lying free on the surface. In many cases they are of a rock type different

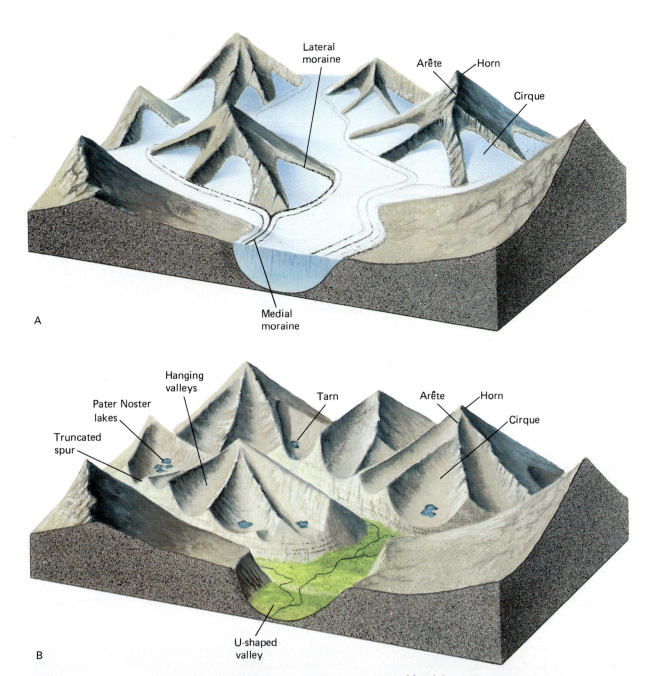

Figure 12.17 Landforms created by alpine glaciers. (A) Mountains covered by alpine glaciers. (B) The same mountains after the glaciers have melted.

from the bedrock in the immediate vicinity. Glaciers carry these boulders, called **erratics**, to their present locations. The origins of glacial erratics can be determined by exploring the terrain in the direction from which the glacier came until the parent rock is found. Some erratics have been carried 500 to even 1000 kilometers from their points of origin; they provide clues to the movements of past glaciers.

Moraines

A **moraine** is a mound or ridge of till deposited by a glacier. Think of a glacier as a giant conveyor belt. If you place a number of suitcases on a conveyor belt, it carries them to the end of the belt and dumps them in a heap. Similarly, a glacier carries sediment and then drops it at its terminus. If a glacier is neither advancing

Figure 12.18 A steep-sided fjord bounded by thousand-meter-high cliffs, Baffin Island, Canada.

nor retreating, its terminus may remain in the same place for years, and the sediment piles up at the terminus in a ridge called an **end moraine** (Fig. 12.19). An end moraine that forms when a glacier is at its greatest advance, before beginning to retreat, is called a **terminal moraine**.

An end moraine deposited by a large alpine glacier may be so high that it would take an hour to climb to its top. A steep moraine can be difficult and dangerous to climb. Till is commonly loose, and large boulders are mixed randomly with rocks, cobbles, sand, and clay. A careless hiker can dislodge boulders and even set off a dangerous landslide.

Terminal moraines show the extent of the most recent continental glaciers. In North America the moraines form a broad, undulating line across the northern United States from Montana to New York. Enough time has passed that soil and vegetation cover moraines (Fig.

Figure 12.19 An end moraine is a ridge of till piled up at a glacier's terminus.

Figure 12.20 This moraine in New York State marks the farthest advance of the latest continental ice sheet.

12.20). Several other types of moraines are described in Table 12.1.

Drumlins

A cluster of about 10,000 elongate hills, called **drumlins**, dots a region in northern Saskatchewan (Fig. 12.21). Each hill looks like an upside-down spoon or a whale swimming through the ground with its back in the air. A drumlin is typically 1 to 2 kilometers long and 15 to 50 meters high. Most drumlins are made of till, although some are stratified drift. In either case, the elongate hill forms as a glacier flows over a mound of drift and streamlines it parallel to the direction of flow.

Figure 12.21 Aerial view of drumlins in northern Saskatchewan. Each drumlin is about 40 meters high. *(Canadian Department of Energy, Mines, and Resources, National Air Photo Library)*

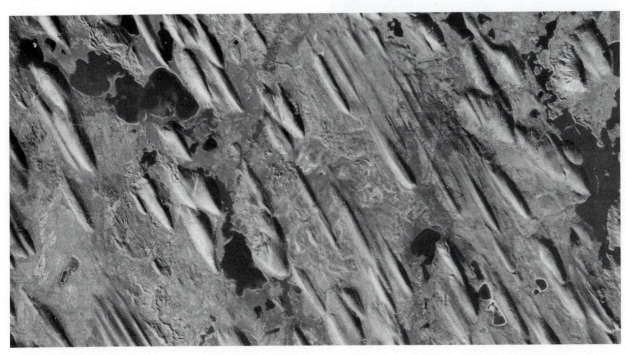

Table 12.1 • Types of Moraines

An **end moraine** forms at the terminus of a glacier. An end moraine that forms when a glacier is at its greatest advance is called a **terminal moraine**.

An end moraine in Glacier National Park, Montana. *(USGS, P. Carrara)*

If the glacier stops retreating and the terminus remains in the same place for a few years or more, a new end moraine, called a **recessional moraine**, forms in the same manner as a terminal moraine. The major differences are that the recessional moraine forms at a later date and upstream from the terminal moraine.

A recessional moraine formed by the Catamount Glacier, British Columbia.

If a glacier retreats at an even rate, the till disperses in a relatively thin layer, forming **ground moraine**. Ground moraine fills old stream channels and other low spots, leveling the terrain. Often this leveling disrupts drainage patterns. Many swamps in the northern Great Lakes region are on young ground moraine.

Boulder-strewn ground moraine in the Canadian Rockies.

Landforms Composed of Stratified Drift

During summer, when snow and ice melt rapidly, streams flow over the surface of a glacier. Many are so deep and wide that hikers cannot jump across them easily. A stream flowing on a glacier commonly runs into a crevasse and plunges into the interior of the glacier. Some water finds its way to the bottom of the ice and flows over bedrock or drift beneath the glacier. Eventually, all of this water flows from the terminus.

Because a glacier erodes so much sediment, a stream flowing from a glacier commonly carries large amounts of silt, sand, and gravel. The stream eventually deposits this sediment downstream from the glacier as **outwash**. Recall from Chapter 10 that a glacial stream often carries such a heavy load of sediment that it becomes braided. A braided glacial stream may fill a valley with a maze of serpentine channels and low-lying bars that shift across the flood plain from year to year. The bars are composed of outwash. If the sediment spreads

An alpine glacier erodes its valley walls as well as its valley floor. Large amounts of debris fall from the steepening valley walls onto the margins of the glacier. As a result of both processes, the edges of the glacier carry large loads of sediment. When the glacier melts, it deposits this sediment near its margins, forming **lateral moraines**.

A lateral moraine formed by an alpine glacier flowing from Mount Sir Sanford, British Columbia.

If two glaciers converge to form a single larger one, lateral moraines from the edges of the glaciers merge into the middle of the larger body of ice. The sediment forms a ridge of till, which appears as a dark stripe on the surface of the ice downstream from the convergence. When the glacier melts, it deposits this till as a **medial moraine** in the middle of the glacial valley.

Three medial moraines appear as streaks of rubble in this glacier on Baffin Island, Canada.

out from the confines of a narrow valley into a larger valley or plain, it forms an **outwash plain**. Outwash plains are also characteristic of continental glaciers. An outwash plain from a continental glacier is a relatively smooth, level landform that often extends beyond the terminal moraines (Fig. 12.22).

As a glacier melts and retreats, streams flow on top of, within, and beneath the ice. They commonly deposit small mounds of sand and gravel called **kames**. A kame can form as a fan or delta at the margin of a melting glacier, or where sediment collects in a crevasse or other depression in the ice. An **esker** is a long, snakelike ridge that forms as the bed deposit of a stream that flowed within or beneath a glacier (Fig. 12.23).

Because kames, eskers, and outwash are stream deposits, they are sorted and show sedimentary bedding. Thus, they are distinguished easily from the unsorted and unstratified till deposited directly from glacial ice.

A large block of ice may be stranded and buried in moraine or outwash as a glacier melts. When the block

Figure 12.22 An outwash plain is deposited by a stream flowing from the terminus of a glacier. *(John S. Shelton)*

melts, it leaves a depression called a **kettle**. Kettles fill with water, forming kettle ponds or lakes. A kettle lake is as large as the stranded ice chunk. It can vary from a few tens of meters to a kilometer or so in diameter, with a typical depth of 10 meters or less.

12.5   The Ice Ages

Geologists have found moraines, drumlins, and striations in bedrock far from mountain glaciers. Once they understood that glaciers had formed these features, they realized that, sometime in the past, a vast continental ice sheet covered high latitudes. A time when alpine glaciers descend into lowland valleys and continental glaciers spread over higher latitudes is called an **ice age** or a **glacial epoch**.

Figure 12.23 Kames, eskers, and kettle lakes are common landforms left by a retreating glacier.

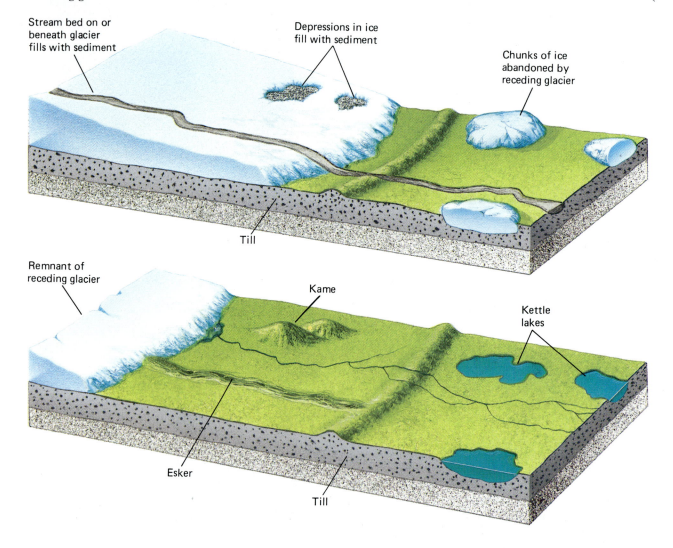

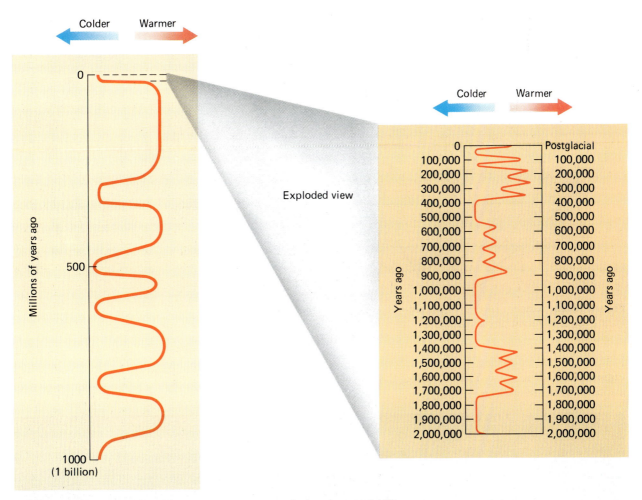

Figure 12.24 Average global temperature variations during the past 1 billion years. The times of lowest temperature are thought to coincide with ice ages. The right-hand scale is an expanded view to show temperature fluctuations during the Pleistocene Ice Age.

Geologic evidence shows that the Earth has been warm and relatively ice free for about 90 percent of the past 2.5 billion years. However, at least five major ice ages have occurred during that time (Fig. 12.24).

The most recent ice age occurred mainly during the Pleistocene epoch and is called the **Pleistocene Ice Age**. It began about 2 million years ago, and many geologists think we are still in this ice age. However, the Earth has not been glaciated continuously during the Pleistocene Ice Age; instead, continental ice sheets and alpine glaciers have grown and then melted away several times. Figure 12.24 also shows fluctuations in average global temperature during the Pleistocene epoch.

During the most recent advance, glaciers reached their maximum extent about 18,000 years ago. At that time, 30 percent of the Earth's land area was ice covered

(Fig. 12.25). The effects of the Pleistocene continental glaciers are discussed in Chapter 21.

Causes of Ice Ages

At least five major ice ages have occurred in the past 2.5 billion years, and the most recent, the Pleistocene Ice Age, was characterized by several glacial advances and retreats (Fig. 12.26). The causes of both the ice ages and the Pleistocene glacial fluctuations were related to global climate changes.

In considering the causes of ice ages, two questions arise:

1. Why have major ice ages occurred at several times in the Earth's history?

2. Why did continental glaciers advance and then retreat several times during the Pleistocene Ice Age?

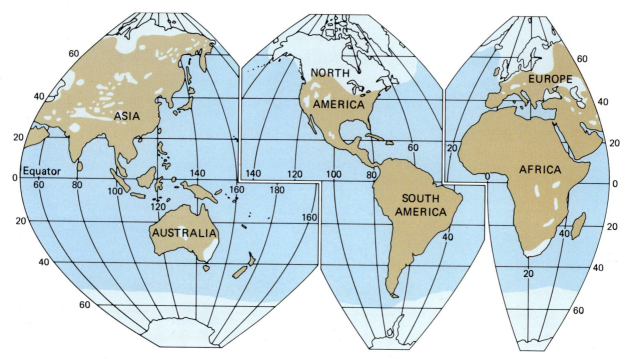

Figure 12.25   Glaciers covered 30 percent of the Earth's continents 18,000 years ago during the most recent glacial maximum.

Figure 12.26   Much of the Earth resembled this part of Greenland during Pleistocene glacial advances.   *(Kevin Killelea)*

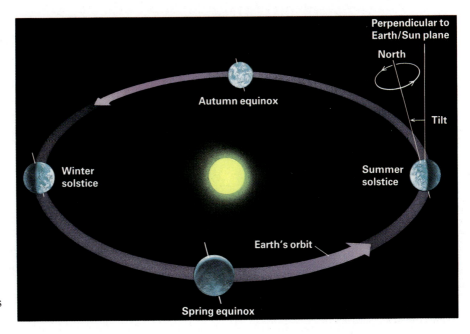

Figure 12.27 Orbital variations may explain the temperature oscillations and glacial advances and retreats during the Pleistocene epoch. Orbital variations occur over time spans of tens of thousands of years.

Scientists have suggested a wide range of explanations for both the major ice ages and for Pleistocene advances and retreats. They include changes in the energy output of the sun, variations in the Earth's orbit around the sun, interactions between the Earth's magnetic field and climate, changes in the positions of the continents due to tectonic motion, changes in deep-ocean currents, surges in Antarctic glaciers, and movements of the pack ice in the Arctic Ocean. Several research teams have suggested a relationship between volcanic activity and climate change. As explained in Chapter 6, volcanoes inject dust into the upper atmosphere, which leads to global cooling.

Two of the most widely discussed theories involve movement of the continents and variations in the Earth's orbit.

Major Glacial Epochs and Plate Tectonics

One explanation for long-term climatic change and the onset of the ice ages comes from plate tectonics. According to this theory, ice ages occur when the continents are positioned close to the polar regions. For example, during the Permian Period (about 250 million years ago), all the continents were gathered into one giant landmass called Pangaea, part of which lay at the South Pole. Glaciation was extensive at this time. As Pangaea broke apart and the continents moved toward the equator, the Permian Ice Age ended.

Cycles of the Pleistocene Ice Age and Orbital Variations

The movements of continents can be used to explain climate changes over periods of tens or hundreds of millions of years. However, these processes are too slow to explain the shorter cycles of glacial fluctuations during the Pleistocene Ice Age.

One explanation for Pleistocene glacial advances and retreats is that the Earth's temperature changes in response to variations in the Earth's orbit and in the orientation of its axis. Three types of variations occur (Fig. 12.27):

1. The shape of the Earth's orbit around the Sun changes on about a 100,000-year cycle. This is known as **eccentricity**.

2. The Earth's axis is currently tilted at about 23.5° with respect to a line perpendicular to the plane of its orbit around the Sun. The **tilt** changes by 1.5° on about a 41,000-year cycle.

3. The Earth's axis, which now points directly toward the North Star, circles like that of a wobbling top. This circling, called **precession**, completes a full cycle every 23,000 years.

Orbital changes do not greatly affect the total amount of solar energy that strikes the Earth, but they do affect the distribution of sunlight with respect to latitude and season. Seasonal changes in the amount of sunlight reaching the Earth can cause an onset of glaciation. One important factor is the summer temperature. In high latitudes, snow falls and ice forms during the winter, even during interglacial periods. If summers are hot and long, the snow and ice melt, but if summers are cool and short, snow and ice persist, leading to glaciation.

Early in the twentieth century, a Yugoslavian astronomer, Milutin Milankovitch, plotted the cycles of the three different variables. He then calculated the combined effects of all three variations on the Earth's climate. The calculations showed that the three orbital variations combine to generate alternating climate changes in the higher latitudes. Moreover, the timing of the calculated climate changes coincided with the known timing of Pleistocene glacial advances and retreats.

SUMMARY

Glaciers sculpt the Earth's surface and create landforms. If snow survives through one summer, it converts to granular ice crystals called **firn**. Firn can metamorphose further to form **glacial ice**. A **glacier** is a thick mass of ice that forms on land and flows plastically. A glacier forms wherever winter snowfall exceeds summer melting. **Alpine glaciers** occur in mountainous regions; a **continental glacier** covers a large part of a continent. A glacier moves by both **basal slip** and **plastic flow**. The upper 40 to 50 meters of a glacier are too brittle to flow, and large cracks called **crevasses** develop in this layer.

In the **accumulation area** of a glacier, more snow falls than melts, whereas in the **ablation area**, melting exceeds accumulation. The **snowline**, or **firn line**, is the boundary between permanent and seasonal snow. The end of a glacier is called the **terminus**.

Glaciers scour both the bottoms and the sides of valleys, giving their valleys a characteristic U shape. Glaciers erode bedrock by **plucking** and by abrasion.

Alpine glaciers create **cirques**, steep-sided depressions eroded into peaks; **arêtes**, thin, sharp ridges separating two glaciated valleys; **horns**, pyramid-shaped peaks formed by the intersection of three or more cirques; and **hanging valleys**, formed by tributary glaciers that hang high above the floor of a larger valley.

Drift is any rock or sediment transported and deposited by a glacier. The unsorted drift deposited directly by a glacier is **till**. Most glacial terrain is characterized by large mounds of till known as **moraines**. **Stratified drift** is sediment transported, sorted, and deposited by glacial streams. An **outwash plain** forms where stratified drift spreads over a broad valley. A **kettle** is a depression created by the melting of a large ice block that was abandoned by a retreating glacier.

Several major **ice ages** occurred during the past 2.5 billion years. The most recent happened during the **Pleistocene Epoch**, when both alpine and continental glaciers created many topographic features that are prominent today.

KEY TERMS

REVIEW QUESTIONS

1. Differentiate between an alpine glacier and a continental glacier. Where are alpine glaciers found today? Where are continental glaciers found today?

2. Distinguish between basal slip and plastic flow.

3. Why are crevasses only about 40 to 50 meters deep, even though many glaciers are much thicker?

4. Describe the surface of a glacier in the summer and in the winter in (a) the accumulation area and (b) the ablation area.

5. How do icebergs form?

6. Describe how glacial erosion can create (a) a cirque, (b) striated bedrock, and (c) smoothly polished bedrock.

7. Describe the formation of arêtes, horns, hanging valleys, and truncated spurs.

8. Distinguish among ground, recessional, terminal, lateral, and medial moraines.

9. How are kames and eskers formed by a receding glacier?

10. Describe four types of topographic features left behind by the continental ice sheets.

DISCUSSION QUESTIONS

1. Outline the changes that would occur in a glacier if (a) the average annual temperature rose and precipitation decreased; (b) the temperature remained constant but precipitation increased; and (c) the temperature decreased and precipitation remained constant.

2. In some regions of northern Canada, both summer and winter temperatures are cool enough for glaciers to form. Speculate on why continental glaciers are not forming in the region.

3. In the 1980s and early 1990s, global temperature was rising. Some climatologists have argued that global warming could lead to increased glaciation. Give a plausible mechanism for this scenario.

4. In some mountain ranges, the tops of mountain peaks are jagged and covered by rubble, whereas the lower elevations of the mountains are rubble free. Give a plausible explanation for these observations.

5. If you found a large boulder lying in a field, how would you determine whether it was an erratic?

6. Imagine that you enountered some gravelly sediment. How would you determine whether it was deposited by a stream or a glacier?

7. Explain how medial moraines prove that glaciers move.

8. Compare and contrast erosion, transport, and deposition of sediment by wind, streams, mass wasting, and glaciers.

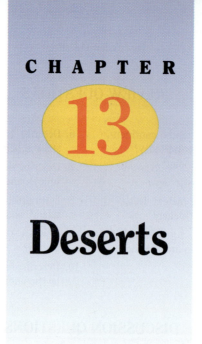
Deserts

*Sand ripples form as wind blows sand over
the surface of a dune in Death Valley.*
(Superstock)

Deserts commonly evoke images of thirsty travelers crawling across an infinity of lifeless sand dunes. These images accurately depict some deserts, but not others. Many deserts are rocky and even mountainous, with colorful cliffs or peaks towering over plateaus and narrow canyons. Cactus, sage, grasses, and other plants may dot the landscape. After a rainstorm, these plants often bloom to produce multicolored flowers. In winter, a thin layer of snow may cover the ground.

Deserts are defined primarily by the amount of rainfall; they generally have an annual precipitation of less than 25 centimeters (10 inches) per year. This limited rainfall supports only sparse and widely spaced vegetation, and in some particularly dry deserts virtually no plants exist.

Animals that live in the desert have adapted to their dry environment. For example, the African oryx, a large antelope, varies its metabolic rate, storing heat during the day and releasing it at night. The North American kangaroo rat feeds on dry seeds, conserving and recycling water, and its urine is highly concentrated. Humans have developed cultural and social adaptations to life in the desert. Traditionally, many desert societies were nomadic. Small groups traveled across the land between water holes with their flocks of camels, sheep, and goats, moving on before they overgrazed the vegetation. Other desert cultures developed extensive irrigation systems and placed farms close to rivers and wells.

13.1   Deserts of the World

If you were to travel widely and visit the great deserts of the Earth, you might be surprised by their geologic and topographic variety. You would see coastal deserts along the beaches of Chile, shifting dunes in the Sahara, deep red sandstone canyons in southern Utah, and stark granite mountains in Arizona. The world's deserts are similar to one another only in that they all receive scant rainfall.

The Effect of Latitude

About 25 percent of the Earth's land surface outside of the polar regions is desert (Fig. 13.1). Most of these arid regions lie at about 30° both south and north of the equator. Why do deserts concentrate at these latitudes?

The Sun shines most directly near the equator, warming the surface air. The warm air absorbs moisture from the equatorial oceans. This warm, moist air rises, forming a low-pressure zone. As the air rises, it cools and the water vapor condenses and falls as rain. For this reason, vast tropical rainforests grow near the equator. The air, which is now drier because of the loss of moisture, flows northward and southward at high altitudes. This air falls back toward the Earth's surface at about 30° north and south latitudes (Fig. 13.2). As air falls, it is compressed, creating a zone of high pressure. This compression heats the air and enables it to hold more water vapor. As a result, water evaporates from the surface into the air. Because the falling air absorbs water, rainfall is infrequent and deserts are common. Thus, two large desert zones, one north and one south of the equator, encircle the globe at about 30° north and south latitudes.

Effect of Topography: Rain Shadow Deserts

When moisture-laden air flows over a mountain range, it rises. As the air rises, it cools and therefore its ability to hold water decreases. As a result, the water vapor condenses into droplets that fall as rain or snow. These conditions cause abundant precipitation on the windward side and the crest of the range. When the air passes the crest and flows down the leeward (or downwind) side, it sinks, as shown in Figure 13.3. This air has already lost much of its moisture. As in the case of sinking air at 30° latitude, the air is compressed as it falls, creating a dry high-pressure zone called a **rain shadow desert** on the leeward side of the range. Figure 13.4 shows the rainfall distribution in California. Note that the leeward valleys are much drier than the mountains to the west.

Continental Coastlines and Interiors

Because most evaporation occurs over the oceans, one might expect that coastal areas would be moist and climates would become drier with increasing distance from the sea. This generalization is often true, but notable exceptions exist.

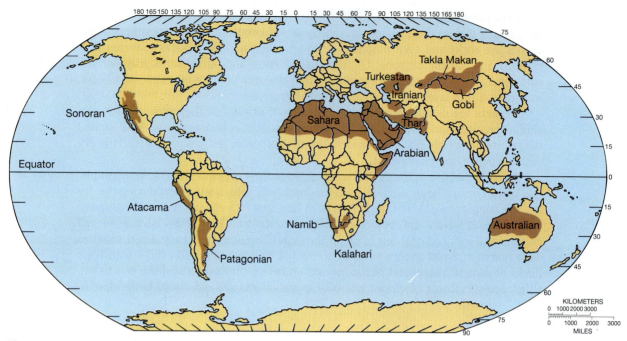

Figure 13.1   The major deserts of the world. Note the global concentration of deserts at 30° north and south latitudes. Most of the deserts are surrounded by semiarid lands.

The Atacama Desert along the west coast of South America is so dry that portions of Peru and Chile have received no rainfall for a decade or more. Cool ocean currents flow along the west coast of South America. When the cool marine air encounters warm land, the air is heated. The moisture in the air can't condense under these conditions and a coastal desert results.

13.2   Water in Deserts

Desert Streams

Large rivers flow through some desert regions. For example, the Colorado River crosses the arid southwestern United States, and the Nile River flows through deserts

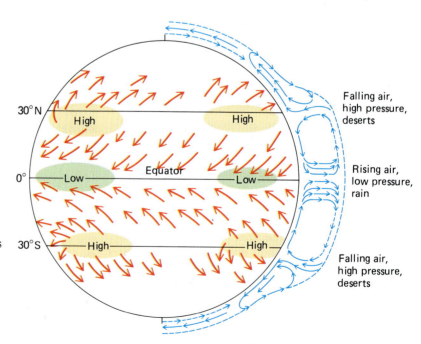

Figure 13.2   The formation of deserts at 30° north and south latitudes. The arrows drawn inside the globe indicate surface winds. The arrows to the right show both vertical and horizontal movement of air on the surface and at higher elevations.

Figure 13.3 Formation of a rain shadow desert. Warm, moist air from the ocean rises as it flows over a mountain range. As it rises it cools, and water vapor condenses to form rain. The dry descending air on the lee side of the range absorbs moisture, creating a desert.

in North Africa (Fig. 13.5). The Colorado and other desert rivers receive most of their water from wetter, mountainous areas bordering the arid lands. Thus, they flow continuously all year long.

In contrast, local desert streams flow for only part of the year. Recall from Chapter 11 that in a moist environment, ground water feeds streams continuously. In the desert, the water table is often so low that water actually seeps out of the stream bed into the ground below. As a result, many desert streams flow for only a short time after a rainstorm or during the spring when winter snows are melting. A stream bed that is dry for most of the year is called a **wash** (Fig. 13.6).

Flash Floods and Debris Flows

Consider what happens when rain falls on a desert. Little vegetation is present to absorb moisture or to protect the soil from erosion. The surface of the land may be covered by large expanses of exposed rock or by tightly compacted soil. Therefore, rain is not absorbed as fast as it falls, and the excess runs over the surface to collect in gullies and washes. During a heavy rainstorm, dry stream beds fill with water so rapidly that **flash floods** occur. A flash flood is one in which flooding occurs suddenly and the water recedes again rapidly. Occasionally, novices to desert camping pitch their tents in

Figure 13.4 Rainfall patterns in California. Note that rain shadow deserts lie east of the mountain ranges. Rainfall is reported in centimeters per year.

Figure 13.5   The Colorado River has cut Grand Canyon through the Arizona desert.

washes, where they find soft sand to sleep on and are sheltered from the wind. However, if a thunderstorm occurs upstream during the night, a flash flood may fill the wash with a wall of water mixed with rocks and boulders. People have been caught by such flash floods and killed. By mid-morning of the next day, the wash may contain only a tiny trickle, and within 24 hours it may be completely dry again. In contrast, in a large river such as the Mississippi, flood waters rise and fall over a period of days or weeks, not hours or minutes.

When rainfall is unusually heavy, the desert soil itself may become saturated enough to flow. The sparse desert vegetation cannot stabilize this muddy slurry. Therefore, viscous, wet mud and water flows downslope, carrying boulders and anything else in its path. This moving mass is called a **debris flow**.

When a flood pours out of the mountains into a valley, the water slows abruptly and most of the sediment is deposited at the mountain front on an alluvial fan. Although alluvial fans form in all climates, they are particularly conspicuous in deserts (Fig. 13.7). A large fan may be several kilometers across and rise a few hundred meters above the surrounding valley floor.

Playas

During the wet season, water enters a desert lake bed by stream and ground-water flow and to a lesser extent by direct precipitation. However, little or no water flows into the lake during the dry season, and water escapes by evaporation and seepage. If the water loss is great enough, the lake dries up completely. An intermittent desert lake is called a **playa lake**, and the dry lake bed is called a **playa** (Fig. 13.8). Recall from Chapters 9 and 10 that water dissolves ions from rock and soil. When this slightly salty water accumulates in a playa lake and the water evaporates, the ions precipitate to deposit salts and other evaporite minerals on the playa. Over years of repetition of this process, economically valuable evaporite deposits may accumulate. The borax deposits of California's Death Valley are examples of playa evaporites.

13.3   Wind Erosion

When wind blows through a forest or across a grassy prairie, the vegetation protects the soil from erosion. In addition, rain commonly accompanies windstorms in temperate climates; the moisture binds soil particles together, so little wind erosion occurs. In contrast, seacoasts, recently glaciated areas, and deserts commonly have little or no vegetation to protect the soil from wind. Therefore, wind erodes soil in these environments.

Figure 13.6   Courthouse Wash, Utah. (See page 248 for a view of the same stream during the wet season.)

GEOLOGY AND THE ENVIRONMENT

Expansion and Contraction of Deserts

Most deserts are surrounded by semiarid regions that receive more moisture than a true desert but less than forests and grasslands. Semiarid lands support small populations of plants and animals. The Sahara is the largest desert on the planet. South of the Sahara lies a semiarid belt called the Sahel. Reports issued in the 1970s and 1980s claimed that the Sahara was advancing steadily southward into the Sahel at a rate of 5 kilometers per year. Scientists argued that human mismanagement of the land was causing the desert advance. However, a new study indicates that the Sahara has not grown continuously, but has alternately expanded and contracted. The desert boundary is clearly visible on satellite photographs as a line of vegetation. Researchers plotted changes in the desert by studying photographs taken between 1980 and 1990. As shown in the figure, the desert expands (blue line) when rainfall declines (red line).

Although land mismanagement may not have increased the size of the Sahara, human activity—for example, overgrazing, extensive cultivation, irrigation, and devegetation for firewood—can reduce the productivity of semiarid lands. One hundred years ago, nomads moved across the Sahara and the Sahel with little regard for national boundaries, traveling with the seasons and abandoning an area after it had been grazed for a short period of time. This constant movement prevented overgrazing. In addition, population levels of the nomadic tribes were stable and low. In recent years, however, their life style has changed. As medical attention and sanitation have improved, the population has expanded dramatically. In addition, enforcement of national borders and civil strife in some countries have curtailed nomadism. As populations have increased and the social structures have changed, increasing pressure has been put on the land to produce food for more people from smaller areas.

Cattle eat the most nutritious grasses and ignore noxious weeds or inedible shrubs. If an area is grazed lightly, grasses remain healthy and compete successfully with weeds and shrubs for water, nutrients, and space. However, if the grasses are overgrazed, their growth is disrupted so seriously that they cannot reseed. This loss of grasses leads to several problems. Less edible plants become dominant, and the value of the range decreases. Loss of grasses leaves the soil susceptible to erosion. When soil is devoid of vegetation and baked in the sun, it becomes so impermeable that water evaporates before it soaks in. Therefore, the water table drops. Because the

deep-rooted bushes depend on underground water, lowering of the water table ultimately kills these plants. The process is accelerated as the grazing cattle pack the ground with their hooves, blocking the natural seepage of air and water through the soil. Firewood gatherers cut the woody shrubs, further destroying the vegetation.

Crops can be grown in deserts if they are irrigated, but extensive irrigation can destroy the land. Recall that all river water is slightly salty. If this water is used to irrigate fields, some of the water evaporates, leaving the salts behind. Over time, the salt content of the soil may increase until the soil becomes too salty to support plants.

Even in the United States, overgrazing of semiarid range land has created a problem. Some critics argue that government policies favor short-term profits for ranchers rather than long-term stability of public range land and that millions of hectares are being destroyed needlessly.

Expansion and contraction of the Sahara Desert from 1980 to 1990. Note that the Sahara has expanded (blue line) when rainfall has decreased (red line). Thus, rainfall, not land management, appears to be the prime factor governing the size of the Sahara. *(Compton J. Tucker, Harold E. Dregne, and Wilbur W. Newcomb, "Expansion and Contraction of the Sahara Desert from 1980 to 1990," Science 253 [July 19, 1991], pp. 299ff)*

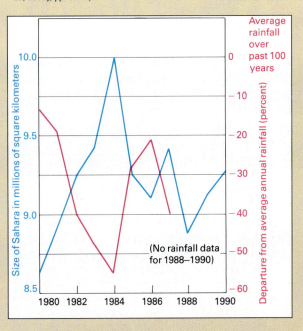

Figure 13.7 An alluvial fan, Canadian Rockies.

Deflation

One can hardly think about wind erosion without evoking an image of a hide-behind-your-camel-type sandstorm in the desert. Wind erosion, called **deflation**, is a selective process. Because air is much less dense than water, wind can move only small particles, mainly silt and sand. (Clay particles usually stick together, and consequently wind does not move them.) Imagine soil, unprotected by vegetation, containing silt and sand mixed with larger pebbles and rocks (Fig. 13.9). When wind blows, it removes only the small particles, leaving the pebbles and rocks to form a continuous cover of stones called **desert pavement** (Fig. 13.10). Once desert pavement forms, the supply of sediment is cut off. Thus, most deserts are rocky and covered with gravel, and flowing dunes are relatively rare.

Only the smallest and lightest particles can be lifted very high or carried great distances by wind. Sand grains, which are relatively large and heavy, are usually lifted less than 1 meter in the air and are transported only a short distance. In contrast, wind carries fine silt in suspension. Skiers in the Alps commonly encounter a silty surface on the snow, blown from the Sahara Desert and carried across the Mediterranean Sea.

Wind erosion also forms depressions called **blowouts**. In the 1930s, a series of intense, dry winds eroded large areas of the Great Plains and created the Dust Bowl. Deflation formed tens of thousands of blowouts, many of which remain today. Some are small, measuring only 1 meter deep and 2 or 3 meters across, but others are much larger (Fig. 13.11). One of the deepest blowouts in the world is the Qattara Depression in western Egypt. It is more than 100 meters deep and 10 kilometers in diameter. Ultimately, the lower limit for a blowout is the water table. If the bottom of the depression

Figure 13.8 Mud cracks pattern the floor of a playa in Utah.

Figure 13.9 Wind erodes silt and sand but leaves rocks behind to form desert pavement.

Wind removes surface sand

Formation of desert pavement complete— No further wind erosion

290

Figure 13.10 Desert pavement is a continuous cover of stones left behind when wind blows silt and sand away.

Figure 13.11 In front of the horse is a hummock left standing in the middle of a large blowout. Soil and vegetation on top of the hummock show the level of the ground surface before wind eroded the blowout. *(N. H. Darton, USGS)*

reaches moist earth near the water table, where water binds the sand grains, wind erosion is no longer effective. If a blowout reaches the water table, a pool forms, creating an oasis.

Abrasion

Moving air by itself is not abrasive enough to erode rocks. But windblown sand and silt are abrasive and are effective agents of erosion. Because wind carries sand only a meter or less above the ground, abrasion concentrates close to the ground. If you see a tall desert pinnacle topped by a delicately perched cap, it was probably not created by wind erosion. The pinnacle is too high. However, if the base of a pinnacle is sculpted, wind may be the responsible agent (Fig. 13.12). Cobbles and boulders lying on the surface often have faces worn flat by windblown sand. Such rocks are called **ventifacts** (Fig. 13.13).

Figure 13.12 Wind abrasion selectively eroded the base of this rock because windblown sand moves mostly near the surface.

Figure 13.13 Windblown sand sculpted the faces on this ventifact. *(Hubbard Scientific Co.)*

GEOLOGY AND THE ENVIRONMENT

Wind Erosion: The Dust Bowl

The semiarid and arid plains of the southwestern United States have experienced periodic droughts for centuries. However, the natural prairie ecosystems were resistant to drought. The native bushes and perennial grasses grow deep roots that collect water and nutrients from lower layers in the soil. Annual plants sprout from seed every spring, grow quickly, and then die in the fall. Annuals generally have shallow roots. During dry years, so little water is available that many of the annuals die, although their seeds remain ready for the next rain. However, the deep-rooted perennials survive drought and protect the soil from wind erosion. In years of high rainfall, the annuals sprout quickly and prevent soil erosion and water runoff. Thus, the soil is protected in both dry and wet years.

Cultivated land is less resistant to changing weather. Generally, farmers till the land before they plant, removing the protective armor of natural vegetation. At this point, the field is vulnerable to erosion. If a drought occurs, the unprotected soil dries up and blows away. If spring rains are too heavy, the exposed soil may be eroded before the sprouts have an opportunity to grow. Many crops, such as corn, cotton, and vegetables, are cultivated in rows, and weeds growing between the rows are removed by herbicides or by further tilling. In short, much of the soil is exposed throughout the growing season. In addition, most crops are shallow-rooted annuals that cannot withstand drought.

During the early 1900s, improper farming practices led to a decline in soil fertility in the southwestern United States. Wind and water erosion had begun to damage the soil. Finally, when a prolonged drought occurred, the seeds failed to sprout, and the dry, exposed soil eroded rapidly. In 1934, a summer wind stripped the topsoil from entire counties and blew some of the dust more than 1500 kilometers eastward into the Atlantic Ocean. In total, 3.5 million hectares of farmland were destroyed, and productivity was reduced on an additional 30 million hectares. The same drought had little effect on parts of the prairie that had not been farmed.

During the 1920s and 1930s, windblown dust had a devastating effect on agriculture in the United States. This photograph, entitled *Buried Machinery,* was taken in South Dakota on May 13, 1936. *(Reprinted with permission of the National Archives)*

13.4 Dunes

A **dune** is a mound or ridge of wind-deposited sand. Dunes are common in regions where glaciers have recently melted, in deserts, and along sandy coastlines. Windblown sand travels by saltation; the grains bounce along like so many ping-pong balls. If the wind blows over a rock or a small clump of vegetation, the downwind, or **lee**, side of the obstacle provides a small, sheltered area where wind slows down. Grains of sand settle out in this protected zone. The growing mound of sand creates a larger windbreak, and more sand is deposited, forming a dune. Dunes commonly grow to heights of 30 to 100 meters, and some giants exceed 500 meters. In places they are tens or even hundreds of kilometers long.

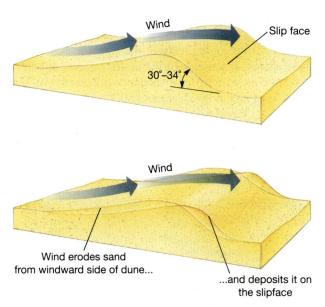

Figure 13.14 A dune migrates in a downwind direction as wind erodes sand from its windward side and deposits it on the slip face.

Most dunes are asymmetrical. Wind erodes sand from the windward side of a dune, and then the sand slides down the sheltered lee side, where it accumulates. In this way, dunes migrate in the downwind direction (Fig. 13.14). The leeward face of a dune is called the **slip face**. Typically the slip face is about twice as steep as the windward face. Wind speed, sand supply, and vegetation all influence the shape and orientation of a dune, as shown in Table 13.1.

Migrating dunes can overrun buildings and highways. For example, U.S. Highway 95 runs across the Nevada desert. Near the town of Winnemucca, dunes advance across the highway several times a year. High-

way crews must remove as much as 4000 cubic meters of sand to reopen the road.

Attempts are often made to stabilize dunes in inhabited areas. One solution is to plant vegetation to reduce deflation and stop dune migration. The main problem with this approach is that desert dunes commonly form in regions that are too dry to support vegetation. Another solution is to build artificial windbreaks to create dunes in places where they do the least harm. For example, a fence traps blowing sand and forms a dune, thereby protecting areas downwind. Fencing is a temporary solution, however, because eventually the dune covers the fence and continues to migrate. In Saudi Arabia, dunes are sometimes stabilized by covering them with tarry wastes from petroleum refining.

Fossil Dunes

When dunes are buried by other sediment and eventually lithified, the resulting sandstone retains the original sedimentary structures of the dunes. For example, Figure 13.15 shows a rock face in Zion National Park in Utah. Notice the sloping sedimentary layering. This rock has not been tilted by tectonic forces. It is a lithified dune, and the dipping beds are the layering of the dune's slip face. The bedding dips in the direction in which the wind was blowing when it deposited the sand. Notice that the planes dip in various directions in different layers. These dip changes indicate changes in wind direction. The layering is an example of **cross-bedding**, described in Chapter 3.

Loess

Most soils contain silt and sand. Wind-transported sand moves close to the ground by saltation and does not travel far. However, wind can carry finer silt for

Figure 13.15 Cross-bedded sandstone in Zion National Park preserves the sedimentary bedding of ancient sand dunes.

Table 13.1 • Types of Dunes

If a dune forms in a rocky area with little sand and sparse vegetation, the dune's center grows higher than its edges. When the dune migrates, the edges move faster because there is less sand to transport. The resulting **barchan dune** is crescent shaped, with the tips of the crescents pointing downwind. Barchan dunes are not connected to one another but instead migrate across the landscape independently.

Barchan dunes of Coral Pinks, Utah.

If sand is plentiful and vegetation sparse, the edges of a dune do not taper as in a barchan, and therefore they move at the same speed as the center. As a result, the dunes migrate as long, parallel ridges called **transverse dunes**. Transverse dunes form perpendicular to wind direction.

Transverse dunes.   *(Hubbard Scientific Co.)*

If vegetation is more plentiful, the plants may control dune shapes. If some disturbance destroys vegetation in a small area, deflation may create a blowout. Wind blows sand out of the depression and deposits it nearby in a **parabolic dune**. Note that parabolic dunes are similar in shape to barchan dunes, except that the tips of parabolic dunes point into the wind, not downwind. The tips of a parabolic dune are often anchored by clumps of vegetation.

These sand ripples are miniature parabolic dunes, showing the role of vegetation in anchoring the tips of the dune.

If wind direction is erratic but prevails from the same general quadrant of the compass, and if the supply of sand is limited, then **longitudinal dunes** form parallel to the wind direction. In portions of the Sahara Desert, longitudinal dunes are 100 to 200 meters high and as much as 100 kilometers long.

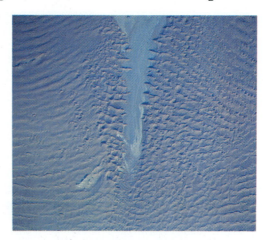

A satellite view of longitudinal dunes in the Namibian Desert.   *(NASA)*

Figure 13.16 Villagers in Askole, Pakistan, have dug caves in these vertical loess cliffs.

hundreds or even thousands of kilometers. Windblown silt can accumulate in thick deposits called **loess** (pronounced "luss"). Loess is porous and uniform and typically lacks layering. Often the individual silt particles are angular and therefore interlock. As a result, even though the loess is not cemented, it typically forms vertical cliffs and bluffs (Fig. 13.16).

The largest loess deposits in the world are found in central China. There, loess beds cover 800,000 square kilometers and reach a thickness of more than 300 meters. The silt was blown from the Gobi and the Takla Makan deserts of central Asia. The particles are so cohesive that people dug caves into the loess cliffs to make their homes. However, in 1920 a great earthquake caused the cave system to collapse, burying and killing an estimated 100,000 people.

Loess is also abundant in North America. The silt accumulated during the Pleistocene Ice Age, when continental ice sheets ground bedrock into fine rock flour. Streams carried this fine sediment away from the melting glaciers and deposited it in vast outwash plains. The outwash zones were cold, windy, and devoid of life, and wind easily picked up and transported the silt. Loess deposits in the United States range from about 1.5 meters to 30 meters thick (Fig. 13.17). Soils formed on loess are generally fertile and make good farmland. Much of the rich soil of the central plains of the United States formed on loess.

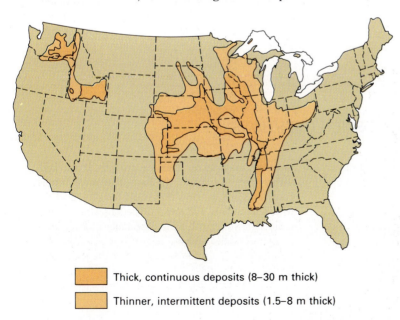

■ Thick, continuous deposits (8–30 m thick)

■ Thinner, intermittent deposits (1.5–8 m thick)

Figure 13.17 Loess covers much of the central plains in the United States.

SUMMARY

Deserts have an annual precipitation of less than 25 centimeters (10 inches). The world's largest deserts occur near 30° north and south latitudes, where warm, dry, descending air absorbs moisture from the land. Deserts also occur in rain shadows of mountains, continental interiors, and coastal regions adjacent to cold ocean currents.

Desert streams are often dry for much of the year but may develop **flash floods** when rainfall occurs. Alluvial fans are common in desert environments. **Playa lakes** are desert lakes that dry up periodically, leaving abandoned lake beds called **playas**.

Wind is an important agent of erosion, transport, and deposition of sediment in environments where little vegetation is present to protect soil. Sparse vegetation is common in deserts, sandy coastlines, and regions recently abandoned by glaciers. Wind selectively erodes silt and sand, leaving larger stones on the surface as **desert pavement**. Wind carries sand grains short distances, at a meter or less above the ground by **saltation**, but can transport silt great distances at higher elevations. Windblown particles are abrasive, but because the heaviest grains travel close to the surface, abrasion occurs mainly near ground level.

A **dune** is a mound or ridge of wind-deposited sand. Most dunes are asymmetrical, with gently sloping windward sides and steeper **slip faces** on the lee sides. Dunes migrate in a downwind direction. Wind-deposited silt is called **loess**.

KEY TERMS

Desert *285*
Rain shadow desert *285*
Wash *287*
Flash flood *287*
Debris flow *288*

Playa lake *288*
Playa *288*
Deflation *290*
Desert pavement *290*
Blowout *290*

Ventifact *291*
Dune *292*
Lee *292*
Slip face *293*
Barchan dune *294*

Transverse dune *294*
Parabolic dune *294*
Longitudinal dune *294*
Loess *295*

REVIEW QUESTIONS

1. Why are many deserts concentrated along zones at 30° latitude in both the northern and the southern hemispheres?
2. List three conditions that produce deserts.
3. Why do flash floods and debris flows occur in deserts?
4. Why are alluvial fans more prominent in deserts than in humid environments?
5. Compare and contrast floods in the desert with those in more humid environments.
6. Why is wind erosion more prominent in deserts and sandy coastlines than in humid regions?
7. Describe the formation of desert pavement.
8. Describe the evolution and shape of a barchan dune, a transverse dune, a parabolic dune, and a longitudinal dune. Under what conditions does each type of dune form?

DISCUSSION QUESTIONS

1. Coastal regions boast some of the wettest and some of the driest environments on Earth. Briefly outline the climatological conditions that produce coastal rainforests versus coastal deserts.
2. Imagine that you lived on a planet in a distant solar system. You had no prior information on the topography or climate of the Earth and were designing a robot spacecraft to land on Earth. The spacecraft had arms that could reach out a few meters from the landing site to collect material for chemical analysis. It also had instruments to measure the immediate meteorological conditions and cameras that could focus on anything within a range of 100 meters. The batteries on your radio transmitter had a life expectancy of 2 weeks. The spacecraft landed and you began to receive data. What information would convince you that the spacecraft had landed in a desert?
3. Deserts are defined as areas with low rainfall, yet water

is an active agent of erosion in desert landscapes. Explain this apparent contradiction.

4. Compare and contrast erosion, transport, and deposition by wind with erosion and deposition by streams and glaciers.

5. Imagine that someone told you that an alluvial fan had formed by wind deposition. What evidence would you look for to test this statement?

6. What type of dune forms under each of the following conditions? (a) Relatively dense vegetation cover, abundant sand supply, and strong wind. (b) Sparse vegetation and scanty sand supply.

7. Imagine that you were looking at a satellite photograph of a distant planet. What deductions could you make if you saw numerous sand dunes?

CHAPTER

14

Coastlines

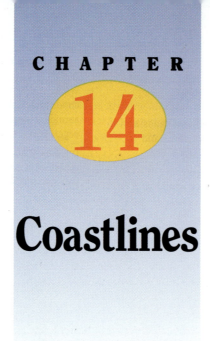

*Sea stacks along the Oregon coast
are wave-eroded remnants of bedrock.*

The seashore has always been an attractive place to live. The ocean, with its great mass of water, resists changes in temperature. Therefore, coastal regions are cooler in summer and warmer in winter than continental interiors. The sea provides food and transportation. In addition, we enjoy the salt air and find the rhythmic pounding of surf soothing and relaxing. Vacationers sail, swim, and fish along the shore. For all these reasons, coastlines have become heavily urbanized and industrialized. In the United States, more than 50 percent of the population, 40 percent of the manufacturing plants, and 65 percent of the electrical power generators are located within 80 kilometers of the oceans or the Great Lakes.

Problems often arise with coastal development because coastlines are among the most geologically active zones on Earth. Convergent plate boundaries occur along many continental coasts. Waves and currents weather, erode, transport, and deposit sediment continuously on all coastlines.

The shallow waters adjacent to continents and oceanic islands are productive biological zones. Many organisms that live in these regions build hard shells or skeletons of calcium carbonate; this material eventually becomes limestone. Thus, living organisms enter into geological cycles.

14.1 Waves

Waves form when wind blows across water. They vary from gentle ripples on a pond to destructive giants smashing against shore during a hurricane. When Hurricane Andrew struck the Florida and Louisiana coasts with 250-kilometer-per-hour winds in August 1992, the storm waves were 5 to 6 meters high. They rolled over low-lying barrier islands and inundated coastal areas. Wind and waves destroyed 100,000 homes, caused nearly $30 billion in damage, and killed 27 people (Fig. 14.1).

In deep water, the size of a wave depends on (1) wind speed, (2) the length of time the wind has blown, and (3) the distance the wind has traveled without being interrupted by land. (Sailors call this distance the fetch.) A gentle breeze blowing at 25 kilometers per hour (16 miles per hour) for 2 to 3 hours across a bay 15 kilometers wide generates waves about 0.5 meter high. In contrast, when a large Pacific storm blows at 90 kilometers per hour (55 miles per hour) for several days and waves roll uninterrupted for 3500 kilometers, they can become 30 meters high, as tall as a ship's mast.

The highest part of a wave is called the **crest**; the lowest is the **trough** (Fig. 14.2). We hardly ever talk about one wave, because waves occur in a series of crests and troughs. The **wavelength** is the distance between successive crests (or troughs).

If you make waves by throwing a rock into a calm lake, the waves travel across the water surface. However, the water does not move at the same speed or in the same direction as the waves. If you are sitting in a boat in deep water, you bob up and down and sway back and forth as the waves pass beneath you, but you do not travel along with the waves. In an ocean or lake wave, water moves in circular paths, as shown in Figure 14.3.

The movement of water continues downward below the wave trough in circles of decreasing size. At a depth equal to about one half the wavelength, the disturbance becomes negligible. Thus, if you dive deep enough, you escape wave motion.

As a wave passes over deep water, even the lowest circles of water motion do not contact the sea floor. But when a wave approaches shore and enters shallow water, its base begins to touch bottom. Friction between the wave and the sea floor slows the wave. Consequently, the wave behind it catches up and the wavelength decreases. In addition, friction slows the base of the wave more than the top (Fig. 14.4). As the crest rides over the trough, the wave steepens until it can no longer support itself. At this point it collapses forward, or breaks, as it rolls into shallower water. As an analogy, think of a skier whose skis catch on a branch hidden under the snow. The skis slow down while the person's upper body continues at constant speed. The skier pitches forward and somersaults into the snow. Once a wave breaks, its water flows toward the beach as a chaotic, turbulent mass called **surf**.

Wave Erosion

Just as stream erosion is most intense during a flood, most coastal erosion occurs during intense storms. A 6-meter-high storm wave strikes shore with 40 times the force of a 1.5-meter-high wave. A giant, 10-meter-high storm wave strikes a 10-meter-wide sea wall with four times the thrust energy of the three main orbiter engines of a space shuttle.

Figure 14.1 High winds from Hurricane Andrew destroyed these houses in Florida City, Florida. *(AP/Wide World Photos)*

A wave striking a rocky cliff drives water into cracks in the rock, compressing air in the cracks. The air and water together may dislodge rock fragments or even huge boulders. Engineers built a breakwater in Wick Bay, Scotland, of car-sized rocks weighing 80 to 100 tons each. The rocks were bound together with steel rods set in concrete, and the sea wall was topped by a steel and concrete cap weighing more than 800 tons. A large storm broke the cap and scattered the upper layer of rocks about the beach. The breakwater was rebuilt, reinforced, and strengthened, but a second storm de-

stroyed this wall as well. On the Oregon coast, a storm wave tossed a 60-kilogram rock over a 25-meter-high lighthouse. After sailing over the lighthouse, it crashed through the roof of the keeper's cottage, startling the inhabitants.

Although flying boulders are spectacular, most wave erosion occurs gradually. Water is too soft to abrade

Figure 14.2 Terminology used to describe waves.

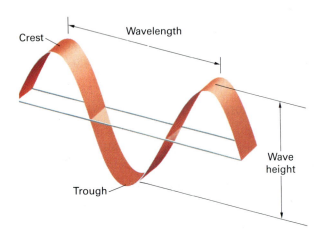

Figure 14.3 The movement of a wave and the movement of water within the wave.

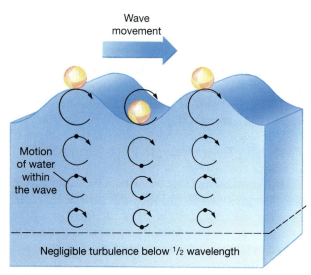

Figure 14.4 When a wave approaches the shore, the circular motion flattens out and becomes elliptical. The wavelength shortens, and the wave steepens until it finally breaks, creating surf. The dashed line shows the increase in wave height as it approaches a beach.

rock, but waves and currents wash sediment over bedrock, acting like liquid sandpaper. Seawater also dissolves soluble rock and carries off ions in solution.

14.2 Currents

An ocean wave moves along the sea surface, but the water in the wave travels in circles, ending up where it started. In contrast, a **current** is a continuous flow of water in a particular direction. Both surface and deepwater currents exist in the central oceans. These currents play important roles in regulating the Earth's climate by transporting warm water toward the poles in some places and cold water toward the equator in others. Geologists are interested primarily in currents adjacent to shore because these currents erode, transport, and deposit sediment and therefore shape coastal landforms.

Longshore Currents

After a wave breaks, the foamy surf flows up the beach and then recedes. If the wave strikes at an angle, water moves along the beach a short distance with each wave. Thus the water zigzags, as shown in Figure 14.5. The

net result produces a **longshore current** that flows parallel to the coast. Longshore currents flow in the surf zone and a little farther out to sea and may travel for tens or even hundreds of kilometers. They transport sand and other sediment parallel to a coastline. Most sediment found on a coast is not produced by erosion at that location, but rather is carried from the interior of the continent by streams and deposited in deltas on the coast. It is then redistributed along the coast by longshore currents.

14.3 Reefs

A **reef** is a wave-resistant ridge or mound built by corals, algae, and other organisms. Because these organisms need sunlight and warm, clear water to thrive, reefs develop in shallow tropical seas where no suspended clay or silt muddies the water (Fig. 14.6). As reef-building organisms die, their offspring grow on their remains. Thus, only the outer and topmost portions of a reef contain living organisms.

The South Pacific and portions of the Indian Ocean are dotted with numerous islands called atolls. An **atoll** is a circular reef that forms a ring of islands around a

Text continued on page 304

California surf.

Figure 14.5 Formation of a longshore current.

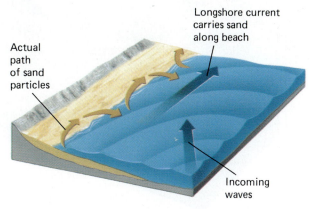

Energy from the Ocean

Energy from Tides   A tide is the cyclical rise and fall of ocean water caused by the gravitational force of the Moon and, to a lesser extent, of the Sun. If islands or narrow-mouthed bays constrict tidal flow, the moving water funnels into a tidal current. Twice a day, seawater flows into the bays with the rising tide, and twice a day it rushes outward (Fig. 1). The energy of these currents can be harnessed by a tidal dam and turbine.

During the 1800s, tidal power was popular in the United States, especially along the coast of Maine with its narrow bays and inlets. At that time, it was an attractive alternative to bulky steam engines that consumed large quantities of wood. However, when fossil fuels became cheap in the 1900s, almost all tidal power plants were abandoned. Today the potential still exists, but even in the most favorable sites tidal dams are not always economical. For example, a 250-megawatt tidal power station was built in France in 1968. Although technologically successful, this facility costs 2.5 times more per unit of energy than a comparable dam on a river.

In addition to being costly, tidal dams can create environmental problems. Tidal bays are often productive es-

Figure 1   Tidal currents can be harnessed to produce electricity.

tuaries where fish come to breed, and they are also popular places for recreation. If these areas are dammed, some of the natural qualities and resources are lost. Nevertheless, proponents of tidal projects point out that when fossil fuels become scarce, tidal energy might become attractive.

Figure 2   A schematic view of one design for harnessing wave energy. An incoming wave raises the water level inside a concrete chamber. Air inside the chamber is compressed and spins a turbine that powers a generator.

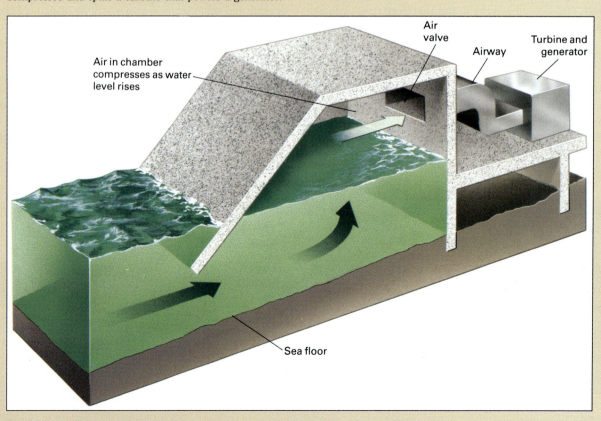

Air in chamber compresses as water level rises

Air valve

Airway

Turbine and generator

Sea floor

Energy from Ocean Waves   Waves strike every coastline in the world, and their combined energy is enormous. As the cost of oil rises and pollution and political availability of petroleum become more troublesome, we ask, "Is it possible and economical to harness wave energy?"

In 1990 engineers designed and built a simple wave generator in Northern Ireland, shown in Figure 2. When water from an incoming wave rises in a concrete chamber, it forces air through a narrow valve. The air spins a rotor that drives an electrical generator. The generator produces electricity for a small coastal village.

To date, the major barrier to widespread use of wave power has been economic. Although the wave energy along an entire coastline is enormous, the potential in any one place is small. Therefore, many wave generators are needed to harness large quantities of energy. At present, the generators are too expensive to be practical.

However, improvements in design and rising oil prices may reverse the economic balance sheet.

Ocean Thermal Power   It is also possible to use the heat of the ocean to produce electricity. In tropical regions, the sea surface is approximately 20°C warmer than subsurface water. The warm water is hot enough to vaporize a pressurized liquid such as ammonia. The gaseous ammonia can drive a turbine just as hot steam drives a turbine in a coal-fired plant. In an ocean thermal generator, the gaseous ammonia is cooled by the subsurface water and reused, as shown in Figure 3. Although the efficiency of such an engine is low, there is so much water in the oceans that a large amount of energy is available. Today the capital costs of ocean thermal power plants are so high that they are uneconomical even though they produce electricity without consuming any fuel.

Figure 3   A schematic view of an ocean thermal power plant. Pressurized ammonia is boiled by the warm surface waters of tropical oceans. The ammonia gas expands against the blades of a turbine, and the spinning turbine drives a generator to produce electricity. The gases are cooled and condensed by colder subsurface water that is pumped into the power plant.

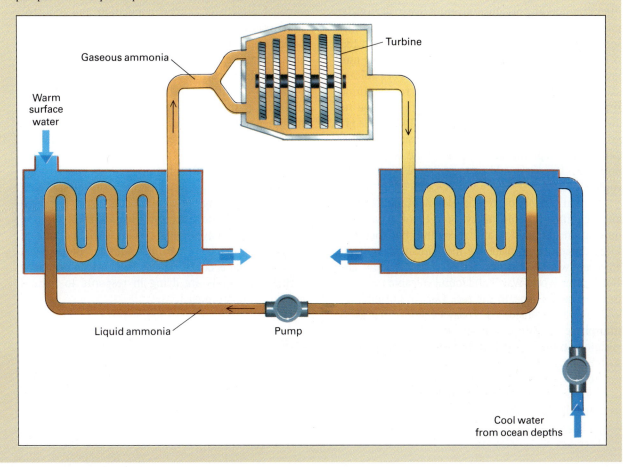

A   B

Figure 14.6   (A) An aerial view of a reef adjacent to the island of Palau in the southwestern Pacific. Note that the waves break on the reef, leaving calmer, shallow water between the reef and the island. (B) An underwater photograph of castle coral, a reef-building organism.   *(Larry Davis)*

calm, protected body of water called a **lagoon**. Atolls vary from 1 to 130 kilometers in diameter. They are surrounded by deep water of the open sea. If corals and other reef-building organisms cannot live in deep water, how did atolls form? Charles Darwin studied this question during his famous voyage on the *Beagle* from 1831 to 1836. He reasoned that to make an atoll, a reef must have formed in shallow water on the flanks of a volcanic island. Eventually the island sank as the reef continued to grow upward, so that the living portion of the reef always remained in shallow water (Fig. 14.7). This proposal was not well received at first because scientists could not accept the idea that volcanic islands sank. Moreover, Darwin could not explain why an island would sink.

Darwin's theory of atoll development mirrors Wegener's continental drift theory. Both theories were well founded on observation and reasoning, but they were rejected because no plausible mechanism was offered. However, when scientists drilled into a Pacific atoll shortly after World War II and found volcanic rock hundreds of meters beneath the reef, Darwin's original hypothesis was reconsidered. Today we know that as the lithosphere beneath a volcanic island cools after the volcano becomes extinct, it becomes denser and sinks isostatically, carrying the island down with it.

Reefs around the world have suffered severe epidemics of disease and predation within the last decade. Studies of fossils show that epidemics and even extinctions have affected reefs periodically for millions of years. However, the past few decades are different from any other interval in the Earth's history because of the phenomenal growth of human population and industry. Human activity has caused the extinction of uncounted species and has decimated the populations of many more. Therefore we ask, are the recent epidemics among reef organisms part of a natural process or a result of human activity? One possible explanation for the epidemics is that sewage provides nutrients for algae and other organisms that smother coral colonies and prevent new ones from forming. In addition, many scientists are concerned that air pollutants are increasing global temperature. Reef-building corals could be adversely affected by a rise in temperature of seawater. Both of these explanations are difficult to test. It is often difficult to locate the sources of chemical compounds in seawater. Some may originate as industrial pollutants, whereas others occur naturally. In addition, global warming remains controversial. Thus, we do not know whether the corals are dying in response to natural causes or human impact.

14.4   Beaches

When most of us think about going to the beach, we think of gently sloping expanses of sand. A **beach** is any strip of shoreline washed by waves and tides. Al-

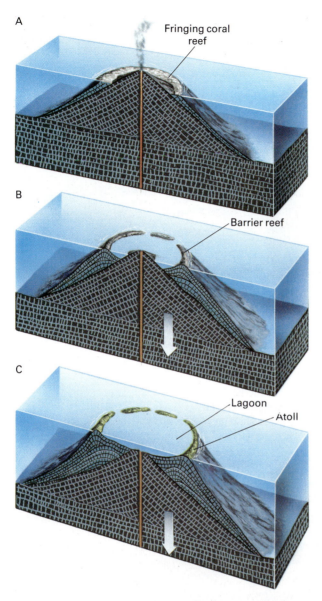

Figure 14.7   An atoll and lagoon develop as a volcanic island sinks.

though many beaches are sandy, others are swampy, rocky, or bounded by sheer cliffs (Fig. 14.8).

A beach is divided into two zones, the **foreshore** and the **backshore**. The foreshore, called the **intertidal zone** by biologists, lies between the high and low tide lines. It is alternately exposed at low tide and covered by water at high tide. The backshore is usually dry but is washed by waves during storms. These storms deposit and erode sediment and dislodge vegetation. Because many plants cannot survive periodic inundation by salt water, the backshore supports a sparse population of salt-resistant grasses. The backshore can be wide or narrow, depending on its topography and the frequency and intensity of storms. Along coasts where the elevation rises steeply, the backshore may be a narrow strip. In contrast, if the coast consists of low-lying plains where hurricanes occur regularly, the backshore may extend several kilometers inland as a zone of shifting coastal dunes partially stabilized by vegetation.

14.5   The Dynamic Nature of Coastlines

A developer builds a resort along a beautiful stretch of sandy beach. Within a few years the beach recedes, becoming narrower and narrower every winter. Then a hurricane rips into the coast and storm waves crash into the buildings.

Coastlines are dynamic geologic environments. Waves and longshore currents erode, transport, and deposit sediment. Sea level rises and falls. Convergent plate boundaries occur along many continental coasts.

Emergent and Submergent Coastlines

An **emergent coastline** forms when the land rises or sea level falls. In an emergent coastline a portion of the continental shelf that was previously under water becomes exposed as dry land. In contrast, a **submergent coastline** develops when coastal land sinks or sea level

Expensive condominiums along the Delaware coast will be flooded if sea level rises.

A

Figure 14.8 (A) A sandy beach on the southern shore of Long Island, New York. (B) A rocky beach on the northern coast of California.

B

Figure 14.9 Emergent and submergent coastlines. If sea level falls or if the land rises, the sediment-laden shallow sea floor is exposed, forming sandy beaches. If coastal land sinks or sea level rises, areas that were once land are flooded. Irregular shorelines develop and beaches are commonly sediment-poor.

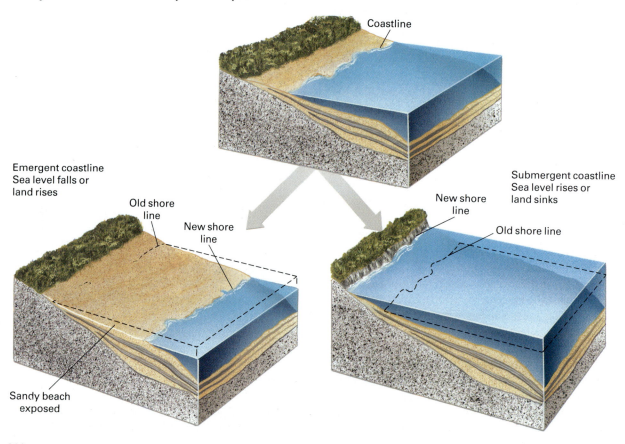

Coastline

Emergent coastline
Sea level falls or
land rises

Old shore
line

New shore
line

Sandy beach
exposed

Submergent coastline
Sea level rises or
land sinks

New shore
line

Old shore line

rises. The sea then floods low-lying areas, and the shore-line moves inland (Fig. 14.9).

Sediment-Rich and Sediment-Poor Coastlines

Many coastal landforms and the overall character of a shoreline depend on the amount of sand and other fine-grained sediment supplied to the coast. If a rich supply of sand is available, sandy beaches and other features built of sand dominate the coast.

Three processes can supply large quantities of sand and other sediment to a coastal area: local erosion of bedrock, transport from an outside source, and coastal emergence.

Some sediment forms by erosion of bedrock or coral reefs along a coast. However, under most conditions local erosion is too slow to supply abundant sediment to a coastline.

Sediment can also be transported from some other place. Rivers carry large quantities of sand, silt, and clay to the sea and deposit it on deltas that may cover thousands of square kilometers. In some regions, glaciers deposited large quantities of till along the coast. Longshore currents erode sediment from these sources and carry it along a coast. For example, much of the sand forming the beaches of the Carolinas and Georgia originated from the Hudson River delta between New York and New Jersey and from glacial deposits on Long Island and southern New England.

In addition to building sandy beaches, coastal processes deposit large quantities of sand and mud below sea level on the continental shelf. If a coastline rises or sea level falls, this vast supply of sediment is exposed to erosion. Thus, emergent coastlines are commonly rich in sediment.

In contrast, submergent coastlines are commonly sediment-poor. In many areas on land, bedrock is exposed or covered by only a thin layer of soil. if this type of terrain is submerged, and if rivers do not supply large amounts of sediment, the coastline is rocky. Submergent coasts are commonly irregular, with many bays and headlands. The coast of Maine, with its numerous inlets and rocky bluffs, is a submergent coastline (Fig. 14.10). Small sandy beaches form in protected coves, but most of the beaches are rocky and bordered by cliffs.

14.6   Sediment-Rich Coastlines

When large quantities of sediment are available, longshore currents and waves erode, transport, and deposit the sediment along the shore. A long ridge of sand or gravel extending out from a beach is called a **spit** (Fig.

Figure 14.10   The Maine coast is a rocky submergent coastline.

14.11). As sediment continues to migrate along the coast, the spit may grow. A well-developed spit may rise several meters above high-tide level and may be tens of kilometers long. If a spit grows until it blocks the entrance to a bay, it forms a **baymouth bar** (Fig. 14.12). The sand on a spit is continuously deposited and eroded by longshore currents and waves. Many seaside resorts are built on spits. Developers often ignore the fact that a spit is a transient and changing landform. If the rate of erosion exceeds that of deposition for a few years in a row, a spit can shrink or disappear completely, leading to destruction of beach homes and resorts.

A **barrier island** is a long, narrow, low-lying island that parallels the shoreline. It looks like a beach or spit

Figure 14.11   An aerial view of a spit on the south shore of Long Island, New York.

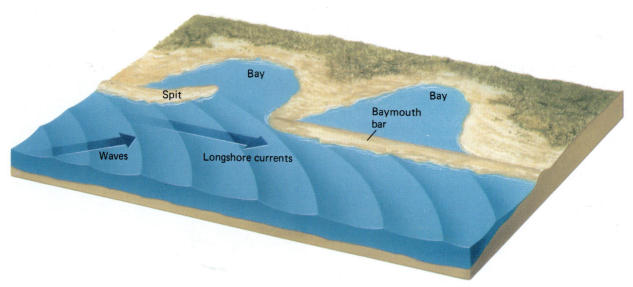

Figure 14.12   A spit forms where a longshore current carries sediment away from the shore and deposits it. If a spit closes the mouth of a bay, it becomes a baymouth bar.

separated from the mainland by a lagoon (Fig. 14.13). A chain of barrier islands extends along the east coast of the United States from New York all the way to Florida. The islands are so nearly continuous that a small boat can travel the entire coast inside the barrier island system and remain protected from the open ocean most of the time.

Barrier islands form in several ways. The two essential ingredients are a large supply of sediment and waves or currents to transport the sediment along the coast. If a coast is shallow for several kilometers outward from shore, storm waves break far out to sea. The breaking waves may carry sediment toward shore and deposit it where their energy diminishes, forming low-lying barrier islands. Barrier islands also form by deposition from longshore currents and are therefore similar to spits. If a longshore current veers out to sea, it slows down and deposits its sediment where it reaches deeper water. If this sediment is further piled up by waves, a barrier island may form. Yet another mechanism involves sea-level change. Underwater sandbars may be exposed as coastlines emerge. Alternatively, sand dunes or beaches may form barrier islands if a coastline sinks.

14.7   Sediment-Poor Coastlines

As explained earlier, submergent coastlines are commonly both irregular and sediment-poor. Erosion and deposition gradually straighten such an irregular coastline. As waves move inland from deeper water, they strike the tip of a headland first and break against the rocks, hurling sediment against the cliffs and eroding them. The part of the wave that strikes the headland

slows down. As a result, both sides bend around the headland. As the waves bend, the wave fronts stretch out and the energy decreases. In turn, when its energy decreases, the wave deposits sediment (Fig. 14.14). Thus, headlands erode and bays fill with sediment. In this way an irregular coastline eventually becomes smooth (Fig. 14.15).

Figure 14.13   An aerial view of a barrier island along the south coast of Long Island, New York.

Figure 14.14 When part of a wave strikes a headland, adjacent parts of the wave slow down and refract. As a result, the point of the headland is selectively eroded.

A

B

C

D

Figure 14.15 (A to C) A three-step sequence in which an irregular coastline is straightened. Erosion is greatest at the points of the headlands, and sediment is deposited inside the bays, leading to a gradual straightening of the shoreline. (D) Muir Beach, California, a sandy beach between rocky headlands.

Figure 14.16 A wave-cut cliff on the California coast.

A **wave-cut cliff** forms when waves erode the lower part of a rock face (Fig. 14.16). Wave erosion may undercut a cliff, and the overhanging rock eventually breaks loose and falls. As the cliff retreats, a flat **wave-cut platform** is left behind at sea level (Fig. 14.17).

If waves cut a cave into a narrow headland, the cave may eventually erode all the way through the headland, forming a scenic **sea arch**. When an arch collapses or when the inshore part of a headland erodes faster than the tip, a pillar of rock called a **sea stack** forms (Fig. 14.18). A sea stack is a temporary landform; eventually it crumbles to rubble.

Figure 14.17 A wave-cut platform on the Oregon coast.

14.8 Fjords and Estuaries

A **fjord** is a deep, long, narrow arm of the sea surrounded by high rocky cliffs (Fig. 14.19). The water in the fjord may be hundreds of meters deep, and often the cliffs drop straight into the sea. Fjords form in moun-

Figure 14.18 A sea stack on the Oregon coast.

Figure 14.19 Inugsuin Fjord on Baffin Island.

Figure 14.20 A satellite view of Chesapeake Bay. *(Chesapeake Bay Foundation)*

tainous coastal terrain where glaciers reach the sea. Whereas a stream can erode only to sea level, a glacier is massive enough to erode bedrock below sea level. When a glacier melts, it leaves a deep fjord. When the Pleistocene ice sheets melted about 10,000 years ago, so much water was added to the oceans that sea level rose by more than 100 meters. Fjords were deepened when the rising seas flooded glacially scoured coastal valleys.

In contrast to a fjord, an **estuary** is a broad river valley submerged by rising sea level or a sinking coast. Estuaries are shallow and have gently sloping beaches. Chesapeake Bay is a major estuary along the Atlantic coast of North America (Fig. 14.20). It formed by submergence of the Susquehanna River valley. It is approximately 100 kilometers long, averages 10 to 15 kilometers wide, and contains numerous bays and inlets. Despite its great size, Chesapeake Bay is only 7 to 10 meters deep near its mouth, and its greatest depth is 50 meters.

Estuaries are the richest protein-producing environments on Earth. Streams wash nutrients into an estuary, and the shallow water provides habitats for many marine organisms. Unfortunately for the organisms, estuaries also are prime sites for industrial activity. Inflowing rivers provide abundant fresh water, and although their shallow entrances sometimes require dredging, estuaries are protected and can be developed into excellent harbors. As a result, many estuaries have become seriously polluted in recent years.

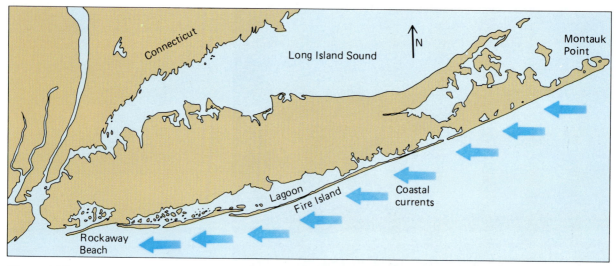

Figure 14.21 A map of Long Island showing barrier islands and the movement of longshore currents.

14.9 Beach Erosion and Human Settlement: The South Shore of Long Island

Long Island extends eastward from New York City and is separated from Connecticut by Long Island Sound (Fig. 14.21). A series of narrow, low-lying barrier islands lies along the southern coast of Long Island. Longshore currents flow westward, eroding sand from the eastern tip of the island and carrying it toward Rockaway Beach and New Jersey. The sand on the beaches and barrier islands is constantly eroded and carried westward, only to be replaced by sand from the eastern end of the island.

Are the beaches of Long Island stable or unstable? The answer to that question depends on our time perspective. Over geologic time, the beach is unstable. The sand at the eastern end of the island will become exhausted and the flow of sand will cease. Then the entire coastline will erode and the barrier islands and beaches will disappear. However, this change will not occur in the near future because of the vast amount of sand still

Figure 14.22 In the absence of industrialization, beaches on Long Island are eroded in winter. The beach sediment is carried back inshore the following summer, and the beach is reestablished.

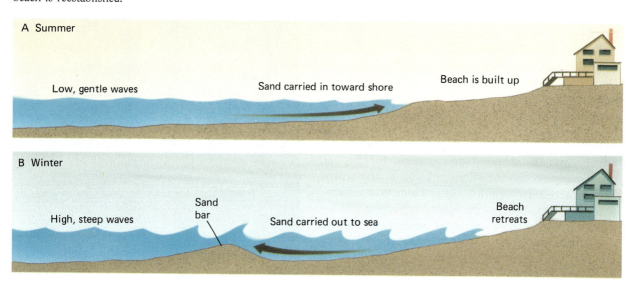

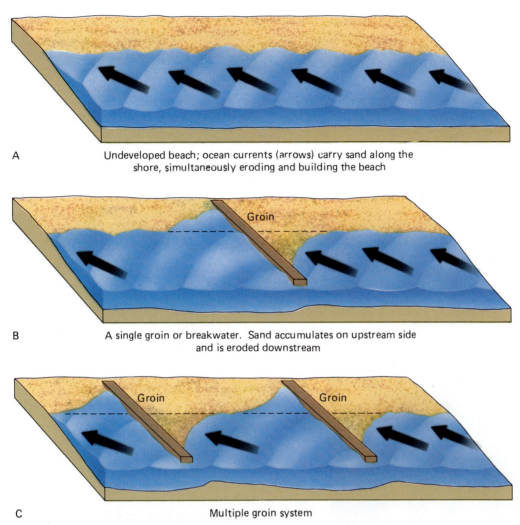

A Undeveloped beach; ocean currents (arrows) carry sand along the shore, simultaneously eroding and building the beach

B A single groin or breakwater. Sand accumulates on upstream side and is eroded downstream

C Multiple groin system

Figure 14.23 The effect of breakwaters on a beach. (A) An undeveloped beach. No net change occurs as sand is simultaneously eroded and deposited by longshore currents. (B) A single groin, or breakwater. Sand accumulates on the upstream side and is eroded downstream. (C) A multiple-groin system.

available at the east end of the island. Thus, the beaches are stable over a time perspective of years or decades. Longshore currents replace sand at the same rate at which they erode it and carry it westward.

If we narrow our perspective ever further and look at a Long Island beach over a season or during a single storm, it alternately shrinks and expands. Over such short times, the rates of erosion and deposition are not equal. Violent winter waves and currents erode beaches, whereas sand accumulates on the beaches during the calmer summer months (Fig. 14.22). In an effort to prevent these seasonal fluctuations and protect their personal beaches, Long Island property owners have built stone barriers called **groins**. If a groin is built

from shore out into the water, it intercepts the steady flow of sand moving from the east and keeps that particular part of the beach from eroding (Fig. 14.23B). But the groin impedes the overall flow of sand. West of the groin the beach erodes as usual, but the sand is not replenished because the upstream groin traps it. As a result, beaches downcurrent from the groin erode away (Fig. 14.24).

So now the land owners living downcurrent from a groin may decide to build a groin and pass the problem farther downstream (Fig. 14.23C). The situation has a domino effect, with the net result that millions of dollars must be spent in attempts to stabilize a system that was naturally stable in its own dynamic manner.

Storms pose another dilemma. Hurricanes occasionally strike Long Island in the late summer or fall, generating storm waves that completely overrun the barrier islands and roll inland, flattening dunes and eroding beaches. When the storms are over, gentler waves and longshore currents carry sediment back to the beach and rebuild it. As the sand accumulates again, salt marshes rejuvenate and the dune grasses grow back within a few months.

These short-term fluctuations are not compatible with human activity, however. People build houses, resorts, and hotels on or near the shifting sands. The

Figure 14.24 (A) Longshore currents on Long Island move from east to west. As a result, sand accumulates on the upstream (east) side of a groin, and the west side is eroded. Here waves lap against the foundation of a house along the eroded portion of the beach. (B) A close-up of the house in part (A).

A

B

owner of a home or resort hotel cannot allow the buildings to be flooded or washed away. Therefore, property owners construct large sea walls along the beach. How do they affect the natural system? When a storm wave rolls across a low-lying beach, it dissipates its energy gradually as it flows over the dunes and pushes the sand inland. The beach is like a master in judo who defeats an opponent by yielding with the attack, not countering it head on. A sea wall interrupts this gradual absorption of wave energy. The waves crash violently against the barrier and erode the base of the sea wall. When the sand on which the wall is built erodes, the wall collapses. It may seem surprising that a reinforced concrete sea wall is more likely to be destroyed permanently than a beach of grasses and sand dunes, yet this is often the case.

14.10   Global Warming and Sea-Level Rise

Sea level has risen and fallen repeatedly in the geologic past, and coastlines have emerged and submerged throughout the history of the planet. It is sobering to realize that coastlines will continue to change in the future. Can sea level rise sufficiently to flood coastal cities and towns?

Both tectonic events and climate change lead to sea-level changes. Tectonic events that cause sea-level fluctuations occur over millions of years and will not affect human settlement in the near future. But climate change can occur much more rapidly.

During the past 40,000 years, sea level has fluctuated by 140 meters, primarily in response to growth and melting of glaciers (Fig. 14.25). The rapid sea-level rise that started about 20,000 years ago began to level off about 7000 years ago. By coincidence, humans began to build cities about 7000 years ago. Thus, civilization has developed at a time when sea level has been relatively constant. However, if a shoreline consists of gently sloping land, a rise of a meter or two may move the beach a few kilometers inland, flooding thousands of square kilometers of land.

Many climatologists predict that the greenhouse effect will raise global temperature by 2° to 3°C in the next few decades. Global warming will melt portions of the huge Greenland and Antarctic ice caps, thus increasing the amount of water in the ocean. Additionally, when the atmosphere warms, seawater warms and expands. This expansion can cause sea level to rise.

During the past century, global sea level has risen 1 to 2.5 centimeters per decade. Many oceanographers

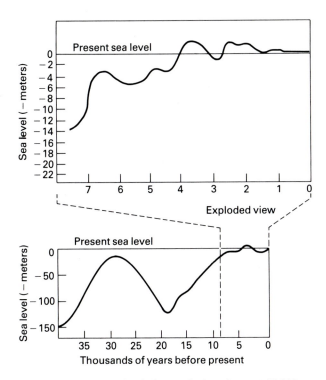

Figure 14.25 Sea-level change during the past 40,000 years. The upper graph shows an exploded view of sea-level change over the past 8000 years. *(J. D. Hansom, Coasts. Cambridge: Cambridge University Press, 1988)*

predict that sea level will continue to rise at this rate. In some regions, problems caused by global sea-level rise have been compounded by local tectonic sinking. In London, where the high-tide level has risen by 1 meter in the past century, a multimillion-dollar series of storm gates was built on the Thames River. A similar system is now planned to protect the city of Venice from further flooding. In North America, some planners are contemplating building gigantic levees to protect coastal cities, and they are visiting the Netherlands to see how the Dutch do it. The cost would exceed that of any construction project in history. People in poor lowland countries cannot afford massive costs. According to one estimate, a 3°C rise in global temperature would flood the Nile delta, displacing 10 million people, and the coastal plain of Bangladesh, displacing 38 million. However, not all scientists agree. One alternative explanation is that global warming will lead to the growth of glaciers. If global temperature increases, more water will evaporate from the oceans. More evaporation leads to more precipitation. Antarctica and Greenland are so cold that they would remain below freezing for most of the year even if the average temperature rose a few degrees. Thus, precipitation would fall as snow at higher latitudes and glaciers would grow. Water would be removed from the oceans and sea level would fall.

SUMMARY

Where waves do not interact with the sea floor, their size depends on (1) wind speed, (2) the amount of time the wind has blown, and (3) the distance that the wind has traveled. The highest part of a wave is the **crest**; the lowest, the **trough**. The distance between successive crests is called the **wavelength**. The water in a wave moves in a circular path. When a wave nears the shore, the bottom of the wave slows and the wave breaks, creating **surf**. Ocean waves erode rock and sediment by hydraulic action, solution, and abrasion and then transport and eventually deposit sediment.

Longshore currents transport sediment along a shore.

A **reef** is a wave-resistant ridge or mound built by corals and other organisms. A **beach** is a strip of shoreline washed by waves and tides.

If land rises or sea level falls, the coastline migrates seaward and abandons old beaches above sea level, forming an **emergent coastline**. In contrast, a **submergent coastline** forms when land sinks or sea level rises. Large quantities of sediment are transported to the sea by rivers. Glacial drift, reefs, and local erosion of rock also add sediment in certain areas. An emergent coastline is commonly sediment-rich because the sediment-rich continental shelf is exposed. A submergent coastline is usually sediment-poor because sediment-poor land surfaces are submerged. A sediment-rich coastline is characterized by sandy beaches, **spits**, **baymouth bars** and **barrier islands**. A sediment-poor coastline is characterized by **wave-cut cliffs**, **wave-cut platforms**, **arches** and **stacks**. Erosion and deposition straighten irregular coastlines. A **fjord** is a submerged glacial valley. An **estuary** is a submerged river bed and flood plain.

Human intervention such as the building of **groins** may upset the natural movement of coastal sediment and alter patterns of erosion and deposition on beaches. Sea level has risen over the past century and may continue to rise into the next.

KEY TERMS

Crest *299*
Trough *299*
Wavelength *299*
Surf *299*
Current *301*
Longshore current *301*
Reef *301*

Atoll *301*
Lagoon *304*
Beach *304*
Foreshore *305*
Backshore *305*
Intertidal zone *305*
Emergent coastline *305*

Submergent coastline *305*
Spit *307*
Baymouth bar *307*
Barrier island *307*
Wave-cut cliff *310*
Wave-cut platform *310*
Sea arch *310*

Sea stack *310*
Fjord *310*
Estuary *311*
Groin *313*

REVIEW QUESTIONS

1. List the three factors that determine the size of a wave.

2. Draw a picture of a wave and label the crest, the trough, the wavelength, and the wave height.

3. Describe the motion of both the surface and the deeper layers of water as a wave passes.

4. Explain how surf forms.

5. What drives longshore currents?

6. Explain the formation of an atoll.

7. List three different sources of coastal sediment.

8. What is an emergent coastline, and how does it form? Are emergent coastlines sediment-rich or sediment-poor? Why?

9. Compare and contrast a beach, a barrier island, and a spit.

10. Explain how coastal processes straighten an irregular coastline.

11. What is a groin? How does it affect the beach in its immediate vicinity? How does it affect the entire shoreline?

12. Explain how warming of the atmosphere could lead to a rise in sea level.

DISCUSSION QUESTIONS

1. Earthquake waves are discussed in Chapter 5. Compare and contrast earthquake waves with water waves.

2. How can a ship survive 10-meter-high storm waves, whereas a beach house can be smashed by waves of the same size?

3. Explain why very large waves cannot strike a beach directly in shallow coastal waters.

4. During World War II, few maps of the underwater profile of shorelines existed. When planning amphibious attacks on Pacific islands, the Allied commanders needed to know near-shore water depths. Explain how this information could be gotten from aerial photographs of breaking waves and surf.

5. Imagine that an oil spill occurs from a tanker accident. Discuss how longshore currents affect the dispersal of the oil.

6. Compare and contrast coastal erosion with stream erosion.

7. We stated in the text that submergent coastlines are commonly sediment-poor. However, the reverse is not necessarily true; sediment-poor coastlines are not necessarily submergent. Outline a sequence of events that would produce a rocky emergent coastline. Outline a sequence of events that would produce a sandy submergent coastline.

8. In Section 14.9 we explained how erosion and deposition tend to smooth out an irregular coastline by eroding headlands and depositing sediment in bays. If coastlines are affected in this manner, why haven't they all been smoothed out in the 4.6-billion-year history of the Earth?

9. Is Chesapeake Bay an emergent or submergent coastline? Briefly outline a plausible sequence of events that could have led to the formation of this coastline.

10. Prepare a three-way debate. Have one side argue that the government should support the construction of groins. Have the second side argue that the government should prohibit the construction of groins. Have the third side defend the position that groins should be permitted but not supported.

11. Prepare another debate on whether government funding should be used to repair storm damage to property on barrier islands.

12. Solutions to environmental problems can be divided into two general categories: social solutions and technical solutions. Social solutions involve changes in attitudes and lifestyles but do not generally require expensive industrial or technological processes. Technical solutions do not mandate social adjustments but require advanced and often expensive engineering. Describe a social solution and a technical solution to the problem of coastline changes on Long Island. Which approach do you believe would be more effective?

13. Imagine that sea level rises by 25 centimeters in the next century. How far inland would the shoreline advance if the beach (a) consisted of a vertical cliff, (b) sloped steeply at a 45° angle, (c) sloped gently at a 5° angle, and (d) were almost flat and rose only 1°?

Erle Kauffman was born in Washington, D.C. on February 9, 1933, and grew up on the Piedmont Plateau in suburban Maryland. His father, a noted conservationist and forester, strongly nurtured his early interests in natural history. In 1951, Erle entered the School of Forestry at the University of Michigan but transferred to geology in his junior year and graduated with his B.Sc. (major in geology, minor in biology) in 1955. Erle remained at Michigan for his M.Sc. (1956) and Ph.D. (1961), while his interests fluctuated between Paleozoic stratigraphy of the Michigan Basin, thrust tectonics of the Rocky Mountains, and, finally, high-resolution stratigraphy, sea-level history, and paleobiology of Cretaceous strata in the Western Interior Basin.

Professor Kauffman joined the Department of Paleobiology at the U.S. National Museum of Natural History, Smithsonian Institution, in 1960, where he became Curator of Fossil Mollusca, the first head of the Smithsonian's Science Diving Program, and Chairman of the Smithsonian Senate of Scientists. He researched a broad variety of topics in paleobiology and stratigraphy. Concurrently, he served as Adjunct Professor of geology at George Washington University, where he helped to build a strong program in paleontology and stratigraphy. Erle Kauffman moved to the Department of Geological Sciences at the University of Colorado in 1980 as chair (1980–1984) and subse-

quently was appointed Director of the Energy and Minerals Applied Research Center (EMARC). He is currently Professor of Geological Sciences at Colorado, in his thirty-seventh year of university teaching, and coordinates a large research program in paleobiology and stratigraphy. Kauffman was Visiting Professor at Oxford University, England (1970–1971), the Universitat Tubingen, Germany (1973–1974), the University of Colorado (1977–1978), and was Fulbright Lecturer to Australia in 1987. He has lectured at over 150 universities, institutes, museums, and corporations worldwide.

Professor Kauffman is a member of 17 professional societies and a fellow of the Geological Society of America and the American Association for the Advancement of Science. He was awarded an honorary doctorate degree from the Universitat Göttigen (1986) and the R. C. Moore Medal from the Society for Sedimentary Geology (SEPM) for innovative contributions to paleontology and stratigraphy. He received a U.S. Government Special Service Award for his work with the Smithsonian Science Diving program (1969), was elected Scientist of the Year by the Rocky Mountain Association of Geologists (1981), and received a Best Paper Award from SEPM (1982) for the paper, "Are These The Oldest Metazoan Trace Fossils?" coauthored with J. R. Steidtmann. In addition to his Fulbright Lectureship, Professor Kauffman has been twice Visiting Lecturer for the American Geological Institute and twice was chosen as Distinguished Lecturer for the American Association of Petroleum Geologists. In over 225 published works, Erle Kauffman has contributed significantly to a broad spectrum of interactive fields in paleobiology, evolution, extinction, stratigraphy, sedimentology, paleoceanography, and, more recently, global change and the modern biodiversity crisis.

What events in your precollege years initiated your interest in science?

From my first cognitive years, I was immersed in natural history. My parents came from opposite ends of the human spectrum. My mother was a concert pianist, a dancer, a writer, and deeply religious. My father was raised in the red clay of a rural Virginia farm; he was a gentle, quiet man who knew how to read the

Earth—an adventurer, a naturalist, a talented writer and storyteller, and an outspoken conservationist. The love of these two different people converged on me from all sides; my father constantly challenged me to look closely at my world, to see life at all scales (and to "leave it be"), and to "read the Earth." We spent a lot of time wandering the Appalachian forests. My mother's music taught me sensitivity and the aesthetic beauty of life. Even now, when I am doing field work in big, lonely places, and "reading the Earth," I can hear her music. I was weaned from the Appalachians at 14, taken to Colorado to work for the next decade of summers as a guide in the Snowmass-Maroon Bells Wilderness. This is the most beautiful place I know, held up with great porphyry stocks and thick piles of Pennsylvanian red beds. The sheer power of that country, and the strange anatomy of the rocks that I viewed from horseback, pulling a packstring over the high passes, set the course of my life. I had to learn how to "read the rocks."

As an undergraduate at Michigan, what influenced your decision to major in geology?

I'd like to say that I chose the University of Michigan over several other schools because of its great geosciences program. But I went there because it had a great track program and it had a great Forestry School. I had decided to follow in my father's footsteps and entered Michigan's Wildlife Biology program in 1951. I took my first geology course the following year—a historical geology survey taught by professor John A. Dorr. Jack was a charismatic lecturer who captivated the class with stories of dinosaur digs and restless magmas. He had a flair that I have rarely encountered in a teacher, and still try to emulate. My lifetime of learning how to understand the natural history of the Earth's surface suddenly took on a new dimension—the natural history of deep time. In short, I was pirated into geology by one great teacher at a pivotal time in my life. I got my first A in college at the end of my sophomore year, and never looked back. Packing 16 to 24 hours of classes into each semester of my junior and senior years, I graduated from Michigan with honors in geology, thus joining the ranks of the "late bloomers."

> The sheer power of that country . . . the strange anatomy of the rocks, set the course of my life.

What advice do you offer undergraduates who are thinking of becoming geology majors?

Geology is the most interdisciplinary of all sciences; in many areas it is also the least precise, as its data are filtered by time, erosion, diagenesis, and metamorphism. These two factors make geology interesting and challenging. Further, it deals with the unfamiliar throughout 3.5 billion years of Earth history. Over 90 percent of geological history occurred under the influence of global environments, characterized by strange beasts, that were very unlike those of today. We thus live in geologically atypical times. Interpreting Earth history is like reading one of those wonderfully complex British murder mysteries; there are lots of strange characters, no simple answers, and more unknowns than knowns. A good geologist today needs to understand the interactions between all dynamic forces in the Earth–Ocean–Atmosphere–Extraterrestrial system to accurately reconstruct even the smallest events in our planetary history from scattered pieces of evidence.

My advice to undergraduates who want to be geologists: (1) Initially seek a broad education in various aspects of the geosciences and peripheral disciplines, and learn how they interact; most of the exciting advances in the geosciences today involve interdisciplinary research aimed at studying the dynamic processes that drive Earth history. (2) Touch the rocks. Spend as much time as possible in the field, not just for class work, but on your own. Discover the many dimensions of geology, and learn how to read the Earth. The greatest joy of any natural science is being able to mentally see beyond the surface, to enjoy not only the look and feel of a mountain, but to know why it is there. (3) Because geological features on Earth are the result of diverse, interactive forces, and because the factual record is so incomplete over deep time, it is imperative for students of the geosciences to develop problem-solving skills and a keen sense of logic early in their careers. Learn how to think about relationships. (4) Communication skills are essential to your success. Learn how to write, illustrate, and carry out in-depth discussions which enhance understanding about your science. Computer and analytical literacy is mandatory. It is important to be able to communicate your science to the lay public as well.

The geosciences are a part of everyday life in our global society, not just in predicting earthquakes, volcanic eruptions, and floods, but also in understanding the devastating effects of the overpopulated human species on global change, and the spiraling loss of nonrenewable resources, natural habitats, and global species diversity. Geologists can provide ancient case histories and predictive models for today's environmental crisis.

What events led to your current research interests in high-resolution stratigraphy, paleobiology, evolution, and extinction theory?

I simply returned to my roots through a circuitous academic path. My intense childhood interest in natural history gave me an early sense of the ecosystem. The most memorable parts of my brief studentship as a wildlife biologist at Michigan were the innovative teaching of a great core of ecologists, especially Marston Bates, who taught me how to look for unifying principles in ecological systems. When I transferred to geology, I put these interests on hold to become a structural geologist, strongly influenced by the charisma of the great E. N. Goodard—father of the geological map of North America. I started a Ph.D. thesis along the eastern side of the Sangre de Cristo Range, in Huerfano Park, Colorado, interpreting the deformational fabric of Cretaceous rocks that formed the soles of huge Laramide thrust sheets. This was a messy and frustrating job. Below the rigid Laramide thrusts, the incompetent Cretaceous mudrocks were horribly deformed. Thin slivers of discrete Cretaceous formations were complexly interwoven beneath these thrusts. This made stratigraphic and structural analysis virtually impossible. To solve this problem, I had to back off and develop a high-resolution lithostratigraphic and biostratigraphic framework for the Cretaceous of the region. Whereas this ultimately allowed solution of the structural problems, in the process I realized that I was actually doing something far more fulfilling. I was once again a natural historian, reconstructing ancient ecosystems by interfacing sedimentologic and paleobiologic data with evidence for climate and sea-level history into a holistic view of a unique marine world—the great Western Interior epicontinental sea of North America. I've subsequently built my career around interdisciplinary physical, chemical, and biological approaches to problem solving and the generation and testing of theory in fields ranging from the patterns of mass extinction, to the prediction of source-rock quality and distribution in sedimentary basins.

Tell us more about your recent investigations.

The main focus of my research is the application of very high-resolution (cm-scale) stratigraphic data to geological problem solving and the testing and generation of theory. We have developed a highly refined system of stratigraphic analysis called HIRES (high-resolution event/cycle stratigraphy). Interactive student-faculty teams, members of which have different specialties, utilize this system to address important problems. High-resolution sedimentologic, geochemical, paleobiological, paleoclimatological, and paleoceanographic data are jointly gathered from a series of sections spanning the interval of interest; data are integrated through computer-aided graphic correlation into a regional matrix of geological time with resolution for correlation at less than 100,000 years per interval. We then utilize this chronostratigraphic and biostratigraphic matrix to study the three-dimensional behavior of geological or biological phenomena and to build predictive models based on our interpretations of these data. Current research projects utilizing this approach are: interpreting the fabric of extinction, survival and recovery of global ecosystems during a mass extinction interval; the evolution and extinction of Jurassic to recent Caribbean reef communities; the characterization and modeling of oceanic anoxic events and their expression in epicontinental seas; rates and patterns of molluscan evolution, Cretaceous sea-level history, and sequence stratigraphy; geological and biological evolution of the Western Interior Seaway in North America, and the passive margin of northern South America; studies in climate and ocean cycles, especially eustatic sea-level changes and Milankovitch climate cyclicity. I have always maintained an interest in descriptive stratigraphy and paleontology to maintain high quality in our data bases.

More recently my interests have turned toward the application of these high-resolution stratigraphic data to predictive modeling in understanding global change and the modern biodiversity crisis—already nearing mass extinction levels. Geology has much to offer here with resolution data and models. Man— *Homo sapiens*—has grossly overpopulated the Earth and ravaged its nonrenewable resources and natural habitats, causing extinction of nearly half the world's species, especially in tropical communities. The destruction of favorable global environments and the loss of biodiversity is occurring at a rate that equals or exceeds that of many ancient mass extinctions. The predictions of geological models for the modern global crisis are frightening; for example, they suggest that the recovery of complex tropical ecosystems (e.g., rainforests) from the impact of the human race may take millions of years. The application of high-resolution event stratigraphy to strata representing the last million years of Earth history is under way in the Caribbean and Tropical Atlantic to assess the relative effects of man and nature on global change. This may be the greatest challenge I will face in my science, and that with the greatest consequences.

In what way have your views about paleobiology and stratigraphy changed over the years?

In my lifetime, both of these fields have changed from largely documentary sciences—the necessary generation of quality data that underlies all interpretation and modeling—to largely interpretive sciences. Stratigraphy has evolved dramatically in its focus. The development of precisely integrated systems of event and cycle chronostratigraphy, assemblage biostratigraphy, more accurate radiometric ages, and refined magnetostratigraphy, has provided more precise chronology for regional correlation and modeling of even short-term events in Earth history. Recently developed seismic methods for examining stratigraphic successions have provided geologists with a better understanding of the interrelationships between sea-level change, sedimentation, and tectonics.

Paleobiology, in particular, has undergone an exciting revolution. It has evolved from a largely descriptive science, focused on systematics and biostratigraphy (the high-quality data base for interpretive science), to a science of interpretation and modeling of adaptive strategies, and evolutionary, ecological, and biogeographic history. From being users of evolutionary theory in the mid-1900s, paleobiologists have become leaders in the generation and testing of new theory. They have pioneered the exciting studies of macroevolution and mass extinction, with its implications to the modern biodiversity crisis. They have given a deep-time perspective to the evolution and extinction of global ecosystems, and a biogeographic perspective to plate tectonic theory. Paleobiology is now prepared to use its paleoenvironmental models for the interpretation of the causes and long-term effects of global change and the modern biodiversity crisis.

The Earth Through Time

◄ *In Paleozoic time, the sea teemed with a wide variety of life.*
(Smithsonian Institution)

Time and Geology

A large vein of light-colored granitic rock cuts across dark metamorphic rocks in the Beartooth Mountains of Wyoming. Cross-cutting relationships indicate that the granitic rocks are younger than the metamorphic rocks which they intrude.

We humans are fascinated by the concept of time. Geology instructors are aware of this interest because they are regularly asked the age of various rock and mineral specimens brought to them by students and returning vacationers. If told that the rock or fossil is tens or hundreds of millions of years old, the collectors are often pleased but also perplexed. "How can this person know the age of this specimen by just looking at it?" they think. If they insist on knowing the answer to that question, they are likely to receive a discourse on the subject of geologic time. It is explained that the rock exposures from which the specimens were obtained have long ago been organized into a standard chronologic sequence based largely on superposition, the stage of evolution indicated by fossils, and actual dating of rocks through the use of the radioactive elements they may contain. The geologist's initial estimate of age is likely to be based on experience. He or she may have spent a few hours kneeling at those same collecting localities and have a background of information to draw on. Thus, the geologist can often recognize particular rocks as being of a certain geologic age. The science that permits him to do this is **geochronology**. In recent decades, the science of geochronology has been greatly refined by methods that use radioactive isotopes to date rocks. But even with these new and precise tools, geologists still rely on concepts for determining the relative age of rock that were developed by such early workers as Nicolaus Steno, James Hutton, William Smith, and Charles Lyell.

15.1 Early Interpreters of Geologic Time

Nicolaus Steno

Nicolaus Steno (1638–1687) was a Danish physician who settled in Florence, Italy. There he became physician to the Grand Duke of Tuscany. Since the duke was a generous employer, Steno had ample time to tramp across the countryside, visit quarries, and examine strata. His investigations of sedimentary rocks led him to formulate such basic principles of historical geology as superposition, original horizontality, and original lateral continuity.

The **principle of superposition** states that in any sequence of undisturbed strata, the oldest layer is at the bottom and successively higher layers are successively younger. It is a rather obvious axiom. Yet Steno, on the basis of his observations of strata in northern Italy, was the first to explain the concept formally. The fact that it is self-evident does not diminish the principle's importance in deciphering earth history. Furthermore, the superpositional relationship of strata is not always apparent in regions where layers have been steeply tilted or overturned (Fig. 15.1). In such instances, the geologist must examine the strata for clues useful in recognizing their uppermost layer. The way fossils lie in the rock and the evidence of mud cracks and ripple marks are particularly useful clues when one is trying to determine which way is up.

The observation that strata are often tilted led Steno to his **principle of original horizontality**. He reasoned that most sedimentary particles settle from fluids under the influence of gravity. The sediment then must have been deposited in layers that were nearly horizontal and parallel to the surface on which they were accumulating (Fig. 15.2). Hence, steeply inclined strata indicate an episode of crustal disturbance after the time of deposition (see Fig. 15.1).

The **principle of original lateral continuity** was the third of Steno's stratigraphic axioms. It pertains to the fact that, as originally deposited, strata extend in all directions until they terminate by thinning at the margin of the basin, end abruptly against some former barrier to deposition, or grade laterally into a different kind of sediment. This observation is significant in that whenever one observes the exposed cross section of strata in a cliff or valley wall, one should recognize that the strata continue laterally for a distance that can be determined by field work and drilling. When geologists stand on a sandstone ledge at one side of a canyon, it is the principle of original lateral continuity that leads them to seek out the same ledge of sandstone on the far canyon wall and then to realize that the two exposures were once continuous.

Today, we recognize that Steno's principles are basic to the geologic specialty known as **stratigraphy**, which is the study of layered sedimentary rocks, including their texture, composition, arrangement, and correlation from place to place. Because stratigraphy enables geologic events as recorded in rocks to be placed

Figure 15.1 Steeply dipping strata grandly exposed in the Himalayan Mountains. It is often difficult to recognize the original tops of beds in strongly deformed sequences such as this. *(D. Bhattacharyya)*

Figure 15.2 Horizontal beds that have not been folded or tilted as a result of crustal movements and therefore have retained the horizontal orientation under which they were deposited. Such strata illustrate Steno's law of original horizontality.

in their correct sequence, it is the key to the history of the Earth.

James Hutton

James Hutton (1726–1797), an Edinburgh physician and geologist, is remembered for his understanding of the Earth's great age and of the processes that have continuously altered the planet's surface since the time of its origin. For Hutton, the Earth was a dynamic, ever-changing place in which new rocks, lands, and mountains arise continuously as a balance against their destruction by weathering and erosion. Hutton took a cyclic view of our planet as opposed to that of many of his contemporaries, who believed that the Earth had changed very little from its beginning to the present time. In addition, Hutton believed that the past history of our globe must be explained by what can be seen to be happening now. This simple yet powerful idea was later to be named **uniformitarianism**.

Perhaps because it is so general a concept, uniformitarianism has been reinterpreted and altered in a variety of ways by scientists and theologians from Hutton's generation down to our own. Some of the ideas about what uniformitarianism now implies would seem strange to Hutton himself. If the term *uniformitarianism* is to be used in geology (or any science), one must understand what is uniform. The answer is that the physical and chemical laws that govern nature are uniform. Hence, the history of the Earth may be deciphered in terms of present observations on the assumption that natural laws are invariant with time. These so-called natural laws are merely the accumulation of all our observational and experimental knowledge. They permit us to predict the conditions under which water becomes ice, the behavior of a volcanic gas when it is expelled at the Earth's surface, or the effect of gravity on a grain of sand settling to the ocean floor. Uniform natural laws govern such geologic processes as weathering, erosion, transport of sediment by streams, movement of glaciers, and the movement of water into wells.

Hutton's use of uniformitarianism was simple and logical. By observing geologic processes in operation around him, he was able to infer the origin of particular features he discovered in rocks. When he witnessed ripple marks being produced by wave action along a coast, he was able to state that an ancient rock bearing similar markings was once a sandy deposit of some equally ancient shore. And if that rock now lay far inland from a coast, he recognized the existence of a sea where Scottish sheep now grazed.

Hutton's method of interpreting rock exposures by observing present-day processes was given the catchy phrase "the present is the key to the past" by Sir Archibald Geike (1835–1924). The methodology implied in the phrase works very well for solving many geologic problems, but it must be remembered that the geologic past was sometimes quite unlike the present. For example, before the Earth had evolved an atmosphere like that existing today, different chemical reactions would have been prevalent during weathering of rocks. Life originated in the time of that primordial atmosphere under conditions that have no present-day counterpart. As a process in altering the Earth's surface, meteorite bombardment was once far more important than it has been for the last 3 billion years or so. Many times in the geologic past, continents have stood higher above the oceans, and this higher elevation resulted in higher rates of erosion and harsher climatic conditions compared with intervening periods when the lands were low and partially covered with inland seas. Similarly, at one time or another in the geologic past volcanism was more frequent than at present.

Nevertheless, ancient volcanoes disgorged gases and deposited lava and ash just as present-day volcanoes do. Modern glaciers are more limited in area than those of the recent geologic past, yet they form erosional and depositional features that resemble those of their more ancient counterparts. All of this suggests that present events do indeed give us clues to the past, but we must be constantly aware that in the past, the rates of change and intensity of processes often varied from those to which we are accustomed today, and that some events of long ago simply do not have a modern analogue.

The eighteenth-century concept of uniformitarianism was not the only contribution James Hutton made to geology. In his *Theory of the Earth,* published in 1785, he brought together many of the formerly separate thoughts of the naturalists who preceded him. He showed that rocks recorded events that had occurred over immense periods of time and that the Earth had experienced many episodes of upheaval separated by quieter times of denudation and sedimentation. In his own words, there had been a "succession of former worlds." Hutton saw a world of cycles in which water sculptured the surface of the Earth and carried the erosional detritus from the land into the sea. The sediment of the sea was compacted into stratified rocks, and then by the action of enormous forces, the layers were cast up to form new lands. In this endless process, Hutton found "no vestige of a beginning, no prospect of an end." In this phrase, we see Hutton's intoxicating view of the immensity of geologic time. No longer could geologists compress Earth history into the short span suggested by the Book of Genesis.

At Scotland's Isle of Arran and at Jedburgh, Scotland, Hutton came across an exposure of rock where steeply inclined older strata had been beveled by erosion and covered by flat-lying younger layers (Fig. 15.3 and 15.4). It was clear to Hutton that the older sequence was not only tilted but also partly removed by erosion before the younger rocks were deposited. The erosional surface meant that there was a gap or hiatus in the rock record. In 1805, Robert Jameson named this relationship an **unconformity**. More specifically, Hutton's Jedburgh exposure was an angular unconformity because the lower beds were tilted at an angle to the upper. This and other unconformities provided Hutton with evidence for periods of denudation in his "succession of worlds." Although he did not use the word *unconformity,* he was the first to understand and explain the significance of this concept.

As is now apparent, Hutton had a grand view of time and of the processes acting on the Earth. He was less interested in correlating strata and putting the pieces of a regional geologic puzzle together. This was to be the contribution of William Smith (1769–1839).

William Smith

William Smith was an English surveyor and engineer who devoted 24 years to the task of tracing out the strata of England and representing them on a map.

Figure 15.3 Angular unconformity at Siccar Point, eastern Scotland. It was here that James Hutton first realized the historical significance of an unconformity. The drawings on page 329 indicate the sequence of events documented in this famous exposure. *(E. A. Hay)*

Small wonder that he acquired the nickname "Strata Smith." He was employed to locate routes of canals, to design drainage for marshes, and to restore springs. In the course of this work, he independently came to understand the principles of stratigraphy, for they were of immediate use to him. By knowing that different types of stratified rocks occur in a definite sequence and that they can be identified by their lithology, the soils they form, and the fossils they contain, he was able to predict the kinds and thicknesses of rock that would have to be excavated in future engineering projects. His use of fossils was particularly significant. Prior to Smith's time, collectors rarely noted the precise beds from which fossils were taken. Smith, on the other hand, carefully recorded the occurrence of fossils and

quickly became aware that certain rock units could be identified by the particular assemblages of fossils they contained. He used this knowledge first to trace strata over relatively short distances and then to extend over great distances his "correlations" to strata of the same age but of different lithology. Ultimately, this knowledge led to the **principle of biologic succession**, which stipulates that the life of each age in the Earth's long history was unique for particular periods, that the fossil remains of life permit geologists to recognize contemporaneous deposits around the world, and that fossils could be used to assemble the scattered fragments of the record into a chronologic sequence.

Smith did not know why each unit of rock had a particular fauna. This was 60 years before the publica-

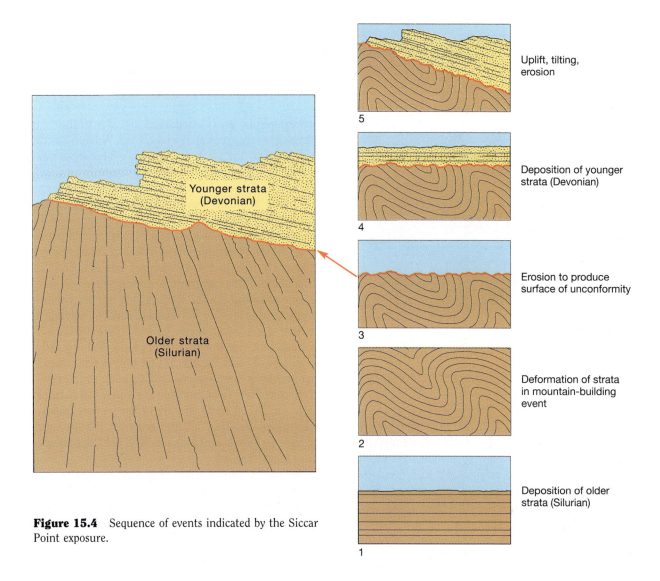

Uplift, tilting, erosion

5

Deposition of younger strata (Devonian)

4

Erosion to produce surface of unconformity

3

Deformation of strata in mountain-building event

2

Deposition of older strata (Silurian)

1

Younger strata (Devonian)

Older strata (Silurian)

Figure 15.4 Sequence of events indicated by the Siccar Point exposure.

tion of Darwin's *Origin of Species*. Today, we recognize that different kinds of animals and plants succeed one another in time because life has evolved continuously. Because of this continuous change, or evolution, only rocks formed during the same age could contain similar assemblages of fossils.

News of Smith's success as a surveyor spread widely, and he was called to all parts of England for consultation. On his many trips, he kept careful records of the types of rocks he saw and the fossils they contained. Armed with his notes and observations, in 1815 he prepared a geologic map of England and Wales that has remained substantially correct even today. His work clearly demonstrated the validity of the principle of biologic succession.

Charles Lyell

In the early nineteenth century, a geologist wrote a book that would both amplify the ideas of Hutton and present under one title the most important geologic concepts of the day. His name was Charles Lyell (Fig. 15.5), and he authored the classic *Principles of Geology*. The first volume of this work was printed in 1830. It grew to four volumes and became immensely important in the Great Britain of Queen Victoria. In these volumes one can recognize Lyell's skill in explaining and synthesizing the geologic findings of contemporary geologists. As his friend Andrew C. Ramsey remarked, "We collect the data, and Lyell teaches us the meaning of them." Lyell's *Principles* became the indispensable handbook

Figure 15.5   Sir Charles Lyell. *(Geological Society of London)*

of every English geologist. In it one finds many of the principles expressed earlier by Hutton regarding the recognition of the relative ages of rock bodies. For example, Lyell discusses the general principle that a geologic feature that cuts across or penetrates another body of rock must be younger than the rock mass penetrated. In other words, the feature that is cut is older than the feature that crosses it. This generalization, called the **principle of cross-cutting relationships**, applies not only to rocks but also to geologic structures like faults and unconformities. Thus, in Figure 15.6, fault *b* is younger than stratigraphic sequence *d*; the intrusion of igneous rock *c* is younger than the fault, since it cuts across it; and by superposition, rock sequence *e* is youngest of all.

Another generalization to be found in Lyell's *Principles* relates to **inclusions**. Lyell logically discerned that fragments within larger rock masses are older than the rock masses in which they are enclosed. Thus, whenever two rock masses are in contact, the one containing pieces of the other will be the younger of the two.

In Figure 15.7A, the pebbles of granite (a coarse-grained igneous rock) within the sandstone tells us that the granite is older and that the eroded granite fragments were incorporated into the sandstone. In Figure 15.7B, the granite was intruded as a melt into the sandstone. Because there are sandstone inclusions in the granite, the granite must be the younger of the two units.

15.2   Correlation

When examining an isolated exposure of a stratum in the bank of a stream or alongside a road, William Smith was aware that the stratum continued laterally across the countryside beneath a cover of loose sediment and soil and that the same layer, or its equivalent, is likely to be found at other localities. Determination of the equivalence of bodies of rock in different localities is called **correlation**. The rock bodies may be equivalent in their lithology (composition, texture, color, and so on), in their age, in the fossils they contain, or in a combination of these attributes. Because there is more than one meaning for the term, geologists are careful to indicate the kind of correlation used in solving a particular geologic problem. In some cases it is only necessary to trace the occurrence of a lithologically distinctive unit, and the age of that unit is not critical. Other problems can be solved only through the correlation of rocks that are of the same age. Such correlations involve time–rock units and are of the utmost importance in geology. They are the basis for the geologic time scale and are essential in working out the geologic history of any region.

The correlation of strata from one locality to another may be accomplished in several ways. If the strata are well exposed at the Earth's surface, as in arid regions, where soil and plant cover is thin, then it may be possible to trace distinctive rock units for many kilometers across the countryside by actually walking along the exposed strata. In using this straightforward method of correlation, the geologist can sketch the contacts between units directly on topographic maps or aerial photographs. The notations can then be used in the construction of geologic maps. It is also possible to construct a map of the contacts between correlative units in the field by using appropriate surveying instruments.

In areas where bedrock is covered by dense vegetation and a thick layer of soil, geologists must rely on intermittent exposures found here and there along the sides of valleys, in stream beds, and in road cuts. Correlations are more difficult to make in these areas but can be facilitated by recognizing the similarity in position of the bed one is trying to correlate with other units in the total sequence of strata. A formation may have changed somewhat in appearance between two localities, but if it always lies above or below a distinctive

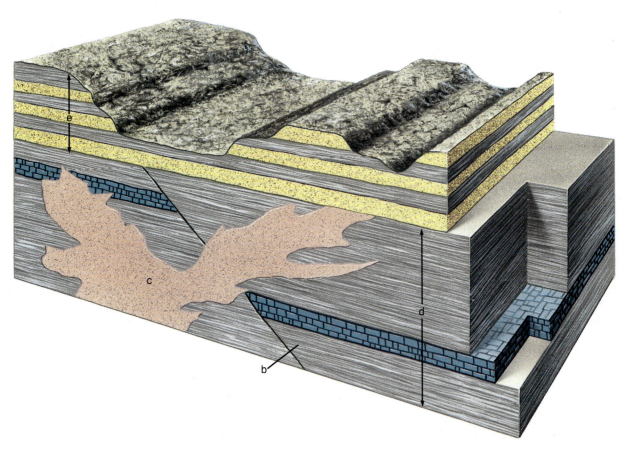

Figure 15.6 An example of how the sequence of geologic events can be determined from cross-cutting relationships and superposition. From first to last, the sequence indicated in the cross section is deposition of *d,* faulting to produce fault *b,* intrusion of igneous rock mass *c,* and erosion followed by deposition of *e.* Strata labeled *d* are oldest, and strata labeled *e* are youngest.

Figure 15.7 (A) Granite inclusions in sandstone indicate that granite is the older unit. (B) Inclusions of sandstone in granite indicate that sandstone is the older unit.

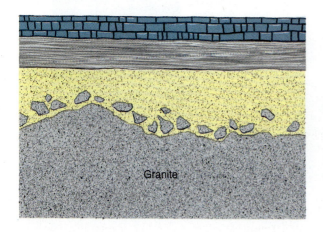

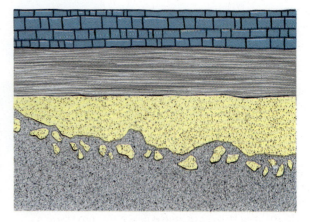

A   B

stratum of consistent appearance, then the correlation of the problematic formation is confirmed (Fig. 15.8).

A simple illustration of how correlations are used to build a composite picture of the rock record is provided in Figure 15.9. A geologist working along the sea cliffs at location 1 recognizes a dense oölitic limestone (formation F) at the lip of the cliff. (An oölitic limestone is a limestone composed largely of small spheres of calcite called oöids.) The limestone is underlain by formations E and D. Months later the geologist continues the survey in the canyon at location 2. Because of its distinctive character, the geologist recognizes the oölitic limestone in the canyon as the same formation seen earlier along the coast and makes this correlation. The formation below F in the canyon is somewhat more clayey than that at locality 1 but is inferred to be the same because it occurs right under the oölitic limestone. Working upward toward location 3, the geologist maps the sequence of formations from G to K. Questions still remain, however. What lies below the lowest formation thus far found? Perhaps years later an oil well, such as that at location 3, might provide the answer. Drilling reveals that formations C, B, and A lie beneath D. Petroleum geologists monitoring the drilling of the well would add to the correlations by matching all the forma-tions penetrated by the drill to those found earlier in outcrop. In this way, piece by piece, a network of correlations across an entire region is built up.

For correlations of rocks of a particular age, one cannot depend on similarities in lithology to establish equivalence. Rocks of similar appearance have been formed repeatedly over the long span of geologic time. Thus, there is the danger of correlating two apparently similar units that were deposited at quite different times. Fortunately, the use of fossils in correlation often prevents mismatching. Fossil correlations are based on the fact that animals and plants have undergone change through geologic time, and therefore the fossil remains of life are recognizably different in rocks of different ages. Conversely, rocks of the same age but from widely separated regions can be expected to contain similar assemblages of fossils.

Unfortunately, there are complications to these generalizations. For two strata to have similar fossils, they would have to have been deposited in rather similar environments. A sandstone formed on a river flood plain would have quite different fossils than one formed at the same time in a nearshore marine environment. How might one go about establishing that the flood plain deposit could be correlated to the marine deposit? In

Figure 15.8 If the lithology of a stratum is not sufficiently distinctive to permit it to be correlated to rocks of a distant locality, the position of the stratum in relation to distinctive rock units above and below may aid in correlation. In the example shown here, the limestone unit at locality A can be correlated to the lowest of the four limestone units at locality B because of its position between the gray shale and the sandstone units.

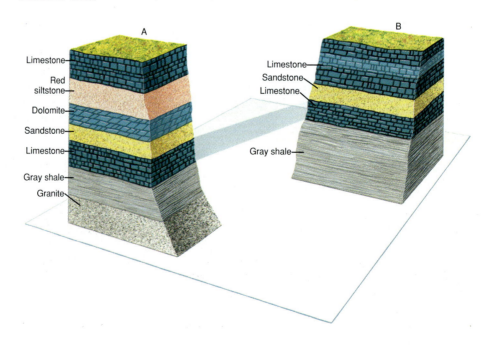

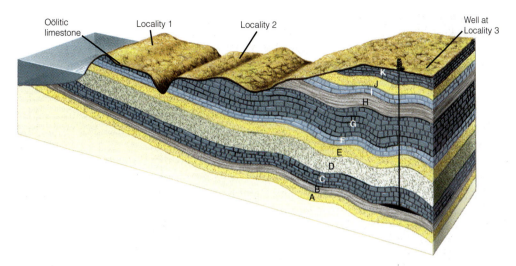

Figure 15.9 An understanding of the sequence of formations in an area usually begins with examination of surface rocks and correlation between isolated exposures. Study of samples from deep wells permits the geologist to expand the known sequence of formations and to verify the areal extent and thickness of both surface and subsurface formations.

some cases this might be done by physically tracing out the beds along a cliff or valley side. Occasionally, one is able to find fossils that actually do occur in both deposits. Pollen grains, for example, could have been wafted by the wind into both environments. Possibly, both deposits occur directly above a distinctive, firmly correlated stratum such as a layer of volcanic ash. Ash beds are particularly good time markers because they are deposited over a wide area during a relatively brief interval of time. Such key beds are exceptionally useful in establishing the correlation of overlying strata. Finally, the geologist may be able to obtain the actual age of the strata using radioactive methods, and these values can then be used to establish the correlation.

15.3 The Standard Geologic Time Scale

The early geologists had no way of knowing how many time units would be represented in the completed geologic time scale, nor could they know which fossils would be useful in correlation or which new strata might be discovered at a future time in some distant corner of the globe. Consequently, the time scale grew piecemeal, in an unsystematic manner. Units were named as they were discovered and studied. Sometimes the name for a unit was borrowed from local geography, from a mountain range in which rocks of a particular age were well exposed, or from an ancient tribe of Welsh people;

sometimes the name was suggested by the kind of rocks that predominated.

Divisions in the Geologic Time Scale

Geologists have proposed the term **eon** for the largest divisions of the geologic time scale. In chronologic succession, the first three eons of geologic time are the **Hadean**, **Archean**, and **Proterozoic** (see Fig. 16–18). The beginning of the Archean corresponds approximately to the ages of the oldest known rocks on Earth. Although not universally used, the term *Hadean* refers to that period of time for which we have no rock record, which began with the origin of the planet 4.6 million years ago. The Proterozoic Eon refers to the time interval from 2500 to 570 million years ago.

The rocks of the Archean and Proterozoic are informally referred to simply as Precambrian. The antiquity of Precambrian rocks was recognized in the mid-1700s by Johann Lehman, a professor of mineralogy in Berlin, who referred to them as the "primary series." One frequently finds this term in the writing of French and Italian geologists who were contemporaries of Lehman. In 1833, the term appeared again when Sir Charles Lyell used it in his formulation of a surprisingly modern geologic time scale. Lyell and his predecessors recognized these "primary" rocks by their crystalline character and took their uppermost boundary to be an unconformity that separated them from the overlying—and therefore younger—fossiliferous strata.

All of the remainder of geologic time is included in the fourth eon, the **Phanerozoic Eon**. As a result of careful study of superposition accompanied by correlations based on the abundant fossil record of the Phanerozoic, geologists have divided it into three major subdivisions, termed **eras**. The oldest is the **Paleozoic Era**, which we now know lasted about 370 million years. Following the Paleozoic is the **Mesozoic Era**, which continued about 170 million years. The **Cenozoic Era**, in which we are now living, began about 66 million years ago.

As shown in Figure 16.18, the eras are divided into shorter time units called **periods**; periods can in turn be divided into **epochs**. Eras, periods, epochs, and divisions of epochs, called **ages**, all represent intangible increments of time. The rocks laid down during a specific time (geochronologic) interval are called time-stratigraphic units. For example, the rocks comprising a geologic **system** were deposited or emplaced during a geologic period. Thus, one may properly speak of climatic changes during the Cambrian Period as indicated by fossils found in rocks of the Cambrian System. Each of the geologic systems is recognized by its distinctive fauna and flora of fossils. The fossils are different in stage of evolution from other fossils in both older and younger systems. **Series** is the time-stratigraphic term used for rocks deposited during an epoch, whereas **stage** represents the tangible rock record of an age (Table 15.1).

Recognition of Time Units

Units of geologic time bear the same names as the time–rock units to which they correspond. Thus, we may speak of the Jurassic System or the Jurassic Period according to whether we are referring to the rocks themselves or to the time during which they accumulated.

Time terms have come into use as a matter of convenience. Their definition is necessarily dependent on the existence of tangible time–rock units. The steps leading to the recognition of time–rock units began with the use of superposition in establishing age relationships. Local sections of strata were used by early geologists to recognize beds of successively different age and, thereby, to record successive evolutionary changes in fauna and flora. (The order and nature of these evolutionary changes could be determined because higher layers are successively younger.) Once the faunal and floral succession was deciphered, fossils provided an additional tool for establishing the order of events. They could also be used for correlation, so that strata at one locality could be related to the strata of various other localities. No single place on Earth contains a complete sequence of strata from all geologic ages. Hence, correlation to standard sections of many widely distributed local sections was necessary in constructing the geologic time scale. Clearly, the time scale evolved piece by piece as a result of the individual studies of many geologists. For some units at the series and stage level, the process continues even today. The fact that the time scale developed in piecemeal fashion is apparent when one reviews its evolution.

Development of the Geologic Time Scale

Most of the geologic systems of the Paleozoic were established by British geologists during the 1800s (Fig. 15.10). Adam Sedgwick (Fig. 15.11) named the **Cambrian** for strata exposed in Wales. The system of rocks takes its name from *Cambria,* which is the Latin name for Wales. An early Celtic tribe known as the *Ordovices* supplied Charles Lapworth with the name for the Ordovician System. Sir Roderick Impey Murchison coined the term **Silurian** for strata that he mapped in southern Wales; and Sedgwick and Murchison, working together, proposed the name **Devonian** for strata near Devonshire, England.

The **Carboniferous System** was named in 1822 by the British geologists William Conybeare and William Phillips for a sequence of strata that contained the coal beds of north-central England. In North America, the Carboniferous has been divided into the **Mississippian** and **Pennsylvanian** systems. The former is characterized by thick limestone sequences, and the latter by coal, dark shales, and siltstones. Finally, the **Permian System** takes its name from post-Carboniferous strata exposed in the province of Perm on the western side of the Ural Mountains in Russia.

The influence of British geologists in providing names for systems is not as evident in the development of Mesozoic nomenclature. The **Triassic**, for example, was a term applied in 1834 by the German geologist Frederick von Alberti for a threefold division of rocks

Table 15.1 • Hierarchy of Time and Time-Stratigraphic Terms

Time Divisions	Equivalent Time-Stratigraphic Divisions
Era	Erathem (rarely used)
Period	**System**
Epoch	**Series**
Age	**Stage**
Chron	**Zone** (Chronozone)

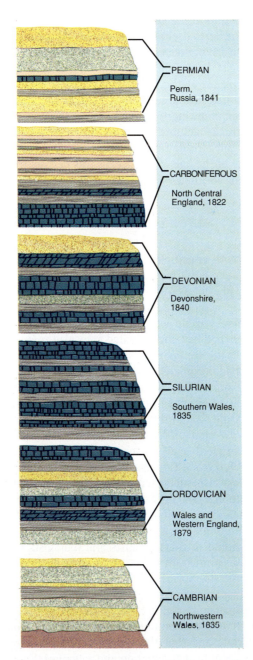

Figure 15.10   The standard geologic time scale for the Paleozoic and other eras developed without benefit of a grand plan. Instead, it developed by the compilation of "type sections" for each of the systems.

Figure 15.11   Adam Sedgwick, one of the foremost geologists of the nineteenth century. A professor of geology at Cambridge University, Sedgwick is best remembered for deciphering the highly deformed system of rocks in northwestern Wales that he defined as the Cambrian System. He also founded the geologic museum at Cambridge which bears his name.   *(Cambridge Museum, Cambridge, England)*

in Germany. Another German geologist, Alexander von Humboldt, proposed the term **Jurassic** for strata of the Jura Mountains. However, in 1795, when he used the term, the concept of systems was not fully developed. As a result, the Jurassic was redefined as a valid geologic system in 1839 by Leopold von Buch. Finally, the term **Cretaceous** (from the Latin *Creta,* meaning chalk) was proposed by the Belgian geologist J. J. D'Omallius d'Halloy in 1822. His selection of the term reflects the prevalence of chalk in the Cretaceous rocks of western Europe.

The derivation of the term **Tertiary**, used for the Tertiary System, takes us back to a time when geology was just beginning as a science. In 1759, the Italian Giovanni Arduino suggested that all rocks could be grouped into four divisions: Primary, Secondary, Tertiary, and Quaternary. Later the term *Tertiary* was defined precisely as the name for a sequence of rocks in France. The term **Quaternary**, also borrowed from an earlier nomenclature, was named for volcanic and sedimentary rocks resting on Tertiary strata in France.

The geologic pioneers who developed the time scale were influenced by conspicuous changes in assemblages of fossils as they compared one exposure of strata to others at distant locations. In many places, they found that such changes frequently occurred above and below an unconformity. The success of their methods is apparent from the fact that, by and large, the systems have persisted and found wide use even to the present day.

15.4   Quantitative Geologic Time

Early Attempts at Quantitative Geochronology

After having constructed a geologic time scale on the basis of relative age, it is understandable that geologists would seek some way to assign actual ages in millions of years to the various periods and epochs. From the time of Hutton, leaders in the scientific community were convinced that the Earth was indeed very old, and certainly it was much older than the approximately 6000 years estimated by biblical scholars from calculations involving the ages of post-Adamite generations. But how old was the Earth? And how might one quantify the geologic time scale?

To geologists of the 1800s, it was apparent that to determine the absolute age of the Earth or of particular rock bodies, they would have to concentrate on natural processes that continue in a single recognizable way and that also leave some sort of tangible record in the rocks. Evolution is one such process, and Charles Lyell recognized this. By comparing the amount of evolution exhibited by marine mollusks in the various series of the Tertiary System with the amount that had occurred since the beginning of the Pleistocene Ice Age, Lyell estimated that 80 million years had elapsed since the beginning of the Cenozoic. He came astonishingly close to the mark. However, for older sequences, estimates based on rates of evolution were difficult, not only because of missing parts in the fossil record, but also because rates of evolution for many taxa were not well understood.

In another attempt, geologists reasoned that if rates of deposition could be determined for sedimentary rocks, they might be able to estimate the time required for deposition of a given thickness of strata. Similar reasoning suggested that one could estimate total elapsed geologic time by dividing the average thickness of sediment transported annually to the oceans into the total thickness of sedimentary rock that had ever been deposited in the past. Unfortunately, such estimates did not adequately account for past differences in rates of sedimentation or losses to the total stratigraphic section during episodes of erosion. Also, some very ancient sediments were no longer recognizable, having been converted to igneous and metamorphic rocks in the course of mountain building. As a result of these uncertainties, estimates of the Earth's total age based on sedimentation rates ranged from as little as a million to over a billion years.

Investigators have also attempted to determine the total age of the Earth's oceans. They reasoned that the ocean basin had filled shortly after the origin of the planet, and thus could be only slightly younger than the age of the Earth itself. The best known of the calculations for the age of the oceans was made in 1901 by the Irish geologist John Joly. From information provided by gauges placed at the mouths of streams, Joly was able to estimate the annual increment of salt to the oceans. Then, knowing the salinity of ocean water and the approximate volume of water, he calculated the amount of salt already held in solution in the oceans. An estimate of the age of the oceans was derived from the following formula:

$$\text{Age of ocean} = \frac{\substack{\text{Total salt in ocean} \\ \text{(in grams)}}}{\substack{\text{Rate of salt added} \\ \text{(grams per year)}}}$$

Beginning with essentially nonsaline oceans, it would have taken about 90 million years for the oceans to reach their present salinity, according to Joly. The figure, however, was off the mark by a factor of 50, largely because there was no way to account accurately for recycled salt and salt incorporated into clay minerals deposited on the sea floors. Vast quantities of salt once in the sea had become extensive evaporite deposits on land; some of the salt being carried back to the sea had been dissolved, not from primary rocks, but from eroding marine strata on the continents. Even though in error, Joly's calculations clearly supported those geologists who insisted on an age for the Earth far in excess of a few million years. The belief in the Earth's immense antiquity was also supported by Darwin, Huxley, and other evolutionary biologists who saw the need for time in the hundreds of millions of years to accomplish the organic evolution apparent in the fossil record.

The opinion of the geologists and biologists that the Earth was immensely old was soon to be challenged by the physicists. Spearheading this attack was Lord Kelvin, considered by many to be the outstanding physicist of the nineteenth century. Kelvin calculated the age of the Earth on the assumption that it had cooled from a molten state and that the rate of cooling followed ordinary laws of heat conduction and radiation. Kelvin estimated the number of years it would have taken the Earth to cool from a hot mass to its present condition. His assertions regarding the age of the Earth varied over 2 decades of debate, but in his later years he confidently believed that 24 to 40 million years was a reasonable age for the Earth. The biologists and geologists found Kelvin's estimates difficult to accept. But how could they do battle against his elegant mathematics when they were themselves armed only with inaccurate dating schemes and geologic intuition? For those geologists unwilling to capitulate, however, new discoveries showed their beliefs to be correct and Kelvin's to be unavoidably wrong.

A more correct answer to the question "How old is the Earth?" was provided only after the discovery of radioactivity, a phenomenon unknown to Kelvin during his active years. With the detection of natural radioactivity by Henri Becquerel in 1896, followed by the isolation of radium by Marie and Pierre Curie 2 years later, the world became aware that the Earth had its own built-in source of heat. It was not inexorably cooling at a steady and predictable rate, as Kelvin had suggested.

Isotopic Methods for Dating Rocks

Atoms and Isotopes

As described in Chapter 2, an atom consists of a minute but heavy nucleus surrounded by rapidly moving, negatively charged **electrons**. Closely spaced positive particles called **protons** are in the nucleus, as well as **neutrons**, which have no charge. The number of protons in the nucleus establishes its number of positive charges and is called its **atomic number**. Each chemical element is composed of atoms having a particular atomic number. Thus, every element has a different number of protons in its nucleus. In naturally occurring elements, the number may range from 1 proton in hydrogen to 92 in uranium (Table 15.2).

The **mass** of an atom is approximately equal to the sum of the masses of its protons and neutrons. (The mass of electrons is so small that it need not be considered.) Carbon-12 is used as the standard for comparison of mass. By setting the atomic mass of carbon at 12, the atomic mass of hydrogen, which is the lightest of the elements, is just a bit greater than 1 (1.008, to be precise). The nearest whole number to the mass of a nucleus is called its **mass number**. Some atoms of the same substance have different mass numbers. Such variants are called **isotopes**. Isotopes are two or more varieties of the same element that have the same atomic number and chemical properties but differ in mass numbers because they have a varying number of neutrons in the nucleus. By convention, atomic mass is noted as a superscript preceding the chemical symbol of an element, and the atomic number is placed beneath it as a subscript. Thus, $_{20}^{40}Ca$ is translated as the element calcium having an atomic number of 20 and a mass number of 40 (see Table 15.2).

Radioactivity

The **radioactivity** discovered by Henri Becquerel was derived from elements such as uranium and thorium, which are unstable and break down or decay to form other elements or other isotopes of the same element.

Table 15.2 • Number of Protons and Neutrons and Atomic Mass of Some Geologically Important Elements

Element and Symbol	Atomic Number (Number of Protons in Nucleus)	Number of Neutrons in Nucleus	Atomic Mass
Hydrogen (H)	1	0	1
Helium (He)	2	2	4
Carbon-12 (C)*	6	6	12
Carbon-14 (C)	6	8	14
Oxygen (O)	8	8	16
Sodium (Na)	11	12	23
Magnesium (Mg)	12	13	25
Aluminum (Al)	13	14	27
Silicon (Si)	14	14	28
Chlorine-35 (Cl)*	17	18	35
Chlorine-37 (Cl)	17	20	37
Potassium (K)	19	20	39
Calcium (Ca)	20	20	40
Iron (Fe)	26	30	56
Barium (Ba)	56	82	138
Lead-208 (Pb)*	82	126	208
Lead-206 (Pb)	82	124	206
Radium (Ra)	88	138	226
Uranium-238 (U)	92	146	238

* When two isotopes of an element are given, the most abundant is starred.

FOCUS ON

Isotopic Dates for Sedimentary Rocks from Interstratified Ash

The geologic time scale was formulated as a relative scale. Since the development of isotopic dating methods, however, geologists have been working to provide numerical ages for the subdivisions of the scale. That effort continues today, with periodic improvements and revisions of the absolute ages that calibrate the time scale (see Fig. 16.18). Because the Phanerozoic time scale is based on fossiliferous strata, an often unattainable but ideal procedure would be to date sedimentary rocks directly by using minerals that crystallized in the environment of deposition. Unfortunately, it is far easier to find radioactive nuclides in the minerals of igneous and metamorphic rocks than in sediments. One rather problematic exception is the sedimentary mineral glauconite,

which can be dated by the potassium–argon method. Glauconite, however, loses argon on burial and thus is only useful in determining minimum ages. Far greater precision in dating sedimentary rocks can be obtained from radioactive isotopes in the minerals of volcanic ash beds that occur interstratified with sedimentary rocks. Based on uranium–lead ratios obtained from zircon crystals in ash layers, recently improved methods have provided numerical ages for key Paleozoic biostratigraphic boundaries with a precision of better than 1 percent. The analyses also provide ages for fossils in the sedimentary rocks and thereby lend important data for use in studies of rates of evolution, paleoecology, and paleogeography.

Ordovician limestone beds separated by a thin layer of altered volcanic ash called bentonite. The altered ash is in the recessed layer beneath the upper massive bed of limestone. It is about 8 inches thick. An age of 453.7 million years was determined for the ash layer by high-resolution U-Pb zircon dating methods.

Reference: R. D. Tucker et al., "Time-Scale Calibration by High-Precision U-Pb Zircon Dating of Interstratified Volcanic Ashes in the Ordovician and Lower Silurian Stratotypes of Britain," *Earth and Planetary Science Letters, 100* (1990), 51–58.

Any individual uranium atom, for example, will eventually decay to lead if given a sufficient length of time. To understand what is meant by "decay," let us consider what happens to a radioactive element such as uranium-238 (^{238}U). Uranium-238 has an atomic mass of 238. The "238" represents the sum of the weights of the atom's protons and neutrons (each proton and neutron having a mass of 1). Uranium has an atomic number (number

of protons) of 92. Such atoms with specific atomic number and weight are sometimes termed **nuclides**. Sooner or later (and entirely spontaneously) the uranium-238 atom will fire off a particle from the nucleus called an **alpha particle**. Alpha particles are positively charged ions of helium. They have an atomic weight of 4 and an atomic number of 2. Thus, when the alpha particle is emitted, the new atom will now have an atomic weight

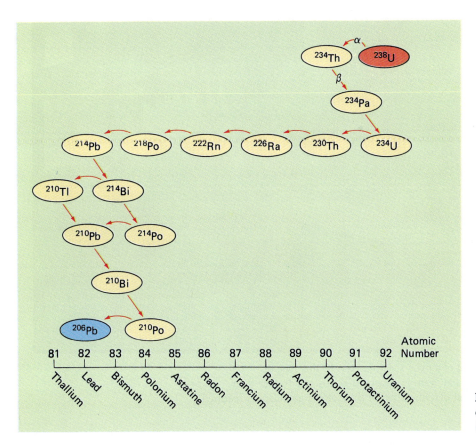

of 234 and an atomic number of 90. The new atom, which is formed from another by radioactive decay, is called a **daughter element**. From the decay of the parent nuclide, uranium-238, the daughter nuclide, thorium-234, is obtained (Fig. 15.12). A shorthand equation for this change is written as follows:

$$^{238}_{92}\text{U} \rightarrow {}^{234}_{90}\text{Th} + {}^{4}_{2}\text{He}$$

Thorium, in turn, is also radiactive and decays to a new daughter. Eventually, a stable isotope, lead-206, forms.

The rate of decay of radioactive isotopes is uniform and is not affected by changes in pressure, temperature, or the chemical environment. Therefore, once a quantity of radioactive nuclides has been incorporated into a growing mineral crystal, that quantity will begin to decay at a steady rate with a definite percentage of the radiogenic atoms undergoing decay in each increment of time. Each radioactive isotope has a particular mode of decay and a unique decay rate. As time passes, the quantity of the original or parent nuclide diminishes, and the number of the newly formed or daughter atoms increases, thereby indicating how much time has elapsed since the clock began its timekeeping. The beginning, or "time zero," for any mineral containing

radioactive nuclides would be the moment when the radioactive parent atoms became part of a mineral from which daughter elements could not escape. The retention of daughter elements is essential, for they must be counted to determine the original quantity of the parent nuclide.

The determination of the ratio of parent to daughter nuclides is usually accomplished with the use of a mass spectrometer, an analytic instrument capable of separating and measuring the proportions of minute particles according to their mass differences. In the mass spectrometer, samples of elements are vaporized in an evacuated chamber, where they are bombarded by a stream of electrons. This bombardment knocks electrons off the atoms, leaving them positively charged. A stream of these positively charged ions is deflected as it passes between plates that bear opposite charges of electricity. The degree of deflection depends on the charge-to-mass ratio. In general, the heavier the ion, the less it will be deflected (Fig. 15.13).

Of the three major families of rocks, the igneous clan is by far the best for isotopic dating. Fresh samples of igneous rocks are less likely to have experienced loss of daughter products, which must be accounted for in

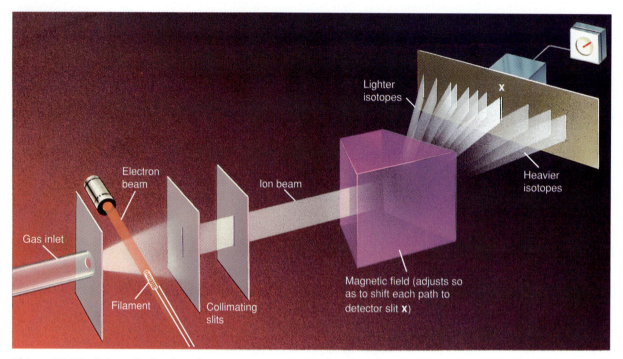

Figure 15.13 Schematic drawing of a mass spectrometer. In this type of spectrometer, the intensity of each ion beam is measured electrically (rather than recorded photographically) to permit determination of the isotopic abundances required for radiometric dating.

the age determination. Igneous rocks can provide a valid date for the time that a silicate melt containing radioactive elements solidified.

In contrast to igneous rocks, the minerals of sediments can be weathered and leached of radioactive components, and age determinations are far more prone to error. In addition, the age of a grain in sandstone does not give an age of the sandstone, but only of the parent rock that was eroded much earlier to produce the grains that were transported, deposited, and lithified to form the sandstone.

Dates obtained from metamorphic rocks may also require special care in interpretation. The age of a particular mineral may record the time the rock first formed or any one of a number of subsequent metamorphic recrystallizations.

Once an age has been determined for a particular rock unit, it is sometimes possible to use that data to approximate the age of adjacent rock masses. A shale lying below a lava flow that is 110 million years old and above another flow dated at 180 million years old must be between 110 and 180 million years of age (Fig. 15.14). Similarly, the age of a shale deposited on the erosional surface of a 490-million-year-old granite mass and covered by a 450-million-year-old lava flow must be between 450 and 490 million years old. The fossils in that shale might then be used to assign the shale to a particular

Figure 15.14 Igneous rocks that have provided absolute radiogenic ages can often be used to date sedimentary layers. (A) The shale is bracketed by two lava flows. (B) The shale lies above the older flow and is intruded by a younger igneous body. (Note: m.y. = million years.)

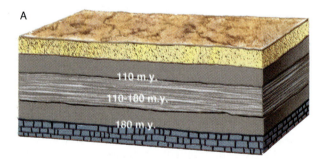

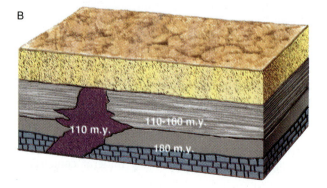

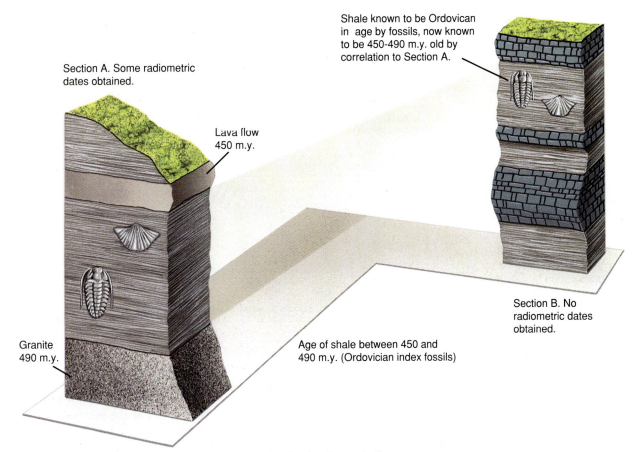

Section A. Some radiometric dates obtained.

Shale known to be Ordovician in age by fossils, now known to be 450-490 m.y. old by correlation to Section A.

Lava flow 450 m.y.

Section B. No radiometric dates obtained.

Granite 490 m.y.

Age of shale between 450 and 490 m.y. (Ordovician index fossils)

Figure 15.15 The actual age of rocks that cannot be dated radiometrically can sometimes be ascertained by correlation.

geologic system or series. Then by correlation, the quantitative age obtained at the initial location (Fig. 15.15, Section A) could be extended to correlative formations at other locations (Fig. 15.15, Section B).

Half-Life

One cannot predict with certainty the moment of disintegration for any individual radioactive atom in a mineral. We do know that it would take an infinitely long time for all of the atoms in a quantity of radioactive elements to be entirely transformed to stable daughter products. Experimenters have also shown that there are more disintegrations per increment of time in the early stages than in later stages (Fig. 15.16), and one can statistically forecast what percentage of a large population of atoms will decay in a certain amount of time.

Because of these features of radioactivity, it is convenient to consider the time needed for half of the original quantity of atoms to decay. This span of time is termed the **half-life**. Thus, at the end of the time constituting one half-life, $\frac{1}{2}$ of the original quantity of radioactive element still has not undergone decay. After

another half-life, $\frac{1}{2}$ of what was left remains or $\frac{1}{4}$ of the original quantity. After a third half-life, only $\frac{1}{8}$ would remain, and so on.

Every radioactive nuclide has its own unique half-life. Uranium-235, for example, has a half-life of 704 million years. Thus, if a sample contains 50 percent of the original amount of uranium-235 and 50 percent of its daughter product, lead-207, then that sample is 704 million years old. If the analyses indicate 25 percent of uranium-235 and 75 percent of lead-207, two half-lives would have elapsed, and the sample would be 1408 million years old (Fig. 15.17).

The Principal Geologic Timekeepers

At one time, there were many more radioactive nuclides present on Earth than there are now. Many of these had short half-lives and have long since decayed to undetectable quantities. Fortunately for those interested in dating the Earth's most ancient rocks, there remain a few long-lived radioactive nuclides. The most useful of these are uranium-238, uranium-235, rubidium-87, and potassium-40 (Table 15.3). There are also a few

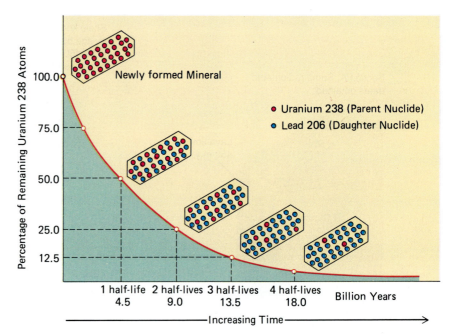

Figure 15.16 Rate of radioactive decay of uranium-238 to lead-206. During each half-life, one half of the remaining amount of the radioactive element decays to its daughter element. In this simplified diagram, only the parent and daughter nuclides are shown, and the assumption is made that there was no contamination by daughter nuclides at the time the mineral formed.

short-lived radioactive elements that are used for dating more recent events. Carbon-14 is an example of such a short-lived isotope. There are also short-lived nuclides that represent segments of a uranium or thorium decay series.

Uranium–Lead Methods Dating methods involving lead require the presence of radioactive nuclides of uranium or thorium that were incorporated into rocks when they originated. To determine the age of a sample

Figure 15.17 Radioactive decay of uranium-235 to lead-207.

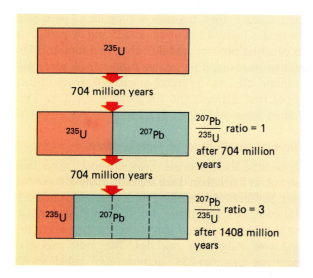

of mineral or rock, one must know the original number of parent nuclides as well as the number remaining at the present time. The original number of parent atoms should be equal to the sum of the present number of parent atoms and daughter atoms. The assumptions are made that the system has remained closed, so that neither parent nor daughter atoms have ever been added or removed from the sample except by decay, and that no daughter atoms were present in the system when it formed. The presence, for example, of original lead in the mineral would cause the radiometric age to exceed the true age. Fortunately, geochemists are able to recognize original lead and make the needed corrections.

As we have seen, different isotopes decay at different rates. Geochronologists take advantage of this fact by simultaneously analyzing two or three isotope pairs as a means to cross-check ages and detect errors. For example, if the $^{235}U/^{207}Pb$ radiometric ages and the $^{238}U/^{206}Pb$ ages from the same sample agree, then one can confidently assume the age determination is valid.

Isotopic ages that depend on uranium–lead ratios may also be checked against ages derived from lead-207 to lead-206. Because the half-life of uranium-235 is much less than the half-life of uranium-238, the ratio of lead-207 (produced by the decay of uranium-235) to lead-206 will change regularly with age and can be used as a radioactive timekeeper (Fig. 15.18). This is called a lead–lead age, as opposed to a uranium–lead age.

The Potassium–Argon Method Potassium and argon are another radioactive pair widely used for dating

Table 15.3 • Some of the More Useful Nuclides for Radioactive Dating

Parent Nuclide*	Half-Life (years)†	Daughter Nuclide	Materials Dated
Carbon-14	5730	Nitrogen-14	Organic materials
Uranium-235	704 million (7.04×10^8)	Lead-207 (and helium)	Zircon, uraninite, pitchblende
Potassium-40	1251 million (1.25×10^9)	Argon-40 (and calcium-40)‡	Muscovite, biotite, hornblende, volcanic rock, glauconite, K-feldspar
Uranium 238	4468 million (4.47×10^9)	Lead-206 (and helium)	Zircon, uraninite, pitchblende
Rubidium-87	48,800 million (4.88×10^{10})	Strontium-87	K-micas, K-feldspars, biotite, metamorphic rock, glauconite

* A nuclide is a convenient term for any particular atom (recognized by its particular combination of neutrons and protons).

† Half-life data from R. H. Steiger and E. Jäger, "Subcommission on Geochronology: Convention on the Use of Decay Constants in Geo- and Cosmochronology," *Earth and Planetary Science Letters 36* (1977), 359–362.

‡ Although potassium-40 decays to argon-40 and calcium-40, only argon is used in the dating method because most minerals contain considerable calcium-40, even before decay has begun.

rocks. By means of electron capture (causing a proton to be transformed into a neutron), about 11 percent of the potassium-40 in a mineral decays to argon-40, which may then be retained within the parent mineral. The remaining potassium-40 decays to calcium-40 (by emission of a beta particle). The decay of potassium-40 to calcium-40 is not useful for obtaining numerical ages because radiogenic calcium cannot be distinguished from original calcium in a rock. Thus, geochronologists concentrate their efforts on the 11 percent of potassium-40 atoms that decay to argon. One advantage of using argon is that it is inert—that is, it does not combine chemically with other elements. Argon-40 found in a mineral is very likely to have originated there following the decay of adjacent potassium-40 atoms in the mineral. Also, potassium-40 is an abundant constituent of many common minerals, including micas, feldspars, and hornblendes. The method lends itself also to the dating of whole-rock samples of solidified lavas. However, like all radiometric methods, potassium–argon dating is not without its limitations. A sample will yield a valid age only if none of the argon has leaked out of the mineral being analyzed. Leakage may indeed occur if the rock has experienced temperatures above about 125°C. In specific localities, the ages of rocks dated by this method reflect the last episode of heating rather than the time of origin of the rock itself.

The half-life of potassium-40 is 1251 million years. As illustrated in Figure 15.19, if the ratio of potassium-40 to daughter products is found to be 1 to 1, then the age of the sample is 1251 million years. If the ratio is 1 to 3, then yet another half-life has elapsed, and the rock would have an age of two half-lives, or 2502 million years.

Potassium–argon dating is widely used in deciphering various types of geologic problems. Geologists are now using the method in studies relating to sea-floor movements. For many years, scientists have been curious about the alignment of the major Hawaiian Islands and the adjacent seamounts. With the advent of the theory of sea-floor spreading, scientists developed the concept that these volcanic islands were built over a relatively fixed "hot spot" deep in the upper mantle (Fig. 15.20). Conduits from the hot spot brought lavas up to the sea floor, where eruptions periodically occurred. Volcanoes that developed over the hot spot were then conveyed along by sea-floor movement, and new volcanoes were produced over the vacated position.

Figure 15.18 Graph showing how the ratio of lead-207 to lead-206 can be used as a measure of age.

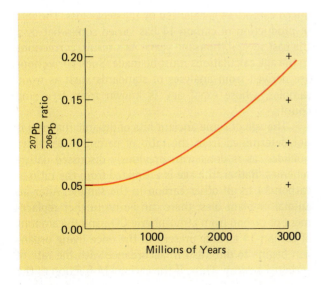

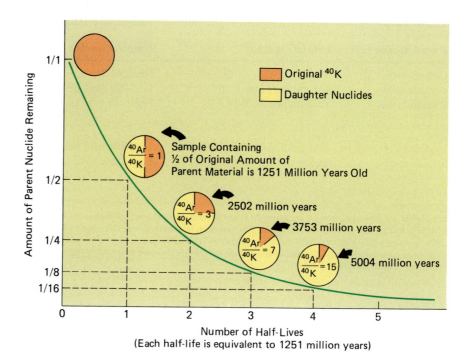

Figure 15.19 Decay curve for potassium-40.

Geologists reasoned that if this process had taken place in the Hawaiian Islands, then potassium–argon radiometric ages should change in sequence along the island–seamount chain.

The Carbon-14 Method   Techniques for age determination based on content of radiocarbon were first devised by W. F. Libby and his associates at the University of Chicago in 1947. The method is an indispensable aid to the archeologic research and is useful in deciphering very recent events in geologic history. Because of the short half-life of carbon-14—a mere 5730 years—organic substances older than about 50,000 years contain very little carbon-14. New techniques, however, allow geologists to extend the method's usefulness back to almost 100,000 years.

Unlike uranium-238, carbon-14 is created continuously in the Earth's upper atmosphere. The story of its origin begins with cosmic rays, which are extremely high-energy particles (mostly protons) that bombard the Earth continuously. Such particles strike atoms in the upper atmosphere and split their nuclei into small particles, among which are neutrons. Carbon-14 is formed when a neutron strikes an atom of nitrogen-14. As a result of the collision, the nitrogen atom emits a proton, captures a neutron, and becomes carbon-14 (Fig. 15.21). Radioactive carbon is being created by this process at the rate of about two atoms per second for every square centimeter of the Earth's surface. The

newly created carbon-14 combines quickly with oxygen to form CO_2, which is then distributed by wind and water currents around the globe. It soon finds its way into photosynthetic plants because they utilize carbon dioxide from the atmosphere to build tissues. Plants containing carbon-14 are ingested by animals, and the isotope becomes a part of their tissue as well.

Eventually, carbon-14 decays back to nitrogen-14 by the emission of a beta particle. A plant removing CO_2 from the atmosphere should receive a share of carbon-14 proportional to that in the atmosphere. A state of equilibrium is reached in which the gain in newly produced carbon-14 is balanced by the decay loss. The rate of production of carbon-14 has varied somewhat over the past several thousand years. As a result, corrections in the age calculations must be made. Such corrections are derived from analyses of standards such as wood samples, whose exact age is known from tree ring counts.

The age of some ancient bits of organic material is not determined from the ratio of parent to daughter nuclides, as is done with previously discussed dating schemes. Rather, the age is estimated from the ratio of carbon-14 to all other carbon in the sample. After an animal or plant dies, there can be no further replacement of carbon from atmospheric CO_2, and the amount of carbon-14 already present in the once living organism begins to diminish in accordance with the rate of carbon-14 decay. Thus, if the carbon-14 fraction of the

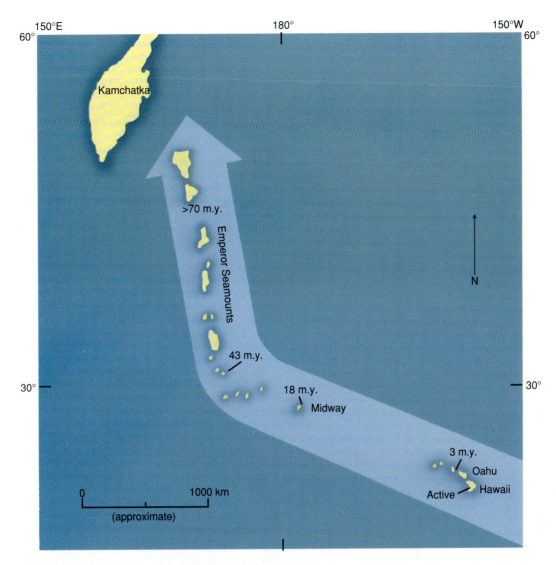

Figure 15.20   Bend in trend of the Hawaiian Island–Emperor Seamount chain was probably caused by change in direction of movement of the Pacific tectonic plate.   *(From J. S. Watkins, M. L. Bottino, and M. Morisawa,* Our Geological Environment. *Philadelphia: W. B. Saunders Company, 1975.)*

total carbon in a piece of pine tree buried in volcanic ash were found to be about 25 percent of the quantity in living pines, then the age of the wood (and the volcanic activity) would be two half-lives of 5730 years each or 11,460 years.

The carbon-14 technique had considerable value to geologists studying the most recent events of the Pleistocene Ice Age. Prior to the development of the method, the age of sediments deposited by the last advance of continental glaciers was surmised to be about 25,000 years. Radiocarbon dates of a layer of peat beneath the glacial sediments provided an age of only 11,400 years. The method has also been found useful in studies of ground-water migration and in dating the geologically recent uppermost layer of sediment on the sea floors.

The Age of the Earth

Anyone interested in the total age of the Earth must decide what event constitutes its "birth." Most geologists assume that "year 1" commenced as soon as the Earth had collected most of its present mass and had developed a solid crust. Unfortunately, rocks that date from those

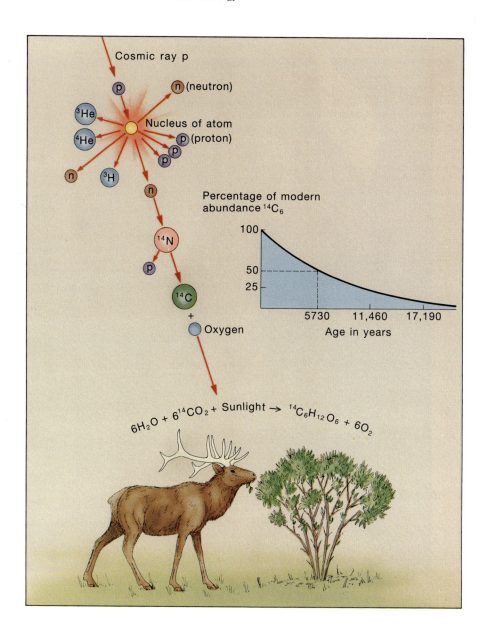

Figure 15.21 Carbon-14 is formed from nitrogen in the atmosphere. It combines with oxygen to form radioactive carbon dioxide and is then incorporated into all living things.

earliest years have not been found on Earth. They have long since been altered and converted to other rocks by various geologic processes. The oldest materials known are grains of the mineral zircon taken from a sandstone in western Australia. The zircon grains are 4.1 to 4.2 billion years old. The zircons were probably eroded from nearby granitic rocks and deposited, along with quartz and other detrital grains, by rivers. Other very old rocks on Earth include 3.7-billion-year-old granites of southwestern Greenland, metamorphic rocks of about the same age from Minnesota, and 3.96-billion-year-old rocks from the Northwest territories of Canada (north of Yellowknife, Canada).

    Meteorites, which many consider to be remnants of a shattered planet or asteroid that originally formed

at about the same time as the Earth, have provided uranium–lead and rubidium–strontium ages of about 4.6 billion years. From such data, and from estimates of how long it would take to produce the quantities of various lead isotopes now found on the Earth, geochronologists feel that the 4.6-billion-year age for the Earth can be accepted with confidence. Evidence substantiating this conclusion comes from returned Moon rocks. The ages of these rocks range from 3.3 to about 4.6 billion years. The older age determinations are derived from rocks collected on the lunar highland, which may represent the original lunar crust. Certainly, the moons and planets of our Solar System originated as a result of the same cosmic processes and at about the same time.

SUMMARY

In no other science does time play as significant a role as in geology. Time provides the frame of reference essential for the interpretation of events and processes on Earth. In the early stages of its development, geology was dependent on the relative dating of rocks and the events they recorded. Nicolaus Steno provided an understanding of the significance of **superposition** in a sequence of strata; and later, Charles Lyell showed how included rock fragments and **cross-cutting relationships** could be used in making relative age determinations. William Smith demonstrated how fossils could be used in correlating strata from one locality to another, and James Hutton helped his nineteenth-century contemporaries visualize the enormous spans of time recorded by thick sequences of rock. A scale of relative geologic time gradually emerged. Initial attempts to decide what the rock succession meant in terms of years were made by estimating the amount of salt in the ocean, the average rate of deposition of sediment, and the rate of cooling to the Earth. However, these early schemes did little more than suggest that the planet was at least tens of millions of years old and that the traditional concept of a 6000-year-old Earth did not agree with what could be observed geologically.

An adequate means of measuring geologic time was achieved only after the discovery of radioactivity at about the turn of the twentieth century. Scientists found that the rate of decay by radioactivity of certain elements is constant and can be measured and that the proportion of parent and daughter elements can be used to reveal how long they had been present in a rock. Over the years, continuing efforts by investigators as well as improvements in instrumentation (particularly of the mass spectrometer) have provided many thousands of age determinations. Frequently, these numerical dates have shed light on difficult geologic problems, provided a way to determine rates of movement of crustal rocks, and permitted geologists to date mountain building or to determine the time of volcanic eruptions. In a few highly important regions, isotopic dates have been related to particular fossiliferous strata and have thereby helped to quantify the geologic time scale and to permit estimation of rates of organic evolution.

The isotopic transformations most widely used in determining absolute ages are uranium-238 to lead-206, uranium-235 to lead-207, thorium 232 to lead-208, potassium-40 to argon-40, rubidium-87 to strontium-87, and carbon-14 to nitrogen-14. Methods involving uranium–lead ratios are of importance in dating the Earth's oldest rocks. The short-lived carbon-14 isotope that is created by cosmic ray bombardment of the atmosphere provides a means to date the most recent events in Earth history. For rocks of intermediate age, schemes involving potassium–argon ratios, those utilizing intermediate elements in decay series, or those employing fission fragment tracks are most useful. A figure of 4.6 billion years for the Earth's total age is now supported by ages based on meteorites and on lead ratios from terrestrial samples.

Improvements in numerical geochronology are being made daily and will provide further calibration of the standard geologic time scale in the future. Some of the time boundaries in the scale, such as that between the Cretaceous and Tertiary systems, are already well validated. Others, such as the boundary between the Paleozoic and Mesozoic, require additional refinement. Additional efforts to further incorporate quantitative ages into sections of sedimentary rocks are among the continuing tasks of historical geologists. The usual methods for determining the age of strata involve the dating of intrusions that penetrate these sediments or the dating of interbedded volcanic layers. Less frequently, strata can be dated by means of radioactive isotopes incorporated within sedimentary minerals that formed in place at the time of sedimentation. At present, the best numerical age estimates indicate that Paleozoic sedimentation began about 570 million years ago, the Mesozoic Era began about 245 million years ago, and the Cenozoic commenced about 66 million years ago.

KEY TERMS

Geochronology *325*
Uniformitarianism *326*
Unconformity *327*
Correlation *330*
Eon *333*
Archean Eon *333*
Proterozoic Eon *333*
Phanerozoic Eon *334*
Paleozoic Era *334*
Mesozoic Era *334*

Cenozoic Era *334*
Period (geologic) *334*
Epoch *334*
Age (geologic) *334*
Series *334*
Stage *334*
Cambrian *334*
Silurian *334*
Devonian *334*
Carboniferous System *334*

Mississippian *334*
Pennsylvanian *334*
Permian System *334*
Triassic *334*
Jurassic *335*
Cretaceous *335*
Quaternary *335*
Electron *337*
Proton *337*
Neutron *337*

Atomic number *337*
Mass number *337*
Isotope *337*
Nuclide *338*
Alpha particle *338*
Daughter element *339*
Half-life *341*

REVIEW QUESTIONS

1. What were the contributions of Adam Sedgwick, Roderick Murchison, Charles Lapworth, and Charles Lyell in the development of the geologic time scale?

2. What types of radiation accompany the decay of radioactive isotopes?

3. How do the isotopes carbon-12 and carbon-14 differ from one another in regard to the following?

   a. number of protons

   b. number of electrons

   c. number of neutrons

4. How does carbon-14 originate in the atmosphere?

5. If a granite intrusion is observed to cut across a thick sandstone stratum, which is older, the granite or the sandstone?

6. Pebbles of basalt within a conglomerate yield an isotopic age of 300 million years. What can be said about the age of the conglomerate? Several miles away, the same conglomerate is bisected by a 200-million-year-old dike. What can now be said about the age of the conglomerate?

7. State the estimated age of a sample of mummified skin from a prehistoric human that contained 12.5 percent of an original quantity of carbon-14.

8. State the effect on the age determination made from a zircon crystal dated by the potassium–argon method if a small amount of argon-40 escaped from the crystal.

9. What is the essential difference between a time-stratigraphic unit and a time unit? Give an example of each.

10. Which of the parent–daughter pairs of isotopes would be most appropriate for dating rocks of the (a) Archean? (b) Jurassic? (c) late Pleistocene?

11. What percentage of Earth history is included in the Phanerozoic Eon?

DISCUSSION QUESTIONS

1. Discuss the contributions to geologic science made by Nicolaus Steno, James Hutton, Charles Lyell, and William Smith.

2. Before about 3 billion years ago, there was little or no oxygen in the Earth's atmosphere. Had James Hutton known this, how might he have changed his statement that "the past history of the globe must be explained by what can be seen to be happening now"?

3. By what methods did geologists attempt to determine the quantitative age of the Earth before the discovery of radioactivity? Why were these methods inadequate?

4. In selecting a mineral for an age determination based on the uranium–lead method, why would the investigator search for an unweathered specimen?

5. Has the amount of uranium in the Earth increased, decreased, or remained about the same over the past 4.5 billion years? What can be stated about the amount of lead?

6. How are dating methods involving decay of radioactive elements unlike methods for determining elapsed time by the funneling of sand through an hourglass?

7. What is the advantage of having both uranium and thorium present in a mineral being used for an isotopic age determination?

8. Minerals suitable for isotopic age determinations are usually components of igneous rocks. How, then, can quantitative ages be obtained for sedimentary rocks?

9. If you were going to determine the paleogeography of a region, why would maps and cross sections depicting time-stratigraphic units be essential?

10. How might you determine the rate of deposition of calcium carbonate if you had a limestone sequence with a layer of altered volcanic ash at its base and at its top?

CHAPTER
16

Life Through Time

The trilobite Isotelus brachycephalus *from strata of Ordovician age near Cincinnati, Ohio. The specimen is 23 centimeters in length.*
(Ward's Natural Science Establishment)

Geologists that investigate organisms of the geologic past call themselves **paleontologists**. Theirs is a science based on the study of **fossils**, the remains or traces of ancient animals and plants. Fossils are a rich source of information about former climatic conditions, marine transgressions and regressions, the chemistry and physical nature of ancient seas, routes taken by drifting continents, the reasons for biologic catastrophes, the response of the biosphere to geologic hazards of long ago, and the characteristics of vanished landscapes. Indeed, we would be severely disadvantaged in our interpretation of the Earth's geologic history were it not for the interpretations of former conditions revealed by fossils. The work often includes comparison of fossils with organisms living today, for this helps one understand how ancient creatures lived, how they interacted with other forms of life, and how they adapted to constantly changing environments.

16.1   Preservation

When one considers the many ways by which organisms are destroyed after death, it is remarkable that fossils are as common as they are. Chemical and bacterial decay, attack by scavengers, and destruction by erosion make the odds against preservation very high. The chances of escaping complete destruction, however, are improved if the organism has a mineralized skeleton or dies in a place where rapid burial occurs. Both of these conditions are found on the floors of the sea, where shelled invertebrates are covered by the continuous rain of sedimentary particles. Although most fossils are found in marine sedimentary rocks, they also occur in terrestrial sediments left by streams and lakes. On occasion, animals and plants have been preserved after being immersed in tar or quicksand, trapped in ice or lava flows, or engulfed by volcanic ash.

The term *fossil* often implies **petrifaction**, literally a transformation into stone. After the death of a shell-bearing organism, scavengers and bacteria ordinarily consume the soft tissue. The empty shell may be left behind, and if it is sufficiently durable and resistant to dissolution, it may remain basically unchanged for a long period of time. Indeed, unaltered shells of marine invertebrates are known from deposits over 100 million years old. In many marine creatures, however, the skeleton is composed of a mineral variety of calcium carbonate called aragonite. Although aragonite has the same composition as the more familiar mineral calcite, it has a different crystal form, which is relatively unstable, and in time changes to the more stable calcite.

Many other processes may alter the shell of a clam or snail and enhance its chances for preservation. Water containing dissolved silica, calcium carbonate, or iron may circulate through the enclosing sediment and be deposited in cavities once occupied by veins, canals, nerves, or tissues. In such cases, the original composition of the bone or shell remains, but the fossil is rendered harder and more durable. This addition of chemical precipitate into pore space is termed **permineralization**. Some forms of petrified wood (Fig. 16.1) have been preserved by permineralization on a very fine scale.

Petrifaction may also involve **replacement** of the original substance of part of a dead animal or plant with mineral matter. Replacement can be a marvelously

Figure 16.1   Petrified wood from the Petrified National Forest in Arizona. This is an example of fine scale permineralization as silica fills cavities inside empty wood cells. The upper specimen is a sawed and polished log, about 46 centimeters in diameter. The lower photograph depicts the surface texture of a similarly petrified log. These fossils occur in the Chinle Formation of Triassic age.

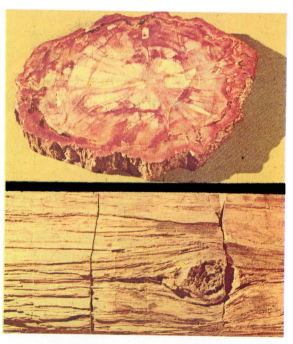

Figure 16.2 This fossil seed fern from rocks of Pennsylvanian age has been preserved by carbonization. The frond is approximately 27 centimeters in length.

Figure 16.3 An internal mold (or steinkern) formed by filling of the spiral cavity of an ancient marine snail.

Figure 16.4 A cast (above) and mold (below) of a trilobite formed in a nodule of calcareous shale. The trilobite (*Calliops*) is 4 centimeters long.

precise process, so that even fine growth lines on shells and delicate structures in bone are preserved.

Another type of fossilization, known as **carbonization**, occurs when soft tissues are preserved as thin films of carbon. As leaves and tissues of soft-bodied organisms such as jellyfish or worms are buried and compressed, most of the oxygen, hydrogen, and nitrogen are lost because of decay. The carbon remains behind as a blackened silhouette (Fig. 16.2).

Any plant or animal may leave an impression of itself if it is pressed into mud. Commonly, among shell-bearing invertebrates, the shell dissolves after burial and lithification, leaving a vacant **mold** bearing surface features of the original shell. If external features (growth lines, ornamentation) of the fossil are visible, the mold is an external mold. Conversely, the internal mold shows features of the inside of a shell, such as muscle scars or supports for internal organs. Many invertebrate shells enclose a hollow space that may be either left empty or filled with sediment (Fig. 16.3). The filling is called a steinkern (stone core). Finally, molds may be subsequently filled, forming **casts** that faithfully show the original form of the shell (Fig. 16.4).

Although it is certainly true that the possession of hard parts enhances the prospects of preservation, organisms having soft tissues and organs are also occasionally preserved. Insects have been found preserved in the hardened resins of conifers and certain other

Figure 16.5 An Eocene insect preserved in amber. The insect is a member of the order Diptera, which includes flies, mosquitoes, and gnats. Although preservation of insects is uncommon, it is likely that they approached modern levels of abundance and diversity early in the Cenozoic Era. *(W. Bruce Saunders)*

trees (Fig. 16.5). X-ray examination of thin slabs of rock sometimes reveal the outlines of tentacles, digestive organs, and vessels of the circulatory systems. Soft parts, including skin, hair, and viscera of ice age mammoths, have been preserved in frozen soil (Fig. 16.6) or in the oozing tar of oil seeps. A striking example of preservation of soft parts was discovered in 1984 at Lindow Moss, England, when an excavating machine uncovered the body of a 2000-year-old human. Named Lindow Man (Fig. 16.7), the Briton had been ritually slaughtered (possibly by Druids), stripped, garroted, and then bled. The body was placed in a pool in the bog, where the skin was preserved and the body flattened by the weight of accumulating peat. Only the upper part of the body was recovered, the lower part having been destroyed by the machine before the remains were detected.

Evidence of ancient life does not consist solely of petrifactions, molds, and casts. Sometimes the paleontologist is able to obtain clues to an animal's appearance and how it lived by examining tracks, trails (Fig. 16.8), burrows, and borings. Such markings are called trace fossils. The tracks of an ancient vertebrate may reveal if the animal that made them walked on two or four legs, on its toes or on the flat of its feet, whether it had an elongate or short body, if it was lightly built or ponderous, and sometimes whether it was aquatic (with webbed toes) or possibly a flesh-eating predator (with sharp claws).

Trace fossils of invertebrate animals are more frequently found than the tracks of vertebrates, and they

Figure 16.6 In the summer of 1977, the carcass of this baby mammoth was dug from frozen soil (permafrost) in northeastern Siberia. The mammoth stood about 104 centimeters tall at the shoulders, was covered with reddish hair, and was judged to be only several months old at the time of death. Dating by the radiocarbon method indicates death occurred 44,000 years ago. *(Klavdija Novikova, Biologopoczvennyj Institut, Vladivostok, Russia)*

Figure 16.7   Preserved torso, arms, and head of the 2000-year-old Lindow man. This example of preservation of soft tissue was found in a peat bog in 1984 at Lindow Moss, England. The lower half of the body was destroyed by the excavating machine.   *(British Museum)*

are also useful indicators of the habits of ancient creatures. One can sometimes infer if the trace-making invertebrate was crawling, resting, grazing, feeding, or simply living within a relatively permanent dwelling. For example, crawling traces are linear and show di-

rected movement (Fig. 16.9A). Shallow depressions that reflect the shape of the animal may be resting traces (Fig. 16.9B). Simple or U-shaped structures approximately perpendicular to bedding are often dwelling traces (Fig. 16.9C). Grazing traces (Fig. 16.9D) occur along bedding planes and are characterized by a systematic meandering or concentric and parallel patterns that represent the animal's effort to cover the area containing food in an efficient manner. Feeding traces are made by animals that consume sediment for the organic nutrients within it. The traces consist of branched or unbranched burrows, as shown in Figure 16.9E.

16.2   The Incomplete Record of Life

The fossil record of life is incomplete. If it were a complete and total record, it would include information on all past forms of life for every increment of time and for every place on Earth. Clearly this is unattainable. Only a limited number of animals and plants have been preserved. Many have been destroyed by geologic events such as metamorphism. Others that were preserved have never been exposed to our view by erosion or drilling. Still others have simply not yet been discovered.

The record is rather more complete for marine life having hard external skeletons and less complete for land life lacking bone or shell. Where burial is rapid and dead organisms are protected from scavengers and agents of decay, the probability of preservation improves. As we have noted, the ocean floor and low-lying land areas where deposition predominates are often

Figure 16.8   Trails made by millipedes about 250 million years ago during the Permian Period. These trace fossils were discovered in the Coconino Sandstone, Grand Canyon National Park. Millipedes are elongated segmented animals with a large number of paired legs. Along with insects and crustaceans, they are members of the Phylum Arthropoda.   *(U.S. Geological Survey, photograph by E. D. McKee)*

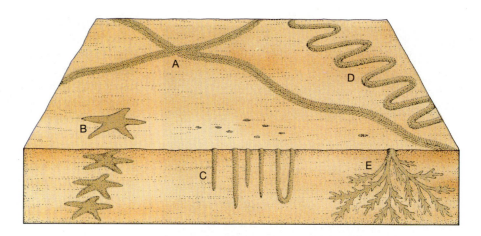

Figure 16.9 Traces that reflect animal behavior: (A) crawling traces, (B) resting traces, (C) dwelling traces, (D) grazing traces, and (E) feeding traces.

favorable places for fossil preservation. Many dinosaur remains, for example, have been discovered in the sands and clays deposited by streams that flowed across the low plains that lay east of the Rocky Mountains during the Cretaceous Period. There were certainly animals and plants living in nearby highland areas as well, but in such places erosion rather than deposition predominated and living things were less likely to escape destruction.

In spite of the many factors that prevent fossilization, the fossil record is remarkably comprehensive. On the basis of this impressive record, paleontologists have been able to piece together a history of past life that is balanced and largely accurate.

16.3   The Rank and Order of Life

The Linnaean System of Classification

Because of the large number of living and fossil animals and plants, random naming would be confusing and inefficient. Realizing this, the Swedish naturalist Carl von Linné, also known as Carolus Linnaeus (1707–1778), established a carefully conceived system for naming animals and plants. The **Linnaean system** uses morphologic structure as the basis of classification and employs binomial nomenclature at the species level. In this scheme, the first, or **generic**, name was used to indicate a general group of creatures that were visibly related, such as all doglike forms. The second, or trivial, name denoted a definite and restricted group—a species. *Canis lupus*, for example, designates the wolf among all canids. Linné went on to recognize larger divisions such as classes and orders. His groupings were based on traits that seemed most basic or natural; the modern nomen-

clature system, however, is based on an attempt to be consistent with evolutionary relationships.

Concepts Involved in Classification

The Species

The **species** is the fundamental unit in biological classification. A group of organisms that have structural, functional, and developmental similarities and that are able to interbreed and produce fertile offspring constitute a species. Because members of one species do not breed with members of different species under natural conditions, species exist in reproductive isolation. Paleontologists recognize that species exist during relatively brief periods of geologic time, that they are derived from ancestral populations, and that they represent the population from which new species arise.

Taxonomy

The species is the basic unit in biologic classification, or **taxonomy**. In this system, the various categories of living things are arranged in a hierarchy that expresses levels of kinship. For example, a **genus** (pl. *genera*) is a group of species that have close ancestral relationships; a **family** is a group of related genera; an **order** is a group of related families; a **class** is a group of related orders; a **phylum** (pl. *phyla*) is a group of related classes; and a **kingdom** is a large group of related phyla. To use an example familiar to all, individual humans are members of the Kingdom Animalia, Phylum Chordata, Class Mammalia, Order Primates, Family Hominidae, Genus *Homo,* and species *Homo sapiens.*

Biologic classification is more than a mere system for cataloging organisms. Because it is based on organic structures, it also depicts the broad outlines of ancestral evolutionary relationships. In a sense, the classification is a depiction of the tree of life.

The conventional system for classifying living and fossil organisms employs five kingdoms (Fig. 16.10). The Monera includes such small and simple organisms as the cyanobacteria (formerly called blue-green algae) and bacteria known as schizomycetes. The Protoctista (formerly Protista) includes mostly one-celled organisms, but some form multicellular colonies. Amoebas, sporozoans (the malarial parasites), and a host of ciliated and flagellated organisms are examples. The Fungi constitute the third kingdom, leaving the kingdoms Animalia and Plantae.

The classification of animals and plants is subject to continuous revision. Current attempts at revision employ evolutionary relationships revealed by molecular structures and sequences. In these newer classifications, all organisms are grouped into three super-kingdoms or domains—the Bacteria, Archaea, and Eucarya—each containing two or more kingdoms.

Because of the incredible variety of living and fossil organisms, scientists have formulated rules of biological nomenclature. With regard to genus and species, the rules stipulate that the generic name (the genus) must be a single capitalized word, that the species name must begin with a lowercase letter, and that these names must be either in Latin or Latinized.

16.4   Organic Evolution

The Roman poet Ovid (43 B.C.E.–A.D. 17) once wrote that "There is nothing constant in the universe, all ebb and flow, and every shape that's born bears in its womb the seeds of change." The words are remarkably relevant when one considers the way life has changed through time. At times, the fossil record indicates that change was startlingly sudden, whereas at other times evidence shows a more gradual change. In either case, older forms have changed or evolved into newer forms to adjust to changes in their environment. This is not to say that the early, relatively simpler forms of life are gone, for many persist along with their often more complex contemporaries.

Darwin's Theory of Natural Selection

As essential component of our present understanding of the way animals and plants evolve is the concept of natural selection. Charles Darwin and his younger

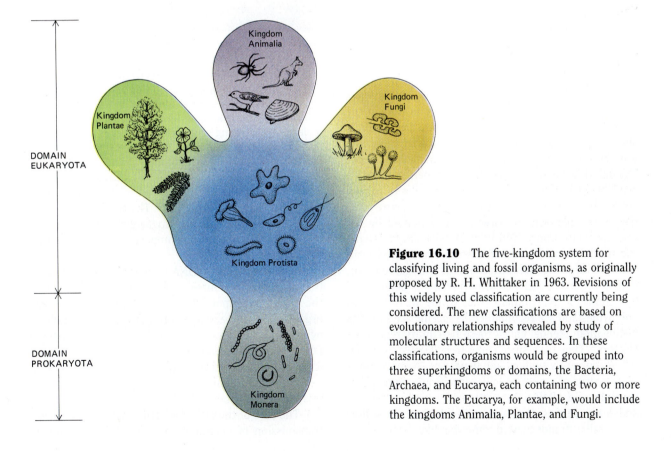

Figure 16.10   The five-kingdom system for classifying living and fossil organisms, as originally proposed by R. H. Whittaker in 1963. Revisions of this widely used classification are currently being considered. The new classifications are based on evolutionary relationships revealed by study of molecular structures and sequences. In these classifications, organisms would be grouped into three superkingdoms or domains, the Bacteria, Archaea, and Eucarya, each containing two or more kingdoms. The Eucarya, for example, would include the kingdoms Animalia, Plantae, and Fungi.

contemporary Alfred R. Wallace jointly proposed this important mechanism for evolution in 1858. Darwin and Wallace observed that living things tend to produce more offspring than can survive. There is simply not enough food, living space, and other vital necessities required for the survival of such potentially prodigious number of offspring. In addition, individuals of the same species vary in morphological and physiological features. Because all of the potentially enormous number of individuals cannot survive, those having variations that provide some advantage in a particular environment are most likely to survive and transmit their traits to the next generation. Less well-fitted variants are likely to be eliminated.

Darwin's *The Origin of Species by Means of Natural Selection* was published in 1859. It appeared at a time when at least part of the European intellectual atmosphere was more liberal and less satisfied with the theologic doctrine that every species had been independently created. Although the ideas of Darwin and Wallace continued to disturb the religious feelings of some, they nevertheless increasingly acquired adherents among nineteenth- and twentieth-century scientists. Darwin did not consider his views impious. He saw a "grandeur in this view of life with its several powers, having been originally breathed by the Creator into a few forms or into one; and that . . . from so simple a beginning endless forms most beautiful and most wonderful have been and are being evolved."

Mendelian Principles of Inheritance

Every theory has its strong and weak points. In Darwin's theory, the weakness lay in an inability to explain the cause of variability in a way that could be experimentally verified. The cause of at least a part of that variability was discovered by a Moravian monk named J. Gregor Mendel while he was conducting experiments on garden peas. Mendel discovered the basic principles of inheritance. His findings, printed in 1865 in an obscure journal, were unknown to Darwin and unheeded by the scientific community until 1900. Mendel had discovered the mechanism by which traits are transmitted from adults to offspring. In his experiments with garden peas, he demonstrated that heredity in plants is determined by "character determiners" that divide in the pollen and ovules and are recombined in specific ways during fertilization. Mendel called these hereditary regulators "factors." They have since come to be known as genes.

As currently understood, genes are chemical units or segments of a nucleic acid—specifically **deoxyribonucleic acid (DNA)**. As suggested by careful chemical and X-ray studies, the DNA molecule is conceived as two parallel strands twisted somewhat like the handrails

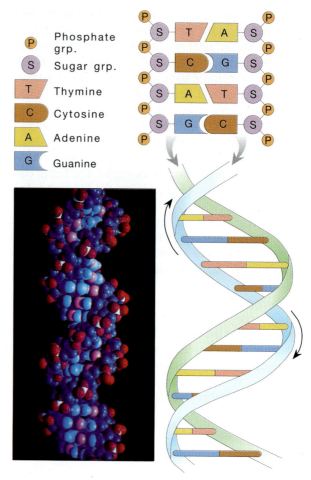

Figure 16.11   Representations of portions of the deoxyribonucleic acid (DNA) molecule. At the lower left is a computer reconstruction by R. J. Feldmann. The drawing depicts the twisted, double-stranded helix, with side rails composed of alternate sugar (deoxyribose) and phosphate molecules. Each rung of the twisted ladder is composed of one pair of nitrogenous bases. Of these, thymine links to adenine, and cytosine to guanine.

of a spiral staircase (Fig. 16.11). The twisted strands are made up of phosphate and sugar compounds and are linked with cross members composed of specific nitrogenous bases.

The importance of DNA is evident when we realize that it indirectly controls the production of proteins, the essential components of many basic structures and organs. Even the activities of organisms are regulated by specific catalytic proteins called enzymes. Without DNA and its products, there would be no life as we know it. Its ability to precisely replicate itself is the basis for heredity, and organic evolution ultimately depends on this remarkable molecule.

The part of the DNA molecule that is active in the transmission of hereditary traits is called a **gene**. In nearly all organisms, genes are linked together to form larger units termed **chromosomes**, the central axis of which consists of a very long DNA molecule comprising hundreds of genes.

Reproduction and Cell Division

Reproduction of an organism may be sexual or asexual or may involve an alternation of sexual and asexual methods. All reproductive methods involve the division of cells. In sexual reproduction there is a union of reproductive or sex cells from separate individuals, whereas in asexual reproduction cells do not unite. The usual methods of asexual reproduction are binary fission, budding, and spore production. **Binary fission** is found in single-celled organisms such as *Amoeba* that divide to form genetically identical daughter organisms. **Budding** occurs in some unicellular as well as multicellular organisms. In this method, the parent organism simply sprouts a bulge or appendage that may either remain attached to the parent or separate and grow as an isolated individual. The third method involving asexual reproduction is the formation of tiny reproductive cells called **spores** by a parent organism. The spores are formed by division of spore parent cells and are shed by the parent organism (called the sporophyte) when they mature. When a spore settles onto a suitable surface, it germinates and grows into a tiny plant (called a gametophyte) that, in turn, produces male and female sex cells. The union of these cells produces a new plant resembling the original sporophyte.

Organisms that reproduce asexually can increase their numbers rapidly whenever conditions are favorable. However, they are not able to develop as much variability among their offspring as do sexually reproducing organisms. To understand the reasons for that variability, we need to know some simple facts about chromosomes. The kind and number of chromosomes are constant for each species and differ between different species. Except in bacteria and blue-green algae (cyanobacteria), chromosomes are located in the nucleus of the cells and occur in duplicate pairs. In humans, for example, there are 46 chromosomes, or 23 pairs. Cells with paired chromosomes are designated **diploid** cells. In all living things, new cells are being produced constantly to replace worn-out or injured cells and to permit growth. In asexual organisms, and in all the body cells of sexual organisms, the process of cell division that produces new diploid cells with exact replicas of the chromosome composition of the parent cells is called **mitosis**.

In most organisms with sexual reproduction, a second type of division, called **meiosis**, takes place when **gametes** (eggs or sperm) are formed. Meiosis may occur in unicellular organisms, but in multicellular forms it takes place only in the reproductive organs (testes or ovaries). During meiosis, two cell divisions occur, which ultimately produce gametes with unpaired chromosomes. These gametes have half the number of chromosomes found in body cells. When sexually reproducing organisms mate, the sperm enters the egg to form a single cell, which can now be called a **fertilized egg**. The egg contains a full complement of chromosomes and genes representing a mix from both parents.

Mutations

Some of the variation we see among individuals of the same species results from the mixing of genes in a fertilized egg. However, if genes never changed, the number of variations they could produce would be limited. For organisms to evolve a truly new variation, a process is required that will change the genes themselves. That process is known as **mutation**. Mutations can be caused by ultraviolet light, cosmic and gamma rays, chemicals, and certain drugs. They may also occur spontaneously. Mutations may occur in any cell, but their evolutionary impact is greater when they occur in sex cells because then succeeding generations are affected.

To understand how mutations occur, it is useful to examine DNA (see Fig. 16.11). The steps on the twisted DNA ladder are composed of two kinds of nitrogenous bases: purines and pyrimidines. Genes owe their specific characteristics to a particular order of these base-pair steps. If the order is disrupted, a mutation may result. For example, during cell division, when twisted strands separate and then recombine, a mishap may position members of a base pair differently from the way they were positioned originally. This alters the gene and may result in an inheritable change.

Thus, three kinds of events cause evolution: mutations, gene recombinations, and natural selection. Mutations are the ultimate source of new and different genetic material. Recombination (mixing) spreads the new traits throughout the population and mixes the new with the old. Natural selection sorts out the varying traits, preserving those that by chance make the organism better fitted to a particular life.

Traits that provide some advantage to an organism are called **adaptations**. The most obvious adaptations have to do with external features of the organism. Among animals, for example, teeth and jaws may be shaped or adapted for dealing with certain kinds of food, or appendages may be fitted for running, swimming, or even flying. In the geologic past, descendants of an ancestral group exhibit diversity that reflects their distinctive adaptations to particular living strategies and environments. The group is said to have undergone an **adaptive radiation**.

The honey creepers of Hawaii provide an illustration of adaptive radiation. These birds, comprising many different species, are believed to have descended from a common ancestor. The most striking differences between the species occur in the sizes and shapes of the beaks, which are adaptively related to the kind of food the birds are dependent on (Fig. 16.12). Some have stout beaks for crushing hard seeds; others have beaks well adapted for seeking out insects in cracks and crevices or for sucking nectar. All of these variations in beak morphology are examples of adaptations, a word that means the acquisition of heritable characteristics that are advantageous to an individual and a population.

On a higher taxonomic level, one finds adaptive radiations among classes, orders, and families of organisms. All the orders of mammals, for example, had a common origin in an ancestral species that lived during the early Mesozoic. By adapting to different ways of life, the descendant forms came to diverge more and more from the ancestral stock, thereby providing the remarkable diversity of mammals that exist today.

Gradual or Punctuated Evolution

Does evolution proceed gradually by means of an infinite number of small advances, as Darwin proposed, or by sporadic advances over a short span of time? Gradual progressive change is called **phyletic gradualism**. Proponents of phyletic gradualism believe that what might be interpreted as a sudden change in the fossil record of an evolving group reflects only a gap in the record resulting from erosion, nondeposition, or lack of preservation.

To show that phyletic gradualism has taken place, one requires a complete sequence of well-dated fossiliferous sediments. Such an uninterrupted sequence occurs in cores of sediment recovered from the ocean floor. In one such sequence, it was shown that four species of the foraminifer *Globorotalia* evolved gradually over a time span of about 8 million years.

Paleontologists who oppose phyletic gradualism argue that the fossil record for the past 3 billion years contains many examples of new groups appearing suddenly in places where there is no evidence of any imperfections in the geologic record. Indeed, they find that

Figure 16.12 The honey creepers of Hawaii provide a good example of adaptive radiation. When the ancestor of today's honey creepers first reached Hawaii, few birds were present. Succeeding generations diversified to occupy various ecologic niches. Their diversity is most apparent in the way their beaks have become adapted to different diets. Some have curved bills to extract nectar from tubular flowers, whereas others have short, sturdy beaks for cracking open seeds, or pointed bills for seeking out insects in tree bark. *Himatione sanguinea*, at the lower right, has a relatively unspecialized beak and appears to be similar to the original honey creeper ancestor.

the slow and stately advance of evolution can be documented only rarely and that evolutionary progress is more often sporadic. Although this concept was suggested more than a half century ago, it has recently received new and strong support from paleontologists such as Stephen J. Gould and Niles Eldridge in their research on Pleistocene snails and Devonian trilobites. Gould and Eldridge have suggested the term **punctuated equilibrium** for evolution that consists of fitful sudden advances that punctuate long episodes of little evolutionary progress.

According to this concept, the punctuation or sudden morphologic change that interrupts equilibrium occurs at the periphery of the geographic area occupied by the population. These segments of the population are referred to as **peripheral isolates**. Gene flow is rapid among the individuals of peripheral isolates, and changes in morphology or physiology leading to speciation may occur within a short span of time. Species do not usually originate in the places where their parental stock exists, but rather in boundary zones where variations can be tested against new environmental situations. Should the parent species suffer extinction or severely decline, the new species may move back into the parental domain or may expand into new territory. With adaptations that are a significant advantage, they are likely to enter a period of rapid evolution. This stage is usually followed by a period of moderate evolutionary rates and stability that lasts until the population begins to decline. Subsequently, yet another new group may begin its evolutionary expansion.

The final chapter on the question of punctuated evolution versus phyletic gradualism has not been written. At present, the proponents of punctuated evolution appear to be more numerous than those of phyletic gradualism. Like most controversies in science, however, the answer need not lie totally in one camp, and it is evident that instances of phyletic gradualism can also be recognized in the fossil record.

Patterns of Evolution

Patterns of evolution are depicted by a diagram called a **phylogenetic tree**, in which various taxa are traced from outermost branches (representing lower taxonomic divisions such as species, genera, or families) through a succession of junctures downward (and backward in time) to the major trunks of the higher taxa. On such a diagram, time is plotted on the vertical axis and biologic change on the horizontal. For the punctuated equilibrium model of evolution (Fig. 16.13A), the phylogenetic tree has short horizontal branches (reflecting sudden morphologic change) that quickly become vertical (to reflect subsequent stability through

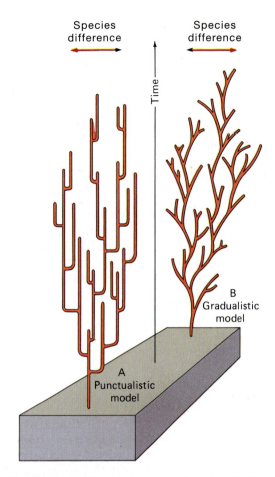

Figure 16.13   Phylogenetic trees depicting (A) the punctualistic model and (B) the gradualistic model of evolution. Morphologic change occurs in sideward directions. Time is depicted by the vertical direction. The short horizontal side branches of the punctualistic model depict sudden change, whereas the inclined branches of the gradualistic model suggest slow and uniform change through time.

time). In contrast, the phyletic gradualism model (Fig. 16.13B) shows gently inclined branches that indicate slow and gradual change through time. Branching occurs by gradual deviation of forms.

The phylogenetic tree serves as a model for depicting particular styles of evolution. As is evident from Figure 16.14, a branch representing a single lineage may simply divide to form two or more lineages. This kind of evolutionary pattern is called **divergence** (Fig. 16.14A). **Convergence** (Fig. 16.14B) occurs when two or more unrelated lineages acquire similar morphologic traits, usually because of adaptations to similar modes of life and environments. A familiar example of convergence is seen in the evolution of superficially similar structures—wings—in birds, bats, and flying reptiles of

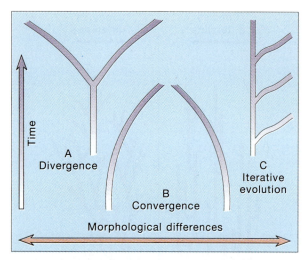

Figure 16.14 Three common evolutionary patterns as indicated by morphologic change among different lineages. (A) The divergence of two lineages. (B) Two lineages evolve toward morphologic similarity as an example of convergence. Presumably, this convergence results from selective pressures to adapt to similar environmental conditions. (C) In iterative evolution, morphologically similar forms evolve repeatedly from a root stock.

the Mesozoic known as pterosaurs. **Iterative evolution** refers to the evolution of similar morphology more than once from the parental lineage (Fig. 16.14C). An example is provided by the coiled chambered mollusks known as cephalopods. Three or more times during the history of this class of invertebrates, irregularly coiled forms evolved from the uniformly coiled parental lineage.

Fossil Evidence for Evolution

The proponents of evolution had a difficult time convincing some of their nineteenth-century contemporaries of the validity of Darwin's theory. One reason was that evolution is an almost imperceptibly slow process. The human life span is too short to witness evolutionary changes across generations of plants or animals. Fortunately, we can overcome this difficulty by examining the remains of organisms left in rocks of successively younger age. If life has evolved, the fossils preserved in consecutive formations should exhibit those changes. Indeed, many examples are known of sequential changes among related creatures through geologic time. The most famous example is provided by fossil horses during the Cenozoic Era (Fig. 16.15). The oldest horses thus far discovered were about the size of fox terriers. Unlike modern horses, they had four toes on their front feet and three on the rear. The many branching lineages of the horse "family tree" in successively higher (hence

younger) formations are a record of evolutionary change. As they evolve, horses show an increase in size, a reduction of side toes with emphasis on the middle toe, and increase in tooth size and complexity, and a deepening and lengthening of the skull.

The change in horse dentition through time provides an interesting example of the important link between environmental change and evolution. During the time that the horse family was evolving, grassland became increasingly widespread. Grass is a harsh food for an animal. The blades contain silica, and because grass grows close to the ground, it is usually coated with abrasive dust. Early members of the horse family lived in forested areas and fed on less abrasive leaves of trees and shrubs. They were browsers and had the low-crowned teeth of leaf eaters. Later members responded to selective processes relating to the more abrasive food and evolved high-crowned teeth with complex patterns of enamel (Fig. 16.16). With teeth better fitted to chewing the harsh grasses, these horses were likely to have been well nourished, lived longer lives, and were able to produce more progeny. Their evolution must have affected the evolution of predators that pursued them and may have even contributed to the evolution of species of grasses resistant to damage by grazing. Evolution is an intricate process in which every animal and plant interacts not only with its physical environment but with its neighbors.

Although fossil remains of horses provide splendid evidence for evolution, hundreds of other examples representing every major group of animals and plants would serve as well. The marine invertebrates known as cephalopods also document progressive evolutionary changes (Fig. 16.17).

16.5   Fossils and Stratigraphy

Establishing Age Equivalence of Strata with Fossils

One of William Smith's major contributions to geology was his recognition that individual strata contain definite assemblages of fossils. Because of the change in life through time, superposition, and the observation that once species have become extinct they do not reappear in later ages, fossils can be used to recognize the approximate age of a unit and its place in the stratigraphic column. (Such a method for judging the age of a rock would not be possible with inorganic characteristics of strata because they frequently recur in various parts of the geologic column.) Further, because organic evolution occurs all around the world, rocks formed

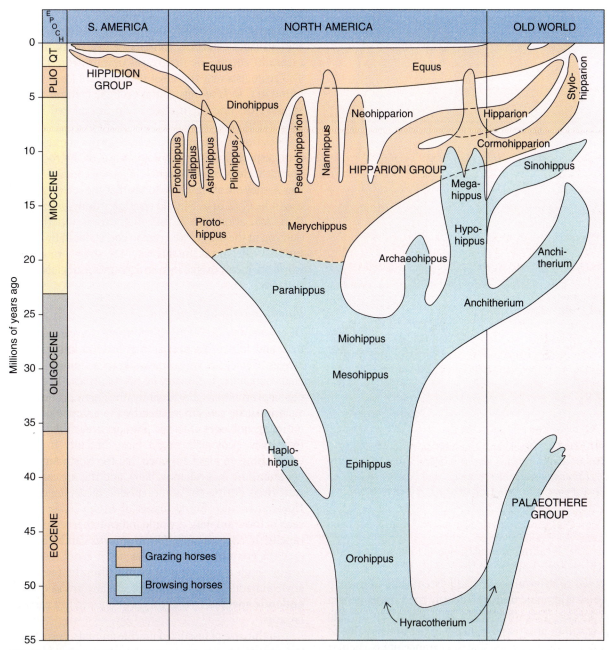

Figure 16.15   Phylogenetic tree of horses showing evolutionary relationships among
the genera and the transition from browsing horses with low-crowned teeth to
grazing horses with high-crowned teeth.   *(After B. J. MacFadden, J. D. Bryand, and P. A. Mueller,
Geology, 19, 1991, 242–245.)*

during the same age in identical environments but diverse localities often contain similar fauna and flora if
there has been an opportunity for genetic exchange.
This fact permits geologists to match chronologically
or correlate strata from place to place. It provides a
means for establishing the age equivalence of strata in
widely separated parts of the globe.

The Geologic Range

Before fossils could be used as indicators of age equivalence, it was first necessary to determine the relative
ages of the major units of rock on the basis of superposition. Geologists began by working out the superpositional sequences locally, and then they added sections

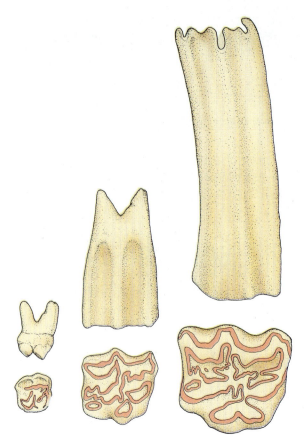

Figure 16.16 Development of high-crowned grinding molars in horses. Enamel shown in black. From left to right, *Hyracotherium, Merychippus,* and *Equus.* *(From H. L. Levin,* Life Through Time. *Dubuque, IA: William C. Brown Co., 1975.)*

from other localities around the world to form a composite chronologic column of units (Fig. 16.18). The next step was to determine the fossil assemblage from each time–stratigraphic unit and identify the various genera and species. This work began well over a century ago and is still in progress. Gradually, it became possible to recognize the oldest (or first appearance of) particular species, as well as their youngest occurrence (last to appear). The interval between the first and last appearance of a species constitutes its **geologic range**.

Identification of Time-Stratigraphic Units

The method for identifying time-stratigraphic units is illustrated in Figure 16.19. Geologists working in Region 1 come upon three time–rock systems of strata designated *O, D,* and *M.* Perhaps years later in Region 2, they again find units *O* and *D,* but in addition, they recognize an older unit—namely, *C,* below and hence older than *O.* Finally, while working in Region 3, they

find a "new" unit, *S,* sandwiched between units *O* and *D.* The section, complete insofar as can be known from the available evidence, consists of five units that decrease in age as one progresses upward from *C* to *M.* The geologists may then plot the ranges of fossil species found in *C* to *M* alongside the geologic column. If they should next find themselves in an unexplored region, they might experience difficulty in attempting to locate the position of this rock sequence in the standard column, especially if the lithologic traits of the rocks had changed. However, on discovering a bed containing species A, they are at least able to say that the rock sequence might be *C, O,* or *S.* Should they later find fossil species B in association with A, they might then state that the outcrop in the unexplored region correlates in time with unit *S* in Region 3. In this way, the strata around the world are incorporated into an accounting of global stratigraphy.

Paleontologic Correlation

Of course, the preceding illustration is greatly simplified, and future discoveries may extend known fossil ranges. Geologists must always keep in mind that the chance inability to find key fossils might lead to erroneous interpretations. Fortunately, the chances of making such mistakes are diminished by the practice of using entire assemblages of fossils. Two or three million years from now, geologists might have difficulty in firmly establishing on fossil evidence that the North American opossum, the Australian wallaby, and the African aardvark lived during the same episode of geologic time. However, if they found fossils of *Homo sapiens* with each of these animals, it would indicate their contemporaneity. In this example, *Homo sapiens* can be considered the **cosmopolitan species**, for it is not restricted to any single geographic location within the terrestrial environment. The aardvark and wallaby are said to be **endemic species**, in that they are confined to a particular area.

In the case of fossilized marine animals, cosmopolitan species have been especially useful in establishing the contemporaneity of strata, whereas endemic species are generally good indicators of the environment in which strata were deposited. Endemic species may slowly migrate from one locality to another. For example, the peculiar screw-shaped marine fossil bryozoan known as *Archimedes* (Fig. 16.20) lived in central North America during the Mississippian Period but migrated steadily westward, finally reaching Nevada by Pennsylvanian time and Russia by the Permian Period.

One must assume a tentative attitude in making correlations based on fossils. The validity of a correlation increases each time the sequence of faunal changes is

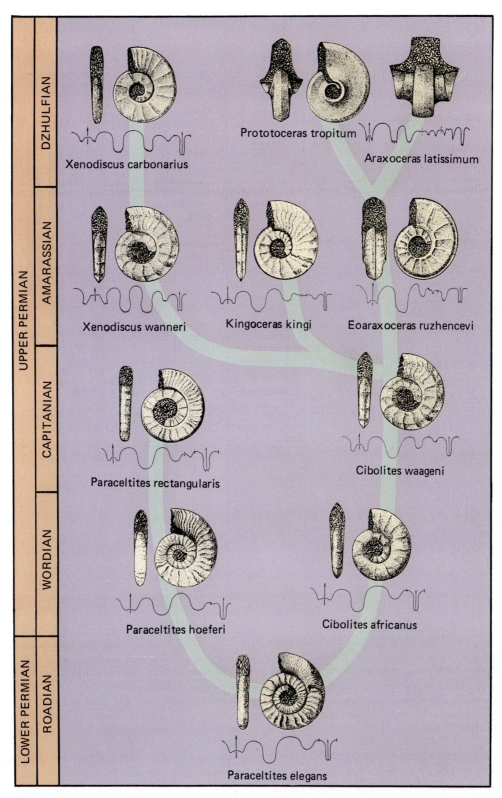

Figure 16.17 An example of progressive evolutionary change in a group of Permian ammonoid cephalopods. According to this interpretation, two evolutionary lineages originated from *Paraceltites elegans,* one terminating in *P. rectangularis* and a second producing *Cibolites waageni.* The latter was the ancestral stock for three additional lineages. The curved lines beneath each drawing are tracings of the suture lines; the terms along the left border are the names of Permian stages. *(From C. Spinosa, W. M. Furnish, and B. F. Glenister,* J. Paleontol, *49[2], 1975, 239–283.)*

363

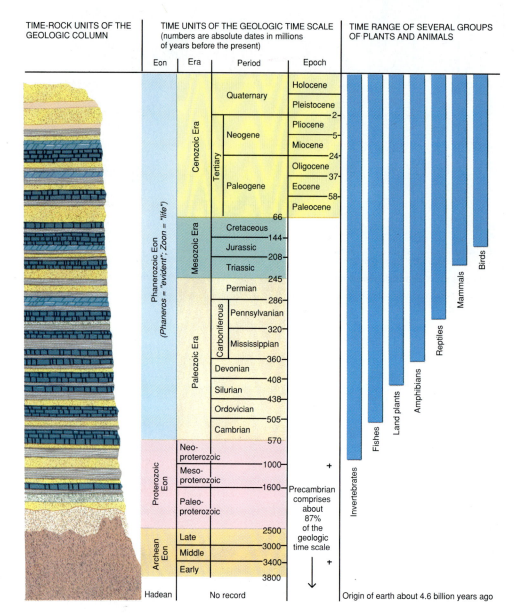

Figure 16.18 Geologic time scale. *(From A. R. Palmer, "The Decade of North American Geology, Geologic Time Scale," Geology, 11, 1983, 503–504. Proterozoic divisions are those recommended by the Subcommission on Precambrian Stratigraphy of the International Union of Geological Sciences; K. A. Plumb, "New Precambrian Time Scale," Episodes, 14(2), 1989, 139–140.)*

found again at different locations around the world. Geologists must also be aware that all changes are not because of evolution, but instead may indicate faunal migrations or shifts in flora that accompanied ancient environmental changes. In this regard, the sudden disappearance of a fossil need not mean that it became extinct, but rather that it moved elsewhere. One might also note that the earliest appearance of a fossil in rocks of a given region might mean it had evolved there;

however, it might also signify only that an older species had come into the new locality.

Guide Fossils. Many fossils are rare and restricted to a few localities. Others are abundant, widely dispersed, and derived from organisms that lived during a relatively short span of geologic time. Such short-lived but widespread fossils are called **guide fossils** because they are especially useful in identifying time–rock units

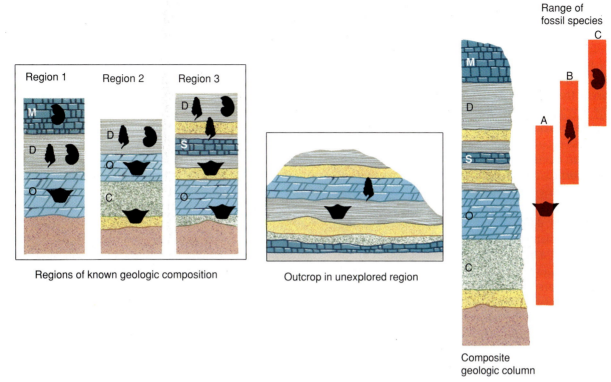

Figure 16.19   Use of geologic ranges of fossils to identify time–rock units.

Figure 16.20   Two specimens of the guide fossil *Archimedes*. Specimen (A) shows
only the screwlike axis of this bryozoan animal. In specimen (B), one can see the
fragile, lacy skeleton of the colony that is attached to the sharp, helical edge. Specimens
are 8 to 10 centimeters in height.

A

B

and correlating them from one area to another. A guide fossil with a short geologic range is clearly more useful than one with a long range. A fossil species that lived during the total duration of a geologic era would not be of much use in identifying the rocks of one of the subdivisions of lesser duration within that era. It is apparent that rate of evolution is an important factor in the development of guide fossils. Simply stated, the **rate of evolution** is a measure of how much biologic change has occurred with respect to time. Figure 16.21 provides an example of a change in the rate of evolution among horned dinosaurs. The data indicate that, although increase in size was a persistent evolutionary trend, the rate of that change diminished with time.

Obviously, different animals have evolved at different rates. Human beings and their ancestors, for example, have had a relatively rapid rate of evolution. In general, lineages with rapid rates of evolution provide greater numbers of guide fossils. Usually, rates of evolution can be related to changes in the environment or to the inherent reproductive or genetic characteristics of the evolving populations.

Although guide fossils are of great convenience to geologists, correlations and interpretations based on assemblages of fossils are often more useful and less susceptible to error or uncertainties caused by undiscovered, reworked, or missing individual species. Also, by using the **overlapping geologic ranges** of particular members of the assemblage, it is often possible to recog-

nize the deposits of smaller increments of time. The advantage of overlapping ranges can be illustrated with the help of Figure 16.22. Here, rather than using actual ranges of species, larger animal categories are used. It is apparent from this chart that a rock containing both stromatoporoids and goniatites can only be considered Devonian, whereas the occurrence of members of only one of the fossil groups would not provide as narrow a limit to the age of the rock.

Zones.   Geologists frequently use the term **biozone** when describing a body of rock that is identified strictly on the basis of its contained fossils. Paleontologic biozones can vary in thickness or lithology and can be either local or global in lateral extent. Without formally naming them, we have already described the two major kinds of biostratigraphic zones: the range zone and the assemblage zone. A **range zone** is simply the rock body representing the total geologic life span of a distinct group of organisms. For example, in Figure 16.23, the *Assilina* range zone is marked by the first (lowest) occurrence of that genus at point *A* and its extinction at Point *B*. Geologists may also designate **assemblage zones** selected on the basis of several coexisting taxa and named after an easily recognized and usually common member of the assemblage. In certain areas, however, even though the guide fossil for the assemblage zone may not be present, the other members permit recognition of the zone. A variation of the assemblage zone is the **concurrent range zone**. It is recognized by the overlapping ranges of two or more *taxa* (sing. *taxon*). (The term **taxon** refers to a group of organisms that constitute a particular taxonomic category, such as species, genus, and family.) For example, the interval between *X* and *Y* in Figure 16.23 might be designated the *Assilina–Heterostegina* concurrent range zone.

Biozones are of fundamental importance in stratigraphy. They are the basic unit for all biostratigraphic classification and correlation. They are also the basis for time-stratigraphic terms because zones are aggregated into stages, stages into series, and series into systems. The boundaries of these units are usually zonal boundaries.

16.6   Fossils as Clues to Ancient Environments

Paleoecology

Although inert and mute, fossils are vestiges of once lively animals and plants that nourished themselves, grew, reproduced, and interacted in countless ways with

Figure 16.21   A study of rate of evolutionary change. In this lineage of horned dinosaurs, the rate of increase in size was initially rapid and subsequently much slower. *(Adapted from E. H. Colbert,* Evolution, *2, 1948, 145–163.)*

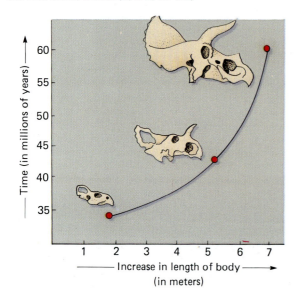

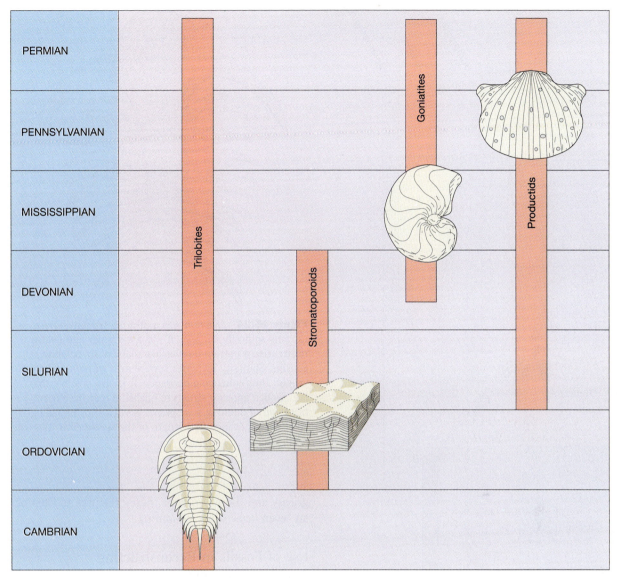

Figure 16.22 Advantage of using overlapping ranges of fossil taxa to recognize smaller increments of geologic time.

other organisms and their physical environment. The study of the interaction of ancient organisms with their environment is called **paleoecology**. Paleoecologists attempt to discover precisely where and how ancient creatures lived and what their habits and morphology reveal about the geography and climate of long ago. This scientific detective work is accomplished in various ways. One can, for example, compare species known only from fossils with living counterparts. The assumption is made that both living and fossil forms had approximately the same needs, habits, and tolerances. One can also examine the anatomy of the fossil and attempt to identify structures that were likely to have developed in response to particular biologic and physical conditions in the environment. As noted earlier, such modifications for living in a certain way and performing particular functions are called **adaptations**. For example, the broad, spiny valve of the brachiopod *Marginifera ornata* (see Fig. 19.33) was an adaptation that served to support the animal on a sea bed composed of soft mud. In another example, reduction of the skeleton to mere spines in the Silurian trilobite *Deiphon* (Fig. 16.24) was probably an adaptation to provide buoyancy in an animal that fed near the surface of the sea. Coiling in living and fossil cephalopods appears to be an adaptation designed to bring the center of buoyancy above the center of

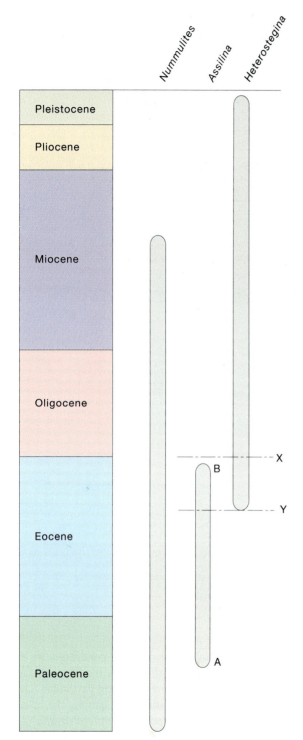

Figure 16.23 Geologic ranges of three genera of foraminifers within the family Nummulitidae. The interval between *A* and *B* is the total range of *Assilina*. The interval between *X* and *Y* could be designated the *Assilina–Heterostegina* concurrent range zone.

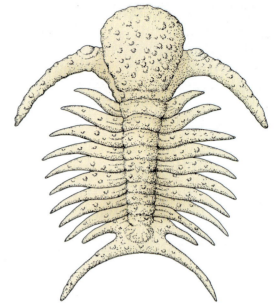

Figure 16.24 *Deiphon* was a Silurian trilobite whose extreme spinosity suggests it was a swimmer and floater rather than a bottom dweller like many of its trilobite cousins. Spinosity increases surface area without adding weight, thus enhancing buoyancy. It is also possible that the swollen anterior structure (glabella) may have been filled with a liquid of low specific gravity, thus providing additional buoyancy. The length of the specimen is 27 millimeters.

gravity and thereby permit these creatures to keep on an "even keel" while swimming.

Use of Fossils in Reconstructing Ancient Geography

The geographic distribution of present-day animals and plants is closely controlled by environmental limitations. Any given species has a definite range of conditions under which it can live and breed, and it is generally not found outside that range. Ancient organisms, of course, had similar restrictions on where they could survive. If we note the locations of fossil species of the same age on a map and correctly infer the environment in which they lived, we can produce a paleogeographic map for that particular time interval. One might begin by plotting on a simple base map the locations of fossils of marine organisms that lived at a particular time. This provides an idea of which areas were occupied by seas and might even suggest locations for ancient coastlines. Figure 16.25 shows major land and sea regions during the middle Carboniferous Period. The locations of marine protozoans called fusulinids are plotted as open circles. Notice that the lofty ranges of the Rocky Mountains were nonexistent, and their present locations were occupied by a great north–south seaway.

Having obtained a fair idea of the marine regions, one might next give attention to the locations of fossils diagnostic of land areas. The fossilized bones of land animals such as dinosaurs or mastodons would suggest a terrestrial environment, as would their preserved footprints. Fossil remains of land plants, including fossilized seeds and pollen, provide excellent evidence of terrestrial paleoenvironments.

By an analysis of the fossils and the nature of the enclosing sediment, it is often possible to recognize deeper or shallower parts of the marine realm or to discern particular land environments such as ancient flood plains, prairies, deserts, and lakes. River deposits

may yield the remains of fresh-water clams and fossil leaves. A mingling of the fossils of land organisms and sea organisms might be the result of a stream entering the sea and perhaps building a delta in the process.

The migration and dispersal patterns of land animals, as indicated by the fossils they left behind, is one important indicator of former land connections as well as mountainous or oceanic barriers that once existed between continents. Today, for example, the Bering Strait between North America and Asia prevents migration of animals between the two continents. The approximately 80 kilometers between the two shorelines, however, is covered by less than 50 meters of water; this might lead one to wonder whether Asia was once connected to North America. The fossil record shows that a land bridge did connect these two continents on several occasions during the Cenozoic Era. The earliest Cenozoic

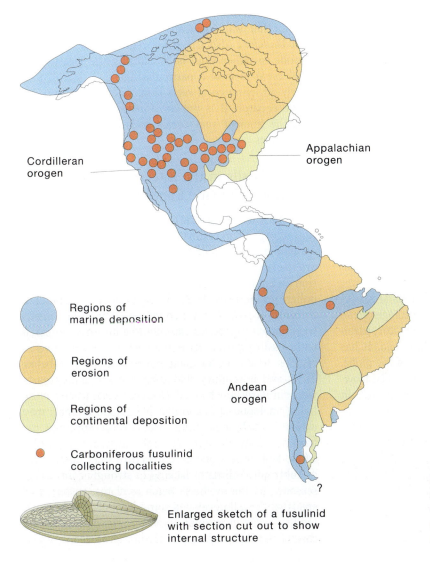

Cordilleran orogen

Appalachian orogen

Andean orogen

Regions of marine deposition

Regions of erosion

Regions of continental deposition

Carboniferous fusulinid collecting localities

Enlarged sketch of a fusulinid with section cut out to show internal structure

Figure 16.25 Major land and sea regions in North and South America during the Carboniferous Period.
(Adapted from C. A. Ross, J. Paleontol, 41(6), 1967, 1341–1354.)

Figure 16.26 A view looking down the steep frontal face of a modern reef. Admiralty Islands, Papua, New Guinea.
(L. E. Davis)

strata on both continents have a fossil fauna uncontaminated by foreign species. Somewhat higher and younger rocks contain fossil remains of animals that heretofore had been found only on the opposite continent. These remains mark a time of land connection. The last such connection may have existed only 14,000 to 15,000 years ago, when Stone Age hunters used the route to enter North America.

Another familiar example of how fossils aid in paleogeographic reconstructions is found in South America. Careful analyses of fossil remains indicate that, in the early Cenozoic Era, South America was isolated from North America; as a result, a uniquely South American fauna of mammals evolved over a period of 30 to 40 million years. The establishment of a land connection between the two Americas is recognized in strata only a few million years old (late Pliocene) by the appearance of a mixture of species formerly restricted to either North or South America. (The migrations were decidedly detrimental to South American species.)

In addition to deciphering the positions of shorelines or locations of land bridges, paleontologists can also provide data that help to locate the equator, parallels of latitude, and pole positions of long ago. It has been observed that in the higher latitudes of the globe, one is likely to find large numbers of individuals. However, these are members of relatively few species. In contrast, equatorial regions tend to develop a large number of species, but with comparatively fewer individuals within each species. Stated differently, the variety or **species diversity** for most higher categories of plants and animals increases from the poles toward the equator. This is probably related to the fact that relatively few species can adapt to the rigors of polar climates. Conversely, there is a stable input of solar energy at the equator, less duress caused by seasons, and a

more stable food supply. These warmer areas place less stress on organisms and provide opportunity for continuous, uninterrupted evolution.

Another way to locate former equatorial regions (and therefore also the polar regions that lie 90° of latitude to either side) is by plotting the locations of fossil coral reefs of a particular age on a world map. Nearly all living coral reefs (Fig. 16.26) lie within 30° of the equator. It is not an unreasonable assumption that the ancient reef corals had similar geographic preferences.

Use of Fossils in the Interpretation of Ancient Climatic Conditions

Among the many climatic factors that limit the distribution of organisms, climate (especially the temperature component of climate) is of great importance. There are many ways by which paleoecologists gain information on ancient climates. An analysis of fossil spore and pollen grains (Fig. 16.27) often provides outstanding evidence of past climatic conditions. Living organisms with known tolerances can often be directly compared with fossil relatives. As noted earlier, corals thrive in regions where water temperatures rarely fall below 18°C, and it is likely that their ancient counterparts were similarly constrained. However, even when a close living analogue is not found, plants and animals may exhibit morphologic traits useful in determining paleoclimatology. Plants with aerial roots, lack of annular rings, and large wood cell structure indicate tropical or subtropical climates lacking in strongly contrasting seasons. Marine mollusks (such as clams, oysters, and snails) with well-developed spinosity and thicker shells tend to occur in warmer regions of the oceans. In the case of particular species of marine protozoans, varia-

A

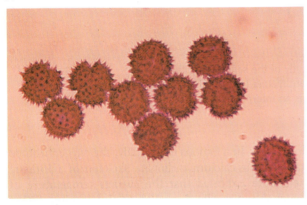

B

Figure 16.27 Pollen grains. (A) Pollen grain of a fir tree. Note the inflated bladders that make the pollen grain more buoyant. The specimen is about 125 microns in its longest dimension. (B) Pollen grains of ragweed. These grains are about 25 microns in diameter.

paleoecologic studies. Today, as in the past, benthic (bottom-dwelling) species have inhabited all major marine environments. As a result of the accumulated data on living foraminifers, inferences about salinity and water depths can be made. Fossil foraminifers have frequently provided the means for recognition of ancient estuaries, coastal lagoons, and nearshore or deep oceanic deposits.

Sometimes, when the overall morphology of a fossil does not provide clues to temperature or climate, the compositions of the skeleton can be used. Magnesium, for example, can substitute for calcium in the calcium carbonate of shelled invertebrates. For particular groups of living marine invertebrates, those living in warmer waters have higher magnesium values than do those residing in colder areas. This knowledge has been used to interpret climate at the times fossil forms were living.

Another method for determining the temperature of ancient seas involves the analysis of oxygen isotopes in the calcium carbonate of marine organisms. When water evaporates from the ocean, oxygen-16, which is lighter, is preferentially removed, whereas the heavier oxygen-18 tends to remain behind. When evaporated moisture is returned to the Earth's surface as rain or snow, water containing the heavier isotope precipitates first, often near the coast, and flows quickly back into the ocean. Inland, the precipitation from the remaining water vapor is depleted in oxygen-18. If the interior of a continent is cold and contains growing glaciers, the glacial ice will lock up the lighter isotope, preventing its return to the ocean and thereby increasing the proportion of the heavier isotope in seawater. As this occurs, the calcium carbonate shells of marine invertebrates such as foraminifers will be enriched in oxygen-18 and thereby reflect episodes of continental glaciation.

tion in the average size of individuals or in the direction of coiling can provide clues to cooler or warmer conditions.

An example of environmentally induced changes in shell coiling directions is provided by the planktonic (floating) foraminifer *Globorotalia truncatulinoides* (Fig. 16.28). In the Pacific Ocean, left-coiling tests (shells) of these foraminifers have dominated in periods of relatively cold climate, whereas right-coiling tests predominated during warmer episodes. Another foraminifer, *Neoglobquadrina pachyderma*, exhibits similar changes in coiling directions, not only in the Pacific Ocean but in the Atlantic as well. Such reversals in coiling may occur quickly and over broad geographic areas. For this reason, they are exceptionally useful in correlation studies.

Aside from morphologic changes within species, entire assemblages of foraminifers are widely used in

Figure 16.28 Shell coiling in the foraminifer *Globorotalia truncatulinoides*. Sketch of both sides of a single left-coiling specimen. (Diameter is about 0.9 millimeters.) *(From F. L. Parker,* Micropaleontology, *8(2), 1962, 219–254.)*

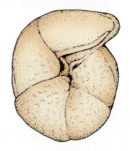
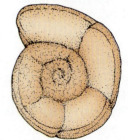

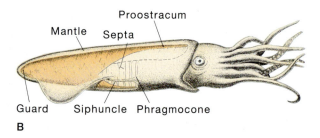

B

Figure 16.29 (A) The usual belemnite fossil consists only of the solid part of the internal skeleton called a guard. The guard here is 9.6 centimeters long. (B) Belemnites are extinct relatives of squid and cuttlefish. Shown here is an interpretation of a belemnite as it would appear if living.

Even in the absence of ice ages, the oxygen isotope method may be useful as a paleotemperature indicator. As invertebrates extract oxygen from seawater to build their shells, a temperature-dependent fractionation of oxygen-18 occurs between the water and the secreted calcium carbonate in the shell matter. Provided there has been no alteration of the shell, it is possible to find the temperature of the water at the time of shell formation. In a famous early study, the oxygen isotope ratio in the calcium carbonate skeleton of a Jurassic belemnite indicated that the average temperature in which the animal lived was 17.6°C, plus or minus 6°C of seasonal variation. Subsequent studies on both Jurassic and Cretaceous belemnites (Fig. 16.29) provided confirmation of the inferred positions of the poles during the Mesozoic and indicated that tropical and semitropical conditions were far more widespread during late Mesozoic time than they have been in subsequent geologic time.

Figure 16.30 The circular pods depicted in this mural of the margins of a Precambrian sea are calcareous algal structures called stromatolites. *(National Museum of Natural History)*

Figure 16.31 *Mawsonites,* a fossil found in the Pound Quartzite of Australia. The formation is late Proterozoic in age. *Mawsonites* is considered to be the mouth end of a jellyfish. *(B. N. Runnegar)*

16.7   An Overview of the History of Life

As yet, no fossil evidence has been found of life on Earth during the planet's first billion years. The oldest direct indications of ancient life are remains of bacteria and primitive algae discovered in rocks over 3.5 billion years old. These early fossils are usually considered evidence that the long evolutionary march had begun. However, they also stand at the end of another long and remarkable period during which living things presumably evoled from nonliving chemical compounds. We have no direct geologic evidence to tell us how and when the transition from nonliving to living occurred. What we do have are reasonable hypotheses supported by careful experimentation. Some of these hypotheses will be examined in Chapter 17, which deals with the Precambrian eons.

For most of Precambrian time, living things left only occasional traces. Here and there paleontologists have been rewarded with finds of bacteria and filamentous or unicellular blue-green algae. Occasionally, filamentous organisms formed extensive algal mats (Fig. 16.30), and these may have produced important quantities of oxygen by photosynthesis. However, life was at the unicellular level until about 1 billion years ago, when the world's first multicellular organisms left their trails and burrows in rocks that would later be called the Torrowangee Group of Australia. In rocks deposited about 0.7 billion years ago, fossil metazoans recognizable as worms, coelenterates, and arthropods have been found—albeit rarely—in scattered spots around the world, including the United States, Australia (Fig. 16.31), Canada, England, Russia, China, and South Africa. Thus, near the end of the Proterozoic, the stage was set for a wide range of Paleozoic plants and animals.

The Evolutionary History of Animals

Some general observations are possible after a brief scanning of the history of life following the Precambrian. The principal groups of invertebrates appear early in the Paleozoic (Fig. 16.32). Less advanced members of each phylum characterized the earlier geologic periods, whereas more advanced members came along later.

Figure 16.32   Geologic ranges and relative abundances of frequently fossilized categories of invertebrate animals. Width of range bands indicates relative abundance. Colored areas indicate where fossils of a particular category are widely used in zoning and correlation.

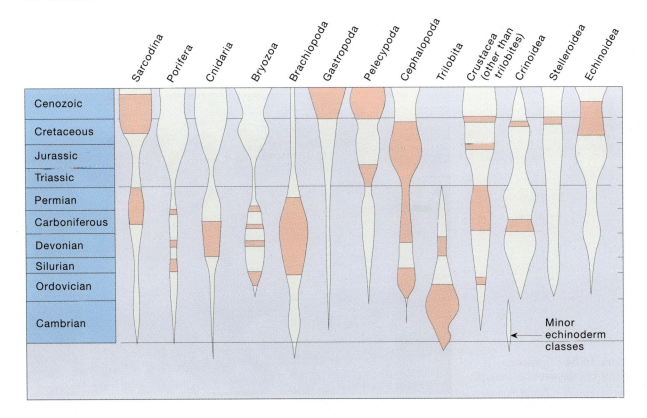

Figure 16.33 A large, straight-shelled cephalopod dominates this scene of an Ordovician sea floor. Also visible are corals, bryozoa, and crinoids. *(National Museum of Natural History)*

Also, most of the principal phyla are still represented by animals today. In our review of fossil creatures, we are not startled by sudden appearances of bizarre or exotic animals and plants. Further, we are able to recognize periods of environmental adversity that caused extinctions. Such episodes were usually followed by much longer intervals of recovery and orderly evolution. One final observation is that there has been a persistent gain in the overall diversity of life through the ages.

After the Precambrian, multicellular organisms evolved rapidly. Marine invertebrates such as trilobites,

brachiopods, nautiloids, crinoids, horn corals, honeycomb corals, and twiglike bryozoans were abundant during the Paleozoic Era (Fig. 16.33). The Paleozoic was also the era when fishes, amphibians, and reptiles appeared and left behind a fascinating record of the conquest of the lands.

Remains of more modern corals, diverse pelecypods, sea urchins, and ammonoids characterize the marine strata of the Mesozoic Era. However, the Mesozoic is best known as the "age of reptiles," when dinosaurs and their kin dominated the continents (Fig. 16.34). No

Figure 16.34 Cretaceous carnivorous dinosaurs in combat. *(American Museum of Natural History)*

less important than these big beasts, however, were the rat-sized primitive mammals and earliest birds that skittered about perhaps unnoticed by the "thunder beasts."

The mollusks were particularly well represented in the marine invertebrate faunas of the Mesozoic, and they have continued in importance into the Cenozoic. However, no ammonoid cephalopods have occurred in this most recent era. Rather, rocks of the Cenozoic are recognized by distinctive families of protozoans (foraminifers) and a host of modern-looking snails, clams, sea urchins, barnacles, and encrusting bryozoa. Because the Cenozoic saw the expansion of warm-blooded creatures such as ourselves, it is appropriately termed the "age of mammals."

The Evolutionary History of Plants

Long before the first animal appeared on Earth, there were plants. Plants also experienced a complex and marvelous evolutionary history. That history began with the origin of the primitive unicellular aquatic algae of the Precambrian. Except for the blue-green algae and bacteria noted previously, the fossil record of this earliest but important part of plant evolution is extremely poor. It improves during the second stage in plant history with the origin in the Silurian Period of seedless (spore-bearing) vascular land plants (Fig. 16.35). These plants

proliferated in the coal forests of the late Paleozoic. A third episode in plant evolution is marked by the advent of nonflowering pollen- and seed-producing plants. Although this was also a late Paleozoic event, the seed plants expanded and diversified during the Mesozoic Era. Finally, in the last period of Mesozoic, plants having both seeds and flowers evolved. Grasses were added to the flowering plant floras during the Cenozoic, and the lands in the age of mammals took on a decidedly modern appearance.

Fossils and the Search for Mineral Resources

One might be surprised to learn that the majority of paleontologists in the world are not based in museums and universities but rather are using their knowledge of fossils in the search for oil and gas. In their investigations, they generally utilize only very small fossils (microfossils) because these are less likely to be broken by drilling tools. During the drilling process, samples of the rock being penetrated are returned to the surface in drilling mud or in rock cores. The rock is then broken down and the microfossils extracted for study. Marine protozoans known as foraminifers (Fig. 16.36) are among the more useful and abundant fossils obtained. Micropaleontologists, specialists in the study of microfossils, prepare subsurface logs that depict the depth

Figure 16.35 Geologic ranges, relative abundances, and evolutionary relationships of vascular land plants.

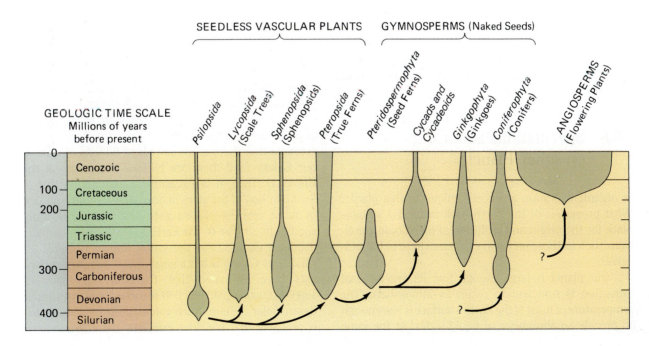

Figure 16.36 Foraminifers obtained from the Monterey Formation of Miocene Age, California.

from which each species was obtained. From the accumulated information on many such paleontologic logs, one is able to identify the geologic period of the formations through which the drill penetrated. Particular units of rock can be recognized by their unique guide fossils. Knowledge of the location of such key fossils relative to known oil-bearing strata is useful when wells are drilled in unproved territory. Indeed, the microfossils permit one to construct a picture of the sequence and orientation of rocks deep beneath the Earth's surface. In addition, the vertical successions of fossil assemblages in well samples often reflect changes in the depth of water in which the sediment was deposited and can therefore be used to infer tectonic conditions within the oil-producing region. Knowledge of the depth of water preferences for microfossils may help in locating porous reef or nearshore sediments that are occasionally saturated with petroleum.

16.8 Speculations about Life on other Planets

Is organic evolution a process unique to this planet? What properties of the Earth have made it a suitable place for the origin and evolution of life? Could conditions such as those on Earth exist elsewhere in the universe?

Our planet is large enough that its gravitational attraction is sufficient to retain an atmosphere. The temperature of most of the Earth's surface is low enough to provide an abundance of liquid water. At the same time, it is high enough for the chemical reactions required for life processes. Our Sun also has a sufficiently long life span to permit time for the emergence and evolution of life. Finally, the Earth has always had all of the chemical elements required for life processes.

Life in Our Solar System

We may first consider the possibilities for extraterrestrial life here in our own Solar System. All evidence to date indicates that because of either size or distance from the Sun, conditions are currently too harsh on neighboring planets to permit the evolution of higher forms of life. The Viking mission to Mars sampled the Martian surface for indications of life but found nothing to suggest the existence of present or past organisms on the red planet.

With regard to our Moon, scientists had predicted that no evidence of life would be found. The Moon lacks such important attributes as an atmosphere and water. The predictions were correct. The Moon is sterile. Satellites of other planets in the Solar System, however, appear to have a greater potential for harboring life. One of these is Europa, a satellite of Jupiter. Data from the Voyager flybys of Europa suggest that the satellite has a liquid water ocean covered by a thin, fractured crust of ice. It is only conjecture, but perhaps these waters harbor simple forms of life. By analogy, here on Earth certain species of algae and diatoms are known to be living in Antarctica in water that is overlain by impressive thicknesses of ice. They are able to photosynthesize even in the feeble light that has penetrated the overlying ice. Perhaps here and there beneath the icy surface of Europa, somewhat analagous conditions and organisms exist.

Life in the Universe

Thus, we see that the Earth is indeed biologically special when compared with other planetary bodies in our Solar System. In contrast, however, it may not be so peculiar at all in the vast realms of the Universe. Astronomers estimate that there are hundreds of billions of stars in our Galaxy (and there are billions of galaxies in the Universe). Thus, it is a reasonable assumption that at least a billion of the stars in our Galaxy would have one or more orbiting planets with size and temperature conditions similar to the Earth's. These are potentially habitable planets. For the entire known Universe (of which our Galaxy is but a small part), reputable scientists have estimated that there are as many as 100,000,000,000,000,000,000 planetary systems similar to our Solar System. Such calculations indicate that it is probable that life does exist out there somewhere.

Amber, the Golden Preservative

To many, the term *fossil* brings to mind remains of ancient organisms that have literally been turned to stone. In this chapter, however, a few rare instances of preservation of actual remains have been described. For such preservations, there is nothing superior to the organic substance known as amber (see Fig. 16.5). Insects are the usual organisms preserved in amber, and the bodies of insects, except for being dessicated, are preserved entirely. In addition to insects, spiders, crustaceans, snails, mammal hair, feathers, small lizards, and even a frog have been discovered in amber. By far the most abundant insects preserved in amber are flies, mosquitoes, and gnats of the order Diptera. (In the film *Jurassic Park*, fictional scientists obtained dinosaur DNA from blood in the belly of a mosquito that had made a meal of dinosaur blood before becoming entombed in amber.)

Paleontologists deplore the fact that during the late nineteenth century thousands of tons of raw amber were melted down for varnish, causing the loss of untold numbers of exquisitely preserved organisms.

Amber is a fossil resin produced by conifers. It is initially exuded from cracks and wounds in trees, and while still soft and sticky it traps and engulfs insects. The insect's struggle to escape can often be recognized in the swirl patterns around appendages. Later, through evaporation of the more volatile components, the soft resin hardens.

Although amber deposits are known in rocks ranging in age from lower Cretaceous to the Holocene, the most famous are those found along the coast of the Baltic Sea southwest of the Gulf of Riga near the seaport of Kaliningrad (formerly Konigsberg). The conifer forests in this area were inundated during a marine transgression in early Oligocene time. Marine sediment buried the trees and their contained amber. Subsequently, chunks of amber eroded from the soft clays were distributed across adjacent areas by streams, wave action, and glaciers. Intrigued by their beauty, neolithic families are known to have gathered the rounded yellow and brown pieces of amber. Such prominent philosophers of antiquity as Aristotle, and later Pliny and Tacitus, described amber's physical and chemical properties. As a semiprecious gem, amber was transported along ancient trade routes from the Baltic region to the Mediterranean. Traders called amber "the gold from the north." Then as now, it was used in the manufacture of beads for jewelry. Amber beads, amulets, necklaces, and bracelets have been recovered from Etruscan tombs and from excavations in Mycenae, Egypt, and Rome. However, for modern ladies adorned in amber jewelry, a word of whimsical caution may be of interest. That fellow gazing intently at your necklace may not be admiring its beauty. He maybe a paleontologist in search of an embalmed mosquito.

Reference: G. O. Poinar, Jr., *Life in Amber* (Stanford, CA: Stanford University Press, 1992).

Indeed, the universe may be rich in suitable habitats for life—but what kind of life? If we assume that such life was formed from the same universal store of atoms and under physical conditions not too dissimilar to those that existed on Earth, then we might very well recognize it as a living thing. But it is highly unlikely that duplicates of humans, cows, or butterflies exist on other planets. There are many variables in the evolutionary interactions of genetics, environment, and time involved in the making of a particular species. To say that the very same mutations, genetic recombinations, and environmental conditions producing a sparrow could occur in precisely the same sequential steps in time on a distant planet seems most improbable.

SUMMARY

Fossils are the remains or traces of life of the geologic past. The processes that are important in fossilization are varied and include the precipitation of chemical substances into pore spaces **(permineralization)**, molecular exchange for substances that were once part of the organisms with inorganic substances **(replace-ment)**, or compression of the animal and plant into a thin film of carbonized remains **(carbonization)**. **Tracks**, **trails**, **molds**, and **casts** are additional varieties of fossils. **Paleontologists**, the specialists that study fossils, are aware that the fossil record is not complete for all forms of life that have existed on Earth.

It is more complete for organisms that had hard parts such as teeth, bone, or shell and for organisms that lived on the floors of shallow seas, where the deposition of sediment is relatively continuous and rapid.

Particular lineages of fossil organisms exhibit change with the passage of time. This observation is part of the evidence for Darwin's conclusion that life has developed from simple to more complex forms (evolution). Because of this change through time, rock layers from different periods can be recognized and correlated on the basis of the kinds of fossils they contain.

Evolution combines natural selection with hereditary mechanisms for change by means of recombinations of hereditary materials and mutations. The hereditary materials are **chromosomes** and **genes**. Chromosomes are located in the nuclei of cells, carry the genes that are composed of DNA, and are responsible for the total characteristics of an organism. A process called **meiosis** segregates the members of homologous pairs of chromosomes and halves the total number for each **gamete**. During fertilization, there is a random union of two gametes from parents of unlike sex that brings together assortments of chromosomes (and genes) from two parents, resulting in the production of individuals with different gene combinations. Changes in genes called **mutations** may occur. Changes may occur also by rearrangement of chromosomes. These changes and redistributions alter the assortment of genes and, hence, also the characteristics that are passed on to succeeding generations.

Adaptation is the process by which organisms become fitted to their environments. Commonly, adaptations involve a combination of characteristics, including not only structure, but physiology and behavior as well. Different vertebrates, for example, display adaptive modifications of limbs for either swimming or running, or of teeth for various kinds of food. The evolution and spread of a single lineage of organisms into several ecologic niches often results in different forms, each of which is distinctly adapted to a particular habitat. This process of diversification from a central stock is called **adaptive radiation**.

Within any large adaptive radiation, one finds many examples of an evolutionary pattern called **divergence**, in which a parent stock splits into two or more distinct lineages. Other evolutionary patterns include **convergence**, in which two morphologically different groups evolve similarities in morphology as they become adapted to similar environments, and **iterative evolution**, wherein organisms of similar morphology evolve more than once from the parental lineage.

Charles Darwin viewed evolution as a gradual stately change in lineages through time. This interpretation, called **phyletic gradualism**, persists today and is probably valid for many groups of organisms. In addition, evolution entails sudden advances that punctuate periods of relative stability. The process is termed **punctuated equilibrium** and occurs when new species arise abruptly from isolated populations. The subsequent expansion of new species from these isolated groups results in the sudden faunal or floral changes often seen in the fossil record.

An important goal for paleontologists is to understand the complex relationships between ancient plants and animals and their habitats. This area of paleontology is termed **paleoecology**. Paleoecology can, in turn, provide information about the distribution of ancient lands and seas, ancient climates, depth of water, natural barriers to migration, and former locations of continents. Fossils are widely employed by commerical paleontologists engaged in the search for fossil fuels. It is likely that this aspect of paleontology will continue in importance as the search for economically essential minerals intensifies. Paleontologists will work in the future not only in such areas of exploration, but possibly also in the detection of once-living things in planets other than the Earth. Statistics suggest the strong probability of present or past life somewhere in the Universe.

KEY TERMS

Fossil *350*	Order *354*	Meiosis *357*	Divergence *359*
Petrifaction *350*	Phylum *354*	Mutation *357*	Convergence *359*
Permineralization *350*	Deoxyribonucleic acid	Adaptive radiation *358*	Iterative evolution *360*
Replacement *350*	(DNA) *356*	Phyletic gradualism *358*	Geologic range *362*
Mold *351*	Gene *356*	Punctuated	Guide fossil *364*
Cast *351*	Chromosome *357*	equilibrium *359*	Biozone *366*
Species *354*	Diploid cell *357*	Peripheral isolate *359*	Species diversity *370*
Family *354*	Mitosis *357*	Phylogenetic tree *359*	

REVIEW QUESTIONS

1. How does sexual reproduction provide for variation in species?

2. What is a fossil? What requirements should fossils meet to be considered guide fossils?

3. What is the total geologic range of each of the following?

   a. trilobites   d. mammals

   b. fishes   e. birds

   c. reptiles   f. angiosperms

4. Beginning with kingdom, list the remaining categories of the taxonomic heirarchy in descending order.

5. Which would probably become better guide fossils: organisms with slow rates of dispersal or organisms with rapid rates of dispersal? Give reasons for your answer.

6. Fossil A occurs in rocks of the Cambrian and Ordovician periods. Fossil B occurs in rocks that range from early Ordovician through Permian in age. Fossil C occurs in Mississippian through Permian rocks.

   a. What is the maximum possible range of age for a rock containing only fossil B?

   b. What is the maximum possible range for a rock containing both fossil A and fossil B?

   c. Which is the better guide fossil, A or C? Support your answer.

7. What are the major stages in the evolutionary history of terrestrial plants?

8. What were the contributions of Darwin and Mendel to our modern concepts of organic evolution?

9. In fossilization by replacement, what are the usual, or most common, replacing substances?

10. What are peripheral isolates? Why are peripheral isolates likely to evolve into new species?

11. Distinguish between phyletic gradualism and punctuated equilibrium. What evidence might you present if you were required to defend the concept of punctuated equilibrium?

12. Provide an example of an adaptive radiation, from either the present world or the geologic past.

DISCUSSION QUESTIONS

1. Evolution is often a response to selective pressures. What is meant by this term? What sort of global or regional environmental changes might result in selective pressures? Is this likely to have happened in the geologic past? Explain.

2. Account for the observation that there are more fossils of marine invertebrates than of any other group of organisms.

3. How has paleontology supported the concept of organic evolution?

4. How may fossils aid in deciphering ancient climates? Give an example.

5. A time–stratigraphic unit contains a different fossil assemblage at one locality than it does at another one located 50 miles away. Suggest a possible cause for the dissimilarity.

6. Does the rate of evolution of a particular group of related organisms affect their usefulness as guide fossils? Substantiate your answer.

7. What is meant by the term evolutionary convergence?

8. Explain why our knowledge of evolution suggests that although there may be life outside the Earth, there will probably not be human beings precisely like ourselves.

9. Why is the determination of a fossil species more subjective than the determination of a still-living (not extinct) species?

10. Using fossils for age correlation is dependent on *a priori* knowledge of their time ranges. How is this information obtained? What might cause the information to be in error?

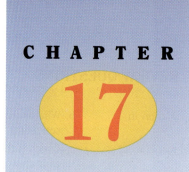

CHAPTER 17

The Oldest Rocks

*Exposure of 3.9 billion year old felsic gneiss from the Slave Province,
Northwest Territories, Canada.*
(Courtesy of S.A. Bowring.)

When nineteenth-century geologists began to determine the relative age of rock units and develop a geologic time scale, they often encountered a primary or basement complex of igneous and metamorphic rocks that lay beneath sedimentary strata. They referred to these older rocks as **Precambrian**, a term proposed in 1835 by Adam Sedgwick. To Sedgwick's contemporaries, correlating and estimating the age of unfossiliferous, deformed Precambrian rocks seemed an impossible task. Nevertheless, Precambrian rocks represent an enormous span of geologic time (Fig. 17.1), form the very cores of continents, and contain rich deposits of iron, gold, and other metals. Within these rocks are found the remains of the Earth's oldest known organisms as well as clues to the primordial conditions under which life began. For these reasons, it was inevitable that a few intrepid geologists would devote their lives to the study of Precambrian rocks.

17.1   Beneath the Cambrian

Among the early investigators of Precambrian rocks in North America, none was more prominent than Sir William Logan, of the Canadian Geological Survey. In the middle 1880s, Logan was able to associate groups of Precambrian rocks in Canada according to their superpositional and cross-cutting relationships. He also inferred that those rocks most severely deformed and metamorphosed were likely to be the oldest. More recent investigations clearly show that Logan frequently erred in attempting to relate degree of metamorphism to age. Older rocks may escape metamorphism, and younger ones may be radically metamorphosed. Yet in southeastern Canada where Logan mapped, one could find an older terrane of gneissic rocks as well as younger sequences of less altered metamorphic and sedimentary rocks. Because of this, it seemed reasonable to think of Precambrian time as divisible into an older **Archean Eon** and a younger **Proterozoic Eon**. All of the remainder of geologic time up to the present day was designated the **Phanerozoic Eon**.

Today, all geologists agree that the only reliable basis for correlation of unfossiliferous Precambrian rocks is through isotopic dating techniques. Only after the absolute ages of rocks in a particular region are known can they be placed in correct chronologic order. Radiometric dating of Precambrian rocks has provided a basis for mapping Precambrian terranes and for deciphering their geologic history. The hundreds of dates already obtained permit a tentative calibration of the Precambrian geologic time table (Table 17.1). The beginning of the Proterozoic is placed at about 2.5 billion years ago. The Archean is generally considered to have begun about 3.8 billion years ago, a date that approximates the age of oldest known crustal rocks.

For the earlier interval that began with the birth of the planet, the term **Hadean Eon** is often used. During the Hadean, the Earth was still very hot, convection was vigorous, and steaming vents and volcanoes blistered the planet's cratered surface. it was during the Hadean that the Earth was enveloped in an atmosphere devoid of free oxygen but rich in carbon dioxide, water, and other volcanic gases. It was also a time during which the Earth, its Moon, and other planets in the Solar System were being vigorously pelted by meteorites and asteroids. Although this episode of bombardment continued for several hundred million years, it has left little

Figure 17.1   Proportions of geologic time encompassed by the Precambrian and its divisions, the Hadean, Archean, and Proterozoic eons.

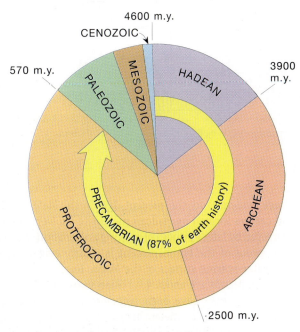

Table 17.1 • Time Divisions for the Precambrian

Time in Billions of Years	Time Divisions Followed in This Book[*]
	Neoproterozoic
1.0	— 1.0 —
	Mesoproterozoic
1.5	
	— 1.6 —
2.0	Paleoproterozoic
2.5	— 2.5 —
	Late Archean
3.0	— 3.0 —
	Middle Archean
3.5	— 3.4 —
	Early Archean
	— 3.8 —
4.0	Hadean
4.5	
	— 4.6 —

*As recommended by the International Union of Geological Sciences.

direct evidence on the present-day Earth. Meteorites and craters formed by Hadean meteorite showers were destroyed by later episodes of melting and erosion. However, proof of the bombardment is clearly evident on the crater-scarred surface of the Moon. Rocks from the lunar highlands are about 4.0 billion years old. At that time, there may have been sufficient heat generated by the vast infall of meteorites to cause melting and consequent resetting of radiometric clocks. Radiometric clocks on the Hadean Earth would have been reset in a similar way.

17.2   Shields and Cratons

Although exposures of Precambrian rocks may occur in the deep canyons of some mountains and plateaus, the most extensive exposures are found in broadly up-warped, geologically stable regions of continents called **shields**. Every continent has one or more shields (Fig. 17.2). Extending across 3 million square miles of northern North America is the great Canadian shield. Because continental glaciers of the last ice age have stripped away the cover of soil and debris, Precambrian rocks of the Canadian shield are extensively exposed at the surface.

To the south and west of the Canadian shield, Precambrian rocks are covered by sedimentary sequences of Phanerozoic age. These younger rocks are mostly flat lying or only gently folded. Such stable regions where basement rocks are covered by relatively thin blankets of sedimentary strata are called **platforms**. The platform of a continent, together with its shield, constitutes that continent's **craton** (Fig. 17.3).

On the basis of differences in the trends of faults and folds, the style of folding, and the ages of component rocks, the Canadian shield has been divided into a number of **Precambrian provinces** (Fig. 17.4). The boundaries of these provinces are often marked by abrupt truncations in structural lineations, or they may be represented by bands of severely deformed rocks of former orogenic belts. In North America, seven Precambrian provinces are recognized: the Superior, Wyoming, Slave, Nain, Hearne, Rae, and Grenville. These provinces were once separate cratonic elements that have been consolidated to form the larger North American craton. They are bound together by belts of deformed, metamorphosed, and intruded rocks that mark the location of collision of the various cratonic elements. For this reason, the belts of deformed rocks are called **collisional orogens**. It has been estimated that all of the Archean and Proterozoic elements of the Canadian shield were consolidated by about 1.8 to 1.9 billion years ago.

17.3   The Archean Crust

Origin of the Oceanic Crust

The Earth's original crust was not granitic. It was a **mafic** crust, meaning that it was dominated by calcium and sodium-rich feldspars, and by dark iron and magnesium silicates. This mafic crust was derived from an even

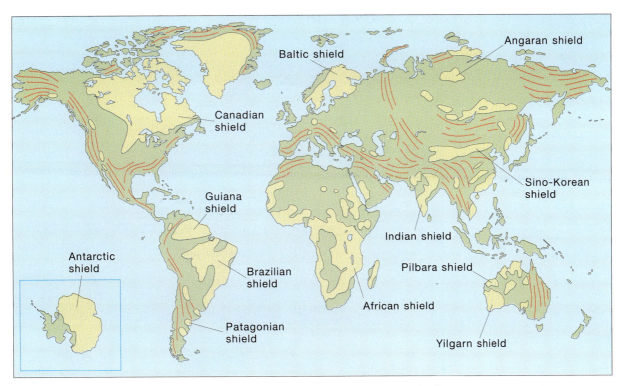

Figure 17.2 Major areas of exposed Precambrian rocks (shown in yellow). Trends of mobile belts are indicated by the red lines.

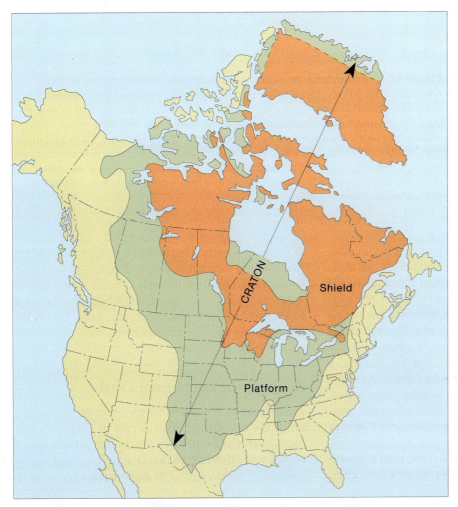

Figure 17.3 North American craton, shield, and platform.

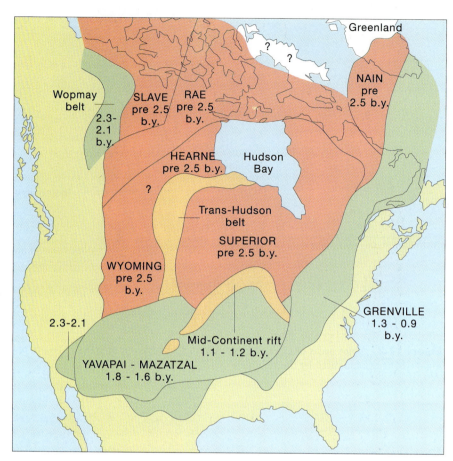

Figure 17.4 Precambrian provinces of North America. Archean provinces (those greater than 2.5 billion years old) are in red. Proterozoic provinces are in green. The Trans-Hudson belt is a zone along which rifting and then closure occurred to weld the Superior province to the Wyoming province. The Hearne and Wyoming provinces may actually be a single province. A major episode of rifting that occurred 1.2 to 1.0 billion years ago is represented by the mid-continent rift zone. The Wopmay belt was developed by rifting followed by closure of an ocean basin. *(Adapted from P. F. Hoffman, Precambrian Geology and Tectonic History of North America, in Geology of North America, Vol. A, Chap. 16 [Boulder, CO: Geological Society of America, 1989].)*

more mafic (hence, **ultramafic**) underlying mantle. To understand the origin of the mafic crust, we must imagine the Earth during its Hadean infancy, when it seethed with heat generated from the decay of short-lived radioactive isotopes, from gravitational forces, and from the infall of meteorites. Even after the final stages of accretion, there was ample heat to cause extensive melting in the mantle. As a result of the melting, a great magma ocean may have covered the Earth's surface. When surface patches of the magma ocean cooled, they solidified as rocks especially rich in iron and magnesium. They were so rich in these elements that they are called ultramafic rocks.

As the early crust was being formed, high temperatures in the upper mantle probably produced convection currents. It appears likely that some form of plate tectonics would then begin, and with plate tectonics there would be a recycling of surface rocks. For example, water-rich melts would rise above subducting plates and be extruded as mafic (not ultramafic) lavas. Eventually, the recycling produced the ordinary rock we know as basalt (Table 17.2).

Origin of the Continental Crust

In contrast to the mafic composition of the oceanic crust, the continental crust is rich in potassium feldspars, quartz, and muscovite. Rocks rich in these minerals are termed **felsic rocks**. One of the oldest known patches of felsic crust (3.96 billion years old) was re-

Table 17.2   Summary of the Characteristics of the Earth's Early Oceanic and Continental Crust

	Oceanic Crust	*Continental Crust*
First appearance	About 4.5 billion years ago	About 4.0 billion years ago
Where formed	Ocean ridges (spreading centers)	Subduction zones
Composition	Ultramafic rocks, basalt	Felsic rocks
Lateral extent	Widespread	Local
How generated	Partial melting of ultramafic rocks in upper mantle	Partial melting of wet mafic rocks in descending slabs

From K. C. Condie, "Origin of the Earth's Crust," *Paleogeography, Paleoclimatology, Paleoecology, 75* (1989), 57–81, with permission.

cently discovered in Canada's Northwest Territories (Fig. 17.5). Southwestern Greenland has similar felsic rocks that are 3.8 billion years old (Fig. 17.6). Also in Greenland, geologists have found exposures of ultramafic rocks that may be remnants of the Earth's surface that was formed before there was any felsic crust. Other patches of extremely old crustal rocks have been found in Antarctica, South Africa, and Australia.

When we compare the ages obtained from the ancient rocks just described to the 4.6-billion-year age of our planet, it is apparent that the Earth had acquired patches of felsic continental crust a few hundred million years after its origin. But how did those original areas form and grow? Many geologists believe that they originated at subduction zones where slabs of sediment-covered crust plunged into the mantle, to be melted and release their felsic components. Because of the lower density of felsic magmas, they would rise to form continental crust, whereas more mafic materials would remain behind in the mantle.

It is likely that the early Archean patches of continental crust were small, perhaps less than about 500

A

B

Figure 17.5   (A) Photomicrograph of one of the 3.96-billion-year-old zircon grains extracted from the Acasta Gneiss, Slave province, Northwest Territories of Canada. The grain is 0.5 millimeters long. Its polished surface has been etched with acid to highlight crystal growth zones. Numbers refer to points selected for analysis. (B) Exposure of the zircon-bearing Acasta Gneiss at the discovery site. *(Sam A. Bowring)*

Figure 17.6 Archean felsic gneiss from the Isua region of southwestern Greenland. *(R. F. Dymek)*

kilometers (310 miles) in diameter. These may have grown by several mechanisms that continue to operate today. One such mechanism may have involved the coming together of two tectonic plates bearing felsic rocks at their encountering margins. Growth may also have occurred as island arcs and microcontinents were added to the host continent along subduction zones. A third method may have involved the welding of offshore sediment onto continental margins. This process begins with the weathering and erosion of preexisting rocks and deposition of the resulting sediment in the ocean adjacent to an upland area. Next, the thick accumulation of sediment is subjected to orogeny caused by an encounter with a converging tectonic plate. The sediment in the prism would have roughly the same bulk composi-

tion as granitic igneous rocks. Such material would be converted to felsic crystalline crustal rocks if subjected to intense metamorphism and melting. The repetition of such cycles of deposition and orogeny, along with suturing events and the incorporation of microcontinents, could account for continuing growth of continental crust. As indicated by the known distribution of 3.0- to 2.5-billion-year-old felsic rocks, very large continents had been assembled by late Archean time.

Figure 17.8 Exposure of greenstone showing well-developed pillow structures, near Lake of the Woods, Ontario, Canada. The original basalts were extruded under water during late Archean time. During the Pleistocene Ice Age, glaciers beveled the surface of the exposures and left behind the telltale scratches (glacial striations). *(K. Schultz)*

Figure 17.7 Greenstone belts of the Superior province. *(After A. M. Goodwin,* Proc. Geol. Assoc. of Canada, *19 [1968], 1–14.)*

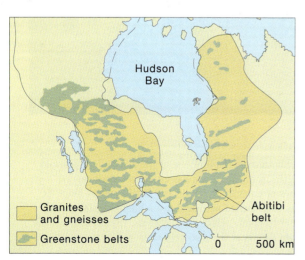

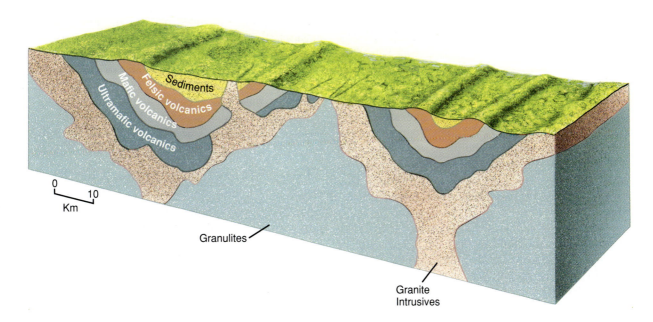

Figure 17.9   Generalized cross section through two greenstone belts. Note their synclinal form and the sequence of rock types from ultrabasic near the bottom to felsic near the top. A late event in the history of the belt is the intrusion of granites. Ultramafic basal layers are particularly characteristic of greenstone belts in Australia and South Africa but do not occur in the Archean of Canada.

Granulites and Greenstones

In general, there are two major groups of Archean rocks. The first is called the **granulite** association and consists largely of gneisses derived from metamorphism of felsic igneous rocks. The second group is called the **greenstone association**. Greenstone rocks occur in synclinal belts (Fig. 17.7) in the Archean of all continents. The Abitibi belt of Canada is one of the largest of these. Like most other greenstone belts, it is composed of basaltic, andesitic, and rhyolitic rocks along with metamorphosed sedimentary rocks derived from the erosion of lavas. As indicated by pillow structures (Fig. 17.8), the lavas were extruded under water.

An intriguing characteristic of greenstone belts is the sequential transition of rock types from ultramafic rocks near the bottom to felsic rocks near the top (Figs. 17.9 and 17.10). The basalt layers above the ultramafic rocks contain chlorite and hornblende, and these green minerals impart the color for which greenstones are named. Proceeding upward in a greenstone succession, one finds felsic volcanics above basalt, followed by shales, graywackes, and conglomerates. The entire greenstone sequence is surrounded by granitic intrusive rocks. In the Superior Province of Canada, the granites were emplaced during a deformational event that occurred about 2.6 to 2.7 billion years ago. That event, called the **Kenoran orogeny**, marks the close of the Archean Eon in Canada.

Archean Tectonics

Because the Earth was hotter during the Archean, plate tectonics probably operated more vigorously than in the Phanerozoic. Volcanism was more extensive, and rates of convection were far greater. The sequence of events

Figure 17.10   A representative greenstone belt stratigraphic sequence from the Precambrian shield of western Australia.   *(Adapted from D. R. Lowe, Ann. Rev. Earth Planet. Sci., 8 [1980], 145–167.)*

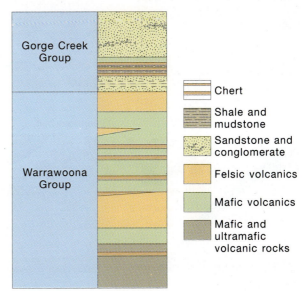

that may have provided for both the granulite and greenstone associations would have begun with the development of volcanic arcs adjacent to subducting plates (Fig. 17.11). Lavas and pyroclastics of the arcs were weathered and eroded to provide sediment to the adjacent trench. In the subduction zones, this wet sediment was carried downward along with oceanic crust and was heated. Partial melting of the mix of materials produced increasingly more felsic magmas, and these worked their way upward and engulfed earlier volcanics. During or shortly after these events, compressional deformation and metamorphism altered the rocks to form the granulite terranes.

Whereas volcanic arcs were the sites for the birth of granulite associations, the back-arc basins were the probable locations for the development of greenstone belts. In the early stages of that development, ultramafic to mafic lavas ascended through rifts in the back-arcs to form the lower layers of the greenstone sequences.

Later extrusions included increasingly more felsic lavas, pyroclastics, and sediments derived from the erosion of adjacent uplands. The final stage of development involved compression of the back-arcs to form the synclinal structures characteristic of greenstone belts. Compression may have occurred simultaneously with the magmatic activity that produced the granitic intrusions which surround the lower level of greenstone belts.

17.4   Archean Sedimentary Rocks

Southern Africa's Barberton Mountain Land and a terrane in western Australia called the Pilbara block contain excellent records of Archean sedimentation. Here one finds large cratons rather than the small protocontinents discussed earlier. These cratons were fully formed and stable as early as 3.0 billion years ago. Because they have been stable since that early time, sedimentary rocks

Figure 17.11   Plate tectonics model for the development of greenstone belts and growth of continental crust. (A) Plates are in motion, driven by convection cells in the upper mantle. Subduction provides for the emplacement of wedges of oceanic crust and for mixing and melting to provide tonalite (felsic) intrusions. Behind the main arc, the back-arc sags by extension and the greenstone volcanic sequence is extruded. (B) Compression has occurred to create the greenstone belts with their synclinal form and also to aggregate small continental patches into a larger continental mass. Later, granites are intruded in and around greenstone belts.   *(Simplified from a model proposed by B. F. Windley, 1977,* The Evolving Continents, *New York, John Wiley & Sons.)*

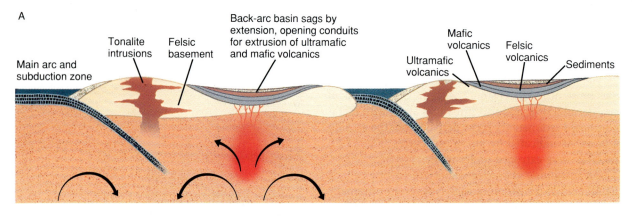

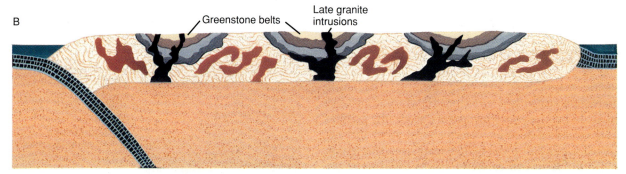

within them have often escaped severe metamorphism that might otherwise have obscured their environments of deposition. In southern Africa, some of these sedimentary rocks contain gold and have therefore been intensely investigated.

In the greenstone belts of Barberton Mountain Land, the rock column consists of a typical greenstone sequence of mafic rocks near the base, grading upward to increasingly more felsic rocks (Fig. 17.12). Sandstones near the top of the sequence contain quartz and orthoclase, indicating nearby granite source areas. By the time these sediments were being deposited, a variety of depositional environments existed. Continental slope and continental rise environments are recorded by graywackes of the Fig Tree Group, and the overlying Moodies Group includes quartz sandstones and shales of shallow marine shelves and deltas.

Also present in rocks of Barberton Mountain Land is a remarkable sequence of river-borne conglomerates and sandstones known as the Witwatersrand Supergroup. The Witwatersrand is notable for several reasons. First, its great areal extent (about 103,000 square kilometers or 40,000 square miles), thickness (8 to 10 kilometers or 5 to 6.3 miles), and nonmarine origin of its sedimentary rocks indicate that large continents were present by 2.7 billion years ago. Second, the Witwatersrand contains placer gold. Indeed, since the discovery of that gold in 1883, half the world's supply of the precious metal has come from river and delta conglomerates of the Witwatersrand.

17.5   Proterozoic Continental Growth

The eon following the Archean is the Proterozoic. It began 2.5 billion years ago and ended only 570 million years ago. It was an eon characterized by a more modern style of plate tectonics and sedimentation. Because Proterozoic rocks are generally less altered than those of the Archean, they are easier to interpret. By Proterozoic time, wide continental shelves and epeiric seas existed. In these environments, shallow-water clastics and carbonates were deposited. Except for the absence of shells of marine invertebrates, the strata had a modern aspect.

During the early Proterozoic, the older cratonic elements of North America were welded together into a large continent called **Laurentia**. The welding occurred along belts of crustal compression called **orogens** (Fig. 17.13). Thereafter, Laurentia continued to grow as sedimentary rocks deposited along once passive continental margins were compressed and accreted onto the continent. Occasionally, small blocks of continental crust would be carried by sea-floor spreading to the margins of continents, providing additional growth.

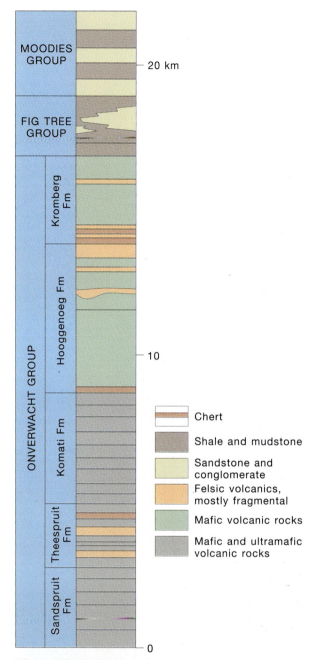

Figure 17.12   Stratigraphic sequence in a greenstone belt in Barberton Mountain Land, South Africa.
(After D. R. Lowe, Ann. Rev. Earth Planet. Sci., 8 [1980], 145–167.)

Consolidation of separate continents into supercontinents followed by breakup of the supercontinents were recurring events during the Proterozoic. The first Proterozoic supercontinent was formed about 1.8 billion years ago. It had broken apart by about 1.3 billion years ago. Once again, however, there was a reassembly, completed by about 1.0 billion years ago. That supercontinent has been dubbed **Proto-Pangaea**. Like those that

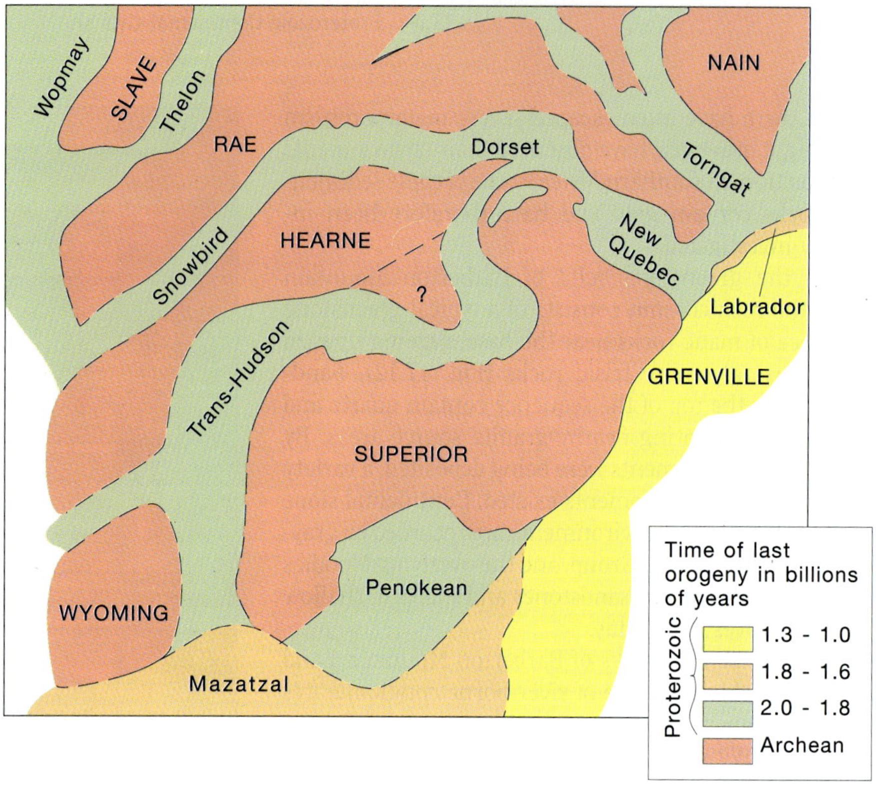

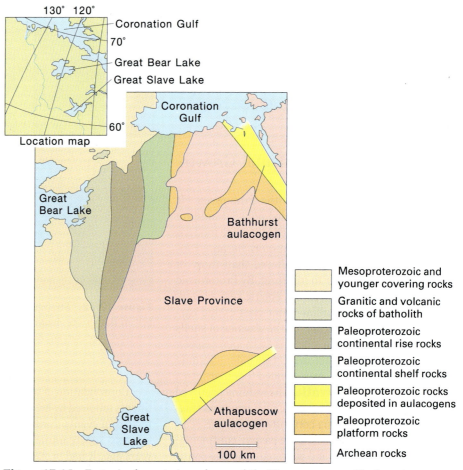

Figure 17.13 Expanded and generalized relationships of Proterozoic orogens to Archean provinces. The Wopmay, Penokean, Labrador, and Mazatzal orogens developed by accretion of sedimentary prisms laid down along the margins of Archean cratons. The other orogens are collisional in nature. *(From P. Hoffman, Precambrian Geology and Tectonic History of North America, Ch. 16 in Geology of North America [Boulder, CO: Geological Society of America, 1989].)*

Time of last orogeny in billions of years

Proterozoic
1.3 - 1.0
1.8 - 1.6
2.0 - 1.8
Archean

Location map

Mesoproterozoic and younger covering rocks

Granitic and volcanic rocks of batholith

Paleoproterozoic continental rise rocks

Paleoproterozoic continental shelf rocks

Paleoproterozoic rocks deposited in aulacogens

Paleoproterozoic platform rocks

Archean rocks

100 km

Figure 17.14 Tectonic elements in and around the Wopmay orogen, Northwest Territories, Canada. *(From P. Hoffman, "Evolution of an Early Proterozoic Continental Margin: The Coronation Geosyncline and Associated Aulacogens of the Northwestern Canadian Shield," Phil. Trans. Royal Soc. London, A., 273 [1973], 547–581.)*

had preceded it, Proto-Pangaea also suffered fragmentation and dispersal of its parts. In the next chapter, we will note how the dispersed fragments of Proto-Pangaea came together to form the supercontinent named Pangaea by Alfred Wegener.

17.6   Paleoproterozoic Events in North America

The Wopmay Orogen

In the Canadian shield, rocks that accreted to Laurentia during the Paleoproterozoic lie primarily in belts that circumvent the older Archean provinces. One such area, which provides a good example of the kinds of geologic events characterizing the Paleoproterozoic in many other parts of the Earth, is the Wopmay orogen along the western edge of Canada's Slave Province in the Northwest Territories (Fig. 17.14). Within the **Wopmay orogen**, there is evidence of the opening of an ocean basin, sedimentation along the resulting continental shelves, and closure of the ocean basin as a result of plate tectonic processes. This sequence of events, well documented in later geologic history, is called a **Wilson cycle** after J. Tuzo Wilson, one of the pioneers of plate tectonic theory.

In the Wopmay orogen, evidence for the initial opening of an ocean basin consists of numerous normal (tensional) faults (Fig. 17.15). Lavas flowed upward through the faults and spread across alluvial sediments that had accumulated in the downfaulted valleys. As the ocean continued to widen, the western margin of the Slave Province became a passive margin. Near shore,

Figure 17.15   Relationship of rock units following Paleoproterozoic opening of an ocean basin along the western margin of the Slave craton.   *(Simplified from P. Hoffman, Geology of North America, Vol. A:447–512 [Boulder, CO: Geological Society of America, 1989].)*

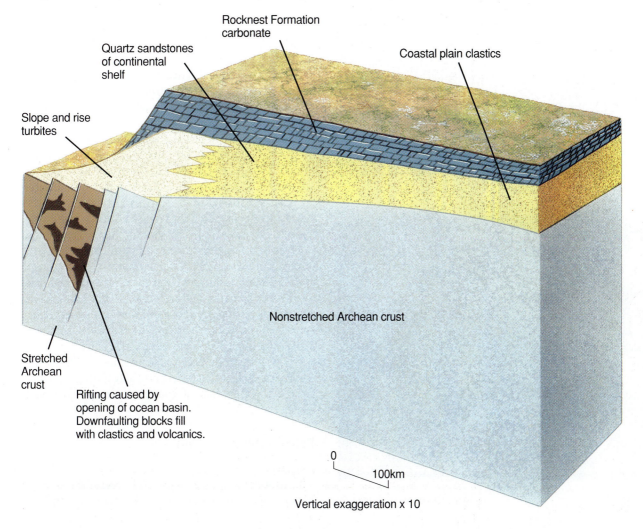

sandstones and carbonates accumulated (Fig. 17.16), whereas farther out to sea, turbidites of the continental rise were deposited. Ultimately, however, this period of relative quiet gave way to the dynamic part of the Wilson cycle that involved ocean closure through plate convergence. In the process, the sedimentary section was compressed, metamorphosed, and raised into a great mountain range.

Figure 17.16   (A) Dolomite and interbedded shales of the Rocknest Formation, Northwest Territories, Canada. (B) This oblique aerial view impressively illustrates the folding of these dolomitic rocks that resulted from east–west crustal shortening and accretion that occurred during the orogenic phase of the development of the Wopmay orogen.   *(P. Hoffman)*

A

B

These events in the Wopmay orogen closely parallel those that occurred during the late Paleozoic evolution of the Appalachian Mountains. The history of both tracts began with deposition of alluvial deposits in fault-bound basins followed by opening of an oceanic tract; progressed to the deposition of continental shelf and rise sediments along passive continental margins; and ended with ocean closure, deformation, and deposition of nonmarine clastic sediments derived from the erosion of mountains produced during the closure episode.

The Trans-Hudson Orogen

The Wopmay is just one of many orogens that developed along the margins of Archean provinces during the Proterozoic. Another is the **Trans-Hudson orogen**, which separates the Superior province from provinces to the north and west (see Fig. 17.4). Rocks of the Trans-Hudson belt record a historical sequence of initial rifting, opening of an oceanic tract, and subsequent closure along a subduction zone. This closure, with the accompanying severe folding, metamorphism, and intrusive igneous activity, welded the Superior plate to the Hearne and Wyoming plates lying to the west.

A Paleoproterozoic Ice Age

In the region north of Lake Huron, rocks of the Paleoproterozoic are called **Huronian**. They consist mainly of coarse clastics that were eroded from older igneous and metamorphic terranes. In addition, the Huronian sequence includes the **Gowganda Formation**, a rock unit that is notable because of its conglomerates and mudstones of glacial origin. Laminations, called varves, in the mudstones represent regularly repeated summer and winter layers. Varves typically form in meltwater lakes adjacent to ice sheets. The varved mudstones in the Gowganda Formation alternate with tillites, indicating periodic advances of the ice sheets into marginal lakes. Cobbles and boulders in the tillite are scratched and faceted, providing ample evidence of the abrasive action of the ice mass as it moved over the underlying bedrock.

Frigid climates and continental glaciation were widespread during the Paleoproterozoic, for rocks similar to the Gowganda are recognized in Finland, southern Africa, and India. These episodes of glaciation may have been the first on Earth, as global temperatures during the Hadean and Archean were probably too high to allow for the accumulation of enough ice to produce vast ice sheets.

Paleoproterozoic Iron

Paleoproterozoic rocks surrounding the western shores of Lake Superior include those of the **Animikie Group**. This group of formations is world famous for the bonanza iron ores they contain. The ore minerals are oxides of iron and thus record the presence of free oxygen in the Earth's atmosphere. By implication, the oxygen-generating process of photosynthesis was probably in vigorous operation by this time. Coarse sandstones and conglomerates deposited in shallow water lie near the base of the Animikie Group. These rocks are overlain by cyclic successions of chert, cherty limestones, shales, and banded iron formations (known by the acronym BIF). Banded iron formations are also known in Archean sequences, but are not nearly as extensive as those of the Proterozoic (Fig. 17.17). Some

Figure 17.17   Banded iron formations. (A) This banded iron formation is exposed at Jasper Knob in Michigan's Upper Peninsula. (B) Banded iron formations occur worldwide in Proterozoic rocks, as suggested by these at Wadi Kareim, Egypt. *(Photograph B courtesy of D. Bhattacharyya)*

A

B

of the Canadian deposits are over 1000 meters thick and extend over 100 kilometers. Within the banded iron sequence, there is a formation known as the **Gunflint Chert**, which contains an interesting assemblage of cyanobacteria and other prokaryotic organisms.

17.7   Mesoproterozoic Events

Keweenawan Sequence

Among the rocks deposited or emplaced during the Mesoproterozoic were those of the Keweenawan sequence (Fig. 17.18). These rocks rest on either crystalline basement or Animikian strata. They extend from the Lake Superior region southward beneath a cover of Phanerozoic rocks for hundreds of kilometers. Keweenawan rocks consists of clean quartz standstones and conglomerates as well as basaltic volcanics. The lava flows are well known for their content of native copper. Holes originally formed as gas bubbles in the extruding lava provided some of the voids in which the copper was deposited. In addition, copper fills small joints and pore spaces in associated conglomerates.

The Keweenawan lavas exceed several thousand meters in thickness. When such prodigious quantities of mafic lava are extruded, it often signals the breaking apart of a continent. The break typically begins with the formation of normal faults that provide the channels along which molten rock flows to the surface. The fault zones associated with the Keweenawan lavas extend from Lake Superior southward beneath sedimentary covering rocks to Kansas. If this system of faults had extended to the edge of the continent, an oceanic tract would have opened within the rift system and the eastern United States would have drifted away. Rifting ceased, however, before such separation could occur.

The Grenville Province

The Grenville Province of eastern Canada is the last Proterozoic province to undergo a major orogeny. Exposures of Grenville rocks extend from the Atlantic coast of Labrador to Lake Huron. Beneath the cover of younger rocks, they continue southwestward all the way to Texas (see Fig. 17.4). Deformation of Grenville rocks (Fig. 17.19) occurred 1.2 to 1.0 billion years ago. That orogeny may have been caused by the collision of another continental mass against the eastern margin of Canada.

17.8   Neoproterozoic Events

The final segment of Proterozoic history is represented in North America by rocks that were deposited in basins and on the craton, or in shelf areas that lay along the

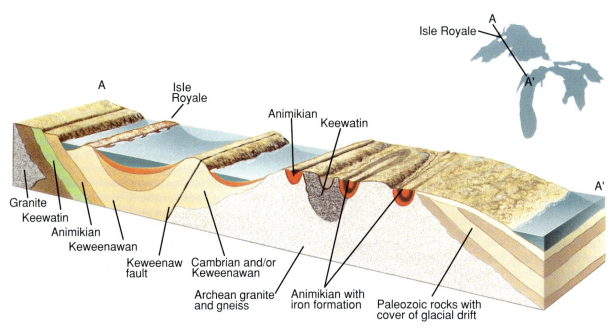

Figure 17.18   Geologic cross section across Lake Superior.   *(Adapted from Michigan Geological Survey,* Bedrock of Michigan, *1968.)*

margins of the continent. An important Neoproterozoic event was extensive rifting along the eastern margin of North America. Lavas upwelled along the tensional faults and flowed into the fault valleys. Erosion of uplifted blocks of crust supplied the coarse clastics that are found interbedded with layers of basalt. Ultimately,

Figure 17.19   Folded sandstones and carbonates of the late Proterozoic Grenville Group, Belmont Township, Ontario, Canada.   *(Geological Survey of Canada)*

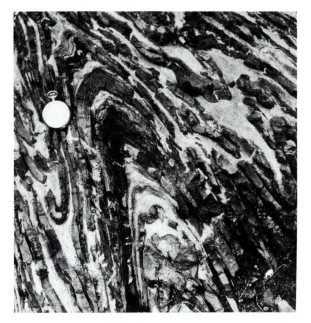

the sea entered the rift valleys as the proto-Atlantic began to form. As the ocean widened, a new continental margin formed east of the old Grenville orogen. The continental shelf and rise of that continental margin would receive the initial deposits of the Paleozoic.

Also in Neoproterozoic sequences, one can find exposures of unsorted boulder beds that are considered by many geologists to be glacial deposits. Such deposits are known in Utah, Nevada, western Canada, Alaska, Greenland, South America, Scandinavia, and Africa. At every locality, the cobbles and boulders exhibit the telltale striations and faceted surfaces that attest to transport by glaciers. This late Proterozoic ice age lasted about 240 million years. It was not the first ice age for the Proterozoic. As described earlier, the Gowganda tillites and correlative formations on other continents reflect widespread glaciation slightly more than 2 billion years ago. Nor would the late Proterozoic glaciations be the last the Earth would witness, for glacial ice advanced again over major portions of the continents during the Ordovician and Silurian, near the end of the Paleozoic, and again during the Quaternary Period of the Cenozoic Era (Fig. 17.20).

17.9   Proterozoic Rocks South of Canada

Although Proterozoic rocks are not as extensively exposed in the United States as in Canada, there are many interesting sites for investigation. Particularly impres-

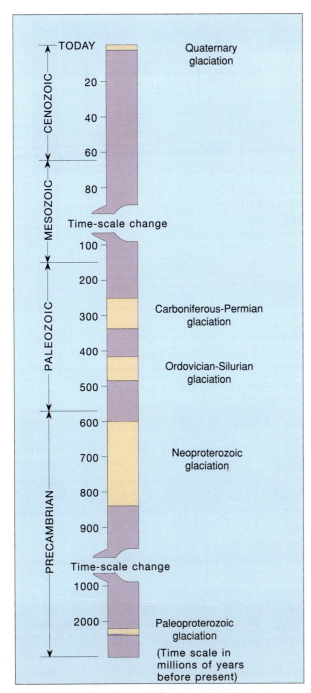

Figure 17.20 In its long history, the Earth has experienced several major episodes of widespread continental glaciation. Each of these major episodes contained shorter intervals of advance and retreat of the glaciers.

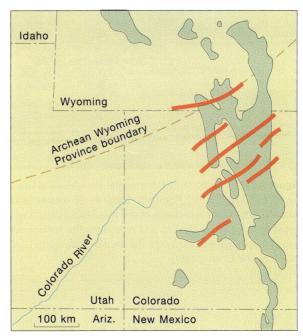

Figure 17.21 Shear zones (red) in Wyoming and Colorado developed when the Archean craton collided with island arc terranes during the Paleoproterozoic. Areas of lower Proterozoic outcrops are shown in green. *(Adapted from K. E. Karlstrom and S. A. Bowring,* Jour. Geology, *96 [1988], 561–571.)*

sive successions of strata are exposed in the Rocky Mountains and Colorado Plateau. The rocks in these regions record a complex geologic history that began over 2.5 billion years ago with the development of an Archean terrane composed of metamorphosed granitic rocks. Remnant patches of lavas and greenstones and a

shear zone (Fig. 17.21) in Wyoming indicate that the old Archean landmass collided with island arc segments about 1.75 billion years ago. Subsequently, huge magnetic bodies (plutons) were emplaced across a broad belt from California to Labrador.

The next event south of the Canadian Shield was a widespread episode of rifting from about 1.4 to 0.8 billion years ago. Large fault-controlled basins developed and provided catchment areas for Proterozoic sedimentary rocks. Among these sedimentary rocks, the **Belt Supergroup** of Montana, Idaho, and British Columbia is of particular interest because of the scenic features with which these rocks are associated. In places, the rocks of the Belt Supergroup are over 12,000 meters thick. They form the dramatic precipices in Waterton Lakes and Glacier National Parks (Fig. 17.22). Belt rocks contain ripple marks and algal structures, which indicate that they were deposited in shallow water along the passive western margin of North America.

Another scenic attraction in which Proterozoic rocks can be viewed is Grand Canyon National Park. Precambrian rocks of the Grand Canyon region consist of two major units. The lower and older unit is the Vishnu Schist, and the upper is the Grand Canyon Supergroup. The Vishnu is a complex of metamorphosed sediments and gneisses that have been invaded by granites. These granites were emplaced about 1.35 billion

Figure 17.22 Undeformed, horizontal limestones of the Belt Supergroup comprise the upper third of Chief Mountain, located just outside of Glacier National Park, Montana. The lower two thirds of the mountain consists of deformed Cretaceous beds. A thrust fault separates the two and represents the surface over which the Proterozoic beds were pushed eastward over the weaker underlying Cretaceous strata. The fault is known as the Lewis thrust.

years ago. A splendid example of a nonconformity can be seen at the contract between the Vishnu Schist and the overlying Grand Canyon Supergroup (Fig. 17.23). The Grand Canyon Supergroup consists of sandstones and shales and is approximately the same age as the Belt Supergroup to the north.

17.10 Proto-Pangaea and Iapetus

The most significant global event near the close of the Precambrian was the gathering of Precambrian continents to form the supercontinent Proto-Pangaea. In the course of this coming together of continents, it is likely that a Southern Hemisphere landmass collided with eastern North America. That collision may have been the cause of the Grenville orogeny. Proto-Pangaea had a relatively short geologic lifespan. Soon after it was fully assembled, tensional fault zones developed and separation began. North America and Greenland moved away from other elements of Proto-Pangaea. The ocean flooded into the rift zone and then slowly widened until an expanding ocean called **Iapetus** was formed.

17.11 The Beginning of Life

The splendid diversity of animals and plants on our planet is a consequence of billions of years of chemical and biological evolution that began during the Precambrian. We would like to have fossil documentation of the earliest events in the progression from nonliving to living things, but that kind of direct evidence is lacking.

Theories for the origin of life are based on our understanding of present-day biology, coupled with evidence obtained from the study of Archean rocks, meteorites, and planetary compositions.

From Simple to Complex

Systematic progressions from simple to complex structures are a manifested characteristic of both living and nonliving systems. Awareness of such progressions has strongly influenced theories on the origin of life. Subatomic particles such as neutrons, protons, and electrons are built into atoms. Atoms in turn combine to form molecules, and molecules may be joined to form sheets, chains, and a variety of complex molecular structures. Perhaps in a somewhat similar fashion, the atoms of elements essential to organisms (carbon, oxygen, hydrogen, nitrogen, sulfur, and phosphorus) are built into organic molecules. At some stage in the progression, simpler molecules might have added components that would gradually transform them into the complex organic molecules essential to all living things. An analogy might be the manner in which small molecules of nitrogenous bases combine with sugar and phosphorous compounds to form the large deoxyribonucleic (DNA) molecule (see Fig. 16.11). In living organisms, DNA is found chiefly in the nucleus of the cell. It functions in the transfer of genetic characteristics and in protein synthesis. Some theorists suggest that such large and complex molecules might have organized themselves into bodies capable of performing specific functions. Such bodies in unicellular organisms are called organelles. Organelles might then have combined to form larger entities that grew, metabolized, reproduced, mu-

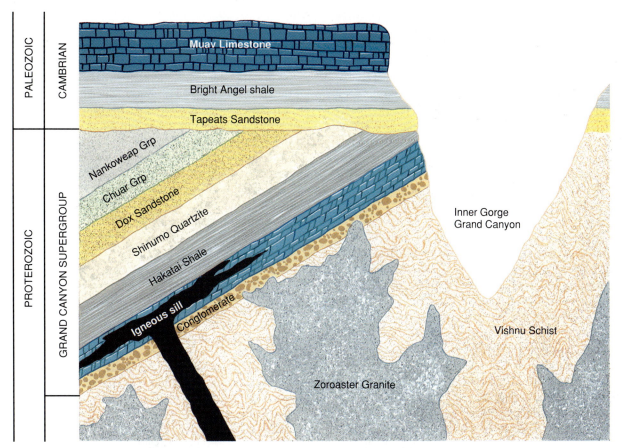

Figure 17.23 Vishnu Schist, Grand Canyon Supergroup, and other rocks in the Grand Canyon of the Colorado River.

tated, and produced mutations. These are the attributes of all living things. Could a progression even remotely similar to this have occurred in the early years of the Archean? We do not know for sure, but the concept is at least feasible.

Preliminary Considerations

The oldest fossil evidence of life is found in 3.5-billion-year-old rocks of Australia's Pilbara shield. Living things, however, probably existed even before the ones that left these traces, as indicated by carbon found in 3.8-billion-year-old banded iron deposits in southern Greenland. The carbon extracted from these ancient sedimentary rocks has ^{13}C to ^{12}C ratios similar to those in present-day organisms.

As noted in Chapter 1, the Earth accreted from a dust cloud about 4.6 billion years ago, and an early atmosphere was subsequently generated by volcanic outgassing. Presumably, the gases emitted were similar to those in volcanic eruptions today in that water and carbon dioxide were dominant, with lesser amounts of

nitrogen, hydrogen, hydrogen sulfide, sulfur dioxide, carbon monoxide, methane, and ammonia. Note the absence of free oxygen in this list. The planet's earliest organisms must have been anaerobic; that is, they did not require oxygen for respiration. The environment in which they lived lacked a protective atmospheric shield of ozone, which is derived from oxygen and which absorbs ultraviolet radiation from the sun. Perhaps early life escaped these lethal rays by living beneath protective ledges of rock or at greater water depths. Some biologists insist that in the early stages of the development of life, ultraviolet radiation may not have been a hindrance but rather served to impel molecules to interact and form complex associations.

The carbon, hydrogen, oxygen, nitrogen, phosphorus, and sulfur necessary to produce life are among the most abundant elements in the Solar System. They were undoubtedly present on the primitive Earth. What had to be accomplished was the enormously intricate organization of those elements into the molecular components of an organism. There are four essential components of life. The first component is protein. **Proteins** are essentially strings of comparatively simple organic molecules called amino acids. Proteins act as building materials and as compounds that assist in chemical reactions within the organism. The second of the basic components is **nucleic acids**, such as DNA, mentioned earlier, and ribonucleic acid (RNA). Organic phosphorous compounds provide a third component of life. They serve to transform light or chemical fuel into the energy required for cell activities. An important fourth ingredient for life is some sort of container, such as a cell membrane. The enclosing membrane provides a relatively isolated chemical system within the cell and keeps the various components in close proximity so that they can interact.

From the previous description, it is apparent that amino acids have an important role in the development of the larger and more complex molecules. They are the building blocks of proteins. Two environmental circumstances in the early years of Earth history may have been important in the natural synthesis of amino acids. Prior to the accumulation of the ozone layer in the Earth's upper atmosphere, ultraviolet rays bathed the Earth's surface. As demonstrated in experiments, ultraviolet radiation is capable of separating the atoms in mixtures of water, ammonia, and hydrocarbons and of recombining those atoms into amino acids. A second form of energy capable of accomplishing this feat is electrical discharge in the form of lightning. Either together or separately, lightning and ultraviolet radiation may have stimulated the production of amino acids in the air, in tidal pools, in the upper levels of the oceans, and wherever suitable physical and chemical conditions existed.

Test Tube Amino Acids and a Step Beyond

Scientists in the mid-nineteenth century had succeeded in manufacturing some relatively simple organic compounds in the laboratory. However, it was not until 1953 that the laboratory synthesis of amino acids and other molecules of roughly similar complexity was announced. Stanley Miller, at the suggestion of Harold Urey, performed the now famous experiment. He infused an atmosphere at that time thought to be like that of the Earth's earliest atmosphere into an apparatus similar to the one shown in Figure 17.24. It was a methane, ammonia, hydrogen, and water vapor atmosphere. As the mixture was circulated through the glass tubes, sparks of electricity (simulated lightning) were discharged into the mixture. At the end of only 8 days, the condensed water in the apparatus had become turbid and deep red. Analysis of the crimson liquid showed that it contained a bonanza of amino acids as well as somewhat more complicated organic compounds that

enter into the composition of all living things. In additional experiments by other biochemists, it was shown that similar organic compounds could also be produced from gases (carbon dioxide, nitrogen, and water vapor) of the preoxygenic atmosphere. The main requirement for the success of the experiments seemed to be the lack, or near absence, of free oxygen. To the experimenters, it now seemed almost inevitable that amino acids would have developed in the Earth's prelife environment. Because amino acids are relatively stable, they probably increased gradually to levels of abundance that would enhance their abilities to join together into more complex molecules.

To come together and form protein-like molecules, amino acids must lose water. This loss can be accom-

Figure 17.24 Stanley Miller and Harold Urey used an apparatus similar to this to replicate what they believed were the conditions of early Earth. An electric spark was produced in the upper right flask to simulate lightning. The gases present in the flask reacted together, forming a number of basic organic compounds. *(Stanley L. Miller, University of California, San Diego)*

plished by heating concentrations of amino acids to temperatures of at least 140°C. Volcanic activity on the primitive crust would be capable of providing such temperatures. However, the biochemist S. W. Fox discovered that the reaction also occurred at temperatures as low as 70°C if phosphoric acid was present. Fox and his co-workers were able to produce protein-like chains from a mixture of 18 common amino acids. He called these structures **proteinoids** and reasoned that billions of years ago they were the transitional structures leading to true proteins. This is not extravagant conjecture, for Fox was able to find proteinoids similar to those he created in his laboratory among the lavas and cinders adjacent to the vents of Hawaiian volcanoes. Apparently, amino acids formed in the volcanic vapors and were combined into proteinoids by the heat of escaping gases.

Hot, aqueous solutions of proteinoids will, on cooling, form into tiny spheres that show many characteristics common to living cells. These **microspheres** (Fig. 17.25), as they are called, have a filmlike outer wall; are capable of osmotic swelling and shrinking; exhibit budding, as do yeast organisms; and can be observed to divide into "daughter" microspheres. They occasionally aggregate linearly to form filaments, as in some bacteria, and they exhibit a streaming movement of internal particles similar to that observed in living cells.

Although complete long-chain nucleic acids have not yet been experimentally produced under prelife conditions, short stretches of ordered sequences of nucleic acid components have now been produced in the laboratory. Indeed, in 1976, Har Gobind Khorana and his associates at the Massachusetts Institute of Technology announced that they had made a functioning artificial gene—one of thousands in the DNA spiral molecule of the bacterium *Escherichia coli*.

Speculations about Earliest Life

There are several good reasons to believe that the earliest organisms originated in the sea. The sea contains the salts needed for health and growth. The waters of the oceans serve as universal solvents capable of dissolving a great variety of organic compounds, and currents in the oceans ceaselessly circulate and mix these compounds. Such constant motion would have favored frequent collisions of the vital molecules and would thereby have increased the probability of their combining into larger bodies. One can imagine the larger bodies adding components, increasing in complexity, and ascending in a thousand infinitesimally small steps from nonliving to transitional things of a biological netherworld, to the first forms of life. Such a sequence may have occurred under conditions lethal to most living organisms. For example, geochemists are now investigating a group of bacteria called thermophiles that thrive at temperatures of 350°C or greater. Such high temperatures occur today around geothermal vents common to the ocean floor along spreading centers. The knowledge that life is possible under such extreme conditions, and that the early Earth was much hotter than it is today, suggests that the earliest forms of life may have been thermophilic microbes (see "Geology and the Environment, The Birthplace of Life").

It is likely that the earliest forms of life would not have evolved a means of manufacturing their own food but rather would have assimilated certain chemical compounds and organic molecules present in the surrounding water medium. Some even would have consumed their developing contemporaries. Today, organisms with this type of nutritional mechanism are termed **heterotrophs**. The food gathered by the ancestral

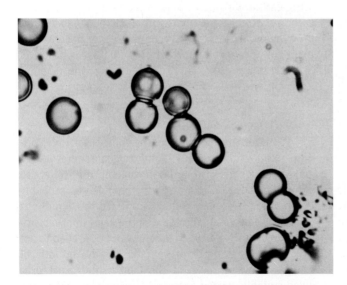

Figure 17.25 Proteinoid microspheres formed in hot aqueous solutions of proteinoids that have been permitted to cool slowly. These spheres are about 1 to 2 μm in diameter. *(Photo by Sidney W. Fox)*

GEOLOGY AND THE ENVIRONMENT

The Birthplace of Life

Since the time of Charles Darwin, scientists have considered the upper sunlit or photic levels of the sea as a favorable place for the origin of life as well as for the evolutionary progression from microbial heterotrophs to photosynthetic autotrophs. There are, however, other environments that deserve serious consideration. One of these lies deep in the abyss, in total darkness, adjacent to scalding waters that jet from hydrothermal vents on the ocean floor. Primitive bacteria termed chemosynthetic autotrophs flourish in this extraordinary environment.

Hydrothermal vents are common along mid-oceanic ridges where molten rock has risen close to the surface of the ocean floor. Seawater percolates downward through fissures until it reaches the zone of hot rocks. There, the waters are heated to temperatures in excess of 350°C, expand, and rise forcefully to the surface to be expelled at hydrothermal vents. In their sojourn beneath the sea floor, the hot waters become charged with chemicals essential for life, including hydrogen sulfide. The bacteria derive vital energy from chemical reactions in the cell that cause hydrogen sulfide to react with oxygen, producing water and sulfur. Thus, early life on Earth may have lived by means of chemosynthesis, the reaction of chemicals without sunlight, rather than photosynthesis, the reaction of chemicals with sunlight. The chemosynthetic bacteria around deep-sea hydrothermal vents comprise the foundation of a food chain that supports a rich and unusual assortment of invertebrates.

If chemosynthesis at hydrothermal vents on the ocean floor played a prominent role in the origin of life, one would hope to find some evidence of this in Archean rocks. Such evidence does exist. It consists of organic compounds within Archean mineral deposits that appear to have originated along deep-sea hydrothermal vent systems.

heterotrophs was externally digested by excreted enzymes before being converted to the energy required for vital function. In the absence of free oxygen, there was only one way to accomplish this conversion—by **fermentation**. There are many variations of the fermentation process. The most familiar reaction involves the fermentation of sugar by yeast. It is a process by which organisms are able to disassemble organic molecules, rearrange their parts, and derive energy for life functions. A very simple fermentation reaction can be written as follows:

$$C_6H_{12}O_6 \rightarrow 2\,CO_2 + 2\,C_2H_5OH + Energy$$
$$\text{Glucose} \qquad \text{Carbon} \qquad \text{Alcohol}$$
$$\text{dioxide}$$

Animal cells are also able to ferment sugar in a reaction that yields lactic acid rather than alcohol.

Eventually, the original fermentation organisms would have begun to experience difficult times. By consuming the organic compounds of their environment, they would eventually have created a food shortage; this scarcity, in turn, might have caused selective pressures for evolutionary change. At some point prior to the depletion of the food supply, organisms evolved the ability to synthesize their needs from simple inorganic substances. These were the first autotrophic organisms. Unlike the heterotrophs, **autotrophs** were able to manufacture their own food. The organisms that had developed this remarkable ability saved themselves from starvation. Their evolution proceeded in diverse directions. Some manufactured their food from carbon dioxide and hydrogen sulfide and were the probable ancestors of today's sulfur bacteria. Others employed the life scheme of modern nitrifying bacteria by using ammonia as a source of energy and matter.

However, more significant than either of these kinds of autotrophs were the **photoautotrophs** (Fig. 17.26), which were capable of carrying on photosynthesis. Their gift to life was the unique capability of dissociating carbon dioxide into carbon and free oxygen. The carbon was combined with other elements to permit growth, and the oxygen escaped to prepare the environment for the next important step in the evolution of primitive organisms. In simplified form, the reaction for photosynthesis can be written as follows:

$$6\,CO_2 + 6\,H_2O \xrightarrow{\text{Sunlight}} C_6H_{12}O_6 + 6\,O_2$$

With the multiplication of the photoautotrophs, billions upon billions of tiny living oxygen generators began to change the primeval nonoxygenic atmosphere to an oxygenic one. Fortunately, the change was probably gradual, for if oxygen had accumulated too rapidly, it would have been lethal to developing early microorganisms. Some sort of oxygen acceptors were needed to act as safety valves and to prevent too rapid a buildup of the gas. Iron in rocks of the continental crust provided suitable oxygen acceptors. Eventually, organisms

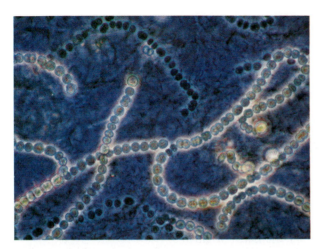

Figure 17.26 Photoautotrophs. These organisms, known as *Nostoc,* use light energy to power the synthesis of organic compounds from carbon dioxide and water. *(Sinclair Stammers/Science Photo Library/Photo Researchers, Inc.)*

evolved oxygen-mediating enzymes that permitted them to cope with the new atmosphere.

After the surficial iron on the Earth's surface had combined with its capacity of oxygen, the gas began to accumulate in the atmosphere and hydrosphere. Solar radiation acted on atmospheric oxygen to convert part

of it to ozone, and ozone in turn formed an effective shield against harmful ultraviolet radiation. Still-primitive and vulnerable life was thereby protected and could expand into environments that formerly had not been able to harbor life. The stage was set for the appearance of aerobic organisms.

Aerobic organisms use oxygen to convert their food into energy. The reaction, which can be considered a form of cold combustion, provides far more energy in relation to food consumed than does the fermentation reaction. This surplus of energy was an important factor in the evolution of more complex forms of life.

Prokaryotes, Eukaryotes, and Symbiosis

There is no unequivocal evidence to date the transition from chemical or prebiotic evolution to organic evolution. The most accurate statement that can be made in this regard is that the transition occurred at some time prior to 3.5 billion years ago. In rocks of that age, paleontologists have discovered the earliest fossil evidence of life. These oldest fossils belong to a category of microorganisms called **prokaryotes** (Fig. 17.27A). Prokaryotes lack definite membrane-bounded organelles and also do not have a membrane-bounded nucleus in which genetic material is neatly arranged into discrete chromosomes. Modern prokaryotes do possess cell

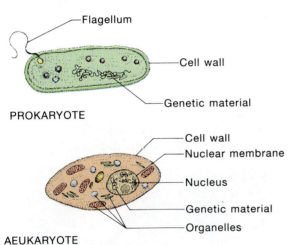

PROKARYOTE

AEUKARYOTE

Figure 17.27 (A) The basic differences between prokaryotes and eukaryotes. The drawings are not at the same scale. Most prokaryotes are only about 1 micrometer in diameter, whereas eukaryotes are well over 10 times that size. (B) An electron micrograph of the prokaryote *Bacillus subtilus,* a living bacterium that is about to complete cell division (magnification ×15,000). *(Figure B from Dr. A. Ryter)*

B

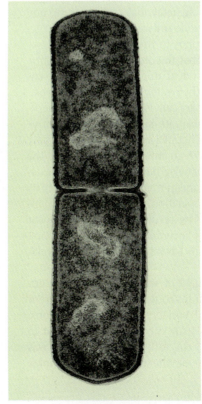

walls, and most are able to move about. The cyanobacteria (formerly called blue-green algae) are prokaryotes that are capable of photosynthesis (Fig. 17.26).

Prokaryotes are asexual and thereby restricted in the level of variability they can attain. In sexual reproduction, as noted in Chapter 16, there is a union of gametes (egg and sperm) to form the nucleus of a single cell, the zygote. The formation of the zygote results in recombination of the parental chromosomes and leads to a multitude of gene combinations among the gametes that give rise to the next generation. In the asexually reproducing prokaryotes, in contrast, a cell divides from the parent and becomes an independent individual containing the identical chromosome complement of the parent cell. Unless mutation intervenes, the number and kinds of chromosomes in individuals produced by asexual reproduction are exactly the same as those in the parent. Thus, the possibilities for variation are more limited than in sexually reproducing organisms. It is probably for this reason that prokaryotes have shown little evolutionary change through more than 2 billion years of Earth history. Nevertheless, such organisms represent an important early step in the history of primordial life.

Evolution proceeded from the prokaryotes to organisms with a definite nuclear wall, well-defined chromosomes, and the capacity for sexual reproduction. These more advanced forms were **eukaryotes** (Fig. 17.27A). Unlike prokaryotes, eukaryotes contain organelles such as chloroplasts (which convert sunlight into energy) and mitochondria (which metabolize carbohydrates and fatty acids to carbon dioxide and water, releasing energy-rich phosphate compounds in the process).

Biologists believe that the organelles in eukaryotic cells were once independent microorganisms that entered other cells and then established symbiotic relationships with the primary cell. For example, an anaerobic, heterotrophic prokaryote such as a fermentative bacterium might have engulfed a respiratory prokaryote and thereby would have an internal consumer of the oxygen that might otherwise have threatened its existence. A nonmotile organism might acquire mobility by forming a symbiotic association with a whiplike organism like a spirochete. The resulting cell would appear to have a flagellum for locomotion. Natural selection would clearly favor such advantageous symbiotic relationships.

With the development of sexually reproducing eukaryotes, genetic variations could be passed from parent to offspring in a great variety of new combinations. This development was truly momentous, for it led to a dramatic increase in the rate of evolution and was ultimately responsible for the evolution of complex multicellular animals.

17.12   The Fossil Record for the Archean

Fossils are rare in Archean rocks. When they are found, they are the remains of prokaryotic microorganisms. The oldest of these now known were discovered by paleobiologist J. William Schopf in 1993. They are the altered remains of strings of connected cells that appear to be cyanobacteria. They occur in rocks that are 3.465 billion years old. The discovery of these microfossils is significant, not only as our earliest known record of life on this planet but as an indicator that oxygen was being generated through photosynthesis a mere billion years after the Earth originated.

Prior to Schopf's 1993 discovery, prokaryotes had been found in cherts of southern Africa's Fig Tree Group. These fossils are 3.1 billion years old. They include a number of tiny, double-walled structures (Fig. 17.28) that resemble modern bacteria. They have been named *Eobacterium,* the "dawn bacterium." In addition, study of thin sections of Fig Tree cherts revealed spherical bodies given the name *Archaeosphaeroides* ("ancient spheres"), also thought to be cyanobacteria. Many beds within the Fig Tree Group contains so much organic matter that they are black in color. On analysis, the dark layers were found to contain organic compounds regarded as the breakdown products of chlorophyll, indicating that photosynthesis was in operation.

More readily apparent indications of the nature and abundance of Archean life consist of structures known as stromatolites. **Stromatolites** are laminar, organic sedimentary structures formed by the trapping of sedimentary particles and precipitation of calcium carbonate in response to the metabolic activities and growth

Figure 17.28   Electron micrographs of rod-shaped *Eobacterium isolatum,* about 0.6 micron in length. Specimen at lower right has been interpreted as a transverse section through a cell.   *(J. W. Schopf)*

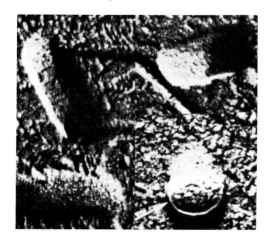

FOCUS ON   *Heliotropic Stromatolites*

Stromatolites are among the most abundant kinds of fossils found in Proterozoic rocks. Today, as during the Proterozoic, the photosynthetic cyanophyta that construct stromatolites depend on sunlight for survival and growth. In a study of modern stromatolites, it has been observed that certain species form columnar laminated growths that are inclined toward the sun to gather the maximum amount of light on their upper surfaces. This tendency among organisms to grow toward the sun is called **heliotropism**. Stromatolites, however, may also exhibit growth responses to currents and rising sea level, and thus must be carefully examined for evidence that a preferred growth orientation is related to sunlight. Stanley Awramic and James Vanyo (paleobiologist and astronomer, respectively) have studied 850-million-year-old stromatolites from Australia's Bitter Springs

Formation for evidence of heliotropism. They believe they found such evidence in species of *Anabaria juvensis*. At the locality studied in central Australia, these stromatolites curve upward in a distinct sine wave form that appears to record growth that followed the seasonal change in the position of the sun above the shallow sea in which the stromatolitic cyanobacteria lived. In their elongate sinuous form, each sine wave represented a year of growth. The stromatolites grew by the addition of a layer or lamina each day, and thus by counting the laminae in the length of a single sine wave, one can obtain the number of days in the year. The results indicated that there were 435 days in the year 850 million years ago when the stromatolites of the Bitter Springs Formation were actively responding to Proterozoic sunlight.

---

of matlike colonies of cyanobacteria and a fewer number of other prokaryotes. The study of present-day stromatolites indicates that they develop when fine particles of calcium carbonate settle on the sticky, filamentous surface of cyanobacterial mats to form thin layers (laminae). The cyanobacterial colonies then grow through the layer and form another surface for the collection of additional fine sediment. Repetition of this process, along with precipitation of calcium carbonate, produces a succession of laminae often forming cabbage-like, columnar, or hummocky masses (Fig. 17.29).

Stromatolites are more common in Proterozoic rocks than in Archean rocks. Warm shelf environments conducive to the growth of marine stromatolites may have been limited during the Archean. Stromatolites are known in rocks as old as those containing the earliest cyanobacteria, such as the Warrawoona Group of Australia (Fig. 17.30). Somewhat younger Archean stromatolites occur in the 3.0-billion-year-old Pongola Group of southern Africa and the 2.8-billion-year-old Bulawayan Group of Australia. The ^{12}C to ^{13}C ratios in the Bulawayan rocks provide evidence of biological fixation of carbon dioxide by photosynthetic organisms.

17.13   The Fossil Record of the Proterozoic

Life of the early Proterozoic was not significantly different from that of the preceding late Archean. Blue-green scums of photosynthetic cyanobacteria (Fig. 17.31)

probably floated on seas and lakes, and anaerobic prokaryotes multiplied in environments deficient in oxygen. Stromatolites, which were relatively uncommon in the Archean, proliferated in the Paleoproterozoic and continued to flourish throughout the remainder of the Phanerozoic Eon. Very likely, the expansion of stromatolites was enhanced by the greater number of epeiric sea and shallow shelf environments evident in the Proterozoic.

By Mesoproterozoic time, evolution had progressed to the point at which eukaryotes begin to appear in the fossil record. You may recall that single-celled eukaryotes contain a true nucleus enclosed within a nuclear membrane and have well-defined chromosomes and cell organelles. Early single-celled eukaryotes gave rise to multicellular forms and were the probable ancestors of the first unicellular animals. Very likely, the expansion of eukaryotes began about 1.4 billion years ago. It was followed in Neoproterozoic time by the early evolution of multicellular animal life.

Proterozoic Prokaryotes

Unicellular prokaryotes were the characteristic organisms of the Archean as well as the Paleoproterozoic. Fossil remains of these relatively primitive organisms were first detected in 1965 by the American paleontologists Elso S. Barghoorn and J. William Schopf in a now-famous rock unit called the **Gunflint Chert**. This rock unit is exposed along the northwestern margin of Lake Superior. Radiometric dating methods provide an age

A   B

Figure 17.29   (A) Modern columnar stromatolites growing in the intertidal zone of Shark Bay, Australia. Metabolic activities of colonial marine cyanobacteria result in the formation of these structures. Fine particles of calcium carbonate settle between the tiny filaments of the matlike colonies and are bound with a mesh of organic matter. Successive additional layers result in the laminations that are the most distinctive characteristic of stromatolites. (B) Fossil stromatolites from Precambrian rocks exposed in southern Africa.   *Photo A courtesy of J. Ross, and B courtesy of J. W. Schopf)*

Figure 17.30   Filamentous prokaryotic microfossils from 3.5-billion-year-old black cherts of the Archean Warrawoona Group, Pilbara shield of western Australia.   *(J. W. Schopf and B. M. Packer)*

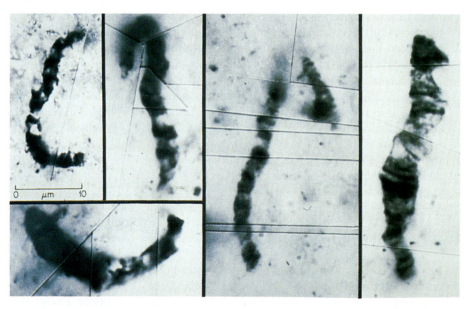

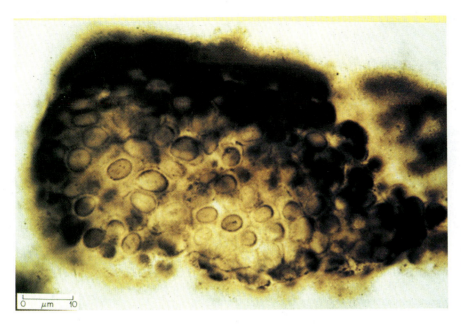

Figure 17.31   A Paleoproterozoic colony of the cyanobacterium *Eoentophysalis* photographed in a thin section of a Belcher Group stromatolitic chert, Belcher Islands, Canada.   *(H. J. Hofmann)*

of 1.9 billion years for the Gunflint Chert. The formation contains an abundant and varied flora of so-called thread bacteria and nostocalean cyanobacteria. Unbranched filamentous forms, some of which are septate, have been given the name *Gunflintia* (Fig. 17.32). More finely septate forms, such as *Animikiea,* are remarkably similar to such living algae as *Oscillatoria* and *Lyngbya*. Other Gunflint fossils, such as *Eoastrion* (literally, the "dawn star"), resemble living iron- and magnesium-reducing bacteria. *Kakabekia* and *Eosphaera* do not resemble any living microorganism, and their classification is uncertain. What does seem certain is that there was an abundance of photosynthetic organisms in the Paleoproterozoic, all actively producing oxygen and thereby altering the composition of the atmosphere. Not only do the Gunflint fossils resemble living photosynthetic organisms, but their host rock contains organic compounds regarded as the breakdown products of chlorophyll.

The Rise of Eukaryotes

The evolution of eukaryotes surely qualifies as one of the major events in the history of life. Eukaryotes possess the potential for sexual reproduction, and this provides enormously greater possibilities for evolutionary change. Unfortunately, the fossil record for the earliest eukaryotes is somewhat ambiguous. This is not surprising when one considers that the first eukaryotes were microscopic unicells whose identifying characteristics (enclosed nucleus, organelles, and so on) are rarely preserved. There is, however, one important clue to the identification of a fossil as a eukaryote. It is based on

the size of cells. Living spherical prokaryotic cells rarely exceed 20 microns in diameter, whereas eukaryotic cells are nearly always larger than 60 microns. Such larger (and hence probable eukaryotic) cells begin to appear in the fossil record by about 1.4 to 1.6 billion years ago.

Another clue to the presence of eukaryotes in the Proterozoic consists of fossils known as acritarchs. **Acritarchs** are unicellular, spherical microfossils with resistant, single-layered walls (Fig. 17.33). The walls may be smooth or variously ornamented with spines, ridges, or papillae. Most paleontologists believe acritarchs were

Figure 17.32   (A) *Eoastrion,* (B) *Eosphaera,* (C) *Animikiea,* and (D) *Kakabekia* from the Gunflint Chert. All specimens are drawn to the same scale. *Eosphaera* is about 30 microns in diameter.

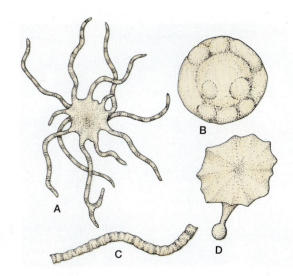

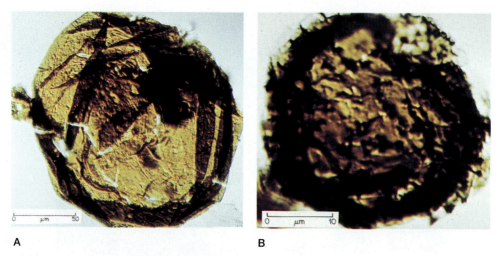

A   B

Figure 17.33   Acritarchs. (A) The striated acritarch *Kildinosphaera lophostriata* from the Chuar Group of Arizona (see Figure 17.23). (B) A spiny acritarch named *Vandalosphaeridium walcotti,* also from the Chuar Group.   *(G. Vidal)*

planktonic algae that grew their thick coats during a resting stage in their life cycle. Some (but not all) resemble the resting stage of unicellular, biflagellate, marine algae known as dinoflagellates. Like dinoflagellates, acritarchs range in size from 20 to 120 μm (microns) and are therefore within the usual size range of eukaryotes.

Figure 17.34   Fossil metazoans from the Conception Group, Avalon Peninsula, Newfoundland. These early metazoans are similar to those of the Ediacaran fauna of Australia. The rocks in which they occur are thought to be late Proterozoic in age. The elongate forms have been interpreted as a form of colonial soft coral, and the circular fossil has been called a jellyfish. There are, however, other interpretations.   *(B. Stinchcomb)*

Fossil microorganisms believed to be acritarchs are known from rocks as old as 1.6 billion years, but they are rare in rocks older than 0.9 billion years. Subsequently, they became more common and more varied, so that there is interest in using them as guide fossils for upper Proterozoic strata. Their period of expansion, however, was rather short-lived, for cooler climates accompanying late Proterozoic glaciation caused widespread extinctions. Only a few of the simple smooth-walled forms survived into the Paleozoic.

Dawn Animals

The most animal-like of the unicellular eukaryotes are protozoans. They are members of the kingdom Protoctista (formerly Protista), and they obtain energy in the animal-like heterotrophic manner by ingesting other cells. Amoebas and ciliates are living examples of protozoans. Protozoan fossils are common in many Phanerozoic rocks, largely because they had evolved shells and skeletal structures by that time, allowing them to be more easily preserved. Most Proterozoic protozoans were naked, soft-bodied creatures, with little chance of preservation. The exception are tiny vase-shaped microfossils about 800 million years old from such widely separated localities as Spitsbergen and Arizona.

The Ediacaran Fauna

The Neoproterozoic fossil record for larger, multicellular animals is considerably better than that for protozoans. Indeed, since the 1960s, important discoveries of large, multicellular late Proterozoic animals have been made in Australia, Africa, South America, Siberia, China,

Figure 17.35   Impression of a soft-bodied discoidal fossil in the Ediacaran Rawnsley Quartzite of southern Australia. This organism has been interpreted as a jellyfish and named *Cyclomedusa*. Some plaeontologists, however, believe it is unrelated to any living organisms. *(B. N. Runnegar)*

Europe, Iran, and North America (Fig. 17.34). The fossils consist of impressions in sedimentary rocks of animals that are clearly metazoans. **Metazoans** are multicellular animals that possess more than one kind of cell and have their cells organized into tissues and organs.

The best known fossils of Neoproterozoic metazoans were first found in 1946 as well-preserved impressions in what were once the sands of an ancient beach. The sands now comprise the Rawnsley Quartzite of the Pound Subgroup. Exposures of the fossil-bearing rock occur in Ediacara Hills in the Flinders Range of southern Australia, and thousands of impressions of large, soft-bodied animals have been recovered from these quartzites. According to their shape, they fall into three groups. There are, for example, discoidal forms such as *Cyclomedusa* (Fig. 17.35) that were initially thought to be jellyfish. Another circular form, *Tribrachidium* (Fig. 17.36), seems to have no modern counterpart and may be a member of an extinct phylum.

The second group of Ediacaran specimens can be called frond fossils. They resemble the living soft corals informally called sea pens (Fig. 17.37). Sea pens look rather like fronds of ferns, except that tiny coral polyps are aligned along the branchlets. The polyps capture and consume tiny organisms that float by. Frond fossils similar to those in the Rawnsley Quartzite are also known from Africa, Russia, and England. In those from England (*Charniodiscus*), the frond is attached to a basal, concentrically ringed disk that apparently served to hold the organism to the sea floor. The disks are frequently found separated from the fronds, which

Figure 17.36   Three members of the Ediacaran fauna. (A) *Pseudohizostomites*, a wormlike form of uncertain affinity. (B) *Tribrachidium*, an unusual discoidal form that appears to have no living relatives. (C) *Parvancorina*, possibly some form of arthropod.   *(Specimen A is an imprint in the Rawnsley Quartzite; B and C are plaster molds made from the original fossils. All specimens courtesy of M. F. Glaessner.)*

Figure 17.37 Diorama of the sea floor in which lived Ediacaran metazoans. The large, frondlike organisms are interpreted here as soft corals known today as sea pens. Silvery jellyfish are seen swimming about. On the floor of the sea one can find *Parvancorina* and elongate wormlike creatures. *(U.S. Natural History Museum, Smithsonian Institution)*

suggests that many of the discoidal fossils common in Ediacaran faunas around the world may be anchoring structures of frond fossils and not jellyfish.

The third group of Ediacaran fossils are ovate to elongate in form and were originally regarded as impressions made by flatworms and annelid worms. Typical of

these fossils is *Dickinsonia* (Fig. 17.38), which attained lengths of up to a meter, and *Spriggina* (Fig. 17.39), a more slender animal with a distinctive crescent-shaped structure at its anterior end.

There is an interesting controversy currently taking place with regard to Ediacaran fossils. Most have long been interpreted as Proterozoic members of such existing phyla as the Cnidaria (which includes jellyfish and corals) and Annelida. Indeed, many paleontologists still retain this opinon. However, that view of the affinities of Ediacaran organisms has been challenged by the innovative German paleontologist Adolf Seilacher. Seilacher believes the resemblance between living sea pens and frond fossils is only superficial, that these animals are sufficiently different from any of today's phyla, and that they should be placed in a new taxonomic category. Seilacher proposes the name Vendoza for the new phylum (after the Vendian, the final geologic period of the Proterozoic).

The Ediacaran fauna survived for about 50 million years following their appearance about 630 million years ago. The fauna are important as a record of the Earth's first evolutionary radiation of multicellular organisms.

One does not find animals with shells among the creatures of the Ediacaran fauna. At present, there is only one known genus of a shell-bearing animal of Ediacaran age. That genus, discovered in Neoproterozoic rocks of Africa, is named *Cloudina* after the American geologist Preston Cloud. *Cloudina* constructed a small, tubular shell (Fig. 17.40) and has been interpreted as a tube-dwelling annelid worm.

Figure 17.38 An exceptionally well-preserved specimen of *Dickinsonia costata* in the Ediacaran Rawnsley Quartzite of southern Australia. This fossil has been interpreted as a segmented worm. Scale divisions are in centimeters. *(B. N. Runnegar)*

Figure 17.39 *Spriggina floundersi* from the Ediacaran Rawnsley Quartzite of southern Australia.

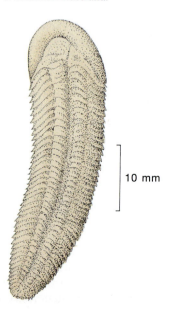

10 mm

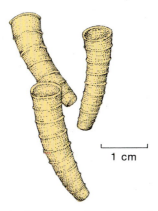

Figure 17.40 *Cloudina,* the earliest known shell-bearing animal. *Cloudina* was first described from the Proterozoic Nama Group of Namibia. It is believed to be a tube-dwelling annelid worm.

Figure 17.41 Trace fossil (horizontal trail) of an Ediacaran metazoan. The trace is in the Rawnsley Quartzite of southern Australia. *(B. N. Runnegar)*

As noted in Chapter 16, fossils need not consist only of petrifactions, molds, or casts (so-called body fossils). Trails, burrows, and other **trace fossils** also provide important clues to ancient life. Traces of Proterozoic burrowing metazoans have been found in North America, Europe, and Australia (Fig. 17.41). In every locality, they occur in rocks deposited after the last Proterozoic episode of continental glaciation. It is interesting that the Proterozoic trace fossils consist of relatively simple, shallow burrows, whereas those of the overlying Cambrian are complex, diverse, and far more numerous. One sees a similar increase in complexity and diversity of body fossils as one passes from the Proterozoic into the Cambrian.

The Changing Proterozoic Environment

The Earth's early atmosphere was largely devoid of free oxygen. Then, about 2 billion years ago, atmospheric oxygen began to accumulate because of increased plant activity (Fig. 17.42). One result of the oxygen buildup was the accumulation on land of considerable amounts of ferric iron oxide, which stained terrestrial rocks a rust-red color. Such so-called **red beds** are considered a valid indication of the advent of an oxygenic environment. Of course, the oxygen level rose slowly and probably did not approach 10 percent of present abundance until the Cambrian Period.

Proterozoic rocks provide evidence for a wide range of climatic conditions, but there is no indication that these climates were particularly unique when compared with climates of the Phanerozoic eras. Thick limestones and dolomites with reeflike algal colonies developed along the Proterozoic equator, where warm tropical conditions prevailed much as they do today. Precam-

brian evaporite deposits in Canada and Australia indicate arid conditions at particular times and places. In the middle and low latitudes, climates were more severe, as indicated by tillites and glacially striated basement rocks.

17.14 Mineral Wealth of the Precambrian

Precambrian rocks have provided enormous quantities of metals, including gold, iron, copper, chromium, zinc, nickel, and uranium. In the past, as today, the availability of Precambrian ores has influenced the health and wealth of nations, standards of living, and even the frequency and outcome of wars. Archean copper, zinc, and iron characteristically occur as metallic sulfides. They formed when seawater penetrated hot submarine lavas, reacted with the lavas to extract metals, and then, on cooling, precipitated the sulfides on the floors of Archean seas. Ores of this kind are called volcanogenic massive sulfide deposits. They have been observed today on the sea floor along the East Pacific rise. Gold is another metal particularly characteristic of some Archean terranes (see *Focus On,* "Riches of Greenstone Belts").

Enormous amounts of copper were once recovered from Proterozoic rocks along the Keweenaw Peninsula of Lake Superior; however, today the easily accessible ores are almost exhausted. The abandoned mines are a testimony of the ultimate fate of even the richest ore deposits. Today, huge tonnages of copper are mined from Proterozoic rocks in Zambia, Africa.

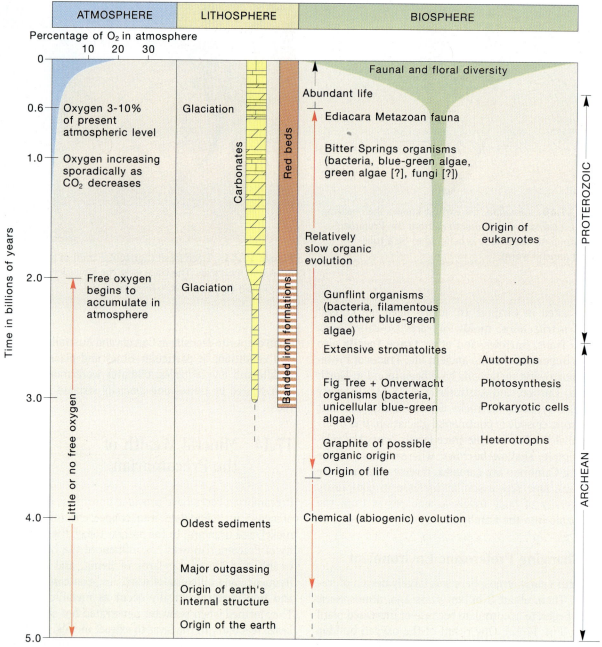

Figure 17.42   Correlation of major events in the history of the biosphere, lithosphere, and atmosphere.

Important deposits of lead and zinc occur in the Proterozoic rocks of Broken Hills, New South Wales (Australia), and Sullivan, British Columbia (Canada). A Proterozoic gabbro intrusion near Sudbury, Ontario, has provided nearly 70 percent of the world's nickel.

Deposits of iron ore are more abundant in rocks of the Proterozoic than of the Archean. The world's largest group of iron mines is in the Canadian shield. Of these, the most famous are the mines that process sedimentary ores in the Lake Superior region. The ores formed where iron-bearing sedimentary rocks were altered and enriched by removal of nonmetallic elements. Other major sources of Proterozoic iron occur in Sweden, the Ukraine, South America, and South Africa.

FOCUS ON

The Riches of Greenstone Belts

At certain locations around the globe, Archean rocks contain important mineral resources. This is especially true of greenstone belts, many of which have contributed significantly to a nation's economic health and supported local economies based on mining and refining ores. It has become apparent that certain kinds of ores occur in particular rock types within the greenstone sequence. As seen in Fig. 17.9, rocks near the bottom of the sequence are ultrabasic volcanics and intrusives. Ores of chromite and nickel, as well as asbestos, are mined from these rocks. In the overlying mafic and felsic volcanics, one finds ores of gold, silver, copper, and zinc. At the top of the greenstone "pile" are sedimentary (or metasedimentary) rocks. Here one may find concentrations of manganese, barite, and iron. Sandstones and conglomerates within the sedimentary section may also contain ancient placer deposits (paleoplacers) that include gold weathered out of older igneous rocks.

The most famous paleoplacer gold deposits in the world are found in the Witwatersrand Supergroup of Late Archean and Proterozoic age in South Africa. As in many of today's placer deposits, the Witwatersrand gold was concentrated in river bars and deltas through the action of water currents. Because particles of gold are heavier than the usual sand and gravel carried by a stream, they tend to be deposited wherever the current has slowed enough to permit their deposition but still has sufficient velocity to carry away lighter rock and mineral grains. Witwatersrand placers are particularly rich where ancient tributaries entered downfaulted, subsiding basins. Since their discovery in 1886, the Witwatersrand gold mines have provided more than half of the world's gold, and they are still producing.

SUMMARY

The Precambrian began 3800 million years ago and ended 570 million years ago. Rocks formed during these 3230 million years of Earth history are most extensively exposed in broadly upwarped, stable regions of the continents known as **Precambrian shields**. Precambrian shields are composed of a number of Precambrian provinces that are distinctive in age and structural characteristics. Margins of provinces often contain belts of deformed rocks called **orogens**.

The original Archean crust was **mafic** in composition and was formed by extrusions of molten rock that originated in the mantle. Later, patches of **felsic** continental crust were produced by partial melting of mafic rocks and by passing the weathered products of mafic source rocks through cycles of melting and metamorphism.

Archean rocks are grouped into **granulite associations**, which consist of gneisses derived from granitic and gabbroic parent rocks, and **greenschist associations**, which consist of mafic volcanics overlain by felsic volcanics and capped by sedimentary rocks. Granulite associations formed along subduction zones, and greenstone belts developed in back-arc basins.

By the beginning of the Proterozoic, many of the smaller crustal elements of the Archean had come together to form a supercontinent known as **Laurentia**. Laurentia and other large cratons underwent further growth by accretion of sediments that had accumulated along continental margins as well as by additions of microplates.

The **Wopmay orogen** of Canada's Northwest Territories is an example of the modern style of tectonics and sedimentation occurring during the Proterozoic. Here one finds the rock record of a cycle of events (a **Wilson cycle**) that included the opening and subsequent closing of an ocean basin.

Two episodes of glaciation occurred during the Proterozoic. The **Gowganda glacial deposits** record a Paleoproterozoic glaciation. The second glaciation occurred near the end of the **Neoproterozoic**.

Proterozoic banded iron deposits are notable for two reasons. They are important mineral resources, and they reflect the buildup of sufficient oxygen in the atmosphere to oxidize iron at the Earth's surface. The development of banded iron formations was followed by deposition of thoroughly oxidized sediments (**red beds**) beginning 2 billion years ago.

The **Keweenawan lavas** of the **Mesoproterozoic** record a period of rifting in the American mid-continental region. The extensive system of faults associated with rifting provided conduits for the extrusion of Keweenawan lavas.

Near the end of the Mesoproterozoic, sedimentary rocks that had accumulated along the eastern margin of North America were compressed and intruded in a mountain-building event known as the Grenville orogeny.

The first forms of life on Earth originated in an environment devoid of free oxygen but which contained the elements needed to make simple organic compounds such as amino acids. Synthesis of more complex molecules followed, and these larger bodies ultimately acquired the organization and behavior of living systems. The first living cells subsisted on the chemical energy they derived by consuming other cells and molecules. They were **heterotrophs** and were followed by organisms (**autotrophs**) that could use the energy of the Sun through **photosynthesis**. Free oxygen accumulated as a byproduct of photosynthesis, making possible the evolution of organisms that use oxygen to burn their food for energy.

Archean fossils consist entirely of **prokaryotes**. During the Mesoproterozoic, the first **eukaryotes** appeared and began their diversification. The final and most dramatic evolutionary event of the Proterozoic was the appearance of metazoans (multicellular animals) near the end of the eon. The fossils of these animals were first discovered in Neoproterozoic rocks exposed in Ediacara Hills of southern Australia. They include large discoidal, frondlike, and elongate animals that have traditionally been assigned to such existing phyla as the Cnidaria and Annelida. Another view is that they are so dissimilar to animals of existing phyla that they should be placed in a new phylum to be named the Vendoza.

Precambrian rocks contain metallic ores that are indispensible to modern civilization. Among these are ores of iron, gold, copper, nickel, zinc, and uranium.

KEY TERMS

Archean Eon *381*
Proterozoic *381*
Hadean Eon *381*
Shield *382*
Platform *382*
Craton *382*
Precambrian province *382*
Orogen *389*
Mafic *382*
Ultramafic *384*
Felsic *384*
Granulite association *387*

Greenstone belt *387*
Kenoran orogeny *387*
Laurentia *389*
Proto-Pangaea *389*
Wopmay orogen *391*
Wilson cycle *391*
Trans-Hudson orogen *392*
Huronian orogeny *392*
Gowganda Formation *392*
Animikie Group *392*
Gunflint Chert *393*
Mesoproterozoic *393*

Neoproterozoic *393*
Keweenawan sequence *393*
Belt Supergroup *395*
Iapetus Ocean *396*
Protein *397*
Nucleic acid *398*
Microspheres *399*
Heterotroph *399*
Fermentation *400*
Autotroph *400*
Photoautotroph *400*
Photosynthesis *400*

Aerobic organism *401*
Prokaryote *401*
Eukaryote *402*
Stromatolite *402*
Cyanobacteria *402*
Acritarch *405*
Ediacaran fauna *406*
Metazoan *407*
Red beds *409*

REVIEW QUESTIONS

1. What is a Precambrian shield? Distinguish between a shield and a craton and between a shield and a platform.

2. Why are rocks of the Archean generally more difficult to correlate than rocks of the Phanerozoic? What method is used to date and correlate most Archean and Proterozoic rocks?

3. What is the economic significance of the Archean Witwatersrand rocks of Africa? Of rocks of the Animikie Group in Canada?

4. What does the presence of banded iron formations (BIF) indicate about changes in the composition of the Earth's atmosphere during the Precambrian?

5. Differentiate between a prokaryotic organism and a eukaryotic organism; between a heterotroph and an autotroph; and between anaerobic and aerobic organisms.

6. What was the tectonic cause of the massive extrusions of basaltic Keweenawan lavas in the Lake Superior region?

7. What are metazoans? When do they appear in the fossil record? With regard to their basic shapes, what are the three major groups of Ediacaran metazoans?

8. What does the heavily cratered condition of the Moon suggest about conditions on Earth during the early stages of the Archean?

9. What is the source of the water that formed the Earth's Precambrian ocean?

10. During the Precambrian, continents often increased in area by means of accretions. Describe this process of continental growth.

DISCUSSION QUESTIONS

1. Differentiate between the terms *mafic* and *felsic*. What is an example of a mafic rock? Of a felsic rock? Continents are basically felsic in composition. Describe how the first continents may have formed from originally mafic source materials during the Archean.

2. It has been stated that temperature gradients near the surface of the Earth during the early Archean were higher than in the Proterozoic. What evidence in the rock record of the Archean suggests that this is true?

3. Describe the typical vertical sequence of rocks found in a greenstone belt. What general trend in rock types is evident?

4. Discuss the possible role of symbiosis in the evolution of eukaryotic organisms from prokaryotic organisms.

5. Account for the laminar structure seen in stromatolites. Did the global expansion of stromatolites have any effect on the Earth's environment during the Precambrian? Explain.

6. What contrasts exist between the Archean and Proterozoic with regard to nature of the rock record, style of tectonics, and kinds of life?

7. Describe the sequence of events recorded by rocks of the Wopmay orogen. Have there been similar sequences of events in other parts of North America during the Precambrian? Explain.

8. What evidence is there for Proterozoic episodes of continental glaciation? When did these glaciations occur? Why is it unlikely that continental glaciations could have occurred during the Archean?

The Paleozoic Era

*The Tapeats Sandsone of Cambrian age exposed along Bright Angel Trail,
Grand Canyon of the Colorado River. The Tapeats sediment represents the
initial shore zone deposit of a transgressing Cambrian sea.*
(Photo by Peter L. Kresan)

Paleozoic rocks have yielded their secrets more readily than rocks of the Precambrian. They are more accessible, less altered, and more fossiliferous. In sequences of Paleozoic rocks around the world, we find clear evidence of episodes of mountain building, dismemberment of landmasses, active plate tectonics, biological catastrophes, and repeated flooding of the continents. This was the era during which continents that had separated during the breakup of Proto-Pangaea were reunited. In the process, they formed the supercontinent Pangaea. The many slow and ponderous collisions of landmasses resulted in orogenies on all the continents that bordered the Atlantic Ocean (Fig. 18.1). The Taconic, Acadian, and Allegheny orogenies produced great mountain chains in eastern North America, whereas the Caledonian and Hercynian orogenies raised mountain ranges in Europe.

18.1 Paleozoic Geography

From paleomagnetic studies we have learned that the Cambrian equator extended across North America from Mexico to Ellesmere Island in the Arctic (Fig. 18.2). During the Ordovician (Fig. 18.3), the equator extended from Baja, California, to Greenland, and thence across the British Isles and central Europe. The location of the ever-moving continents relative to the equator explains why we find salt and gypsum deposits as well as limestones containing fossils of warm-water invertebrates in early Paleozoic rocks of Arctic Canada.

Although **Proto-Pangaea** had begun to break apart near the end of the Proterozoic, both paleomagnetic and fossil evidence suggests that the continents were still relatively close together at the beginning of the Paleozoic. However, by late Cambrian, narrow oceanic tracts flooded into rift zones and widened gradually. The rifted margins of the separating plates became collecting areas for continental shelf and rise sediments derived from the continents. Subsequent closures of the oceans deformed these sediments. We can now observe them in the folded strata of the Appalachians and other ranges formed during the Paleozoic.

Six large continents had become isolated by the breakup of Proto-Pangaea. Their ultimate consolidation to form the Pangaea supercontinent recognized by Alfred Wegener began in late Ordovician time. It was then that the ancient continent of Baltica pushed into North America, forming a chain of mountains called the Caledonides. The next major collison occurred about 300 million years ago when the Northern Hemisphere continents joined Gondwanaland. The largely eroded and partly covered Hercynian Mountains of Europe are the vestiges of this later event.

In North America, the collision with Gondwanaland produced the Allegheny orogeny. Strange as it may seem to us now, the mountains of our southern Appalachians were the result of a collision with northwestern Africa. Similarly, the eroded and largely buried mountains of the Ouachita fold belt in Arkansas and Oklahoma were formed during the collision with the South American segment of Gondwanaland.

As the Paleozoic was coming to a close, Siberia was joined to western Europe. Sediments trapped in the closing ocean were folded and thrust upward to form the Ural Mountains. The clustering of once separate continents was now nearly complete. Great mountain ranges mark the suture zones where continents had come together, and a vast ocean called Panthalassa spanned nearly 300 degrees of longitude around the globe.

Text continued on page 418

Figure 18.1 Pangaea with mountain belts formed during the Paleozoic.

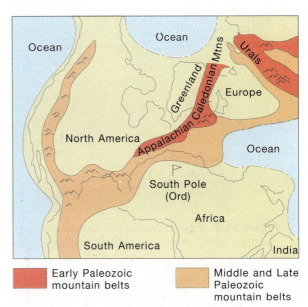

🟥 Early Paleozoic mountain belts	🟧 Middle and Late Paleozoic mountain belts

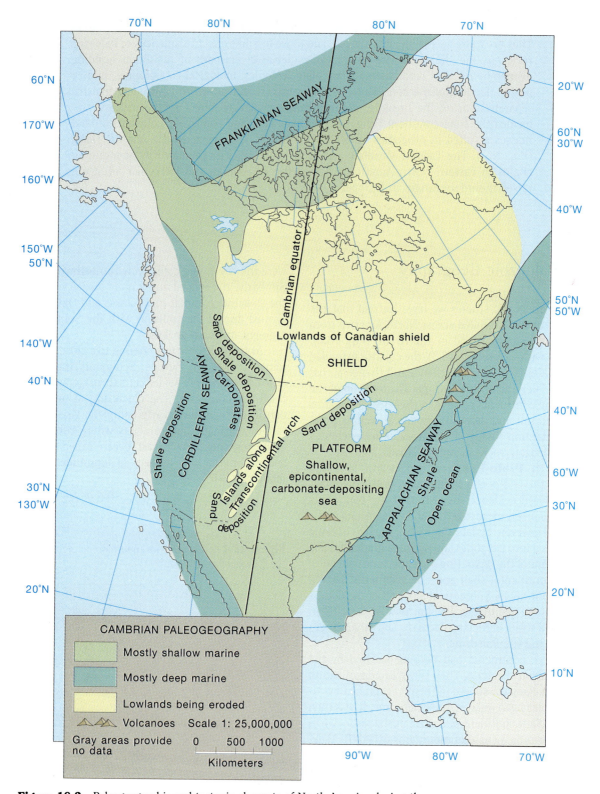

Figure 18.2   Paleogeographic and tectonic elements of North America during the Cambrian period, showing position of the Cambrian paleoequator.

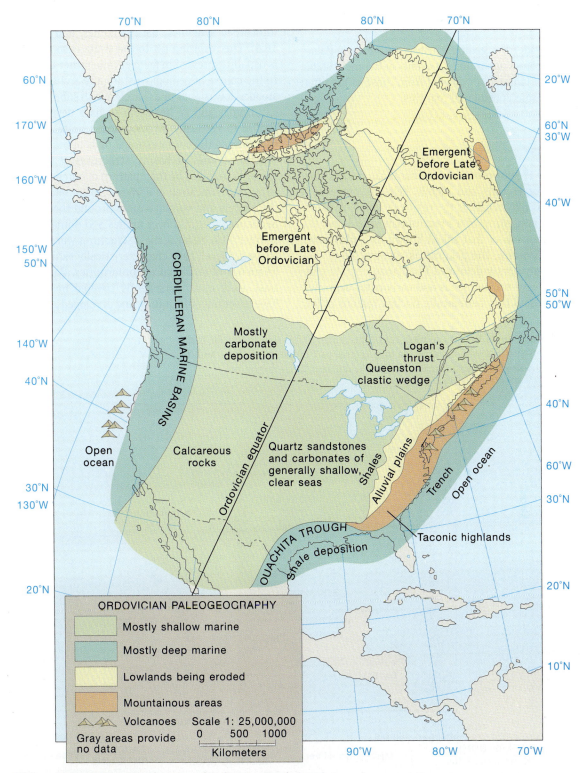

Figure 18.3 Paleogeographic map of Ordovician North America.

18.2   The Continental Framework

As we consider the geologic history of the early Paleozoic, it becomes apparent that events of the more stable interiors of landmasses differed from those of the often more dynamic continental margins.

Platforms

In Chapter 17, we noted that continents can be described in terms of cratons and mobile belts or orogens. A craton, you may recall, is the relatively stable part of a continent consisting of a Precambrian shield and the buried extension of that shield known as a **platform**. In North America, the platform is constructed of flat-lying or only gently deformed Phanerozoic strata consisting of wave-washed sands and carbonates deposited in shallow seas that periodically flooded regions of low

relief. Here and there, these strata are gently warped into broad synclines, basins, domes, and arches (Fig. 18.4). The resulting tilt to the strata is so slight that it is usually expressed in feet per mile rather than degrees. In the course of geologic history, the arches and domes stood as low islands or as barely awash submarine banks.

Domes and arches seem to have developed in response to vertically directed forces quite unlike those that formed the compressional folds of mountain belts. Whether domes or basins, structures of the platform can be recognized by their pattern of outcropping rocks. Erosional truncation of domes exposes older rocks near the centers and younger rocks around the peripheries. Sequences of strata over arches and domes tend to be thinner. Also, because these structures were periodically above sea level, they characteristically exhibit a greater frequency of erosional unconformities. Basins, in contrast, were more persistently covered by inland seas, have fewer unconformities, and developed greater thick-

Figure 18.4   Map of the central platform of the United States showing major basins and domes. Structural section below the map crosses part of the Ozark dome and the Illinois basin. The basins and domes developed at different times during the Phanerozoic.

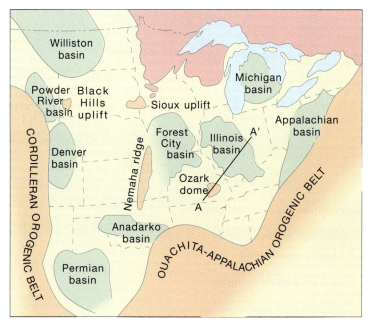

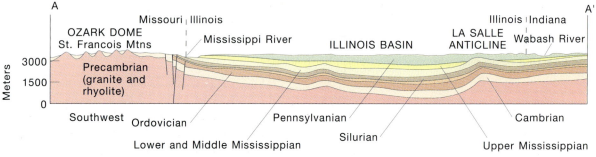

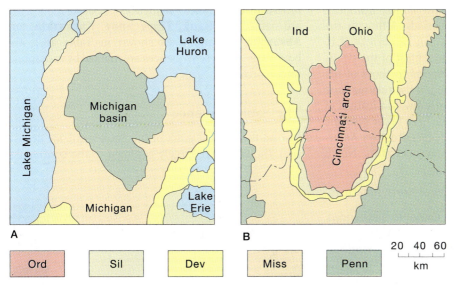

Figure 18.5 (A) In an erosionally truncated basin such as the Michigan basin, the youngest beds are centrally located. (B) In a domelike structure such as the Cincinnati arch, the oldest beds are located in the center.

ness of sedimentary rocks. In erosionally truncated basins, younger rocks are centrally located and older strata occur successively farther toward the peripheries (Fig. 18.5).

Orogens

The North American craton is bounded on four sides by great elongate tracts that have been the sites of intense deformation, igneous activity, and earthquakes. One or more of such orogens are present on all continents (Fig. 18.6). Most are located along present or past margins of continents, such as the North American Cordilleran belt. As described in Chapter 4, the passive margin of a continent may experience an early stage in which thick sequences of sediments accumulate along the continental shelf and rise. This stage may be followed by subduction of oceanic lithosphere along continental

margins and may terminate in continent-to-continent collisions, as in the classic Wilson cycle. As we review the history of eastern North America later in this chapter, we will describe the geologic evidence for this sequence of events along the Appalachian orogen.

18.3 Events on the Craton

It is convenient to discuss the geologic history of North America in terms of its cratonic and orogenic regions. It is also advantageous to subdivide the Paleozoic on the bases of advances (transgressions) and retreats (regressions) of the shallow seas (epeiric seas) that periodically covered parts of the craton. Regressions exposed the old sea floors to erosion and produced unconformities that are used as stratigraphic markers.

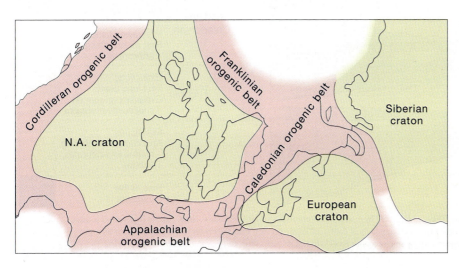

Figure 18.6 Cratons and mobile belts of North America and Europe.

In Chapter 15, we discussed the largely European basis for defining the geologic periods. Although the cratonic rocks of North America can be correlated to those of Europe by means of fossils, most of the stratigraphic boundaries do not correspond to those of the European geologic systems. Indeed, if the relative time scale had been worked out in North America instead of Europe, there would probably have been fewer geologic periods in the Paleozoic. To remedy this situation, American geologists divide the Paleozoic rocks of the North American craton into "sequences" of deposition. They are named the **Sauk**, **Tippecanoe**, **Kaskaskia**, and **Absaroka** sequences. As shown in Table 18.1, the boundaries of these sequences are marked by widespread unconformities.

The Sauk Sequence

During the earliest years of the Sauk sequence deposition, the seas were largely confined to the continental

Table 18.1 • Cratonic Sequences of North America*

Geologic Time		Cratonic Sequences (← Center of craton Margin of craton →)		Orogenic Events	Biological Events	Ice Ages
	Cenozoic		Tejas	Himalayan Alpine	Age of mammals	
				Laramide	Massive extinctions	
Mesozoic	Cretaceous		Zuni	Sevier	First flowering plants Climax dinosaurs and ammonites	
				Nevadan		
Mesozoic	Jurassic				First birds Abundant dinosaurs and ammonites	
Mesozoic	Triassic			Sonoma	First dinosaurs First mammals Abundant cycads	
Late Paleozoic	Permian		Absaroka	Sonoma	Massive extinctions (including trilobites) Mammal-like reptiles	
Late Paleozoic	Pennsylvanian			Alleghenian	Great coal forests Conifers First reptiles	
Late Paleozoic	Mississippian		Kaskaskia	Antler	Abundant amphibians and sharks Scale trees Seed ferns	
Late Paleozoic	Devonian			Antler	Extinctions First insects First amphibians First forests First sharks	
Early Paleozoic	Silurian		Tippecanoe	Acadian–Caledonian	First jawed fishes First air-breathing arthropods	
Early Paleozoic	Ordovician			Taconic	Extinctions First land plants Expansion of marine shelled invertebrates	
Early Paleozoic	Cambrian		Sauk		First fishes Abundant shell-bearing marine invertebrates Trilobites	
	Neoproterozoic				Rise of the metozoans	

65 mya; 248 mya; 408 mya; 570 mya

*The green areas represent sequences of strata. They are separated by major unconformities, indicated in yellow. Note that the rock record is most complete near cratonic margins just as the time spans represented by unconformities are greatest near the center of the craton. Major biologic, orogenic, and glacial events are added for reference. (Cratonic sequence model after Sloss, L. L. 1965. *Bull. Geol. Soc. Amer.* 74:93–114.)

Figure 18.7 The Upper Cambrian Galesville Sandstone exposed in a quarry near Portage, Wisconsin. The mature sandstone is being quarried for use in the manufacture of glass and in premixed cement.

margins, so that most of the craton was exposed and undergoing erosion. No doubt it was a bleak and barren scene, for vascular land plants had not yet evolved. Uninhibited by protective vegetative growth, weathering and erosional forces gullied and dissected the surface of the land. The old Precambrian igneous and metamorphic rocks were deeply weathered and must have formed a thick, sandy "soil." Eventually, marine waters began to spread across the craton and deposit the initial beds of the Sauk sequence.

The craton was not a level, monotonous plain during Sauk time, but had distinct upland areas composed of Precambrian igneous and metamorphic rocks. During times of marine transgression, these uplands became islands in early Paleozoic seas and provided detrital sediments to surrounding areas. The absence of marine sediments over upland tracts provides evidence of their existence and extent. One of the largest of the highland regions was the transcontinental arch (see Fig. 18.2), which extended southwestward across the craton from Ontario to Mexico.

By late Cambrian time, seas spread across the southern half of the craton from Montana to New York. A vast apron of clean sand (Fig. 18.7) covered the sea floor for many miles behind the advancing shoreline (Fig. 18.8). This sandy aspect, or facies, of Cambrian

Figure 18.8 Upper Cambrian lithofacies map. *(Simplified and adapted from* Stratigraphic Atlas of North America and Central America, *Shell Oil Company, Exploration Department.)*

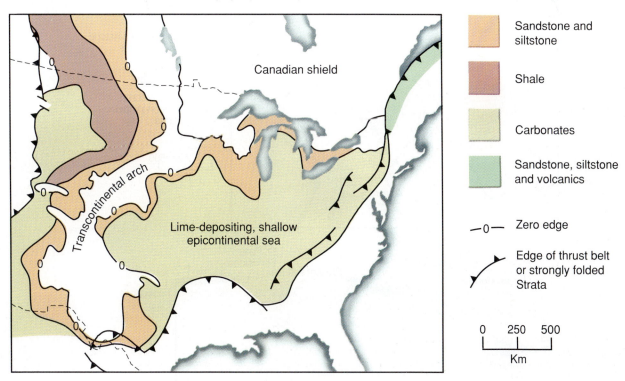

421

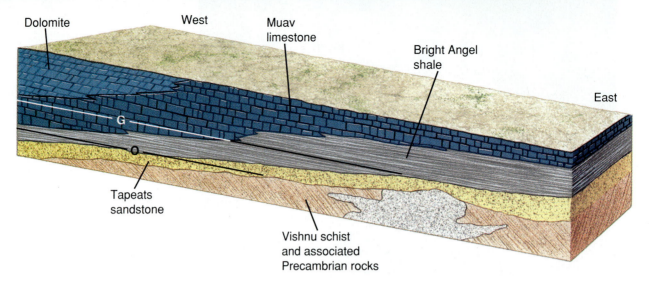

Figure 18.9 East–west section of Cambrian strata exposed along the Grand Canyon. The line labeled "O" indicates the top of the lower Cambrian *Olenellus* trilobite zone. The line labeled "G" indicates the location of the middle Cambrian *Glossopleura* trilobite zone. The section is approximately 200 kilometers long, and the section along the western margin is about 600 meters thick. *(Adapted from E. D. McKee,* Cambrian Stratigraphy *of the Grand Canyon Region [Washington, DC: Carnegie Institute, 1945, Publ. 563].)*

Figure 18.10 Lower Tippecanoe lithofacies map and cross section along line A to A'. In the centrally located white area, Tippecanoe strata have been removed by erosion. Note the extensive unconformity between the Sauk and the Tippecanoe that is depicted on the cross section.

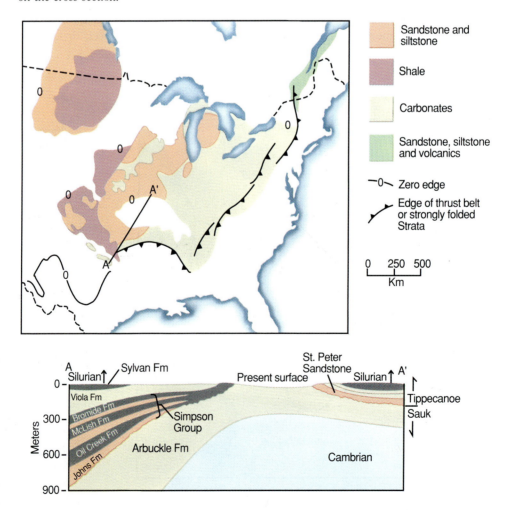

deposition was replaced toward the south by carbonates. Here the waters were warm, clear, and largely uncontaminated by clays and silts from the distant shield. Marine algae flourished and contributed to the precipitation of calcium carbonate. Invertebrates, although present, did not contribute to the volume of sediment to the degree they did in later periods.

In the Cordilleran region, the earliest deposits were sands, which graded westward into finer clastics and carbonates. An excellent place to study this Sauk transgression is along the walls of the Grand Canyon of the Colorado River. In this region one finds the Lower Cambrian Tapeats Sandstone (see chapter opening photograph) as an initial strandline deposit above the old Precambrian surface. The Tapeats can be traced both laterally and upward into the Bright Angel Shale, which was deposited in a more offshore environment (Fig. 18.9). Next is the Muav Limestone, which originated in a still more seaward environment. As the sea continued its eastward transgression, the early deposits of the Tapeats were covered by clays of the Bright Angel Formation, and the Bright Angel was in turn covered by limy deposits of the Muav Limestone. Together these formations form a typical transgressive sequence, recognized by coarse deposits near the base of the section and increasingly finer (and more offshore) sediments near the top.

Cambrian rocks of the Grand Canyon region not only provide a glimpse of the areal variation in depositional environments as deduced from changing lithologic patterns, but they also illustrate that particular formations are not usually the same age everywhere they occur. Detailed mapping of the Tapeats, Bright Angel, and Muav units combined with careful tracing and correlation of trilobite occurrences indicate that deposition of the highest facies (Muav) had already begun in the west before deposition of the lowest facies (Tapeats) had stopped in the east. Correlations based on the guide fossils provided reliable evidence that the Bright Angel sediments were largely early Cambrian in age in California and mostly middle Cambrian in age in the Grand Canyon National Park area. This example illustrates the **principle of temporal transgression**, which stipulates that sediments deposited by advancing or retreating seas are not necessarily of correlative geologic age throughout their areal extent.

The episode of carbonate deposition that was so characteristic of the southern craton continued into the early Ordovician almost without interruption. Then, near the end of the early Ordovician, the seas regressed, leaving behind a landscape underlain by limestones that was subjected to deep subaerial erosion. That erosion produced a widespread unconformity that geologists use as the boundary between the Sauk and the Tippecanoe sequences (Fig. 18.10).

Tippecanoe Sequence

The second major transgression to affect the platform occurred when the Tippecanoe sea flooded the region vacated as the Sauk sea regressed. Again the initial deposits were great blankets of clean quartz sands. Perhaps the most famous of these Tippecanoe sandstones is the St. Peter Sandstone (Fig. 18.11), which is nearly pure quartz and is thus prized for use in glass manufacturing. Such exceptionally pure sandstones are usually not developed in a single cycle of erosion, transportation, and deposition. They are the products of chemical and mechanical processes acting on still older sandstones. In the St. Peter Formation, waves and currents of the transgressing Tippecanoe sea reworked Upper Cambrian and Lower Ordovician sandstones and spread the resulting blanket of clean sand over an area of nearly 7500 square kilometers. As described in Chapter 3, sandstones can be described by their texture and composition as either mature or immature. These textural and compositional traits are, of course, related to the processes that produced the sandstone. Sandstones with a high proportion of chemically unstable minerals and angular,

Figure 18.11   The St. Peter Sandstone is the massive bed directly beneath the overhanging ledge of horizontally bedded dolomite of the Joachim Formation. Both formations are middle Ordovician in age. The St. Peter is the initial deposit of the transgressing Tippecanoe sea. The brown stains are iron oxide. This exposure is 30 miles southwest of St. Louis, Missouri.

Figure 18.12 Fossiliferous limestone deposited in the Tippecanoe sea. This specimen, from near Danville, Missouri, is about 20 centimeters in longest dimension. The most abundant fossils are brachiopods. On close examination, one can also find many twiglike branches of colonial bryozoans, marine snails, and parts of trilobites and echinoderms.

Figure 18.13 Block diagram and stratigraphic section of Niagara Falls. The Lockport Dolomite forms the resistant lip of the falls. The rocks dip gently to the south in this area, and where harder dolomite layers such as the Lockport Dolomite intersect the surface, they form a line of bluffs known as the Niagaran Escarpment.
(Map after E. T. Raisz)

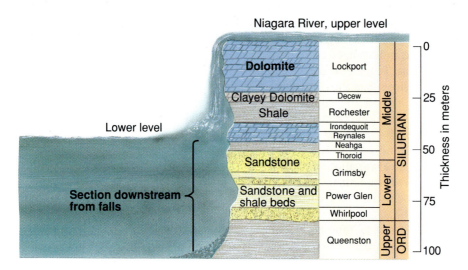

Figure 18.14   Niagara Falls, formed where the Niagara River flows from Lake Erie into Lake Ontario. The classic section of the American Silurian System is exposed along the walls of the gorge below the falls.   *(Guido Alberto Rossi/The Image Bank)*

poorly sorted grains are considered immature, whereas aggregates of well-rounded, well-sorted, highly stable minerals (such as quartz) are considered mature. The St. Peter Sandstone is so pure as to be geologically unusual and perhaps should be termed an ultramature sandstone.

The sandstone depositional phase of the Tippecanoe was followed by the development of extensive limestones, often containing calcareous remains of brachiopods, bryozoans, echinoderms, mollusks, corals, and algae (Fig. 18.12). Some of these cabonates were chemical precipitates, many were fossil fragment limestones (bioclastic limestones), and some were great organic reefs. Frequently, the deposited calcite underwent chemical substitution of some of the calcium by magnesium and in the process was converted to the carbonate rock dolomite.

In the region east of the Mississippi River, the dolomites and limestones were gradually supplanted by shales. As we shall see, these clays are the peripheral sediments of the Queenston clastic wedge which lay far to the east. The geologic section exposed at Niagara Falls is a classic locality to examine these rocks (Figs. 18.13 and 18.14).

Near the close of the Tippecanoe sequence, landlocked, reef-fringed basins developed in the Great Lakes region (Fig. 18.15). During the Silurian, evaporation of these basins caused the precipitation of salt and gypsum on an immense scale. These deposits indicate the existence of excessively arid conditions. In the salt-bearing Salina Group, for example, evaporites have an aggregate thickness of 750 meters. The precipitation of this amount of salt and gypsum, if theoretically confined to a single continuous episode of evaporation, would require a column of seawater 1000 kilometers tall. Because such a column of water has never existed on Earth, it is likely that the evaporating basins had connections to the ocean through which replenishment could occur. One setting in which this might have occurred would be in a basin that had its opening to the sea restricted by a raised sill (or possibly also by a submerged bar). Evaporation within the basin would have produced heavy brines, which would have sunk to the bottom and been prevented from escaping because of the sill or bar (Fig. 18.16). This type of basin is called a **barred basin**.

The Kaskaskia Sequence

The Sauk and Tippecanoe sequences were deposited during the Cambrian, Ordovician, and Silurian geologic periods. With the withdrawal of the Tippecanoe sea, the craton experienced erosion that produced a widespread

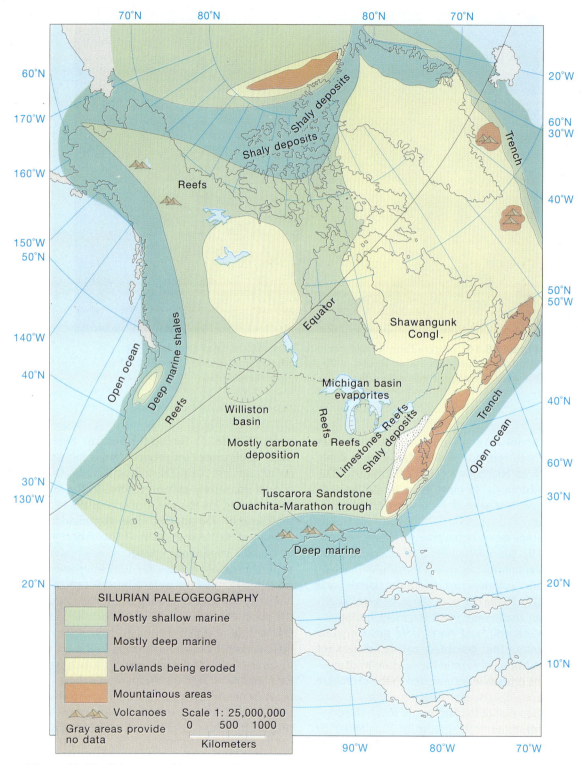

Figure 18.15 Paleogeographic map of Silurian North America.

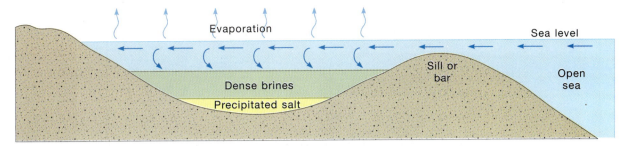

Figure 18.16   Cross section illustrating a model for the deposition of evaporites in a barred basin. Seawater is able to flow more or less continuously into the basin over the partially submerged barrier. It is evaporated to form dense brines. Because of their density, the brines sink and are thus unable to return to the open sea because of the barrier. When the brine has become sufficiently concentrated, salts are precipitated.

unconformity. Gradual flooding of that erosional surface by the Kaskaskia sea (Fig. 18.17) was the initial event of the Devonian. Except for minor regressions, the sea persisted over extensive areas of the craton until the end of the Mississippian Period. Erosion of the layers of sediment deposited on the Kaskaskia sea floor began, only to be interrupted by the advance of the final Paleozoic inundation, the Absaroka sea.

Kaskaskia sedimentation begins with deposition of well-sorted quartz sands. The most famous of these formations is the Oriskany Sandstone of New York and Pennsylvania. Because of its purity, the Oriskany Sandstone is quarried for use in the manufacture of glass. With the continued advance of the Kaskaskia sea, limy sediments were deposited over the Oriskany, and corals began to build reef structures. In areas where water circulation was restricted, salt and gypsum were deposited. During the middle and late Devonian of the region east of the Mississippi Valley, carbonate sedimentation gave way to shales. The change to clastic deposition was a consequence of mountain building associated with the Acadian orogeny in the Appalachians. As will be described shortly, highlands formed during this orogeny were rapidly eroded, and clastics were transported westward to form an extensive apron of sediments that are coarser and thicker near their eastern source area and give way to thin but widespread black shales in the east-central region of the United States.

In the far western part of the craton, Middle and Upper Devonian rocks are largely limestones, although there are shales as well. In a depressed area known as the Williston basin (from South Dakota and Montana northward into Canada), extensive reefs were developed. Arid conditions and restricted circulation in reef-enclosed basins resulted in the deposition of impressive thicknesses of gypsum and salt. The reefs provided per-

meable structures into which petroleum migrated, resulting in some of Canada's richest oil fields.

During the late Devonian, mountains existed along the eastern margin of North America. A clastic wedge was built along the western side of the mountains. Conglomerates and sandstones were the characteristic deposits of the wedge, but farther to the west, suspended muds were carried into the shallow sea that covered the platform. These muds were deposited as a thin but extraordinarily widespread sheet of dark shale. There are many local names for the shale, but the formal name most commonly used is **Chattanooga Shale**. Because the shale is both widespread and easy to recognize, it has served as an excellent marker in regional correlation.

During the passage from Devonian to Mississippian time, highland source areas that had provided the Chattanooga Shale were worn low, and the quantity of muddy sediment decreased. Carbonates then became the most abundant and widespread kind of sediment in the epeiric seas of the platform (Fig. 18.18). Cherty limestones (Fig. 18.19), limestones containing the remains of countless crinoids (Fig. 18.20) and other invertebrates, and limestones composed of myriads of oöids (Fig. 18.21) formed extensive beds across the central and western parts of the North American craton.

The crinoidal, oölitic, and other varieties of limestone deposited in the Kaskaskia sea record one of the most extensive inundations of North America since Ordovician time. Geologists refer to the broad blanket of carbonates as "the great Mississippian lime bank."

The Absaroka Sequence

When the Kaskaskia seas had finally left the craton at the end of Mississippian time, the exposed terrain was

Text continued on page 430

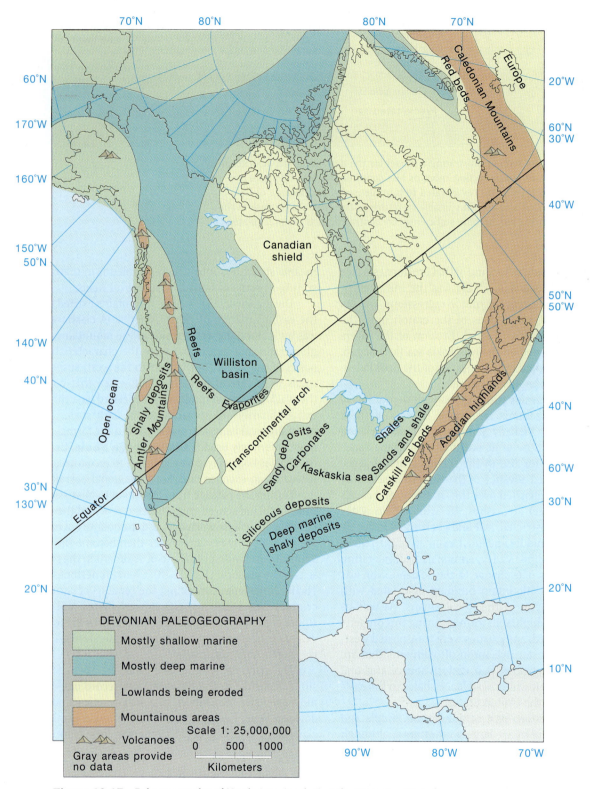

Figure 18.17   Paleogeography of North America during the Devonian Period.

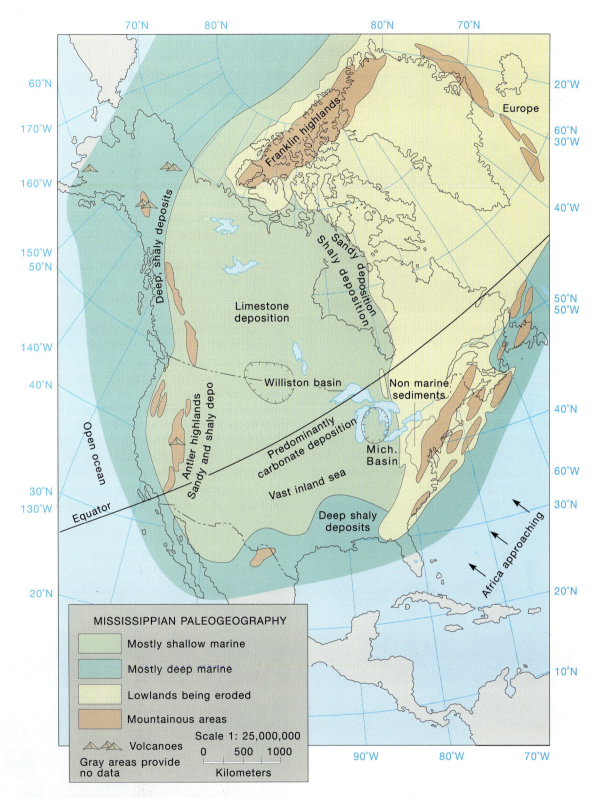

Figure 18.18   Paleogeography of North America during the Mississippian Period.

Figure 18.19 White nodular chert in tan limestone. Fern Glen Formation west of St. Louis, Missouri.

subjected to erosion that resulted in one of the most widespread regional unconformities in the world. Entire systems of older rocks over arches and domes were lost to erosion. The unconformity provides a criterion for separating those strata equivalent to the Carboniferous of Europe into the Mississippian and Pennsylvanian sys-

tems. It is appropriate to use these two names in North America, not only because of the unconformity but also because the rocks above the erosional hiatus differ markedly from those below. Indeed, the overlying Pennsylvanian strata are a consequence of quite different tectonic circumstances.

It was not until near the beginning of middle Pennsylvanian time that the seas were able to encroach onto the long-exposed central region of the craton. The deposits of this seaway are those of the Absaroka sequence. In general, the Pennsylvanian section of rocks near the eastern highlands are thicker, and virtually all are continental sandstones, shales, and coal beds (Fig. 18.22). This eastern section of Pennsylvanian rocks gradually thins away from the Appalachian belt and changes from predominantly terrestrial to about half marine rocks and half nonmarine rocks. Still farther west, the Pennsylvanian outcrops are largely marine limestones, sandstones, and shales.

One of the most notable aspects of Pennsylvanian sedimentation in the middle and eastern states is the repetitive alternation of marine and nonmarine strata. A group of strata that records the variations in depositional environments caused by a single advance and retreat of the sea is called a **cyclothem**. A typical cyclothem in the Pennsylvanian of Illinois contains ten units (Fig. 18.23). Units one through five are continental deposits, the uppermost of which is a coal bed (Fig. 18.24). The strata deposited above the coal bed represent an advance of the sea over an old forested area. In Missouri and Kansas, at least 50 cyclothems are recognized, and

A

Figure 18.20 (A) Specimen of crinoidal limestone (Burlington Formation) in which crinoid fragments (mostly stem plates) stand out in relief because of weathering. (B) Reconstruction of crinoids growing on the floor of the Kaskaskia (Mississippian) sea. *(Latter photo courtesy of the U.S. National Museum)*

B

Figure 18.21   Oöids. Diameter of field is 2 centimeters.

many of these extend across thousands of square kilometers.

It is apparent that cyclothems are the result of repetitive and widespread advance and retreat of seas. But what was the cause of such oscillations? One explanation involves periodic regional subsidence of the land to a level slightly below sea level, so that marginal seas could spill onto swampy lowlands. A relatively short time later, subsidence might cease, and sediments could be built above sea level to extend the shoreline seaward and reestablish continental conditions. Alternatively, the reestablishment of dry lands may have resulted from temporary regional uplifts. Finally, worldwide (eustatic) changes in sea level related to continental glaciation might have produced repetitive marine invasions and regressions. We have ample evidence that large regions of Gondwanaland were covered with glacial ice from late Mississippian through Permian time. When the ice sheet grew in size, sea level would be lowered because of water removed from the ocean and precipitated as snow onto the continent. During warmer episodes, meltwater returning to the ocean would raise sea level. These sea-level oscillations could result in the shifting back and forth of shorelines that produced cyclothems. Additional support for this idea comes from recognition of cyclothems of the same age on continents other than North America, as this indicates that the cause of the cyclicity was global.

Ordinarily, cratonic areas of continents are characterized by stability. The southwestern part of the North American craton provides an exception to this general rule, for during the Pennsylvanian this was a region of

mountain building. The resulting highlands are generally called the Colorado mountains (also called the ancestral Rockies) and the Oklahoma mountains. These mountains and related uplifts appear to have resulted from movements of crustal blocks along large, nearly vertical reverse faults. The Colorado mountains included a range that extended north–south across central Colorado (the Front Range–Pedernal uplifts) and a segment curving from Colorado into eastern Utah (the Uncompahgre uplift). The central Colorado basin lay between these highland areas (Fig. 18.25). A separate range, the Zuni–Fort Defiance uplift, extended across northeastern Arizona on the southwestern perimeter of the Paradox basin. To the east of the Colorado mountains lay the southeastward-trending Oklahoma mountains. Eroded stumps of this once rugged range form today's greatly reduced Arbuckle and Wichita mountains. Remains of the Amarillo mountains are now buried beneath younger rocks and are known principally from exploratory drilling for oil.

Judging from the tremendous volume of sediment eroded from the Colorado Mountains, it is likely that they had attained heights of more than 1000 meters above surrounding basins. They also were subjected to repeated episodes of uplift during the late Pennsylvanian and into the Permian. Erosion of the mountains eventually exposed their Precambrian igneous and metamorphic cores. As erosion of the mountains continued, deposits of red arkosic sediments were spread across adjacent basins. A small part of this massive accumulation of conglomerates and sandstones is dramatically exposed just a few miles west of Denver (Fig. 18.26) and in the steeply dipping "flatirons" near Boulder, Colorado.

Geologists have long puzzled over the cause of the cratonic disturbances that created the Colorado mountains, but plate tectonics has provided a hypothesis for this unusual event. It seems likely that the collision of Gondwanaland with North America along the site of the Ouachita orogenic belt generated stress in the bordering region of the craton to the north. Crustal adjustments to relieve the stresses resulted in the deformations that produced the mountains and intervening basins.

As the Paleozoic drew to a close, the epeiric seas that had occupied much of the western craton became increasingly more restricted. Thick and extensive salt beds in Kansas provide testimony to the gradual restriction and evaporation of these seas. The last Permian sea occupied the western part of Texas and western New Mexico. In this region, a remarkable association of lagoon, reef, and open basin sediments were deposited (Figs. 18.27 and 18.28). In this region, several

Text continued on page 436

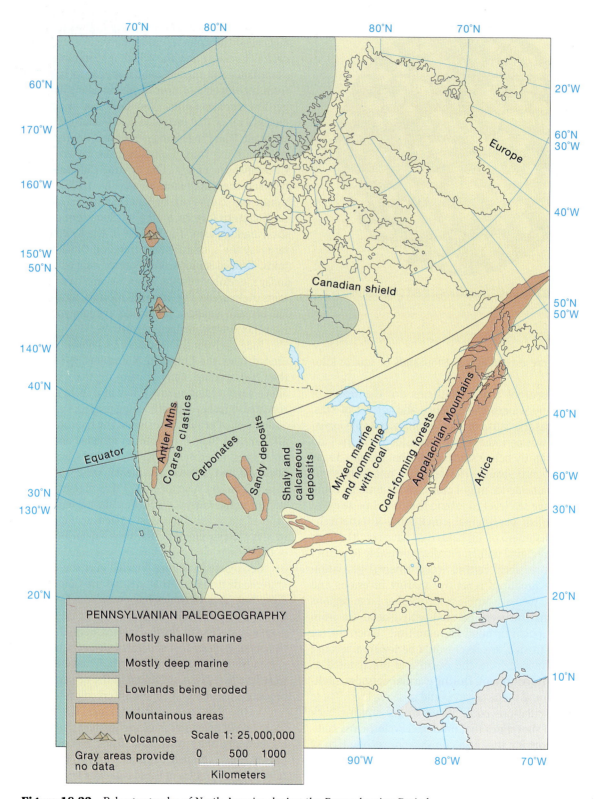

Figure 18.22   Paleogeography of North America during the Pennsylvanian Period.

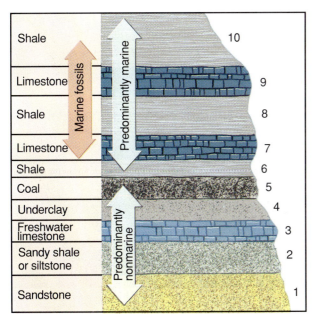

Shale		10
Limestone		9
Shale		8
Limestone		7
Shale		6
Coal		5
Underclay		4
Freshwater limestone		3
Sandy shale or siltstone		2
Sandstone		1

Marine fossils

Predominantly marine

Predominantly nonmarine

Figure 18.23 An ideal coal-bearing cyclothem showing the typical sequence of layers. Many cyclothems do not contain all ten units, as in this illustration of an idealized sequence. Some units may not have been deposited because changes from marine to nonmarine conditions may have been abrubt, and/or units may have been removed by erosion following marine regressions. The number 8 bed usually represents maximum inundation and, correlated with the same bed elsewhere, provides an important correlative stratigraphic horizon.

Figure 18.24 Part of an Illinois cyclothem. The lowermost layer is the coal seam (cyclothem bed 5), followed upward by shale (bed 6) near the geologist's hand, limestone (bed 7), shale (bed 8), another limestone (bed 9), and the upper shale (bed 10). Part of another sequence caps the exposure. This cyclothem is part of the Carbondale Formation. *(D. L. Reinertsen and the Illinois Geological Survey)*

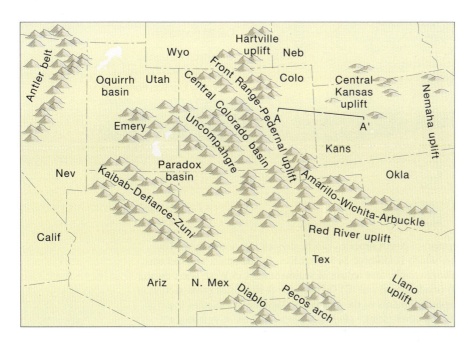

Figure 18.25 Location of the principal highland areas of the southwestern part of the craton during Pennsylvanian time.

Figure 18.26 Coarse arkosic red beds of the Pennsylvanian Fountain Formation exposed around the famous Red Rocks Amphitheatre, a few miles southwest of Denver.

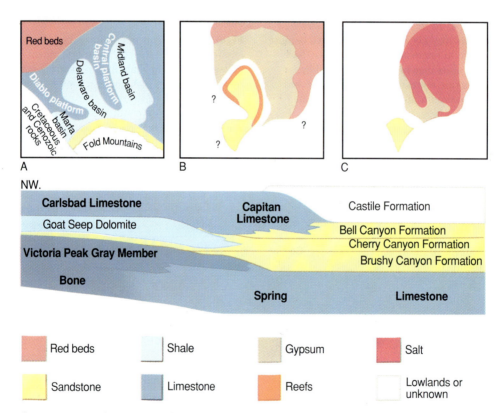

Figure 18.27 The Permian of North America can be divided into four stages. The oldest is the Wolfcampian, which is followed by Leonardian, Guadalupian, and Ochoan. A, B, and C show the paleogeography and nature of sedimentation in West Texas during the (A) Wolfcampian, (B) Guadalupian, and (C) Ochoan. (D) A simplified cross section of Leonardian and Guadalupian sediments of the Guadalupe Mountains indicates the relationship of the reef to the other facies. *(After many sources, but primarily P. B. King, U.S. Geological Survey Professional Paper 215, 1948.)*

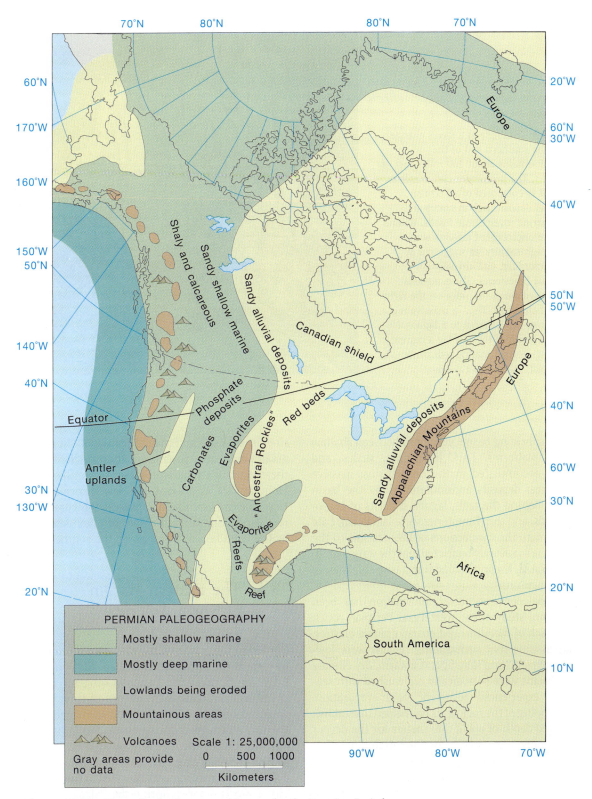

Figure 18.28   Generalized paleogeographic map for the Permian Period.

Figure 18.29   Example of a facies change in rocks at the southern end of the Guadalupe Mountains about 100 miles east of El Paso, Texas. The prominent cliff is called El Capitan. It is composed of a Permian limestone reef deposit, rich in corals and associated marine invertebrate fossils. Behind the reef there once existed lagoons with abnormally high salinity, as suggested by evaporites, dolomites, and virtually unfossiliferous limestones. These lagoonal beds are exposed along the ridge on the right side of the photograph. Thus, a massive reef limestone facies lies adjacent to a lagoonal facies. To the south of the reef lay a marine basin in which normal marine sediments were deposited.   *(National Park Service)*

irregularly subsiding basins developed between shallow submerged platforms. Dark limestones, shales, and sandstones accumulated in the deep basins; massive carbonate reef deposits accumulated along the basin edges; and, behind the reefs, lagoon deposits of evaporites, redbeds, and thin limestones were laid down. Late in the Permian, the connections of these basins to the south became so severely restricted that the waters gradually evaporated, leaving behind impressive thicknesses of salt and gypsum.

Much paleoenvironmental information has been obtained from study of the West Texas Permian rocks. From the lack of medium- and coarse-grained clastics, one may assume that surrounding regions were low-lying. The gypsum and salt suggest a warm, dry climate in which basins were periodically replenished with seawater and experienced relatively rapid evaporation. Careful mapping of the rock units has helped to establish an estimated depth of about 500 meters for the deeper basins and only a few meters for the intrabasinal platforms. The basin deposits are dark in color and rich in organic carbon as a likely consequence of accumulation under stagnant, oxygen-poor conditions. Perhaps upwelling of these deeper waters may have provided a bonanza of nutrients on which phytoplankton and reef-forming algae thrived. Along with the algae, the reefs contain the skeletal remains of over 250 species of ma-

rine invertebrates. Today, because of their relatively greater resistance to erosion, these ancient reefs form the steep El Capitan promontory in the Guadalupe Mountains of West Texas and New Mexico (Fig. 18.29). Here one can examine the forereef composed of broken reef debris that formed a sort of talus deposit caused by the pounding of waves along the southeastern side.

18.4   The Eastern Margin of North America

Bordering the craton on the east from Newfoundland to Georgia is the Appalachian orogen. (Its southwestward extension is called the Ouachita orogen.) The Appalachian orogen contains the record of three major orogenic events and many minor disturbances. Study of the rocks deformed during these events has been the basis for classic theories of mountain building as well as new concepts that view the Appalachian region as a collage of microcontinents accreted to the eastern margin of North America during the Paleozoic.

The elongate basin in which sediments of the Appalachian region were deposited during the Paleozoic (see Fig. 18.2) can be divided into a western belt that received mostly shales, limestones, and sandstones and an eastern belt containing graywackes, volcanics, and siliceous shales, which are often metamorphosed and intruded by granite masses of batholithic proportions. Today these rocks underly the Blue Ridge and Piedmont Appalachian provinces, whereas rocks of the western belt occur beneath the Valley and Ridge province and Appalachian plateaus (Fig. 18.30).

At the beginning of the Paleozoic, the eastern margin of North America was a passive margin, much as it is today. Shallow-water deposits were spread across the continental shelf, with deeper-water sediment carried seaward to the continental rise. Initial deposits along the shelf were largely quartz sands and clays derived from the stable interior, which was still largely emergent in earliest Cambrian time. Gradually, however, the seas spread inland and sandy strandlines migrated westward. The early sandstone deposits of the shelf were covered by a thick sequence of carbonate rocks that formed a shallow bank or platform extending from Newfoundland to Alabama. Mudcracks and stromatolites attest to the deposition of these carbonates (now dolomites) at or near sea level.

Figure 18.30 Physiographic provinces of the eastern United States.

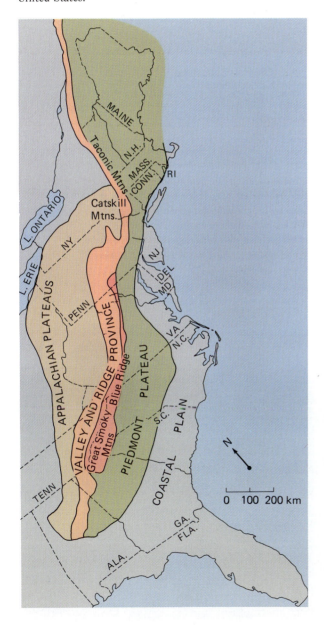

Early Paleozoic Unrest

This rather quiet depositional scenario of the Cambrian and early Ordovician changed dramatically during the middle Ordovician (Fig. 18.31). At that time, carbonate sedimentation ceased, the carbonate platform was downwarped, and huge volumes of black shales and immature sands spread westward over the dolomites. The Ordovician section in New England contains such coarse detritus as well as volcanic pyroclastics and interbedded lava flows, which indicate that a subduction zone with an accompanying volcanic chain had formed along the eastern margin of North America. Apparently, the Proto-Atlantic (or Iapetus) Ocean that had opened during the Proterozoic and Cambrian was beginning to close again. That closure resulted in the subduction–volcanic arc complex of the Paleozoic and ultimately carried the continents that were to compose Pangaea into contact with each other.

The pulses of orogenic activity that had begun in the early Ordovician were followed by several more intense deformational events in middle and late Ordovician that constitute the **Taconic orogeny**. The Taconic orogeny was caused by the collision of another tectonic plate, probably part of ancestral western Europe, with the eastern coast of North America. Later orogenic episodes resulted from further compression between the opposing plates. The effects of the Taconic orogeny are most apparent in the northern part of the Appalachian belt. Caught in the vise between closing lithospheric plates, sediments were crushed (Fig. 18.32), metamorphosed, and shoved northwestward along a great thrust fault known as **Logan's thrust**. Continental rise deposits are carried over and across about 48 kilometers of shelf and shield rocks by Logan's thrust. Today we are able to view the remnants of this activity in the Taconic Mountains of New York. Ash beds, now weathered to a clay called bentonite, attest to the violence of volcanism. Masses of granite now exposed in the Piedmont help to record the great pressures and heat to which the

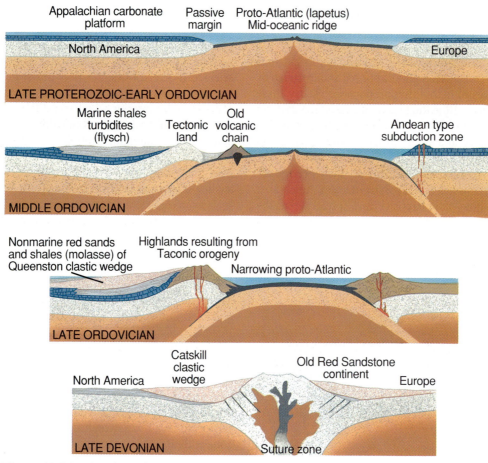

Figure 18.31 Hypothesis depicting the role of plate movements in the evolution of the northern Appalachians. Cambrian and Lower Ordovician carbonates were deposited on the passive margin of North America. With the development of plates with opposing movements in the Ordovician, highland areas were produced that shed clastics into adjacent basins. Devonian clastic sediments had as their source the mountains produced by the closure of the Iapetus Ocean and convergence of North America and Europe. *(Based in part on concepts expressed in J. M. Bird and J. F. Dewey, Bull. Amer. Jour. Sci, 81 [1972], 1031–1059; R. D. Hatcher, Jr., Amer. J. Sci., 274 [1972], 135–147.)*

Figure 18.32 These Ordovician shale beds exposed at Black Point, Port au Port, Newfoundland, are intensely folded as a result of the Taconic orogeny of late Ordovician time. *(Geological Society of Canada, Ottawa)*

sediments of the subduction zone complex were subjected. But even if the igneous rocks had never been found, geologists would know mountains had formed, for the great apron of sandstones and shales that outcrop across Pennsylvania, Ohio, New York, and West Virginia must have had their source in the rising Taconic ranges (Fig. 18.33).

From a feather edge in Ohio, this barren wedge of rust-red terrestrial clastics called the **Queenston clastic wedge** becomes coarser and thicker toward the ancient source areas to the east (Fig. 18.34). Streams flowing from those mountainous source areas were laden with sand, silt, and clay. As they emerged from the ranges, they spread their load of detritus in the form of huge deltaic systems which grew and prograded westward into the basin, often covering earlier shallow marine deposits. The deltaic aspect of many of these deposits accounts for the alternate name of Queenston delta for the Queenston clastic wedge. It has been estimated that over 600,000 cubic kilometers of rock were eroded to produce the enormous volume of sediment in the Queenston clastic wedge. This indicates that the

Taconic ranges may have exceeded 4000 meters (13,100 feet) in elevation.

During the Silurian, the locus of most intense orogenic activity seems to have shifted northeastward into the Caledonian belt. Meanwhile, erosion of the Taconic highlands continued. Early Silurian beds are often coarsely clastic, as evidenced by the Shawangunk Conglomerate of New York (see Fig. 18.15). These pebbly strata give way upward and laterally to sandstones such as those found in the Cataract Group of Niagara Falls. Silurian iron-bearing sedimentary deposits accumulated in the southern part of the Appalachian region. The greatest development of this sedimentary iron ore is in central Alabama, where there are also coal deposits of Pennsylvanian age. The coal is used to manufacture coke, which is needed in the process of smelting. Limestone, used as blast furnace flux, is also available nearby. The fortunate occurrence of iron ore, coal, and limestone in the same area accounts for the formerly important steel industry of Birmingham, Alabama.

Extending across the southern margin of the North American craton is the Ouachita–Marathon depositional

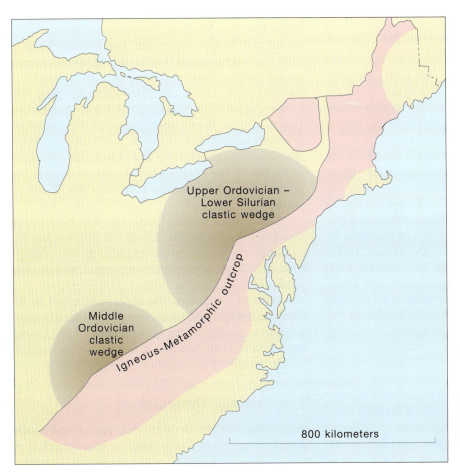

Figure 18.33 Great wedges of clastic sediments spread westward as a result of erosion of mountain belts developed during the early Paleozoic. *(After J. B. Hadley, in* Tectonics of the Southern Appalachians. *V.P.I., Dept. Geol. Sci. Memoir 1, 1964.)*

Upper Ordovician – Lower Silurian clastic wedge

Igneous-Metamorphic outcrop

Middle Ordovician clastic wedge

800 kilometers

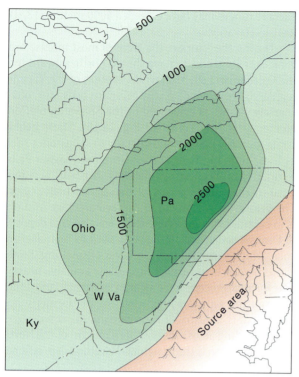

Figure 18.34 Isopach map illustrating regional variation in thickness of Upper Ordovician sedimentary rocks in Pennsylvania and adjoining states. *(After M. Kay,* Geol. Soc. Am. Mem., *48 [1951].)*

trough (see Fig. 18.15). Although over 1500 kilometers long, only about 300 kilometers of its folded strata are exposed. Additional information about the distribution of early Paleozoic rocks within the belt is derived mostly from oil well drilling and geophysical surveys.

Nearly 10,000 meters of Paleozoic sediment fill the Ouachita–Marathon trough. One can recognize deep-water deposits of the former continental rise along the southern perimeter of the belt and shallow-water deposits of the continental shelf along the northern margin. Subsidence was most rapid during the late Paleozoic, as indicated by the thicker section of late Paleozoic rocks.

Paleozoic rocks of this region are noted for their exceptionally siliceous and cherty character. Very likely, the silica was derived from the submarine weathering of volcanic ash ejected from volcanoes that lay to the south of the Ouachita–Marathon orogen.

The Caledonides

Scotland's ancient name, still used in poetry, is Caledonia. Eduard Suess used the name for the Scottish remnants of an early Paleozoic mountain range, the Caledonides. It is also the basis for the name of the Caledonian

orogenic belt which extends along the northwestern border of Europe (see Fig. 18.6).

The Caledonian and Appalachian belts have had a generally similar history. This is not surprising, for both are really part of one greater Appalachian–Caledonian system. Both elongate depositional sites evolved as a result of a cycle of ocean expansion and contraction, such as is schematically illustrated in Figure 18.31. The cycle began with a late Precambrian to middle Ordovician episode of spreading, as the Iapetus Ocean widened to admit new oceanic crust along a spreading center. On the continental shelves and rises of the separating blocks, thick lenses of sediment accumulated. This depositional phase is marked by the development of two distinct groups of rocks. Subduction zone rocks include more than 6000 meters of volcanics, graywackes, and shales. A shelf facies of clean sandstones and fossiliferous limestones comprises the second group of rocks. Here and there along the margins of the Caledonian trough there are Upper Silurian fresh-water shales containing fossil remains of early fish.

The closure of the Iapetus Ocean and crumpling of the Caledonian marine basins began in middle Ordovician time, when subduction zones developed along the margins of the formerly separating continents. This event is recognized from the volcanic rocks that occur in the Canadian Maritime Provinces, northwestern England, northeastern Greenland, and Norway. Little by little, the Iapetus Ocean closed until the opposing continental margins converged in a culminating mountain-building event termed the **Caledonian orogeny**. The orogeny reached its climax in late Silurian to earliest Devonian time. It was most intense in Norway, where Precambrian and lower Paleozoic sedimentary rocks are drastically metamorphosed and thrust-faulted. Southwestward, in the British Isles, the effects were not as severe, although mountainous terrains also developed there.

South of the unruly Caledonian belt there is evidence of a landmass that during late Silurian and Devonian received sandy detritus from the growing Caledonides. This region, on which were deposited clays and sands of the Old Red Sandstone, is often referred to as the Old Red Continent. The deposits here are in the form of a thick, clastic wedge very similar to the Queenston clastic wedge on the opposite side of the Iapetus Ocean. In fact, the two clastic wedges are approximate mirror images of one another (see Fig. 18.31).

Late Paleozoic Orogenies

Beginning with the Devonian, the eastern and southeastern margins of North America experienced their

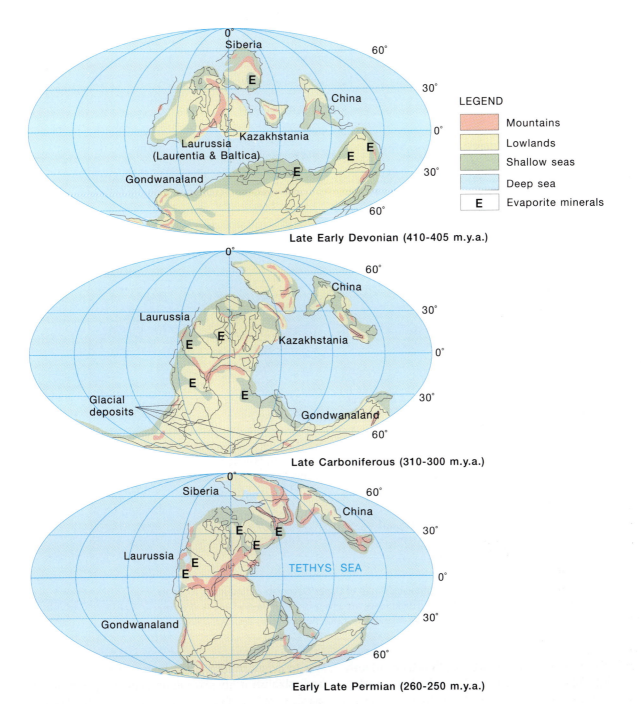

Figure 18.35   Global paleogeography of the late Paleozoic. *(From C. R. Scotese, R. K. Bambach, C. Barton, R. Van Der Voo, and A. M. Ziegler, J. Geol., 87(3) [1979], 217–277.)*

culminating and most intense episodes of orogenesis. The crumpling of the Appalachian and Ouachita depositional tracts was a consequence of the reassembly of the formerly separated continents into Pangaea. The northern part of the Appalachian–Caledonian belt had taken the shock of a collision with Europe during the Silurian and Devonian. That collision was the cause of the Caledonian orogeny in Europe and the Acadian orogeny in northeastern North America. As indicated in Figure 18.35, the southern half of the elongate Appalachian depositional basin as well as the Ouachita trend were crushed into mountainous tracts when they were

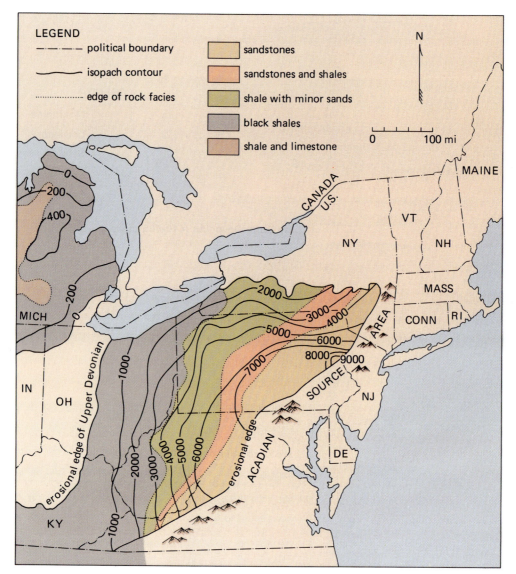

Figure 18.36 Isopach and lithofacies map of Upper Devonian sedimentary rocks in the northeastern United States. Thicknesses are given in feet. *(Modified from W. D. Sevon, Geol. Soc. Am. Special Paper, 201 [1985], 79–90; and W. G. Ayrton, Pa. Geol. Survey Rpt., 39(4) [1963], 3–6.)*

struck by the northern margin of Gondwanaland (South America and northwestern Africa), causing deep-seated deformations such as those that produced the Colorado and Oklahoma mountains.

The effects of the Devonian deformation, called the **Acadian orogeny**, are clearly seen in a belt extending from Newfoundland to West Virginia. Here one finds thick, folded sequences of turbidites interspersed with rhyolitic volcanic rocks and granitic intrusions. The intensity of the compression that affected these rocks is reflected in their metamorphic minerals, which indicate that mineralization occurred at temperatures exceeding 500°C and pressures equivalent to burial under 15,000 meters of rock. The overall result of the Acadian orogeny

was to demolish forever the marine depositional basin and establish in its place mountainous areas in which erosion was the prevalent geologic process. Here and there in isolated basins among the mountains, Devonian nonmarine sediments were deposited. However, the greatest volume of erosional detritus was spread outward from the highlands as a great wedge of terrigenous sediment called the Catskill clastic wedge (Figs. 18.36 and 18.37).

The Catskill Clastic Wedge

For many reasons, the Devonian rocks of the **Catskill clastic wedge** (also called the Catskill delta) have ex-

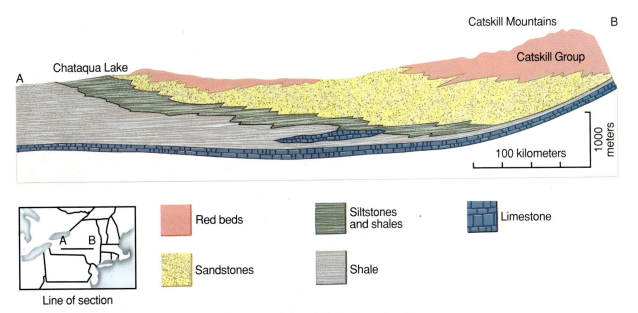

Figure 18.37   East–west section across the Devonian Catskill clastic wedge, New York. Note that continental red beds interfinger with nearshore marine sandstones, and these in turn grade toward the west into offshore siltstones and shales. Continental deposits prograded upon the sea, pushing the shoreline progressively westward. *(Based on several classic studies by G. H. Chadwick and G. A. Cooper, completed between 1924 and 1942.)*

cited the interest of geologists for well over a century. This apron of sediments provides an ideal area for the examination of facies from a varied assortment of both marine and nonmarine depositional environments (Fig. 18.38). Catskill facies characteristically exhibit rapid lateral changes from sandstones to shales. Such relationships form traps for petroleum and are the reason for the thousands of wells drilled into Catskill strata. The flat slabs of red sandstones have also provided vast tonnages of flagstones for buildings in eastern cities. But for historical geologists, the most pervasive reason for studying Catskill rocks is for the clues they provide to the locations and time of occurrence of phases of the Acadian orogeny. Most of these clues come from the study of subsidiary clastic wedges within the larger Catskill complex. The locations of these wedges indicate that the Acadian orogeny was caused by the oblique convergence of a displaced continental fragment named the **Avalon terrane** against the irregular eastern margin of the North American craton.

In regard to Catskill facies, the older term, Catskill delta, is not entirely appropriate. These sediments were not deposited as large deltas of one or two major streams. Study of the directional properties of the fluvial sandstones indicates deposition from many small streams, all flowing westward out of the Acadian highlands. In many areas, marine processes reworked the sediment supplied by streams so quickly that even small deltas did not develop.

Unlike the earlier Queenston rocks, the Catskill sediments accumulated at a time when land plants were abundant and provided a green mantle for the alluvial plains and hills. Fossil remains of these plants (Fig. 18.39) are most often seen in the sediments deposited along former stream valleys. They are fossils of tropical plants that record generally warm and humid conditions.

Post-Acadian Events

Mississippian strata crop out in the Appalachian region from Pennsylvania to Alabama. Nonmarine shales and sandstones predominate, indicating that erosion of the mountainous tracts developed during the Acadian orogeny was still in progress. Some of the finer clastics were spread westward onto the craton to form the vast deposits of Lower Mississippian black shales mentioned earlier. Particularly coarse sediments, including conglomerates, were deposited as part of the **Pocono Group**. Pocono sandstones (Fig. 18.40) form some of the resistant ridges of the Appalachian Mountains in Pennsylvania. Westward, the Pocono section thins and changes imperceptibly into marine siltstones and shales. Evidently, the depositional framework consisted of another great clastic wedge complex of alluvial plains that sloped westward and merged into deltas that were being built outward into the epicontinental seas. The plains and deltas, standing only slightly above sea level, were

A

Figure 18.38   Sedimentary rocks of the Catskill clastic wedge. (A) Sandstone of the Devonian Catskill Formation filling an abandoned stream channel (scale divisions are 10 centimeters). (B) Tidal flat deposits near Altoona, Pennsylvania.   *(W. D. Sevon)*

B

Figure 18.39   Fossil plant root traces in mudstones of the Catskill clastic wedge (Oneonta Formation), near Unadilla, New York.   *(W. D. Sevon)*

backed by the rising mountains of the Appalachian fold belt. A large part of the coarser clastics had settled before reaching the southern portions of the epicontinental seas, where marine limestones are the most prevalent rocks.

Pennsylvanian rocks of the Appalachians are characterized by cross-bedded sandstones and gray shales that were deposited by rivers or within lakes and swamps. Coal seams are, of course, prevalent in the Pennsylvanian System of the eastern United States and reflect the luxuriant growths of mangrove-like forests. This was an ideal environment for coal formation. Vegetation that accumulated in the poorly drained swampy areas was frequently inundated and killed off. Immersed in water or covered with muck, the dead plant material was protected from rapid oxidation; however, it was attacked by anaerobic bacteria. These organisms broke down the plant tissues, extracted the oxygen, and released hydrogen. What remained was a fibrous sludge with a high content of carbon. Later, such peatlike layers

Figure 18.40 Exposure of the Pocono Group (Mississippian) in a highway cut in western Pennsylvania. The section includes thick, cross-bedded fluvial sandstones as well as a few dark, coaly shales. The cut is across the axis of a syncline that was formed during the Allegheny orogeny. *(J. D. Glaser)*

were covered with additional sediments—usually siltstones and shales—and then compressed and slowly converted to coal.

The culminating deformational event in the southern Appalachians has been termed the **Allegheny orogeny** (also called the Appalachian orogeny). This great episode of mountain building probably began during the Pennsylvanian Period and continued throughout the remainder of the Paleozoic. It affected a belt that extended for over 1600 kilometers, from southern New York to central Alabama. Like the earlier Acadian orogeny, the Allegheny deformations were the result of closure of an oceanic tract (the proto-Atlantic, or Iapetus) and the convergence of continents. Whereas the Acadian even had been governed by convergence of North America and Europe, the Allegheny orogeny occurred when northwestern Africa came into contact with southeastern North America (see Fig. 18.35).

The results of the Allegheny orogeny were profound and included Permian compression of early continental shelf and rise sediments as well as strata deposited along the bordering tract of the craton. The great folds now visible in the Valley and Ridge province were developed during this orogeny (Fig. 18.41). Less visible at the surface but no less impressive are enormous thrust faults formed along the east side of the southern Appalachians. The folds are asymmetrically overturned toward

the northwest, and the fault surfaces are inclined southeastward, suggesting that the entire region was moved forceably against the central craton. Not unexpectedly, erosion of the rising mountains produced another great clastic blanket of nonmarine sediments.

The Ouachita Deformation

The deformation of the Ouachita orogenic belt was caused by the collision of the northern margin of Gondwanaland (northern South America and perhaps part of northwestern Africa) with the southeastern margin of the North American craton (see Fig. 18.36). That deformation began rather late in the Paleozoic, for the rock record indicates that from early Devonian to late Mississippian time the region experienced slow deposition interrupted by only minor disturbances. Carbonates predominated in the more northerly shelf zone, whereas cherty rocks known as **novaculites** accumulated in the deeper marine areas (Fig. 18.42).

Novaculites are hard, even-textured, siliceous rocks composed mostly of microcrystalline quartz. They are formed from bedded cherts that have been subjected to heat and pressure. Arkansas novaculite is used as a whetstone in sharpening steel tools. The pace of sedimentation increased dramatically toward the end of the Mississippian, when a thickness of over 8000 meters of

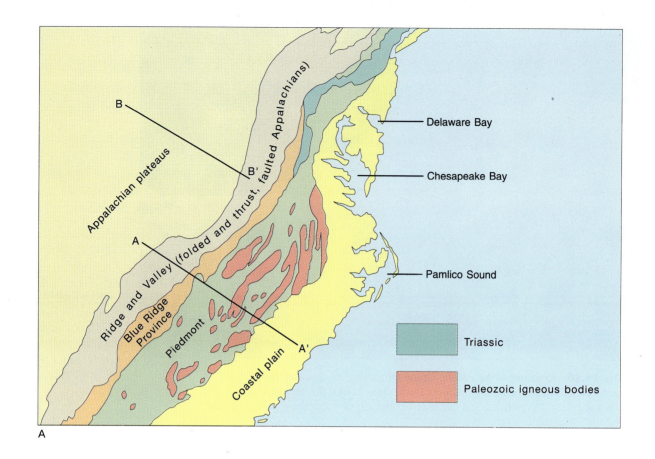

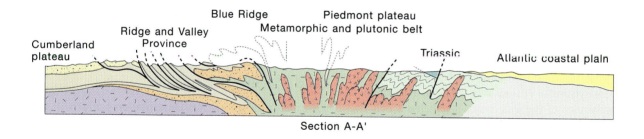

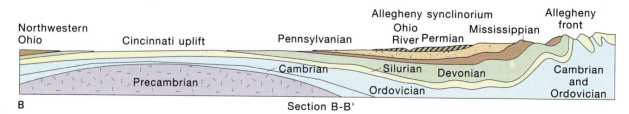

Figure 18.41 (A) Map of physiographic provinces of the Appalachian region. (B) Two cross sections showing the structural relationships of the Appalachian Mountains.
(Photo A from Tectonic Map of the United States, *Geological Society of America, 1944; B from P. B. King,* Bull. Am. Assoc. Petr. Geol., *1950.)*

Figure 18.42 Tightly folded beds of Arkansas novaculite. *(C. G. Stone)*

graywackes and shales was spread into the depositional basin. The flood of clastic deposition continued into the Pennsylvanian, forming a great wedge of sediment that thickened and became coarser toward the south, where mountain ranges were being formed (Fig. 18.43). Radiometric dating of now deeply buried basement rocks from the Gulf Coastal states indicates that these rocks were metamorphosed during the late Paleozoic and were the likely source for the Pennsylvanian clastics spread into the belt. These coarse sediments document several

pulses of orogenesis that ultimately produced mountains along the entire southern border of North America. In the faulting that accompanied the intense folding, strata of the continental rise were thrust northward onto the rocks of the shelf. By Permian time, stability returned, and strata of this final Paleozoic period are relatively undisturbed.

Since the time of active mountain building, erosion has leveled most of the highlands, leaving only the Ouachita Mountains of Arkansas and Oklahoma and the

Figure 18.43 Geologic section of Pennsylvania rocks across Oklahoma and Texas, showing the thick wedge of sediment shed from mountain ranges to the south. *(From Stratigraphic Atlas of North and Central America, Shell Oil Company, Exploration Department.)*

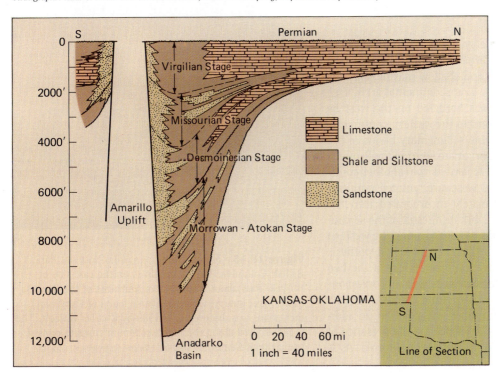

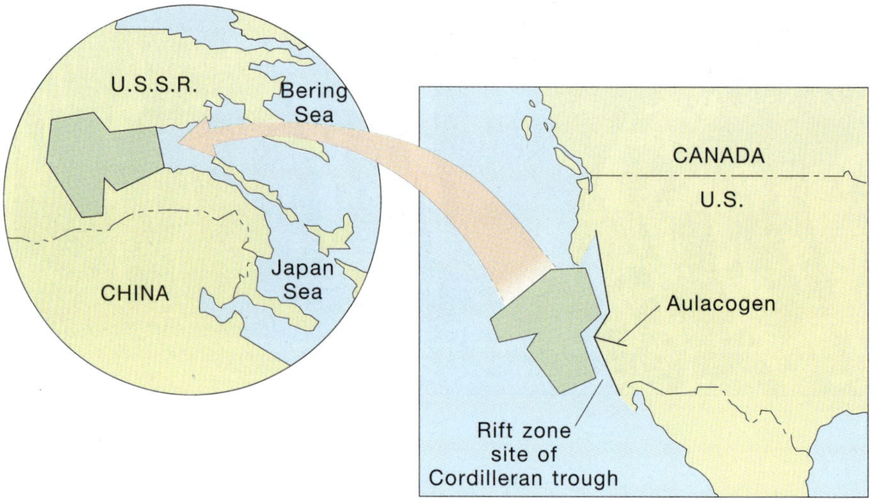

Figure 18.44 Conceptual drawing indicating the development of a passive margin trough along the western edge of North America by detachment of a landmass that ultimately was incorporated into the Siberian Platform of Russia. *(Concept from J. W. Sears and R. A. Price,* Geology, *6 [1978], 227–270.)*

Marathon Mountains of southwest Texas as remnants of once lofty ranges. Although the Ouachita belt has been traced over a distance of nearly 2000 kilometers, only about 400 kilometers are exposed. Thus, the actual configuration of the belt has been determined by the examination of millions of well samples and other data obtained during drilling activities associated with petroleum exploration.

18.5 Paleozoic History of Western North America

During much of early Paleozoic time, large areas of the Cordilleran region of western North America lay beneath sea level. A coastal plain and shallow marine shelf extended westward from the Transcontinental arch. This relatively level expanse was bordered on the west by deep marine basins adjacent to the western continental margin. Many of the marine sedimentary basins of the Cordillera appear to have originated during the Proterozoic when North America separated from another continental block that drifted westward, rotated counterclockwise, and was ultimately incorporated into the Siberian platform (Fig. 18.44). As a result of the separation, a passive margin was developed along the western border of North America. Evidence for this hypothesis consists of similarities in Proterozoic rocks of the two now widely separated segments, as well as correspondence in the alignment of structural trends when the two blocks are placed in proper juxtaposition. In addition, there is geophysical evidence for the presence of a failed rift or aulocogen in southern California. As is

Figure 18.45 Mount Burgess in Yoho National Park, British Columbia. The mountain is on the east side of Burgess Pass about 3 kilometers southeast of the Burgess Shale fossil quarry, where a diverse fauna of Cambrian metazoans was discovered in 1909 by Charles Walcott. The rocks exposed along the flanks of Mount Burgess consist primarily of Cambrian shales. *(Image Bank/Wm. A. Logan)*

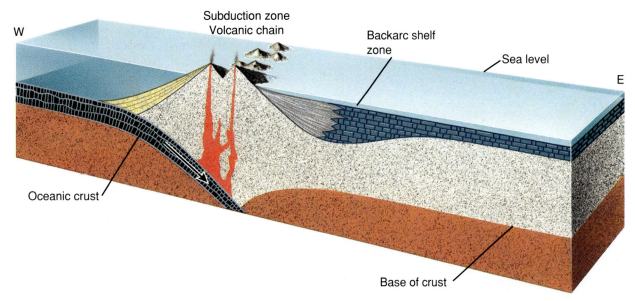

Figure 18.46   Interpretive cross section of conditions across the Cordilleran region during early Paleozoic time.

characteristic of aulocogens, this one is perpendicular to the possible line of separation between the two landmasses. One notes further that the Belt Supergroup of Montana, Idaho, and British Columbia, as well as the Uinta Series of Utah and the Pahrump Series of California, were all deposited in fault-controlled basins formed during an episode of Proterozoic rifting.

Some of the grandest sections of Cambrian rocks in the world are exposed in the Canadian Rockies of British Columbia and Alberta. The formations have been erosionally sculpted into some of Canada's most magnificent scenery (Fig. 18.45). Lower Cambrian rocks include ripple-marked quartz sandstones that were probably derived from the Canadian shield. By middle Cambrian time, the seas had transgressed farther eastward, and shales and carbonates became more prevalent. An interesting section of these Middle Cambrian rocks is exposed along the slopes of Kicking Horse Pass in British Columbia near the border with Alberta. One of the units in this section, the Burgess Shale, has excited the keen interest of paleontologists around the world because it contains an extraordinary fauna of complex, soft-bodied animals. We will examine the importance of the Burgess Shale fauna in Chapter 19.

Patterns of sedimentation in the Cordilleran region during the Ordovician and Silurian indicate that the western margin of the continent was no longer a passive margin, as was the case in the Neoproterozoic and earliest Cambrian. The Pacific plate was now moving against North America, and a subduction zone with an associated volcanic chain had formed (Fig. 18.46). A subduc-

tion zone complex containing graywackes and volcanics developed in the trench; whereas east of the volcanoes, siliceous black shales and bedded cherts accumulated. Dissolution of particles of volcanic ash and the siliceous shells of marine microorganisms provided the silica required for the formation of chert beds.

By Devonian time, the advance of the Pacific plate brought a volcanic arc to the western margin of North America. Convergence of the arc against the continent resulted in a disturbance known as the **Antler orogeny** (Fig. 18.47) and produced thrust faulting on a massive

Figure 18.47   Location of the Antler orogenic belt.

Figure 18.48 The Mississippian Bragdon Formation of the eastern Klamath Mountains of California. The Bragdon is composed of siltstones, sandstones, and felsic ash shed from a volcanic island arc that formed next to North America during late Paleozoic time. *(M. Miller)*

scale. Continental rise and slope deposits were thrust westward by as much as 80 kilometers and now rest on shallow-water sediments of the former continental shelf.

The Antler orogeny continued actively into the Mississippian and Pennsylvanian. Late Paleozoic deposits derived from the erosion of the Antler highlands include an immense volume of coarse clastics and volcanic rocks that were swept eastward from the source region (Fig. 18.48). Over 2000 meters of sandstones, shales, and ash beds are found in the Klamath Mountains of northern California. Volcanic rocks in western Idaho and British Columbia attest to a continuation of vigorous volcanism from the Mississippian through the Permian. Crustal deformation along the west side of the Cordilleran belt is indicated by an extensive area of angular unconformities between Permian and Triassic sequences. These Permian–Triassic disturbances of the Cordilleran region have been named the **Cassiar orogeny** in British Columbia and the **Sonoma orogeny** in the southwestern United States. Like the earlier Antler orogeny, the Sonoma event was probably caused by the collision of an eastward-moving island arc against the North American continental margin in west-central Nevada. Oceanic rocks and remnants of the arc were thrust into the edge of the continent and became part of North America.

Permian conditions in the shelf area east of the Antler uplift were quieter than those to the west. The region was occupied by a shallow sea during much of the Permian Period. Within that sea several large platform deposits accumulated. One of these was the Kaibab Limestone, an imposing formation that forms the vertical cliffs along the rim of the Grand Canyon. Beneath and eastward of the Kaibab Limestone, there are red beds that reflect deposition on coastal mud flats and flood plains. Sand dunes were prevalent nearby, as indicated by the massive, extensively cross-bedded Coconino Sandstone.

While the Kaibab was being deposited in the southwest, a relatively deep marine basin was developing to the north in the general area now occupied by Wyoming, Montana, and Idaho. In this basin, sediments of the Phosphoria Formation were deposited. Although the Phosphoria includes beds of cherts, sandstones, and mudstones, it takes its name from its many layers of dark phosphatic shales, phosphatic limestones, and phosphorites. **Phosphorite**, a dark gray concretionary variety of calcium phosphate, is mined from the formation and used in the manufacture of fertilizers and other chemical products. The unusual concentration of phosphates may have been produced by upwelling of phosphorus-rich seawater from deep parts of the basin. The metabolic activities of microorganisms may have assisted in the precipitation of phosphate salts.

18.6 Aspects of Paleozoic Global Climate

The Earth during the Paleozoic was characterized by latitudinal and topographic variations in climate, just as exist today. During episodes of extensive marine inundations of the continents, climates would have been relatively mild; during those periods when continents stood high and great mountain chains affected atmospheric circulation, climates became more diverse and extreme. There were also factors that would have made the climates of long ago different from those of today. During the early periods of the Paleozoic, the Earth turned faster and the days were shorter. Tidal effects were stronger, and there was no green covering of vascular plants to absorb the Sun's radiation. There may even have been differences in atmospheric composition and solar radiation that have left no clues in the rock record.

GEOLOGY AND THE ENVIRONMENT

The Ecological Effects of the Big Freeze in North Africa

We twentieth-century residents of the Earth are accustomed to reading about ecologic problems in our newspapers almost daily.

We are ourselves to blame for many of these problems, including those caused by pollution, acid rain, deforesting, and modification of natural water systems. Yet environmental and ecological crises have occurred many times in the geologic past, long before there were humans to generate them. For example, consider the effects on life of the Ordovician Ice Age described in this chapter. We have evidence that the South Pole at this time was located in what is now a barren area between Algeria and Mauritania. Evidence for an enormous ice sheet in this region occurs as glacial grooves and striations, outwash plains, moraines, and meltwater channels that extend across thousands of square kilometers.

There must have been a truly enormous buildup of continental ice, and that accumulation would surely have resulted in cooler temperatures in ocean regions far to the north. Eustatic and climatic changes would have had a pervasive effect on life. Indeed, paleontologists have long recognized that the late Ordovician was marked by extinctions of large numbers of marine invertebrates, including entire families of bryozoans, corals, brachiopods, sponges, nautiloid cephalopods, trilobites, crinoids, and graptolites. Twenty-two percent of all families of known animals were wiped out. It was one of the greatest episodes of mass extinction in the geologic record.

It seems likely that the demise of these families of organisms resulted from general cooling of tropical seas as well as the draining of epeiric and marginal seas as sea level was lowered. The loss of the shallow-water environments in which early Paleozoic invertebrates thrived, and the crowding of species into remaining habitats, would explain this Ordovician ecological disaster.

At the beginning of the Cambrian, climate was probably somewhat cooler than the average for the first half of the Paleozoic. The Earth was still recovering from the ice age that occurred near the end of the Proterozoic. Soon, however, conditions became warmer. Broad, shallow seas gradually flooded the interiors of continents, and North America, Europe, and even Antarctica were along or near the early Paleozoic equator, as indicated by paleomagnetic studies. The paleomagnetic data is compatible with the depositional record of widespread limestones, reefs, and evaporites.

There were, however, severe as well as equable climates during the early Paleozoic, as indicated by glacially striated bedrock and tillites in the region of the present Sahara Desert. The presence of glacial deposits in northern Africa coincides with paleomagnetic data indicating a nearby location of the Ordovician South Pole.

Gondwanaland experienced yet another episode of continental glaciation during the late Paleozoic (see Fig. 17.20). Tillites (Fig. 18.49) and glacially scoured bedrock (Fig. 18.50) are present at many places in South America, South Africa, Antarctica, and India (Fig. 18.51).

Figure 18.49   Upper Carboniferous tillite at base of slope (beneath hammer) composed of Precambrian basalt. Tillite is consolidated and unsorted glacial debris. Kimberley, South Africa.   *(Warren Hamilton, U.S. Geological Survey)*

Figure 18.50 Glacial striations formed in early Permian time on Proterozoic quartzite in Australia. The ice flowed to the right, parallel to the pen. Note the downflow smoothing and crescent-shaped "chatter" marks with their concave sides facing upflow (upper right). *(Warren Hamilton, U.S. Geological Survey)*

Figure 18.51 The distribution of glacial tillites (blue triangles), coal (red circles), and evaporites (irregular green areas) during the Permian, about 250 million years ago. *(From G. E. Dewey, T. S. Ramsey, and A. G. Smith, J. Geol, 82(5) [1974], 539.)*

There are indications of four or more glacial advances, suggesting a pattern of cyclic glaciation similar to that in North America and Europe during the Pleistocene. The orientation of striations chiseled into bedrock by moving ice indicates that the ice moved northward from centers of accumulation in southwestern Africa and eastern Antarctica. During the warmer interglacial stages and in outlying less frigid areas, *Glossopteris* and other plants tolerant of cool, dry climates grew in profusion and provided the materials for thick seams of coal.

Late Paleozoic climates of the northern hemisphere differed significantly from those of Gondwanaland. As we have seen, the South Pole was located in Africa, providing cool conditions over much of Gondwanaland. The North Pole was over open ocean, and the late Paleozoic equator extended northward across southern Europe (see Fig. 18.23). Coal beds containing the fossils of tropical plants and coral reefs indicate generally milder conditions north of Gondwanaland. Dune-deposited red sandstones associated with evaporites attest to arid conditions in some regions. It is likely that the northern regions of Pangaea lay within the zone of northeasterly trade winds. As these winds traveled westward and encountered mountains, they would have been forced to ascend, form clouds, and produce precipitation on the windward side of the ranges. On reaching the far or lee side of the mountains, much of the moisture in the air would have been lost, and arid conditions would be the expected result, at least until moisture was again replenished as winds continued westward over epeiric seas.

18.7   Rock and Mineral Resources of the Paleozoic

Paleozoic sequences in North America contain an abundance of economically important Earth materials. Among these are all kinds of sedimentary, metamorphic, and igneous building stones; limestones used in the manufacture of cement; quartz sandstones used in the manufacture of glass products; salt obtained from Silurian and Permian sections; and phosphates from the Phosphoria Formation of Montana, Utah, Idaho, and Wyoming. Sedimentary iron ores occur from Newfoundland southward along the Appalachians to Alabama. Mountain building, with its attendant intrusive and volcanic igneous activity, was frequently accompanied by the emplacement of metallic ores. Paleozoic orogenies produced ores of copper, silver, gold, zinc, and tin.

A

B

Figure 18.52   (A) A remote-controlled continuous mining machine with a flooded bed scrubber for dust control in removing coal in a West Virginia underground mine. (B) Large stripping shovel in the pit of an Illinois surface mine.   *(Peabody Coal Co.)*

Among the fossil fuels, coal is an important resource found in post-Devonian rocks. The Appalachian and Illinois basins have yielded enormous tonnages of coal (Fig. 18.52). Single sequences of Pennsylvanian strata may include many coal beds, as in West Virginia, where 117 different layers are recognized. In the anthracite district of eastern Pennsylvania, orogenic compression has partially metamorphosed coal into an exceptionally high-carbon, low-volatile variety prized for its industrial uses. The coal industry, which has deteriorated for the past few decades, is currently experiencing an upsurge as a consequence of decreasing supplies of petroleum and natural gas.

Commercial quantities of oil and gas have been recovered from Ordovician through Pennsylvanian rocks in the Appalachian states as well as Ohio,

FOCUS ON

Glass from Sand

Glass, a product formed by melting and cooling quartz sand and minor amounts of certain other common materials, is so familiar to us that we sometimes forget its importance. In fact, it would be difficult to imagine a world without glass. There would be no goblets, bottles, mirrors, or windows; no glass laboratory retorts, tubes, or beakers; no electric light bulbs or neon signs; no glass lenses for cameras, microscopes, telescopes, or motion picture projectors; and no fiber optics for use in viewing the inside of the human body. We see glass used everywhere in our buildings, electronic equipment, and medical devices. We not only see it, we see through it.

Glass has a long and colorful history. Natural glass, such as obsidian, was used by prehistoric humans to make arrowheads and knife blades. Human-made glass was made by Egyptians as long ago as 3000 B.C.E. At Ur in Mesopotamia, glass beads nearly 4500 years old have been found in archaeological excavations. The Greeks and Romans learned the art of making glass ornaments, bottles, vases, and trinkets that were prized throughout the ancient world. During the Middle Ages, Europeans produced beautiful stained-glass windows for cathedrals. Blood-red stained glass was made by adding copper compounds to the molten silica. The Europeans knew that cobalt gave the glass a rich, deep-blue color; manganese turned it purple; and antimony provided golden yellows. Iron oxides were added to color glass green or brown.

Today, the sands used in making glass have a silica content as high as 95 to 99.8 percent. As mentioned in this chapter, two sources of such exceptionally pure sands are the Ordovician St. Peter Sandstone of Missouri and Illinois and the sandstone members of the Devonian Oriskany Formation of West Virginia and Pennsylvania. These clean, pure sandstones were deposited in shallow marine environments where wave action could remove clay impurities and concentrate grains of sand-size quartz. Both sandstones are composed of grains that have been eroded from older sandstones, and this reworking has also contributed to their extraordinary purity.

In the manufacture of window glass, the mixture prepared for melting consists of about 72 percent silica from quartz grains. Certain metallic oxides such as soda (Na_2O) and lime (CaO) serve to lower the temperature required for melting. That temperature for soda–lime–silica glass is 600 to 700°C. Once molten, the "batch" (as it is called) is cooled, poured, rolled, or blown into the shapes desired. Compounds of boron and aluminum can be added to provide heat-resistant glass, such as in the Pyrex® brand of cookware. Fine cut glass and so-called crystal are a silica–lead–soda glass known not only for its brilliance but musical tone as well. By varying the kinds and amounts of metallic oxides, glass can be produced for a great variety of special uses and products. There is little doubt that it will always be an essential part of life in this age of technology.

Figure 18.53 Drilling for oil in upper Paleozoic rocks of eastern Kansas.

Oklahoma, and Texas. Devonian reefs within the Williston basin of Alberta and Montana have provided enormous amounts of petroleum. Devonian petroleum has also been produced in the Appalachians. Indeed, in 1859, the first U.S. oil well was drilled into a Devonian sandstone. (Oil was struck at a depth of only 20 meters.) Carboniferous formations of the Rocky Mountains, midcontinent (Fig. 18.53), and Appalachians also contain oil reservoirs. However, wells drilled into reefs and sandstones of the Permian basin of West Texas have yielded the greatest amount of oil from upper Paleozoic formations of the western United States.

The stores of metallic and nonmetallic economic minerals in Paleozoic rocks have contributed significantly to the power and welfare of nations. It must be remembered, however, that these vitally important resources are nonrenewable, exhaustible, and often limited in extent. Every effort should be made to conserve the ever-depleting supply of ore minerals and fossil fuel so future generations can enjoy an adequate standard of living.

SUMMARY

At the beginning of the Paleozoic, the **Proto-Pangaean** supercontinent had begun to pull apart, and ocean tracts formed between the separating landmasses. **Iapetus Ocean** was an early Paleozoic Proto-Atlantic that was widening at the beginning of the Paleozoic but was too close during later Paleozoic time as a result of the convergence of tectonic plates bearing North America, western Europe, and Gondwanaland. The reassembly of landmasses ultimately produced the supercontinent Pangaea.

During earliest Paleozoic time, most of North America was above sea level and undergoing erosion. Initial sites of marine sedimentation were along rifted, passive margins. Soon, however, the seas spilled out of these marginal basins and inundated the continental platform. Sediment deposited on the floors of the shallow seas that covered the platform record four major cycles of transgression and regression, as well as many minor cycles. The rocks of the major cycles compose the Sauk, **Tippecanoe, Kaskaskia,** and **Absaroka cratonic sequences**. In general, Paleozoic sections on the platform are thicker in basins and thinner over domes. Loss of sections of rock because of unconformities also characterizes domes. Stratigraphic sequences called **cyclothems** record repeated alternations of marine and nonmarine conditions during the Pennsylvanian Period. Near the close of the Paleozoic, the shallow sea that had covered the platform withdrew. Evaporite and red beds in the Permian basins of the southwestern United States provide the final record of Paleozoic sedimentation on the platform.

The relative calm that characterized the platform during the Paleozoic was not at all characteristic of the eastern margin of North America. As tectonic plates were moving against one another to form Pangaea, this region experienced one orogenic event after another. The first was the Taconic orogeny, followed by the **Caledonian orogeny**, whose effects can now be seen in northwestern Europe. Then came the **Acadian orogeny** and, finally, the **Allegheny orogeny**. Each mountain-building event was accompanied by vigorous erosion that produced great aprons of detritus known as **clastic wedges**. The Taconic orogeny produced the **Queenston clastic wedge**; the Caledonian's clastic wedge bears the odd name of **Old Red Continent**; and the Acadian orogeny resulted in the **Catskill clastic wedge**. Nonmarine clastic sediments of the Dunkard and Monongahela series form the clastic wedge for the Allegheny orogeny.

In the American West, movement of the Pacific plate against the continent resulted in the formation of a subduction zone with an associated volcanic arc. Following the development of the subduction zone, there were several mountain-building events in the Cordillera, including the late Devonian–Mississippian **Antler orogeny**.

In general, the range of climatic conditions during the Paleozoic was not significantly different from that of the recent geologic past. Cold climates are indicated by both Ordovician and Permian glacial deposits in Southern Hemisphere continents. Warmer climates prevailed where we find Paleozoic reefs and associated beds of richly fossiliferous limestones. There were also periods of aridity, as indicated by thick sections of evaporites in the Silurian of Michigan and the Permian of West Texas.

KEY TERMS

Proto-Pangaea *415*

Platform *418*

Sauk cratonic sequence *420*

Tippecanoe cratonic sequence *423*

Principle of temporal transgression *423*

Kaskaskia cratonic sequence *425*

Chattanooga Shale *427*

Absaroka cratonic sequence *427*

Cyclothem *430*

Taconic orogeny *437*

Logan's thrust *437*

Queenston clastic wedge *439*

Caledonian orogeny *440*

Acadian orogeny *442*

Catskill clastic wedge *442*

Pocono Group *443*

Allegheny orogeny *445*

Novaculite *445*

Antler orogeny *449*

Cassiar orogeny *450*

Sonoma orogeny *450*

Phosphorite *450*

REVIEW QUESTIONS

1. What are the names and locations of the major orogenic events of the Paleozoic, and when did each occur?

2. What is the significance of the Tommotian stage in the standard geologic time scale? How do multicellular organisms found as fossils below the Tommotian differ from those above?

3. What geologic evidence indicates that cratonic arches and domes were often above sea level during the Paleozoic? How might a dome be distinguished from a basin on a geologic map?

4. What is a barred basin? Why is a barred basin a particularly effective place for the precipitation of evaporites? Where might such basins have existed during the Paleozoic?

5. What is a clastic wedge? Under what geologic circumstances do clastic wedges form? Are they primarily marine or nonmarine?

6. With what orogenic events are the Queenston and Catskill clastic wedges associated?

7. What is the paleoenvironmental significance of extensive red sandstones and shales in association with thick salt and gypsum accumulations?

8. What is a cyclothem? What may have been the cause of this type of cyclic deposition during the Pennsylvanian?

9. When and where were the Colorado mountains (ancestral Rockies) formed?

10. Name three metallic and three nonmetallic mineral resources that are obtained in commercial quantities from Paleozoic rocks.

DISCUSSION QUESTIONS

1. It is often inferred that rates of erosion proceeded more rapidly during the Cambrian and Ordovician than in subsequent geologic time. What is the basis for this inference?

2. Why are rocks such as the Tapeats Sandstone, Bright Angel Shale, and Muav Limestone not precisely the same age at widely separated locations?

3. Why is it logical to divide the Paleozoic rocks of the platform of North America into cratonic sequences?

4. With regard to plate movements, what was the cause of the Caledonian and Acadian orogenies? What similarities exist between clastic wedges developed as a result of these events?

5. With regard to plate movements, what was the cause of the Ouachita orogeny? Where might one travel to examine the rocks and structures formed during the Ouachita orogeny?

6. What evidence suggests the occurrence of two ice ages during the Paleozoic? Where were the continental glaciers centered? What evidence supports the location you have indicated?

CHAPTER
19

Life of the Paleozoic

Reconstruction of some of the arthropods preserved in the Cambrian Burgess Shale of British Columbia. The organisms provide a view of marine life as it existed 530 million years ago.

The rich fossil record of the Paleozoic stands in startling contrast to that of the earlier Precambrian. We have noted that the Precambrian was a world of bacteria, algae, and fungi. Unquestioned multicellular animals are found only in uppermost Proterozoic strata, such as those at Ediacara Hills in Australia. Except for a few tiny tubular creatures, latest Precambrian animals had not acquired the ability to secrete shells.

In passing from late Proterozoic to early Cambrian time, the record of life is much improved. Scattered around the shallow margins of the continents, isolated groups of invertebrates had established themselves. Soon, epeiric seas began to creep across the cratons, providing a multititude of habitats and opportunities for the diversification and expansion of life.

Certainly the improvement in the fossil record evident in Cambrian strata can be attributed to the spread of shell-building abilities among invertebrates. During the Cambrian, shell-building brachiopods and trilobites were abundant, but the most rapid expansion and diversification of shell-building groups began later, in early Ordovician time. This proliferation in shell building was probably related to the advantages afforded by shells in providing protection against predators as well as support for soft tissues. Shells also gave bottom dwellers the increased mass necessary to maintain their positions on the sea floor.

The evolution of plant life during the Paleozoic was no less impressive than that of animal life. Beginning with unicellular aquatic forms, plants evolved the adaptations necessary for survival on land and produced great forests of ferns, scale trees, and, ultimately, even conifers.

19.1   Plants of the Paleozoic

The history of plants has an obscure beginning among the bacteria and algae of the Precambrian. Indeed, the Precambrian sedimentary iron ores are presumed to have been produced by bacterial activity. However, as

Figure 19.1   Hummocky stromatolites of Cambrian age exposed along the banks of the Black River in Missouri.

has been noted, fossils of such tiny, soft-bodied creatures are very rare.

An exception to the generally poor fossil record of the earliest plants is provided by stromatolites, which are abundant in upper Proterozoic rocks and continue to be found in limestones of all younger ages. During the Cambrian Period, stromatolitic reefs were widespread (Fig. 19.1) but became more restricted in the Ordovician, perhaps because of the rise of marine invertebrates that grazed on the algae responsible for producing stromatolites.

Another relatively common group of early Paleozoic marine plant fossils are the **receptaculids** (Fig. 19.2). Receptaculids are lime-secreting green algae of the Family Dasycladacea. In general appearance, they remind one of the seed-bearing central area of a large sunflower; this has caused amateur fossil collectors to designate them "sunflower corals." They are, however, neither sunflowers nor corals. Nor are they sponges, a designation once considered because of their spicule-like rods and circulatory passages. Although most frequently found in Ordovician rocks, members of this group also occur sparsely in Silurian and Devonian strata.

Invasion of the Lands

The appearance of our landscape today owes much to its verdant covering of forests and grasslands. At the beginning of the Paleozoic, however, there were no trees

Figure 19.2 *Fisherites,* a receptaculid from the Ordovician Kimmswick Formation of Missouri. Sometimes called sunflower corals, receptaculids are the fossil remains of a type of green algae and not in any way related to corals (10 × 16 centimeters).

clothing the continents. Most living organisms were confined to the seas. It is likely that some species of algae invaded fresh-water lakes and streams, and some may even have evolved means of moisture control independent of their environment. These algae moved into damp areas on land and began the evolutionary journey toward complex land plants. The algal group most likely to have begun this progression were probably the **chlorophytes** or green algae. A relationship between chlorophytes and land plants is indicated by the presence of the same kind of green pigment in both groups and by the fact that both produce the same kind of carbohydrate during photosynthesis. Also, some chlorophytes have become adapted to fresh water—a necessary accomplishment for the vegetative invasion of the continents. Today some chlorophytes can even survive in moist areas on land.

The vegetative invasion of the continents required the evolution of a vascular system. Vascular plants have tubes and vessels that convey fluids from one part of the plant to another. The importance of this system for a land plant is apparent when we remember that surficial moisture above ground is an undependable source of water in most environments. There is nearly always a greater persistence of moisture in the pore spaces of soil. But no light for photosynthesis is available underground. The vascular system of a plant permits part of it to exist underground, where there is water but no light, and another part of it to grow where water supplies are uncertain but sunlight for photosynthesis is ensured.

The transition from aquatic plants to land plants was apparently difficult, for it was late in coming. The first *unquestioned* remains of vascular plants are found in rocks of middle Silurian age. These are **psilophytes**. They were small plants (less than 30 centimeters tall), consisting of horizontal stalks or rhizomes that grew just beneath the surface of the ground in moist soil. Extending vertically from the rhizomes (Fig. 19.3) were short slender stems bearing smaller branches and spore sacs. Delicate strands of wood cells and a rudimentary vascular system have been recognized in psilophyte fossils.

The psilophytes paved the way for the evolution of large trees. Because of the evolution of wood, plants were able to stand tall against the pull of gravity and the force of strong winds. By late in the Devonian, there were forests of lofty, well-rooted, leafy trees (Fig. 19.4). One of these forests, located near Gilboa, New York, has left many fossils as ample evidence of its existence. Some of the Gilboa trees were over 7 meters tall. Impressive as these trees were, however, they were to be dwarfed by some of their Carboniferous descendants.

When one surveys the entire history of vascular plants, three major advances become apparent. Each

Figure 19.3 A psilophyte from the Lower Devonian of the Gaspé Peninsula (plant at left about 16 centimeters tall).

Figure 19.4   Restoration of a middle Devonian forest in the eastern area of the United States. (A) An early lycopod, *Protolepidodendron*. (B) *Calamophyton*, an early form of the horsetail rush. (C) Early tree fern, *Eospermatopteris*.
(Field Museum of Natural History; painting by C. R. Knight)

involved the development of increasingly more effective reproductive systems. The first advance led to seedless, spore-bearing plants, such as those that were ubiquitous in the great coal-forming swamps of the Carboniferous. The second saw the evolution of seed-producing, pollinating, but nonflowering plants (gymnosperms). This was a late Paleozoic event. The evolution of plants with both seeds and flowers (angiosperms) came late in the Mesozoic Era.

Among the moisture-loving plants of the Carboniferous were the so-called scale trees, or **lycopsids**. Today, they are represented by their smaller survivors, the club mosses, of which the ground pine *Lycopodium* is a member. Small size was not particularly characteristic

of the late Paleozoic lycopsids. The forked branches of *Lepidodendron* (Fig. 19.5) reached 30 meters into the sky. The elongate leaves of the scale trees emerged directly from the trunks and branches. After being released, they left a regular pattern of leaf scars. In *Lepidodendron*, the scars are arranged in diagonal spirals, whereas species of *Sigillaria* have leaf scars in vertical rows.

Another dominant group of plants that grew side by side with the Carboniferous lycopsids were the **sphenopsids**. Living sphenopsids include the scouring rushes and horsetails. Fossil sphenopsids, such as *Calamites* (Fig. 19.6) and *Annularia* (Fig. 19.7), possessed slender, unbranching, longitudinally ribbed stems with

Figure 19.5   Fossilized bark of the Carboniferous tree *Lepidodendron* (scale rows are about 1 centimeter wide).
(Photograph courtesy of Ward's Natural Science Establishment, Inc., Rochester, NY)

Figure 19.6   *Calamites,* a sphenopsid (plants shown are about 3 to 5 meters tall).

Figure 19.7   *Annularia,* an abundant sphenopsid of Pennsylvanian age.

a thick core of pith and rings of leaves at each transverse joint. At the top, a cone bore the spores that would be scattered in the wind.

True ferns (Fig. 19.8) were also present in the coal forests. Many were tall enough to be classified as trees. Like the lycopsids and sphenopsids, they reproduced by means of spores carried in regular patterns on the undersides of the leaves.

Seed plants made their debut during the late Paleozoic. They probably arose from Devonian fernlike plants. These plants, appropriately dubbed "seed ferns," had fernlike leaves, but unlike true ferns they reproduced by means of seeds.

One of the most widely known seed fern groups is *Glossopteris,* which was restricted to the Southern Hemisphere during the Carboniferous and Permian (Fig. 19.9). Most species of *Glossopteris* were plants with thick, tongue-shaped leaves. Because of certain anatomic traits and because of their association with glacial deposits, *Glossopteris* and associated plants are thought to have been adapted to cool climates.

Fossils of seed plants are present in the Northern Hemisphere also. Cordaites (primitive members of the conifer lineage) were abundant, some towering 50 meters (Fig. 19.10). Their branching limbs were crowned with clusters of large, straplike leaves. These and somewhat more modern cone-bearing conifer plants spread

widely during the Permian, perhaps as a consequence of drier climatic conditions. The first ginkgoes made their appearance during the Permian. Today, only a single species, *Ginkgo biloba,* remains as a survivor of this once flourishing group.

19.2   Invertebrates of the Paleozoic

People are accustomed to thinking of the ocean as the birthplace of life. Thus, it is no surprise to learn that the Earth's earliest animals lived in the sea. The only fossils ever found of animals that lived during the end of the late Proterozoic, the Cambrian, and the Ordovician have been those of ocean dwellers. As is the case today, these animals varied in their choices of habitat as well as in the kinds of food they required. In the

Figure 19.8   *Pecopteris,* a true fern from the Pennsylvanian of Illinois (note the penny for scale).

Figure 19.9   Fossil *Glossopteris* leaf associated with coal deposits and derived from glossopterid forests of Permian age. This fossil was found on Polarstar Peak, Ellsworth Land, Antarctica.   *(U.S. Geological Survey, J. M. Schopf, and C. J. Craddock)*

Figure 19.10   Reconstruction of a cordaitean plant of the Carboniferous.

geologic past, as in the present, there were basically three modes of life. Organisms that could live on the sea floor or burrow in sea-floor sediment are termed benthic, those that floated are termed planktic, and those that swam are termed nektic. In addition, many animals in the sea spend a phase of their lives in one of these modes and then change to accommodate themselves to another. In fact, the mobile larvae of a multitude of otherwise sedentary (stationary) "critters" are dispersed in this way.

As indicated by the Ediacaran fossils described in Chapter 17, multicellular marine organisms were already present near the end of the Precambrian. Those organisms, however, were soft-bodied and are only infrequently preserved as imprints in once soft sediment. When animals had achieved the ability to secrete hard parts, their chances of being preserved improved dramatically. The oldest assemblage of fossils to accomplish this feat occurs in rocks of the latest Precambrian (Yudomian Stage) and earliest Cambrian (Tommotian Stage) of Siberia, Sweden, Antarctica, and England. The fossils are the remains of small, shelly animals that include snails, sponges (known from spicules), and a variety of tube- and cap-shaped creatures whose biologic classification is uncertain (Fig. 19.11). Unlike the majority of later shelly invertebrates that built their shells of calcium carbonate, these early shell-bearers secreted calcium phosphate for their skeletons. As a result, they are found in slabs of limestone and dolomite when the rock is etched in weak acid.

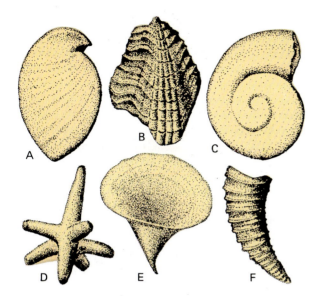

Figure 19.11   Late Precambrian and Early Cambrian shell-bearing fossils from Siberia. (A) *Anabarella,* ×20, a gastropod; (B) *Camenella,* ×18, affinity uncertain; (C) *Aldanella,* ×20, a gastropod; (D) sponge spicule, ×30; (E) *Fomitchella,* ×45, affinity uncertain; (F) *Lapworthella,* ×20.   *(After S. L. Matthews and V. V. Missarzhevsky,* J., *Geol. Soc. London, 131 [1975], 289–304.)*

Following the Tommotian Stage, rocks are clearly dominated by more familiar fossils such as trilobites and brachiopods. Their robust shells provided a fine fossil record. But there were also many soft-bodied creatures. Some of these comprise the early Cambrian Chengjian fossils of China and the middle Cambrian Burgess Shale fauna of Canada.

Window into the Past: Fossils of the Burgess Shale

High on a ridge near Mount Wapta, British Columbia, there is an exposure of Burgess Shale that contains one of the most important faunas in the fossil record (Fig. 19.12). The fossils in the Burgess Shale are reduced to shiny, jetlike impressions on the bedding planes of the blackrock (Fig. 19.13). Most of them are the remains of animals that lacked shells. Altogether, they form an extraordinary assemblage (Fig. 19.14) that includes four major groups of arthropods (trilobites, crustaceans, and members of the taxonomic groups that include scorpions and insects), as well as sponges, onycophorans (see Fig. 19.12), crinoids, sea cucumbers, chordates, and many species that defy placement in any known phyla.

The special significance of the Burgess Shale fauna lies in the perspective that it provides on life at the beginning of the Paleozoic. It reveals that the Cambrian seas teemed with diverse and complex animals and that marine life at the beginning of the Paleozoic had already reached a quite advanced stage of evolution. The discovery of the Burgess fossils further demonstrates that reconstructions of ancient communities based on the fossils of shelled animals (often all that are preserved) are likely to be seriously incomplete. Much like marine communities today, there were more animals with soft parts in the Burgess fauna than those having hard parts, and these soft-bodied animals were more varied than the shelled ones. Thus, the Burgess fauna is a more valid representation of life as it really was than is usually found for other fossil faunas.

The Burgess fauna is significant for yet another reason. Sharing the sea floor with many other unique and sometimes bizarre Burgess creatures were small, elongate animals that are currently believed to be the earliest known chordates. Chordates are animals that have, at some stage in their development or throughout their lives, a notochord (internally supportive rod) and a dorsally situated nerve cord. Chordates in which the notochord is supplanted by a series of vertebrae are termed vertebrates. These earliest known chordates have been assigned the generic name *Pikaia* in honor of Mount Pika near the Burgess fauna discovery site. Fossils of *Pikaia* (Fig. 19.15) exhibit two features upon which their designation as chordates is based. The animals had a dorsal, longitudinal supportive structure known as a **notochord**, as do all other chordates at some stage in their life history. In addition, one can discern the outlines of muscles arranged in a series of V-shaped segments along the sides of the animal as is characteristic of the musculature in fish. These muscles, working with the rigid but flexible notochord, provided a sinuous, fishlike body motion required for swimming. Like *Pikaia,* humans are chordates. For this reason, we are keenly interested in this little animal that appears to be the earliest representative of our phylum.

The discovery of the Burgess Shale fossil locality is an interesting story in itself. The initial discovery of a single fossil from the formation was made in August 1909 by Charles D. Walcott. Walcott's field crew included his wife, Marrella, who was an avid fossil collector, and his two sons. As they were riding along a trail south of Mount Wapta, Mrs. Walcott's horse dislodged a slab of black shale with a silvery impression on its surface. The slab caught Walcott's eye, and he dismounted to examine it closely. It contained the remains of a crustacean that Walcott later named *Marrella.* Unfortunately, the discovery came at the end of the field season, and exposures at higher elevations from which the rock had come were becoming snow covered. As a result, the Walcotts were forced to leave the site and return the following year, when they traced the fossil to its source

Text continued on page 465

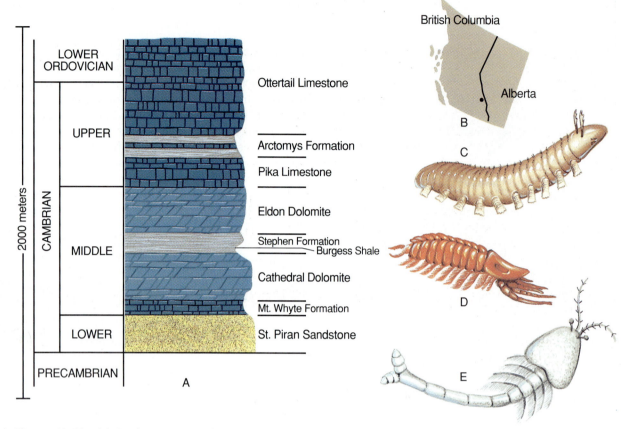

Figure 19.12 (A) Cambrian stratigraphic section at (B) Kicking Horse Pass, British Columbia. According to a now famous story, this is where a slab of fossiliferous shale was kicked over by a pack horse and caught the eye of paleontologist Charles Walcott, who followed the trail of rock debris up the side of the mountain to discover the Burgess Shale beds with their rich fauna of Middle Cambrian fossils. Among these were (C) *Aysheaia,* an onycophoran believed to be intermediate in evolutionary position between segmented worms and arthropods. Among the thousands of specimens were trilobite like arthropods such as (D) *Leanchoilia superlata* and (E) *Waptia fieldensis.*

Figure 19.13 *Hyolithes carinatus.* This photograph illustrates the nature of preservation of Burgess Shale fossils. *Hyolithes* had a tapering shell surmounted by a lid or operculum which could be closed for protection. The lateral blades on either side may have served as props. *Hyolithes* is tentatively considered a mollusk. The length of the shell is 2 centimeters. *(Smithsonian Institute; photograph by Chip Clark)*

in the Burgess Shale. Quarrying was begun and continued during the 1912, 1913, and 1917 field seasons.

Altogether, about 60,000 specimens were collected and stored in the U.S. National Museum of Natural History, where Walcott served as secretary of the Smithsonian Institute. In the 1960s, these fossils were reexamined by Harry B. Whittington, who then joined palentologists of the Geological Survey of Canada in reopening the quarry and assembling a second collection. Whittington devoted the next 15 years to an incisive and critical scrutiny of the Burgess Shale fauna. He was able to describe and depict the anatomy of the Burgess animals with great clarity. During the study, it became apparent that earlier interpretations of many Burgess animals as ancestors to existing taxonomic groups were incorrect and that Burgess fauna included creatures so different as to warrant placement in new phyla.

Early Paleozoic Diversification

As indicated by the highly structured and diversified Burgess Shale community, as well as other communities known mostly from shelled animals, marine life during the early Paleozoic had evolved a variety of lifestyles and adapted to a multitude of habitats. There were **epifaunal** animals that lived on the surface of the sea floor, as well as **infaunal** creatures that burrowed beneath the surface. There were borers as well as burrowers, attached forms and mobile crawlers, swimmers, and floaters. Some were filter-feeders that strained tiny bits of organic matter or microorganisms from the water; others were sediment-feeders that passed the mud of the sea floor through their digestive tracts to extract the nutrients within. There were animals that grazed on algae that covered parts of the ocean bottom, and carnivores that consumed these grazers. A host of scavengers processed organic debris and aided in keeping the seas suitable for life. Some classes of invertebrates maintained a single mode of life, whereas groups within other classes adapted to several different lifestyles. Snails, for example, included scavengers, herbivores, and carnivores. With the passage of geologic time, many evolutionary changes and extinctions were to occur among the invertebrates, but these changes tended to occur *within* the taxonomic groups that had been established during the early Paleozoic.

Unicellular Animals

The postulated biochemical steps leading to the first unicellular animals were discussed in Chapter 17. The first unicellular organisms were plants, but among modern unicells are creatures that defy classification as either animal or plant. Fortunately, there are also some single-celled forms that can with confidence be

Figure 19.14   The Burgess Shale diorama at the U.S. National Museum of Natural History. The reconstruction is based on actual fossil remains of organisms. It depicts a benthic marine community of the middle Cambrian. A steep submarine escarpment forms the background. Slumping along this wall contributed to the preservation of the Burgess fauna by burying the organisms. Along the wall one can see green and pink vertical growths of two types of algae. The large purplish creatures that resemble stacks of automobile tires are sponges *(Vauxia)*. The blue-colored animals are trilobites *(Olenoides)*. The brown arthropods with distinct lateral eyes are named *Sidneyia*. Climbing out of the hollow on the sea floor are crustaceans called *Canadaspis*. The yellow animals swimming toward the right above the sea floor are *Waptia. Opabinia* has crawled out of the left side of the hollow. Burrowing worms are visible in the vertical cut at the bottom.   *(National Museum of Natural History; photograph by Chip Clark)*

Figure 19.15   Reconstruction of *Pikaia,* the earliest known member of our own phylum, the Chordata. Note the rod along the animal's back that appears to be a notochord (length about 4.0 centimeters).

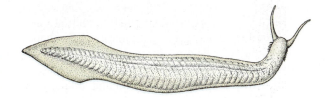

Figure 19.16   Fusulinid limestone of Permian age from the Sierra Madre of western Texas.   *(W. D. Hamilton)*

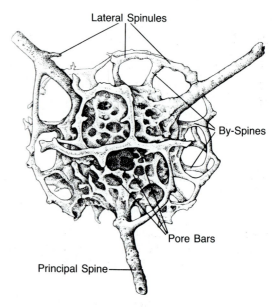

Figure 19.17   A late Ordovician radiolarian from the Hanson Creek Formation of Nevada. It is unusual to find well-preserved radiolarians in rocks of the early Paleozoic.   *(From J. B. Dunham and M. A. Murphy, J. Paleo., 50(5) [1976], 883.)*

identified as animals. Two groups of creatures in the last category that are also capable of building shells or supportive structures are the **foraminifers** and **radiolarians**.

Foraminifers build their tiny shells by adding chambers singly, in rows, in coils, or in spirals. Some species construct the shell, or test, of tiny particles of silt; others secrete tests composed of calcium carbonate. The test is characteristically provided with holes through which extend "tentacles" of protoplasm for feeding. It is from these holes, or foramina, in the test that foraminifers take their name. Foraminifers range from the Cambrian to the present, but they are rare and poorly preserved in lower Paleozoic strata. They became more numerous and diverse by Carboniferous time. The increase is strikingly evident among the foraminifers known as **fusulinids**. The tests of fusulinids superficially resemble grains of wheat (Fig. 19.16). Although many different species appear similar in external appearance, they have complex internal structures that are quite distinctive at the species level.

Radiolarians (Fig. 19.17) are also single-celled planktic organisms that have been present on Earth at least since the Paleozoic began. Like the foraminifers, they have threadlike pseudopodia that project from an ornate, lattice-like skeleton of opaline silica or a proteinaceous substance.

Cup Animals: Archaeocyathids

Archaeocyatha means "ancient cups." It is an appropriate name for a somewhat enigmatic group of Cambrian organisms that constructed conical or vase-shaped skeletons out of calcium carbonate (Fig. 19.18). **Archaeocyathids** hold two paleontologic records. They are the earliest abundant reef-building animals on Earth, and they are members of a separate phylum that suffered extinction. Although quite abundant during the early Cambrian, they had died out by the end of that period. Gregarious in habit, they carpeted shallow seas near shorelines. The reefs formed by archaeocyathids can be studied today in North America, Siberia, Antarctica, and Australia.

Pore-Bearers

Among the many stationary animals to colonize the early Paleozoic sea floor were the sponges. Sponges are members of the phylum **Porifera**. They appear to have evolved from colonial flagellated unicellular creatures

Figure 19.18   An archaeocyathid.

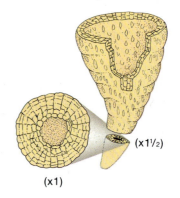

(x1½)

(x1)

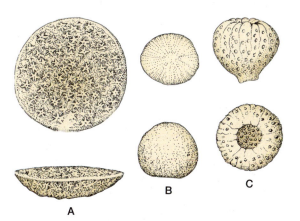

Figure 19.19 Early Paleozoic (Silurian) sponges. (A) *Astraeospongium* (it takes its name from the starlike spicules). (B) *Microspongia.* (C) *Astylospongia.*

and thus provide insight into how the transition from unicellular to multicellular animals may have occurred.

Sponges are a relatively conservative branch of invertebrates that have a long history. Cambrian representatives of all but one modern class of Porifera are known as fossils. Some fossil sponges, such as *Protospongia* from the Cambrian; *Astraeospongium, Microspongia,* and *Astylospongia* (Fig. 19.19) from the Silurian; and the siliceous sponge *Hydnoceras* from the Devonian are all well-known guide fossils (Fig. 19.20).

Sponges have always been predominantly marine creatures, although a few modern species live in fresh water. Spicules formed in the walls and cell layers of sponges are distinguishing characteristics of the phylum and provide both protection and support. The spicules may be composed of silica, calcium carbonate, or a proteinaceous material termed spongin. Naturally, the mineralized spicules are more commonly preserved and are frequently found by geologists examining rocks that have been disaggregated for study. Spicules are also important in the classification of sponges. For example, the Desmospongea consist entirely or in part of spongin, which may be reinforced with siliceous spicules. Hexactinellida develop siliceous spicules of distinctive shape, and Calcarea are characterized by spicules made of calcium carbonate. The Sclerospongea are a minor class of tropical marine sponges with skeletons composed of siliceous spicules, spongin fibers, and a basal layer of calcium carbonate.

Although sponges vary greatly in size and shape, their basic structure (Fig. 19.21) is that of a highly perforated vase modified by folds and canals. The body is attached to the sea floor at the base, and there is an excurrent opening, or osculum, at the top. The wall consists of two layers of cells. Facing the internal space is a layer of collar cells (choanocytes), and on the outside

Figure 19.20 The Devonian siliceous sponge *Hydnoceras.* The specimen is about 15 centimeters in height. *(J. Keith Rigby, Brigham Young University)*

is a protective wall of flat cells that resemble the bricks of a worn masonry pavement. Between these two layers one finds a gelatinous substance called mesenchyme. Here amoeboid cells go about the work of secreting the spicules. Sponges lack true organs. Water currents moving through the sponge are created by the beat of flagellae. These currents bring in suspended food particles, which are ingested by the collar cells. In a simple sponge, water enters through the pores, flows across sheets of choanocytes in the central cavity, and passes out through the osculum.

A group of early Paleozoic sclerosponges of particular interest because of their reef-building capabilities is the **stromatoporoids** (Fig. 19.22). These organisms constructed fibrous, calcareous skeletons of pillars and thin laminae that can nearly always be found in reef-associated carbonate rocks of the Silurian and Devonian. Apparently, stromatoporoids grew profusely and in close association with corals, brachiopods, and other invertebrate reef dwellers.

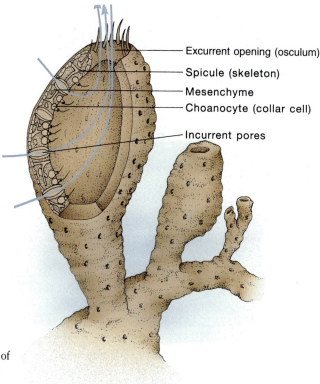

Excurrent opening (osculum)
Spicule (skeleton)
Mesenchyme
Choanocyte (collar cell)
Incurrent pores

Figure 19.21 Schematic diagram of a sponge having the simplest type of canal system. The path of water currents is indicated by arrows.

Corals and Other Cnidaria

Sea anemones, sea fans, jellyfish, the tiny *Hydra*, and reef-forming corals are all representatives of the phylum **Cnidaria**, which is known for the great diversity and beauty of its members (Fig. 19.23). The cnidarian body wall is composed of an outer layer of cells, the ectoderm; an inner layer, the endoderm; and a thin, noncellular

Figure 19.22 Polished slab of limestone containing the stromatoporoid named *Stromatoporella,* from Devonian rocks of Ohio.

23010   1

intermediate layer, the mesoglea. In the endoderm are found primitive sensory cells, gland cells that secrete digestive enzymes, flagellated cells, and nutritive cells to absorb nutrients. A distinctive feature of cnidarians is the presence of stinging cells, which, when activated, can inject a paralyzing poison. These specialized cells, called cnidocytes, are a unique characteristic of all cnidaria.

Body form in cnidaria may be either polyp or medusa (Fig. 19.24). The medusoid form is seen in the jellyfish, which resembles an umbrella in shape. Jellyfish have a concave undersurface that contains a centrally located mouth; for this reason, it is designated the oral surface. In jellyfish such as the living *Aurelia,* which is common along the eastern shore of the United States, the mouth is surrounded by four oral arms. All jellyfish have tentacles, and in *Aurelia* these are located around the margin of the umbrella.

In the polyp form, as exemplified by *Hydra*, there is a circle of tentacles above the saclike body. Corals and sea anemones are among the frequently encountered cnidaria with this polyp form. In stony corals (class Anthozoa), the polyp secretes a calcareous cup, in which it lives. The animal may live alone or may combine with other individuals to form large colonies. The cup, or

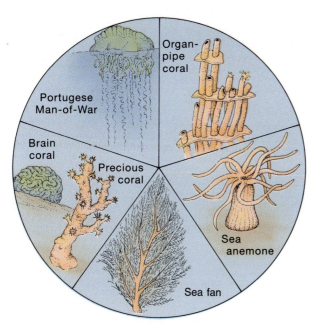

Figure 19.23   Diversity among the cnidarians.   *(From H. L. Levin,* Life Through Time *[Dubuque, IA: William C. Brown Co., 1975].)*

colonial forms, such as the honeycomb and chain corals. The rugose corals and tabulate corals became extinct at the end of the Paleozoic and were followed in the Mesozoic by anthozoans of the order Scleractinia. In all scleractinian corals, septa are inserted between the mesenteries in multiples of six.

The fossil record for Cnidaria begins with fossil jellyfish impressions discovered in upper Precambrian rocks. The phylum is still poorly represented in Cambrian rocks. In the succeeding Ordovician the record improves dramatically, for the lime-secreting anthozoans begin to expand and diversify. The first stony corals were the tabulates, recognized by their simple, often clustered or aligned tubes divided horizontally by transverse tabula. These were tabulate corals, such as the common honeycomb coral *Favosites* (Fig. 19.25). Tabulate corals were the principal reef formers during the Silurian. They declined after the Silurian, but their role in reef building was assumed by the rugose corals. Both solitary and compound colonies of rugosids (Fig. 19.26) are abundant in the fossil record of the Devonian and Carboniferous, but this group declined and became extinct during the late Permian.

Reefs built on the North American platform by Paleozoic corals were extensive and often exceeded 250 meters in total thickness. Most of these reefs lie beneath younger rocks and were discovered during oil explorations. Because of their porosity, these buried Paleozoic reefs became sites for the accumulation of petroleum.

Moss Animals: Bryozoa

Bryozoans are minute, bilaterally symmetric animals that grow in colonies that frequently appear twiglike. The individuals, called zooids, are housed in a capsule,

theca, may be divided by vertical plates called **septa**, which serve to separate layers of tissue and provide support. As the animal grows, it also secretes horizontal plates termed **tabulae**. Corals are identified according to the nature of their septa, tabulae, and other skeletal features. For example, after the development of an initial embryonic set of six protosepta, the Paleozoic **rugose corals**, or Rugosa, insert new septa at only four locations during growth. In other Paleozoic corals, septa are absent or poorly developed, so that tabulae are the most important features in classification. These are the Tabulata. **Tabulate corals** include many interesting

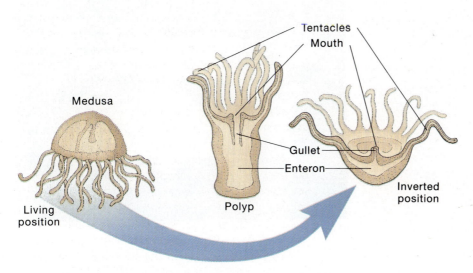

Figure 19.24   Comparison of polyp and medusa forms in cnidarians.

Figure 19.25 The common tabulate coral *Favosites*. An individual polyp resided in each of the tubes. Tiny pores visible in the walls of the tubes allowed for inter-connections of tissues between adjacent polyps.

Figure 19.26 Devonian rugose corals. (A) The solitary horn coral *Zaphrenthis* with clearly visible radiating septa in the hornlike theca. (B) The compound (colonial) rugose coral *Lithostrotionella*. (C) A polished slab of the compound coral *Hexagonaria*. Water-worn fragments of this coral are found along the shore of Lake Michigan at Petoskey, Michigan, and this accounts for its being called Petoskey stone. Although not a rock, Petoskey stone is the designated state rock of Michigan. (D) Reconstruction of compound and solitary rugose corals on the floor of a Devonian epeiric sea. *(Diorama photograph courtesy of the National Museum of Natural History)*

A

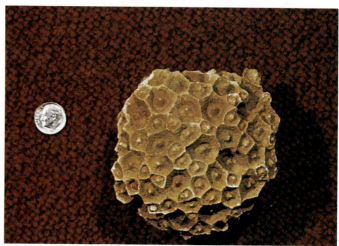

B

C

D

470

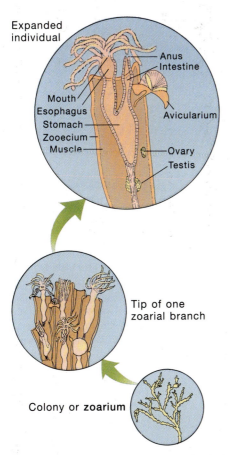

Figure 19.27   Relationship of bryozoan individuals to the zoarium. *(From L. H. Levin,* Life Through Time *[Dubuque, IA: William C. Brown Co., 1975].)*

or zooecium (Fig. 19.27), that appears as a pinpoint depression on the outside of the colony, or zoarium. The zooid has a complete, U-shaped digestive tract with mouth surrounded by a tentacled feeding organ called the **lophophore**.

Today, there are more than 4000 living species of bryozoans, but nearly four times that number are known as fossils. Their earliest unquestioned occurrence is from Lower Ordovician strata of the Baltic, but they did not become abundant until middle Ordovician and Silurian time. In rocks of these ages, their remains sometimes make up much of the bulk of entire formations. Like the corals, bryozoa contributed to the framework of reefs. One of the commonest of all the early Paleozoic bryozoans was *Fistulipora,* different species of which develop massive, incrusting, or arborescent colonies. The genus *Hallopora* (Fig. 19.28A) is one of many branching twig bryozoans. Star-shaped patterns on the surface of *Constellaria* provided the inspiration for its generic name. Late Paleozoic bryozoans include varieties that constructed lacy, delicate, fanlike colonies (Fig. 19.28B). Associated with these so-called **fenestel-lid** colonies were the bizarre Mississippian corkscrew bryozoans known by the generic name *Archimedes* (Figs. 19.28C and D).

Brachiopods

Brachiopods are probably the most abundant, diverse, and useful fossils readily found in Paleozoic rocks. They are characterized by a pair of enclosing valves, which together constitute the shell of the animal (Fig. 19.29A). They resemble clams in this regard, but in symmetry of the valves and soft-part anatomy, brachiopods are quite different from pelecypods. Brachiopod valves are almost always symmetric on either side of the midline, and the two valves differ from each other in size and shape. The valves of a clam are right and left, whereas those of a brachiopod are dorsal and ventral. The valves may be variously ornamented with radial ridges or grooves, spines, nodes, and growth lines. Calcium carbonate is the usual hard tissue of brachiopods, although the valves of some families are composed of mixtures of chitin and calcium phosphate. This is particularly true of the group called **inarticulates** (Figs. 19.30 and 19.29C).

In the **articulate** brachiopods, the valves are hinged along the posterior margin and are prevented from slipping sideways by teeth and sockets. The less common inarticulates lack this definite hinge, are held together by muscles, and characteristically have simple spoon-shaped or circular valves. Although brachiopod larvae swim about freely, the adults are frequently anchored or cemented to objects on the sea floor by a fleshy stalk (pedicle) or by spines. Some simply rest on the sea floor. One of the more conspicious soft organs of brachiopods is the lophophore, a structure consisting of two ciliated, coiled tentacles whose function is to circulate the water between the valves, distribute oxygen, and remove carbon dioxide (see Fig. 19.29). Water currents generated by cilia on the lophophore move food particles toward the mouth and short digestive tract.

Brachiopods still live in the seas today, although in far fewer numbers than during the Paleozoic. Cambrian brachiopods included many chitinous inarticulate types. These increased in diversity during the Ordovician but then declined. A few species of only three families remain today.

The articulates also first appeared in the Cambrian but became truly abundant during the succeeding Ordovician Period, when there was a great expansion of all sorts of shelled invertebrates. Across the floors of early Paleozoic epeiric seas, large and small aggregates of these filter-feeders could be found. Today, their skeletons compose much of the volume of thick formations

A

Figure 19.28   Paleozoic bryozoans. (A) The branching twig bryozoan *Hallopora* from the Ordovician of Kentucky. (B) *Fenestella,* a lacy bryozoan from Devonian limestones at the Falls of the Ohio River. (C) *Archimedes,* with part of the spirally encircling frond of lacy bryozoan colony attached and visible. (D) The central axis of *Archimedes.*

B

Figure 19.29   Living positions of articulate and inarticulate brachiopods. (A) The articulate living brachiopod *Magellania* attached to the sea floor by pedicle and with lophophore barely visible through the gape in the valves. (B) Interior of brachial valve showing ciliated lophophore. (C) The inarticulate brachiopod *Lingula,* which excavates a tube in bottom sediment and lives within it. The pedicle secretes a mucus at the end that glues the animals to the tube.

C

D

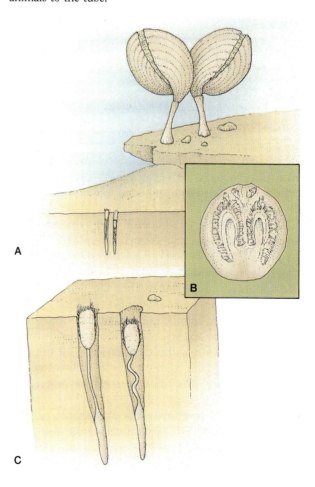

A

B

C

Figure 19.30 The inarticulate brachiopod *Obolus* from the Cambrian Conasauga Formation near Coosa, Georgia. The specimens are about 0.6 centimeter in diameter.

and provide the stratigrapher with essential markers for correlation. Among the articulate brachiopods that were particularly abundant during the early Paleozoic were the strophomenids, orthids, pentamerids, and rhynchonellids (Fig. 19.31). During the Devonian Period of the late Paleozoic, spiriferid brachiopods became particularly abundant. Spiriferids take their name from the calcified internal spirals that supported the lophophore (Fig. 19.32). The most characteristic brachiopods of the Carboniferous and Permian were the productids. Productids were distinctively spinose (Fig. 19.33) with large, inflated ventral valves. They were so numerous during the Carboniferous that the period might well be dubbed the age of productids.

Mollusks

A stroll along almost any seashore will provide evidence that mollusks are today's most familiar marine invertebrates. The shells of most members of this phylum have been readily preserved and provide enjoyment to collectors and conchologists today. Such well-known animals as snails, clams, chitons, tooth shells, and squid are included within the phylum **Mollusca**. Although most mollusks possess shells, some, such as the slugs and octopods, do not. The various classes of Mollusca differ considerably in external appearance, yet they have fundamental similarities in their internal structure. There is a muscular portion of the body called the foot that functions primarily in locomotion. In cephalopod forms, the foot is modified into tentacles. A fleshy fold called the mantle secretes the shell. In aquatic mollusks, respiration is accomplished by means of gills. Well-developed organs for digestion, sensation, and circulation attest to the advanced stage of evolution that mollusks have attained. The most abundant mollusks in the fossil record are the pelecypods, gastropods, and cephalopods.

The **Pelecypoda** (also termed the Bivalvia) are a class of mollusks that includes clams, oysters, and mussels. Pelecypods are generally characterized by layered gills, a muscular foot (Fig. 19.34), bilobed mantle, and absence of a definite "head." Although members of the class made their first appearance in the Cambrian, they did not become notably abundant until Pennsylvanian and Permian, when they populated the shallow seas of the time.

Gastropods first appear in Lower Cambrian strata. Earliest forms constructed small, depressed conical shells. During later Cambrian and Ordovician time, gastropods with the more familiar coiled conchs became commonplace (Fig. 19.35). It appears that coiling in a plane was not unusual intitially but was supplanted by spiral coiling. Gastropods of the Ordovician and Silurian had shapes similar to those of living species. One presumes that their soft parts were also similar, with distinct head, mouth, eyes, tentacles, and a vertically flattened foot to provide for gliding movement. By Pennsylvanian time, gastropods had become abundant and diverse. During this period, the oldest known air-breathing (pulmonate) gastropods appeared. The succeeding Permian Period ended with widespread extinctions among marine invertebrates, and many families of gastropods were decimated.

The **cephalopods** may very well be the most complex of all the mollusks. Today, this marine group is represented by the squid, cuttlefish, octopods, and the lovely chambered nautilus. The nautilus in particular provides us with important information about the soft anatomy and habits of a vast array of fossils known only by their preserved conchs. In the genus *Nautilus* (Fig. 19.36), one finds a bilaterally symmetric body, a prominent head with paired image-forming eyes, and tentacles developed on the forward portion of the foot. Water is forcefully ejected through the tubular "funnel" to provide swift, jet-propelled movement.

Text continued on page 477

Figure 19.31   Some early Paleozoic brachiopods: (A) *Rafinesquina* (A), *Strophomena*
(B), and *Leptaena* (C) are all strophomenids. (D) *Hebertella,* is an Ordovician orthid;
(E) *Lepidocyclus* is an Ordovician rhynconellid; (F) *Pentamerus,* is a Silurian
pentamerid ($\times\frac{1}{2}$).

A

B

C

Figure 19.32 Devonian spiriferid brachiopods. (A) *Mucrospirifer.* (B) *Platyrachella.* (C) A spiriferid brachiopod with shell broken to reveal the internal spiral supports for the lophophore.

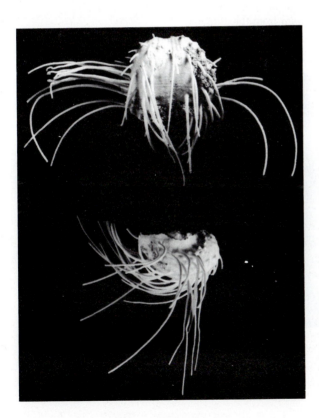

Figure 19.33 Ventral (upper photograph) and side (lower photograph) views of the Permian spinose productid brachiopod *Marginifera ornata* from the Salt Range of West Pakistan. Valves (not including spines) are about 2 centimeters wide. *(R. E. Grant, U.S. Geological Survey)*

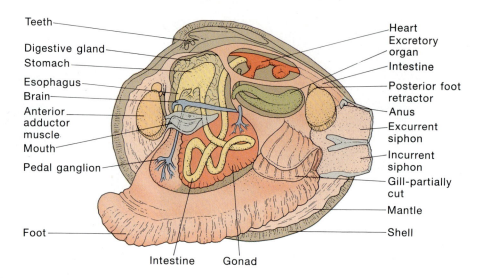

Teeth
Digestive gland
Stomach
Esophagus
Brain
Anterior adductor muscle
Mouth
Pedal ganglion
Foot
Intestine
Gonad

Heart
Excretory organ
Intestine
Posterior foot retractor
Anus
Excurrent siphon
Incurrent siphon
Gill-partially cut
Mantle
Shell

Figure 19.34 Internal anatomy of a clam.

A

B

Figure 19.35 Two representative early Paleozoic gastropods. (A) *Taeniospira* from the Upper Cambrian Eminence Formation of Missouri. (B) *Clathospira* from the Ordovician Platteville Formation of Minnesota.

A

B

Figure 19.36 *Nautilus,* a modern nautiloid cephalopod. (A) Conch of a nautiloid sawed in half to show large living chamber, septa, and septal necks through which the siphuncle passed. (B) Living animal photographed at a depth of about 300 meters.

(W. Bruce Saunders)

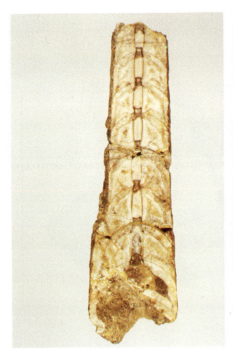

Figure 19.37 Variation in conch shape among early Paleozoic nautiloid cephalopods. Both of these specimens are from the Silurian of Bohemia. (A) A sawed and polished section of the straight conch of *Orthoceras potens* showing septa and siphuncle. (B) Sawed and polished section of *Barrandeoceras,* exhibiting a coiled form. Specimen (A) is 22.5 centimeters in length; (B) has a diameter of 18 centimeters.

A   B

At first glance, the conchs of cephalopods resemble snail shells, but the resemblance is only superficial. Although we have mentioned an exception to this rule, the gastropod shell generally coils in a spiral, whereas the cephalopod conch characteristically coils in a plane. More importantly, the planispirally coiled conch is divided into a series of chambers by transverse partitions, or septa. The bulk of the soft organs reside in the final chamber. Where the septa join the inner wall of the conch, suture lines are formed. These lines are enormously useful in the identification and classification of cephalopods. For example, cephalopods placed in the subclass **Nautiloidea** have straight or gently undulating sutures, whereas the **Ammonoidea** have more complex sutures.

The oldest fossils to be classified as cephalopods were small, conical conchs discovered in Lower and Middle Cambrian rocks of Europe. The class gradually increased in number and diversity and became ubiquitous inhabitants of Ordovician and Silurian seas. Indeed, by Silurian time, a great variety of conch forms—from straight to tightly coiled (Fig. 19.37)—had developed. Some of the elongate forms exceeded 4 meters in length. The first signs of a decline in nautiloid populations can be detected in Silurian strata. After Silurian time, the group continued to dwindle until today only a single genus, *Nautilus,* survives.

During the Devonian, the first ammonoid cephalopods appeared. These were the **goniatites**, characterized by angular and generally zig-zag sutures without

any additional crenulations (Fig. 19.38). The goniatites persisted throughout the late Paleozoic and gave rise to the ceratites and ammonites of the Mesozoic.

Arthropods

The **arthropod** phylum is enormous. It includes such living animals as lobsters, spiders, insects, and a host of

Figure 19.38 Goniatite ammonoid cephalopod exhibiting zig-zag sutures. Diameter is 4.6 centimeters.

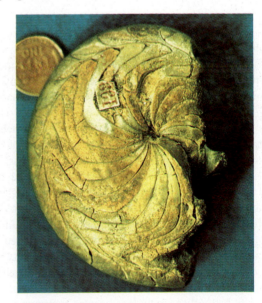

FOCUS ON   *The Eyes of Trilobites*

"The eyes of trilobites eternal be in stone, and seem to stare about in mild surprise at changes greater than they have yet known," wrote T. A. Conrad. Indeed, trilobites may have been the first animals to look out on the world, for they possessed the most ancient visual systems known. The eyes of trilobites include both simple and compound types. A simple eye appears as a single, tiny lens, resembling a small node. A few receptor cells probably lay beneath each simple lens. Simple eyes are rare among trilobites. Far more abundant are compound eyes composed of a large number of individual visual bodies, each with its own biconvex lens. Each lens is composed of a single crystal of calcite. One of the properties of a clear crystal of calcite is that in certain orientations, objects viewed through the crystal produce a double image. Trilobites, however, were not troubled by double vision because the calcite lenses were oriented with the principal optic axis (the light path along which the double image does not occur) normal to the surface of the eye.

In some compound trilobite eyes, termed holochroal, the many tiny lenses are covered by a continuous, thin, transparent cornea. Beneath this smooth cover there may be as many as 15,000 lenses. In other trilobites, such as *Phacops* and *Calliops* (see Fig. 19.41 and accompanying figure), there are discrete individual lenses each covered by its own separate cornea and separated from its neighbors by a cribwork of exoskeletal tissue. Such compound eyes are termed schizochroal.

Trilobite eyes can sometimes provide clues useful in determining the habits of certain species. Eyes along the

The large schizochroal right eye of this well-preserved, silicified cephalon of the trilobite *Calliops*. Light directed toward the back of the specimen is emerging through some of the lensal pits.

anterior margin of the cephalon, for example, may indicate active swimmers. Eyes located on the ventral side of the cephalon may indicate that the trilobite was a surface-dweller. In the majority of trilobites, the eyes are located about midway on the cephalon in a position appropriate for an animal that crawled on the sea bottom or occasionally swam above it. Trilobites that either lacked eyes or were secondarily blind appear to have been adapted for burrowing in the soft sediment on the ocean floor.

---

other animals that possess chitinous exterior skeletons, segmented bodies, paired and jointed appendages, and highly developed nervous system and sensory organs. Members of Arthropoda that have left a particularly significant fossil record are the **trilobites**, **ostracods**, and **eurypterids**.

Trilobites (Fig. 19.39) were swimming or crawling arthropods that take their name from division of the dorsal surface into three longitudinal segments, or lobes. There are, for example, a central axial lobe and two lateral (pleural) lobes (Fig. 19.40). There was also a transverse differentiation of the shield into an anterior cephalon, a segmented thorax, and a posterior pygidium. The skeleton was composed of chitin strengthened by calcium carbonate in parts not requiring flexibility. As in many other arthropods, growth was accomplished by molting. Although many trilobites were sightless, the majority had either single-lens eyes or compound eyes

composed of a large number of discrete visual bodies. The earliest trilobites had small pygidia and a large number of thoracic segments. These characteristics have led paleontologists to speculate that trilobites may have evolved from annelid worms sometime during the Precambrian.

If one considers the entire fossil record for only the Cambrian, the trilobites are clearly the most abundant and diverse. More than 600 genera are known from the Cambrian. They first appear in rocks above those of the Tommotian Stage of the lowermost Cambrian.

The earliest trilobites apparently were bottom-dwelling, crawling scavengers and mud processors. Some preferred limy bottoms of cratonic and shelf areas, and others, such as *Paradoxides* of the middle Cambrian, inhabited the muddy or silty floors of deep-water tracts. As a result of such environmental preferences, it has been possible to delineate trilobite faunal provinces in

Figure 19.39 A gallery of trilobites. (A) The middle Cambrian guide fossil *Ogygopsis klotzi* from Mount Stephens, British Columbia, near the site of the famous Burgess Shale quarry (length 6.0 centimeters). (B) *Ellipsocephalus* from the middle Cambrian of Bohemia (average length 2.8 centimeters). (C) *Isotelus* from Ordovician limestones in New York state (length 3.8 centimeters). (D) *Dalmanites* from Silurian beds in Indiana (length 3.2 centimeters).

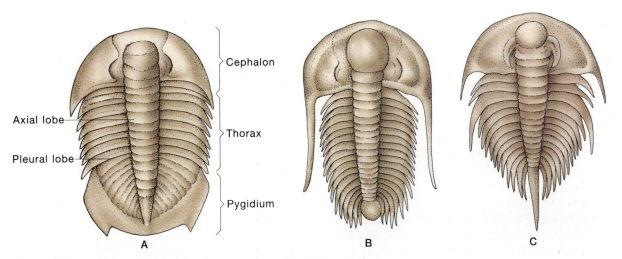

Figure 19.40 Three well-known Cambrian trilobites. (A) *Dikelocephalus minnesotensis* (Upper Cambrian), (B) *Paradoxides harlani* (Middle Cambrian), and (C) *Olenellus thompsoni* (Lower Cambrian).

479

North America and Europe that suggest that these two continents were once close together. (For example, the Lower Cambrian *Olenellus* assemblage is found in exposures in Pennsylvania, New York, Vermont, Newfoundland, Greenland, and Scotland.)

The optimum time for trilobites was reached in the late Cambrian. After that they began to decline, perhaps in response to predation from cephalopods and fishes; however, they remained fairly abundant throughout the Ordovician and Silurian. Extinctions during the middle and late Devonian decimated the majority of trilobite groups, but the large-eyed group known as proparians (Fig. 19.41) persisted and now serve as guide fossils. The largest trilobites ever found lived during the Devonian. One giant reached a length of over 70 centimeters. After the temporary increase in diversity during the Devonian, trilobites continued to wane until they became extinct near the close of the Paleozoic. Nevertheless, they were not at all biologic failures, for they had been important animals in the oceans for over 300 million years.

Arthropod companions to the trilobites in the Paleozoic seas were the small, bean-shaped ostracods (Fig. 19.42). At first glance, ostracods appear so different from trilobites as to cause one to question their classification within the same phylum. The ostracods have a bivalved shell vaguely suggestive of some sort of tiny clam. However, this bivalved carapace encloses a segmented body from which extend seven pairs of jointed appendages. Adult animals are about 0.5 to 4 millimeters in length. The valves are composed of both chitin and calcium carbonate and are hinged along the dorsal margin.

Ostracods first appeared early in the Ordovician, and some limestones of this age are almost completely composed of their discarded carapaces. *Leperditia* and *Euprimitia* are common Ordovician genera. Ostracods continue in relative abundance to the present day. They occur in both marine and fresh-water sediments. Be-

Figure 19.41   Side view of the trilobite *Phacops rana* showing the discrete visual bodies of the large compound eyes. Length is 3.2 centimeters.

Figure 19.42   Ostracods *(Leperditia fabulites)* in the Ordovician Plattin Limestone, Jefferson County, Missouri. Ostracods are the smooth, bean-shaped fossils. These average about 1 centimeter in length and are thus very large ostracods. Two strophomenid brachiopods are also present on the surface of the limestone.

cause of their small size, they are brought to the surface in wells drilled for oil, and along with other microfossils, such as foraminifers and radiolaria, are used by exploration geologists in correlating the strata of oil fields.

Eurypterids (Fig. 19.43) are a group of arthropods that, because of their rarity, are less useful in stratigraphic studies than are either trilobites or ostracods. Nevertheless, they are among the most impressive of early Paleozoic marine invertebrates. Although many were of modest size, some were nearly 3 meters long and, had they survived, would have been suitable subjects for a Hollywood monster film. On their scorpion-like bodies were five pairs of appendages and a fearful-looking pair of pincers. Some were also equipped with a tail spine. Eurypterids ranged across portions of the sea floor and brackish estuaries from Ordovician until Permian time, but they were especially abundant during the Silurian and Devonian.

Spiny-Skinned Animals

Just as modern seas abound with starfish, sea urchins, and sea lilies, so were the oceans of the early Paleozoic populated with members of the phylum Echinodermata

Figure 19.43   The fossil eurypterid *Eurypterus lacustris* from the Bertie Waterlime of Silurian age, Erie County, Pennsylvania. This well-preserved specimen is about 28 centimeters long.   *(Ward's Natural Science Establishment)*

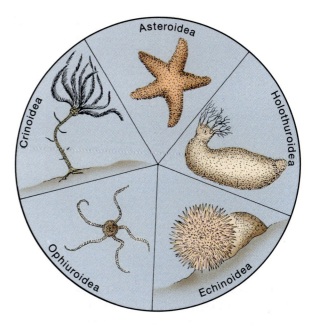

Figure 19.44   Living echinoderms.   *(From H. L. Levin,* Life Through Time *[Dubuque, IA: William C. Brown, 1975].)*

(Fig. 19.44). **Echinoderms** are animals with mostly five-way symmetry that masks an underlying primitive bilateral symmetry. They are well named "the spiny skinned," for spines are indeed present in many species. A unique characteristic of the phylum is the presence of a system of tubes—the water vascular system—which functions in respiration and locomotion (Fig. 19.45). Members of the phylum are exclusively marine, typically bottom-dwelling, and either attached to the sea floor or able to move about slowly.

Of considerable interest is the evolutionary relationship between echinoderms and vertebrates. There are so many allied embryologic parallelisms among primitive chordates and echinoderms that some zoologists speculate that both may have arisen from similar ancestral forms. For example, the larvae of echinoderms closely resemble those of the living protochordates called acorn worms. Also, in embryologic development, the mesodermal layer of cells, as well as certain other elements of the body, arises in the same way. Biochemistry has provided additional evidence for a relationship between echinoderms and chordates by revealing chemical similarities associated with muscle activity and oxygen-carrying pigments in the blood.

Among the many classes of the phylum Echinodermata, the Asteroidea (starfish), Ophiuroidea (brittle stars), Echinoidea (sea urchins), Edrioasteroidea, Crinoidea (sea lilies), Blastoidea, and Cystoidea are the most

abundant and useful in geologic studies. The mostly attached and sessile crinoids, blastoids, and cystoids were particularly characteristic of the Paleozoic Era, whereas the more mobile asteroids and echinoids were more frequent in later eras.

Echinoderms appear to have evolved late in the Proterozoic, although fossil remains of Proterozoic forms are few and often enigmatic. The Ediacaran fauna, for example, includes a globular fossil named *Arkarua* that has five rays on its surface and may be related to a group of echinoderms called edrioasteroids.

Figure 19.45   Partially dissected starfish showing elements of the water vascular system and other organs.

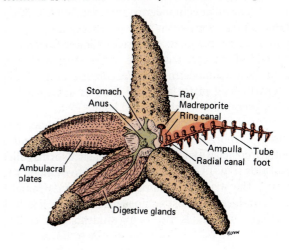

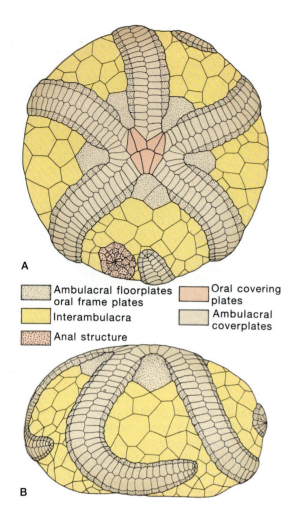

A

Ambulacral floorplates oral frame plates

Interambulacra

Anal structure

Oral covering plates

Ambulacral coverplates

B

Figure 19.46   *Edrioaster bigsbyi*, a middle Ordovician edrioasteroid. Specimen is 45 millimeters in diameter. (A) Oral surface. (B) Lateral view of the globoid fossil. *(From B. M. Bell,* J. Paleo., *51(3) [1977], 620.)*

Figure 19.47   The spiraled, spindle-shaped early Cambrian echinoderm *Helicoplacus*. The specimen is about 2.6 centimeters long.   *(After J. W. Durham and K. E. Caster,* Science, *140 [1963], 820–822.)*

Edrioasteroids (Fig. 19.46) are considered by many paleontologists to be ancestral to starfish and sea urchins.

By early and middle Cambrian time, several classes of rather peculiar echinoderms appeared, but this initial radiation was apparently not very successful. None of these became abundant, and three became extinct before the end of the Cambrian. One of the most bizarre of these early echinoderms was *Helicoplacus* (Fig. 19.47), a form having plates and food grooves arranged in a spiral around its spindle-shaped body.

The stemmed or stalked echinoderms first occur in middle Cambrian strata but do not become abundant until Ordovician and Silurian time. Stalked forms called **cystoids** (Fig. 19.48) are the most primitive among this group. The striking pentamerous (five-way) symmetry that is evident in most echinoderms is often less well developed in cystoids. Beginning students of paleontol-

ogy often recognize cystoids by the characteristic pores that occur in rhomboid patterns on the plates of the calyx and compose part of the water vascular system. Although cystoids range from Cambrian to late Devonian, they are chiefly found in Ordovician and Silurian rocks.

Unlike some of the cystoids, stalked echinoderms known as **blastoids** (Fig. 19.49) have a beautifully symmetric arrangement of plates. The five radial areas (ambulacral areas) are prominent, bear slender branches or brachioles along their margins, and have a well-developed and unique water transport system. The blastoids first appeared in Silurian time, expanded in the Mississippian, and declined to extinction in the Permian.

Crinoids, like most cystoids and blastoids, are composed of three main parts: the calyx (which contains the vital organs), the arms, and the stem with its rootlike holdfast (Fig. 19.50). The arms bear ciliated food grooves and, like the brachioles of blastoids, serve to move food particles toward the mouth. Crinoids are found in rocks that range in age from Ordovician to Holocene. In some areas, Ordovician, Silurian, and Carboniferous rocks contain such great quantities of disaggregated plates of crinoids that they are named crinoidal limestones. So abundant are crinoid remains in Mississippian rocks that the period has been dubbed the age of crinoids (Fig. 19.51).

Although one group of crinoids survived the hard times at the end of the Paleozoic and gave rise to those

Figure 19.48   A well-preserved specimen of the Silurian cystoid *Caryocrinites ornatus* from the Lockport Shale of New York.   *(From J. Sprinkle, J. Paleo., 49(6) [1975], 1062–1073.)*

Figure 19.49   Calyx of the blastoid *Pentremites symmetricus* from the Upper Mississippian near Floraville, Illinois. Height of calyx is 2.6 centimeters.

Figure 19.50   Crinoid in living position on sea floor.

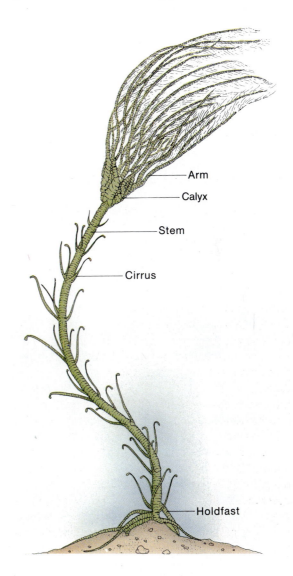

still living today, most died out before the end of the Permian. The blastoids also died out before the end of the Paleozoic. Free-living echinoderms such as echinoids (Fig. 19.52) and starfish were locally abundant during the late Paleozoic but not as numerous or diverse as they would become in post-Paleozoic time.

Graptolites

Just as the early Paleozoic sea floors bustled with the activities of a myriad of invertebrates, so also did the overlying waters. However, many of these planktonic creatures lacked skeletons and so are poorly known. One group, the **graptolites** (Fig. 19.53), had preservable chitinous skeletons that housed colonies of tiny individual animals.

The graptolites made their appearance at the very end of the Cambrian but were rare until the Ordovician, when they became quite abundant and diverse.

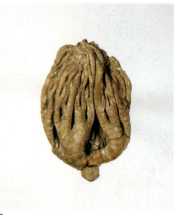
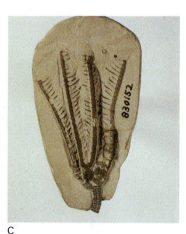

A   B   C

Figure 19.51   (A) Diorama depicting crinoids living on the floor of a Mississippian epeiric sea. (B) Calyx and arms of the Mississippian crinoid *Taxocrinus* (height 5.1 centimeters). (C) *Scytalocrinus* (height 4.5 centimeters). Both fossil crinoids are from the Keokuk Formation of Mississippian age, near Crawfordsville, Indiana.   *(Diorama photograph courtesy of the National Museum of Natural History)*

Figure 19.52   The large Mississippian echinoid *Melonechinus* from the St. Louis Limestone, St. Louis, Missouri. Average diameter of specimens in 12 centimeters.   *(Photograph by John Simon)*

Unfortunately for paleontologists, they did not survive beyond the Mississippian, and the nature and functions of their soft organs can only be surmised. The structure of the exoskeleton is relatively simple. The tiny cups, or thecae, are arranged along a branch, or stipe. The stipes may be solitary or formed into a system of two or more branches. The entire colony is referred to as the rhabdosome and is supported or attached at one end by a thin filament—the nema. Where the lower end of the nema reaches the base of a stipe, there is a conical sicula, which may have served as the theca for the first individual. From the sicula, subsequent members of the colony add their thecae by budding. Most graptolites were planktic organisms. Some attached themselves to floating debris by the threadlike nema.

Graptolites occur as streaks of flattened, carbonaceous matter in fine-grained rocks. Recently, however, uncompressed graptolites have been studied that reveal an unexpected relationship to primitive chordates. For example, both graptolites and the living protochordates known as pterobranchs secreted tiny enclosed tubes, and in both the unique structure of the enclosing wall of the sheath is so similar that a relationship is virtually certain.

During the early Paleozoic, graptolites were at times so abundant that they have been visualized as forming sargasso-like floating masses. They were apparently carried about by ocean currents and thus achieved a worldwide distribution, which has enhanced their use as index

Figure 19.53   Part of a stipe of the Ordovician graptolite *Orthograptus quadrimucronatus* (×15). The specimen has been bleached to reveal internal features. The cuplike thecae are clearly visible on either side. One can also see the inverted cone of the sicula. The nema would rise from the pointed end of the sicula.

(From S. R. Herr, J. Paleo., 45(4) [1971], 628–632.)

fossils. Their use, however, is limited in that they are seldom found preserved except in fine-grained, usually carbonaceous, shaly sediments. Perhaps these occurrences came about as masses of graptolite colonies floated into areas of toxic conditions, died, and settled to the sea floor. It is also possible that such accumulations resulted from fortuitously favorable preservation.

Conodonts

One group of well-known but enigmatic microscopic fossils may have affinities to a prechordate animal. However, their true biologic relationships are still being hotly debated. These problematic fossils are **conodonts**. Some are platelike or cone-shaped (Fig. 19.54), and all are composed of calcium phosphate. Most conodonts are about 1 millimeter in length. At their base one finds cavities or attachment scars where they were presumably fastened to a supportive part of a parent organism. Conodont elements occur mostly as single, disarticulated fossils, but occasionally paired groupings are discovered that are suggestive of a feeding or filtering mechanism.

Although their true natures are unknown, conodonts are still excellent guide fossils. They occur in a variety of marine sedimentary rocks and range from latest Proterozoic to Triassic age. There are now about 140 conodont-based stratigraphic zones. Their uncertain origins do not detract from their value in stratigraphy. Nevertheless, paleontologists are keenly interested in discovering their true affinities. Fortunately, the search has been narrowed by two or three discoveries of conodonts found in close association with soft

Figure 19.54   Scanning electron micrographs of Ordovician conodonts. These simple, recurved, cone-shaped forms have a deeply excavated base resembling a pulp cavity. Specimens are about 1 millimeter long.   *(D. Hearns)*

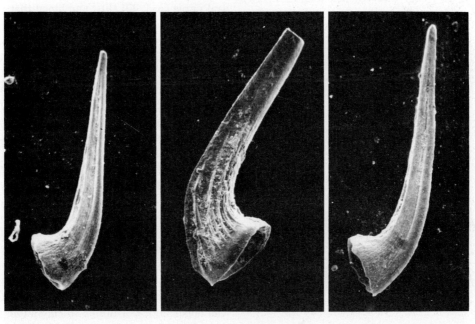

remains of the animals that apparently contained them. Mississippian rocks near Edinburgh, Scotland, provided one of the most recent of these discoveries. The conodont-bearing fossil was an elongate (about 40 millimeters long), segmented animal with ray-supported fins. Conodonts were found clustered at the anterior end just behind a depression resembling the animal's mouth. The conodonts were arranged symmetrically in an apparatus-like grouping. The investigators are convinced they have found a true conodont animal, although once again its precise affinities are obscure.

19.3   Continental Invertebrates

Because of the greater hazards of postmortem destruction of continental (terrestrial) invertebrates, their fossil record is not as complete as that for marine invertebrates. Nevertheless, as plants invaded the continents, animals were able to follow. Arthropods, with their sturdy legs and protective exoskeletons, were probably the marine animals best preadapted for coming ashore. Early invaders may have lived within the wet, nutritious plant debris washed up along the beaches of early Paleozoic seas. Their fossil record begins with possible millipede trace fossils in Ordovician rocks. During the Devonian, the record for arthropods improves, with discoveries of wingless insects, including mites, springtails, spiders, and centipedes. Carboniferous strata contain a slightly better record of insects, including giant dragonflies with wingspans of over 70 centimeters (2 feet). Cockroaches (Fig. 19.55) that reached lengths of 10

Figure 19.55   Reconstruction of a primitive Pennsylvanian cockroach. Length is 10 centimeters.

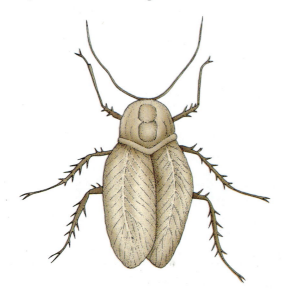

centimeters crept about among the rotting vegetation. Eurypterids, although not common, persisted on the continents in lakes and streams throughout the late Paleozoic. Several species of advanced spiders have been recovered from Devonian and Carboniferous strata in North America, Germany, and Great Britain. In Nova Scotia, an interesting collection of land snails has been collected from hollows within the preserved stumps of trees. Scorpions and centipedes have been found in abundance in a concretionary Pennsylvanian siltstone exposed along Mazon Creek in northern Illinois.

Of all the land invertebrates, it is apparent that the arthropods and gastropods have been the most persistently successful. Many paleontologists believe this may be because of attributes already evolved by their aquatic ancestors. As protection against desiccation, arthropods had evolved relatively impervious exoskeletons. Snails derived similar benefits from their shells. Both groups included very active animals with sufficient mobility to seek out food aggressively.

19.4   Vertebrates of the Paleozoic

The Rise of Fishes

A momentous biologic event occurred in early Paleozoic time. It was the birth of the chordate line. In our earlier discussion of the Burgess Shale fauna, we briefly described chordates as animals that have (at least at some stage in their life history) a stiff, elongate supporting structure. In addition, all chordates have a dorsal, central nerve cord, gill slits, and blood that circulates forward in a main ventral vessel and backward in the dorsal. The supportive structure in primitive chordates is called a notochord and has been studied by generations of biology students in such animals as the lancelet *Branchiostoma* (Fig. 19.56). In taxonomic hierarchy, vertebrates are simply those chordates in which the notochord is supplemented or replaced by a series of cartilaginous or bony vertebrae.

As we have mentioned previously, the ancestors of the vertebrate lineage may lie somewhere among the echinoderms. The theoretic evolutionary progression that was to lead to vertebrates may have begun with sedentary, filter-feeding animals that had exposed cilia located along their arms, somewhat in the fashion of crinoids. From such a beginning there may have evolved filter-feeders with cilia brought inside the body in the form of gills. In a subsequent stage, the organisms may have become free-swimming, gilled animals, in superficial appearance not unlike small, simple fishes or frog tadpoles. *Pikaia,* from the Middle Cambrian Burgess

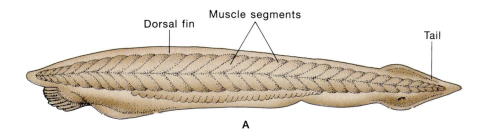

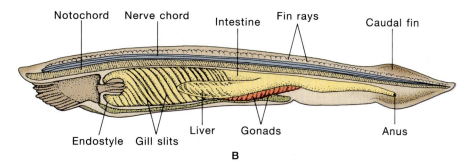

Figure 19.56
Branchiostoma, a member
of the subphylum
Cephalochordata. (A) External
view; (B) longitudinal section
(length 3 to 5 centimeters).

Figure 19.57 The basic evolutionary radiation of fishes
that began during the early Paleozoic and produced the
sharks and bony fishes during the Devonian.
(From E. H. Colbert, Evolution of the Vertebrates *[New York: John Wiley,
1969]. With permission of the author.)*

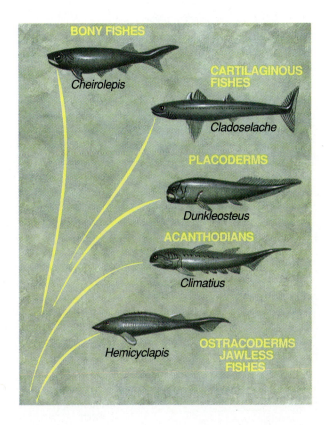

Shale, appears to have been such an animal. *Pikaia* (see
Fig. 19.15) or animals like it may be at the base of the
evolutionary sequence leading to vertebrates. Presumably, we may call those earliest vertebrates fishes.

The oldest fossil remains of fishes consist of small
pieces of external plates found in Upper Cambrian deposits of Wyoming. Their distinctive surface patterns and
internal structure clearly indicate that these plates were
part of the body covering of armored jawless fishes.
Somewhat younger scales and plates of jawless fishes
are found in Ordovician rocks of Colorado, Spitzbergen,
and Greenland.

The vertebrates that we loosely call fishes are actually divided into at least five distinct taxonomic classes
(Fig. 19.57). There are the jawless fishes, or Agnatha;
two groups of archaic jawed fishes, the Acanthodii and
Placodermi; the cartilaginous fishes, or Chondrichthyes;
and the familiar Osteichthyes with their highly developed bony skeletons. It is the first three of these categories—namely, the Agnatha, Acanthodii, and Placodermi—that frequented fresh-water and saltwater
bodies of the early Paleozoic.

The agnaths still exist today as the inelegant hagfish
and lamprey. However, these specialized survivors are
quite unlike the jawless fishes of the Ordovician and
Silurian. Early Paleozoic agnaths are collectively termed
ostracoderms. The name means "shell skin" and refers
to the bony exterior that was a distinctive trait of many
of these fishes. Ostracoderms comprised a rather diverse
group that included the unarmored forms *Thelodus*
and *Jamoytius,* as well as such armored creatures as

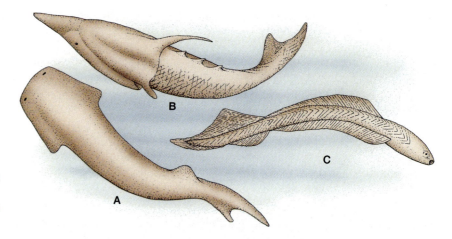

Figure 19.58 The early Paleozoic ostracoderms, (A) *Thelodus,* (B) *Pteraspis,* and (C) *Jamoytius,* drawn to the same scale.

Pteraspis (Fig. 19.58). The purpose of the bony armor that is the hallmark of many of the ostracoderms is still being debated. A widely held view is that the armor provided protection, although the fossil record shows few creatures of the same age and habitat that could have preyed on the ostracoderms. Nevertheless, it does not seem unreasonable that some of the smaller ostracoderms might have provided an occasional meal for a large cephalopod or eurypterid. Another theory stipulates that the dermal armor was not primarily for protection but rather was a device for storing seasonally available phosphorus. According to this idea, phosphates could be accumulated as calcium phosphate in the dermal layers during times of greater availability and then used during periods when supply was deficient. This cache of phosphorus may have enabled early vertebrates to maintain a suitable level of muscular activity.

Hemicyclaspis (Fig. 19.59) is one of the most widely known of the ostracoderms. It is recognized by its large, semicircular head shield, on the top of which can be found four openings: two for the upward-looking eyes, a small pineal opening, and a single nostril. The pineal opening may have housed a light-sensitive third eye. The mouth was located ventrally in a position suitable for taking in food from the surface layer of soft sediment. Along the margin of the head shield were depressed areas that may have had a sensory function. Internally,

Figure 19.59 The ostracoderm *Hemicyclapsis.* The creature was about 15 centimeters long. *(After A. E. Stensio, Br. Mus., 1932, p. 3, fig. 15.)*

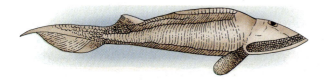

Figure 19.60 Origin of jaws. (A) Gill arches as they might exist in a primitive jawless fish. (B) Early jawed fish with a complete gill slit behind jaws. (C) A jawed fish (like a shark) with the first gill slit reduced to a spiracle. *(Modified from E. L. Cockrum and W. J. McCauley, Zoology [Philadelphia: W. B. Saunders, 1965].)*

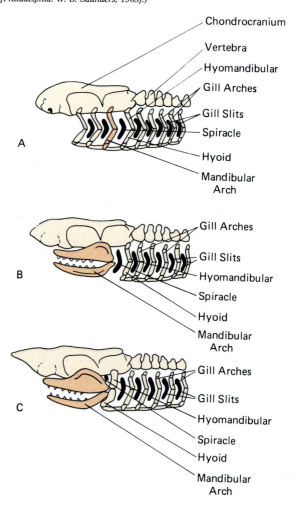

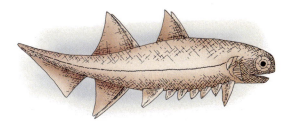

Figure 19.61 The early Devonian acanthodian fish *Climatius*. *(After A. S. Romer,* Vertebrate Paleontology *[Chicago: University of Chicago Press, 1945].)*

the head contained a regular segmental arrangement of nerves, blood vessels, and gill pouches. These and other features of the soft internal anatomy have been discerned by careful dissection of exceptionally well-preserved specimens.

The ostracoderms continued into the Devonian but did not survive beyond that period. For the most part they were small, rather sluggish creatures restricted to mud-straining or filter-feeding modes of life. They were to be gradually replaced by fishes that had developed bone-supported, movable jaws. The evolution of the jaw was no small accomplishment, for it enormously expanded the adaptive range of the vertebrates. Fishes with jaws were able to bite and grasp. These new abilities led to more varied and active ways of life and to new sources of food not available to the agnaths.

The evolutionary development of the vertebrate jaw involved a remarkable transformation in which an older structure was modified to perform a role that was entirely different from its original function. Those older structures were supports called gill arches that were located between the gill slits (Fig. 19.60). As the mouth extended posteriorly, the first two of these arches were sequentially eliminated. The third set of gill arches was gradually modified into jaws. Classic anatomic studies have clearly shown the similarities in jaw architecture and neurology to elements of the gill arches and the nerves that serve these supports. In the earliest fishes, the upper jaw was attached to the skull by ligaments only. A full gill slit occurred immediately behind the jaw. In more advanced fishes, the upper jaw was anchored to the skull by the upper part of the next gill arch, the hyomandibular. The pair of gill slits that had once existed behind the jaw was squeezed into a smaller space and ultimately developed into the spiracle openings found in members of the shark family today. The spiracle was a structure destined to become part of the ear apparatus in higher vertebrates.

The oldest fossil remains of jawed fishes are found in nonmarine rocks of the late Silurian. These fishes, called **acanthodians** (Fig. 19.61), became most numerous during the Devonian and then declined to extinction in the Permian. Acanthodians were archaic jawed fishes and were quite distinct from the great orders of modern fishes. Another group of archaic fishes includes the **placoderms**, or "plate-skinned" fishes. They too arose in the late Silurian, expanded rapidly during the Devonian, and then in the latter part of that period began to decline and be replaced by the ascending sharks and bony fishes.

There was considerable variety among the placoderms. The most formidable of these plate-skinned fishes was a carnivorous group called arthrodires. *Dunkleosteus* (Fig. 19.62) was a Devonian arthrodire whose

Figure 19.62 The gigantic armored skull and thoracic shield of the formidable late Devonian placoderm fish known as *Dunkleosteus*. *Dunkleosteus* was over 10 meters (about 30 feet) long. The skull shown here is about 1 meter tall. It is equipped with large bony cutting plates that functioned as teeth. Each eye socket was protected by a ring of four plates, and a special joint at the rear of the skull permitted the head to be raised and thereby provided for an extra large bite. *Dunkleosteus* ruled the seas 350 million years ago.
(National Museum of Natural History; photograph by Chip Clark)

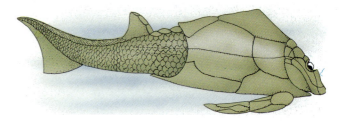

Figure 19.63 The Devonian antiarch fish *Pterichthyodes.*

(From A. S. Romer, Vertebrate Paleontology *[Chicago: University of Chicago Press, 1945], p. 54, fig. 38.)*

length exceeded 9 meters and whose huge jaws could be opened exceptionally wide to engulf even the largest of available prey. Other placoderms, called antiarchs (Fig. 19.63), had the heavily armored form and mud-grubbing habits of their predecessors, the ostracoderms.

Among the late Silurian acanthodians and placoderms were the ancestors of the bony and cartilaginous fishes that were to dominate the marine realm in succeeding periods. The ascendancy of these more modern kinds of fishes began with a veritable evolutionary explosion during the Devonian. Two important categories of fishes, the cartilaginous chondrichthyians and the bony osteichthyians, made their debut in the Devonian. Today, the cartilaginous fishes are represented by sharks,

rays, and skates. Among the better known of the late Paleozoic sharks were species of *Cladoselache* (Fig. 19.64A). Remains of this shark are frequently encountered in the Devonian shales that crop out on the south shore of Lake Erie. During the late Carboniferous, a group represented by *Xenacanthus* (Fig. 19.64B) managed to penetrate the fresh-water environment. A third group of Paleozoic cartilaginous fishes were the bradyodonts. These had flattened bodies like modern rays and blunt, rounded teeth for crushing shellfish. Apparently, modern sharks arose from cladoselachian ancestors but retained their archaic traits until the Jurassic.

Because of the role of bony fishes in the evolution of tetrapods (four-legged animals) and because they are

A

Figure 19.64 Models of (A) the Devonian marine shark *Cladoselache,* and (B) the Pennsylvanian fresh-water shark *Xenacanthus.*

B

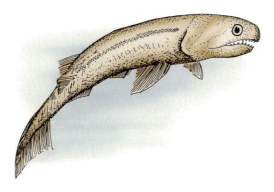

Figure 19.65 *Cheirolepis,* the ancestral bony fish that lived during the Devonian Period.

Figure 19.66 *Dipterus,* a Devonian lungfish.

the most numerous, varied, and successful of all aquatic vertebrates, their evolution is of particular importance. Bony fishes may be divided into two categories, namely the familiar ray-fin, or Actinopterygii, and the lobe-fins, or Sarcopterygii.

As implied by their name, ray-fin fishes lack a muscular base to their paired fins, which are thin structures supported by radiating bony rays. Unlike the Sarcopterygii, they do not possess paired nasal passages that open into the throat. The ray-fins began their evolution in Devonian lakes and streams and quickly expanded into the marine realm. They became the dominant fishes of the modern world. The more primitive Devonian bony fishes are well represented by the genus *Cheirolepsis* (Fig. 19.65). From such fishes as these evolved the more advanced bony fishes during the Mesozoic and Cenozoic.

The second category of bony fishes, the Sarcopterygii, are characterized by sturdy, fleshy lobe-fins and a pair of openings in the roof of the mouth that led to clearly visible external nostrils. Such fish were able to rise to the surface and take in air that was passed on to functional lungs. Lungs and fins do not seem to occur together in modern fishes, but in late Paleozoic fishes the combination was not uncommon. Studies of living examples of Sarcopterygii indicate that lungs probably began their evolution as saclike bodies developed on the ventral side of the esophagus and then became enlarged and improved for the extraction of oxygen. (In modern fishes, the lung has been converted to a swim bladder, which aids in hydrostatic balance.)

Two major groups of lungfishes lived during the Devonian. They are designated the **Dipnoi** and **Crossopterygii**. The Dipnoi, represented in the Devonian

by *Dipterus* (Fig. 19.66), were not on the evolutionary track that was to lead to tetrapods. They are, nevertheless, an interesting group, which includes the living fresh-water lungfish of Australia, Africa, and South America. Their restricted presence south of the equator suggests that Gondwanaland was the probable center of dispersal for dipnoans. Dipnoi means "double breather." The name was suggested by the observation that living species are able to breathe by means of lungs during dry seasons. At such difficult times, they burrow into the mud before the water is gone. When the lake or stream is dry, they survive by using their accessory lungs, and when the waters return they switch to gill respiration.

Because of the arrangement of bones in their muscular fins (Fig. 19.67), the pattern of skull elements

Figure 19.67 Comparison of the limb bones of a crossopterygian fish (upper right) and an early amphibian. *(From H. L. Levin,* Life Through Time *[Dubuque, IA: William C. Brown, 1975].)*

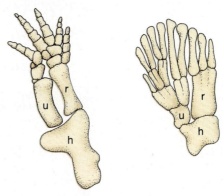

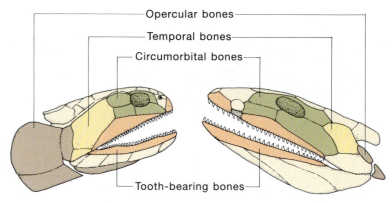

Figure 19.68 Comparison of skulls and lower jaws of a crossopterygian (left) and the Devonian amphibian *Ichthyostega*. *(From H. L. Levin,* Life Through Time *[Dubuque, IA: William C. Brown, 1975].)*

(Fig. 19.68), and the structure of their teeth (Fig. 19.69), fossil Crossopterygii are considered the ancestors of the amphibians. A rather advanced fish that exemplifies Devonian crossopterygians is *Eusthenopteron* (Fig. 19.70). In this genus, the paired fins were short and muscular. Internally, a single basal limb bone, the humerus (or femur for the pelvic structure), articulated with the girdle bones and was followed by two bones, the ulna and radius, for the pectoral fins and the tibia and fibula for the posterior fins.

Of course, the robust skeleton and sturdy limbs of the crossopterygians did not evolve because fishes had miraculously taken on a desire for life on land. These were adaptations for moving to and remaining in water during periods of drought. The air-gulping fishes found land a hostile environment and occasionally dragged themselves onto the land only as a means of reaching another pond or fresher body of water.

During the Devonian, two distinct branches of crossopterygians had evolved. One of these, the rhipidistians, led ultimately to amphibians, and the other led to salt-water fishes called coelacanths. Coelacanths were thought to have undergone extinction during the late Cretaceous, but their survival down to the present day has been documented by several catches of the coelacanth *Latimeria* (Fig. 19.71) near Madagascar.

Amphibians

It required tens of millions of generations to convert the crossopterygian fishes into animals that could live comfortably on solid ground. Even so, the conversion was not complete, for amphibians continued to return to water to lay their fishlike, naked eggs. From these eggs came fishlike larvae, which, like fish, used gills for respiration.

A number of changes accompanied the shift to land dwelling. A three-chambered heart developed to route

Figure 19.70 The Devonian crossopterygian lungfish *Eusthenopteron* has sturdy fins whose structure foreshadowed that of four-footed land animals that were to be their descendants. *(St. Louis Science Center diorama)*

Figure 19.69 Cross section of a crossopterygian tooth clearly exhibits the distinctive pattern of infolded enamel. This is also a characteristic of the teeth of the amphibian descendants of crossopterygian fishes. *(From H. L. Levin,* Life Through Time *[Dubuque, IA: William C. Brown, 1975].)*

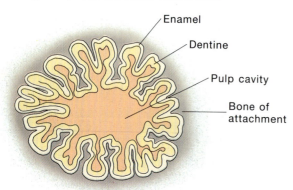

- Enamel
- Dentine
- Pulp cavity
- Bone of attachment

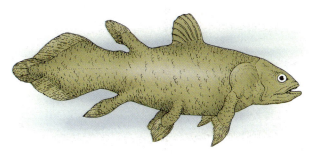

Figure 19.71 *Latimeria,* a surviving coelacanth living in the ocean near Madagascar and southeast Africa. *Latimeria* is a large fish, nearly 2 meters in length.   *(From J. L. Levin, Life Through Time [Dubuque, IA: William C. Brown, 1975].)*

the blood more efficiently to and from the more efficient lungs. The limb and girdle bones were modified to overcome the constant drag of gravity and better hold the body above the ground. The spinal column, a simple structure in fishes, was transformed into a sturdy but flexible bridge of interlocking elements. To improve hearing in gaseous rather than fluid surroundings, the old hyomandibular bone, used in fishes to prop the braincase and upper jaw together, gradually evolved into an ear ossicle—the stapes. The fish spiracle (a vestigial gill slit) became the amphibian eustachian tube and middle ear. To complete the auditory apparatus, a tympanic membrane (ear drum) was developed across a prominent notch in the rear part of the amphibian skull.

The fossil record for amphibians begins with a group called **ichthyostegids** (Fig. 19.72). As suggested by their name, these creatures retained many features of their fish ancestors. The amphibians that followed the ichthyostegids fall mostly within a group collectively termed the **labyrinthodonts**. The labyrinthic wrinkling and folding of tooth enamel inherited from the rhipidistian Sarcopterygii provided the inspiration for the labyrinthodont name (see Fig. 19.69). During the Carboniferous, large numbers of labyrinthodonts wallowed in swamps and streams, eating insects, fish, and

one another. A labyrinthodont that exemplifies the culmination of their lineage is *Eryops* (Fig. 19.73). Eryops was a bulky, inelegant creature with a flattish skull typical of labyrinthodonts and bony nodules in the skin for protection against some of its more vicious contemporaries. The labyrinthodonts declined during the Permian, and only a relatively few survived into the Triassic.

Reptiles

The evolutionary advance from fish to amphibians was no trivial biologic achievement, yet the late Paleozoic was the time of another equally significant event. Among the evolving land vertebrates were some that had developed a way to reproduce without returning to the water. They accomplished this feat by evolving enclosed eggs in which the embryonic animal was allowed to pass through larval and other developmental stages before being hatched in an essentially adult form. This enclosed egg liberated the tetrapods from their reliance on water bodies and has been hailed as a major milestone in the history of vertebrates. The animals that first evolved the so-called amniotic egg were amphibians. Evolutionary processes had provided them with a method for protecting developing young from predation and desiccation.

The oldest known remains of reptiles were discovered in early Pennsylvanian swamp deposits of Nova Scotia. Named *Hylonomus,* these reptiles were only about 24 centimeters (9.5 inches) from head to tail and had a body form resembling that of a salamander. Their skeletal remains are found in holes within the stumps of Pennsylvanian scale trees. In slightly younger late Pennsylvanian rocks of Kansas, additional early reptiles have been recovered from claystones and siltstones of former mudflats. The most common of these reptiles is called *Petrolacosaurus. Petrolacosaurus* had a longer neck and longer legs than *Hylonomus.* Details of its skull indicate that it was an early member of the lineage that would eventually lead to dinosaurs.

The small and relatively primitive reptilian groups of the Pennsylvanian provided the stock from which a

Figure 19.72   The skeleton of *Ichthyostega* still retains the fishlike form of its crossopterygian ancestors.   *(From L. H. Levin,* Life Through Time *[Dubuque, IA: William C. Brown, 1975].)*

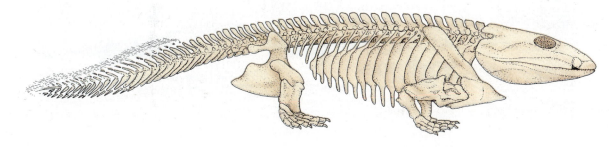

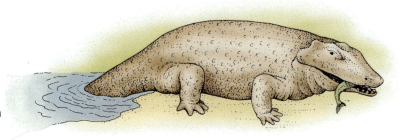

Figure 19.73 The large Permian amphibian *Eryops.* Length is about 2 meters.

succession of more specialized Permian reptiles evolved. The most spectacular of these were the **pelycosaurs**, several species of which sported erect "sails" supported by rodlike extensions of the vertebrae. The pelycosaurs were a varied group. Some, such as *Edaphosaurus,* were plant eaters, whereas others, as indicated by their great jaws and sharp teeth, ate flesh. *Dimetrodon* (Fig. 19.74) was one such predator. There have been many attempts to explain the function of the pelycosaurian sail. The most reasonable explanation is that the sail-fin acted in temperature regulation by serving sometimes as a collector of solar heat and at other times as a radiator giving off surplus heat. Because pelycosaurs had several mammalian skeletal characteristics, they are thought to be the early mammal-like reptilian group from which advanced mammal-like reptiles known as **therapsids** arose. In this regard, it is interesting that they were attempting a kind of body temperature control.

For paleontologists, therapsids are among the most fascinating of fossil vertebrates. They were widely dispersed during Permian and Triassic time. There is a fabulous record of these reptiles and their contemporaries in the Karoo basin of South Africa and more modest discoveries in Russia, South America, Australia, India, and Antarctica. Therapsids were predominantly small to moderate-sized animals that displayed at least the beginnings of several mammalian skeletal traits. There were fewer bones in the skull than generally found in reptile skulls, and there was a mammal-like enlargement of the lower jaw bone (dentary) at the expense of more posterior elements of the jaw. A double ball-and-socket articulation had evolved between the skull and neck. Teeth showed a primitive but distinct differentiation into incisors, canines, and cheek teeth. The limbs were in more direct vertical alignment beneath the body, and the ribs were reduced in the neck and lumbar region for greater overall flexibility. These features are apparent in the Permian therapsid *Cynognathus* (Fig. 19.75). Therapsids continued from the Permian through the Triassic but became extinct early in the Jurassic. However, before they died out completely, they gave origin to the early mammals.

Figure 19.74 Permian reptiles. The sailback reptiles with the larger skulls and teeth are *Dimetrodon. Edaphosaurus,* a herbivorous form, is just right of center. The smaller lizardlike reptiles are *Casea.* *(Chicago Field Museum of Natural History; painting by C. R. Knight)*

Figure 19.75   Mammal-like reptiles. The scene depicts three carnivorous forms *(Cynognathus)* about to attack a plant-eating reptile *(Kannemeyeria).*   *(Chicago Field Museum of Natural History; painting by C. R. Knight)*

19.5   Extinctions at the Close of the Paleozoic

The fossil record indicates that there has been a persistent increase in the numbers of different species through geologic time. It also shows that this trend has been upset or even reversed at times as a result of environmental adversity. Extinctions of large numbers of previously successful animals and plants may have resulted from their inability to adapt to changing conditions. Early geologists were able to identify such times of biologic crisis and used extinctions to define further the termination of the Paleozoic and the Mesozoic eras.

As discussed in the previous chapter, there were drastic geographic changes as the Paleozoic drew to a close. Pangaea was experiencing mountain building and regional uplift, there was widespread volcanic activity, and seas had withdrawn from the extensive tracts they once covered. Climates became more severe, and a protracted glacial age occurred in the south. There is evidence of marked seasonal changes and aridity in many areas of the globe. As one or another group of animals gradually diminished in response to these changes, they contributed to the demise of others farther along in the food chain. Then, as now, the balance in the biologic environment could be easily upset, causing far-reaching waves of extinction.

Among the invertebrate groups that became extinct near the end of the Paleozoic were the trilobites, eurypterids, rugose and tabulate corals, productid brachiopods, many families of bryozoans, and blastoids. Entire families of other invertebrate taxa were lost as well. The archaic jawed fishes (acanthodians) became extinct during the Permian, whereas amphibians, reptiles, and several groups of plants declined conspicuously. The misfortunes of the invertebrates may have been associated with the drastic reduction of the inland seas, whereas the amphibians may have suffered in the competition with evolving reptiles. Among those reptiles were a few hardy groups that were able to meet the new challenges and become the founders of the reptilian communities that were to dominate the Mesozoic terrestrial scene.

SUMMARY

The fossil record of the Paleozoic is immensely better than that of the preceding Proterozoic. Shell building is one important reason for this improvement in the fossil record of invertebrates.

The earliest Paleozoic plants were mainly bacteria and algae, with the most obvious larger fossils being **stromatolites** and, by Ordovician time, the **receptaculids**. By middle Silurian, the first land plants, the **psilophytes**, made their appearance. It is probable that they evolved from **chlorophytes**. Psilophytes were followed in the Devonian and Carboniferous by **lycopsids**, **sphenopsids**, and **seed ferns**. In temperate Southern Hemisphere continents, a flora dominated by *Glossopteris* developed. Permian forests included conifers and ginkgoes.

The ability to secrete hard parts had already begun in earliest Cambrian time, and this adaptation led to a rapid radiation of marine invertebrates, with **archaeocyathids**, **brachiopods**, and **trilobites** becoming abundant. By Ordovician time, however, all major invertebrate phyla had become common. The best record of soft-bodied invertebrates are those of the middle Cambrian **Burgess Shale fauna**, which includes many previously unknown arthropods, echinoderms, sponges, and cnidarians, as well as an animal *(Pikaia)* believed to be the earliest known chordate. Among the other important invertebrate groups of the Paleozoic were **foraminifers** (fusulinids expanded during the Pennsylvanian and Permian), sponges, corals (**tabulate** and **rugose**), bryozoa, mollusks, echinoderms (especially **blastoids** and **crinoids**), and **graptolites**. The enigmatic **conodonts** serve as excellent index fossils for Paleozoic rocks.

The first uncontested fossils of vertebrate animals are found in rocks of Ordovician age. They are remains of jawless fishes known as **ostracoderms**. Archaic jawed fishes appear late in the Silurian but do not begin to dominate the marine realm until Devonian time. This was also the time when cartilaginous (chondrichthyian) and bony (osteichthyan) fishes began their rise to dominance. A special group of bony fishes called crossopterygians arose from osteichthyan stock. These fishes could breathe air by means of accessory lungs and possessed muscular fins that provided short-distance overland locomotion. The bones within the fins of crossopterygians resembled those of primitive amphibians. Other skeletal traits clearly suggest that these Devonian fishes were the ancestors of the first amphibians—the **ichthyostegids**. From a start provided by ichthyostegids, amphibians called **labyrinthodonts** underwent a successful adaptive radiation that lasted until the close of the Triassic. Long before their demise, however, they provided a lineage from which evolved the reptiles.

The transition from amphibian to reptile entailed a significant breakthrough in evolution. It involved the development of the amniotic egg, a biologic device that liberated land vertebrates from the need to return to water bodies to reproduce. The first known reptiles are Pennsylvanian in age and include such forms as *Hylonomus* and *Petrolacosaurus*. During very late Pennsylvanian and Permian, reptiles underwent an elaborate evolutionary radiation that produced the fin-back reptiles and mammal-like reptiles, as well as several other major reptilian groups.

The end of the Paleozoic is marked by a major episode of extinction. Near the end of the Permian period, nearly half of the known families of animals disappeared. The decimation of marine animals was particularly dramatic and included the loss of fusulinids, spiny productid brachiopods, rugose corals, two orders of bryozoans, and many taxa of stemmed echinoderms, including blastoids. Among the vertebrates, over 70 percent of amphibian and reptilian families perished. Fortunately, a few groups survived the Permian crisis and were able to continue their evolution during the Triassic. The cause of the Permian biologic crisis is not definitely known. Many believe it was the result of a high stand of Permian continents, reduction of epeiric seas, mountain building, and sharp climatic changes resulting from all of these factors.

KEY TERMS

Receptaculid *458*
Chlorophyte *459*
Psilophtyte *459*
Lycopsid *460*
Sphenopsid *460*
Notochord *463*
Epifaunal *465*
Infaunal *465*
Foraminifer *466*
Radiolarian *466*
Archaeocyathid *466*
Porifera *466*
Stromatoporoids *467*

Cnidaria *468*
Theca *469*
Septa *469*
Tabulae *469*
Rugose coral *469*
Tabulate corals *469*
Bryozoan *469*
Lophophore *471*
Fenestellid bryozoan *471*
Brachiopod *471*
Mollusca *473*
Pelecypod *473*
Gastropod *473*

Cephalopod *473*
Nautiloidea *477*
Ammonoidea *477*
Goniatite *477*
Trilobites *478*
Ostracod *480*
Eurypterid *480*
Echinoderm *480*
Cystoid *482*
Blastoid *482*
Crinoid *482*
Graptolite *483*
Conodont *485*

Ostracoderm *487*
Acanthodians *489*
Placoderm *489*
Dipnoi *491*
Crossopterygii *491*
Ichthyostegid *493*
Labyrinthodont *493*
Pelycosaur *494*
Therapsid *494*

REVIEW QUESTIONS

1. What group of algae were the probable ancestors of the first land plants? What evidence supports this interpretation?

2. The receptaculid *Fisherites* was once considered a type of sponge but is now believed to be the fossil remains of algae. Does this new interpretation affect the value of *Fisherites* as a guide fossil? If not, why?

3. Why was the evolution of a vascular system an essential precursor to the colonization of the continents by land plants?

4. What is the evolutionary significance of the Burgess Shale fossil named *Pikaia?*

5. To what geologic system would rocks containing the following fossils be assigned? (a) arcaeocyathids, (b) fusulinids, (c) *Archimedes*

6. The phylum Cnidaria was formerly termed Coelenterata. Why is Cnidaria an appropriate name for animals in this phylum?

7. How do the sutures of nautiloids differ from those of goniatites?

8. Describe the principal kinds of plants that grew abundantly in the swamps and forests of North America during the Pennsylvanian Period.

DISCUSSION QUESTIONS

1. In terms of location, climatic preference, and reproduction, how did *Lepidodendron* differ from *Glossopteris?*

2. What does the fauna of the Burgess Shale reveal about the bias in the fossil record of early Paleozoic life?

3. Distinguish between the terms *chordate* and *vertebrate*. What phylum of invertebrates has been considered ancestral to the earliest chordates? Why?

4. List three characteristics of therapsid reptiles that may have provided advantages over earlier amphibians and reptiles.

5. Discuss the origin and derivation of components of the vertebrate jaw as seen in the evolution of jawed fishes.

6. Discuss the possible causes for extinctions that occurred worldwide near the end of the Ordovician and Permian Periods.

The Mesozoic Era

The Egyptian Temple, Capitol Reef National Monument, Utah. The massive rock at top is Shinarump Formation, beneath which are banded beds of Moenkopi Formation.
(U.S. Geological Survey, L.C. Huff photograph)

The mass extinction of animals and plants that occurred at the close of the Permian did not go unnoticed by the founders of geology. The sudden disappearance of many major groups of fossils was used to mark the close of the Paleozoic sequence. Overlying younger strata contained different and often more progressive organisms, yet not as modern as those that live today. For these, it seemed appropriate to propose a middle chapter in the history of life. Geologists named that chapter the Mesozoic Era. The Mesozoic lasted 160 million years. It ended with another biological crisis marked by the extinction of dinosaurs and scores of other animals, both on land and in the sea.

During the Mesozoic, many new families of plants and animals evolved. It was the era in which two new vertebrate classes, the birds and mammals, appeared. Flowering plants evolved during this era. This was also the time in which the supercontinent Pangaea, which had formed during the Paleozoic by the joining of ancestral continents, was dismembered. The process of fragmentation and drift ultimately produced the present physical geography of the planet (Fig. 20.1).

20.1 The Breakup of Pangaea

The dismemberment of Pangaea occurred in four stages. The first stage began in the Triassic with rifting and volcanism along normal fault systems. Such events characterize regions of the lithosphere that are subjected to tensional stresses. In this instance, the forces were associated with the separation of North America and Gondwanaland. As the rifting progressed, the eastern border of North America parted from the Moroccan bulge of Africa (see Fig. 20.1). Oceanic basalt flowed from fault zones on the floor of the widening Atlantic Ocean. The largely tensional structures, as well as the volcanic rocks and sediments in the Triassic System of both the eastern United States and Morocco, exhibit many striking similarities. These now widely separated regions were on opposite sides of the same axis of sea-floor spreading.

The second stage in the breakup of Pangea involved the rifting and opening of narrow oceanic tracts between southern Africa and Antarctica. This rift zone extended northeastward between Africa and India and developed a branch that separated northward-bound India from the yet to be dismembered Antarctica–Australia landmass. The rifting was, as usual, accompanied by the extrusion of great volumes of basaltic lava.

In stage three of the breakup of Pangaea, the Atlantic rift began to extend itself northward. Clockwise rotation of Eurasia tended to close the eastern end of the Tethys Sea, a forerunner of the Mediterranean. By the end of Jurassic time, an incipient breach began to split South America away from Africa. The cleft worked its way up from the south, creating a long seaway that would widen to become the South Atlantic. Australia and Antarctica remained together, but India had moved well along on its journey to Laurasia. By late Cretaceous time, South America had completely separated from Africa. The Mid-Atlantic Rift propagated northward and divided into two branches, one on each side of Greenland. The western branch separated North America from Greenland, and the eastern branch separated Greenland from Europe. During the following Cenozoic Era, Antarctica and Australia parted company.

20.2 North America During the Mesozoic

Eastern North America

Triassic Events

At the beginning of the Mesozoic, conditions in the eastern part of North America were much the same as they had been near the end of the Paleozoic. The rugged Appalachian ranges that had been raised during the Allegheny orogeny were undergoing vigorous erosion. Then, during the late Triassic and early Jurassic, North America began to experience the tensional forces that precede separation of continents. As a result, a series of fault-bounded troughs developed along the eastern coast from Nova Scotia to North Carolina (Figs. 20.2 and 20.3). Swiftly flowing streams carried arkosic gravels and sands into the downfaulted basins (Fig. 20.4). These sediments, now consolidated as arkosic conglomerates and sandstones, constitute the **Newark Group** of late Triassic and early Jurassic age.

In addition to arkosic alluvial deposits, the Newark Group includes shales deposited in lakes where drainage in the basins had become impounded. The lake deposits contain the remains of fresh-water crustaceans and fish.

Text continued on page 503

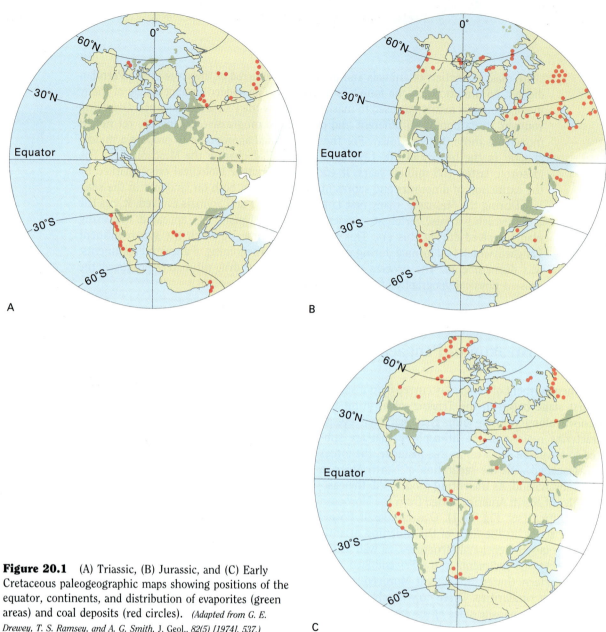

Figure 20.1 (A) Triassic, (B) Jurassic, and (C) Early Cretaceous paleogeographic maps showing positions of the equator, continents, and distribution of evaporites (green areas) and coal deposits (red circles). *(Adapted from G. E. Drewey, T. S. Ramsey, and A. G. Smith, J. Geol., 82(5) [1974], 537.)*

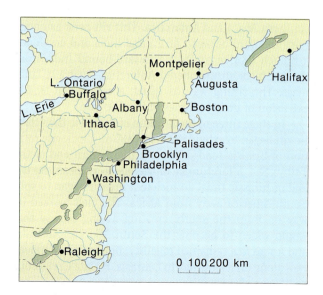

Figure 20.2 Outcrop areas of Triassic rocks in eastern North America. Green areas show troughlike deposits of Late Triassic age.

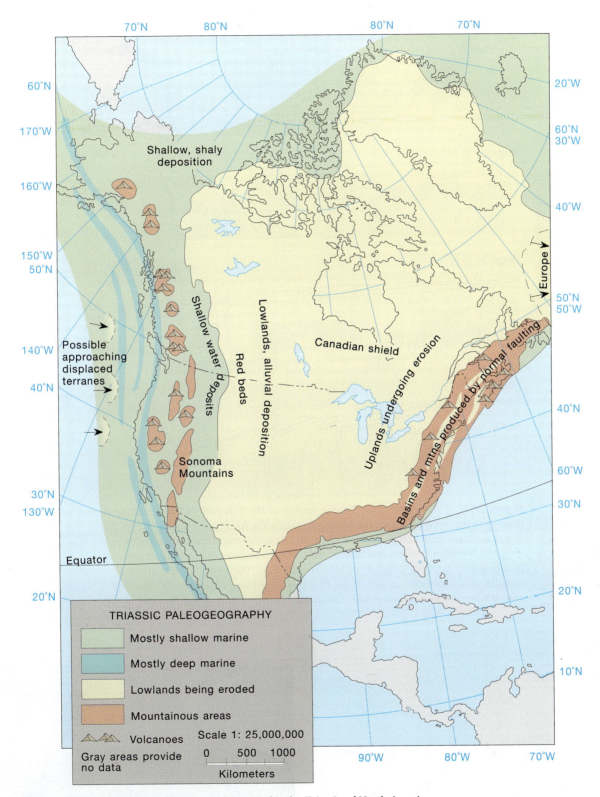

Figure 20.3   Generalized paleogeographic map for the Triassic of North America.

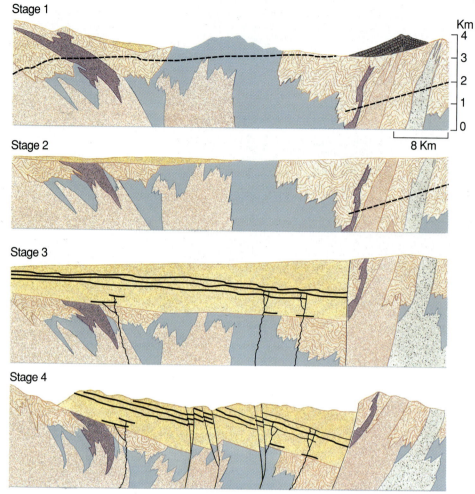

Figure 20.4   Four stages in the Triassic history of the Connecticut Valley. (Stage 1) Erosion of the complex structures developed during the Allegheny orogeny. (Stage 2) Mountains have been eroded to a low plain, and Triassic sedimentation has begun. (Stage 3) Newark sediments and basaltic sills, flows, and dikes accumulate in troughs resulting from faulting. (Stage 4) In Late Triassic time, the area is broken into a complex of normal faults as part of the Palisades orogeny.   *(Modified from the classic 1915 paper by J. Barrell, Central Connecticut in the Geologic Past,* Conn. Geol. Nat. Hist. Bull. *No. 23.)*

Figure 20.5   Slab of Newark Group sandstone containing the cast of a footprint made by a three-toed Triassic reptile. The footprint is about 20 centimeters long. The linear ridges in the slab are casts of mud cracks.

Figure 20.6 Palisades of the Hudson River. *(Palisades Interstate Park Commission)*

Ripple marks, mud cracks, raindrop impressions, and even the footprints of early members of the dinosaur lineage (Fig. 20.5) occur on the surfaces of Newark beds. Lavas, rising along steeply dipping fault surfaces, flowed out on the newly deposited sediment, and volcanoes spewed ash onto the surrounding hills and plains.

Three particularly extensive lava flows and an imposing sill are included within the Newark Group in the New York–New Jersey area. The exposed edge of a vertically jointed sill forms the well-known Palisades of the Hudson River (Fig. 20.6).

Jurassic Events

During the Jurassic, the Gulf of Mexico had only begun to open. Its opening was part of the initial stage of a Wilson cycle. In addition to its restricted condition, the seaway lay only 15 degrees north of the equator. Thus, conditions were ideal for the precipitation of evaporites. Salt and gypsum accumulated to thicknesses exceeding 1000 meters (Fig. 20.7).

The Jurassic evaporite beds are the source of the salt domes of the Gulf Coast. Salt domes are economically important structures associated with the entrapment of petroleum and natural gas. When compressed by a heavy load of overlying strata, salt flows plastically. As it moves upward in the direction of lesser pressures, the salt bends and faults overlying strata, producing traps for oil as it does so.

Evaporative conditions abated later in the Jurassic, and several hundred meters of normal marine limestones, shales, and sandstones accumulated in the alternately transgressing and regressing seas of the Gulf embayment. In the United States, these rocks are deeply buried beneath a thick cover of Cretaceous and Cenozoic sediments (Fig. 20.8).

Cretaceous Events

The Cretaceous was a time of widespread marine inundations of low-lying continental areas. Early in the period, the Atlantic coastal plain, which had been undergoing erosion since the beginning of the Mesozoic, began to subside. At about the same time, the Appalachian belt that lay to the west was elevated. Waters of the Atlantic flooded onto the subsiding coastal plain. In time, a wedge of sediment that thickened seaward had accumulated. Today, the thinner, western border of the wedge is exposed in New Jersey, Maryland, Virginia, and the Carolinas. But the thickest section of Cretaceous sediment was deposited farther east, along the present continental shelf.

To the south, the Florida region was a shallow submarine bank during the Cretaceous. The oldest strata consist of limestones, but later in the period sands and clays were carried into the area by streams flowing from the southern Appalachians. As these source areas were worn down, carbonate deposition resumed.

503

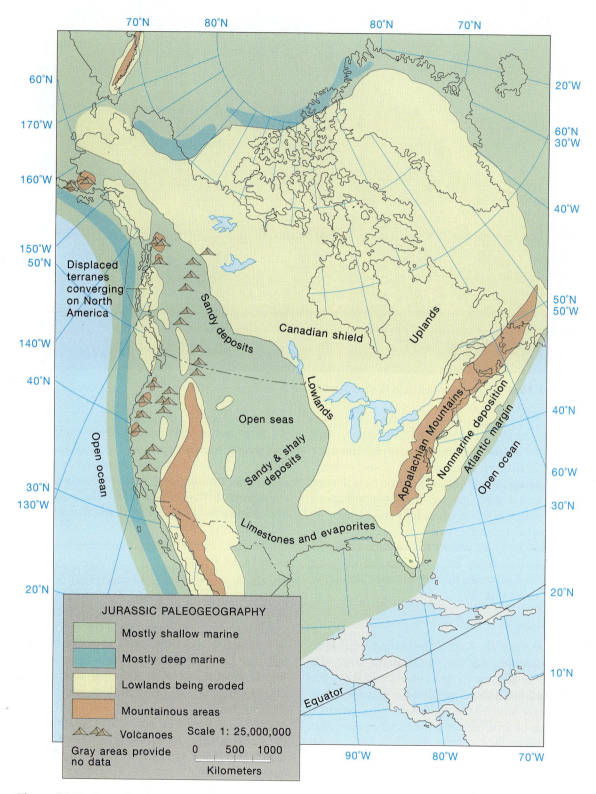

Figure 20.7 Generalized paleogeographic map for the Jurassic of North America.

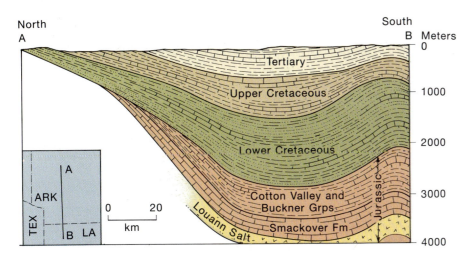

Figure 20.8 Jurassic rocks in the Gulf Coast region are deeply buried beneath Tertiary and Cretaceous rocks. *(Adapted from* Stratigraphic Atlas of North and Central America, *Shell Oil Co.)*

The Cretaceous is noteworthy as a time when carbonate reefs were developed extensively. Among the invertebrates that contributed their skeletons to these reefs were a group of pelecypods called **rudists** (Fig. 20.9). The shells of rudists form the basic framework of many Cretaceous reefs. Because of their high porosity and permeability, rudistid reefs are reservoir rocks for some of the world's greatest oil accumulations.

The late Cretaceous was a time of maximum coverage of continents by epeiric seas. A broad, shallow sea occupied a tract from the Gulf of Mexico to the Arctic Ocean (Fig. 20.10). Among the sediments of this vast inland sea, chalk was particularly prevalent. Chalk is a white, soft, fine-grained variety of limestone composed mainly of the microscopic calcareous platelets (called coccoliths) of golden-brown algae. It is common among beds of Cretaceous age in many parts of the world (Fig. 20.11). Indeed, the Cretaceous System takes its name from *creta,* the Latin word for chalk.

Western North America

As the eastern border of North America was undergoing tension caused by the separation of Europe and Africa, compressional forces prevailed in the west. While the newly formed Atlantic Ocean was widening, North America moved westward, overriding the Pacific plate. Thus, deformation of western North America is related to events in the east. It has been shown, for example, that the pace of tectonic activity in the North American Cordillera was most intense during the time when seafloor spreading was most rapid in the Atlantic.

Accretionary Tectonics

During the Mesozoic, a steeply dipping subduction zone existed along the western margin of North America. It was not a simple subduction zone, however, for the advancing Pacific plate carried considerably more than

ocean basalts and sea-floor sediments. Entire sections of volcanic arcs and fragments of distant continents were carried to North America's western margin as well. These foreign fragments, now incorporated in the Cordilleran belt, constitute **displaced** or **allochthonous terranes**. They are recognized by their distinctive age, rock associations, fossils, and other features that identify the part of the world from which they came. More than 50 allochthonous terranes have been recognized in our western mountain ranges. They reveal that the growth

Figure 20.9 *Eoradiolites davidsoni,* a Cretaceous rudistid pelecypod from Texas. Rudistids appeared in the Jurassic and proliferated during the Cretaceous when they were important reef-making organisms. They bear a close resemblance to some Paleozoic corals. This specimen is 8.6 cm tall.

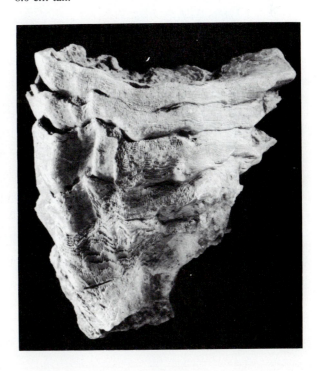

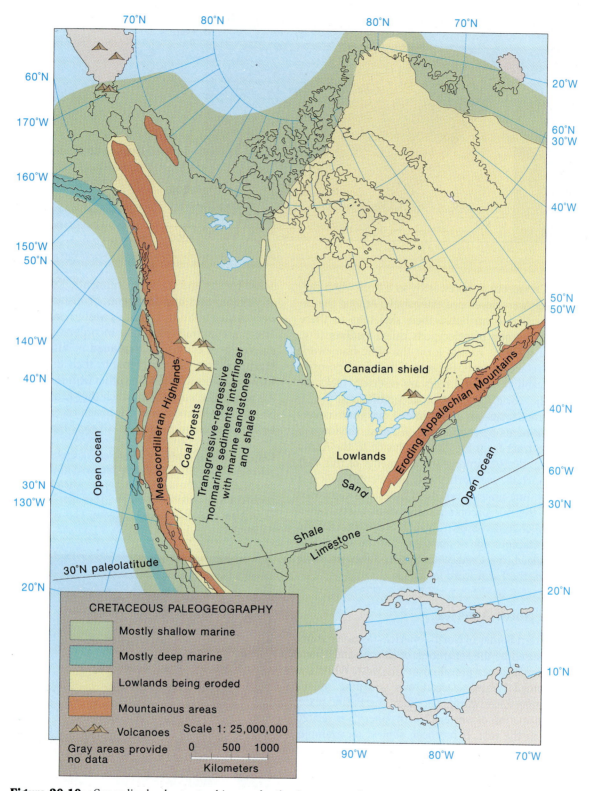

Figure 20.10   Generalized paleogeographic map for the Cretaceous of North America.

B

Figure 20.11 (A) Sea cliffs composed of chalk along the Dorset coast of England. (B) Close-up view of the chalk. The dark nodules are flint, a gray or black variety of chert.

A

of North America occurred not only by accretion of materials along subduction zones, but also by incorporation of large chunks of continental crust formed elsewhere and conveyed to this continent by sea-floor spreading.

Triassic Sedimentation

In general, the Cordilleran region during the Mesozoic consisted of a western belt containing thick volcanic and siliceous deposits and a wider eastern tract that lay adjacent to the stable interior of the continent. Nearly 800 meters of clastic sediments and volcanics exposed

in Nevada and southeastern California attest to the dynamic conditions in the western zone.

The first big Mesozoic orogenic event of the Cordilleran region occurred at or near the Permian–Triassic boundary, when an eastward-moving volcanic arc collided with the Pacific margin of North America. As a result of the collision, the island arc was welded onto the western edge of North America, adding about 300 kilometers to the width of the continent. In the course of this smashup, oceanic rocks were thrust eastward onto the eroded structures of the old Antler orogen (Fig. 20.12). The larger of the thrust sheets moved scores

Figure 20.12   Tectonic conditions and paleogeography along the Cordilleran region during the Late Permian and Early Triassic.

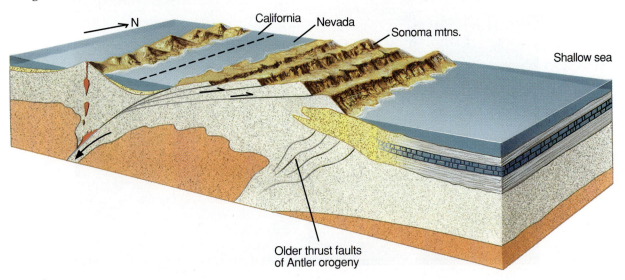

Older thrust faults
of Antler orogeny

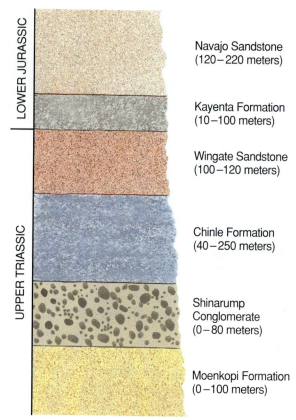

Figure 20.13   Generalized geologic section of Upper Triassic and Lower Jurassic sedimentary rocks of central Utah.

of kilometers eastward and broke into scores of overlapping smaller sheets. Each of these brought a repeated stratigraphic sequence to the surface.

In the quieter western zone of the Cordillera, the Triassic section begins with sandstones and limestones of shallow seas. By the Triassic, however, the epeiric seas regressed, leaving the former sea floor exposed to the forces of erosion. Upper Triassic beds rest on the unconformity resulting from that erosion. These Upper Triassic formations consist largely of continental deposits transported by rivers flowing westward across an immense alluvial plain. Lowermost is the sandy **Moenkopi Formation**, followed by the pebbly **Shinarump Formation**, whose gravels appear to have been derived from nearby highlands. Above the beds of the Shinarump are the vividly colored shales and sandstones of the **Chinle** and **Kayenta** formations (Fig. 20.13). These alternate with sand dune accumulations of the **Navajo Sandstone** (Fig. 20.14) and the **Wingate Sandstone** (Fig. 20.15). One can view these rocks in the walls of Zion Canyon in southern Utah. They display sweeping cross-bedding, such as is characteristic of sands transported by wind.

The Painted Desert of Arizona is sculpted mostly in rocks of the Chinle Formation. The formation is known throughout the world for the petrified logs of conifers it contains. Each year, thousands of tourists examine these logs, now turned to colorful agate, in Petrified Forest National Monument (Fig. 20.16). Appar-

Figure 20.14   Panoramic view of Triassic and Jurassic formations in Zion National Park, Utah. The lowermost slope is formed by shales of the Moenkopi Formation. Above that slope is a prominent ledge of Shinarump sandstones. The next slope above the Shinarump is the one formed in Chinle Formation. The towering cliffs at the top are Navajo and Carmel formations.   *(U.S. National Park Service)*

Figure 20.15 The erosional features in the foreground are sculpted from Wingate Sandstone, beneath which lies the red Chinle Formation. the cliffs in the far background are also Wingate Sandstone. Colorado National Monument near Grand Junction, Colorado. *(R. J. Weimer)*

ently, when Triassic floods occurred, uprooted trees were left on sandbars or trapped in log jams and then covered with sediment. Percolating solutions of underground water subsequently converted the wood to silica.

Orogenies

Most of the mountain-building activity in the Cordillera during the Mesozoic was the result of the continuing eastward subduction of oceanic lithosphere beneath the western margin of North America (Fig. 20.17). That subduction resulted in shifting phases of deformation that began in the far western part of the Cordillera and then proceeded eastward to finally reach the margin of the craton. The folding, faulting, and igneous activity associated with the western tract is referred to as the **Nevadan orogeny**. In this far western belt, graywackes, siliceous oceanic sediments, and volcanics that had been swept into the subduction zone were compressed, complexly folded, and metamorphosed to form a jumbled mass of rocks termed a mélange (Fig. 20.18). The Franciscan fold belt of California (Fig. 20.19) is a good example of a mélange.

In addition to the deformation of the sedimentary section, enormous volumes of granodiorite were generated above the subduction zone. The melts intruded overlying rocks repeatedly during the Jurassic and Cretaceous.The Sierra Nevada (Fig. 20.20), Idaho, and Coast Range batholiths record the vast scale of this Nevadan magmatic activity (Fig. 20.21).

Somewhat before the Sierra Nevada batholith was emplaced during the Cretaceous, another series of deformations began east of the present Sierra Nevada. The new orogeny has been named the **Sevier**, and it had a distinctly different style of deformation. Strata were sheared from underlying Precambrian rocks and broken along parallel planes of weakness to form multiple, low-angle thrust faults (see Fig. 20.17). The French word **décollément** (which means "unsticking") is used to describe this kind of structure, in which older rocks are

Text continued on page 512

Figure 20.16 Petrified logs, Petrified National Monument, Arizona. The petrified logs and wood fragments are *Araucarioxylon*. They have been weathered from the Chinle Formation of Triassic age. *(Lehi F. Hintze)*

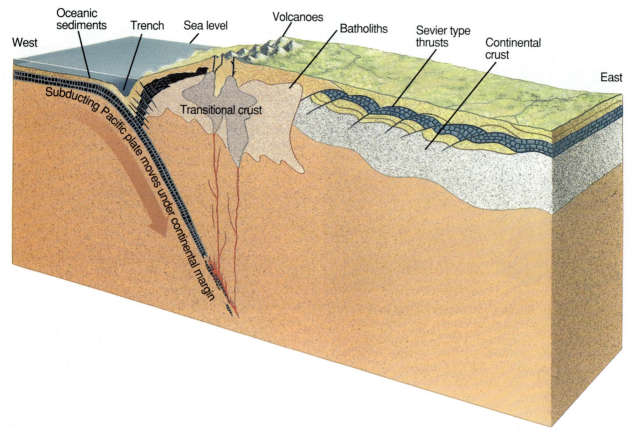

Figure 20.17   An advanced stage in the evolution of the North American Cordillera, with structures developing as a consequence of underthrusting of the continent by the Pacific oceanic plate. Note the multiple, imbricated, low-angle thrusts on the east side of the section.   *(The diagram is simplified from J. F. Dewey, and J. M. Bird, J. Geophys. Res., 75(14) [1970], 2638.)*

Figure 20.18   Mélange that is within the Franciscan fold belt of northern California. The exposure is north of Crescent City, California. Width of photographed area is 2.4 meters.

Figure 20.19 Ophiolites of the Franciscan Formation exposed at Lime Point, Marin Peninsula, California. The west side of the Golden Gate Bridge tower rests on rocks of the Franciscan Formation. *(U.S. Geological Survey)*

Figure 20.20 Granodiorite of the Sierra Nevada batholith grandly exposed in Yosemite Valley, California.

Figure 20.21 Mesozoic batholiths in west–central North America.

thrust on younger in multiple, nearly parallel crustal slabs.

Although the term *Sevier* is usually reserved for deformational events in the Nevada–Utah region, a similar style of deformation occurred to the north in Montana, British Columbia, and Alberta. Many of the major mountain ranges in this region were produced as Proterozoic and Paleozoic strata were thrust eastward. The most famous of these great faults is the Lewis thrust (Fig. 20.22), along which Proterozoic rocks of the Belt Supergroup were carried 65 kilometers to the east.

Much of the thrust faulting and magmatic activity had subsided by late Cretaceous and early Tertiary time. Deformational events then shifted eastward to the region where the Rocky Mountains of New Mexico, Colorado, and Wyoming are now located. These more eastward disturbances are called the **Laramide orogeny**. The structures produced by the Laramide orogeny characteristically consist of broadly arched domes, basins, and anticlines. Often, the strata of the domes and anticlines are draped over central masses of Precambrian igneous and sedimentary rocks. The exposed, uptilted edges of resistant beds that surround the cores stand as mountains. It has been suggested that the domes and anticlines are related to deep-seated faults produced by drag when the eastward-moving subducted Pacific plate scraped along the sole of the craton.

Most of the structures of the present Rocky Mountains are the result of Laramide deformations. The landscapes we see today, however, are the result of repeated episodes of Cenozoic erosion and uplift. Erosion, acting on the geologic structures already present, sculpted the final scenic design.

Figure 20.22 Northeast–southwest cross section showing the Lewis and Eldorado thrust faults at a location about 66 kilometers south of Glacier National Park. Formations shown in green are Proterozoic in age. Tan color indicates Paleozoic and Mesozoic rocks. *(After M. R. Mudge and R. L. Earhart,* U.S. Geological Survey Prof. Paper, *No. 1174, 1980.)*

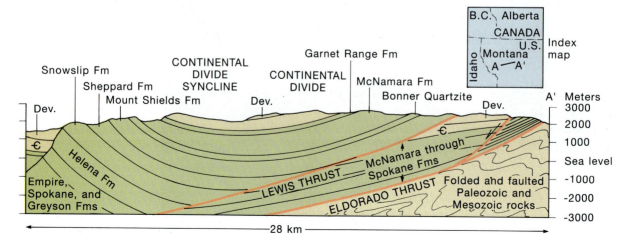

Figure 20.23 Cross stratification in the Navajo Sandstone, Checkerboard Mesa, Zion National Park, Utah. *(Lehi F. Hintze)*

Jurassic Sedimentation

During the Jurassic, while a melange was being formed in the subduction zone along the western margin of the continent, less dramatic events were occurring inland. Early Jurassic deposits consist of clean sandstones such as the **Navajo Sandstone**. Large-scale cross-bedding is well developed in the Navajo (Fig. 20.23), and thin beds of fossiliferous limestones and evaporites occur locally. It is likely that the sands of the Navajo were deposited largely as dunes along coastal regions of the inland sea (Fig. 20.24).

Following the deposition of the Navajo Formation, the entire west-central part of the continent was flooded by a wide seaway that extended well into central Utah (Fig. 20.25). This great embayment has been named the **Sundance Sea**. Within the Sundance Sea, sands and silts of the Sundance Formation were deposited. The formation is well known for the fossils of Jurassic marine reptiles it contains.

Huge volumes of sediment were brought to the Sundance Sea from the Cordilleran highlands that lay to the west. Eventually, the basin was filled and what had been an inland sea was transformed into a vast, swampy plain across which meandering rivers built broad flood plains. These deposits compose the **Morrison Formation**, famous for its rich content of bones from over 70 species of dinosaurs.

Cretaceous Sedimentation

During the Cretaceous, the Pacific border of North America was a land of lofty mountains undergoing rapid erosion and subjected to repeated episodes of uplift, deformation, and igneous activity. Regions to the east, however, were occupied by epeiric seas. The seas were most extensive during the late Cretaceous, when an embayment from Canada was joined by one from the Gulf of Mexico to effectively divide North America into two separate landmasses. Chalk, fossiliferous limestones, and limy clays are the most common deposits

Figure 20.24 Paleogeographic map for the early Jurassic of the western United States, showing general extent of sea and land as well as paleolatitudes. Arrows indicate wind directions. *(From K. O. Stanley, W. M. Jordan, and R. H. Dott,* Bull. Am. Assoc. Petrol. Geol., *55(1) [1971], 13.)*

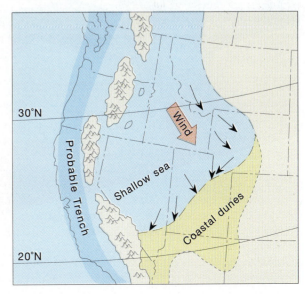

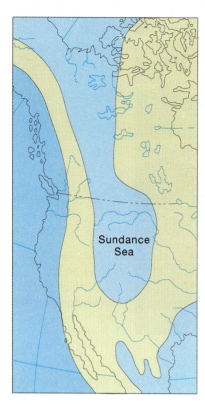

Figure 20.25 Region in western North America inundated by the Middle Jurassic Sundance sea. (Land areas are shown in tan, marine areas in blue.)

of this great inland sea, and these beds yield abundant fossils of Cretaceous mollusks as well as the skeletons of extinct marine reptiles and aquatic birds. West of the sea, a low-lying coastal plain received terrigenous clastics from the Cordilleran highlands.

In the context of plate tectonics, the depositional basin receiving the terrigenous clastics supplied by streams flowing from the Cordillera is termed a back-arc basin (Fig. 20.26). The back-arc basin of the western interior of North America was immense. It extended from the Arctic to the Gulf of California and in places was over 1600 kilometers wide. It was bounded on the west mostly by fold and thrust belts such as the Sevier and on the east by the craton. Many of the rock formations in the back-arc basin yield commercial quantities of oil, natural gas, and coal. Exploration for these resources has provided hundreds of well logs and stratigraphic columns that permit geologists to map Cretaceous rocks of the piedmont, coastal plains, shorelines, and marine areas. The picture that has emerged shows conglomerates and sandstones immediately adjacent to the Cordilleran highlands. These interfinger with coal-bearing continental deposits that similarly interfinger with marine rocks farther to the east (Fig. 20.27). Marine and nonmarine phases shifted repeatedly with advances and regressions of the Cretaceous epeiric seas. As is usually true, the transgressive phases are recorded by

Figure 20.26 Conceptual cross section indicating major tectonic features present during the Cretaceous across the western United States.

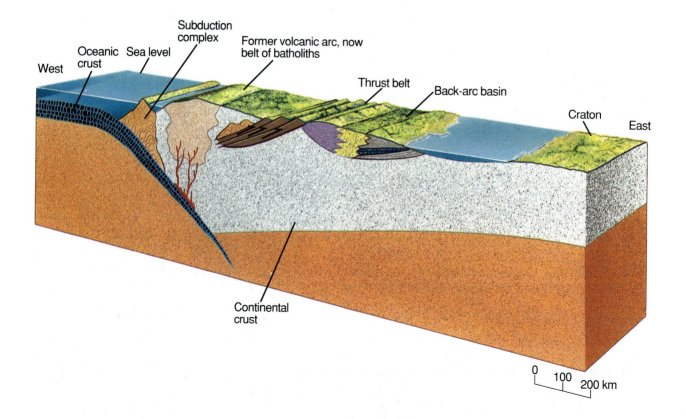

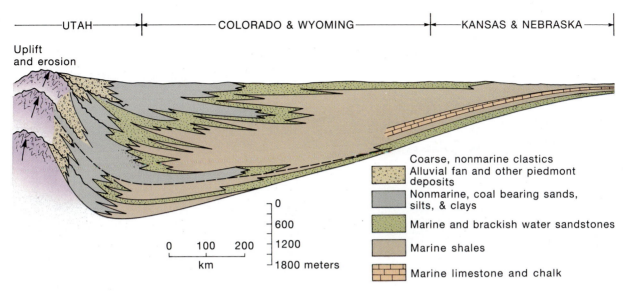

Figure 20.27   Generalized cross section illustrating the pattern of deposition of Cretaceous rocks in the back-arc basin of the western interior of the United States.
(Simplified from P. B. King, Evolution of North America *[Princeton, NJ: Princeton University Press, 1977].)*

sandstones, such as those of the late Cretaceous **Dakota Group** (Fig. 20.28). The brown layers of the Dakota are exposed in many places along the eastern front of the Rocky Mountains, where the uptilted beds form prominent ridges called *hogbacks* (Fig. 20.29). In many parts of the Great Plains, the Dakota sandstones provide ground water for agricultural and domestic use.

Cretaceous rocks of Wyoming and Colorado also include extensive beds of bentonite, a clay formed by the alteration of volcanic ash. Volcanoes in the Idaho region during the late Jurassic and Cretaceous explo-

sively ejected tremendous volumes of ash that were carried into adjacent states by westerly winds. The ash beds, subsequently converted to bentonites, yield radiometric ages and are of great value in correlation.

The Cretaceous Period closed with a withdrawal of the inland seas that had covered so much of the continent. The withdrawal was contemporaneous with the Laramide orogeny that produced the Rocky Mountains. As the seas regressed, coal-bearing deltaic and other continental sediments were gradually spread across the old sea floor.

Figure 20.28   The north side of the impressive Interstate-70 road cut west of Denver, Colorado. The rocks visible in this view are part of the Dakota Group. They are dipping steeply toward the east (right). Sandstones of the Dakota Group serve as an important source of artesian ground water beneath the Great Plains. They have also yielded oil and gas in eastern Colorado, Nebraska, and Wyoming.

Figure 20.29   Prominent north–south trending hogback formed by the eastward-dipping Dakota Sandstone of Cretaceous age, a few miles south of Golden, Colorado. The Dakota Sandstone at this location is a beach deposit of the Cretaceous sea. Well-developed ripple marks and occasional dinosaur footprints are preserved in the sandstone. *(U.S. Geological Survey)*

20.3   Mesozoic Mineral Resources

Uranium Ores

Rocks of the Mesozoic System contain a wealth of important mineral resources. Notable among these are the nuclear fuels. In the United States, uranium ores are derived chiefly from continental Triassic and Jurassic rocks of New Mexico, Colorado, Utah, Wyoming, and Texas. The most abundant mineral is **carnotite** (Fig. 20.30), a yellow, earthy oxide of uranium that usually occurs within the pore spaces of sandstones, as en-

Figure 20.30   The bright yellow uranium mineral carnotite disseminated through and encrusting brown Shinarump sandstone. Specimen is 9 centimeters long.

crustations or as replacements of fossil wood. In some instances, petrified logs have provided amazing concentrations of not only uranium but also vanadium and radium. These are striking exceptions, however, for most of the ores are of very low concentration.

Fossil Fuels

Although nuclear power plants are now common, the nations of the world still rely heavily on fossil fuels. Mesozoic rocks contain significant amounts of these critical resources. The Jurassic, for example, contains workable deposits of coal in Siberia, China, Australia, Spitzbergen, and North America. Cretaceous coal underlies more than 300,000 square kilometers of the Rocky Mountain region (Fig. 20.31). Much of this coal is prized because of its low content of environmentally offensive sulfur.

In addition to coal, Mesozoic rocks supply large quantities of oil and gas to energy-hungry industrial nations. The oil provinces of the Middle East and North Africa probably contain more oil than the combined reserves of all other countries. Middle East petroleum comes primarily from thick sections of Jurassic and Cretaceous sedimentary rocks that accumulated in the central part of the Tethys seaway. Other areas of petroleum production from Mesozoic rocks occur in the western United States, Alaska, Arctic Canada, the Gulf Coastal states, western Venezuela, southeast Asia, beneath the North Sea, and beneath the eastern offshore area of Australia. Notwithstanding the size of some re-

Cretaceous Epeiric Seas Linked to Sea-Floor Spreading

During the Cretaceous, continents were extensively inundated by inland seas. Marine sedimentary rocks in North America, Europe, Australia, Africa, and South America indicate that about a third of the land area now emergent was under water. Because all the continents experienced marine transgressions at about the same time, the cause must have been eustatic rise in sea level and not the result of subsidence of the land. However, there is little evidence of glaciation during the time that submergence was at its peak. It is therefore unlikely that meltwater from ice caps or continental glaciers was the cause of the eustatic rise in sea level. An alternative would be for crustal materials to take up space in the ocean and thereby cause a rise in sea level. This might occur if there were an increase in the rate at which oceanic crust is formed at mid-ocean ridges.

Geologists have been able to estimate the volume of new ocean crust produced during 10-million-year intervals as far back as the beginning of the late Cretaceous. These studies indicate that the rate at which ocean crust was being produced along mid-ocean ridges was exceptionally rapid. Newly formed ocean crust is hotter than older crust and therefore occupies more volume. This is evident on the present ocean floor, where the surface of newly formed crust stands 2.8 kilometers below sea level, as contrasted to 5 kilometers for older, cooler oceanic crust. Also, when spreading rates are high (producing high-volume crust), the corresponding subduction of cooler, low-volume crust is also greater. Taken together, these sea-floor spreading processes decrease the space for water in the ocean basins, causing sea level to rise and resulting in marine transgressions across formerly emergent regions of the continents. In North America, the rise in sea level was responsible for epeiric seas that advanced southward from Alberta, Canada, and joined an embayment that spread northward from the Gulf of Mexico. At its peak, that seaway was over 1600 kilometers wide and divided our continent into two separate landmasses.

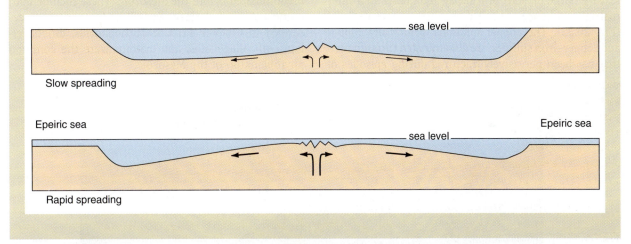

cent discoveries in rather forbidding parts of the world, it appears likely that the present rates of oil and gas consumption will cause exhaustion of the world's resources in less than a century. Therefore, oil and gas must be replaced by other energy sources in the near future, especially if we are to conserve these valuable materials for the chemical industry. The time is approaching when petroleum will be too valuable to burn.

Metallic Mineral Deposits

Metalliferous deposits were formed widely throughout the active orogenic belts of the Mesozoic world. A variety of metals now found in the Rocky Mountains, the Pacific coastal states, and British Columbia were emplaced during the batholithic intrusions that accompanied Jurassic and Cretaceous mountain building. Among these ore deposits are the gold-bearing quartz veins known as the "Mother Lode." The California gold rush of 1849 was a consequence of the discovery of gold-bearing gravels eroded from the Mother Lode.

The copper, silver, and zinc veins of the Butte, Montana, and Coeur d'Alene, Idaho mining district were emplaced as a result of Cretaceous igneous activity. One belt of copper-bearing rocks extends from Denver into northeastern Arizona. These deposits and similar ones

Figure 20.31   Coal beds in the Mesaverde Formation (Upper Cretaceous) exposed in a road cut on Highway 6-50 at Castlegate, Utah.

in Utah occur in igneous rock having a porphyritic texture (large crystals of feldspar and quartz in a finer matrix) and hence are called porphyry copper deposits (Fig. 20.32). The copper minerals, principally copper and copper–iron sulfides, are disseminated through rock that has been pervasively fractured during episodes of

explosive volcanic activity. The many fractures provided easy passage for hot, mineralizing solutions from which the ore was precipitated. Porphyry copper deposits occur not only in western North America but also in South America, the Soviet Union, Iran, the former Yugoslavia, and the Phillipines.

20.4   The Mesozoic Biosphere

Mesozoic Marine Plants

Because plants that live in the oceans are suspended in water, they do not require the vascular and support systems that characterize land plants. Most marine plants are therefore unicellular, although they may grow in impressive colonies and aggregates. The major groups of marine plants are part of the vast realm of floating organisms called plankton; because they are plants, they are further designated **phytoplankton**. The geologic record of the important groups of phytoplankton is shown in Figure 20.33. As indicated in the chart, phytoplankton that did not possess mineralized coverings predominated before the Mesozoic. These include both cyanobacteria and algae, as well as the acritarchs described in Chapter 17. Coccolithophorids, dinoflagellates, silicoflagellates, and diatoms were the common phytoplankton of the Mesozoic.

Figure 20.32   The open pit copper mine at Bingham Canyon, Utah. The copper minerals at Bingham Canyon are disseminated throughout porphyritic igneous rock.

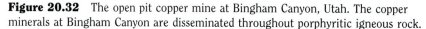

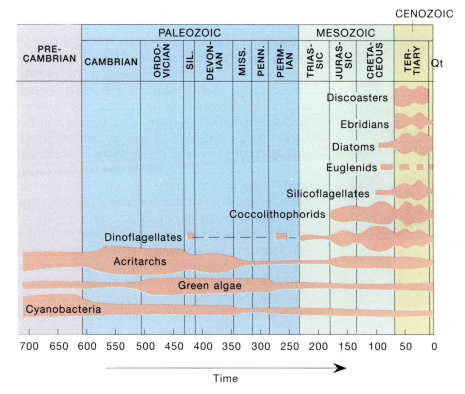

Figure 20.33 Geologic distribution and abundances of phytoplankton. *(From H. Tappan and A. R. Loeblich, Jr.,* Geol. Soc. Am. Special Paper, *127 [1970], 257.)*

Dinoflagellates are unicellular organisms with a cellulose wall and two flagella. Only the cysts of the adult organism (Fig. 20.34), which are highly resistant to decay, are known as fossils. These tiny cysts, however, are useful in stratigraphic correlation.

Coccolithophorids (Fig. 20.35) are calcium carbonate secreting organisms. Typically, the coccolithophorid secretes complex discoidal structures termed **coccoliths**, and these form a calcareous armor around the spherical cell. The coccoliths are particularly useful fossils for determining the age and correlation of Cretaceous and Tertiary limestones as well as for the sediments obtained by coring into the sea floor. Many thick and extensive chalk formations are composed largely of the skeletal remains of coccolithophorids.

Siliceous phytoplankton includes silicoflagellates and diatoms. In silicoflagellates, the cell is supported from within by a spherical or discoidal structure com-

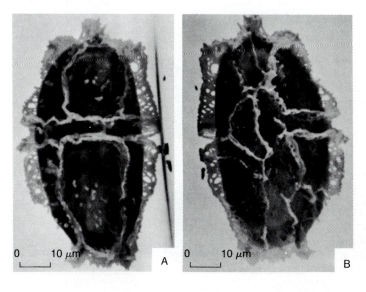

Figure 20.34 Fossil dinoflagellate cyst, *Prionodinium alveolatum,* from the Cretaceous of Alaska. *(From H. A. Leffingwell and R. P. Morgan, J. Paleontol, 51(2) [1977], 292.)*

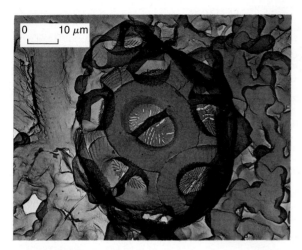

Figure 20.35 Scanning electron micrograph of a coccosphere covered with coccoliths. *(From J. M. McCormick and J. V. Thiruvathukal,* Elements of Oceanography, *2nd ed. [Philadelphia: Saunders College Publishing, 1981]. Photography courtesy of S. Honjo.)*

posed of hollow rods of opaline silica. Silica is also the skeletal material in diatoms (Fig. 20.36). The siliceous covering of the diatom cell is called the frustule, and it may be circular, cylindrical, triangular, or a variety of other shapes. Together with silicoflagellates, diatoms are most abundant in the sedimentary rock called diatomite as well as in siliceous oozes of the deep sea. In the past, a proliferation of diatoms was often associated with volcanic activity. Apparently, silica supplied to seawater as fine volcanic ash stimulated diatom productivity.

Land Plants

Although the Mesozoic Era is often dubbed the Age of Reptiles, an expert on fossil plants might argue that the title Age of Cycads is equally appropriate. The **cycads**, members of the class Cycadophyta (Fig. 20.37), are seed plants in which true flowers have not yet evolved. Jurassic cycads were tall trees with rough columnar branches marked by the leaf bases of leathery, pinnate leaves. Although highly successful throughout the Jurassic and the first half of the Cretaceous, by the late Cretaceous cycads began to decline, and only a few have survived to the present day. One such survivor is the sago palm, often used as a house plant.

In the previous chapter, it was noted that there were three important episodes in the evolution of land plants. The first stage led to the development of the spore-bearing plants. The second involved the evolution of such nonflowering, pollinating seed plants as cycads, ginkgoes (Fig. 20.38), seed ferns, and conifers. The third stage in plant history is marked by the advent of plants having enclosed seeds and flowers. Such plants are known as **angiosperms**. Angiosperms diversified widely during the Cretaceous, and wooded areas included stands of birch, maple, walnut, beech, sassafras (Fig. 20.39), poplar, and willow. Before the end of the Cretaceous, angiosperms had surpassed the nonflowering plants in both abundance and variety. Flowering trees, shrubs, and vines expanded across the lands and, except for the absence of grasses, gave the landscape a modern appearance.

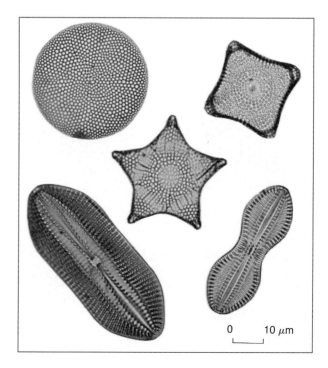

Figure 20.36 Diatoms. These modern diatoms were collected off the coast of Crete. *(Naja Mikkelsen, Scripps Institution of Oceanography)*

Figure 20.37   A living South African cycad. Note the immense seed cones.   *(W. H. Hodge/Peter Arnold, Inc.)*

The evolution of flowering plants was, either directly or indirectly, related to the evolution of Mesozoic insects, birds, reptiles, and mammals. Angiosperms produced a variety of seeds, nuts, and fruits needed for the survival of the plant embryo but that also provided for the survival of animals that used these plant products as food. The relation of insects to flowering plants is a familiar one. By encouraging insect visits, angiosperms use insects to deliver their pollen. This method provides a more efficient means of pollen dispersal than the more

Figure 20.39   Fossil leaf of a Cretaceous sassafras tree from the Dakota Sandstone, Elsworth, Kansas. The slab is about 12.5 centimeters (5 inches) wide.

Figure 20.38   *Ginkgo biloba,* the ginkgo or maidenhair tree (so called because the leaves resemble those of the maidenhair fern). Note the naked, fleshy seeds that characterize female trees. The ginkgo represents the oldest genus of living trees. Fossils over 200 million years old are nearly indentical to the living forms.
(W. H. Hodge/Peter Arnold, Inc.)

random wind pollination, which requires each plant to disperse millions of pollen grains rather than a few dozen.

Invertebrates of the Mesozoic

At the end of the Paleozoic, many families of marine invertebrates had either declined or suffered extinction. The recovery from the episode of Permian extinctions was slow, as indicated by the limited nature of early Triassic faunas. However, after a few groups had become well established, marine invertebrates expanded dramatically. The pelecypods became increasingly prolific from the middle Triassic on and quickly surpassed the brachiopods in colonizing the sea floor. Oysters were among the most abundant pelecypods, and such genera as *Gryphaea* (Fig. 20.40), *Exogyra* (Fig. 20.41), and the oddly shaped rudistid pelecypods (Fig. 20.42) were particularly widespread.

In the shallow, warm seas of the Jurassic and Cretaceous, corals built extensive reefs. Corals of the Mesozoic are called **scleractinids**. Scleractinids provided the basic framework of the reefs, and the reefs in turn provided

A

B

Figure 20.40 (A) The pelecypod *Gryphaea* is found in shallow marine Jurassic and Cretaceous strata around the world. It is a member of the oyster family and is characterized by a large left valve that is arched up and over the smaller right valve. (B) Fossils of *Gryphaea* exposed by weathering of the Tununk Shale Member of the Mancos Shale of Cretaceous age. *(Tununk Shale photograph courtesy of Lehi Hintze)*

Figure 20.41 *Exogyra arietina* from the Cretaceous Del Rio Clay of Texas. *Exogyra* differs from *Gryphaea* in having the beak of the larger left valve twisted to the side so as not to overhang the right valve. Both genera occur in enormous numbers in shell banks of Cretaceous age along the Gulf and Atlantic coastal plains.

Figure 20.42 The late Cretaceous rudist pelecypod *Coralliochama orcutti* from Baja California. (A) Cluster of specimens from the reef deposit. (B and C) Two single specimens. *(From L. Marincovich, Jr., J. Paleontol, 49(1) [1975], 212–223.)*

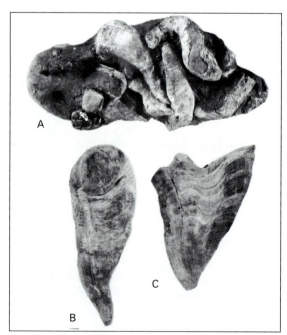

habitats for a host of other kinds of marine invertebrates and fishes. Persisting groups of brachiopods clung to the reef structures, as did pelecypods, gastropods, bryozoans, echinoderms (Fig. 20.43), and crustaceans. Among the latter, modern types of crayfish, lobsters, crabs, shrimp, and ostracodes were abundant. At some localities barnacles grew in profusion.

Those cousins of the crinoids, the echinoids (Fig. 20.44), became far more diverse and abundant in the Mesozoic than they had been in the preceding era. Some early Cretaceous formations in the Gulf Coast states contain prodigious remains of these spiny creatures. Collectors in Europe prize the silicified echinoids found in Cretaceous chalk beds. In addition to echinoids, starfish and serpent stars (Fig. 20.45) were also abundant in Mesozoic seas.

Among the molluscan groups of the Mesozoic, none is more useful in precise regional correlation than the ammonoid cephalopods. You may recall that two orders

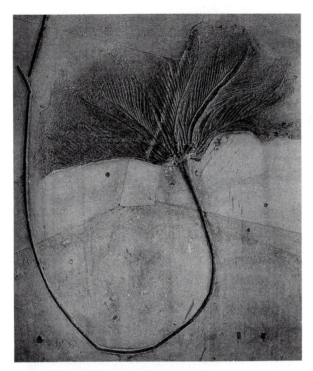

A

B

Figure 20.43 Large, well-preserved specimen of the crinoid *Pentacrinus subangularis* from Lower Jurassic strata near Holzmaden, Germany. The extended arms span about 1.2 meters. *(Washington University collection, photograph by John Simon)*

Figure 20.44 (A) The regular fossil echinoid *Cidaris* from Jurassic marine strata of Germany. (B) *Hemiaster,* a fossil irregular echinoid from the Cretaceous of Mississippi. Specimens are about 6 centimeters in diameter.

Figure 20.45 Two members of the echinoderm class Stelleroidea. (A) An ophiuroid (brittle star or serpent star) from the Triassic of England. Slab is about 10 centimeters wide. (B) An imprint of an asteroid (starfish) in a California Cretaceous sandstone. The starfish is about 15 centimeters across.

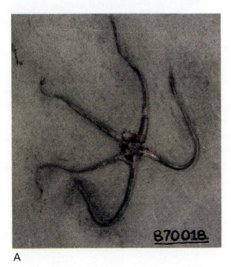

A

B

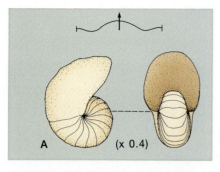

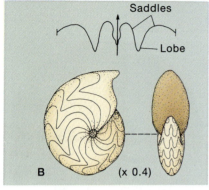

Saddles

Lobe

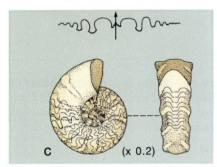

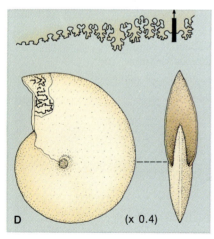

Figure 20.46 Sutures of cephalopods. (A) Nautiloid cephalopod (arrow at mid-ventral line pointing toward conch opening). (B) Ammonoid cephalopod with goniatitic sutures. (C) Ammonoid with ceratitic sutures. (D) Ammonoid cephalopod with ammonitic sutures. *(From W. H. Twenhofel and R. R. Shrock, Invertebrate Paleontology [New York: McGraw-Hill, 1935]. Copyright © 1935 McGraw-Hill Book Co.)*

Figure 20.47 Cretaceous ammonoid cephalopods from the Mancos Shale of New Mexico. (A) *Hoplitoides sandovalensis* (vertical diameter of 9.3 centimeters). (B) Apertural view of another individual of same species. (C) Part of outer whorl of *Tragodesmoceras socorroense* (maximum height of 16 centimeters).

(W. A. Cobban, U.S. Geological Survey)

of cephalopods arose during the Paleozoic Era: the Nautiloidea, which have smooth sutures, and the Ammonoidea, which have wrinkled or pleated sutures. **Sutures** of cephalopods are lines formed on the inside of the conch where the edge of each chamber's partition or **septum** meets the inner wall. Wrinkled septa are simply a reflection of septa that, like the edges of a pie crust, are pleated. Based on the complexity of the suture patterns (Fig. 20.46), **ammonoids** can be divided into three groups: goniatites, which lived from Silurian to Permian time; ceratites, which were abundant during the Permian and Triassic; and ammonites (Fig. 20.47). Ammonites, although found in the rocks of all three Mesozoic periods, were most prolific during the Jurassic and Cretaceous.

As one examines the complex suture patterns of ammonoids, it is difficult not to wonder about the purpose or function of pleated septal margins. Most paleontologists favor the theory that, like the corrugated steel panels used in buildings, fluted septa gave the conch the strength needed to prevent collapse because of external water pressure. Nautilods are able to resist external water pressure because their conch wall is thicker than in ammonoids.

The rich variety of Mesozoic ammonoids indicates their success in adapting to a variety of environments and living strategies. Near the end of the Cretaceous, however, the entire diverse assemblage began to decline. By the end of the Cretaceous, this once highly successful group had become extinct. Their close relatives, the nautiloids, survive.

Another group of common Mesozoic cephalopods were the squidlike **belemnites**. The shell of a belemnite was internal. The posterior part resembled a bullet with the pointed end at the rear of the animal. The forward part was chambered. As indicated by the carbonized imprints of a few well-preserved fossils, ten tentacles surrounded the mouth. Most belemnite shells were less than a half meter long, but some attained lengths of over 2 meters.

Single-celled protozoans such as radiolarians (Fig. 20.48) and foraminifers (Fig. 20.49) were also abundant in Mesozoic seas. Radiolarians secrete opaline silica in the construction of their open, delicate skeletons. Along with diatoms, their remains form extensive deep-sea deposits of siliceous ooze.

The shells (called tests) of foraminifers are generally more durable than those of radiolarians. They have left an imposing Mesozoic fossil record and one of great importance in stratigraphic correlation. Because of their small size and strong tests, large numbers of fossil foraminifers can be obtained from small pieces of rock

Figure 20.48 Scanning electron photomicrographs of Jurassic radiolaria from the Coast Ranges of California. Top row, left to right: *Paronaella elegans, Crucella sanfilippoae,* and *Emiluvia antiqua.* Bottom row, left to right: *Tripocyclia blakei, Parvicingula santabarbaraensis,* and *Parvicingula hsui.* *(From E. A. Pessagno, Jr.,* Micropaleontology, *23(1) [1077], 56, 113, selected from Plates 1–12.)*

Figure 20.49   A living planktonic foraminifer similar to those that became abundant during the Cretaceous and Cenozoic. The fine rays are cytoplasm that is extruded through the pores of the tiny calcium carbonate shell. The shell is approximately 85 μm in diameter. *(Manfred Kage/Peter Arnold, Inc.)*

brought to the surface during the drilling of oil wells. Not only are these fossils useful in tracing stratigraphic units from well to well, but they are often sensitive indicators of the water temperature, depth, and salinity at the time of deposition.

Vertebrates of the Mesozoic

Although marine invertebrate faunas changed abruptly in passing from the Paleozoic into the Mesozoic, there was considerable continuity among land animals. The labyrinthodont amphibians continued into the Triassic before becoming extinct, and the mammal-like reptiles were also able to cross the era boundary. The most progressive of the mammal-like reptiles, the ictidosaurs, lived on as contemporaries of the first mammals.

Many new groups of reptiles appeared in the Triassic. Among these were the first true turtles. The Triassic was also the period during which many lineages of marine reptiles appeared. Most interesting of all, however, were reptiles known as **archosaurs**. The Archosauria are a large and important group of reptiles that include dinosaurs, flying reptiles (pterosaurs), crocodiles, and thecodonts. Thecodonts have a distinguished place in vertebrate history, for they are the ancestors of the dinosaurs.

Thecodonts

Early **thecodonts**, as exemplified by *Hesperosuchus* (Fig. 20.50), were small, agile, lightly constructed animals with long tails and short forelimbs. They had evolved the habit of walking on their hind legs. This bipedal mode was an important innovation. Bipedalism

permitted thecodonts to move about more speedily than their sprawling ancestors. Because their forelimbs were not used for support, they could be employed for catching prey; even more important, they could be modified for flight. Thus, the thecodonts were the ancestors not only of dinosaurs but of flying reptiles as well.

Not all thecodonts were nimble, bipedal sprinters; some reverted to a four-footed stance and evolved into either armored land carnivores or aquatic reptiles called **phytosaurs** (Fig. 20.51). Occasionally in the history of life, initially unlike organisms from separate lineages

Figure 20.50   *Hesperosuchus* from the Triassic of the southwestern United States. Adult *Hesperosuchus* was about 4 feet long.

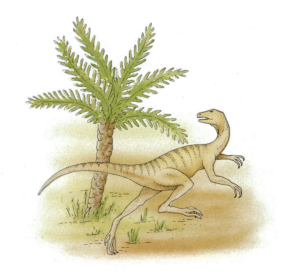

Figure 20.51   *Rutiodon,* a Triassic phytosaur.

gradually become more alike. The evolutionary process responsible for the trend toward similarity in form is called **convergence**. Phytosaurs and crocodiles provide a good example of evolutionary convergence. Indeed, the most visible distinction between the two groups is the position of the nostrils, which are at the end of the snout in crocodiles but just in front of the eyes in phytosaurs.

Dinosaurs

Of all the reptiles that have ever lived on this planet, few are more fascinating than the dinosaurs (Fig. 20.52).

Figure 20.52   Evolution of the dinosaurs.   *(From E. H. Colbert,* Evolution of the Vertebrates *[New York: John Wiley & Sons, 1969]. Used with permission of the author, artist Lois Darling, and publisher.)*

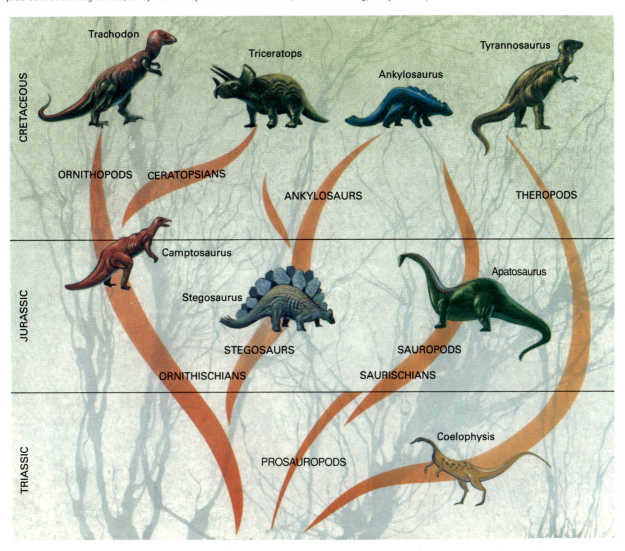

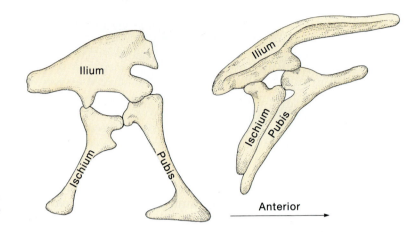

Figure 20.53 Basis for the division of dinosaurs into two groups. On the left is the arrangement of pelvic bones in the *Saurischia;* on the right, the arrangement in *Ornithischia.* These views show one side only; the bones are duplicated on the other side.

They are the most awesome, familiar, and popular of all prehistoric beasts. These headliners of the Age of Reptiles compose not one order, but two, each having evolved separately from the thecodonts. The two orders are the Saurischia (lizard-hipped) and ornithischia (bird-hipped). As suggested by these names, the arrangement of bones in the hip area provides the criterion for the twofold division. The reptile pelvis is composed of three bones on each side. The uppermost is the ilium, which is firmly clamped to the spinal column. The bone extending downward and slightly backward is the ischium. Forward of the ischium is the pubis. In the saurischians, the arrangement of the three pelvic bones is triradiate, as it was in their thecodont ancestors. In the ornithischians, however, the pubis is swung backward so that it is parallel to the ischium, as in birds (Fig. 20.53).

The earliest dinosaurs were nearly all saurischians. Most were relatively light, nimble, carnivorous bipeds known from fossils discovered in Triassic beds of both South and North America as well as China. *Coelophysis*

(Fig. 20.54), for example, was a small, hollow-boned early saurischian found in the Chinle Formation of New Mexico. These birdlike reptiles, called coelurosaurs, continued into the Jurassic and Cretaceous. *Ornithomimus* (Fig. 20.55) must have looked very much like an ostrich, with its long neck, toothless jaws, and small head. Because it lacked teeth, paleontologists speculate that it lived on the eggs laid by its contemporaries.

The larger carnivorous saurischians (including coelurosaurs) are called **theropods**. Of more dramatic dimensions than the coelurosaurs were the large carnosaurians such as *Allosaurus* (Fig. 20.56) of the Jurassic and the Cretaceous dinosaurs *Deinonychus* (Fig. 20.57) and *Tyrannosaurus* (Fig. 20.58). The last-named beasts attained lengths of over 13 meters and weighed in excess of 4 metric tons. Carnosaur hindlimbs were robust and muscular. Great curved claws for tearing flesh protruded from each of three toes, whereas a nearly functionless fourth toe bore a smaller claw. The forelimbs were small but powerful. As befits an animal that must kill with

Figure 20.54 The late Triassic carnivorous dinosuar *Coelophysis*. This reptile was about 2.5 meters long. *(From David Peters, A Gallery of Dinosaurs and Other Early Reptiles [New York: Alfred A. Knopf, Inc.]. Copyright © 1989, with permission.)*

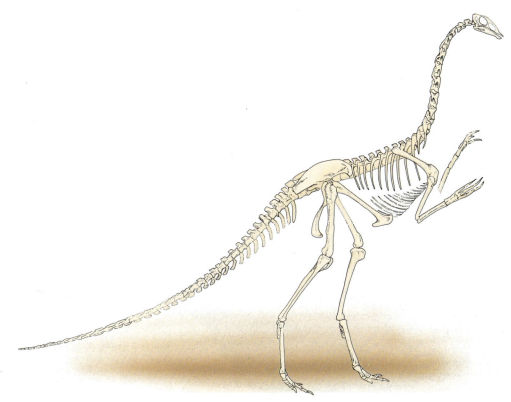

Figure 20.55 *Ornithomimus,* an ostrich-like Cretaceous dinosaur that was capable of swift running. *(From H. F. Osborn,* Bull. Am. Mus. Nat. Hist., *35 [1917], 733–777.)*

Figure 20.56 Skull of the carnivorous late Jurassic theropod *Allosaurus.* Note the long jaws and dagger-like teeth, which *Allosaurus* used effectively in dealing with its prey of other reptiles. The skull consisted largely of a framework of sturdy bony arches to which powerful muscles were attached. *(After H. F. Osborn,* Bull. Am. Mus. Nat. Hist., *35 [1917], 733–771.)*

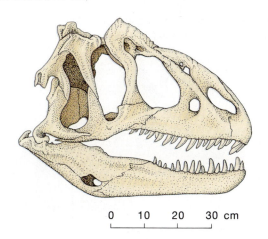

```
 0    10    20    30 cm
```

its jaws and teeth, the head of the carnosaur was large. Doubly serrated teeth, up to 6 inches long, lined the powerful jaws. These were truly spectacular predators.

The saurischian group included herbivorous **sauropods** as well as flesh-eating theropods. The ancestry of the sauropods can be traced to the late Triassic protosauropod known as *Plateosaurus* (Fig. 20.59). From this smaller, partially bipedal form, the more typical giant sauropods evolved at the beginning of the Jurassic and survived right up to the end of the Cretaceous. They are the animals that people first think of when they hear the word *dinosaur.* The best-known sauropods were enormous, long-necked, long-tailed beasts that had returned to the four-legged stance to support their tremendous bulk. A well-known Jurassic representative whose remains have been found in the Morrison Formation of Colorado is *Apatosaurus* (Fig. 20.60). This favorite of schoolchildren (formerly known as *Brontosaurus*) measured almost 20 meters in length and weighed over 30 metric tons. Yet *Apatosaurus* was a relative lightweight when compared to *Supersaurus* and *Ultrasaurus,*

Figure 20.57   The Cretaceous dinosaur *Deinonychus.* This reptile was about 8 feet long. It possessed a large skull, and the margins of the jaws were set with serrated, saber-like teeth.

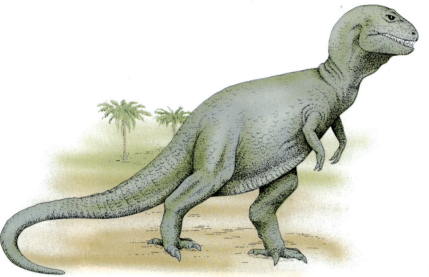

Figure 20.58   *Tyrannosaurus,* a late Cretaceous theropod.

Figure 20.59   *Plateosaurus,* the late Triassic ancestor of the giant sauropods.

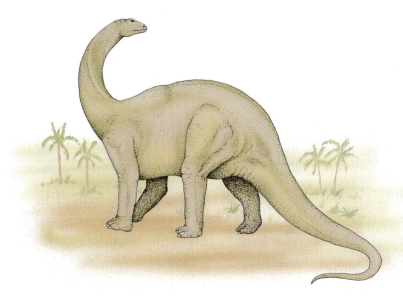

Figure 20.60   The enormous Jurassic sauropod *Apatosaurus* (formerly called *Brontosaurus*).

whose weights probably exceeded 80 tons. *Seismosaurus,* the Earth shaker, was over 35 meters (115 feet) long. Like its smaller relative *Diplodocus* (Fig. 20.61), it had a neck and tail that were unusually long even for a sauropod.

For many years paleonotologists speculated that, even with their pillar-like legs, these largest of land animals could not have supported their own weight continuously. It was therefore surmised that they dwelt in the buoyant waters of lakes and streams. However, there is little evidence to support this theory. It is likely that sauropods roamed the Jurassic forests and uplands much like modern elephants.

Large size afforded certain advantages to the sauropods. Predators often avoid encounters with huge animals. In addition, great size in reptiles also serves to slow changes in body temperature. The ratio of surface area to mass for an animal decreases as size increases. Consequently, a large animal has a proportionately smaller surface for heat loss, and just as a large pot of water loses its heat more slowly than a small pot, so does a large animal lose its heat more slowly than a small animal.

The other major dinosaur line, the Ornithischia, evolved near the end of the Triassic and thrived throughout the remaining Mesozoic. Ornithischians were plant

Figure 20.61   Mounted skeleton of the lengthy Jurassic sauropod *Diplodocus.* From head to tail, *Diplodocus* measured 27 meters (88 feet) long.   *(National Museum of Natural History)*

FOCUS ON   *Changing Views of the Dinosaurs*

Until a few decades ago, paleontologists engaged in the study of dinosaurs devoted most of their efforts to the discovery and description of dinosaur skeletal remains. Assumptions about a dinosaur's behavior were often based on comparisons with crocodiles, large land turtles, or creatures like the Komodo dragon (a giant lizard) of Indonesia. As a result, most dinosaurs were seen as cold-blooded, solitary, and weak-minded. This view of dinosaurs began to change in 1969, when John H. Ostrom presented evidence that some dinosaurs were very likely warm-blooded. Other paleontologists quickly joined Ostrom with additional new arguments for endothermy in dinosaurs, as well as fresh ideas about dinosaur behavior, physiology, and ecology. Among these investigators was a tall, laconic dinosaur hunter named John (Jack) R. Horner.

In 1978, Horner was shown a fossil site in the Cretaceous Two Medicine Formation of western Montana. On close examination, it became evident that the fossil location included a hatchery of hadrosaurian ornithopod dinosaurs, complete with nests, clutches of eggs, hadrosaur embryos, and nestlings. There was evidence that hadrosaur babies were nurtured by their parents, that they lived within the social structure of large herds, that they were warm-blooded, and that, in general, they behaved more like birds than living reptiles.

Excavations within the Two Medicine Formation revealed that hadrosaurs had hollowed out bowl-shaped nests in soft fluvial sediments and that within each nest they had laid about 20 eggs in neatly arranged circles. Impressions in the sediment suggest that the eggs were covered with decaying vegetation so that fermentation would provide warmth. Nests were about 7.5 meters (24.6 feet) apart from one another, a distance approximating the length of the adult parent. Horner found two lines of evidence suggesting that the parent hadrosaurs, named *Maiasaura* (from the Greek word for good mother lizard), nurtured their young. He observed that

many of the nests contained the bones of juveniles that were about a meter long. Thus, babies that were only about 30 centimeters long at the time they hatched stayed in their nests, where food was brought to them until they had grown sufficiently to fend for themselves. In addition, the teeth of these juveniles exhibited distinct signs of wear, suggesting they had hatched earlier and had been feeding for some time while still in the nest. The interpretation that newly hatched babies stayed in the nest until they tripled their size also supported the interpretation that *Maiasaura* was warm-blooded. A baby crocodile would require three years to triple its size. It is inconceivable that *Maiasaura* babies would have stayed in their nests three years. Like today's ostriches, they probably attained their meter-long size in just a few months. Such rapid growth occurs only in endothermic vertebrates.

Reference
J. R. Horner and J. Gorman, *Digging Dinosaurs* (New York, Workman Publishing, 1988).

---

eaters. The teeth in the forward part of the jaws were replaced by a beak suitable for cropping vegetation. The group included both quadrupedal and bipedal varieties, with the bipedal condition considered more primitive. Even the most advanced quadruped ornithischians had such short forelimbs that their descent from bipedal forms seems certain.

The bipedal group of ornithischians is known as **ornithopods**. Their evolutionary history began in the Triassic with relatively small species that lived primarily

on dry land. A representative large Jurassic ornithopod is *Camptosaurus* (Fig. 20.62). This was a bipedal dinosaur of medium size with a heavy tail, short forelimbs, and long hind legs. The articulation of the jaw was arranged to bring the teeth together at the same time, an arrangement frequently seen in herbivores down through the ages. Leaves and stems were cropped by the forward, beaklike part of the jaws and passed backward to the cheek teeth for chopping and chewing. From camptosaurid-like ancestors, the larger Cretaceous or-

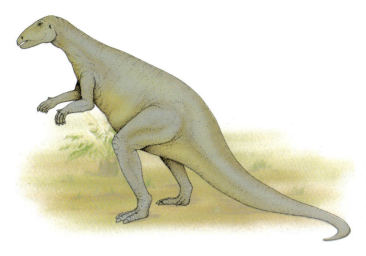

Figure 20.62 *Camptosaurus,* a small to medium-sized (2 to 3 meters long) herbivorous dinosaur. Although there were no teeth in the beaklike forward part of the jaw, the remainder of the jaw was equipped with a tight mosaic of row upon row of teeth suitable for grinding plant food.

nithopods developed. Among these was *Iguanodon* (Fig. 20.63), one of the first dinosaurs to be scientifically described. *Iguanodon,* sometimes called the thumbs-up dinosaur because of the horny spike that substituted for a thumb, is thought to have been a gregarious animal that may have moved about in herds. Evidence for this comes from a Belgian coal mine in which 29 of these individuals were found together as a result of having fallen into an ancient fissure.

By Cretaceous time, ornithopods had moved into a variety of terrestrial environments. A particularly successful group were the trachodonts or hadrosaurs. Some

of the hadrosaurs appear to have been aquatic and inhabited lakes and streams. The evidence for this consists of skin impressions showing that the feet were webbed. The vertically flattened tail of these reptiles may have been used in swimming. Other hadrosaurs, however, seem to have lived in drier areas of coastal plains and flood plains. Discoveries of nesting sites in what were the foothills of the ancestral Rocky Mountains indicate they may have moved to higher ground for nesting.

In hadrosaurs, the forward part of the face was broad, flat, and toothless, sometimes resembling the bill of a duck (Fig. 20.64); hence the name *duckbill* dino-

Figure 20.63 *Iguanodon* (in rear) was a herbivorous ornithischian dinosaur that lived during the early Cretaceous. Approaching from the right is the carnivorous dinosaur *Baryonx.* The boy in the painting is for scale. *(From David Peters,* A Gallery of Dinosaurs and Other Early Reptiles *[New York: Alfred A. Knopf, Inc.] Copyright © 1989, with permission.)*

Figure 20.64   Mounted skeletons of two Cretaceous duckbill dinosaurs (trachodonts) on display at the American Museum of Natural History in New York.

saurs for members of this group. Behind the toothless forward part of the jaw were lines of lozenge-shaped teeth that appear to have been well adapted for chewing coarse vegetation.

An interesting peculiarity of some hadrosaurs was the development of bony skull crests containing tubular extensions of the nasal passages (Figs. 20.65 and 20.66). The function of these bizarre structures is of special interest to paleontologists. It has been suggested recently that the cranial crests were used to catch the attention of sexual partners of the same species and could be further employed as vocal resonators for promoting an individual's success in obtaining a breeding partner.

Not all ornithopods had skull crests, and in some the crests lacked nasal tubes. One crestless ornithopod that certainly qualifies as a legitimate "bone head" was *Pachycephalosaurus* (Fig. 20.67). The skull in pachycephalosaurians consisted mostly of solid bone with only a small space to accommodate the unimpressive brain.

The best known of the quadrupedal ornithischians are the **stegosaurs** (Fig. 20.68). Stegosaurs had two pairs of heavy spikes mounted on the tail that were used for defense. However, their more identifying feature was the row of pentagonal plates that stood upright along the back. Scientists have debated the purpose of these plates for many years. Probably the plates functioned in the regulation of body temperature by serving as body heat dissipaters. Indeed, the arrangement, size,

Figure 20.65   (A) Skull of the hadrosaur *Lambeosaurus* with its peculiar hatchet-shaped crest. (B) *Corythosaurus* had a helmet-shaped crest. These crests may have functioned as vocal resonators for dinosaur bellowing. The skulls are approximately 0.75 meters (30 inches) long.   *(National Museum of Natural History)*

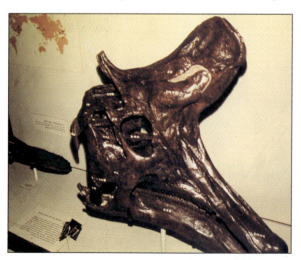

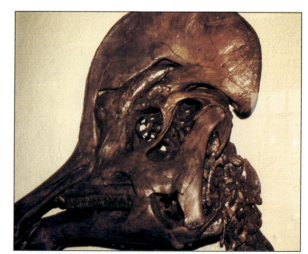

A   B

GEOLOGY AND THE ENVIRONMENT

Volcanic Ash: A Recurrent Geologic Hazard

In A.D. 79, hot gas and ash erupted violently from Mount Vesuvius and overwhelmed the Roman cities of Pompeii and Herculaneum. Excavations of these once thriving cities have uncovered paved streets, temples, dwellings, and even the remains of human victims of the eruption. Some of the bodies entombed in ash left perfect molds which, when filled with plaster, provided casts that show even the tortured expressions on the faces of the hapless victims. Similar fiery catastrophes have occurred many times in the geologic past. One particularly interesting eruption resulted in sudden death to a mighty herd of dinosaurs.

In northern Montana just east of the Rockies, paleontologist John R. Horner found a bonanza of dinosaur bones scattered over an area about one quarter of a mile wide by one and a quarter miles long. Horner and his team estimated that there were over 30 million pieces of bone in this area, representing the skeletal remains of about 10,000 dinosaurs. A layer of volcanic ash lay directly on top of the bed containing the extraordinary concentration of bones. Horner further observed that many of the elongate bones (such as limb bones) were uniformly aligned in an east–west direction and that smaller bones were concentrated at one end of the bone layer. Finally, he noticed that the bones were smoothly broken. They were not splintered in the way that bone breaks in a recently killed vertebrate.

From these observations, Horner was able to reconstruct a scenario for events at this location during one small moment of the geologic past. He envisioned a

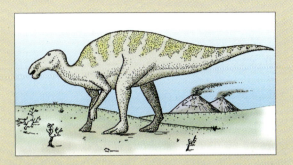

great herd of duckbill dinosaurs browsing peacefully for a time on the flood plain of a river flowing eastward from the Rockies. The duckbills were members of the genus *Maiasaura* (see figure). Nearby, one or more of the many volcanoes abundant in this region during the Cretaceous erupted, engulfing the surrounding area in scorching, lethal gases and suffocating ash. The terrified dinosaurs in the herd had no way to escape. Like the people of Pompeii, they died in agony. All that remained was an ash-covered killing field littered with the partially buried bodies of *Maiasaura*. Much later, after the bones had become mineralized by ground-water solutions, a turbulent flood passed over the area, transporting and aligning the bones and carrying the smaller ones to the far end of the area. The flood would account for the absence of complete *Maiasaura* skeletons, and the smooth fractures on the dinosaur bones suggest that the deluge occurred long after the volcanic event.

Reference
J. R. Horner and J. Gorman, *Digging Dinosaurs* (New York: Workman Publishing, 1988).

and shape of the plates, and their probable rich supply of blood vessels, would favor the thermoregulatory suggestion, although the plates may have served as camouflage or protection as well.

During the Cretaceous, stegosaurs were succeeded by the heavily armored **ankylosaurs**. These bulky, squat ornithischians were completely covered by closely fitted bony plates across the entire length of their 6-meter backsides. The head was small and armored. Small, weak teeth were present in the jaws of some

ankylosaurs, but *Nodosaurus* (Fig. 20.69) was completely toothless.

The fourth group of quadrupedal ornithischians are the **ceratopsians**. These beasts take their name from horns that grew on the face of all but the earliest forms. Typically, ceratopsians possessed a median horn just above the nostrils, and in some species an additional pair projected from the forehead. The head was quite large in proportion to the body and displayed a shieldlike bony frill at the back of the skull roof that served as

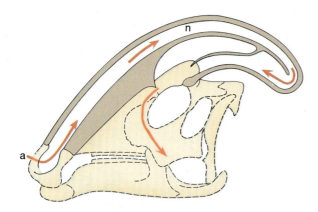

Figure 20.66 Internal structure of the skull crest of *Parasaurolophus cyrtocristatus.* The superficial bone of the left side has been removed to expose the left nasal passage *(n)*. Air enters nostrils at *a,* moves up and around partition in crest, and from there moves down and back to internal openings in the palate. *(From J. A. Hopson,* Paleobiology, *1 [1975], 1.)*

Figure 20.67 Reconstruction of the head of *Pachycephalosaurus,* the "bone-head" dinosaur.

protection for the neck region and as a place of attachment for powerful jaw and neck muscles. All ceratopsians possessed jaws that had the shape of a parrot's beak (Fig. 20.70). Judging from the scars that mark the shield bones of some ceratopsians, they were often attacked by the great carnosaurs. No doubt they frequently emerged the victor. During the late Cretaceous, ceratopsians moved eastward from Asia across the land connection to North America and inhabited the region that lay to the west of the epicontinental sea.

Were Dinosaurs Warm-Blooded? Since the late eighteenth century, when dinosaurs were first studied scientifically, they have been regarded as reptiles and hence ectothermic or cold-blooded. Ectothermic animals have little or no ability to maintain a uniform body temperature by physiologic processes. Some ectotherms, however, may regulate their body temperature to a certain degree by seeking either sun or shade in

Figure 20.68 *Stegosaurus.* Restudy of the fossil remains of this dinosaur indicates that the plates along the back may have been arranged in a single line rather than in two alternating rows, as has been frequently depicted.

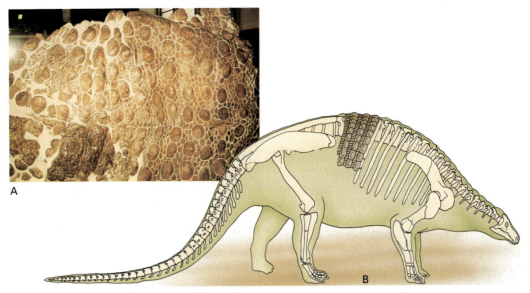

Figure 20.69   The Cretaceous ankylosaur *Nodosaurus*. This heavily armored dinosaur was about 5 meters long. Part of its armored carapace is shown in the photograph.
(Drawing from R. S. Lull, Amer. Jour. Sci., 5th ser. *1 [1921], 97–126.)*

response to temperature needs. In living reptiles, the pineal gland may play a role in directing this type of behavior. In extinct reptiles, certain anatomical features such as the sail in *Dimetrodon* or the plates on the back of *Stegosaurus* may have served to catch the Sun's rays or dissipate body heat.

In contrast to largely ectothermic animals, endothermic animals such as mammals and birds maintain

a constant body temperature by physiologic production of heat internally and the radiation of excess heat away from the body. Like other animals, mammals produce heat by oxidizing food. However, when body temperature rises, special physiologic mechanisms regulated by the hypothalamus (part of the brain) help to dissipate the heat. These mechanisms include expansion of the blood vessels in the skin, perspiring, or (in furry ani-

Figure 20.70   The Cretaceous ceratopsian dinosaurs *Styracosaurus* (A) and *Triceratops* (B). To an advancing enemy, *Styracosaurus* presented a many-horned threat.   *(Figure A from David Peters,* A Gallery of Dinosaurs and Other Early Reptiles *[New York: Alfred A. Knopf, Inc.]. Copyright © 1989, with permission.)*

A

B

mals) panting. When temperatures fall, other mechanisms—such as restriction of blood vessels in the skin and shivering—minimize heat loss.

In 1968, paleontologist Robert Bakker proposed that dinosaurs were warm-blooded, or endothermic, and thus could no longer be classified as reptiles. Bakker knew that birds and dinosaurs are similar in general anatomical design. If, like birds, dinosaurs were truly endothermic animals, then it would appear logical to restructure the classification of vertebrates by removing dinosaurs from the class Reptilia and erecting a new taxonomic class that would include both dinosaurs and birds. The class Aves would be eliminated. In the new classification, there would be four classes of terrestrial vertebrates: the class Amphibia (ectothermic amphibians), the class Reptilia (ectothermic reptiles such as lizards and snakes), the class Dinosauria (endothermic dinosaurs and birds); and the class Mammalia. In a very broad interpretation of this proposed reclassification of vertebrates, dinosaurs still "live" today—as birds.

Bakker supported his theory of dinosaurian endothermy with several interesting lines of evidence. One relates to the way dinosaurs stood and walked. Today's lizards and salamanders are ectotherms and most have a sprawling stance, with their limbs directed more or less to the side. In contrast, mammals and birds stand and walk more erect. The limbs are held directly beneath the body. Dinosaur stance resembles that of mammals and birds, and hence, by correlation, dinosaurs were endothermic. But is this an entirely valid correlation? Dinosaur posture may be only an evolutionary solution for supporting their enormous weight.

A second observation considered by some to favor dinosaur endothermy relates to the microscopic structure of bones. Dinosaur bones are richly vascular, like the bones of mammals. Most reptiles have less vascular bones, indicating a poorer supply of blood to the bone tissue. The correlation, however, is not absolute. Some of the bones in living crocodiles have considerable bone vascularity, and crocodiles are primarily ectothermic. Perhaps the high amount of vascular bone in the dinosaurs evolved in response to requirements related to growth rates or size and was not related to endothermy.

Isotope analysis of bone also provides a test of endothermy. In ectothermic animals there are large differences in the oxygen isotope content of bones of the extremities as compared to bones of the body core. Endothermic vertebrates do not have this disparity. Analysis of bone from Cretaceous dinosaurs shows isotope variability similar to that in warm-blooded vertebrates.

Remains of dinosaurs have been discovered at Jurassic and Cretaceous paleolatitudes that would normally have had temperate or even cool climates. Their occurrence at such northern locations provides yet another argument in favor of their being warm-blooded.

Among the many additional arguments for dinosaur endothermy is one that seeks to make a correlation between predator to prey proportions in mammals (endotherms) as opposed to living reptiles (ectotherms). Today's endothermic communities consist of about 3 percent predators and 97 percent plant-eating prey. It clearly takes a lot of food to fuel the energy requirements of an endothermic predator such as a lion or wolf. In the ectothermic community, 33 percent of the animals are predators and 66 percent are prey. Determining the proportion of predators to prey among dinosaurs is rather tricky because the fossil record cannot be as precise as data obtained from living communities. At the present time, the evidence is contradictory.

In the debate about dinosaur warm-bloodedness, it should be remembered that endothermy is often a matter of degree. Even if it were demonstrated that dinosaurs were not warm-blooded in the usual sense, this does not mean that some did not achieve a measure of body temperature regulation.

Aerial Reptiles of the Mesozoic

Again and again in the history of life, the descendants of a small group of animals that were initially adapted to a narrow range of ecologic conditions have, by means of evolutionary processes, radiated into peripheral environments. As the new groups diverged from their ancestral lineage, they changed in ways that made them better suited to their new surroundings. This process, known as **adaptive radiation**, is well demonstrated by the Mesozoic reptiles. The adaptive radiation originated with the stem reptiles of the late Carboniferous and ultimately produced the enormous variety of large and small, herbivorous and carnivorous, dry land and aquatic animals of the Mesozoic. The radiation did not end with terrestrial vertebrates, however, for during the Mesozoic, reptiles invaded the marine environment and even managed to overcome the pull of gravity for short periods of time.

The first reptiles to attempt to conquer the air were not active flyers, but rather were gliders similar to present-day flying lizards and so-called flying squirrels. For such animals, gliding provided an easy way to move from tree tops to the ground or from branch to branch. The earliest of such gliders were discovered in rocks of Permian age. They have been given the name *Coelurosauravus*. The skin membranes that served as wings for *Coelurosauravus* were supported along the sides of the body by greatly elongated ribs. Rib-supported wings are not unusual in gliding reptiles. The Triassic lizard *Icarosaurus* (named after Icarus, the ill-fated son of Daedalus) also evolved a similar support for its wings. It is also a characteristic of the present-day *Draco* lizard of the Orient.

A distinct different structure for flying appeared in the Triassic reptile *Longisquama*. The scales in this bizarre creature were as long as its body, and along the forward edge of its forelimbs they were nearly seven times as long. The long scales could be spread out to either side to make an effective wing. Paleontologists believe that feathers could have evolved from such scales.

The Triassic beds that contained the remains of *Longisquama* also yielded a reptile that was not a passive glider but an active flyer. Its name is *Sharovipteryx*. Because this reptile could truly fly and maneuver in the air, it is considered the first known **pterosaur**. In *Sharovipteryx,* the skin membranes extended between the elbows and the knees and from the rear legs to the tail (Fig. 20.71A).

A discovery from Jurassic beds in Russia confirmed the generally accepted theory that active, wing-flapping

Figure 20.71 Two flying reptiles from Central Asia. (A) *Sharovipteryx* is a Triassic form that is particularly interesting because it appears to represent an intermediate stage between primitive forms and more highly adapted flyers of the Jurassic and Cretaceous (B). The larger form is *Sordes pilosus,* of late Jurassic age. The hairy covering of *Sordes pilosus,* clearly visible in the fine limestone that enclosed the fossil, provides evidence that these creatures were warm blooded.

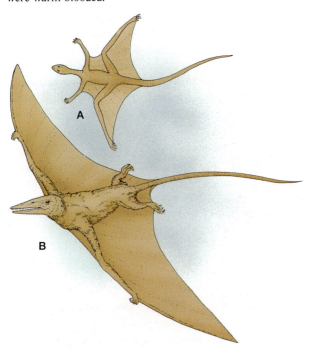

reptilian flyers must have been able to maintain a constant high internal temperature. Without that kind of metabolism, cold air would have reduced their power of exertion and they would have stalled. The actual evidence was found in a well-preserved pterosaur that had a covering of soft hair. Appropriately, it was named *Sordes pilosus* (Fig 20.71B), meaning "the hairy devil."

The pterosaurs may appear to us as ugly, graceless creatures, but their existence from late Triassic until late Cretaceous attests to their adaptive success. The Jurassic and Cretaceous pterosaurs that are most familiar typically had rather large heads and eyes and long jaws, which in most forms were lined with thin, slanted teeth. The bones of the fourth finger were lengthened to help support the wing, whereas the next three fingers were of ordinary length and terminated in claws. The wing was a sail made of skin stretched between the elongate digit, the sides of the body, and the rear limbs. There were two general groups of pterosaurs: the more primitive were the rhamphorhynchoids, which had long tails, and the more advanced were tailless pterodactyloids. The latter group is exemplified by *Pteranodon* (Fig. 20.72), species of which had an astonishing wingspan of over 7 meters. The body of *Pteranodon* was about the size of that of a goose. The skeleton was lightly constructed, as is fitting for an aerial vertebrate. The animals probably soared along much like oversized sea birds, snapping up various sea creatures in their toothless jaws. Relative to body size, pterosaurs had somewhat larger brains than some of their land-dwelling relatives. Perhaps this was a result of the higher level of nervous control and coordination needed for flight.

The first prize for large size among pterosaurs must certainly go to *Quetzalcoatlus northropi* (Fig. 20.73) from Upper Cretaceous beds of western Texas. This giant, named after an Aztec god that took the form of a feathered serpent, had an estimated wingspan of 15.5 meters (50.8 feet). *Quetzalcoatlus northropi* ranks as the largest known flying vertebrate ever to soar over the Earth's surface.

A Return to the Sea

The marine habitat is one in which the archosaurs were not notably successful. Only one archosaurian group— the sea crocodiles—were able to invade the oceanic environment. Other reptile groups, such as the ichthyosaurs, plesiosaurs, mosasaurs, and sea turtles, however, were highly successful in adapting to life in the sea. Many fed upon the ammonoids and fishes that had preceded them in populating the seas. During the Cretaceous, the most modern of the bony fish, the teleosts, made their appearance.

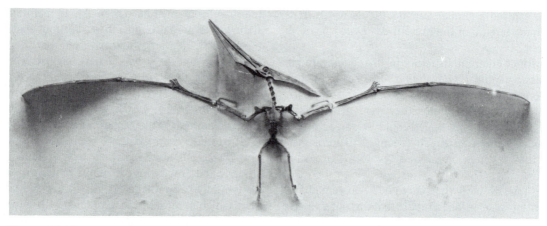

Figure 20.72   Restored and mounted skeleton of the great flying reptile *Pteranodon ingens* from Cretaceous chalk deposits of Kansas. *Pteranodon* had a wingspan of over 7 meters.

Not unexpectedly, the invasion of the marine environment required many modifications in form and function. Paddle-shaped limbs and streamlined bodies evolved to allow efficient movement through the water. Because these reptiles were unable to abandon air breathing and reconvert to gills, their lungs were modi-

Figure 20.73   the giant flying reptile *Quetzalcoatlus* discovered in Upper Cretaceous strata of West Texas. This reptile had a wingspan of 15.5 meters. The flying reptile *Pterodaustro* stands in front of *Quetzalcoatlus.*   *(From David Peters,* A Gallery of Dinosaurs and Other Early Reptiles *[New York: Alfred A. Knopf, Inc.], Copyright © 1989, with permission.)*

fied for greater efficiency. In those sea-going reptiles that were unable to lay their eggs ashore, reproductive adaptations provided for birth of the young while at sea.

Marine reptiles that had paddle-shaped limbs for locomotion were already present during the Triassic Period. One group, the **nothosaurs**, were just beginning to take on adaptations that would be perfected in their likely descendants, the pleosiosaurs. The nothosaurs were joined in the Triassic by a group of mollusk-eating, flippered reptiles known as **placodonts** (Fig. 20.74). These bulky animals had distinctive pavement-type teeth in the jaws and palate, which they used for crushing the shells of the marine invertebrates upon which they fed.

By far the best known of the paddle swimmers were the **plesiosaurs** (Fig. 20.75). Their earliest remains are found in Jurassic strata. Sometimes nicknamed "swan lizards," they had short, broad bodies and large, many-boned flippers. In some species the neck was extraordinarily long and was terminated by a smallish head. Slender, curved teeth, suitable for ensnaring fish, lined

Figure 20.74   The skull of *Placodus,* a bulky, mollusk-eating, paddle-limbed marine reptile of Triassic age. The forward teeth were modified for plucking shellfish from the sea floor, and the cheek teeth for crushing their shells.

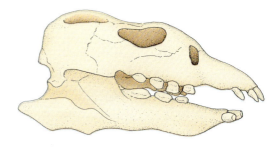

Figure 20.75   Cretaceous plesiosaurs. *(From a painting by David Peters, with permission.)*

the jaws. *Elasmosaurus,* a well-known Cretaceous plesiosaur, attained an overall length in excess of 12 meters.

In addition to the long-necked types, there were many plesiosaurs characterized by short necks and large heads. It is likely that the short-necked plesiosaurs were aggressive divers. *Kronosaurus,* a giant, short-necked form from the Lower Cretaceous of Australia, had a skull size that probably exceeds that of any known reptile. It was 3 meters long.

The most fishlike in form and habit of all the marine reptiles were the Triassic and Jurassic **ichthyosaurs** (Fig. 20.76). In many ways, they were the reptilian counterparts of the toothed whales of our present-day oceans. Ichthyosaurs had fishlike tails in which vertebrae extended downward into the lower lobe, boneless dorsal fins to help prevent sideslip and roll, and paddle limbs for steering and braking. The head was a suitably pointed entering wedge for cutting rapidly through the water. A ring of bony plates surrounded the large eyes and may have helped protect them against water pressure.

Clearly, these were active predators with good vision and the ability to move swiftly through the water.

The **mosasaurs**, a group of giant monster lizards, were a highly successful group of Cretaceous sea dwellers. A typical mosasaur (Fig. 20.77) had an elongate body and porpoise-like flippers. The creatures propelled themselves through the water by the sculling action of their long, vertically flattened tails and the rhythmic undulations of their long bodies. The lower jaw had an extra hinge at midlength which greatly increased its flexibility and gape. Mosasaurs were primarily fish eaters, but some frequently dined on large mollusks. Conchs of cephalopods have been found with puncture marks that precisely match the dental pattern of their mosasaurian foes.

Perhaps less spectacular than the mosasaurs, but far more persevering, were the sea turtles. In this group, also, we find a trend toward increased size. The Cretaceous turtle *Archelon,* for example, reached a length of nearly 4 meters. As an adaptation to their aquatic habitat, the carapace of the marine turtle was greatly reduced and the limbs were modified into broad paddles.

As briefly indicated earlier, the marine crocodiles (Fig. 20.78) were the only archosaurian group that entered the sea. They became relatively common during the Jurassic, but only a few remained by the early Cretaceous. It is likely that they did not fare well in competition with the mosasaurs.

The Birds

From the time of Darwin, naturalists have been aware of the structural similarities between birds and reptiles.

Figure 20.76   Restoration of an ichthyosaur, a marine reptile similar to a modern porpoise in form and habits.

Figure 20.77   Mounted skeleton of a Cretaceous mosasaur. These giant marine lizards attained lengths of over 9 meters (30 feet).   *(National Museum of Natural History)*

These similarities have prompted the statement that birds are only glorified reptiles that have gained wings and feathers and lost their teeth. However, such statements depreciate the marvelous attainments of birds, not the least of which are their superior powers of flight and high level of endothermy. Both of these attributes are related to the evolution of feathers. Developmentally, feathers are homologous with reptilian scales. In their earliest development, they may not have functioned in flight but instead for insulation, camouflage, or display.

There is little doubt that birds evolved from small Triassic theropods that were already birdlike in their bipedal stance and in the structure of their forelimbs and hindlimbs, shoulder girdle, and skull. Earlier in this chapter we noted a proposed new classification of vertebrates in which the close relationship between dinosaurs and birds is indicated by the inclusion of both groups within the same taxonomic class (the Dinosauria).

Figure 20.78   Toothy skull of the Jurassic marine crocodile *Geosaurus.* Length of skull is about 45 centimeters.

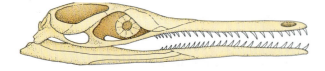

The first undisputed fossil bird to be discovered, *Archaeopteryx,* is the perfect evolutionary link between small, bipedal theropods and modern birds. *Archaeopteryx* (Fig. 20.79) was about the size of a crow. With the exception of its distinctly fossilized feathers (arranged on the forelimbs as in modern flying birds), its skeletal features were largely reptilian. The jaws bore teeth, and the creature had a long, lizard-like tail that bore feathers. Unlike the wings of modern birds, in which the bones of the digits coalesce for greater strength, the primitive wings of *Archaeopteryx* retained claw-bearing free fingers for climbing and grasping. The light-weight sternum lacked a keel, indicating that the heavy muscles needed for sustained flight were lacking. Because of this, some paleontologists believe *Archaeopteryx* was a structurally primitive evolutionary side branch and that it did not give rise to more advanced fliers. Support for this view came in 1986 with the discovery of two crow-sized skeletons of birds that not only were 75 million years older than *Archaeopteryx* but also possessed keel-like breast bones. Although feather impressions were not found with the skeletons, the tiny bones of the forearms possessed a series of nodes that may have been the sites for the attachment of feathers. This seemingly more advanced yet older vertebrate has been given the name *Protoavis.*

Small, delicate, hollow-boned animals are not readily preserved, and the fossil record for birds—especially Mesozoic birds—is not good. The cretaceous provides

A   B

Figure 20.79   The Jurassic bird *Archaeopteryx*. (A) The skeleton of *Archaeopteryx* in
the Solnhofen Limestone of Germany. The Solnhofen beds formed from lime mud
deposited on the floor of a tropical lagoon. The specimen resides in the Berlin Museum
of Natural History.   *(John D. Cunningham/Visuals Unlimited)*   (B) Restoration of *Archaeopteryx*.
(From a painting by Rudolph Freund, courtesy of the Carnegie Museum of Natural History)

only a partial glimpse of bird evolution. Toothed birds
otherwise resembling terns and gulls are occasionally
found in the Cretaceous deposits of the inland chalky
seas. Some Cretaceous birds, such as *Hesperornis*, became excellent swimmers. They retained feathers and
other birdlike characteristics but lost their wings and
relied on their webbed feet for swimming (Fig. 20.80).
Marine sediments provided the rapid burial needed to
preserve these aquatic birds. On land, preservation of
Mesozoic birds was a rare occurrence.

The Mammalian Vanguard

While the Mesozoic reptiles were having their heyday,
small, furry animals were scurrying about in the under-
growth and unwittingly awaiting their day of supremacy.
These shrewlike creatures were the primitive mammals.
On the basis of rare and often minuscule remains, they
are known from all three systems of the Mesozoic.
Among the earliest of the mammals were the **morganucodonts**, jaws, teeth, and skull fragments of which have
been recovered in Upper Triassic rocks of south Wales.

It is evident from these rather scrappy remains that
morganucodonts still retained many vestiges of reptilian
structures. However, the articulation of the jaw to the
skull was mammalian, and the lower jaw was function-
ally a single bone, as in mammals. As indicated by the
shapes of teeth and their patterns of cusps, Mesozoic
mammals included plant and insect eaters, small predators, and species that had a lifestyle similar to that of
rodents. One group, the **eupantotheres**, is of particular
interest, for in the Cretaceous they gave rise to marsupials and placentals.

For the mammals, the Mesozoic was a time of evolutionary experimentation. For 100 million years, they
effectively and unobtrusively lived among the great reptiles while simultaneously improving their nervous, circulatory, and reproductive anatomy. Equipped with an
exceptionally reliable system for control of body temperature, they were able to thrive in cold as well as warm
climates. As the reptile population declined near the
end of the era, mammals quickly expanded into the
many habitats vacated by the saurians.

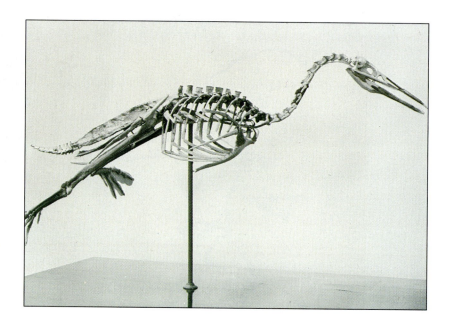

Figure 20.80 Skeleton of the Cretaceous diving bird *Hesperornis*. This bird was over a meter tall when standing. Propulsion was provided by its large webbed feet. A notable primitive trait of this bird was the retention of teeth. *(U.S. Natural History Museum)*

20.5   A Second Time of Crisis

Just as the end of the Paleozoic was a time of crisis for animal life, so also was the conclusion of the Mesozoic. Primarily on land but also at sea, extinction overtook many seemingly secure groups of vertebrates and invertebrates. In the seas, the ichthyosaurs, plesiosaurs, and mosasaurs perished. The ammonoid cephalopods and their close relatives, the belemnites, as well as the rudistid pelecypods, disappeared. Entire families of echinoids, bryozoans, planktonic foraminifers, and calcareous phytoplankton became extinct. On land the most noticeable losses were among the great clans of reptiles. Gone forever were the magnificent dinosaurs and soaring pterosaurs. Turtles, snakes, lizards, and crocodiles, and the New Zealand reptile *Sphenodon* (the tuatara lizard) are the only reptiles that survived the great biologic crash. Overall, in the late Cretaceous, extinctions eliminated about one fourth of all known families of animals.

The question of what caused the decimation in animal life at the end of the Mesozoic continues to intrigue paleontologists. Scores of theories, some scientific and many preposterous, have been offered. Those that have the most credibility attempt to explain simultaneous extinctions of both marine and terrestrial animals and seek a single or related sequence of events as a cause. In general, the theories tend to fall into two broad categories. The first of these categories relies on some sort of extraterrestrial interference, such as an encounter with an asteroid or comet. Because it is difficult to distinguish between the effects of a meteorite impact and those of a comet, scientists employ the term **bolide** to designate any large extraterrestrial body that strikes the Earth's surface.

The second group of hypotheses proposes that events that occurred here on Earth were the cause of extinctions. Such events are not called catastrophic because their effects were distributed over a relatively long span of time.

Extraterrestrial Causes of Extinctions

Asteroid Impact

Since the time that geologists first became aware of the extinctions that mark the Cretaceous–Tertiary boundary, there have been attempts to place the blame for those extinctions on meteorites, comets, or lethal cosmic radiation. Tangible evidence to support those ideas, however, was lacking. This situation changed in 1977 when geologist Walter Alvarez discovered a thin layer of clay at the Cretaceous–Tertiary boundary outside of the town of Gubbio, Italy (Fig. 20.81). Alvarez sent samples of the clay to his father Luis, a physicist, who had the clay analyzed. The result of the analysis was startling. The samples contained approximately 30 times more of the metallic element iridium than is normal for the Earth's crustal rocks. Where could this high concentration of iridium have come from? Iridium is probably present in the Earth's core and perhaps the mantle, but how could the metal from so deep a source find its way into a clay layer at the Cretaceous–Tertiary boundary?

Figure 20.81 The coin marks the location of the iridium-rich layer of clay that separates Cretaceous and Tertiary rocks near Gubbio, Italy. The gray Cretaceous limestone below the coin contains abundant fossil coccolithophores, but few of these phytoplankton remain in the Tertiary beds above the iridium-rich clay layer. *(Lawrence Berkeley Laboratory, University of California)*

While volcanism is a possibility, iridium also occurs in extraterrestrial objects such as asteroids (or their fragments, meteorites). The father and son Alvarez team favored this extraterrestrial origin for the iridium in the clay layer and proposed that an iridium-bearing asteroid had crashed into the Earth at the end of the Cretaceous. The explosive, shattering blow from the huge body (presumed to have been over 10 kilometers in diameter) would have thrown dense clouds of iridium-bearing dust and other impact ejecta into the atmosphere. Mixed and transported by atmospheric circulation, the dust might have formed a lethal shroud around the planet, blocking the Sun's rays and thereby causing the demise of marine and land plants upon which nearly all other forms of life ultimately depend. As the dust settled, it would have formed the iridium-rich clay layers found at Gubbio and subsequently in Denmark, Spain, New Zealand, North America, Austria, Haiti, the former Soviet Union, and in sediment layers beneath the Atlantic and Pacific oceans.

In addition to the iridium-rich clay layer found at various sites around the world, there are other kinds of evidence for the impact of a large extraterrestrial body at the end of the Cretaceous. One of these is the widespread occurrence of **shocked quartz** (Fig. 20.82) at the Cretaceous–Tertiary boundary. The mineral grains are recognized by distinctive parallel sets of microscopic planes (called shock lamellae) that were produced when high-pressure shock waves, such as those emanating from the impact of a large meteorite, travel through quartz-bearing rocks. Often in the same stratum containing grains of shocked quartz one also finds tiny glassy spher-

Figure 20.82 A shocked-quartz grain from the Cretaceous–Tertiary boundary layer. Note the parallel intersecting sets of microscopic planes called shock lamellae. These intersecting lamellae are similar to those observed in quartz from rock fragments clearly known to have been subjected to meteorite impact. Maximum diameter 0.3 millimeter. *(Jim S. Alexopoulos)*

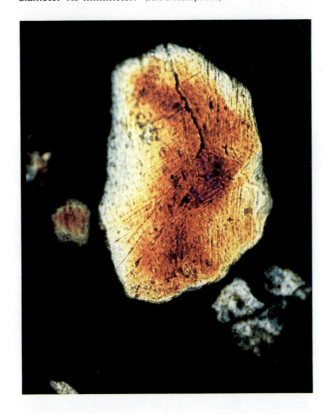

FOCUS ON   *Crater Linked to Demise of Dinosaurs*

If we accept the hypothesis that a huge bolide smashed into our planet at the end of the Cretaceous (with catastrophic effects on life), where then is the crater produced by that lethal event? The most probable candidate for the crater appears to have been located near the town of Chicxulub, Mexico, in the northern part of the Yucatán Peninsula (see accompanying map). At this location, magnetic and gravity surveys, as well as cores and logs of oil wells, reveal a buried, circular, crater-like structure about 180 kilometers in diameter. Andesitic rock exists in the central core of the structure. The andesite has an isotopic and chemical composition similar to that of small particles of glass called tektites that are abundant in the Cretaceous–Tertiary (K/T) boundary layer at many locations in the Caribbean region. Tektites are thought to form by melting and rapid cooling during meteorite impact. Thrown into the air during impact, they can be transported worldwide by wind systems.

Further evidence of bolide impact is found in core samples of rocks penetrated during the drilling of oil wells in and around the Chicxulub structure. Prominent among these rocks are coarse breccias that occur both above and interbedded with andesite. The breccias appear to be part of a blanket of severely fragmented rock produced by a massive impact. Abundant grains of shocked quartz (see Fig. 13.62) and feldspar occur in clasts of the breccia.

Assuming the Chicxulub structure is indeed the crater formed by bolide impact, does its age correlate with the extinctions that were widespread at the end of the Cretaceous? Isotopic age determinations indicate an af-

firmative answer to this question. Melt rocks subjected to $^{40}Ar/^{39}Ar$ analyses provide an age of 65.2 ± 0.4 million years. In addition, these same melt rocks acquired remanent magnetism indicating they had solidified during an episode of reverse geomagnetic polarity known to exist at the time of deposition of the K/T boundary layer.

If the Chicxulub structure is indeed an impact crater as the predominance of evidence suggests, it is the largest ever blasted out of the Earth. The bolide that produced the crater would have an estimated diameter of more than 10 kilometers. As it splashed down it would have produced a wave over 34 kilometers high that would flood land areas adjacent to the Caribbean and Gulf of Mexico. Rock would be vaporized, and the atmosphere would fill with dust, water vapor, and carbon dioxide from melted limestones. These conditions would have caused great stress on life of the late Cretaceous.

ules thought to represent droplets of molten rock thrown into the atmosphere during the impact event. Finally, sediments at the Cretaceous–Tertiary boundary often include a layer of soot that may be the residue of vegetation burned during widespread fires caused by extraterrestrial impact.

If a huge comet or asteroid (a **bolide**) did strike our planet about 65 million years ago, one would hope to find traces of the crater it produced. Recently, a buried impact structure of appropriate age, shape, and size was discovered in the Yucatán Peninsula of Mexico. (See the Commentary, "Crater Linked to Demise of Dinosaurs.")

The bolide-impact hypothesis seems a tidy way to account for the extinction of dinosaurs and many of

their plant and animal contemporaries. Like all hypotheses, however, it must stand the test of scrutiny by the scientific community. Were the Alvarezes correct in assuming that the iridium was derived from a bolide? And, if such an event did occur, was it sufficient to cause geologically instantaneous extinctions or merely to contribute to lethal environmental processes already in operation?

In regard to the first question, geologists have found evidence that iridium in clays like that at Gubbio may have its source in the Earth's mantle, from which it can move to the surface by way of conduits and blast into the atmosphere as iridium-rich volcanic ash and dust. Volcanism was indeed very prevalent during the late stages of the Cretaceous.

If we assume the impact hypothesis is correct, then it is still difficult to explain why some animals perished from the proposed blast, yet small mammals, birds, lizards, crocodiles, bony fishes, many deciduous plants, and certain mollusks survived the hard times and proliferated during the Cenozoic.

Terrestrial Causes of Extinctions

Several hypotheses explain the extinction of plants and animals near the end of the Cretaceous as a consequence of environmental changes that upset the ecologic balance needed for survival. Earlier in this chapter we noted the presence of vast epeiric seas during the Cretaceous. These warm inland seas were very favorable for marine life and helped to moderate climates on the continents as well. Under these conditions, plants and animals increased in variety and abundance. These favorable conditions, however, were soon to change. Studies of the stratigraphic sequence across the Cretaceous–Tertiary boundary indicate a general lowering of sea level at the end of the Mesozoic. That change in sea level spelled disaster to animals and plants living in shallow coastal areas and to the phytoplanton on which the marine food web depends. Without the moderating effects of vast tropical seas, continental regions would also have experienced harsher climatic conditions and more extreme seasonality. Unable to adjust to the new conditions, many groups would have met their end. Their demise would result in waves of extinctions among species lower in the food chain.

A somewhat similar scenario might result from an episode of extensive volcanism. A volcanic event that occurred at the right time and on a suitably immense scale was the extrusion of phenomenal volumes of basaltic lava in northwestern India. These now solidified lavas (called the Deccan traps) form India's Deccan Plateau. Vent and fissure eruptions in India released large quantities of carbon dioxide, possibly causing lethal temperature and climatic changes.

Whatever the final outcome of the controversy over a terrestrial or extraterrestrial cause for extinctions, the fact remains that hard times near the end of the Mesozoic doomed the dinosaurs, pterosaurs, ammonites, and over three fourths of the known species of marine plankton. The extinctions, however, were not always simultaneous or sudden. Particular groups died out over an interval of 0.5 to 5 million years. This observation leads many paleontologists to reject extinction caused by a sudden extraterrestrial event. Perhaps the real cause of the great dying will be found in a combination of the two categories of hypotheses. Fossil evidence indicates that extinctions were underway in the late Cretaceous. The debilitated and diminished survivors of environmental hard times may have been dealt a *coup de grace* by a bolide.

SUMMARY

Geographic change was a significant characteristic of the 160 million years of Mesozoic history, as large segments of Pangaea separated from the supercontinent. North America lost its connection to Gondwanaland, and by late in the Jurassic rifts had developed between the Gondwanaland segments of Australia and Antarctica. These two continents, like northeastern North America, Greenland, and northwestern Europe, retained a hold on one another until early in the Cenozoic.

At the beginning of the era, most of North America was emergent. In the east, crustal tension produced large-scale block faulting and volcanism. In downfaulted valleys, arkosic sandstones, conglomerates, and lake shales accumulated. While this was occurring, the ancestral Gulf of Mexico began to take form. Thick sequences of limestones, evaporites, and calcareous shales were deposited in the Gulf and Atlantic coastal region during the Cretaceous and Jurassic.

Continental red beds characterize the Triassic of the Rocky Mountain region, whereas marine sedimentation continued along the Pacific margin. During the Jurassic, this far-western belt was deformed as North America began to override a subduction zone at the leading edge of the eastward-moving Pacific plate. The advancing Pacific plate carried volcanic arcs, oceanic plateaus, and microcontinents to the western margin of North America, where they became incorporated into the Cordillera as a tectonic collage of displaced terranes. Growth of a continental margin in this way is called accretionary tectonics.

In general, deformation of the North American Cordillera during the Mesozoic began along the Pacific coast and moved progressively eastward. By the late Cretaceous, inland seas that had occupied the western interior of the continent had withdrawn. Majestic mountain ranges stood in their place. Crustal unrest continued to affect the western states even during the early epochs of the Cenozoic.

Rocks of Mesozoic age have supplied the world with a variety of important mineral resources. These include both fossil and nuclear fuels and such metals as copper, zinc, silver, chromite, gold, lead, and mercury. Most of the ore minerals were emplaced during orogenesis. They are concentrates of materials that had been conveyed to orogenic belts by sea-floor spreading.

In the Mesozoic seas, **coccoliths** and **diatoms** flourished, as did such invertebrate groups as **ammonoids**, **belemnites**, oysters, and other pelecypods, echinoderms, corals, and **foraminifers**. On land, seed ferns and conifers were common in Triassic and Jurassic forests. Flowering plants expanded during the Cretaceous, and with them a multitude of modern-looking insects. The changing composition of plant populations was matched by innovations among terrestrial vertebrates, as dinosaurs, pterosaurs, and birds evolved from the small bipedal **thecodonts** of the Triassic.

The dinosaurs were the ruling reptiles of the Mesozoic. Both carnivorous and herbivorous varieties occupied a variety of habitats. Based on a basic difference in pelvic structure, two orders of dinosaurs, the **Saurischia** and **Ornithischia**, are recognized. Certain of the dinosaurs as well as the pterosaurs may have been endothermic, as indicated by bone structure, posture, and predator–prey relationships. The reptilian dynasty extended to the oceans as well, where **ichthyosaurs**, **plesiosaurs**, and **mosasaurs** competed successfully with the most modern of fishes, the **teleosts**. The Mesozoic is also noteworthy as the era during which two new vertebrate classes appeared: mammals and birds. The birds, with true feathers for insulation and flight, evolved from small carnivorous dinosaurs (theropods). The oldest remains that are unquestionably those of a true bird are of *Archaeopteryx* from the Jurassic. *Protoavis,* from Triassic strata, is currently being evaluated as a possibly older bird.

Mammals made their debut during the Triassic. In general, most of these primitive mammals remained small and inconspicuous. One group, the **eupantotheres**, gave rise to marsupial and placental mammals during the Cretaceous.

Like the Paleozoic, the Mesozoic Era closed with an episode of extinctions. Many groups of both marine and terrestrial reptiles succumbed, as did the ammonoids and belemnites. Several of the major taxonomic classes survived, but entire families within those classes were exterminated. Still other groups of Mesozoic organisms experienced rapid evolutionary development and thereby were able to keep pace with environmental changes.

Geologists are not agreed on the cause of the terminal Cretaceous extinctions. There is, however, an impressive body of evidence that a large meteorite or asteroid struck the Earth about 65 million years ago, and the effects of the impact may have caused or contributed to the mass extinction. Others argue that the extinctions can be attributed to extensive volcanic activity, withdrawal of epeiric seas, and changes in climate.

KEY TERMS

Newark Group *499*	Sundance Sea *513*	Ammonoids *525*	Adaptive radiation *538*
Rudist *505*	Morrison Group *513*	Belemnite *525*	Pterosaur *539*
Allochthonous terranes *505*	Dakota Formation *515*	Archosaur *526*	Nothosaur *540*
Moenkopi Formation *508*	Carnotite *516*	Thecodont *526*	Placodont *540*
Chinle Formation *508*	Phytoplankton *518*	Phytosaur *526*	Plesiosaur *540*
Navajo Sandstone *508*	Coccoliths *519*	Convergence *527*	Ichthyosaur *541*
Wingate Sandstone *508*	Cycad *520*	Sauropod *529*	Mosasaur *541*
Nevadan orogeny *509*	Angiosperm *520*	Ornithopod *532*	Eupantotheres *543*
Sevier orogeny *509*	Scleractinian *521*	Stegosaur *534*	Morganucodont *543*
Decollément *509*	Sutures (of	Ankylosaur *535*	Bolide *545*
Laramide orogeny *512*	cephalopods) *525*	Ceratopsian *535*	Shocked quartz *546*

REVIEW QUESTIONS

1. What is the total duration of the Mesozoic Era? Which period of the Mesozoic was the longest?

2. Describe the source area for sediments of the Morrison Formation. What evidence indicates that the Morrison Formation was *not* a marine formation?

3. During which period of the Mesozoic were epeiric seas most extensive? When were they most limited?

4. At what time in geologic history did the modern Atlantic Ocean begin to form? Which region of the Atlantic was the last to open?

5. What nonmetallic mineral resources are obtained from Mesozoic rocks? What metallic ores? How might sea-floor spreading be related to the origin of metallic ore deposits in mountain ranges?

6. What kind of deformation is particularly characteristic of the Sevier orogeny?

7. During what geologic period did most of the ranges of the Rocky Mountains undergo their major deformation?

8. When were large volumes of evaporites deposited in the Gulf Coast region? Do any of these evaporites form oil traps?

9. How do coccolithophores and diatoms differ in morphology and composition? Which has a role in the formation of chalk?

10. How did the terrestrial plants of the Cretaceous differ from earlier Mesozoic plants?

11. What Paleozoic marine invertebrates are absent or diminished in the Mesozoic?

12. What two classes of vertebrates appear for the first time in the Mesozoic?

13. Formulate a single-sentence description of each of the following Mesozoic animals.

   a. pterosaur f. theropod

   b. ichthyosaur g. eupantothere

   c. sauropod h. mosasaur

   d. ornithischian i. morganucodont

   e. teleost j. saurischian

14. What reptile groups managed to survive the extinctions that occurred at the end of the Cretaceous?

DISCUSSION QUESTIONS

1. In what way(s) might prodigious phytoplankton productivity affect atmospheric composition and global climate?

2. What attributes of ammonoid fossils have resulted in their having great value as guide or index fossils?

3. What evidence might one seek in attempting to establish that a particular group of Mesozoic reptiles was endothermic?

4. What is the meaning of evolutionary convergence? Cite two examples of living vertebrates that exhibit convergence with extinct reptiles.

5. What attributes present in Mesozoic mammals contributed to their survival during the biological crisis that decimated the reptiles at the end of the Cretaceous?

6. What evidence at the boundary between the Cretaceous and Tertiary systems supports the bolide-impact hypothesis for the extinction of the dinosaurs? What evidence tends to dispute the bolide-impact hypothesis as a cause of dinosaur extinctions?

CHAPTER 21

The Cenozoic Era

The landscapes of the modern world acquired their present form during the Cenozoic Era.
Even the spectacular scenery of the Colorado Plateau has been sculpted since about
middle Tertiary time. This view of the Colorado River was taken from Dead Horse Pass, Utah.
(Lehi Hintze)

The Cenozoic is the latest era of geologic time. It is the era during which the continents and their landscapes acquired their present form. Many of the animals and plants of the present world have been shaped and modified by Cenozoic events.

There are two sets of terms used for the periods of the Cenozoic Era (Table 21.1). European geologists prefer a time table that divides the era into **Paleogene, Neogene,** and Quaternary periods, arguing that this is better suited to the Cenozoic sequence in Europe. American geologists generally prefer using the terms **Tertiary** and **Quaternary** for the Cenozoic, and these terms are used in this book.

The Cenozoic was a time of considerable tectonic plate motion and sea-floor spreading. It has been estimated that approximately 50 percent of the present ocean floor has been renewed along mid-ocean ridges during the past 66 million years. Much of this new ocean floor was emplaced in the expanding Atlantic and Indian oceans. As this widening progressed, the Americas moved westward. The region that is now California came into contact with the northward-moving Pacific plate and thereby produced the San Andreas fault system. South America moved against the Andean trench and actually bent and displaced it. Orogenic and volcanic activity was vigorous along the western backbone of the Americas and resulted in the formation of the Isthmus of Panama, which today links North and South America. Greenland separated from Scandinavia along the North Atlantic rift, destroying the land connection between Europe and North America. Far to the south, Australia parted company with Antarctica and moved to its present location. A branch of the Indian Ocean rift split Arabia from Africa and in the process created the Gulf of Aden and the Red Sea (Fig. 21.1). The most dramatic crustal events of the Cenozoic, however, were the collisions of Africa and India with Eurasia. These magnificent smashups produced the Alps and Himalayas.

The interior regions of continents stood relatively high during much of the Cenozoic. As a result, marine transgressions were limited. Climatic zones were more sharply defined than in earlier periods, and a gradual cooling trend is apparent from studies of fossil plants, as well as pollen grains derived from those plants.

21.1   Before the Ice Age

Eastern United States

The eastern margin of North America was relatively quiet during the Cenozoic. Erosion continued in the Appalachians. As these uplands were being reduced by erosion, there were repeated, gentle uplifts. Revitalized by each uplift, streams sculpted new generations of ridges and valleys. The most recent of the uplifts was accompanied by tilting of the Atlantic coastal plain and adjacent parts of the continental shelf. Clastic sediments carried out of the highlands by streams were deposited on the plains and shelf areas to the east. The Cenozoic section of marine rocks is thinner near the Appalachian source area and becomes thicker and less clastic eastward (Fig. 21.2). In the region where Florida now lies, terrigenous clastics were less available. Carbonates accumulated along a series of subsiding coralline platforms that were similar to the Bahama Banks today (Fig. 21.3). In the final stages of the Tertiary, the northern end

of this tract was uplifted, forming much of the land area of Florida.

Gulf Coast

The best record of Cenozoic strata in North America is found in the Gulf Coastal Plain (Fig. 21.4). The rocks of eight major transgressions and regressions occur in this region. The Paleocene transgression brought marine waters as far north as southern Illinois. In the cycles of transgression and regression, near-shore deltaic sands were often deposited above offshore shales. The resulting alternation and interfingering of permeable sands and impermeable shales provided hundreds of traps for oil and gas.

"A wedge of sediment that thickens seaward" is an appropriate description of the Cenozoic deposits of the Gulf Coast. Seismic data suggest that the thickness of Tertiary sediments beneath the northern part of the Gulf region may exceed 10,000 meters. The floor of the Gulf must have subsided periodically to provide space for this immense volume of sediment.

Table 21.1 • Geochronologic Terminology Used for Divisions of the Cenozoic Era

ERA	PERIOD		EPOCH
CENOZOIC	QUATERNARY		RECENT
			PLEISTOCENE
	TERTIARY	NEOGENE	PLIOCENE
			MIOCENE
		PALEOGENE	OLIGOCENE
			EOCENE
			PALEOCENE

Rocky Mountains and High Plains

While marine sedimentation was occurring along the eastern coastal regions, terrestrial deposition prevailed in the Rocky Mountain region. The exception was a small internal sea of Paleocene age that occupied western North Dakota. There is no evidence that this sea, called the Cannonball Sea (Fig. 21.5), had a connection to the Gulf of Mexico or the Arctic. It is therefore thought to be a remnant of the last Cretaceous epeiric sea.

Late Cretaceous and early Tertiary phases of the deformations described in the previous chapter were largely responsible for the major structural features of the Western Cordillera. However, the present topography of this region is due primarily to erosion following uplifts that began during the Miocene. Erosional debris was often trapped in intermontane basins and provides a record of the early Tertiary epochs. Later uplifts and erosion produced the spectacular relief of the Rockies and caused the detritus of the older basins and newly uplifted mountains to be spread over the plains that lay to the east. The result was the creation of a vast apron of nonmarine Oligocene through Pliocene sands, shales, and lignites that lie beneath the western high plains. Ash beds are interspersed with these sediments and attest to periodic episodes of volcanic pyrotechnics.

Earliest Tertiary sediments deposited in intermontane basins of the Cordillera consist of gray siltstones, carbonaceous shales, lignite, and coal. These rocks are well represented in the Fort Union Formation, a unit famous for its content of low-sulfur coal. The coal beds indicate swampy conditions in this region during the Paleocene. Lower Tertiary formations lying above the Fort Union Formation are seen by thousands of tourists

Figure 21.1   The Red Sea, shown here in a photograph taken from the Gemini spacecraft, formed about 30 million years ago when the Arabian Peninsula broke away from Africa. The seaway continues to widen today. The view is toward the south. The wedge of land in the lower left is the Sinai Peninsula, bordered on the right by the Gulf of Suez and Egypt and on the left by the Gulf of Aqaba and Saudi Arabia. *(NASA)*

Figure 21.2   Cross section of Tertiary strata across the New Jersey coastal plain. *(From* Stratigraphic Atlas of North and Central America, *Shell Oil Company)*

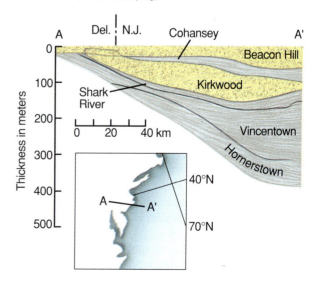

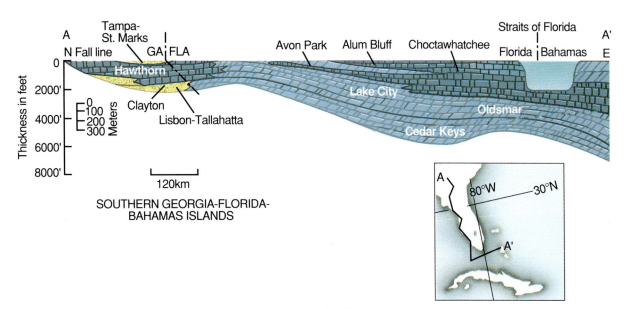

Figure 21.3   Cross section of Tertiary strata across the trend indicated on the location map from southern Georgia to the Bahamas.   *(From* Stratigraphic Atlas of North and Central America, *Shell Oil Company)*

Figure 21.4   Cross section of Tertiary strata across the Gulf coastal plain and Gulf of Mexico.   *(From* Stratigraphic Atlas of North and Central America, *Shell Oil Company)*

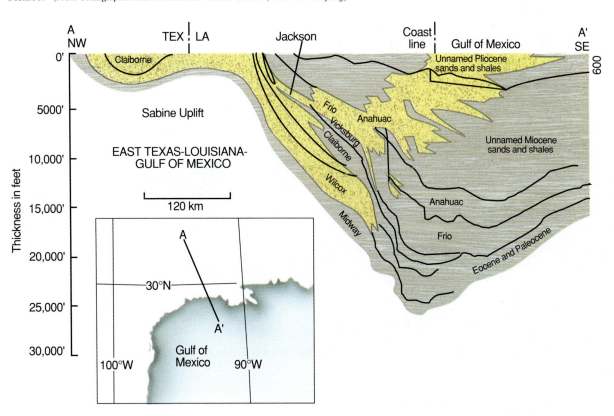

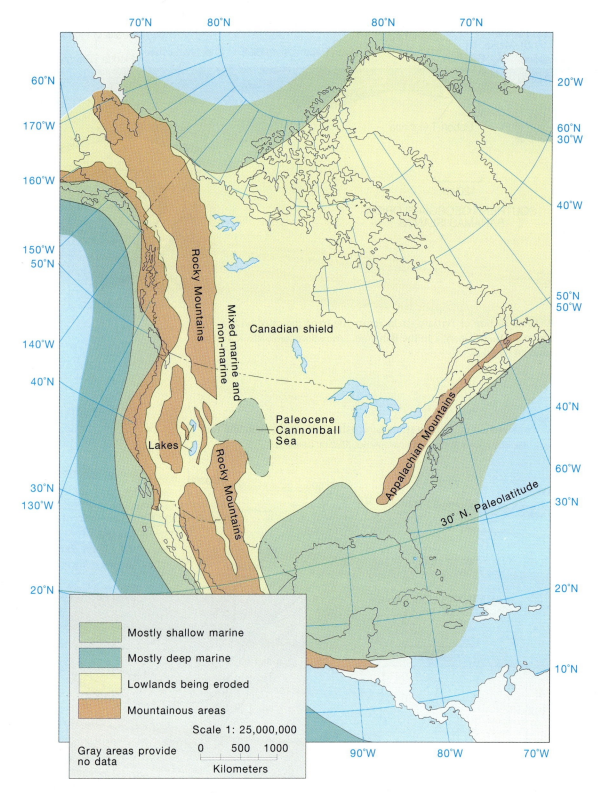

Figure 21.5   Paleogeographic map of North America during the early Tertiary.

Figure 21.6 Cedar Breaks National Monument in Utah. Here the Eocene Cedar Breaks Formation has been eroded into steep ravines, pinnacles, and razor-sharp divides.

Figure 21.7 Cenozoic basins of Colorado, Utah, and Wyoming containing important oil shale deposits. Map boundary is the Upper Colorado River drainage basin. *(Simplified from D. A. Rickert, W. J. Ulman, and E. R. Hampton, eds. 1979. Synthetic Fuels Development, U.S. Geologic Survey publication.)*

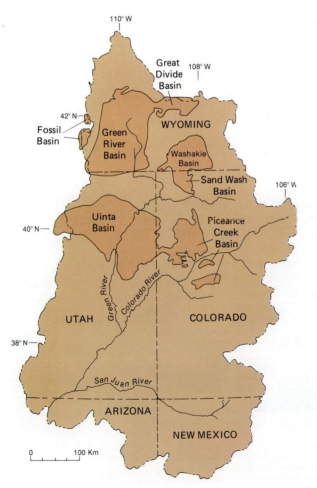

each year as they visit the spires and pinnacles carved by erosion into the Cedar Breaks Formation at Bryce Canyon and Cedar Breaks, Utah (Fig. 21.6).

Also during the early Tertiary, water from streams draining the Uinta, Wind River, and Big Horn Mountains filled intermontane basins, creating large lakes (Fig. 21.7). One such lake occupied the Green River basin of southwestern Wyoming. Its sediments comprise the Green River Formation, noted for its well-preserved fossil fishes (Fig. 21.8) and fossil plants. In addition, the shales of the Green River are rich in waxy hydrocarbons. They are called **oil shales** and can be processed to yield up to 200 liters of petroleum per ton.

The Uinta basin south of the Green River basin is notable because it is structurally the deepest of those in the Colorado Plateau. It received a particularly thick succession of Tertiary sediments (Fig. 21.9). These formations can still be observed in exposures in the interior of the basin, where they are generally flat lying. Along the northern edge of the basin, the beds form steeply dipping hogbacks.

During the late Eocene and Oligocene, ash from volcanoes blanketed the region that today includes Yellowstone National Park and the San Juan Mountains. For paleontologists, however, the more interesting rocks of this time are flood plain deposits of the White River Formation. This formation contains entire skeletons of Tertiary mammals in extraordinary number, variety, and excellence of preservation. If one recalls newsreels of recent floods, it is not difficult to understand how Oligocene mammals became buried in the sediment of overflowing streams. The badlands of South Dakota have been sculpted into the silt and ash beds of the White River Formation (Fig. 21.10).

Figure 21.8 Fossil remains of an Eocene fresh-water fish *(Diplomystis)* from the Green River Formation, Wyoming (length of fish is 8 centimeters).

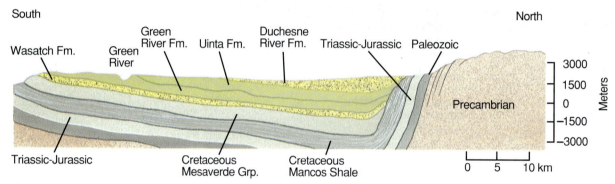

Figure 21.9 North–south cross section through the Uinta basin. Tertiary units are above the Mesaverde Group.

Figure 21.10 Poorly consolidated, horizontal claystone and shale beds of the Oligocene Cedar Breaks Formation form the picturesque pinnacles of the South Dakota badlands. Lacking the protection afforded by vegetation, the soft sediments are susceptible to severe erosion during infrequent but torrential rains. *(R. Arvidson)*

GEOLOGY AND THE ENVIRONMENT

The Good and Bad of Oil Shale

The Eocene oil shale that occurs in parts of Wyoming, Utah, and Colorado is rich in an oil-yielding compound known as kerogen. When heated to 480°C, kerogen in the shale decomposes to yield vaporized hydrocarbons. The vapor can then be condensed to form a thick oil that, when enriched with hydrogen, can be refined into gasoline and other products in much the same way as ordinary crude oil. The retort in which the shale is heated resembles a giant pressure cooker that fuels itself with the gases generated during heating.

Many oil shales yield from one half to three quarters of a barrel of oil per ton of rock. If one were to mine only the shale layers that are thicker than 10 meters, they would provide an impressive 540 billion barrels of oil. If this oil were to be produced during the next decade, it would reduce appreciably the United States' dependency on foreign crude oil, and that is the good news. Production of oil shale, however, requires vast amounts of water and poses a serious threat to the environment. Thus, Americans are faced with the question of how to balance the harmful effects resulting from the use of an important resource with the critical need for that resource.

Two environmental problems have prevented full-scale mining and retorting of oil shales. The first is the waste disposal problem. In the process of being crushed and retorted, the shale expands to occupy about 30 percent more volume than was present in the original rock. Geologists call this the popcorn effect. Where can this great volume of light, dusty material be placed? The second problem relates to air quality, for the processing of huge tonnages of oil shale is likely to release large amounts of dust into the atmosphere. Unfortunately, oil shales occur in arid regions. There is little water available for use in processing the rock to contain the dust and provide for revegetation of the land. Until these problems are solved, production of oil from oil shale will be slow. Perhaps the delay will work to our advantage, for we may need this oil for processed goods a century from now, when the internal combustion engine may be as rare as the horse and buggy is today.

By Miocene time, climates in North America had become somewhat cooler. There were expanding areas of grasslands populated by camels, horses, rhinoceroses, deer, and other grazing animals. In the central and southern Rockies, Miocene formations include beds of volcanic ash as well as interspersed lava flows attesting to explosive volcanic activity. The gold deposits of Cripple Creek, Colorado, are mined from veins associated with one of these volcanoes. Regional uplift of the Rockies also began in the Miocene and was accompanied by increased rates of erosion. Great volumes of terrigenous detritus eroded from the mountains and filled intermontane basins, and when the basins could hold no more, the surplus was carried eastward by streams to contribute to the construction of the Great Plains. Crustal uplifts continued in the Rockies throughout the remaining epochs of the Cenozoic, resulting in some of the most spectacular scenery of the American west. In northwestern Wyoming, the lofty Teton Range (Fig. 21.11) was elevated along great normal faults with displacements of up to 6000 meters.

Basin and Range Province

The region now occupied by the Basin and Ranges (Nevada, Arizona, New Mexico, and part of southern California) had been folded and overthrust during the Mesozoic. Structurally, it had the form of a broad regional arch during the early Tertiary. Beginning in the Miocene, the arch subsided between great tensional faults that developed on both the west and east sides. Similar faults with general north–south alignment developed in the interior of the region. The uplifted blocks formed linear mountain ranges that became sources of sediment for the adjacent downdropped blocks (Fig. 21.12 and 21.13). The cause of all the large-scale tensional faulting that produced the Basin and Ranges is uncertain. One hypothesis is that the faults were caused by uparching and crustal stretching that occurred when westward-moving North America overrode part of the oceanic plate and spreading center being subducted along the west coast. When the subducted spreading center reached the region of the present Basin and Ranges, it

Figure 21.11   The jagged Teton Range has been cut by water and ice erosion from a crustal block uplifted along a nearly vertical fault. Movement along the fault began in the Pliocene and continued sporadically through the Pleistocene.   *(Eva Moldovanyi)*

Figure 21.12   The California and Nevada Nopah Range of the Basin and Range Province.

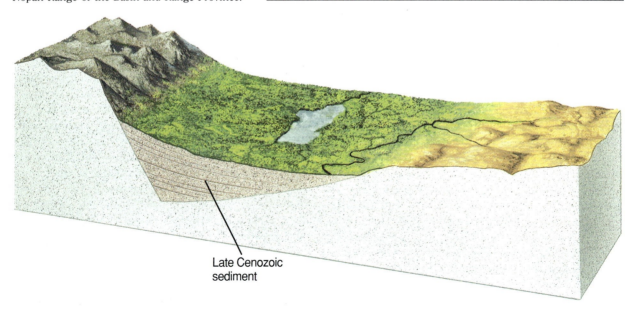

Late Cenozoic sediment

Figure 21.13   Block diagram showing general relationship of Basin and Range topography to normal faulting.

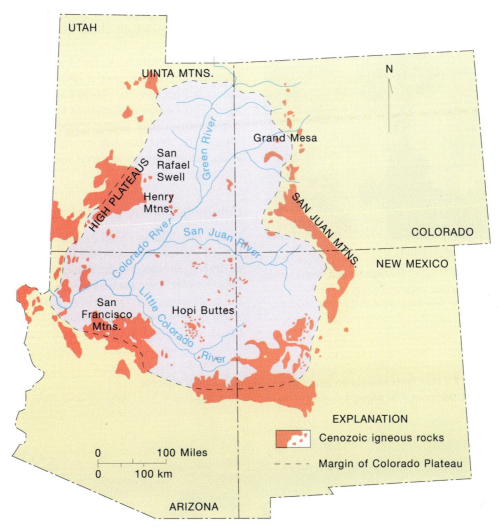

Figure 21.14   Location of volcanic and other centers of igneous activity of Cenozoic age in the Colorado Plateau and its periphery.   *(From C. B. Hunt,* Cenozoic Geology of the Colorado Plateau, *U.S. Geological Survey Professional Paper 279, 1956.)*

caused uplift, stretching of the crust, and tensional faulting.

Colorado Plateau

The Colorado Plateau (Fig. 21.14) is one of the most magnificent regions of uplift on Earth. Somehow this immense area escaped deformation during the Mesozoic, for its Paleozoic and Mesozoic strata are flat lying (Fig. 21.15). It is as if the plateau formed a buttress against surrounding forces. Folding occurred only around its perimeter. The plateau was raised repeatedly during the Pliocene, and often the uplifts were accompanied by the development of nearly vertical faults that provided conduits for the upward escape of lava and ash. The San Francisco Mountains near Flagstaff,

Figure 21.15   Canyon of the Dolores River, Colorado Plateau south of Grand Junction, Colorado.

Figure 21.16 Vertical aerial photograph of a large cinder cone in the San Francisco volcanic field of northern Arizona. The solidified flow issuing from the cone is 7 kilometers long and more than 30 meters thick. *(U.S. Geological Survey)*

Arizona, are a group of impressive recent volcanoes and cinder cones built above the level of the plateau (Fig. 21.16).

Without question, the best-known feature resulting from the linked processes of uplift and erosion on the Colorado Plateau is the Grand Canyon of the Colorado River. This awesome monument to the forces of erosion is more than 2600 meters (8500 feet) deep in some places. The river has cut its way through the entire Phanerozoic sequence and into underlying Precambrian crystalline rocks.

Columbia Plateau and Cascades

Unlike the Colorado Plateau, which is constructed of layered sedimentary rocks, the Columbia Plateau in the northwestern corner of the United States is built largely of layer upon layer of solidified lava (Fig. 21.17). During the late Tertiary and Quaternary, basaltic lavas erupted along deep fissures in the region. Glowing blankets of liquid rock spread from the fissures and buried more

Figure 21.17 Lava flows exposed on either side of Summer Falls, Columbia Plateau. *(U.S. Department of the Interior, Bureau of Reclamation, Columbia Basin Project)*

than 500,000 square kilometers of existing topography. In some places the low-viscosity lava flowed 170 kilometers from the source fissures. The combined thickness of the flows exceeds 2800 meters.

West of the Columbia Plateau lies an uplifted belt that was also the site of extensive volcanic activity. Here, however, the more viscous melts produced the mountains of the Cascade Range. Volcanism in this region began about 4 million years ago. The fact that it has continued to the present was made dramatically obvious to Americans during the violent eruption of Mount St. Helens in 1980 (Fig. 21.18).

Mount St. Helens is not the only famous volcano in the Cascades, nor is it the only peak having periodic eruptions. As recently as 1914, Mount Lassen extruded lava and ash for a year and then exploded violently on May 19, 1915, producing a **nuée ardente.** Other well-known Cascade peaks include Mount Baker, Mount Hood, and Mount Rainier. The last major eruption of Mount Rainier was about 2000 years ago, although minor disturbances occurred frequently in the late 1800s. Crater Lake in Oregon (Fig. 21.19) was formed when the top of the volcanic cone Mount Mazama collapsed into its subsiding lava column to form a caldera.

The volcanoes of the Cascades are surface manifestations of an ongoing collision between the American plate and the small Juan de Fuca plate of the eastern Pacific. The Juan de Fuca plate plunges eastward under

Figure 21.18 Aerial view of erupting Mount St. Helens volcano taken at about 11:30 A.M. on May 18, 1980. Clouds of steam and ash are being blown toward the northeast from the prominent plume, which reached about 20,000 meters in altitude. The white linear features at the base are logging roads, and the dark patches are stands of mature trees. The May 18 eruption produced an amount of ash roughly equivalent to that ejected during the A.D. 79 eruption of Mount Vesuvius that buried Pompeii. *(Photograph by Austin Post, courtesy of U.S. Geological Survey)*

Figure 21.19 View of the west wall of Crater Lake, Oregon, with Wizard Island rising from the caldera. *(L. E. Davis)*

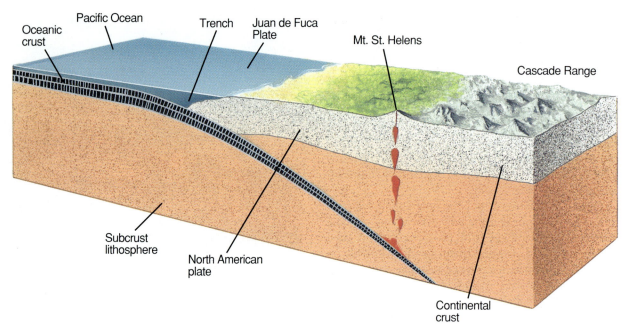

Figure 21.20   As the small Juan de Fuca plate plunges beneath Oregon and Washington, molten rock is generated and rises to supply the volcanoes of the Cascade Range.

the coasts of Oregon and Washington. Molten rock generated as the plate moves downward rises to supply the lavas of the Cascade volcanoes (Fig. 21.20).

Sierra Nevada and California

South of the Cascades lie the Sierra Nevada. These majestic ranges were folded and intruded during the Jurassic Nevadan orogeny. For most of the Tertiary, the peaks were undergoing rapid erosion. Eventually, their granitic basement lay exposed at the surface. Then, in Pliocene time, the entire Sierra Nevada block was raised along normal faults bounding its eastern side and tilted westward. The high eastern front was lifted an astonishing 400 meters, whereas the depressed western margin formed the California trough. As these movements were underway, rejuvenated streams and powerful valley glaciers began to sculpt the magnificent landscapes of the present-day Sierras.

West Coast Tectonics

During most of the Cenozoic, the western edge of North America lay adjacent to an eastward-dipping subduction zone. This subduction was, in one way or another, responsible for the batholiths, compressional structures, volcanism, and metamorphism that accompanied Mesozoic and early Tertiary orogenies. The oceanic plate that

was being fed into the subduction zone has been named the **Farallon plate**. During the Cenozoic, the Farallon plate was being consumed at the subduction zone faster than it was receiving additions at its spreading center. As a result, most of the plate and part of the East Pacific rise that generated it were gobbled up at the subduction zone along the western edge of California (Fig. 21.21). Today the Juan de Fuca plate near Oregon and Washington, and the Cocos plate off the coast of Mexico, are remnants of the once more extensive Farallon plate.

With the loss of the Farallon plate near California, the North American plate was brought into direct contact with the Pacific plate, and a new set of plate motions began. Before making contact with the west coast, the Pacific plate had been moving toward the northeast. As a result, when contact was made, the Pacific plate did not plunge under the continental margin but rather slipped along laterally, giving rise to the San Andreas fault. No longer did California have an Andean-type subduction zone (Fig. 21.22). Strike-slip movements along the San Andreas were to be the new style of tectonics.

Birth of the Alps and Himalayas

During most of the Cenozoic, a marine tract known as the Tethys seaway lay to the south of Europe, separating Eurasia from Gondwanaland. Tectonic upheavals along

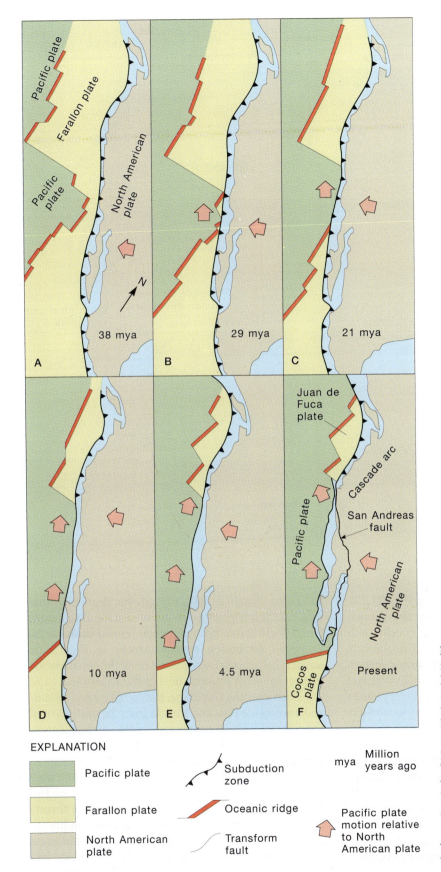

Figure 21.21 Schematic model of the interaction of the Pacific, Farallon, and North American plates for six time intervals during the Cenozoic. Note how the Farallon plate was largely subducted by late Cenozoic time, leaving only remnants to the north (Juan de Fuca plate) and to the south (Cocos plate). The San Andreas and associated faults were caused by right-lateral movements beginning about 29 million years ago. *(Adapted from T. H. Nilsen, San Joaquin Geological Society Short Course No. 3, 1977. Data source from T. Atwater, Bull. Geol. Soc. Am., 81[12]:3513–3536, 1970.)*

EXPLANATION

Pacific plate

Farallon plate

North American plate

Subduction zone

Oceanic ridge

Transform fault

mya Million years ago

Pacific plate motion relative to North American plate

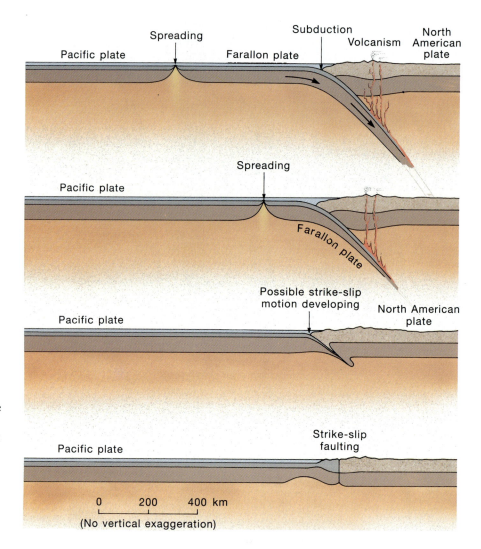

Figure 21.22 Sequence of cross sections of California and its offshore area illustrating the subduction of the Farallon plate. In this model, the Pacific plate is considered fixed as the spreading center (East Pacific Rise) encounters the continental margin. *(After T. Atwater,* Bull. Geol. Soc. Am., *81[12]:3513–3536, 1970.)*

this seaway produced the Alps, Himalayas (Fig. 21.23), Atlas, Apennines, Carpathians, Caucusus, and Pamirs. Crustal disturbances began during the Eocene with large-scale folding and thrust faulting. These events occurred at a time when the northward-moving African block encountered the western underside of Europe and crumpled the strata that now compose the Pyrenees and Atlas mountains. Then, with a sort of scissor-like movement, the Alpine region began to be squeezed. Crests of folds formed elongate islands and shallow banks in the Tethys. Between these elevated islands, siliceous shales, cherts, and graywackes accumulated.

By Oligocene time, compression from the south caused enormous recumbent folds to rise as mountain ranges out of the old seaway and to slide forward over lands that lay to the north of the Tethys. The folds, severed along their undersides by thrust faults, were pushed one on top of the other. North of these contorted

and rising structures lay a topographic depression that received the terrigenous detritus eroded from the mountains. These terrestrial sediments, termed **molasse** by European geologists, resemble rocks of the clastic wedges that accompanied formation of the Appalachians in eastern North America during the Paleozoic.

Even after the Oligocene, the compressions continued. During the Pliocene, thrusts from the south carried the older folded belts northward over the molasse deposits and crumpled the Jura folds, which today form the northern front of the Alps. The thrusting was followed by spasmodic uplifts that continue to the present day.

That part of the Tethys east of the Alps was no less active during the Cenozoic. Volcanism, folding, thrusting, and emplacement of plutons began early in the era and increased during the Miocene. Great elongate tracts of the former sea floor were squeezed into folds and shoved southward. Much of the early deformation oc-

Figure 21.23 Shivling Peak in the Himalayas. The mountain has been eroded by glaciers into a glacial horn. *(D. Bhattacharyya)*

curred at or near sea level, but ultimately the sea floor was forced upward to become land. A broad, subsiding trough formed along the northern edge of newly arrived peninsular India. On this lowland more than 5000 meters of continental sediments were deposited. In the final epochs of the Cenozoic, regional (epeirogenic) uplifts brought the plateaus and ranges to their spectacular elevations. The uplifts appear to have been caused by continued northward movement of the Indian block beneath the Asian plate.

When the Mediterranean Dried Up

Prior to the Miocene, that part of the Tethys now called the Mediterranean was linked to the Atlantic Ocean by two narrow, shallow straits. At the time the straits were in existence, the Antarctic ice sheet expanded, and this resulted in a loss of ocean water. Sea level was lowered by as much as 50 meters. Because the straits that connected the Mediterranean to the Atlantic were less than 50 meters deep, the Mediterranean became a landlocked sea. Without replenishment from the Atlantic, the sea began to evaporate. Eventually, it dried up completely, forming a sort of Mediterranean "Death Valley." This arid state of affairs, however, ended about 5.5 million years ago when the Strait of Gibraltar opened and allowed Atlantic waters to cascade spectacularly into the Mediterranean basin. Evidence that the Mediterranean was once a dry basin includes a layer of evaporites penetrated during drilling into the Mediterranean sea floor, as well as seismic cross sections that show gorges cut by streams that flowed into the once empty basin from northern Africa.

21.2 The Great Pleistocene Ice Age

The final two epochs of the Cenozoic are the Pleistocene and Holocene (or Recent). They represent only about 2 million years of geologic time, but for humans they are of great interest. It was during the Pleistocene that primates of our own species evolved and rose to a dominant position. In addition, during the Pleistocene, more than 40 million cubic kilometers of snow and ice were dumped on about a third of the land surface of the Earth. Such a vast cover of ice and snow had profound effects, not only on the glaciated terrains themselves but also on regions far away from the glaciers. Climatic zones in the Northern Hemisphere were shifted southward, and arctic conditions spread across northern Europe and the United States. Mountains were sculpted by great valley glaciers like those illustrated in Chapter 12. While the ice accumulated in high latitudes and high elevations, rainfall increased in lower latitudes with beneficial effects on plant and animal life. Even as late as the beginning of the Holocene, present arid regions in Africa were well watered, fertile, and populated by nomadic tribes. People of middle and late Pleistocene time hunted along the fringes of the ice sheets, for there was abundant game and spoilage of meat was inhibited by the frigid conditions.

Although one cannot avoid being impressed by the magnitude of the Pleistocene great ice age, similar widespread continental glaciations had occurred in the Precambrian, Ordovician, Permian, and possibly in the Oligocene and Pliocene as well.

Pleistocene and Holocene Chronology

The Pleistocene Epoch began about 1.8 million years ago. According to the International Union of Geological Sciences, the epoch ended 10,000 years ago. The lower boundary of the Pleistocene series can often be recognized in shallow-water marine sediments by the occurrence of key mollusks and foraminifers. When examining cores of deep-sea sediments, one can frequently recognize the basal Pleistocene oozes by the extinction point of fossils called discoasters. **Discoasters** (Fig. 21.24) are calcareous, often star-shaped microfossils formed by golden-brown algae related to coccolithophorids. In continental deposits, the fossil remains of the modern horse (*Equus*), the first true elephants, and certain other mammals are used to identify deposits of the lower Pleistocene.

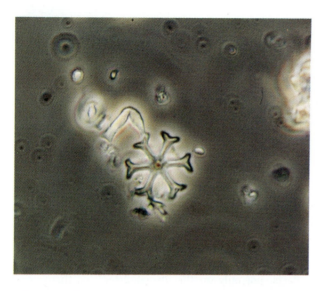

Figure 21.24 Discoaster *(Discoaster challengeri)* seen through a light microscope at a magnification of ×1200.

Before the mid-1970s, geologists considered that the Pleistocene included four distinct glacial stages with intervening interglacial stages. Oxygen isotope and faunal analyses of continuous sections of Pleistocene deep-sea deposits now indicate that there have been many additional periods of severe cold during the Pleistocene. Over a 3-million-year span, there have been 30 episodes of frigid conditions. Antarctica (Fig. 21.25) has been frozen in ice for at least the past 15 million years. In addition, episodes of global cooling did not end with the close of the Pleistocene Epoch. Historical records and carbon-14 dating of terminal moraines indicate that cold spells recurred periodically during the Holocene.

Figure 21.25 The Antarctic ice sheet near Victoria Land, Antarctica. *(P. B. Larson and the U.S. Geological Survey)*

Climates on our planet are in delicate balance with atmospheric, geographic, and such astronomic variables as sunspot activity, which influences the amount of solar energy reaching the Earth. One well-documented period of cooler conditions in the Northern Hemisphere occurred between A.D. 1540 and 1890, when temperatures were often 2° to 4°F cooler than today. These four centuries are called the **little ice age**. Exceptionally cold conditions during the little ice age extended across most of Europe and in the United States as far south as the Carolinas. Sea ice intruded far to the south of its present Arctic margin. The cold caused loss of harvests with resulting famine, food riots, and warfare. In the middle of the fourteenth century, excessive cold contributed to the demise of a once flourishing Norse colony in Greenland.

Environmental Impact of the Ice Age

The glaciations of the Pleistocene changed the face of the Earth in many ways. As the ice fronts advanced and retreated, there was a corresponding decline and rise in sea level with consequent alternate exposure and inundation of coastal plains and other low-lying terrains. There was isostatic rebound of regions once covered by ice, changes in the locations of major streams, and the formation of immense lakes, including the Great Lakes of the United States. Before the Pleistocene, the region now occupied by the Great Lakes was dry land. Lobes of the great continental glacier moved into these generally low-lying areas and scoured them deeper. As the glaciers retreated, their meltwaters flowed into the depressed areas to form the lakes (Fig. 21.26). Niagara Falls, between Lake Erie and Lake Ontario, developed when the retreating ice of the Wisconsin glacial stage uncovered an escarpment formed by the southwardly dipping Lockport Limestone. Weak shales beneath the limestone are continuously undermined, causing southward retreat of the falls (see Fig. 18.14).

Another large system of ice-dammed lakes covered a vast area of North Dakota, Minnesota, Manitoba, and Saskatchewan. The largest of these lakes has been named Lake Agassiz in honor of Jean Louis Rodolphe Agassiz, the great French naturalist who initially insisted on the existence of the Ice Age. Today, rich wheatlands extend across what was once the floor of the lake.

Other lakes developed during the Pleistocene, occupying basins that were not near ice sheets. These were formed as a consequence of the greatly increased precipitation and runoff that characterized regions south of the glaciers. Such water bodies are called **pluvial lakes** (which comes from the Latin *pluvia,* meaning rain). Pluvial lakes were particularly numerous in the northern part of the Basin and Range Province of North

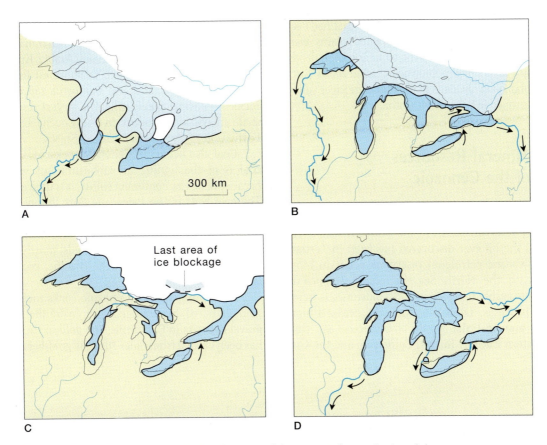

Figure 21.26 Four stages in the development of the Great Lakes as the ice of the last glacial advance moved away. *(After J. L. Hough,* Geology of the Great Lakes. *Urbana: Univ. of Illinois Press, 1958, figs. 56, 69, 73, and 74.)*

America, where faulting produced more than 140 closed basins. So-called pluvial intervals, when lakes were most extensive, were generally synchronous with glacial stages, whereas during interglacial stages, many lakes shrank to small saline remnants or even dried out completely. Lake Bonneville in Utah was one such lake. It once covered more than 50,000 square kilometers and was as deep as 300 meters in some places. Parts of Lake Bonneville persist today as Great Salt Lake, Utah Lake, and Sevier Lake.

A particularly spectacular event associated with the formation of Pleistocene lakes occurred in the northwestern corner of the United States. Lobes of the southwardly advancing ice sheet repeatedly blocked the Clark Fork River, and the impounded water formed a long, narrow lake extending diagonally across part of western Montana. The fresh-water body, called Lake Missoula, contained an estimated 2000 cubic kilometers of water. With the recession of the glacier, the ice dam broke, and tremendous floods of water rushed out catastrophically across eastern Washington, causing severe erosion and depositing huge volumes of gravel, boulders, and cobbles. The dissected region is appropriately termed the **channeled scablands** (Fig. 21.27).

The glacial conditions of the Pleistocene also had an effect on soils. In many northern areas, fertile topsoil was stripped off the bedrock and transported to more southerly regions, which are now among the world's most productive farmlands. Because of the flow of dense, cold air coming off the glaciers, winds were strong and

Figure 21.27 Location of the channeled scablands.

persistent. Fine-grained glacial sediments that had been spread across outwash plains and flood plains were picked up and transported by the wind and then deposited as thick layers of windblown silt called **loess**. Such deposits blanket large areas of the Missouri River valley, central Europe, and northern China.

21.3   Mineral Resources of the Cenozoic

Most of the known accumulations of oil have been found in Tertiary reservoir rocks. In the United States, about half of all the oil ever discovered has been in Tertiary rocks. Paleocene petroleum reservoirs occur in Libya and beneath the North Sea. Petroleum trapped in Eocene permeable limestones and sandstones is being pumped in Texas, Louisiana, Iraq, the Soviet Union, Pakistan, and Australia. Strata of Oligocene age yield oil in western Europe, Burma, California, and the Gulf

Coast states of the United States. Miocene sandstones yield oil on every continent except Australia. They are extraordinarily productive in Saudi Arabia, Kuwait, California, Texas, and Louisiana. Offshore oil fields in the Gulf Coast (Fig. 21.28) and California tap oil held in rocks of Pliocene age. Finally, the largest reserves of oil shale in the world occur in the Eocene Green River Formation of the western United States.

Coal also occurs in Cenozoic rocks. Most Cenozoic coal is of the lignitic or subbituminous variety, but because of its low content of sulfur, it is used extensively. The coal beds of the Paleocene Fort Union Formation are mined in the Dakotas, Wyoming, and Montana. They represent the largest recoverable fossil fuel deposits in the United States. Scattered coal beds, mostly of Eocene age, are exploited along the Pacific coast of the United States.

Metals mined from Cenozoic rocks include placer gold gathered by dredging and hydraulic mining of gravel deposits in California. The gold in these deposits has been eroded from older Jurassic gold-bearing quartz

Figure 21.28   Drilling for offshore oil. Shown here is a huge semisubmersible drilling rig.   *(Reading and Bates Drilling Company)*

veins in the Sierra Nevada. About 60 percent of the world's supply of tin is also supplied by Cenozoic stream deposits. Manganese, an essential additive in the manufacture of steel, is mined extensively from lower Tertiary rocks of the Soviet Union. These deposits comprise 75 percent of the world's richest manganese ores. Another metal used in making certain types of steel is molybdenum. One deposit in Climax, Colorado, supplies 40 percent of the world's molybdenum. Other metals, such as copper, silver, lead, mercury, and zinc, are frequently found associated with Tertiary intrusive rocks of western North America and the Andes and with Tertiary orogenic belts along the western margins of the Pacific Ocean. Similar Cenozoic rocks were formed in southern Europe and Asia as a consequence of the deformation of the Tethys belt.

Nonmetallic materials from Cenozoic rocks are also important resources. Diatomite, the white, porous rock formed from the silica coverings (frustules) of diatoms, is mined in great quantities in California. Cenozoic building stone, clay, phosphates, sulfur, salt, and gypsum are quarried in North America, Europe, the Middle East, and Asia.

21.4 Life of the Cenozoic

Because biological developments of the Cenozoic have occurred so recently and because Cenozoic fossils are topmost in the stratigraphic column, we have more facts about this era than about the far lengthier ones that preceded it. Armed with this more adequate data, paleontologists are better able to compare biological evolution with environmental and paleogeographic changes in the Cenozoic. Continental fragmentation clearly stimulated biological diversification and resulted in distinctive faunal radiations on separated landmasses. Among the many interesting evolutionary developments, none seem more fascinating than the changes experienced by the primates, which by late Tertiary time had produced species considered to be the ancestors of humans. In the Pleistocene Epoch, our own species, *Homo sapiens,* evolved.

Plants

Response of Mammals to the Spread of Prairies

Although they are the most recent group to evolve, the flowering plants are now the most widespread of all vascular land plants. Angiosperm floras did not explode on the lands until mid-Cretaceous (Fig. 21.29). There were few spectacular floral innovations during the Cenozoic. Rather, this was a time of steady progress toward the development of today's complex plant populations. The Miocene is particularly noteworthy as the epoch during which grasses appeared and grassy plains and prairies spread widely over the lands. In response to the proliferation of this particular kind of forage, grazing mammals began their remarkable evolution.

Numerous evolutionary modifications among herbivorous mammals can be correlated to the development of extensive grasslands. Especially evident were changes in the dentition of grazing mammals. Grasses are abrasive materials. Many contain silica, and because they grow close to the ground, they are often coated with fine particles of soil. To compensate for the wear that results from chewing grasses, the major groups of grazers evolved high-crowned cheek teeth that continue to grow at the roots during part of the animals' lives. To provide space for these high-crowned teeth, the overall length of the face in front of the eyes increased. Enamel, the most resistant tooth material, became folded, so that when the tooth was worn, a complex system of enamel ridges was formed on the grinding surface (see Fig. 16.16). Incisors became aligned into a curved arc for nipping and chopping grasses.

In the open plains environment, it was more difficult to escape detection by predators. As a result, many grazers evolved modifications that permitted speedy flight from their enemies. Limb and foot bones were lengthened, strengthened, and redesigned by the forces of selection to prevent strain-producing rotation and

Figure 21.29 Relative proportions of genera in Cretaceous to Recent floras.

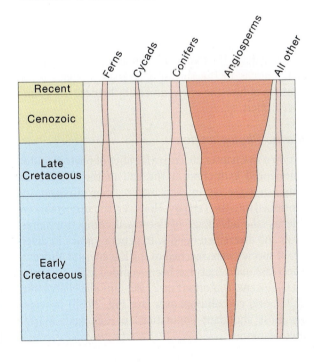

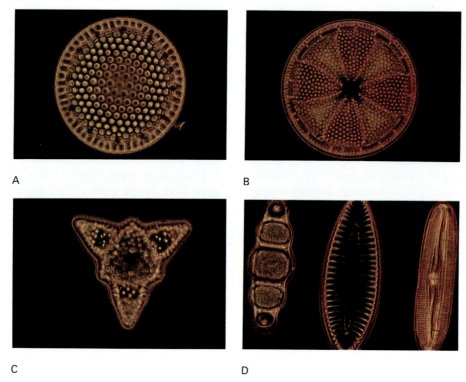

Figure 21.30   Cenozoic diatoms exhibiting various frustule shapes, including discoidal, triangular, and spindle shaped. (A) *Arachnoidiscus*. (B) *Actinoptychus*. (C) *Triceratum*. (D) From left to right, *Biddulphia, Surirella,* and *Pinnulria* (×50).

permit rapid fore-and-aft motion. To achieve greater speed, the ankle was elevated, and, like sprinters, animals ran on their toes. In many forms, side toes were gradually lost. Hoofs developed as a unique adaptation for protecting the toe bones while the animals ran across hard prairie sod. Some grazers called ungulates evolved a four-chambered stomach in response to selection pressures favoring improved digestive mechanisms for breaking down the tough grassy materials. The response of mammals to the spread of grasses provides a fine example of how the environment influences the course of evolution.

Marine Phytoplankton

As noted in Chapter 20, entire families of phytoplankton experienced extinction at the end of the Cretaceous. Only a few species in each major group survived and continued into the Tertiary. However, the survivors were able to take advantage of decreased competitive pressures and diversified rapidly. In general, peaks in diversity of species were reached in the Eocene and Miocene. A decrease in diversity has been recorded in the intervening Oligocene. Diatoms (Fig. 21.30), dinoflagellates (Fig. 21.31), and coccolithophores (see Fig. 20.35) provided the most abundant populations of Cenozoic marine phytoplankton.

Invertebrates

The invertebrate animals of the Cenozoic seas included dense populations of foraminifers, radiolarians, mollusks of all kinds, corals, bryozoans, and echinoids. The fauna had a decidedly modern aspect. Once successful groups such as ammonites and rudistid pelecypods were no longer present.

Figure 21.31   Dinoflagellate as seen with the aid of a scanning electron microscope. From Paleocene sedimentary rocks of Alabama, ×450. *(Standard Oil Company of California, photographer, W. Steinkraus)*

Figure 21.32   Diversity of form in Cenozoic foraminifers. Magnification ×50.

Foraminifers

Foraminifers were enormously prolific and diverse during the Cenozoic (Fig. 21.32) and included large numbers of benthic as well as planktic forms. Among the larger forms, coin-shaped nummulitic foraminifers thrived in the Tethys seaway as well as in warm ocean waters of the western Atlantic. The shells of these organisms have accumulated to form thick beds of nummu-

litic limestone. The ancient Egyptians quarried this rock and used it to construct the Pyramids of Gizeh. Even the famous Sphinx was carved from a large residual block of nummulitic limestone. The incredible numbers and variety of foraminifers have permitted their extensive use in correlating Cenozoic strata in oil fields of the Gulf Coast, California, Venezuela, the East Indies, and the Near East.

Corals

Corals (Fig. 21.33 and 21.34) grew extensively in the warmer waters of the Cenozoic oceans. Fossil solitary corals are commonly found in shallow-water deposits of early Tertiary age in Europe and the United States Gulf Coastal region. Reef corals were most extensively developed in parts of the Tethyan belt, West Indies, Caribbean, and Indo-Pacific regions. Careful comparison of coral species in Cenozoic rocks on either side of the Isthmus of Panama has given geologists clues to when the Atlantic and Pacific oceans were connected across this present-day barrier.

Mollusks

Cenozoic shells of mollusks look very much like those likely to be found along coastlines today. Pelecypods and gastropods were the dominant classes of Cenozoic mollusks (Fig. 21.35). Their range of adaptation was truly outstanding. Arcoids, mytiloids, pectinoids, cardioids, veneroids, and oysters were particularly abundant pelecypods. Although the climax of cephalopod evolution had passed with the demise of the ammonites, nautiloids similar to the modern pearly nautilus lived in Tertiary seas, as they do today. Shell-less cephalopods, such as squid, octopi, and cuttlefish, were also well

Figure 21.33   *Fungia*, a solitary coral that takes its name from its resemblance to the underside of the cap of a mushroom. It has a geologic range from the Miocene to the present and is common today in the seas of the Indo-Pacific region. This specimen is about 9.0 centimeters in diameter.

Figure 21.34   The living coral *Montastrea cavernosa*, with polyps extended for feeding.   *(Charles Seaborn)*

Figure 21.35   Common Cenozoic pelecypods (A, B, C, D) and gastropods (E, F). (A) *Cardium.* (B) *Pecten.* (C) *Ostrea.* (D) *Mya.* (E) *Turritella.* (F) *Crepidula,* a small marine snail having a horizontal shelf that extends across the posterior part of the shell.

represented in the marine environment, although their fossil record is understandably sparse.

Echinoderms, Bryozoans, Crustaceans, and Brachiopods

Many other invertebrates have continued successfully through the Cenozoic. Echinoderms, mainly free-mov-

ing types, were particularly prolific. Members of the phylum Bryozoa are common in Tertiary rocks and are still very abundant in many parts of the ocean today. It was during this final era that the modern crustaceans became firmly established both in fresh-water bodies (Fig. 21.36) and in the oceans. Brachiopods declined in abundance and diversity during the Cenozoic. Fewer

Figure 21.36   The fresh-water shrimp *Bechleja rostrata* from the Eocene Green River Formation.   *(Rodney M. Feldmann. From M. Feldmann, et al., J. Paleontol, 55[4]:788–799, 1981.)*

than 60 genera survive today. They consist mostly of terebratulids (Fig. 21.37A), rhynchonellids, and such inarticulate brachiopods as *Lingula* (Fig. 21.37B).

Vertebrates

Fishes and Amphibians

Bony fishes that evolved to the highest level of ossification and skeletal perfection are the **teleosts**. Teleosts have achieved an enormous range of adaptive radiation during the Cenozoic. That radiation has produced such varied forms as perches, bass, snappers, seahorses, sailfishes, barracudas, swordfishes, flounders, and others too numerous to mention. The Green River strata of Wyoming are well known for their content of beautifully preserved Eocene teleosts (see Fig. 21.8).

In addition to the bony fishes, the cartilaginous sharks were at least as common in the Tertiary as they are today. *Carcharodon,* an exceptionally large shark, had teeth as big as a man's hand and was more than 12 meters long.

Amphibians throughout the Cenozoic have resembled modern forms. All have been small-bodied, smooth-skinned creatures not at all like their Paleozoic ancestors. Frogs, toads, and salamanders were relatively abundant. The first frogs appeared during the Triassic and by Jurassic time were already completely modern in appearance. Thus, they have continued almost unchanged for more than 200 million years.

Figure 21.37   (A) Cluster of present-day articulate terebratulid brachiopods, *Terebratulina septentrionalis.* (B) *Lingula,* a persistent primitive inarticulate brachiopod with a thin shell of proteinaceous material and a long, fleshy, musular pendicle.

A

B

C

Figure 21.38 Surviving
reptiles. (A) monitor lizard, (B)
hog-nose snakes emerging from
their shells, (C) *Alligator
mississippiensis,* (D) three-toed
box turtle. *(B from Animals Animals
© 1992 Zia Leszczynski; C from Ed
Reschke; D from L. Stone/The Image
Bank)*

D

Reptiles

By the beginning of the Cenozoic, the dinosaurs had disappeared from the lands, as had the flying reptiles from the air and several groups of marine reptiles from the seas. The reptiles that have managed to survive and continue to the present include the tuatara, which inhabits islands off the coast of New Zealand, as well as turtles, crocodilians, lizards, and snakes (Fig. 21.38). The tuatara, formally known as *Sphenodon,* resembles a large lizard. It is the sole survivor of a group of ancient reptiles known as rhynchocephalians that evolved and diversified during the Triassic.

Cenozoic turtles are the descendants of a lineage that can be traced back into the late Permian. Turtles are readily recognized by everyone because of their distinctive adaptations. The most apparent of these adaptations is the shell. In turtles, the ribs have expanded differentially to form a broad carapace. On the underside, a bony growth called the plastron provides a similar covering. Both carapace and plastron are covered with a horny sheath. The jaws in turtles are toothless and covered by a beak that is used effectively in slicing through plants or meat.

Crocodilians also began their evolution during the Triassic and were very successful contemporaries of the dinosaurs throughout the Mesozoic. Modern crocodilians include the broad-snouted alligators, the narrow-snouted crocodiles, and the very narrow-snouted gavials.

Both lizards and snakes belong to an order of reptiles known as the Squamata. The Squamata are by far the most varied and numerous of living reptiles. Lizards are the ancestors of snakes. In fact, snakes are essentially modified lizards in which the limbs are lost, the skull bones modified into a highly flexible and mobile structure for engulfing prey, and the vertebrae and ribs greatly multiplied to provide the elongate form.

As evidence of their tetrapod ancestry, certain primitive snakes retain vestiges of rear limb and pelvic bones. Poisonous snakes evolved during the Miocene. They are characterized by specialized teeth or fangs for the injection of poison. One type of poison (neurotoxin) affects parts of the nervous system that control breathing and heart action, whereas a second type (hemotoxin) destroys red blood cells and causes disintegration of small blood vessels. It has been suggested that the evolution of poisonous snakes and the feeding behavior and skull structure of all snakes may be tied to the diversification of the Cenozoic mammals that served as prey.

Birds

The Cenozoic record for birds is generally poor. Birds are rarely preserved. The fragmentary record indicates, however, that birds have been essentially modern in basic skeletal structure since the beginning of the Cenozoic. Distinctive skeletal features of birds include the fusion of bones of the "hand" to help support the wing, development of a vertical plate or keel on the sternum for attachment of the large muscles leading from the breast to the wings, and fusion of the pelvic girdle and vertebrae to provide rigidity during flight. Other characteristics include a body covering of feathers, light and porous bones, jaws in the form of a toothless horny beak, a four-chambered heart, and constant body temperature. The avian fauna is composed of a rich variety of families, including song birds such as robins, upland birds (pheasants), forest birds (owls), oceanic birds (albatrosses), wading birds (plovers), flightless aquatic birds (penguins), and flightless land birds (ostriches).

The fossil record for large terrestrial flightless birds is somewhat better than for small flying varieties. *Diatryma* (Fig. 21.39), an Eocene representative of this group, stood more than 2 meters tall. In New Zealand, huge moas lived until relatively recent time. Some were over 3 meters tall and laid eggs having a 2-gallon capacity. Perhaps the most famous of all Cenozoic ground birds was the dodo (*Didus ineptus*), which lived on the island of Mauritius east of Madagascar until about A.D. 1700, when they were exterminated by sailors searching for provisions. The moas of New Zealand were also exterminated by humans. Maori tribespeople killed them for food. The African ostrich, the South American rhea, and the emus and cassowaries of Australia are surviving flightless land birds.

Figure 21.39 *Diatryma* from the Eocene (Wasatch Formation) of Wyoming. This large, flightless bird was about 7 feet tall.

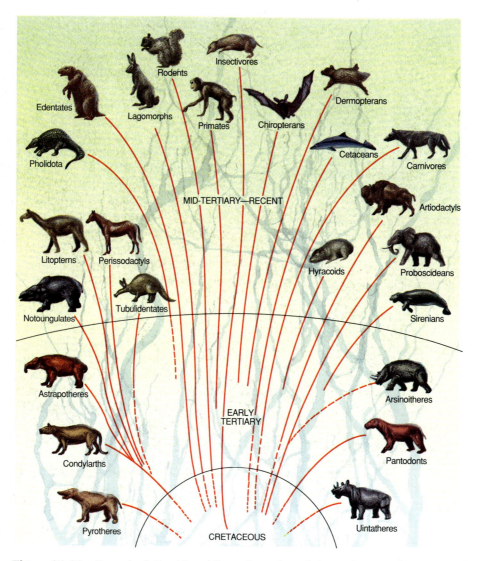

Figure 21.40   General relationships of the major orders of placental mammals.

Mammals

During the Cenozoic, mammals (Fig. 21.40) came to dominate the earth in much the same way as reptiles had done during the preceding Mesozoic Era. As we noted in Chapter 20, however, the evolution of mammalian traits had already begun among the therapsids of the Permo-Triassic. The Karoo beds of Africa, for example, contain bones of near-mammals that had almost made the transition from reptile to mammal.

Just as birds are easily recognized by the possession of feathers, so are mammals by their possession of hair. It is sometimes only present as a few whiskers, but it is present. Like feathers, hair functions as an insulating layer. Mammals are, of course, also recognized by mammary glands. However, neither body hair nor mammary glands are particularly helpful to the paleontologist, who must recognize mammalian remains on the basis of skeletal characteristics. Among these, the lower jaw is particularly useful. It consists of a single bone, the dentary, on either side. In reptiles and birds, there are always several. Another mammalian trait is the bony mechanism that conveys sound across the middle ear. In mammals, there is a chain of three little bones rather than a single one, as in lower tetrapods. Two of these ossicles, the incus and malleus, were derived from bones of the reptilian jaw. Unfortunately, these delicate ossicles are so small that they are rarely preserved.

Typically, mammals have seven cervical, or neck, vertebrae, regardless of the length of the neck. The skull is usually recognizable by the expanded brain case, and teeth are nearly always of different kinds, serving different functions in eating. As an aid to their endothermic metabolism, mammals have a well-developed secondary palate that separates the oral cavity from the nasal passages. This makes simultaneous breathing and feeding possible. Without the secondary palate, an infant mammal could not suckle.

Paleontologists speculate that more rigorous climatic conditions during the Permo-Triassic favored selection of the mammalian traits of warm-bloodedness and postnatal care. All evidence indicates that the first mammals were diminutive creatures, and small animals lose heat rapidly. A high rate of heat loss could have been partly compensated by insulating fur and an ample supply of food. If the earliest mammals laid eggs, as do primitive mammalian monotremes today, then the newly hatched young might also have had to cope with the heat problem. They were very likely kept warm by snuggling next to the soft fur that may have formed an incubation patch on the female. Perhaps at the same time, they were nourished by secretions from glands that preceded the development of true mammary glands.

Earliest Mammals.   The mammalian fossil record begins with rare and difficult-to-find small jaw fragments, tiny teeth, and scraps of skull bone. The fossils, however, are sufficient to indicate that the earliest members of our own taxonomic class were very small, that they evolved from mammal-like reptiles such as those previously described from the late Paleozoic, and, judging from tooth patterns, they were insect eaters. Evidence from brain casts made of their inner cranial walls also suggests that the parts of their brains dealing with smell and hearing were particularly well developed, as in nocturnal animals. Perhaps the most significant and ironic fact about these tiny creatures is that they came on the scene at about the same time as the dinosaurs that they were destined to succeed. For 140 million years of dinosaur dominance, they thrived quietly as if waiting for just the right conditions for their own biological triumph. Then, following the terminal Cretaceous extinctions of dominant reptilian groups, mammals expanded rapidly. By Paleocene time, 18 mammalian orders had appeared, and throughout the early Tertiary, there was an intricate interplay of appearance of new groups and extinctions of older groups that ultimately produced more than 30 taxonomic orders of mammals. The list is far too long to treat each group, but a brief review of some of the more interesting forms suggests the spectacular variety of the Cenozoic mammalian fauna.

Monotremes.   The most primitive of all living mammals are called **monotremes**. These relics of an older time still lay eggs in the reptilian manner. However, unlike reptiles, monotremes have primitive mammary glands and provide their newly hatched offspring with nourishment, if only for a brief period. The oldest known remains of monotremes (a few distinctive teeth) were recently found in Cretaceous rocks.

The platypus of Australia and Tasmania and two species of spiny anteaters of New Guinea and Australia (Fig. 21.41) are the only three species of monotremes still in existence.

Figure 21.41   Two present-day monotremes. (A) The duck-bill platypus of Australia and Tasmania. (B) The spiny anteater of Australia.   *(A from E. R. Degginger)*

A   B

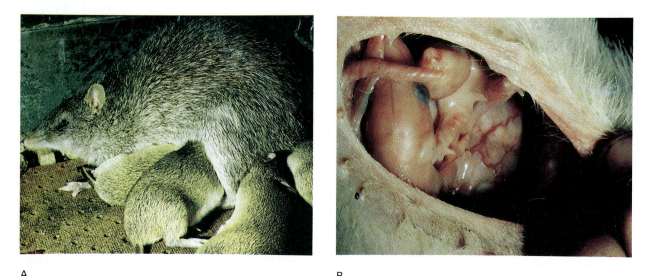

A   B

Figure 21.42   The short-nosed bandicoot, a common marsupial of Australia. (A) Advanced offspring that have left the pouch return for nursing. (B) A fetal bandicoot attached to a teat in the mother's pouch.

Figure 21.43   Adaptations of marsupials for various habitats is evident in their diversity. (The Tasmanian wolf is now thought to be extinct.)

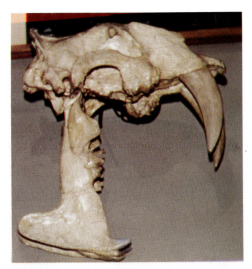

Figure 21.44 *Thylacosmilus,* a Pliocene South American carnivorous marsupial comparable to the placental saber-toothed cats. The bladelike upper canine teeth were about 18 centimeters long and were protected when the jaws were closed by a deep flange of bone in the lower jaw.

Marsupials. **Marsupials** are mammals that nurture their young in a special pouch, or marsupium (Fig. 21.42). Today, they are represented by such animals as kangaroos, wallabies, wombats, phalangers, bandicoots, koalas of Australia, and opossums of the New World (Fig. 21.43). Compared with the placentals, it is apparent that they are a dwindling group.

Marsupials had their greatest success in Australia and South America. Both continents were more or less isolated during most of the Tertiary, and marsupials were able to evolve without excessive competition from placentals. Although a few middle Tertiary Australian mammal sites have recently been found, the fossil record of Australian marsupials is not good until Pleistocene time. An older and more complete record exists in South America, where marsupials evolved from opossum-like ancestors and produced an array of mostly flesh-eating types. *Thylacosmilus* (Fig. 21.44), a large Pliocene form, was similar in appearance to the North American saber-toothed cats. South American marsupials fared poorly in competition with placental immigrants from the north when North and South America became connected in the early Pleistocene.

Placental Mammals. Placental mammals appear during the Cretaceous as small, unspecialized insectivores. Modern moles are members of the order **Insectivora**, but the tiny shrew (Fig. 21.45) is more representative of the kind of animal from which other orders of Cenozoic placentals evolved. The descendants of the insectivores include the edentates, bats, primates, ro-

dents, flesh-eating mammals, a host of herbivores, and various marine mammals.

Armadillos, tree sloths, and South American anteaters are **edentates** that have not become extinct. Fossil species of armadillos have been found in rocks as old as Paleocene. Among the extinct edentates are the glyptodonts, which survived up until the Pleistocene. *Glyptodon* (Fig. 21.46) was a walking fortress with a spike-covered knob on its tail for bludgeoning pesky predators. Quite unlike the glyptodonts were the ungainly Cenozoic ground sloths (Fig. 21.46).

Rodents have been exceptionally successful Cenozoic mammals. Today they probably outnumber all other mammals and have invaded an extraordinary variety of habitats. There are burrowing rodents such as the marmot, aquatic muskrats, desert-dwelling jerboas, and arboreal squirrels. Beavers are also well represented and include Pleistocene giants that were as large as bears.

Bats were also evolving during the Cenozoic. Their teeth have been discovered in Paleocene strata of France. A well-preserved skeleton was recovered from lower Eocene strata in Wyoming and indicates that by that time bats were already very similar to their living relatives. Bats are the only mammals to have achieved true flight. They have greatly elongated finger bones for the support of the membrane that forms the wing.

The history of flesh-eating placental mammals began in the Cretaceous with the advent of small, weasel-like animals called **creodonts** (Fig. 21.47). They were soon joined by members of the order **Carnivora**, which have common ancestry with the Creodonta. It appears that toward the end of Eocene time, with the coming of speedier and more progressive plant eaters, the creodonts gradually lost ground to the Carnivora, which by late Tertiary included bears, raccoons, weasels, hyenas,

Figure 21.45 The common tree shrew most resembles the ancient insectivores that gave rise to the primates. *(Warren Barst/Tom Stock & Associates)*

Figure 21.46 Restoration of a scene during the middle Pleistocene in Argentina. The heavily armored animals are glyptodonts. On the left is a giant ground sloth. *(Copyright © Chicago Museum of Natural History, painting by C. R. Knight, with permission)*

dogs, and cats. Wild dogs were doing very well long before humans came along to make companions of them. The modern genus *Canis,* in the form of the dire wolf, was much in evidence during the Pleistocene. The most famous of the extinct cats are the so-called stabbing cats, exemplified by *Smilodon* (Fig. 21.48). This robust flesh eater preyed on the larger herbivorous animals of the Pleistocene. A second line of cat evolution

produced the biting cats, which resembled modern leopards and pumas. The skeletons of these carnivores indicate they were strong, speedy, and agile predators.

As is true of carnivores today, Cenozoic carnivores were an essential element in the evolutionary process. To survive, they had to equal or better the speed and cunning of the herbivores on which they fed. The herbivores, in turn, responded to the carnivore threat by

Figure 21.47 The Eocene creodont, *Patriofelis.* *(Smithsonian Institution)*

Figure 21.48 The saber-toothed cat *Smilodon.*

evolving adaptations for greater speed and defense. Then, as now, predators were not villains but necessary constituents of the total biological scheme.

Not all the Carnivora of the Cenozoic were land dwellers. Some, such as seals, sea lions, and walruses, gathered their food in or at the edge of the sea. As indicated by their sharp, pointed teeth, early Cenozoic seals and sea lions ate fish as they do today. The walrus had (in addition to tusks) broad, flat teeth for crushing the shells of mollusks. Seals, sea lions, and walruses probably evolved from semiaquatic mammals similar to otters, but the transitional forms are as yet not discovered.

In adaptation to the marine environment, **cetaceans** (whales and porpoises) have made the most complete adjustment. Whales first appeared quite suddenly about 50 million years ago as already fully developed oceanic creatures. For this reason, paleontologists suspect that cetaceans underwent an extraordinarily rapid evolution. The first whales are represented by *Basilosaurus* of the Eocene (Fig. 21.49). The structure of the skull in this toothed whale provides some evidence that the line was derived from early Cenozoic terrestrial flesh eaters.

Modern whales arose from the group that included *Basilosaurus* and divided into two lineages: the toothed whales and the whale-bone whales. Toothed whales, which made their debut during the Oligocene, today include porpoises, killer whales, and sperm whales like Melville's famous Moby Dick. The titanic blue whale, right whale, and Greenland whale are all representatives of the second cetacean group. These plankton-feeding giants first appeared in the Miocene. Instead of teeth, they possess ridges of hardened skin that extend downward in rows from the roof of the mouth. The ridges are fringed with hair that serves to entangle the tiny invertebrates on which these cetaceans feed—thus, the paradox of the largest of all animals feeding on some of the smallest. The great blue whales far exceed in size even the largest dinosaurs, for some have attained lengths of 30 meters and weigh over 135 metric tons.

Figure 21.49 The Eocene whale *Basilosaurus.* The tendency toward increase in size was already evident in this whale, which was over 20 meters long.

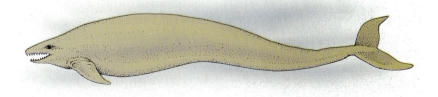

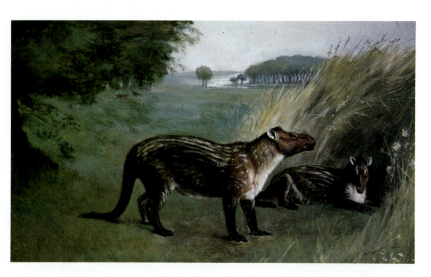

Figure 21.50 The early Tertiary archaic herbivorous mammal *Phenacodus* was a member of a group of primitive plant eaters called condylarths. Like most of the early Tertiary herbivores, *Phenacodus* walked on all five toes in a method termed plantigrade. Speed was not essential to this forest-dwelling browser. *(National Museum of Natural History, Smithsonian Institution)*

The largest category of Cenozoic plant eaters is that comprising the **ungulates**. Simply defined, ungulates are animals that walk on hoofs and feed on plants. The earliest ungulates were members of a group called condylarths. *Phenacodus* (Fig. 21.50), a representative condylarth, had simple primitive teeth and five toes, each terminated with a small hoof. The animal walked mostly in a flat-footed fashion, for these ungulates had not yet evolved the ability to walk or run on their toes for greater speed. This type of stance, termed **plantigrade**, was characteristic of early Tertiary ungulates, including the huge six-horned *Uintatherium* (Fig. 21.51). Primitive plantigrade and relatively small-brained plant eaters are often termed **archaic ungulates**.

The modern and more familiar ungulates fall into two categories: the **perissodactyls** and the **artiodactyls**. The perissodactyls include the modern horses, tapirs, and rhinoceroses and the extinct chalicotheres and titanotheres (Fig. 21.52). They seem to have originated from condylarth ancestors and reached the peak of their evolutionary history during the Miocene Epoch. Since that time, they have declined steadily.

Perissodactyls have certain distinctive characteristics. For example, the number of toes on each foot is usually odd, and the axis of the foot, along which the weight of the body is primarily supported, lies through the third toe. There is a tendency toward reduction of the lateral toes. In the modern horse, only the single, central toes remains. Perissodactyls are clearly digitigrade, meaning that they run or walk on their toes to attain a longer stride and greater speed.

The oldest known perissodactyl, *Hyracotherium* (Fig. 21.53), has been found in upper Paleocene and Eocene strata of both North America and England. Its ancestors were probably condylarths. Familiarly known as eohippus, *Hyracotherium* has the distinction of being

the earliest member of the horse family. This little "dawn horse" was not much larger than a fox. It had four hoofed toes on the front feet (Fig. 21.54) and three on the hind feet. The back curved like that of a condylarth, and the dentition was primitive, with canines still present and the premolar teeth not similar to the molars, as they are in more recent perissodactyls. In addition, the molars were bluntly cusped for browsing and had not yet developed the ridged oral surface and high crowns that characterize modern grazers (see Fig. 16.16).

From ancestors such as *Hyracotherium*, the horse family evolved, at least until the Miocene, in a rather straightforward manner (see Fig. 16.15). The skeletal remains clearly show progressive increases in size, length of legs, height of crowns of cheek teeth, and brain size. The curved back became straightened, and the middle toe was strengthened and emphasized at the expense of lateral toes (see Fig. 21.54). The premolars came to resemble the molars, and their grinding surfaces developed increasingly complicated patterns of resistant enamel ridges. *Mesohippus*, an Oligocene horse about twice the height of *Hyracotherium*, showed the beginnings of many of these trends but still retained relatively low-crowned teeth best adapted for browsing.

With the spread of grass-covered prairies during the Miocene, the family tree of horses became more complicated as more conservative species stayed behind in the forests while others took advantage of the more open prairie environment. The descendants of these grassland dwellers led ultimately to the modern horse, *Equus*. During the Pleistocene, several species of *Equus* thrived, including *Equus occidentalis*, a horse about the size of a small pony, and the huge *Equus giganteus*, which rivaled modern draft horses of today. Pleistocene horses spread across the length of North America,

Figure 21.51   Eocene mammals. (A) *Uintatherium* is the large six-horned and tusked animal in the upper right. Other animals on this restoration are (B) the small, fleet rhinoceros *Hyrachus;* (C) *Trogosus,* a grawing-toothed mammal; (D) *Mesonyx,* a hyena-like flesh eater; (E) *Stylinodon,* a gnawing-toothed mammal; (F) three early members of the horse lineage *(Orohippus);* (G) a saber-toothed mammal, *Machaeroides;* (H) *Patriofelis,* an early carnivore; and (I) *Palaeosyops,* an early titanothere. Restorations are based on skeletal remains from the middle Eocene Bridger Formation of Wyoming.   *(National Museum of Natural History, J. H. Matternes mural, with permission)*

Figure 21.52   Generalized diagram of the evolution of perissodactyls.

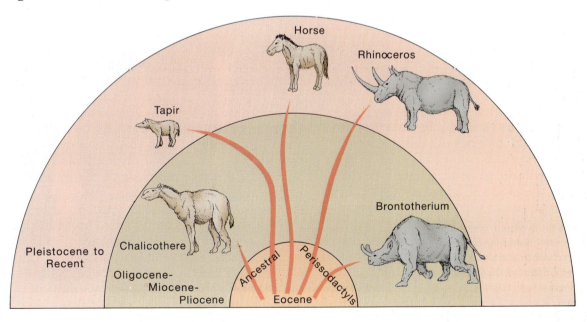

Figure 21.53 *Hyracotherium,* the small Eocene horse more popularly known as eohippus, the "dawn horse." This tiny horse was only about a half meter in length. *(American Museum of Natural History)*

Eurasia, and Africa. Only a few thousand years ago, horses suffered extinction in North America. (The continent was later restocked with the progeny of domestic horses that had escaped from the Spanish explorers in the sixteenth century.) The cause of the extinction of horses in North America is something of a mystery. Some believe it was brought on by contagious disease, whereas others speculate that the cause was overkill by prehistoric human hunters.

Among the other surviving perissodactyls are the tapirs and rhinoceroses. Tapirs retain the primitive condition of four toes on the forefeet and three on the rear, as well as low-crowned teeth. They are forest-dwelling, leaf-eating animals whose fossil record begins in the Oligocene. More impressive to humans are the rhinoceroses, which began their evolution during the Eocene as small, swift-running creatures and eventually produced such giants a *Baluchitherium* (Fig. 21.55). *Baluchithe-*

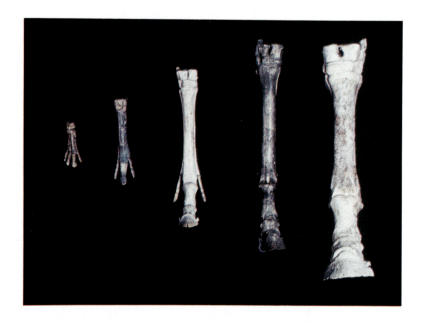

Figure 21.54 Evolution of the lower foreleg in horses, beginning at far left with *Hyracotherium* and ending with the modern horse *Equus* at the far right. *(National Museum of Natural History)*

Figure 21.55 The giant Oligocene to early Miocene hornless rhinoceros *Baluchitherium*.

rium is the largest land mammal thus far discovered. It stood 5 meters tall at the shoulders.

The perissodactyls that did not survive into modern times were the **titanotheres** and **chalicotheres**. The most familiar titanothere of the Cenozoic was ponderous *Brontotherium* (Fig. 21.56G), readily remembered because of the pair of hornlike processes that grew over the snout. Chalicotheres differed from all other Cenozoic perissodactyls in having three claws rather than hoofs on their feet. These odd creatures (Fig. 21.57A) were rather similar to a horse in the appearance of the head and torso. The fore legs were longer than the hind legs, and the back sloped rearward. They lived from Eocene into the Pleistocene, at which time they became extinct.

The even-toed ungulates, or artiodactyls, have been far more successful than the perissodactyls in terms of survival, variety, and abundance (Fig. 21.58). Modern artiodactyls include pigs, deer, hippos, goats, sheep, cattle, and camels. The center of artiodactyl evolution was not North America, as had been the case with the odd-toed ungulates, but rather Eurasia and Africa. They were already present during the Eocene and by late Tertiary time had clearly achieved numerical and varietal superiority over the other herbivores.

Most artiodactyls have an even number of toes on each foot, either four or two. Most of the weight of the animal is carried on the two middle toes, which form a symmetrical pair and thus appear cloven. Unlike advanced perissodactyls, the molar and premolar teeth are dissimilar. Among the Tertiary artiodactyls that became extinct, the **oreodonts** and **entelodonts** are particularly interesting. Oreodonts (see Fig. 21.57H) were short, stocky grazers that roamed the grassy plains of North America in enormous numbers. Entelodonts, some of which were as large as buffalo, were repulsive-looking, hoglike beasts. Their trademarks were curious

bony processes that grew along the sides of the skull and jaws (Fig. 21.57F).

The major radiation of artiodactyls began in the Eocene and ultimately produced the three major surviving categories: swine, camels, and ruminants. Of these, the swine family has probably remained the most primitive. They have kept their four toes, even though most of the weight is carried by the two middle digits. The swine group includes both pigs and the more lightly constructed and primarily South American peccaries. Hippopotami are the only modern amphibious artiodactyls. They are relatives of the pig family, having arisen from a group of Miocene piglike animals called anthracotheres.

People are often surprised to learn that much of the evolutionary development of the camel lineage—**Tylopoda**—occurred in North America. The geologic history of camels and llamas began in Eocene time with tiny creatures about the same size as a small goat. As they evolved through the Tertiary, they lost their side toes and increased the length of their legs and neck. The long-necked and long-limbed *Oxydactylus* (Fig. 21.59) of the Miocene probably browsed on leaves in much the same manner as the giraffe. By Pleistocene time, there were numerous modern-looking camels and llamas in North America. The llamas moved southward and took up residence in the highlands and plains of South America. Camels migrated across the Bering land bridge to Eurasia and Africa, where they established themselves in arid regions. Those that were left behind in North America mysteriously became extinct.

The **ruminants** are the most varied and abundant of modern-day artiodactyls. The ruminants take their name from the rumen, or the first of four compartments in their multichambered stomach. For the most part, they are cud chewers. The earliest ruminants were small,

Figure 21.56 A restoration of Oligocene mammals based primarily on skeletal remains from the White River Formation of South Dakota and Nebraska. The flora is based on nearly contemporaneous plant fossils from the Florissant beds of Colorado. (A) *Trigonias,* an early rhinoceros. (B) *Mesohippus,* a three-toed horse. (C) *Aepinocodon,* a remote relative of the hippopotamus. (D) *Archaeotherium,* an enteledont. (E) *Protoceras,* a horned ruminant. (F) *Hyracodon,* small, fleet rhinoceros. (G) The giant titanothere *Brontotherium.* (H) Oreodonts named *Merycoidodon.* (I) *Hyaenodon,* an Oligocene carnivore. (J) *Poëbrotherium,* an ancestral camel. *(National Museum of Natural History, J. H. Matternes mural, with permission)*

Figure 21.57 Mural depicting an assemblage of early Miocene mammals. (A) The chalicothere *Moropus.* (B) The small artiodactyl *Merychyus.* (C) *Daphaenodon,* a large wolflike dog. (D) *Parahippus,* a three-toed horse. (E) *Syndyoceras,* an antelope-like animal. (F) *Dinohyus,* a giant, piglike enteledont. (G) *Oxydactylus,* a long-legged camel. (H) *Stenomylus,* a small camel. *(National Museum of Natural History, J. H. Matternes mural, with permission)*

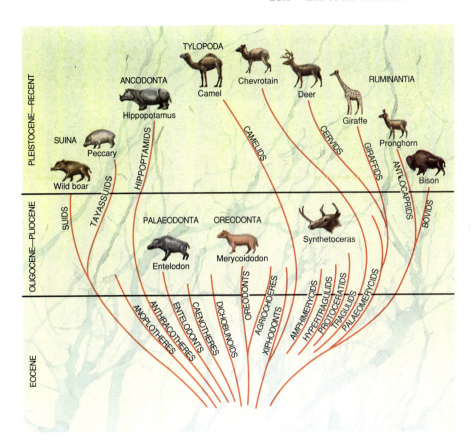

Figure 21.58   The evolutionary radiation of the even-toed ungulates, or artiodactyls. *(From E. H. Colbert,* Evolution of the Vertebrates. *New York: John Wiley & Sons, 1969. Used with permission of the author, artist Lois Darling, and the publisher.)*

Figure 21.59   Skeleton of the Miocene camel, *Oxydactylus.* This small, graceful camel was only one of a varied group of camels that populated the grasslands of North America. *(Denver Natural History Museum)*

delicate, four-toed animals called **tragulids**. They are represented today by the skinny-legged and timid mouse deer of Africa and Asia. One of the most distinctive tragulids was *Synthetoceras,* a Pliocene form with a Y-shaped horn on its snout (Fig. 21.60B).

Ruminants such as sheep, cattle, giraffes, and deer are called **pecorans**. Deer are primarily browsers that have made the forests their principal habitat ever since their first appearance in the Oligocene. Early members of the deer family were small, hornless browsers. Subsequent evolution involved increases in size and the development of antlers. The culmination of these trends is represented by the Pleistocene *Megaloceras,* whose antlers were more than 3 meters from tip to tip (Fig. 21.61).

The bovids, a group that includes cattle, sheep, and goats, are presently the most numerous of ruminants. Miocene strata provide the earliest fossil record of bovids. The bison are particularly interesting to Americans. Seven species of bison lived in North America during the Pleistocene, some of which were truly giants with horns that measured 2 meters from end to end.

A discussion of Cenozoic mammals would not be complete without a brief mention of elephants and their older relatives, the mammoths and mastodons. Because these animals have a trunk, they are collectively termed

Figure 21.60 A variety of early Pliocene mammals. (A) *Amebelodon,* the shovel-tusked mastodon. (B) *Synthetoceras.* (C) *Cranioceras.* (D) *Merycodus,* an extinct prong-horned antelope. (E) *Epigaulis,* a burrowing, horned rodent. (F) *Neohipparion,* a Pliocene horse. (G) The giant camel *Megatylopus* and smaller *Procamelus.* (H) *Prosthennops,* an extinct peccary. (I) The short-faced canid *Osteoborus.* *(National Museum of Natural History, J. H. Matternes mural, with permission)*

proboscideans. The trunk is the elephant's principal means of bringing food to its mouth. Other animals as tall as elephants reach food on the ground easily because of their long necks. However, proboscideans, with short, muscular necks to support their massive heads, have evolved their own unique anatomic solution to food gathering. Paleontologists are able to follow the development of the trunk in early proboscideans by noting the position of the external nasal openings at the front of the skull. Those openings recede toward the rear of the skull in sequential stages of trunk development. Another elephantine trademark is the tusks, which evolved by elongation of the second pair of incisors.

The early fossil record of proboscideans includes a trunkless, tapir-like animal named *Moeritherium* (Fig. 21.62). Bones of this ancestral proboscidean have been found in upper Eocene and lower Oligocene beds near Lake Moeris in Egypt. It is likely that the moeritheres arose from early Tertiary condylarths. From an ancestry represented by the moeritheres, proboscideans separated into two branches. One led toward a group of

Miocene and Pliocene animals called **dinotheres** (Fig. 21.63). The tusks of dinotheres were distinctive in that they were present only in the lower jaws and curved downward and backward, an orientation presumably useful for uprooting plants and digging for roots and tubers.

The other branch of the proboscidean family produced mastodons and elephants. Mastodons were browsers that lived mostly in forested areas. They take their name from their large cheek teeth that, in many species, bore a pair of blunt cusps shaped like human breasts (Greek *mastos,* breast, plus *odons,* teeth). Most mastodons had tusks in both jaws. The term **mammoth** is loosely applied to Ice Age elephants of North America, Europe, and Africa. Their cheek teeth have multiple cross ridges formed of infolded enamel to provide a grinding mill for harsh grasses. Mammoths include the famous woolly mammoths (Fig. 21.64) drawn by our own ancestors on the walls of their caves. The great imperial mammoth reached heights of 4.5 meters and ranged widely across California, Mexico, and Texas.

Figure 21.61 *Megaloceros,* the giant Irish "elk," was actually a deer whose remains are frequently found in the peat bogs of Pleistocene age in Ireland. From tip to tip, the antlers were 3.6 meters in breadth.

21.5   Demise of the Pleistocene Giants

At the time of maximum continental glaciation (approximately 11,000 years ago), the Northern Hemisphere supported an abundant and varied fauna of large mammals, comparable to that which existed in Africa south of the Sahara several decades ago (Fig. 21.65). The fauna included giant beavers, mammoths, mastodonts, elk, most species of perissodactyls, many even-toed forms, and ground sloths. From all available evidence, most of these great beasts maintained their numbers quite well during the most severe episodes of glaciation but experienced rapid decline and extinction in the period around 8000 years ago. A favored theory proposes that the extinctions were the result of human overkill.

Early humans had developed the ability to hunt in highly organized social groups and skillfully bring down large numbers of big animals on open prairies and tundras. Human predators may have killed then, as they do now, in excess of their needs. Unlike such predators as wolves, early human hunters probably did not seek out the weak or sick to kill but brought down the best animals in the herds. With this decimation of particular species of herbivores, the other, nonhuman predators would have necessarily also suffered. In support of this idea, paleontologists note that most Pleistocene extinctions involved large terrestrial animals. Marine genera, protected by the sea from human predation, continued to thrive. Small mammals also survived; although frequently hunted, they were difficult to exterminate because of their more rapid breeding rate, their greater numbers, and the probability that they were killed only one at a time. The big, gregarious herbivores were easily accessible, reproduced more slowly, and entire herds could be driven off cliffs or into ravines, thus providing opportunities for slaughter of hundreds at the same time.

Another line of evidence favoring overkill stresses the observation that an early wave of extinction began in Africa and southern Asia, where predatory humans first became prevalent. The better-known extinctions of the colder north did not occur until much later, when humans had moved into this region after having first

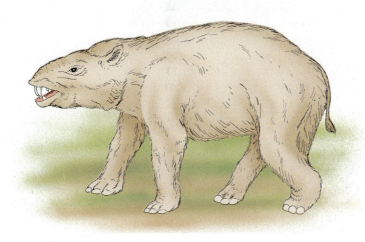

Figure 21.62 The first proboscideans (members of the group that includes elephants) were moeritheres, named after this typical genus *Moeritherium,* which lived in Egypt during the late Eocene.

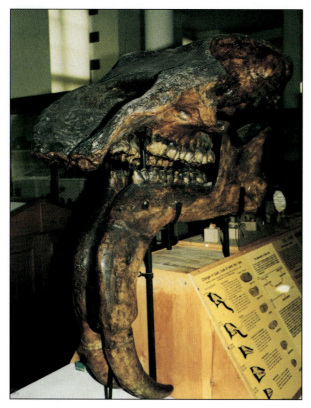

Figure 21.63 Skull of *Dinotherium,* a Miocene proboscidean. Length of skull is about 1.2 meters. This specimen is on display in the Sedgwick Museum of Cambridge University.

devised the means to better clothe and shelter themselves.

21.6 Human Origins

Primates

Evolution during the late epochs of the Cenozoic produced a mammal capable of shaping and controlling its own environment. That mammal was the primate of our own species, *Homo sapiens.* Unlike previous vertebrates, this remarkable creature has profoundly changed the surface of the planet, modified the environment in ways both beneficial and destructive, and so manipulated the populations of other creatures as to change the entire biosphere.

We have examined the characteristics of Cenozoic herbivores, flesh eaters, and rodents, but what traits should an animal have to qualify as a primate? This is not an easy question to answer, for primates have remained structurally generalized compared with most other orders of placental mammals. They retain the primitive number of five digits, have teeth that are not specialized for dealing with either grain or flesh, and have never developed hoofs, horns, trunks, or antlers. In the course of their evolution from shrewlike insectivores, their principal changes have involved progressive enlargement of the brain and certain modifications of the hand, foot, and thorax that were related to their life

Figure 21.64 The woolly mammoth. In the late Pleistocene, these magnificent animals lived along the borders of the continental glaciers. Their remains have been found frozen in the tundra of northern Siberia. *(Courtesy and copyright of the Chicago Museum of Natural History; painting by C. R. Knight, with permission)*

Figure 21.65 An assemblage of animals that lived in Central Alaska about 12,000 years ago, during the late Pleistocene. Fossil remains of this period are abundant and indicate a fauna in which grazing animals predominated, with the remaining fauna composed of browsers and predators. *(National Museum of Natural history, J. H. Matternes mural, with permission)*

in the trees and their manner of obtaining food. These adaptations were not insignificant, however, for they enabled the human primate to shape a mode of life qualitatively different from that of any other animal.

The fictional pig Snowball in George Orwell's book *Animal Farm* remarked that "the distinguishing mark of man is the hand, the instrument with which he does all his mischief." Although this is clearly a biased point of view, it is correct in its assessment of the importance of the primate grasping hand with its opposable thumb. This characteristic not only permitted primates a firm grip on their perches but also allowed them to grasp, release, and manipulate food and other objects. The forearms retained their primitive mobility, in which rotation of the ulna and radius upon one another permitted the hands to be reversed in position.

The development of the grasping, mobile hand was accompanied by improvement in visual attributes. The eyes of primates became positioned toward the front of the face so that there was considerable overlap of both fields of vision, resulting in an improved ability to judge distances.

Other evolutionary modifications of primates were related to changes in the eyes and limbs. To protect the eyes, a postorbital bar evolved. As the eyes were positioned more closely together, the snout was reduced so that the face became flatter. In response to a brachiating habit (swinging from branches), forelimbs and hindlimbs diverged in form and function and an inadvertent predisposition toward upright posture developed.

The Prosimian Vanguard

The primates are divided into the suborders Prosimii, which includes tree shrews, lemurs (Fig. 21.66), and tarsiers, and the order Anthropoidea, which includes monkeys, apes, and humans. Table 21.2 provides the common names for members of these suborders. In general, evolution has progressed from Prosimii to Anthropoidea.

The fossil record for primates begins with the appearance of a creature named *Purgatorius,* known only from a few teeth discovered in the Hell Creek Formation at Purgatory Hill in Montana. These finds indicate that the earliest primates were contemporaries of the last of the dinosaurs, at least in the tropical latest Cretaceous environments of North America. The fossil record improves somewhat in the early epochs of the Cenozoic. *Plesiadapis* (Fig. 21.67), found in Paleocene beds of both the United States and France, has the distinction of being the only genus of primate (other than that of humans) that has inhabited both the Old World and the New World. The presence of this **prosimian** on the now widely separated continents is one of many clues indicating that the northern continents were not yet completely separated by the widening Atlantic in the Paleocene. *Plesiadapis* was a rather distinctive and specialized primate and represents a sterile offshoot of the primate family tree. The incisors were rodent-like and were separated from the cheek teeth by a toothless gap or **diastema**. Fingers and toes terminated in claws

Figure 21.66 The ring-tailed lemur of Madagascar. *(Frans Lanting/Minden Pictures)*

rather than nails. The rodent-like characteristics and habits of the plesiadapiformes may have contributed to their extinction by late Eocene time. About that time, rodents were having their own radiation. Rodents are highly successful as gnawing, seed-eating animals and are able to reproduce at a rate with which no primate can compete.

General trends in prosimian evolution during the Eocene Epoch (37 to 55 million years ago) involved reduction in the length of the muzzle, increase in brain size, shifting of the eye orbits to a more forward position, and development of a grasping big toe. These trends are evident in the fossil remains of *Cantius* from the Wind River basin of Wyoming. In the Wyoming stratigraphic sequence, one can trace the evolution of successive species of *Cantius* from lower to upper beds. Near the top of the section, a new genus, *Notharctus* (Fig. 21.68), occurs as the apparently direct descendant of *Cantius*. *Notharctus* has been known from other localities for over a century. In general form, it appears to be just

Table 21.2 • A Simplified Classification of the Order Primates

Order	Suborder	Superfamily	Common Names of Representative Forms	
Primates	Prosimii	*TUPAIOIDEA*	Tree shrew	
		LEMUROIDEA	Lemur	
		LORISOIDEA	Bush baby, Slender loris	
		TARSIOIDEA	Tarsier	
	Anthropoidea	*CEBOIDEA*	Howler monkey, Spider monkey, Capuchin, Common marmoset, Pinche monkey	
		CERCOPITHECOIDEA	Macaque, Baboon, Wanderloo, Common langur, Proboscis monkey	
			Family	
		HOMINOIDEA	HYLOBATIDAE	Gibbon, Siamang
			PONGIDAE	Orangutan, Chimpanzee, Gorilla
			HOMINIDAE	Humans

Figure 21.67 The Paleocene rodentlike prosimian *Plesiadapis*.

the kind of lemur-like animal from which monkeys and apes are derived.

The Early Anthropoids

The next step in the primate evolutionary advance was the appearance of the first **anthropoids**. Discoveries in the Fayum region of Egypt have provided a wealth of information about the early anthropoids. The fossils there occur in several upper Oligocene horizons and include well over 100 specimens. Although many of the skull fragments and teeth retain subtle vestiges of prosimian ancestry, none of the remains are from prosimians. They are fossils of primates that have reached what may be called the monkey stage of organization. One primate discovered at Fayum is *Aegyptopithecus*, a relatively robust arboreal primate with monkey-like limbs and tail, a brain larger than that of *Notharctus*, and eye orbits rotated to the front for stereoscopic vision. The fossil of the species *Aegyptopithecus zeuxis* is dated as 33 to 34 million years old, indicating that the prosimian–anthropoid transition had taken place by Oligocene time.

Fossils of the earliest known New World monkeys are known from upper Oligocene and lower Miocene strata of South America. Although they appear superficially similar to some Old World forms, New World monkeys evolved quite independently from prosimian ancestors and without genetic contact with their Old World cousins.

Because evolution is a continuing process, with each animal truly a transitional link between older and younger species, it is a difficult and often arbitrary task to assign the fragment of a fossil jaw or tooth to either the monkey or the ape category. One of many clues used in attempting to make this distinction is the pattern of

Figure 21.68 The Eocene lemur *Notharctus*.

cusps on certain molar teeth. In Old World monkeys, specific molars have four cusps. Among apes (and humans), these same molars have five cusps, with an intervening Y-shaped trough. Molars belonging to *Aegyptopithecus* display the Y-5 pattern (Fig. 21.69), which is one reason anthropologists consider the genus to be the possible very early ancestor of Miocene apes. It seems safe to state that the Oligocene Epoch may have witnessed not only the transition from prosimians to anthropoids, but also the differentiation of apes from monkeys.

During the Miocene, plate tectonics had a significant influence on the evolution of primates. Africarabia (Africa and Arabia) was drifting northward and ultimately collided with Eurasia. (The Arabian portion of the continent then rotated counterclockwise to eventually produce the Red Sea.) As a result of these movements, east–west circulation of tropical currents across the former Tethys seaway was prevented, and East Africa became cooler and dryer. Extensive grass-covered savannas replaced former areas of dense forests. The evolution of Miocene primates was strongly affected by these changes.

Among the new players that came on the scene during the Miocene were a group called **dryomorphs** (formerly called dryopithecines). The dryomorphs take

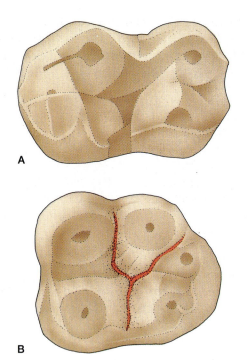

A

B

Figure 21.69   General pattern of cusps on the molars of Old World monkeys. (A) The lower molar of a baboon showing a cusp at each corner. (B) The lower molar of a chimpanzee showing the "lazy Y-5" pattern characterized by a Y-shaped depression that separates five cusps.

their name from *Dryopithecus fontani,* a species first discovered in France in 1856. The dryomorphs varied in size and appearance and include forms bearing different generic names from France, Spain, Greece, Hungary, Turkey, India, Pakistan, and Africa. Many anthropologists consider the important species *Proconsul africans*

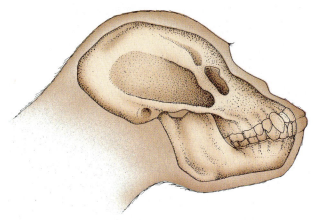

Figure 21.70   Skull of *Proconsul* (a dryomorph) from Lake Victoria, Kenya.

to be either a very early dryomorph or the immediate dryomorph ancestor. *Proconsul africans* was discovered by Mary and Louis Leakey in 1948 on an island in Lake Victoria, Kenya. The skull, jaws, and teeth of this small primate are decidedly apelike rather than monkey-like (Fig. 21.70). The slightly more advanced middle Miocene dryomorph descendants of *Proconsul* are probably ancestral to both modern African apes and the first hominids, the australopithecines. This means that *Proconsul* is distantly ancestral to humans as well.

The study of Miocene primates clearly indicates an absence of orderly sequential change along a single trend. The family tree of primates is a complex of parallel and diverging branches. For this reason, the attempt to trace the ascent of humans up through the many bifurcations and dead ends is an exciting but difficult task.

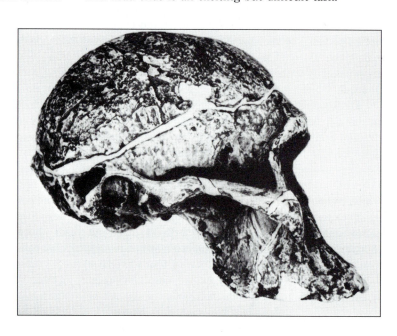

Figure 21.71   *Australopithecus africanus* from the Transvaal of South Africa.   *(Photograph of Wenner-Gren Foundation replica by David G. Gantt)*

The Australopithecine Stage and the Emergence of Hominids

The story of the emergence of humans (hominids) begins with Raymond Dart's 1924 discovery of fossil remains of an immature primate in a limestone quarry in South Africa. Dart named the fossil *Australopithecus africanus* (Fig. 21.71). In succeeding years, many additional skeletal fragments of species of *Australopithecus* have been found in upper Pliocene or lower Pleistocene deposits of Africa, Java, and China. Fossil sites in East Africa have become increasingly important because of the veritable bonanza of hominid bones they have yielded. The new collections of fossils have given paleoanthropologists an unsurpassed record of human evolution over the past 3 million years. Some of the East African sites, such as Olduvai Gorge (Fig. 21.72), are now famous as a result of the lifelong research programs of Mary Leakey and the late Louis Leakey.

Their efforts, and those of their son Richard, have provided fossils of relatively heavy-bodied, robust **australopithecines** as well as lighter, so-called gracile types. Among the latter are remains of creatures that may well belong within our own genus.

Volcanoes located along the east African rift system formed a fiery array during the Cenozoic. As a result, many of the fossil sites have interspersed layers of volcanic ash and lava flows. This fact has been of great benefit to the paleoanthropologists, for samples of ash can often be dated by the potassium–argon method. Dates from two succeeding ash beds would then provide a close estimate of the age of fossils found in the intervening layers of sediment.

Among the oldest undisputed hominids of East Africa are those discovered at Laetoli in Tanzania and Hadar in Ethiopia (see Fig. 21.72). These localities have yielded remains of human ancestors that roamed Africa between 3 and 4 million years ago. Here have been

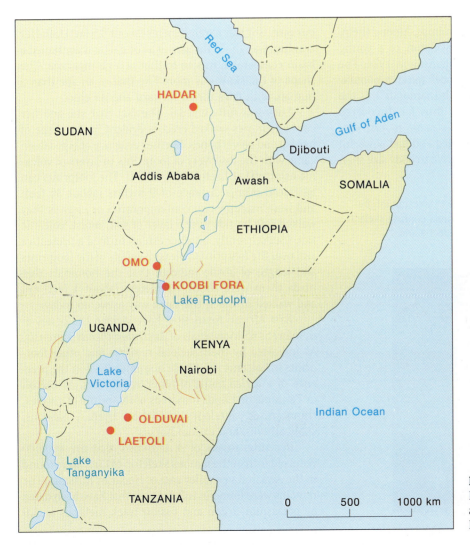

Figure 21.72 Location of the Hadar, Laetoli, Olduvai, Omo, and Koobi Fora (East Lake Rudolf) paleontologic sites.

Figure 21.73 A male and female *Australopithecus afarensis* leave telltale footprints in deposits of wet volcanic ash as they stroll across the East African landscape about 3.5 million years ago. The lady has been named Lucy by her discoverer. She was 42 inches tall and weighed about 80 pounds. Tooth structure suggests a diet of plants. *(American Museum of Natural History, with permission)*

found the remains of the oldest hominids. Among them, none is more famous than the skeleton of a young female discovered by Donald C. Johanson in 1974. The individual, informally dubbed "Lucy" (after a popular Beatles song, "Lucy in the Sky with Diamonds"), is the most complete Pliocene hominid thus far discovered (Fig. 21.73). Lucy, whose scientific name is *Australopithecus afarensis,* was well muscled and had powerful arms that were slightly longer relative to body size than

Figure 21.74 Footprints made in wet volcanic ash by *Australopithecus afarensis.* The footprints confirm skeletal evidence that the species had a fully erect stance. *(Peter Jones)*

our own. She was barely a meter (about 3.5 feet) tall. Her cranium was about the size of a softball and contained a brain comparable in size to that of a chimpanzee. The shape of the skull was more like that of an ape than a human, with jaws thrust forward and no chin.

Lucy was fully bipedal. The fact that she and her contemporaries walked erect is evident in the shape of her leg and pelvic bones. Further evidence of bipedalism came with the discovery of footprints in layers of volcanic ash (tuff) at Laetoli (Fig. 21.74). The footprints can be traced over a distance of 9 meters (about 30 feet) and were made by two contemporaries of Lucy as they walked side by side over a layer of soft, moist, volcanic ash. (Subsequently, another ashfall formed a protective seal over the footprints.) Measurements of the footprints and length of stride indicate that the hominids that made them were about 1.2 meters (about 4 feet) tall and walked in much the same way as we do today.

East African fossil sites have yielded the bones of as many as 65 individuals of *Australopithecus afarensis.* As indicated by their skeletal remains and other footprints in volcanic ash, these hominids lived in the presence of a varied fauna of other vertebrates. At Laetoli in Tanzania one finds hundreds of footprints of giraffes, rhinoceroses, antelopes, ostriches, hyenas, baboons, chalicotheres, and extinct horses.

Recently, fossils of *A. afarensis* that are 400,000 years older than the famous skeleton of Lucy were found in the Awash River valley east of Hadar in Ethiopia. The remains were found in tuff deposits that have been dated by potassium–argon and fission track methods as 4.0

million years old (middle Pliocene). As such, they represent the oldest hominid fossils known. The femur was that of a male, about 4.5 feet tall, who clearly had a bipedal posture and locomotion. The absence of any significant differences between the skull and other bones in this creature and those of Lucy suggests that, as a species, *A. afarensis* was stable over a remarkable span of time. Some paleontologists see this as support for the punctuational pattern of evolution (see Chapter 16) rather than the more traditional view of evolution by slow and gradual steps.

Although debate about the precise evolutionary position of these fossils is currently lively, many paleontol-ogists consider Lucy and her kin to be the direct ancestors of the genus *Homo,* as well as the ancestor of species of *Australopithecus* (Fig. 21.75).

Two other East African fossil localities that have yielded species of both *Homo* and *Australopithecus* are the Omo and Koobi Fora sites (see Fig. 21.72). The Omo digs are located along the Omo River in remote southwestern Ethiopia. Here one finds a nearly continuous section of sediments and volcanics that span 2.2 million years. The Koobi Fora locality is on the east side of Kenya's great alkaline Lake Rudolf (Fig. 21.76). It has been investigated by Richard Leakey and Glynn Isaac. Altogether, the Omo, Koobi Fora, Olduvai, Hadar,

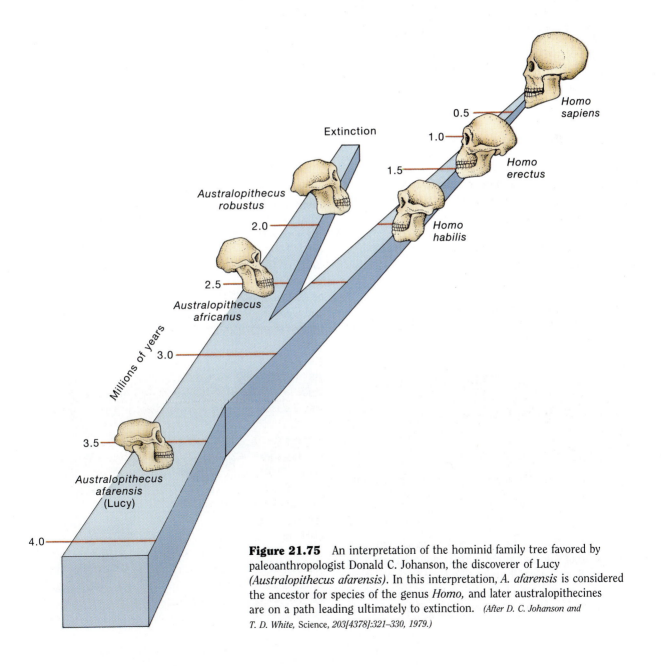

Figure 21.75   An interpretation of the hominid family tree favored by paleoanthropologist Donald C. Johanson, the discoverer of Lucy *(Australopithecus afarensis).* In this interpretation, *A. afarensis* is considered the ancestor for species of the genus *Homo,* and later australopithecines are on a path leading ultimately to extinction.   *(After D. C. Johanson and T. D. White,* Science, *203[4378]:321–330, 1979.)*

Figure 21.76 The Koobi Fora fossil site in East Africa where the skeletal remains of many australopithecines, as well as the skull designated KNM-ER-1470, were discovered. In addition to the hominid skeletal remains unearthed at Koobi Fora, paleontologists found pebble choppers and flake tools associated in deltaic deposits with the bones of a hippopotamus. The position of both the tools and hippo bones suggests that this was a butchering site. Apparently one day about 1.8 million years ago, some australopithecines came across the body of a hippopotamus that had recently died. They feasted on its carcass, using the sharp edges of flaked chert to get at the meat.

(Kay Behrensmeyer)

and Laetoli localities have given us sufficient skulls, jaws, teeth, and other bones to indicate that the australopithecines were not an entirely homogeneous group. However, certain characteristics appear in nearly all of the specimens unearthed. From the structure of their pelvic girdles, it is known that they stood upright in a fashion more human-like than apelike. Australopithecine dentition was essentially human, although the teeth were more robust than in modern humans. In contrast to these human-like characteristics, australopithecine cranial capacity was comparable to that of modern large apes and reached a volume of 600 or 700 cubic centimeters. However, even with a brain only about half the size of *Homo sapiens,* they were able to make several kinds of crude implements from horn, teeth, and bone.

Recent investigations by anthropologists working in Kenya and Ethiopia have provided evidence that there may have been several human-like lineages coexisting in Africa between 1 and 3 million years ago. One of these lineages included *Australopithecus africans* and the more recent *Australopithecus habilis* (some prefer

Homo habilis) from Olduvai Gorge. The possible existence of a second lineage became apparent in 1972, when a team led by Richard Leakey unearthed a skull at the Koobi Fora site in beds that are about 2.0 million years old. The skull (Fig. 21.77), assigned the code name KNM-ER-1470, has a cranial capacity of about 800 cubic centimeters, much larger than that of any *Australopithecus* and close to the cranial capacity of later hominids already assigned to the genus *Homo*. Although KNM-ER-1470 may not have quite reached the *Homo* stage of evolution, it was tending in that direction. The fossil indicates that large-brained, human-like creatures may have been present on Earth much earlier than previously suspected.

Yet another group of primates at the australopithecine stage had smaller brains but were larger than the *Australopithecus africanus* line. This group is exemplified by *Paranthropus* and may have evolved into a still heavier creature appropriately named *Gigantopithecus*. Such species as *Australopithecus boisei* (once called *Zinjanthropus boisei*) and *Australopithecus robustus* (formerly *Paranthropus robustus*) represent dead-end

Figure 21.77 Lateral view of a nearly complete skull, designated KNM-ER-1470, from the Koobi Fora fossil site in Kenya. The creature may be an early representative of the genus *Homo*.

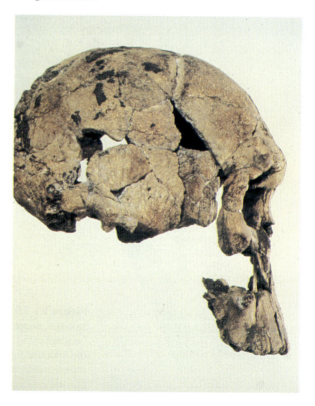

side branches that failed to give rise to any known descendants.

The *Homo Erectus* Stage

The next stage in hominid evolution is represented by *Homo erectus,* formerly known as *Pithecanthropus erectus* from Java and *Sinanthropus pekinensis* (Fig. 21.78) from China. Louis Leakey discovered a skull cap of *Homo erectus* in Bed II of Olduvai Gorge. Bed II had been dated by potassium–argon techniques and found to be about 750,000 years old. These finds indicate that *Homo erectus* was probably widely dispersed in Asia and Africa during the middle Pleistocene. *Homo erectus* has been considered the first true species of humans and is generally regarded as having evolved from a species of *Australopithecus.*

The bones of the axial skeleton and limbs of *Homo erectus* were quite similar to those of modern humans. In this regard, these hominids were more advanced than the australopithecines. For example, the pelvic bones of *Australopithecus* indicate that, although they were fully bipedal, they walked with their feet turned outward in a sort of half-running, rolling gait. *Homo erectus,* on the other hand, was an excellent walker.

Although the postcranial skeleton of *Homo erectus* was quite modern, the skull was not. Cranial capacity ranged from about 775 cubic centimeters in earlier forms to nearly 1300 cubic centimeters in specimens of more recent age. Brain capacity in modern *Homo sapiens* ranges from 1200 to 1500 cubic centimeters, and thus the brain size of *Homo erectus* overlaps at least the lowermost range of modern peoples. Pithecanthropines represent a stage in the evolution of hominids during which relatively rapid increases in brain size had begun. No doubt the expansion of the brain involved not only the reshaping of the cranium but also the enlargement of the birth canal.

The skull of *Homo erectus* was massive and rather flat, with heavy supraorbital ridges over the eyes (see Fig. 21.78). The forehead sloped, and the jaws jutted forward at the tooth line in a condition termed **prognathus**. A definite jutting chin was lacking, and the nose was probably broad and flat. These are primitive traits. However, except for being somewhat more robust, the dental arcade in *Homo erectus* was essentially modern.

Homo erectus appears to have made good use of his larger brain. Bones found at living sites indicate that these hominids were good hunters. Simple implements like scrapers and axes were fashioned from flint. However, we do not know if *Homo erectus* spoke a language, wore clothes, or built dwellings.

Homo Sapiens

The Neanderthals

From the *Homo erectus* stage of the middle Pleistocene (Fig. 21.79), it was only a short step to late Pleistocene *Homo sapiens neanderthalensis* (Fig. 21.80). The initial

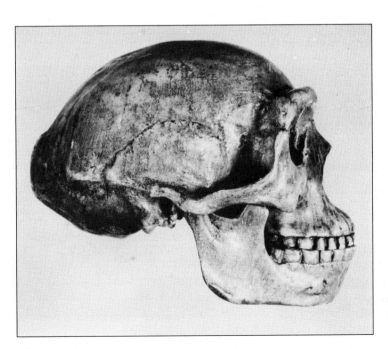

Figure 21.78 Replica of *Homo erectus,* previously known as *Sinanthropus pekinensis.*
(Photograph of Wenner-Gren Foundation replica by David G. Gantt)

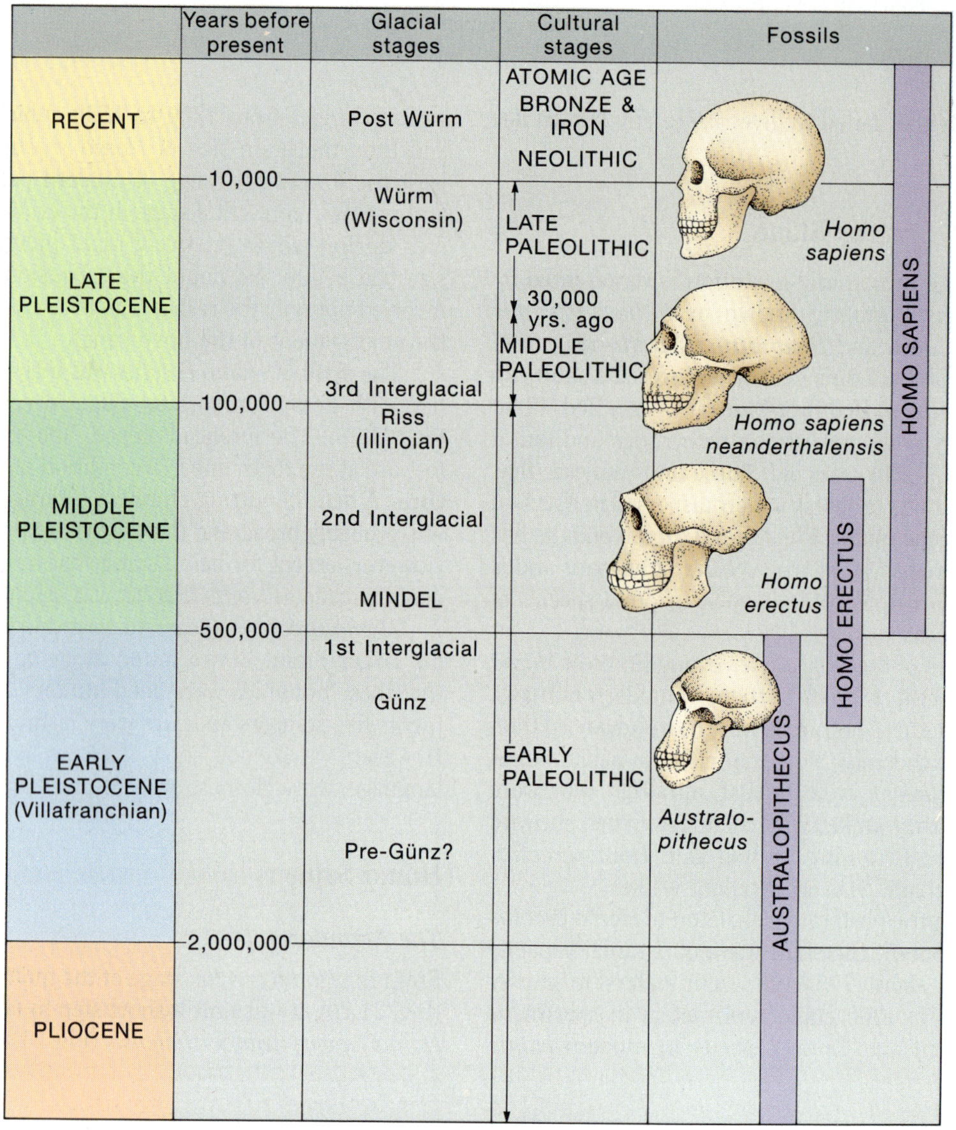

	Years before present	Glacial stages	Cultural stages	Fossils		
RECENT		Post Würm	ATOMIC AGE BRONZE & IRON NEOLITHIC			
LATE PLEISTOCENE	—10,000—	Würm (Wisconsin)	LATE PALEOLITHIC	*Homo sapiens*		
		3rd Interglacial	↕ 30,000 yrs. ago MIDDLE PALEOLITHIC		HOMO SAPIENS	
	—100,000—	Riss (Illinoian)		*Homo sapiens neanderthalensis*		
MIDDLE PLEISTOCENE		2nd Interglacial				HOMO ERECTUS
		MINDEL		*Homo erectus*		
	—500,000—	1st Interglacial				
		Günz	EARLY PALEOLITHIC		AUSTRALOPITHECUS	
EARLY PLEISTOCENE (Villafranchian)				*Australopithecus*		
		Pre-Günz?				
	—2,000,000—					
PLIOCENE						

Figure 21.79 Chronologic chart of Pleistocene fossil humans.

Figure 21.80 Reconstruction of a Neanderthal family group. *(National Museum of Natural History)*

specimen of this early human was found in the Neander Valley near Dusseldorf, Germany. Many subsequent discoveries indicate that the Neanderthal people ranged all across the Old World. With their heavy brows and prognathous jaws, neanderthaloids have become the very personification of the "cave man." Indeed, the face seems a brutish carryover from the middle Pleistocene. In most other features, however, the neanderthaloids were quite modern. Below the neck, their skeletons matched our own, and their brain size equaled or exceeded that of present humans. The cartoon depiction of a neanderthaloid as a bent-kneed, in-toed, bull-necked brute with a curved back is false. Reexamination of the skeleton on which the original restoration was based revealed that it was that of an old man who had been severely afflicted with osteoarthritis.

Most of the so-called classic neanderthaloids were sturdy people of small stature that had adapted to life in the cold climates near the edge of the ice sheet. The relatively short limbs and bulky torso may have been an advantage in helping them to conserve body heat. Neanderthals often lived in caves and successfully hunted many contemporary cold-tolerant mammals, including cave bears, mammoths, woolly rhinoceroses, reindeer, bison, and fierce ancestors of modern cattle known as aurochs. They were not at all devoid of culture and manufactured a variety of stone spear points, scrapers, borers, knives, and saw-edged tools. In addition, Neanderthals made ample use of fire and apparently could ignite one at will in the hearths they excavated in the floors of caves.

The use and control of fire had many advantages for Pleistocene hominids. It provided light in caves that otherwise would have been avoided because of the near helplessness of humans in deep darkness. It gave warmth to creatures that had evolved in the tropics but had moved northward into colder realms, and it gave protection at night from predators. During the frigid winters of the Ice Age, fire provided the means to thaw meat that had become quickly frozen after a kill. Conceivably, cooking may have been a simultaneous result of the thawing process. Cooked foods provided the added advantages of promoting easier digestion and destroying harmful microorganisms.

In addition to their use of fire, there is evidence that Neanderthals constructed shelters of skins, sticks, and bones in areas where caves may not have been available. Perhaps as a hunting ritual, these people killed cave bears and stacked the skulls carefully in chests constructed of stones. That they also pondered the nature of death and believed in an afterlife is indicated by their custom of including artifacts in graves.

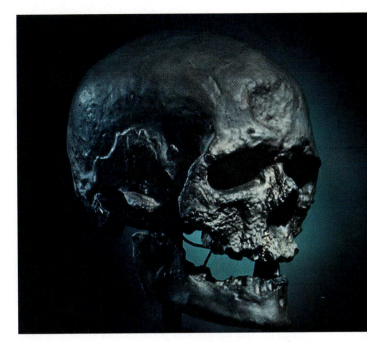

Figure 21.81   The badly corroded skull of a Cro-Magnon human. As in present-day humans, the face of Cro-Magnon was vertical rather than projecting, and there was a prominent chin.

Cro-Magnon

About 34,000 years ago, during the fourth glacial stage, humans closely resembling modern Europeans moved into regions inhabited by the Neanderthals and in a short span of time completely replaced them, probably by tribal warfare and by competition for hunting grounds. The new breed, designated *Homo sapiens sapiens* but informally dubbed **Cro-Magnon** (Fig. 21.81), were taller than their predecessors, had a more vertical brow, and had a decided projection to the chin. In short, Cro-Magnon's bones were modern, and anthropologists have recognized definite Cro-Magnon skull types among today's western and northern Europeans, as well as North Africans and native Canary Islanders.

The modern types of *Homo sapiens* did not appear suddenly on the evolutionary scene. Changes in dentition and supporting facial architecture were not an "overnight" occurrence. Transitional forms between Neanderthal and Cro-Magnon do exist (Fig. 21.82) and include fossil skulls from Palestine, South Africa, Germany, and Czechoslovakia. These forms vary among themselves but generally show less robust brow ridges, the beginnings of the modern chin, and less pronounced muscle markings than are evident in typical Neanderthal specimens. The transitional forms seem to have evolved outside Europe and, after they had reached the

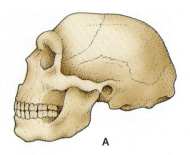 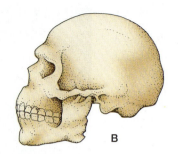 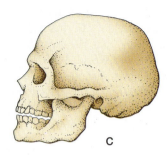

Figure 21.82 Comparison of the skulls of (A) *Homo sapiens neanderthalensis;* (B) a neanderthal/Cro-Magnon transitional form from a rock shelter on the slope of Mount Carmel; and (C) Cro-Magnon. The Mount Carmel skull is both intermediate in form and age between neanderthal and Cro-Magnon.

Cro-Magnon stage, migrated into Europe and Asia during a temporary regression of the ice sheets about 35,000 years ago.

The cultural traditions of the Neanderthals appear to have been continued and further developed by Cro-Magnon. The variety and perfection of stone and bone tools were increased. Finely crafted spear points, awls, needles, scrapers, and other tools are found in caves once inhabited by these people. These caves also retain splendid paintings and drawings on their walls and ceilings. Carvings and sculptures of obese women (Fig. 21.83), probably used in fertility rites, were produced from fragments of bone or ivory, as were small, elegant engravings and statues of mammoths and horses.

Figure 21.83 Prehistoric art of late Pleistocene *Homo sapiens.* (A) Venus figure originally found in Austria ("The Venus of Willendorf"). (B) Thong-stropper used either to work hide thongs or to straighten arrow shafts. The tool is made from an antler and is intricately carved. *(Musée de l'Homme, Paris.)*

 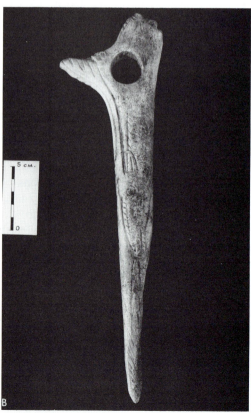

FOCUS ON   *Population Growth*

One of the many lessons of historical geology is that the advent of *Homo sapiens* represents only one recent and momentary event along the sinuous, branching, 500-million-year evolution of vertebrates. We humans are linked by similarities of structure and body chemistry to lungfish struggling out of stagnating Devonian pools, to the small, shrewlike precursors of the primate order, and to the axe-carrying hunters that wandered along the margins of great ice sheets. The descendants of those and other Pleistocene hunters have emerged as the dominant species of higher life presently on this planet. Like other animals, *Homo sapiens* has been shaped by the combined powers of genetic change and environment. However, our species has quickly become a pervasive force in modifying the very physical and biological environment from which it was shaped. Humanity has come to rely heavily on science to improve its lot but has had great difficulty in finding ways to manage the resulting

technology and to control the burgeoning problems arising from too few resources for too many humans.

About 10,000 years ago (a mere moment to a geologist), there were more than 6 million humans on the earth. By the year A.D. 1, the number had jumped 50-fold to 300 million. In 1970, the figure stood at 3.6 *billion*. If we project current rates of increase into the future, there will be 7.5 billion people on earth by the year 2000. Only a decade into the twenty-first century, global population is likely to exceed 10 billion. Can our fragile planet sustain so enormous a population? Few would doubt that the survival of *Homo sapiens* will depend on our ability to control population growth and to solve the many often related resource and environmental problems. Should we fail to solve these problems, other animals may replace our species and add their own chapters to the history of life.

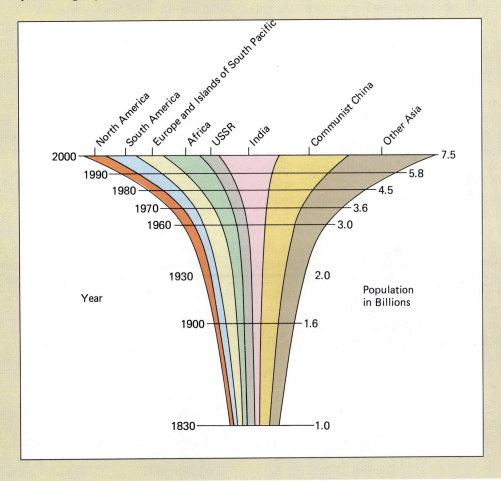

Cro-Magnon people enjoyed wearing body ornaments and frequently fashioned necklaces from pieces of ivory, shells, and teeth. Burial of the dead became a truly elaborate affair. Hunters were buried with their weapons and children with their ornaments. This apparent concern for an afterlife, and the sense of self-awareness that resulted in art and complex ritual, suggests that the beginning of the age of the philosopher had arrived.

Through most of his early history, *Homo sapiens* was a wandering hunter and gatherer of wild edible plants. However, about 10,000 to 15,000 years ago, near the beginning of the Holocene Epoch, tribes began to domesticate animals and cultivate plants. They learned to grind their tools to unprecedented perfection and to make utensils of fired clay. With more reliable sources of food, permanent settlements were developed, and individuals, spared the continuous demands of searching for food, were able to build and improve their cultures. Languages improved, and symbols developed into forms of writing. The era of recorded history had begun.

SUMMARY

The drift of continents to their present locations, the formation of the Alps and Himalayas, the onslaught of a major ice age, and the appearance of *Homo sapiens* are among the important events of the Cenozoic Era. Along the eastern margin of North America, repeated cycles of uplift and erosion sculpted the Appalachians to their present topography. Deposition of thick sequences of sandstones, shales, and limestones occurred along the Atlantic and Gulf coastal regions. Florida was a shallow, carbonate bank until uplifted late in the era.

In western North America, compressional forces that had produced the structures of the Cordillera lessened, and instead there were vertical crustal adjustments. Erosional debris from the mountains filled the basins between ranges and spread eastward to form the clastic wedge of the Great Plains. Crustal movements with associated faulting produced the Basin and Ranges and resulted in the uplift of the Sierra Nevada Mountains. Intense volcanism accompanied these crustal movements. That activity continues today along sectors of the west coast that lie adjacent to subduction zones, such as the Cascades, which are fed by lavas from the subducting Juan de Fuca plate. Also in the far west, the Farallon plate and its spreading center were overridden by westward-moving North America. As a result, the formerly convergent plate boundary was changed to a shear boundary.

Climate was on a fluctuating cooling trend during the Cenozoic. The culmination of this trend was the Pleistocene Ice Age. Repeated glaciations depressed continental regions, changed drainage systems, and were responsible for the development of the Great Lakes.

Biological events in the Cenozoic include the appearance of grasses and consequent spread of prairies, the proliferation of such marine invertebrates as mollusks and foraminifers, and the formation of coral reefs along warm coastal areas and around volcanic islands of the Pacific. The period of reptilian dominance had ended, and mammals increased rapidly in both numbers and diversity. The first and earliest radiation produced the archaic mammals, and by the middle of the era a rich array of progressive predators and herbivores had spread across the continents. Among the prominent groups were perissodactyls, artiodactyls, and proboscideans. Many of the large herbivores that lived during the Pleistocene, however, became extinct near the end of the epoch. The possibility of overkill by early human hunters may account for the demise of many Pleistocene giant mammals.

Primate evolution at the prosimian stage was already under way during the Paleocene and Eocene, and by Miocene time anthropoid dryomorph ancestors of hominids had appeared. Australopithecines were present by about 4 millon years ago, followed by *Homo erectus* in the middle Pleistocene and late Pleistocene Neanderthal and Cro-Magnon peoples.

KEY TERMS

Paleogene *551*
Neogene *551*
Oil shale *555*
Farallon plate *562*
Molasse *564*
Discoaster *565*
Little ice age *566*
Pluvial lake *566*

Teleost *573*
Marsupial *579*
Insectivora *579*
Edentate *579*
Creodont *579*
Carnivora *579*
Ungulate *582*
Archaic ungulate *582*

Perissodactyl *582*
Artiodactyl *582*
Chalicothere *585*
Oreodont *585*
Entelodont *585*
Ruminant *585*
Tragulid *587*
Dinothere *588*

Mammoth *588*
Prosimian *591*
Anthropoid *593*
Dryomorph *593*
Australopithecine *595*
Cro-Magnon *601*

REVIEW QUESTIONS

1. What Eurasian mountain ranges resulted from the compression and upheaval of large areas of the Tethys seaway?

2. What epochs of the Cenozoic Era are included within the Tertiary? What epochs comprise the Paleogene?

3. What is the economic importance of the Fort Union Formation? The Green River Formation?

4. What land bridge, important in the migration of ice age mammals, resulted from glacial lowering of sea level?

5. What is the explanation for the gradual rise in land elevations that has occurred within historic time around Hudson Bay, the Great Lakes, and the Baltic Sea?

6. What was the effect of geographic separation of landmasses on the evolution and distribution of Cenozoic mammals?

7. During what epoch of the Cenozoic did grasslands first become extensive? How did Cenozoic herbivores adjust to this event?

8. List the general evolutionary changes apparent in the teeth and skeleton of horses during their Cenozoic history.

9. What groups of reptiles survived the extinctions at the end of the Mesozoic and are still living today?

10. Early arboreal primates were characterized by close-set eyes and grasping hands. Other than for mobility in trees, what function might these adaptations have had?

11. What advantage is there in a fossil hominid site that consists of fossil-bearing strata interspersed with layers of volcanic ash?

12. What human-like traits were present in *Australopithecines?* What apelike traits were present?

DISCUSSION QUESTIONS

1. Discuss the manner in which each of the following major physiographic features developed.

   a. Mountains of the Basin and Range Province

   b. Great Plains

   c. Columbia and Snake River Plateau

   d. Cascade Range

2. The Cenozoic stratigraphic record for the Gulf Coast region indicates that there were eight major transgressions and regressions. What characteristics of a stratigraphic sequence might indicate that it was deposited during a transgression? A regression?

3. What evidence indicates that the Gulf of Mexico experienced steady subsidence during the Cenozoic?

4. What might be the effect on humans if all the ice now present as continental ice sheets were to melt?

5. What advantages might be inherent in the ruminant type of digestion? Name three living groups of ruminants.

6. Discuss the adaptive radiation of Cenozoic mammals and give examples of the habitats in which they were successful.

7. What properties of chert and flint account for their use in weapons and tools of early humans?

8. Discuss the kinds of environmental problems that may occur as a result of severe global overpopulation.

Conversation with PAUL HOFFMAN

Paul F. Hoffman was born in 1941 in Toronto, Canada. He studied at McMaster University (B.Sc. 1964) and The Johns Hopkins University (Ph.D. 1970) before joining the Geological Survey of Canada as a Research Scientist. He has also been a visiting lecturer at Franklin & Marshall College and the University of California at Santa Barbara, a Fairchild Distinguished Scholar at the California Institute of Technology, a visiting professor at the University of Texas at Dallas and the Lamont–Doherty Geological Observatory of Columbia University, and an adjunct professor at Carleton University. In 24 summers of field work in the northern Canadian shield, his work on Precambrian sedimentary basins and related orogenic belts provided evidence that the movement of tectonic plates has occurred throughout most of Earth history. Since 1985, he has been working on a synthesis of the Precambrian geology of North America.

He serves on committees of the International Lithosphere Program, the Canadian Lithoprobe Project, the International Commission on Stratigraphy, the Circum-Pacific Map Project, the Geological Society of America, and the Massachusetts Institute of Technology. He has made distinguished lecture tours for the American Association of Petroleum Geologists, the Canadian Institute of Mining and Metallurgy, and the Geological Association of Canada and has won

best paper awards from the Society of Economic Paleontologists and Mineralogists, the American Association of Petroleum Geologists, the Canadian Society of Petroleum Geologists, and the Geological Society of Washington. A recipient of the Past-Presidents' Medal from the Geological Association of Canada and the Bownocker Medal for Research in Earth Sciences from The Ohio State University, Hoffman was elected a Fellow of the Royal Society of Canada in 1981.

How did your interest in geology begin?

I first became interested in geology through Saturday classes for primary school students in various subjects in natural science at the Royal Ontario Museum in Toronto. At the age of 10, I joined the mineralogy and geology group at the museum. We went on field trips to collect minerals and fossils in the Paleozoic rocks around Toronto, Ontario. As a result of those field trips and mineral collecting, I decided to be a geologist by the age of 12.

I went to McMaster University in Hamilton, Canada. Initially, I went to McMaster thinking that I would go into mineralogy because mineral collecting had been my love. What I found out at the university was that studying mineralogy was mainly an indoor pursuit and involved lots of mathematics, not the kind of thing that I was interested in or very good at.

My interest in mapping began the summer of my freshman year when I worked as a field assistant on a geological mapping party in northwestern Ontario. We spent 4½ months living out of canoes and mapping a large area of the northwestern Ontario Archean terrane.

I spent every summer of the next 24 years mapping in the Northwest Territories with the Geological Survey of Canada. The outcrop exposures are much better there than in the southern Canadian shield, so the geology is more challenging and rewarding. After I graduated from McMaster, I decided to do a Ph.D. thesis on the Canadian shield in the Northwest Territories. The Geological Survey of Canada funded my thesis project, which was done at The Johns Hopkins University.

Although you started out as a field geologist doing field mapping on a quadrangle level, you've become best known for your global scale tectonic models and interpretations. Tell us more about these.

Well, they were *large* quadrangles (10,000 square kilometers) and I've always been interested in the large-scale tectonics of the Canadian shield. My Ph.D. thesis project was basically designed to compare a Precambrian orogenic belt with the Appalachians, which is a Phanerozoic orogenic belt. Since The Johns Hopkins University is close to the Appalachians, we went on a lot of field trips there.

After reading about younger orogenic belts, I thought that the best way to begin a study of the Appalachians would be to start with the sedimentary rocks because they're the best preserved and because the layers of sediments record the evolution of the mountain belt. In contrast, in most previous studies of the Precambrian, geologists had gone to the inner parts of these deeply eroded belts and looked at the metamorphic rocks first, which I felt were more difficult to work with. From my work as an undergraduate field assistant, I knew that several small basins of well-preserved sediments existed.

I started with a basin in the east arm of Great Slave Lake about 1000 miles north of Montana. I had the idea that the area to the north, which gave older radiometric ages, was a stable foreland and that the area to the south, which gave younger radiometric ages, was the internal part of the ancient mountain belt. In the Appalachians there is a pattern where the succession of sediments laid down on a subsiding continental margin are followed by sediments that were derived from the rising mountains formed as that continental margin collided with another continent. When I started out, I wasn't thinking in terms of plate tectonics, so I wanted to see whether there was the same pattern of sedimentation as in the Appalachians and also whether the paleocurrent indicators, the directional indicators of sediment transport, showed a reversal as they do in the Appalachians. There the early paleocurrents are directed toward the ocean, and the subsequent paleocurrents are directed toward the interior of the continent away from the rising mountain belt.

> At its best, field geology rewards the intellectual, the athletic, and the aesthetic spirits simultaneously.

What I found in the Precambrian of the Great Slave Lake was a sedimentary succession very similar to the Appalachians and a similar reversal in the paleocurrents, but the paleocurrent directions weren't what I had expected. Instead of going toward the south initially and then switching toward the north, they were going toward the west initially and then toward the east. So my thesis postulated that there had been a Precambrian ocean and later a mountain belt hidden to the west in the area of Great Slave Lake that is now covered by Devonian sediments. But I thought that relics of the belt might be exposed in the Canadian shield to the north, so I predicted that it would be there. When I joined the Geological Survey in 1969, I proposed going to that more northerly area to see whether my prediction panned out. The area is called the Wopmay orogen, named after a famous bush pilot.

My prediction turned out to be correct. But that begged the question as to why in the east arm of Great Slave Lake the fold-belt trends northeast–southwest, with more intense deformation to the south than to the north. And why was it that the paleocurrents there seemed to be parallel to the structural trends rather than transverse to them, as in the Appalachians? In trying to wrestle with these problems, I stumbled on a paper by two Russian geologists outlining the idea of "aulacogens," subsided rifts that extend into the interior of continents. Thus, I proposed that the east arm of Great Slave Lake was a Precambrian aulacogen. This subsequently turned out to be a very popular model. In fact, I believed it for a number of years. But now I think that it's quite wrong. The Great Slave Lake area is in fact of collisional origin as I had originally thought, but the paleocurrents of the sedimentary fill I was looking at are related to the slightly younger belt to the west.

What is the difference between working for the Geological Survey and working in a university?

One of the differences of working with the Survey is that your specialty tends to be regionally defined rather than defined by discipline. When I started working on the Wopmay orogen, I found that my compelling interest was to study the various parts of the orogen from the standpoint of stratigraphy, sedimentology, and structure to volcanology, igneous petrology, and metamorphic geology as I systematically examined its different zones. So rather than studying only the sediments of many regions as a university specialist would, I ended up studying all the different rock types and different zones of the orogen within only one general region.

Tell us more about your work for the Survey and your current research.

The geologic evidence in the Northwest Territories convinced me and others that plate tectonic processes were operating in the Precambrian. This became apparent to me at a very early stage; my thesis work and projects in the late 1960s showed that there was very little difference between the rocks, structures, and organization of orogenic belts of the Precambrian and those of the Phanerozoic. If Phanerozoic orogens were products of plate tectonics, it followed logically that the Precambrian ones must be also. I think that in the early 1970s, geologists were starting to come around to this view.

Anyway, I mainly worked in this one general region because it is extremely well exposed and offers a great range of structural levels. Most of my work up until 1983 was on the margins of a small microcontinent called the Slave microcontinent. I was operating in an area only about 600 by 600 kilometers studying the tectonic, volcanic, and metamorphic activity that resulted from a pair of collisions between three microcontinents, one on the east margin and one on the west margin of the Slave. My reasoning was that to convince people of the reality of plate tectonics 2 billion years ago, I had to study orogens of that age in as much detail as Phanerozoic belts because most other Precambrian belts were very poorly known relative to Phanerozoic ones.

I had always felt that many of the other orogenic belts in the shield could also be interpreted in terms of plate tectonics and that their histories would be interrelated, but we didn't really have a good way of dating them until the late 1970s and early 1980s when Tom Krogh of the Royal Ontario Museum advanced the uranium–lead dating method. This method allowed one to date these Precambrian rocks with very high precision, and that's when the thing really started to become exciting. In about 1983, the Geolog-

ical Society of America began to publish a series of volumes compiling, summarizing, and synthesizing everything that was known about the geological evolution of the North American plate. Because I had always had an interest in the Canadian shield as a whole, I asked to be the principal compiler for the Precambrian of Canada. A year later, the editor-in-chief for the North American overview volume asked me to prepare a chapter in that volume outlining the Precambrian evolution of North America as a whole. So at that point I threw myself into the North American synthesis. It was about this time that the explosion in uranium–lead age determination in the Precambrian began. We also started to get colored digital data sets for gravity and magnetic anomalies on the scale of the continent. These allowed you to see the continuation of the Precambrian structures underneath the sedimentary cover of the Great Plains.

Because a number of things were coming together at once, it made it an excellent time to take a new look at the entire Precambrian structure of North America. What ultimately emerged from the synthesis of North America is that you can't understand North America just by looking at North America alone; you have to know what's going on in the rest of the world. North America was part of the supercontinent called Pangaea in the late Paleozoic and early Mesozoic time. Similarly, North America was joined with other continents at various times in various configurations during the Precambrian. So, it was a natural extension of the North American Precambrian synthesis to start looking at what occurred in the Precambrian on a global scale. My current greatest interest now is with Gondwana, the ancestral supercontinent that gave birth to Africa, South America, Antarctica, India, and Australia. Gondwana formed at the end of the Precambrian as a result of a series of collisions known collectively as the Pan-African orogenic episode.

What's your encore after describing the origin of the North American craton? What do you think you will be doing in the future?

The evolution of the North American craton shows us that it has been involved in the aggregation and dispersal of as many as six successive supercontinents over the past 2.6 billion years. If the notion of episodic supercontinents is valid, then one should see evidence on many continents of contemporaneous collisional mountain building, pointing to the aggregation of supercontinents, and contemporaneous rifting, signifying supercontinental breakup. The supercontinent "cycle" also predicts changes in global climate, the chemistry of ocean sediments, and changes in continental paleolatitudes which can be detected from remanent magnetism in rocks. So my next goal is to compare the Precambrian geology of all the continents. After North America, the best-known Precambrian continent is Gondwana. It is not as well studied as North America, but it has the advantage of being very large, almost a supercontinent in itself. Unlike North America, which originated about 1.8 billion years ago, Gondwana was first aggregated only 0.6 billion years ago, almost at the end of the Precambrian. However, several of the older, smaller cratons within Gondwana contain collision zones that are very close in age to those in North America and to the other northern continents. Ultimately, I hope that piecing together past continental connections will lead to a better understanding of global Precambrian crustal development. Obviously, we will never fully succeed in this aim, but I suspect we will learn a lot that we can't anticipate at this time.

What would you tell a student interested in studying geology and possibly interested in pursuing it as a profession?

Each science is so diverse that you can find opportunities regardless of your interests and abilities. It has changed greatly since I was a student 25 years ago. Then it was said that it was not an experimental sci-

ence because of the problem of scaling phenomena of such great size and duration. Now, computerized numerical simulations of geologic processes of all scales using geophysical modeling are commonplace. The advent of satellite observations has given us a useful global perspective on geologic problems. There has also been a healthy breakdown of barriers between disciplines and a growing interest in the interactions between the solid Earth, biosphere, hydrosphere, atmosphere, and solar system. Concern over the future environment is encouraging Earth scientists to see their value to society beyond the traditional service to resource industries.

Despite increased technology, there is still a place for those romantics, like myself, who are attracted to the out-of-doors and the use of a vivid imagination and a fit body. Contrary to what is often heard, "low-tech" field work is intensely scientific—it requires you to constantly synthesize observations, to make conjectures that fit your observations, and to test those conjectures through predictions that can be proven wrong through new observations. In contrast, much time in "high-tech" science is spent trying to get the equipment to work. At its best, field geology rewards the intellectual, the athletic, and the aesthetic spirits simultaneously. It also has the advantage of having enough physical discomfort to discourage the uncommitted!

What good is an understanding of geology to someone who doesn't plan to become a geologist?

There is intense public interest these days in various issues that are centrally related to geology, such as environmental pollution, earthquakes, landslides, climate change, resource depletion, and biological extinctions. Current issues affecting society and its future are much more interesting and meaningful if one has some knowledge of how the Earth works. Moreover, it is crucial in this age of litigation to understand the nature of scientific inquiry and the strengths and limitations of scientific evidence.

Geology provides a means of understanding geography, the oceans, volcanoes, mountain ranges, rivers, and plains, as well as the people who live there. Travel becomes a deeper experience when what one sees conjures up images of times past. Geology, the science of Earth's history, converts one's perception of the planet from a snapshot to a movie.

◄ *The volcano, Maat Mons, On Venus.*
(NASA/JPL)

Geologic Resources

An offshore oil drilling platform in the Gulf of Mexico.
(Oryx Energy)

Many animals use tools. Apes use sticks as back scratchers and occasionally as clubs. Woodpecker finches on the Galapagos Islands of South America use small twigs to prod insects from trees. No other species uses tools to the extent that humans do, however, and no other animal controls fire.

Tools and fire are essential to civilization. Prehistoric people made wooden tools, and wood was the major fuel of antiquity. Prehistoric people also used geologic resources such as flint and obsidian to make weapons and hide scrapers. Later, people learned to mine and refine metals and to extract coal and petroleum from the Earth, and today we depend on those resources.

Modern geologic resources fall into two categories:

1. *Energy resources:* Petroleum, coal, and natural gas are called **fossil fuels** because they formed from the remains of plants and animals that lived in the geologic past. Uranium is the basic fuel for nuclear reactors.

2. *Mineral resources:* About 30 metals are commercially important. Some, such as iron, lead, copper, aluminum, silver, and gold, are familiar to all of us. Others, such as vanadium, titanium, and tellurium, are less well known but are vital to industry.

Many nonmetals are also mined for a variety of uses. Sand and gravel are mined for road building and the manufacture of concrete. Limestone is quarried to produce cement. Granite, limestone, marble, and other rocks are quarried for use as building stone. Phosphorus and potassium are essential fertilizers. Sulfur is used widely in the chemical industry.

ENERGY RESOURCES

22.1 Fossil Fuels

Coal

In forests and grasslands, dead plants fall to the ground. In most environments, oxygen and water mix with this plant litter. The oxygen supports organisms that decompose the litter. The water flows downward through the soil and removes decay products.

In some swamps, however, plants grow and die so rapidly that plant litter accumulates faster than it decomposes. Newly fallen vegetation covers the older, partially rotted material, so atmospheric oxygen cannot penetrate into the deeper layers. Therefore, decomposition stops before it is complete, leaving brown, partially decayed plant matter called **peat**. Commonly, peat is then buried by mud deposited in the swamp. Burial compresses the peat, squeezing out the water and other volatiles. Plant matter is composed mainly of carbon, hydrogen, and oxygen. During burial, most of the hydrogen and oxygen escapes as gases and the proportion of carbon increases. The result is **coal**, a combustible rock composed mainly of carbon (Fig. 22.1). Several types of coal are distinguished by their hardness, carbon content, and heat value (Table 22.1).

Petroleum

Organic matter eroded from land is carried to the oceans by streams and deposited with mud in shallow coastal waters. As plants and animals living in the sea die and settle to the bottom, they add more organic matter to the mud. This organic-rich mud is then buried by younger sediment. Burial increases temperature and pressure. These conditions convert the mud to shale and the organic material to liquid petroleum (Fig. 22.2). Typically petroleum forms in the range of 50° to 100°C. (As a comparison, lukewarm tea is about 50°C, and water boils at 100°C.)

Initially, the liquid oil is dispersed throughout the shale. But dispersed oil in shale does not constitute a reservoir. A commercial petroleum **reservoir** forms when oil flows from the shale and concentrates in other rock. A commercial reservoir develops in three phases.

Formation of Source Rock

Organic material accumulates in clay-rich mud. When the mud is buried and heated, the solid organic matter converts to liquid oil at about the same time that the

A Litter falls to floor of stagnant swamp

B Debris accumulates, barrier forms, decay is incomplete

C Sediment accumulates, organic matter is converted to peat

D Peat is lithified to coal

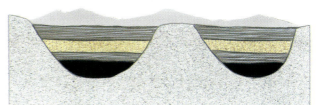

Figure 22.1 Formation of coal deposits.

Table 22.1 • Classification of Different Types of Coal

Type	Color	Water (%)	Other Volatiles and Noncombustible Compounds (%)	Carbon (%)	Heat Value (Btu/pound)
Peat	Brown	75	10	15	3,000–5,000
Lignite	Dark brown	45	20	35	7,000
Bituminous (soft coal)	Black	5–15	20–30	55–75	12,000
Anthracite (hard coal)	Black	4	1	95	14,000

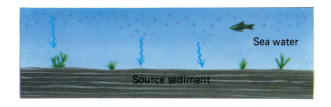

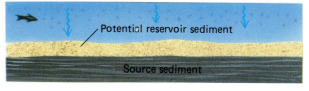

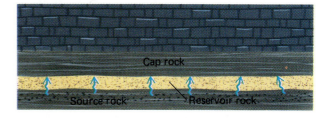

Figure 22.2 The first stages in the development of petroleum reserves.

mud converts to shale, which is called the **source rock**. Oil cannot be recovered from shale because shale is relatively impermeable; that is, liquids do not flow through it rapidly.

Oil Migration

If conditions are favorable, petroleum is forced out of the source rock and migrates to a nearby layer of sandstone or limestone with open spaces called pores between the grains. If liquids can flow readily through porous rock, the rock is said to be permeable.

Concentration of Oil in a Trap

Petroleum, being less dense than water and much less dense than rock, rises through permeable rock. An **oil trap** is any barrier to the upward migration of oil or gas. When oil or gas seeps into the rock below a trap, it accumulates there and forms a reservoir. Most reservoirs form in permeable sandstone or limestone. Many types of traps accumulate oil. In one common type, a dome of impermeable **cap rock**, such as shale, covers the permeable reservoir rock (Figs. 22.3A and D). The cap rock prevents the petroleum from rising farther. In other types of traps, rock is faulted or layered so that impermeable caps cover the reservoir (Figs. 22.3B and C).

It is important to emphasize that a petroleum reservoir is not an underground pool or lake of oil. Instead, it is permeable, porous rock saturated with oil, more like an oil-soaked sponge than a bottle of oil.

The oil pumped from the ground, called **crude oil**, is a gooey, viscous, dark liquid made up of thousands of different chemical compounds. Oil refineries convert crude oil to propane, gasoline, jet fuel, heating oil, motor oil, road tar, and other useful materials. Some of the compounds in oil are used to manufacture plastics, medicines, and many other products.

Natural Gas

Natural gas, or methane (CH_4), forms when crude oil is heated above 100° to 150°C during burial. Many oil wells contain natural gas that floats above the heavier liquid petroleum. In other instances, the lighter, more mobile gas escapes and is trapped elsewhere in a separate reservoir.

Text continued on page 618

Figure 22.3   Formation of oil traps. (A) Petroleum rises into a dome in the rock. (B) A trap is created when rock fractures and slips along the fracture, moving impermeable shale above the oil reservoir. (C) Horizontally bedded shale overlies inclined, petroleum-bearing sandstones. (D) A salt dome. Layers of salt, emplaced at great depth, rise as pressure accumulates. This rising bubble of salt deforms the overlying strata.

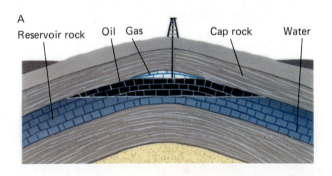

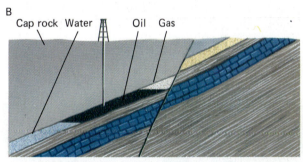

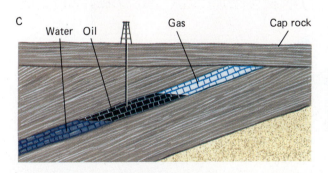

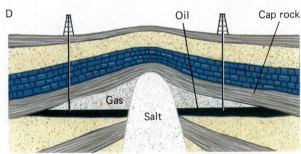

GEOLOGY AND THE ENVIRONMENT

Energy Strategies for the United States

In the United States, petroleum consumption has increased by a factor of 2.5 since 1950. In contrast, domestic oil production peaked in 1970 and has declined for the last 2 decades. These two opposing trends have led to a steady increase in oil imports (Fig. 1). Today we import about 45 percent of our petroleum at an annual cost of $25 billion. This cash outflow has a negative impact on our national economy.

In addition, much of the imported oil is purchased from the Middle East, which is vulnerable to political disruptions. Twice in the 1970s, the major oil-producing nations reduced oil production and raised the price, causing economic problems in oil-importing nations. In 1990 Iraq invaded Kuwait and threw global petroleum markets into turmoil. The price of petroleum skyrock-

eted from below $20 per barrel to $40 per barrel and then fell again after the brief Gulf War.

In 1989, before the Gulf War, the United States spent between $15 billion and $54 billion to safeguard petroleum supplies and shipping lanes in the Middle East. This wide range in estimates reflects difficulties in assigning costs to military operations. The Gulf War in 1990–1991 added another $30 billion to the cost of protection. Using the lowest estimates, these figures add about $23.50 to the cost of each barrel of petroleum. Thus, the cost of defending oil imports more than doubled the cost of imported oil. Such military expenditures are not added to the cost of gasoline at the retail level, but they must be paid for nevertheless.

In short, the United States no longer has an abundant, low-cost energy resource. . . . Indeed, the U.S. energy picture is growing closer to Europe and Japan, which lack cheap domestic energy sources and

Figure 1 The sources of oil used in the United States. The left side of the graph shows what has happened in the past. The right side shows what will happen in the future if current trends continue. Notice that imports grew from near zero in 1950 to about 45 percent of consumption in 1990 due to increased consumption and decreased internal production. If these trends continue, our dependence on imports will rise steadily. *(From "Improving Technology: Modeling Energy Futures for the National Energy Strategy," Energy Information Administration paper SR/NES/90-01, 1991)*

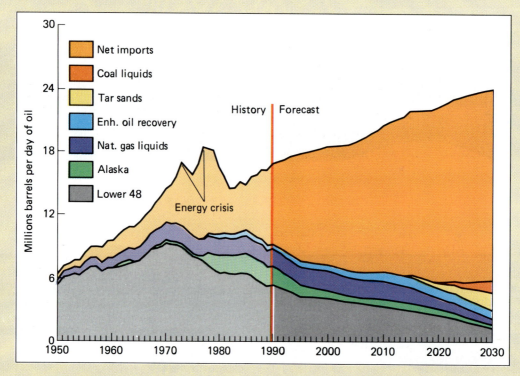

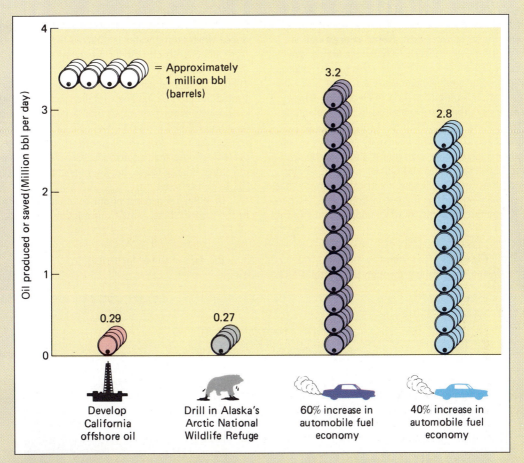

Figure 2   Petroleum savings from conservation versus drilling in the Arctic National Wildlife Refuge (ANWR) and in California offshore fields.   *(From Union of Concerned Scientists, Nucleus 12, no. 3 [1990])*

have to work hard to function amidst high energy costs. The U.S. has a long way to go considering that it uses around twice as much oil per person as European nations do.[1]

Most people agree that the current situation is undesirable, but they disagree about solutions. Three options are possible: (1) Drill for more oil in the United States, (2) exploit other energy sources, and (3) conserve energy.

Drill for More Oil in the United States   Existing U.S. oil fields are being depleted, production is declining, and discovery rates per foot of drilling have decreased by half since 1950. But what about finding new oil fields? Let us examine one case in particular.

In 1980 Congress set aside an 18-million-acre wildlife refuge in northeastern Alaska called the Arctic National Wildlife Refuge (ANWR). After the conclusion of

the Gulf War in 1991, President Bush advocated drilling for oil in the refuge. The argument for this proposal is that our lifestyle, economy, and national security depend on petroleum, and therefore it is imperative to drill for domestic oil wherever it exists. The argument against drilling in ANWR is that the refuge is one of the finest wildlife sanctuaries in the world and that the government must protect what little is left of such places. The U.S. Department of the Interior estimates that 3.2 billion barrels of oil lie beneath ANWR. Under normal operating conditions, the oil reserves would be extracted over 30 years. Production would peak at 500,000 barrels per day and average about 270,000 barrels per day. Today we import about 8 million barrels per day. Therefore, the average contribution from ANWR would result in a 3 percent reduction of imports. Calculations show that production of other reserves would result in similarly small reductions of imports. Thus, we cannot obtain petroleum independence simply by exploiting fragile environments (Fig. 2). In 1991 Congress voted to maintain protection of ANWR and prohibit drilling.

---

[1] Christopher Flavin, "Conquering U.S. Oil Independence," *Worldwatch*, January-February, 1991.

continued

Exploit Non-Fossil-Fuel Energy Sources   Several non-fossil-fuel energy sources exist. **Solar energy** can be harnessed in several ways. Many buildings are designed and oriented to collect sunlight. In others, solar collectors trap the Sun's radiation and use it to heat water. Solar cells convert sunlight directly to electricity. **Biomass energy** is the energy obtained by using plants as fuels. In 1991, more homes in the United States were heated with wood than with electricity generated by nuclear power plants. More than half of the household trash in the United States and Canada is paper, which can be used as fuel. The largest electrical generator that operates solely on trash is located in Rotterdam, Netherlands. It generates 550 megawatts, enough to supply the homes of 250,000 people. **Hydroelectric energy** is produced when falling water is directed through a turbine to produce electricity. About 5 percent of the world's energy is supplied by hydroelectric generators (Fig. 3). In 1990, 13,000 wind generators produced electricity in the United States (Fig. 4). Their total capacity was 1000 megawatts, about the output of a large nuclear reactor. As discussed in Chapter 11, the heat of the Earth's interior can be used to produce **geothermal energy**. At present, geothermal resources satisfy only a tiny portion of human energy needs, but there is considerable potential to be exploited.

In 1990, 7.6 percent of our total energy consumed in the United States came from nuclear fission, 3.5 per-

Figure 3   The Glen Canyon dam. Water retained by the dam is stored for irrigation, and water falling through the dam produces electricity.

22.2   Fossil Fuel Sources and Availability

Coal is forming today in some swamps. Oil and gas are forming in organic-rich muddy sediments in many marine basins and coastal areas. However, because the processes are extremely slow—much slower than current rates of consumption—fossil fuels will eventually be depleted. To estimate the time remaining before humans will exhaust the Earth's fossil fuels, we must estimate the amount left in the ground and future rates of consumption.

One method of estimating future fuel consumption is to use past patterns to predict future trends. Figure 22.4 shows energy consumption in the United States from 1949 through 1989. Notice that the curves are not smooth. Both total energy consumption and petroleum consumption increased steadily from 1948 to 1973, doubling from 1948 to 1970. If oil consumption were to continue to double every 23 years, 660 years from now we would consume as much petroleum annually as there

Figure 4   When the West was being pioneered, windmills like this one were used to pump water. Modern windmills produce electricity.

cent from hydroelectric power, and lesser amounts from solar, geothermal, wind, and biomass sources. Although these numbers are small compared to energy provided by fossil fuels, greater potential exists. We have already discussed the problems and promise of nuclear energy. The state of California has established a program to exploit renewable energy sources, including wind, solar, hydroelectric, and geothermal power, for more than 40 percent of its electricity.

Although these renewable energy sources produce electricity, none can be used directly in automobiles. Eventually transportation may be dominated by electric cars, trains, and trolleys, but the changes will not occur quickly or cheaply.

Conservation   A new nuclear power plant or mass transit system can be built in about 10 years if no major problems arise. If you and your neighbor carpool to work rather than commute separately, you can cut your transportation fuel consumption and air pollution in half starting tomorrow. Conservation is the cheapest and

quickest way to reduce dependence on foreign oil. Two conservation strategies exist. **Social solutions** involve personal choices and sacrifices. If you choose to carpool rather than drive your own car, you inconvenience yourself by coordinating your schedule with your neighbor, driving or walking to a meeting place, and waiting if your friend is late. However, if everyone carpooled or used mass transit, highways would be less congested and there would be fewer traffic jams and more parking places. **Technical solutions** involve using more efficient machinery. For example, a car that runs 40 miles on a gallon of gasoline can carry you as quickly and comfortably as an old gas hog that gets 12 miles per gallon. In 1989 people in the United States used 14.5 million barrels of oil per day *less* than they would have if they had retained the technology of 1973. Further improvements are possible. A 40 percent increase in average automobile fuel economy would save 2.8 million barrels of oil a day, about one third of the current import level.

is water in the world's oceans. Calculations of this sort assure us that we cannot go on doubling our fossil fuel consumption every 22 years.

However, the rapid increase in oil consumption between 1948 and 1970 was not sustained. In the mid-1970s, oil-producing countries began to realize how valuable and limited their fuels were, so they raised prices. As prices went up, consumers used less fuel. People drove fewer miles and purchased smaller, more fuel-efficient automobiles. They added insulation to buildings and saved energy in many other ways. Conser-

vation resulted in a decline in energy use between 1973 and 1975. Then, from 1975 to 1990, energy consumption fluctuated but generally increased.

Future Availability of Petroleum

Many predictions have been made regarding the future availability of petroleum. Although they differ as to the exact time limit, they all share the conclusion that sometime in the next 10 to 35 years, petroleum supply will

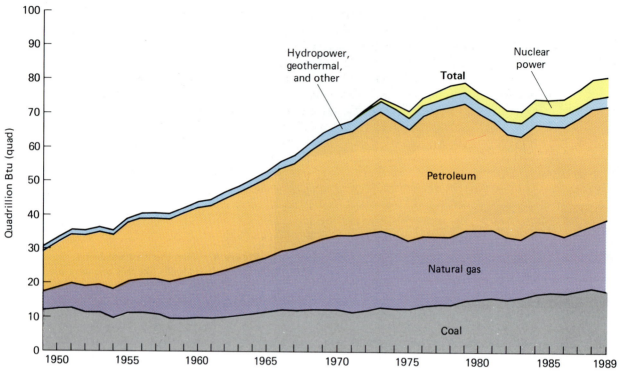

Figure 22.4   Energy consumption in the United States from 1949 through 1989.
(Annual Energy Review 1989, *Energy Information Administration*)

be insufficient to meet demand. The United States produced 145 billion barrels of petroleum in the 100 years from the start of commercial drilling to 1990. About 27 billion barrels of recoverable oil remain in known fields.[1] How much oil lies undiscovered? The discovery rate of new wells has declined in recent years. For every meter of drilling, we now find half as much oil as we did in the 1950s. According to a report published in 1989 by the United States Geological Survey, geologists expect that an additional 55 billion barrels of oil will be discovered in the united States. Thus, as of 1989, known plus estimated reserves in the United States equaled approximately 82 billion barrels of oil. At the current consumption rate of 5.4 billion barrels per year, this is a 15-year supply. Currently, imports account for about 45 percent of the petroleum consumed in the United States. If the flow of oil is not impeded by political instability or war, and if that percentage remains constant, imports will extend the domestic supply 35 years, or to about the year 2025. People in their twenties in the mid-to-late 1990s will see, within their lifetimes, a restructuring of our society resulting from a permanent petroleum shortage.

Humans and their immediate ancestors have lived for about 2 million years. Thus, the "petroleum age" will appear as a short episode in the history of our species.

Future Availability of Natural Gas

Natural gas reserves in the United States will last from 32 to 77 years if consumption remains constant (Fig. 22.5). The large range reflects uncertainties regarding advances in extraction technology. On thing is certain, however: Consumption is increasing. Natural gas is extracted as a nearly pure compound; it releases no sulfur and other pollutants when it burns. As a result, gas is becoming increasingly popular and will be even more desirable if pollution laws become stricter. If natural gas consumption increases, reserves will be depleted faster than predicted in Figure 22.5.

Future Availability of Coal

Large reserves of coal exist in many parts of the world. Figure 22.6 shows that widespread availability of this fuel is expected until at least the year 2200. However, coal cannot be used directly in conventional automobiles, in most home furnaces, or in many industries. One solution to this problem is the conversion of coal

---

[1] *Annual Energy Review 1990*, Energy Information Administration.

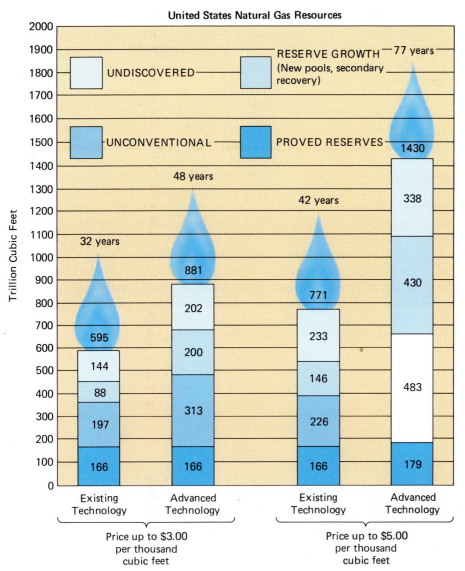

Figure 22.5 Future availability of natural gas, given four different possibilities of price and extraction technology. Above each column is the number of years our gas reserves will last, given the conditions outlined in the column. All predictions assume that consumption rates remain constant. *(American Association of Petroleum as reported in* Geotimes, *November 1989)*

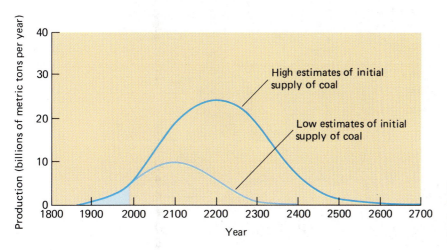

Figure 22.6 Past and predicted world coal production based on two different estimates of initial supply. *(Adapted from M. King Hubbert)*

to liquid and gaseous fuels, which has been accomplished both in laboratories and in industry. This process is not used commercially simply because it is too costly to compete with petroleum. However, when petroleum becomes scarce and prices rise, conversion of coal to liquid and gas will probably occur on a large scale.

Secondary Recovery from Oil Wells

On the average, more than half of the oil in a reservoir is left behind after a well has "gone dry." Conventional pumping cannot recover this oil. In the United States, more than 300 billion barrels of this type of oil remain in oil fields. One technique for recovering it is to force superheated steam into old wells at high pressure. The steam heats the oil and increases its ability to flow into the well. Of course, a great deal of energy is needed to heat the steam, so this type of extraction is not always cost effective or energy efficient. Another process involves pumping detergent into the reservoir. The detergent carries oil to the well. The petroleum is then recov-

ered and the detergent recycled. Not all of the 300 billion barrels can be recovered by these techniques, but if the price of fuel rises, such **secondary recovery** of oil will become economical.

Oil Shale

Some shales and other sedimentary rocks contain large amounts of a waxy, solid organic substance called **kerogen**, the precursor of liquid petroleum. Kerogen-bearing rock is called **oil shale**. If oil shale is mined and heated properly, the kerogen converts to petroleum. In the United States, oil shales contain the energy equivalent of 2 to 5 trillion barrels of petroleum, enough to fuel the nation for 400 to 900 years at current rates of consumption (Fig. 22.7). However, low-grade oil shales require more energy to mine and convert the kerogen to petroleum than is generated by burning the oil, so they will probably never be used for fuel. Oil from higher-grade oil shales in the United States would supply this country for about 75 years if consumption remained

Figure 22.7   The locations of major oil shale fields in the United States.   *(Adapted from U.S. Geological Survey Circular 2223, 1965)*

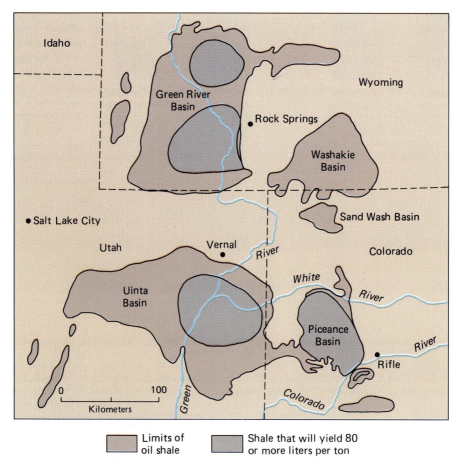

Limits of oil shale

Shale that will yield 80 or more liters per ton

at current levels. Worldwide deposits are not as rich, so the global situation is not as promising.

When oil prices skyrocketed from $22 per barrel in 1978 to $45 per barrel in 1981, major oil companies began to explore the development of oil shale and built experimental recovery plants. However, when prices plummeted a few years later, most of this activity came to a halt. In 1991, the cost of crude oil was about $20 per barrel and gasoline cost a little over $1 per gallon at the pump. We can expect renewed interest in oil shale when the wholesale price of oil doubles.

22.3   Environmental Problems Resulting from Production of Fossil Fuels

All steps in fossil fuel extraction and refining are potential causes of environmental problems.

Coal Mining

Coal is mined in both underground tunnels and surface mines. Underground **tunnel mines** do not directly disturb the surface of the land. However, many abandoned mines collapse, and occasionally houses have fallen into the holes. In addition, the bedrock dug out to get at the coal must go somewhere. This material, called **mine spoil**, is often deposited as banks or mounds of loose, unvegetated rock near mines. Mine spoil erodes easily, and the muddy runoff silts nearby streams and destroys aquatic habitats.

An underground tunnel mine.

Tunnel mines are often more expensive and less efficient than **surface mines** (Fig. 22.8). To dig a surface mine, huge power shovels scrape off soil and bedrock to expose the coal. Mine spoil is piled up on the surface. In the United States, the Surface Mining Control and Reclamation Act requires that mining companies restore mined land so that it is useful for the same purposes as before mining began. In addition, a tax is levied to reclaim land that was mined and destroyed before the law was enacted. However, many citizens claim that the government has not adequately enforced the law and that 6000

Figure 22.8   (A) An aerial view of a coal strip mine, the Decker Mine in Montana. *(E. N. Hinrichs/USGS)*   (B) The Navajo coal strip mine in New Mexico.   *(H. E. Malde/USGS)*

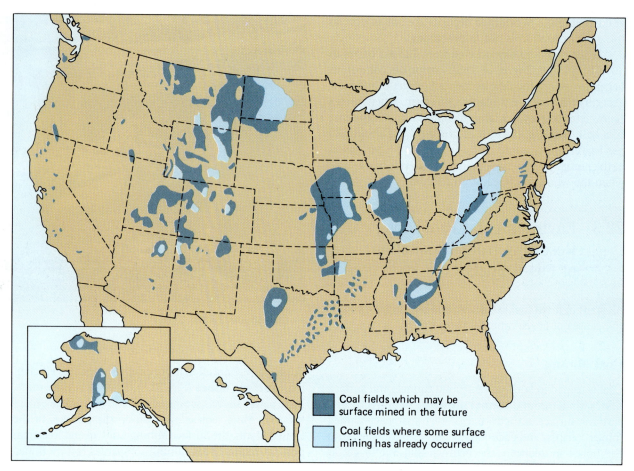

Figure 22.9 Areas in the United States where economically attractive coal deposits are found. Lightly shaded areas indicate coal fields where some surface mining has already occurred. Dark shading indicates coal fields that may be surface mined in the future.

surface mine sites have not been restored in compliance with the law (Fig. 22.9).

Most coal contains sulfur. When the coal is exposed, some of this sulfur reacts with water and air to produce sulfuric acid (H_2SO_4). If pollution control is inadequate, the sulfuric acid runs into streams and ground water below the mine or mill. This type of pollution is called **acid mine drainage**. In addition, when sulfur-rich coal burns, the sulfur reacts to form hydrogen sulfide and sulfuric acid, which in turn create acid precipitation.

Petroleum Extraction, Transport, and Refining

Relatively little environmental damage occurs when oil wells are drilled in temperate and arid regions. An oil well occupies only a few hundred square meters of land. Wells are even drilled in people's backyards, with minimal environmental problems. As these relatively acces-

sible reserves are being depleted, however, oil companies have begun to extract petroleum from more fragile environments, such as the ocean floor and the Arctic tundra.

Rich offshore oil reserves exist on continental shelves in many parts of the world, including the coast of southern California, the Gulf of Mexico, and the North Sea in Europe. To obtain the oil, engineers must first build platforms on pilings driven into the ocean floor. Drilling rigs are then mounted on these steel islands (Fig. 22.10). Despite great care, accidents occur during drilling and extraction of oil. Broken pipes, excess pressure, and difficulty in capping a new well have all led to blowouts, spills, and oil fires. When these accidents occur on shore, they can destroy small areas of farmland but do not affect entire ecosystems. When accidents occur at sea, however, millions of barrels of oil can be dispersed throughout the waters, poisoning marine life and disrupting ocean ecosystems. Significant oil spills have occurred in virtually all offshore drilling areas.

Petroleum from Alaska's northern coast is piped through the Alaska pipeline to Valdez on the southern coast of Alaska (Fig. 22.11). There it is pumped into tankers that carry it to refineries. In March 1989, the supertanker *Exxon Valdez* was navigating around floating ice when it ran aground near Valdez, spilling 260,000 barrels of petroleum into Prince William Sound (Fig. 22.12). The thick, black crude oil covered hundreds of kilometers of coastline and killed thousands of birds and sea mammals. It temporarily disrupted commercially important fisheries and damaged the local tourism industry. It is unclear how long it will take for the ecosystem to return to normal.

Crude petroleum must be **refined** to produce gasoline, propane, diesel fuel, motor oil, and chemicals. During refining, the crude oil is heated under pressure to break apart its large molecules and form useful products such as high-octane gasoline. A large refinery covers hundreds of acres and processes thousands of barrels of petroleum every day (Fig. 22.13). Each stage in the refining process is a potential source of air pollution. Although all refineries use expensive pollution control equipment, these devices are never completely effective.

After an oil well has been drilled, the expensive drill platform is removed and replaced by a pumper, such as this one on the Wyoming prairie.

Figure 22.10   An offshore oil drilling platform.   *(Schlumberger Inc.)*

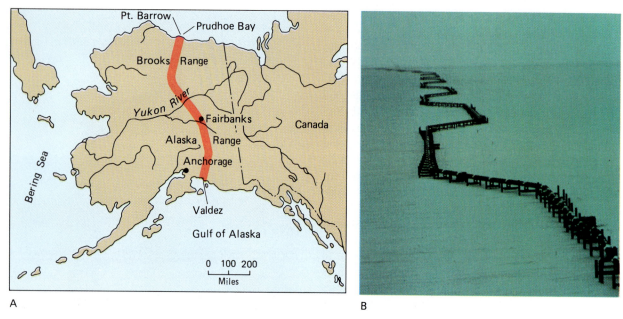

Figure 22.11 (A) The route of the Alaska Pipeline. (B) The Alaska Pipeline is built in zigzag fashion so it will not rupture when the permafrost soil expands and contracts. *(Alyeska Pipeline Company)*

Figure 22.12 The *Exxon Valdez* after it ran aground and began to spill oil. The slick appears off the ship's bow. *(Wide World Photos)*

As a result, toxic and carcinogenic chemicals escape into the atmosphere.

Mining and Refining of Oil Shale

Oil shale can be mined either in surface mines or by underground methods. In either case, environmental disruptions similar to those created by coal mining are created routinely. But additional problems specific to shale development also occur. When oil shale is broken into small pieces and exposed to the atmosphere, it absorbs water and swells. Thus, the volume of rock actually increases even though oil is removed from it. (As an analogy, take a potato, slice it into thin slices, and fry the slices to make potato chips. Now place the potato chips in a pile. Which occupies more volume, the pile of potato chips or the potato?) After the kerogen is removed, the expanded shale must be dumped somewhere, and the piles of rock are unsightly and erode easily.

Another problem inherent in shale development is water consumption. Approximately two barrels of water are needed to produce each barrel of oil from shale. Oil shale occurs most abundantly in western North America, where the climate is semiarid. In this region, water is also needed for agriculture, domestic consumption, and industry. Proponents of energy development claim that when oil prices rise and oil shale becomes

Figure 22.13   An oil refinery.

economical, new dams can be built to conserve water, and industries can move elsewhere. Moreover, most of the ranchers in the region use water inefficiently and much of the land is marginally productive. Therefore, according to this argument, it might be in the national interest to use that water instead for oil shale development. But the law allocates water rights according to history of use. The original ranchers and farmers passed down the water rights to later generations or sold them with the land. It would be unfair and perhaps unconstitutional to change water laws and harm individuals, even if the nation as a whole would benefit.

22.4   Nuclear Energy and Uranium Reserves

Fossil fuels are **nonrenewable resources**; when they are used up, we will have to do without them. What other sources of energy are available? Nuclear energy is one. Other energy sources are listed in the essay, "Energy Strategies for the United States."

Modern nuclear power plants use **nuclear fission** to generate heat (Fig. 22.14). A certain isotope of uranium, U-235, is the major fuel. When the nuclei of

Figure 22.14   A nuclear power plant. The domed structure on the left is the reactor; the larger structure on the right is the cooling tower.

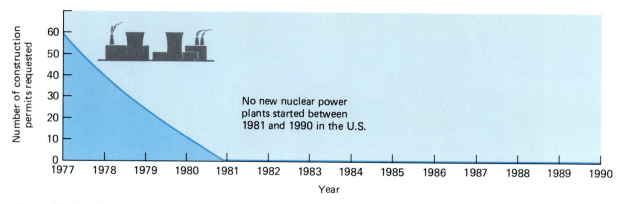

Figure 22.15   The numbers of construction permits requested by the nuclear power industry in recent years. The rapid decline reflects smaller-than-predicted growth in electric consumption and the fact that nuclear power is no longer economical.   *(Annual Energy Review 1991, Energy Information Administration)*

U-235 are bombarded with neutrons, they break apart (hence the word *fission,* which means splitting). Large quantities of heat are released during fission. The heat is used to generate steam to drive turbines, which in turn generate electricity. During fission, uranium nuclei break apart to produce several daughter products, some of which are also radioactive.[2]

Over the past 40 years, optimism about nuclear power has waxed and waned. Many hazards and economic problems were overlooked initially, and proponents of nuclear power wrote statements such as "we can look forward to universal comfort, practically free transportation, and unlimited supplies of materials." By the early 1980s, this rosy forecast had changed. Construction of new reactors had become so costly that electricity generated by nuclear power was more expensive than that generated by coal-fired power plants. In addition, public concern about accidents and disposal of radioactive waste became acute. As a result, the growth of the nuclear power industry came to an abrupt halt. Between 1981 and 1991, no new orders were placed for nuclear power plants in the United States and 117 planned power plants were canceled (Fig. 22.15). However, nuclear power may regain importance if fossil fuels become expensive and alternative energy sources are not developed.

In 1990, nuclear fuels generated 7.6 percent of the energy used in the United States. The current world's uranium reserves are about 6 million tons. With these reserves, the present rate of nuclear power generation could be maintained for about 100 years without mining low-grade uranium ore.

Nuclear Fusion

Nuclear fusion occurs when nuclei of light elements combine to form heavier nuclei. Fusion generates tremendous amounts of energy. Our Sun generates energy by fusion of hydrogen nuclei to form helium nuclei. Controlled nuclear fusion is a potential source of energy that, on a human time scale, is limitless. No useful fusion reactor has yet been developed.

MINERAL RESOURCES

22.5   Ore

If you walked outside, picked up any rock at random, and sent it to a laboratory for analysis, the report would probably show that the rock contains measurable quantities of iron, gold, silver, aluminum, and other valuable metals. However, the concentrations of these metals are so low in most rocks that the cost of extracting them would be much greater than the income gained by selling them.

A **mineral deposit** is a local enrichment of one or more minerals. **Ore** is any natural material sufficiently enriched in one or more minerals to be mined profitably. Table 22.2 shows that the concentrations of elements in ore may be as much as 100,000 times those in ordinary rock. The **mineral reserves** of a region are the known supply of ore in the ground. Reserves are depleted when ore is mined, but reserves may also increase in two ways. First, new mineral deposits may be discovered. Second, increases in price or improvements in mining

---

[2] For a review of radioactive decay, refer to Chapter 4.

Table 22.2 • Comparison of Concentration of Specific Elements in Earth's Crust with Concentration Needed to Operate a Commercial Mine

Element	Natural Concentration in Crust (% by weight)	Concentration Required to Operate a Commercial Mine (% by weight)	Enrichment Factor
Aluminum	8	24–32	3–4
Iron	5.8	40	6–7
Copper	0.0058	0.46–0.58	80–100
Nickel	0.0072	1.08	150
Zinc	0.0082	2.46	300
Uranium	0.00016	0.19	1,200
Lead	0.00010	0.2	2,000
Gold	0.0000002	0.0008	4,000
Mercury	0.000002	0.2	100,000

or refining technology may convert subeconomic mineral deposits into ore.

Many known mineral deposits are enriched in one or more minerals but are not quite rich enough to mine at a profit. For example, a deposit containing 30 percent iron is not ore because it would not be profitable to extract the metal from the rock. But if technology improved so that the iron could be refined more cheaply or if the price of iron increased, then the deposit would suddenly become ore.

As an example of the changing nature of reserves, in 1966 geologists estimated that global reserves of iron were about 5 billions tons.[3] At that time, world consumption of iron was about 280 million tons per year. Assuming consumption continued at the 1966 rate, the global iron reserves identified in 1966 would have been exhausted in 18 years (5 billion tons/280 million tons per year = 18 years), and we would have run out of iron ore in 1984. Something must have changed, because iron ore remains available today and is relatively inexpensive. The critical change resulted from development of inexpensive methods for processing iron ores of lower grade. Thus, deposits that were uneconomical to mine in 1966, and therefore not counted as reserves, are now ore. In fact, as shown in Figure 22.16, the reserves of many elements have increased dramatically in recent years. Furthermore, many regions, especially in less developed countries, have not been explored thoroughly, so new deposits probably will be discovered.

---

[3] B. Mason, *Principles of Geochemistry*, 3rd ed. (New York: John Wiley & Sons, 1966), Appendix III.

An aerial view of the Bingham open-pit copper mine. *(Agricultural Stabilization and Conservation Service/USDA)*

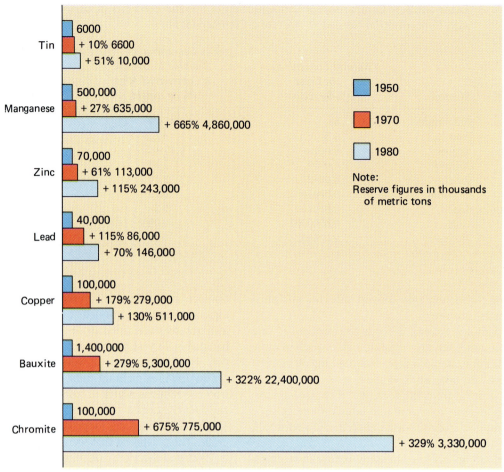

Figure 22.16   The change in the reserves of selected ores from 1950 to 1980.

22.6   How Ore Deposits Form

One of the primary professional objectives of many geologists is to find new ore deposits. Successful exploration requires an understanding of the processes that concentrate elements to form ore. For example, platinum concentrates in certain types of igneous rocks. Therefore, if you were exploring for platinum, you would not look in shale or limestone.

Magmatic Processes

Cooling magma does not solidify and crystallize all at once. Instead, high-temperature minerals crystallize first. Lower-temperature minerals crystallize later, when the temperature drops (see Chapter 3).

Solid minerals are denser than liquid magma. Consequently, crystals that form first sink to the bottom of a magma chamber in a process called **crystal settling** (Fig. 22.17). These crystals form a layer that commonly consists of a single mineral or a mixture of minerals with similar melting points. If the minerals contain valuable metals, an ore deposit may form. The largest ore deposit formed in this way is the Bushveldt intrusion of South Africa. It is about 375 by 300 kilometers in area—roughly the size of the state of Maine—and about 7 kilometers thick. Large quantities of chromium and platinum are mined from the Bushveldt.

Hydrothermal Processes

Magma and underground rock contain large amounts of hot water. The water contains the same dissolved ions found in seawater. This mixture of hot water and dissolved ions is a **hydrothermal solution**. (The word *hydrothermal* is derived from the roots *hydro* for water and *thermal* for hot.)

Hydrothermal solutions are corrosive and can dissolve metals such as copper, gold, lead, zinc, and silver from hot rock or magma. Recall that most rocks contain low concentrations of many metals. For example, gold makes up 0.0000002 percent of average crustal rock; copper makes up 0.0058 percent, and lead 0.0001 per-

GEOLOGY AND THE ENVIRONMENT

Pollution from Metal Mining: A Case History

The largest complex of toxic waste sites in the United States was produced not by a chemical factory or petroleum refinery, but by the mines and smelters in Butte and Anaconda in Montana. A short case history of this region highlights pollution problems that result from mining and refining metals.

Remember that most ore is not pure metal but rock containing metal-rich minerals. Some of the raw ore from the mines in Butte contained as little as 0.3 percent copper, with even smaller concentrations of manganese, zinc, arsenic, lead, cadmium, and a variety of other metals. When ore is extracted from a mine, the waste rock is simply piled up somewhere. Such piles of crushed rock are prone to erosion, and the muddy runoff silts streams and destroys aquatic habitats. But that is only the beginning of the problem, because the smelting process generates additional wastes.

In the boom days in Butte, during the late 1800s and early 1900s, metal smelting was inefficient, so large quantities of arsenic, cadmium, zinc, lead, copper, and other metals were discarded. Metal ores commonly contain large amounts of sulfur, which was discarded with the other wastes. Furthermore, Butte's copper ore contains gold, which the early miners attempted to extract by dissolving the copper ore in mercury, and the waste mercury was dumped on the piles as well. Today in the Butte–Anaconda area there are about 25 square kilometers (about 10 square miles) of mine and mill spoils averaging 15 meters (about 50 feet) high, or about as tall as a five-story building.

These wastes are a serious environmental threat because both the sulfur and the metals are poisonous. For example, sulfur compounds react in water to form sulfuric acid. Sulfuric acid is soluble in water and readily disperses into streams and ground-water reserves. In one study, workers in a cadmium recovery plant suffered a higher incidence of lung cancer than the public at large. In another study, skin cancer was prevalent among a group of people who drank well water contaminated with arsenic. Metal contamination is also harmful to the environment, causing a reduction in the growth rate of plants, of animals that eat the plants, and of microorganisms in the soil.

Sulfur and metal poisons have leached into nearby streams and ground-water reserves and have migrated up to 150 kilometers downstream from Butte and Anaconda. Windblown dust from the spoils has contaminated entire counties. No one really knows how to clean up such a mess or how much it will cost. Even a partial clean-up will cost tens to hundreds of millions of dollars.

---

cent. Although the metals are present in the rock in low concentrations, hydrothermal solutions percolate slowly through vast volumes of rock, dissolving and accumulating large amounts of the metals. The metals are then deposited when the solutions encounter changes in temperature, pressure, or chemical environment (Fig. 22.18). In this way, hydrothermal solutions scavenge metals from average crustal rocks and then deposit them locally to form ore.

If the metals precipitate in fractures in rock, a hydrothermal **vein** deposit forms. Ore veins range from less than a millimeter to several meters in width. They can be incredibly rich. Single gold or silver veins have yielded several million dollars' worth of ore.

The same hydrothermal solutions that flow rapidly through open fractures to form rich ore veins may also soak into large volumes of country rock around the fractures. If metals precipitate in the rock, they may create a very large but much less concentrated **disseminated ore deposit**. Because they may form from the

Figure 22.17 Concentration of minerals by crystal settling in a magma chamber.

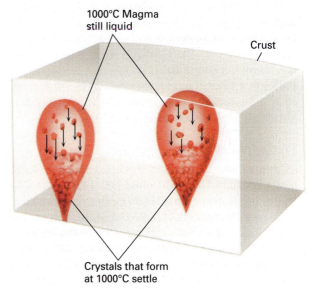

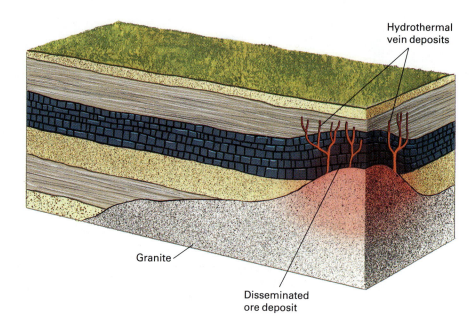

Figure 22.18 Formation of hydrothermal deposits.

same solutions, vein deposits and disseminated deposits are often found together. The history of many mining districts is one in which early miners dug shafts and tunnels to follow the rich veins. After the veins were exhausted, later miners used huge power shovels to extract low-grade ore from disseminated deposits surrounding the veins.

Sedimentary Processes

Two types of sedimentary processes form ore deposits: sedimentary sorting and precipitation.

Sedimentary Sorting: Placer Deposits

Gold occurs naturally as a pure metal and is denser than any other mineral. Therefore, if a mixture of gold dust, sand, and gravel is swirled in a glass of water, the gold falls to the bottom first. Differential settling also occurs in nature. Many streams carry silt, sand, and gravel with an occasional small grain of gold. The gold settles first when the current slows down. Thus, grains of gold concentrate near bedrock or in coarse gravel, forming a **placer deposit** (Fig. 22.19).

Precipitation

As ground water percolates through rock, it dissolves minerals and carries off dissolved ions. In most environments, this water eventually flows into streams and then to the sea. Some of the dissolved ions, such as sodium and chloride, make seawater salty. In deserts, however, lakes develop with no outlet to the ocean. Water flows into the lakes but can escape only by evaporation. As

the water evaporates, the dissolved ions concentrate until they begin to precipitate.

You can perform a simple demonstration of evaporation and precipitation. Fill a bowl with warm water and add a few teaspoons of table salt. The salt dissolves

Panning for gold.   *(Montana Historical Society)*

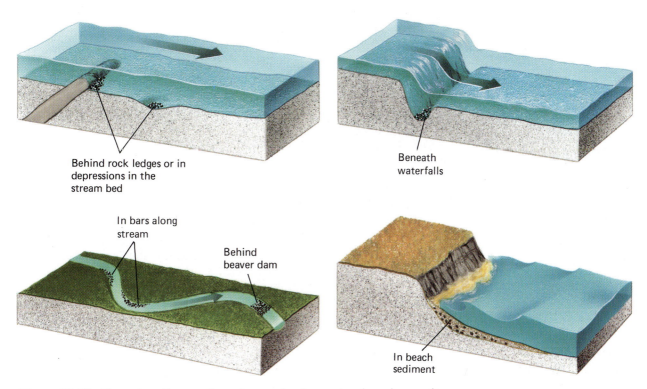

Figure 22.19   Placer deposits occur in environments where water slows down and sediment is deposited.

and you see only a clear liquid. Set the bowl aside for a day or two until the water evaporates. The salt precipitates and encrusts the sides and bottom of the bowl.

Salts that form by evaporation are called **evaporite deposits**. Evaporite minerals include table salt, borax, sodium sulfate, and sodium carbonate. These salts are used in the production of paper, soap, and medicines and for the tanning of leather.

Several times during the history of the Earth, shallow seas covered large portions of North America and all other continents. At times, those seas were so poorly connected to the open oceans that water did not circulate freely between them and the oceans. Consequently, evaporation concentrated the dissolved ions until salts began to precipitate as **marine evaporites**. Periodically, storms flushed new seawater from the open ocean into the shallow seas, providing a new supply of salt. In this way, thick marine evaporite beds formed. Nearly 30 percent of North America is now underlain by these deposits. Table salt, gypsum (used to manufacture plaster and sheetrock), and potassium salts (used in fertilizer) are mined extensively from marine evaporites.

Weathering Processes

In environments with high rainfall, the abundant water dissolves and removes most of the soluble ions from the soil and rock. The insoluble ions left behind form **residual deposits**. Both aluminum and iron have very low solubility in water. **Bauxite**, our principal source of aluminum, forms in this manner, and in some instances iron also concentrates enough to become ore.

22.7   Future Availability of Mineral Resources

The Romans discovered veins of nearly pure copper on the island of Cyprus. Those deposits have been mined out and are now gone. In modern times, many rich and easily accessible ores, such as the high-grade iron deposits of the Mesabi Range in the north-central United States, are being used rapidly. These ore deposits are nonrenewable. But our technological life will not end with the exhaustion of rich deposits because less concentrated ore is still available.

On the other hand, we cannot extract metals economically from very low-grade ore deposits. Gold, silver, and a few other elements exist in their pure forms in the crust, but most elements combine naturally with others. Thus, pure iron is rarely found in the crust. The iron in iron ore is chemically bonded to oxygen (Fig. 22.20). To extract iron, the ore must be dug up and crushed and the iron minerals separated from the other

Figure 22.20 A banded iron formation. The red bands are iron minerals; the dark-colored layers are chert. *(Ward's Natural Science Establishment, Inc.)*

Figure 22.21 The energy required to mine and refine aluminum and copper ores as a function of the metal concentration of the ore. Notice that there is very little change in energy requirements for very concentrated ores, but as the metal concentration drops off, the energy requirements rise very rapidly. However, almost no mines contain 70 percent aluminum or 20 percent copper. Most mining is already being done in the steep section of the curve, so considerably more energy will be needed as lower-grade ores are exploited.

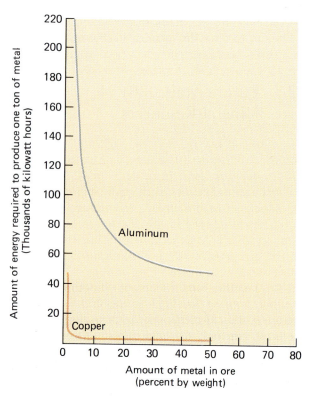

minerals in the rock. Finally, the iron is chemically separated from oxygen to obtain metallic iron. Each step, especially the chemical separation, requires energy. Low-grade ore requires more energy to process than high-grade ore because more rock must be dug, transported, and crushed to obtain a given amount of metal, and more energy is needed to separate the metal from the ore.

Figure 22.21 shows the energy required to extract and refine 1 ton of copper and 1 ton of aluminum from different concentrations of ore. Notice that relatively little energy is needed for rich ores. As concentrations decrease, the energy consumption rises rapidly.

A 1984 article in *Science* discussed the future availability of 67 commercially important elements.[4] The authors predicted that 33 will still be abundant by the year 2100. Table 22.3 shows that 19 metals, including iron, aluminum, chromium, and nickel, are on this list.

Even if sufficient ore is not available, shortages can be alleviated. As the prices of ores rise, recycling

---

[4] H. E. Goeller and A. Zucker, "Infinite Resources: The Ultimate Strategy," *Science, 223* (1984), 456.

Figure 22.22 A granite quarry near Barre, Vermont.

Table 22.3 • Predictions of Future Supplies of Elements Used in Industry

Availability	Metals	Nonmetals
Limitless supply	Sodium, magnesium, calcium	Nitrogen, oxygen, neon, argon, xenon, krypton, chlorine, bromine, silicon
Abundant supply	Aluminum, gallium, iron, potassium	Sulfur, hydrogen, carbon
Abundant only if research continues to improve extraction	Lithium, strontium, rubidium, chromium, nickel, cobalt, platinum, palladium, rhodium, ruthenium, iridium, osmium	Boron, iodine
Might be in short supply by the year 2100	Gold, silver, mercury, copper, zinc, lead, arsenic, tin, molybdenum, plus 22 other less familiar metals	Helium, phosphorus, fluorine

Source: H. E. Goeller and A. Zucker, "Infinite Resources: The Ultimate Strategy," *Science, 223* (1984), 456.

becomes economical, and engineers and chemists look for cheaper substitutes for expensive materials such as gold, silver, copper, lead, and zinc.

22.8 Nonmetallic Resources

When we thnk about becoming rich from mining, we usually think of finding gold. As undramatic as it seems, more money has been made mining sand and gravel than gold.

Concrete is an essential and versatile building material. Reinforced with steel, it is used to build roads, bridges, and buildings. Concrete is a mixture of cement, sand, and gravel. Sand and gravel are mined from stream and glacial deposits, sand dunes, and beaches. Cement is made by heating a mixture of crushed limestone and clay.

Many buildings are faced with stone, usually granite or limestone, although marble, sandstone, and other rocks are also used. Stone is mined from **quarries** cut into bedrock (Fig. 22.22).

Phosphorus and potassium are valuable fertilizers extracted from sedimentary rocks.

Look at the room around you and think about how many utensils and building materials are manufactured from geologic resources. The "lead" in your pencil is a mixture of graphite and clay. Your coffee cup may be ceramic and therefore made of clay; if not ceramic, it is probably plastic, manufactured from coal or petroleum. The inside walls of your building are probably lined with wallboard or plaster, both of which are made of gysum. The exterior of the building may be brick, which is baked clay, or faced with granite or limestone. You may be wearing jewelry of a semiprecious stone such as turquoise or topaz, or perhaps even a gem such as a ruby or diamond.

SUMMARY

Fuels and minerals are the two types of geologic resources. **Fossil fuels** include oil, gas, and coal. If oxygen and flowing water are excluded, plant matter decays partially to form **peat**. Peat converts to **coal** when it is buried and subjected to elevated temperature and pressure. **Petroleum** forms from the remains of organisms that settle to the ocean floor and are incorporated into **source rock**. The organic matter converts to liquid oil when it is buried and heated. The petroleum then migrates to a **reservoir**, where it is retained by an **oil trap**.

Oil and gas shortages will probably occur in the early twenty-first century, whereas coal reserves will be plentiful for 200 years or more. Additional supplies of petroleum can be recovered by secondary extraction from old wells and from **oil shale**.

Metal ores and coal and oil shale are mined in **tunnel mines** or **surface mines**. Both disrupt the land and may create **acid mine drainage**. Petroleum drilling in temperate ecosystems is usually not environmentally disruptive, but spills in fragile environments damage ecosystems, and tanker accidents pollute the ocean and beaches.

Nuclear power is expensive, and questions about safety and disposal of nuclear wastes have not been answered to everyone's satisfaction. As a result, no new nuclear power plants have been ordered since 1981. Inexpensive uranium ore will be available for a century or more.

Ore is a rock or other material that can be mined profitably. **Mineral reserves** are the estimated supply of ore in the ground.

Four major kinds of geologic processes concentrate elements to form mineral deposits and ore:

1. **Magmatic processes** form ore during the solidification of magma. **Crystal settling** is one example.

2. **Hydrothermal processes** involve transportation of dissolved ions by hot water. Minerals precipitate from those solutions in fractures and pores in rock to form **hydrothermal mineral deposits**.

3. Two types of **sedimentary processes** concentrate minerals. Dense minerals that concentrate by settling out of flowing water form **placer deposits**. Precipitation from lake water or seawater forms **evaporite deposits**.

4. **Weathering** removes easily dissolved elements and minerals, leaving behind **residual deposits**.

The future availability of ore reserves depends on the quantity and concentration of mineral deposits, the availability of energy, and problems of pollution.

Nonmetallic resources include sand and gravel for concrete, limestone for cement, building stone, and phosphorus and potassium for fertilizers.

KEY TERMS

Fossil fuel *613*	Secondary recovery *622*	Nuclear fission *627*	Disseminated ore
Peat *613*	Kerogen *622*	Nuclear fusion *628*	deposit *631*
Coal *613*	Oil shale *622*	Mineral deposit *628*	Placer deposit *632*
Reservoir *613*	Tunnel mine *623*	Ore *628*	Evaporite deposit *633*
Source rock *615*	Surface mine *623*	Mineral reserve *628*	Marine evaporites *633*
Oil trap *615*	Acid mine drainage *624*	Crystal settling *630*	Residual deposit *633*
Cap rock *615*	Nonrenewable resources *627*	Hydrothermal solution *630*	

REVIEW QUESTIONS

1. Name the two different categories of geologic resources.

2. Explain how coal forms. Why does it form in some environments but not in others?

3. Explain the importance of source rock, reservoir rock, cap rock, and oil traps in the formation of petroleum reserves.

4. Discuss problems in predicting the future availability of fossil fuel reserves. What is the value of the predictions?

5. Discuss the prospects for the availability of petroleum in the next 10, 20, and 40 years. What uncertainties are inherent in these predictions?

6. Discuss two sources of petroleum that will be available after conventional wells go dry.

7. What is ore? What are mineral reserves? Describe three factors that can cause changes in estimates of mineral reserves.

8. If most elements are widely distributed in ordinary rocks, why should we worry about someday running short?

9. Explain crystal settling.

10. Discuss the formation of hydrothermal ore deposits.

11. Discuss the formation of marine evaporites.

12. Explain why the availability of mineral resources depends on the availability of energy, on other environmental issues, and on political considerations.

13. Compare the environmental impact of underground mines with that of surface mines.

DISCUSSION QUESTIONS

1. If you were searching for petroleum, would you search primarily in sedimentary rock, metamorphic rock, or igneous rock? Explain.

2. Which of the following materials would you expect to find on the Moon: iron, gold, aluminum, coal, or petroleum? Defend your answer.

3. Imagine that you were a space traveler and were abandoned on an unknown planet in a distant solar system. What clues would you look for if you were searching for fossil fuels?

4. Is an impermeable cap rock necessary to preserve coal deposits? Why or why not?

5. Prepare a class debate. Have one side argue that we are quickly approaching an energy crisis that will debilitate our society, and have the other side argue that society will adjust to changes in energy supply without disruption.

6. What factors can make our metal reserves last longer? What factors can deplete them rapidly?

7. It is common for a single mine to contain ores of two or more metals. Discuss how geologic processes might favor concentration of two metals in a single deposit.

8. Compare the depletion of mineral reserves with the depletion of fossil fuels. How are the two problems similar, and how are they different?

9. Imagine that you plan to write a novel about a society that lives on the Earth in the year 2100. In your story, a nuclear war has occurred and major industries have been destroyed. Write a brief description of the types of materials that would be available to your characters and explain how they would extract and process them.

10. List ten objects that you own. What resources are they made of? How long will each of the objects be used before it is discarded? Will the materials eventually be recycled or deposited in the trash? Discuss ways of conserving resources in your own life.

11. Explain why the environmental effects of extracting fossil fuels are likely to increase as fuel reserves are exhausted gradually.

12. Who pays for the cost of reclaiming surface mines? What costs arise from surface mines if they are not reclaimed, and who pays these expenses? Which costs are more immediately apparent to the average citizen?

Planets and Their Moons

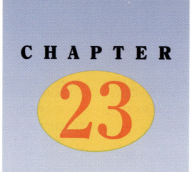

Color reconstruction of a volcanic eruption on Io, one of the moons of Jupiter. Planetary astronomers calculate that Jupiter's gravity is strong enough to stress rocks in Io's interior, generating heat and volcanic activity.
(NASA/JPL)

Twenty-five years ago scientists knew more about the compositions and structures of distant stars than they knew about the planets and moons in our own Solar System. As explained in Chapter 19, stars emit their own light. Spectral analysis of starlight provides information about composition, temperature, and other important parameters. Planets and moons, on the other hand, emit no light. They merely reflect sunlight, so spectral analysis provides little information about them. Our closest neighbor in space, Venus, is particularly difficult to study because it is surrounded by dense clouds.

Thus much of our knowledge about the planets and moons has been obtained from spacecraft launched within the past few decades. Several important missions are in space now, and the data they are relaying to Earth will modify some of the information presented in this chapter.

23.1   The Origin of the Solar System

Recall from Chapter 1 that the Solar System formed about 5 billion years ago from a cloud of dust and gas rotating slowly in space. More than 90 percent of the matter in the cloud gravitated toward its center to form the protosun, the earliest form of the Sun. The protosun was heated by the gravitational collapse of the cloud, but it was not a true star because it did not yet generate energy by nuclear fusion. Heat from the protosun warmed the inner region of the disk-shaped cloud. Then, as the gravitational collapse became nearly complete, the disk cooled. Gases condensed to form small aggregates, much as raindrops or snowflakes form when moist air cools.

Pressure inside the protosun then became great enough for fusion to begin, and the modern Sun was born. At about the same time, matter surrounding the Sun continued to coalesce, forming planets. Solar radiation boiled off most of the light elements, such as hydrogen and helium, from the inner planets, leaving mainly rock and metal behind. Thus, Mercury, Venus, Earth, and Mars are mostly solid spheres and are called the **terrestrial planets** (Fig. 23.1). The outer **Jovian planets**, Jupiter, Saturn, Uranus, and Neptune, are far enough from the Sun to have retained large amounts of hydrogen, helium, and other light elements. All have relatively small solid cores surrounded by swirling liquids and gaseous atmospheres. The outermost planet, Pluto, is anomalous and is discussed further in Section 23.9. The properties of the planets are summarized in Table 23.1.

23.2   Mercury: A Planet the Size of the Moon

Mercury has a radius of 2000 kilometers, less than four tenths that of Earth. It is the closest planet to the Sun and therefore orbits faster than any other planet. Each Mercurial year is only 88 Earth days long.[1] Mercury rotates slowly on its axis, completing only three rotations for each two revolutions around the Sun. Because it is so close to the Sun and its days are so long, temperature on its sunny side reaches 450°C, as hot as the inside of a self-cleaning oven and hot enough to melt tin or lead. In contrast, temperature on its dark side drops to −175°C, nearly cold enough to liquefy gaseous oxygen if any existed on the planet. The lack of an atmosphere is partly responsible for these extremes of

[1] Unless otherwise noted, all indications of time in this chapter are given in Earth days and years.

Figure 23.1   The terrestrial planets and the larger moons of the Solar System at the same scale. *(NASA)*

Table 23.1 • Comparison of the Nine Major Planets

Planet	Distance from Sun (Millions of Kilometers)	Radius (Compared to Radius of Earth = 1)	Mass (Compared to Mass of Earth = 1)	Density (Compared to Density of Water = 1)	Composition of Planet	Density of Atmosphere (Compared to Earth's Atmosphere = 1)	Number of Satellites
Terrestrial Planets							
Mercury	58	0.38	0.06	5.4		One billionth	0
Venus	108	0.95	0.82	5.2	Rocky with metallic core	90	0
Earth	150	1	1	5.5		1	1
Mars	229	0.53	0.11	3.9		0.01	2
Jovian Planets							
Jupiter	778	11.2	318	1.3	Liquid hydrogen surface with liquid metallic mantle and solid core	Dense and turbulent	16
Saturn	1420	9.4	94	0.7			19
Uranus	2860	4.0	15	1.3 }	Hydrogen and helium outer layers with solid core }	Similar to Jupiter except that some compounds that are gases on Jupiter are frozen on the outer planets }	15
Neptune	4490	3.9	17	1.7			8
Most Distant Planet							
Pluto	5910	0.17	0.0025	2.0	Rock and ice	0.00001	1

temperature because there is no wind to carry heat from one region to another.

Little was known about the surface of mercury before the spring of 1974, when the spacecraft *Mariner 10* passed within a few hundred kilometers of the planet and began relaying information to Earth. The first photographs revealed a cratered surface similar to that of the Moon (Fig. 23.2). The craters on both Mercury and the Moon formed during intense meteorite bombardment early in the history of the Solar System. Why are Mercury and the Moon so heavily cratered, and Earth is not? Earth was also pockmarked by the same episode of meteorite impacts. If you could go back about 4 billion years, its surface would appear similar to that of Mercury today. However, tectonic activity and erosion have erased the Earth's craters. The preservation of those features on Mercury tells us that tectonic activity and erosion have not occurred on Mercury during the past 4 billion years.

Flat plains on Mercury are probably vast lava flows that formed early in its history when the interior of the planet was hot enough to produce magma. However, Mercury is so small that its interior has cooled and tectonic activity has ceased. It is so close to the Sun that its atmosphere has boiled off into space, so no wind, rain, or flowing water has eroded its surface.

In 1991 radar images of Mercury revealed highly reflective regions at the poles. One possible explanation is that the poles consist of ice. If polar ice caps exist, then scientists will have to modify their inferences of temperatures on Mercury.

23.3 Venus: The Greenhouse Planet

Of all the planets in our Solar System, **Venus** most closely resembles Earth in size, density, and distance from the Sun. Therefore, astronomers once thought

Figure 23.2 A close-up of Mercury. The photograph shows an area 580 kilometers from side to side. *(NASA)*

that the environment on Venus might be similar to that on Earth and that life might be found there. Until recently it was impossible to study the surface of Venus, for it is obscured by a thick, dense atmosphere with an opaque cloud cover (Fig. 23.3). Today our knowledge of Venus has been greatly increased by data obtained from spacecraft. It now seems certain that no life exists on Venus because its environment is too harsh. Its surface is nearly as hot as that of Mercury, hot enough to destroy the complex organic molecules necessary for life.

The Atmosphere of Venus

Although Venus is slightly smaller than Earth and its gravitational force is less, its atmosphere is 90 times more dense than the Earth's. Thus, atmospheric pressure at the surface of Venus is equal to the pressure 1000 meters beneath the sea on our planet. The Venusian atmosphere is more than 97 percent carbon dioxide, with small amounts of nitrogen, helium, neon, sulfur dioxide, and other gases. Drops of concentrated sulfuric acid rain fall from sulfurous clouds.

Why is Venus's atmosphere so different from ours? We can imagine that early in the history of the Solar System, Venus and Earth were similar to one another because they formed from the same cloud of dust and gas. A space traveler would have viewed the two planets as sisters. Rivers may have flowed over the surface of Venus, and oceans may have filled its low-lying basins. On Earth, some limestone precipitates chemically. The same process may have occurred on Venus, forming beds of limestone beneath Venusian seas.

However, there is one important difference between the two planets: Venus is closer to the Sun than Earth

is, and it therefore receives more solar heat. This small difference in solar radiation led to chemical changes in the atmosphere of Venus. These changes made Venus very hot. Recall from Chapter 15 that carbon dioxide gas and water vapor both absorb infrared radiation and thereby warm the surface of a planet in a process called the greenhouse effect. Carbon dioxide exists in many forms: as a gas in the atmosphere, dissolved in water, and combined with other compounds to form limestone. Water commonly exists as a solid (ice), liquid, or gas.

Three processes may have combined to produce a runaway greenhouse effect on Venus:

Figure 23.3 The solid surface of Venus is obscured by a turbulent cloud cover. *(NASA/JPL)*

1. Carbon dioxide dissolved in Venusian seawater escaped into the atmosphere when the water was heated.

2. The heating also accelerated chemical weathering of limestone, which released more carbon dioxide.

3. Finally, as the primordial Venusian oceans warmed, more water evaporated from them. Because both carbon dioxide and water vapor absorb infrared energy, greenhouse warming occurred. As the temperature increased, all three of these processes speeded up. Most of the water reacted with sulfur dioxide to form sulfuric acid. Thus, the atmosphere of Venus became hot, acidic, and rich in carbon dioxide. Greenhouse warming heated the surface of the planet until eventually it reached its current temperature of about 450°C.

The Surface and Geology of Venus

Astronomers use radar to penetrate the Venusian atmosphere and produce photo-like images of its surface. The most spectacular images have come from the *Magellan* spacecraft, which began relaying data in September 1990.

Recall from our discussion of Mercury that dense swarms of meteorites bombarded the Solar System early in its history. Scientists have sampled rocks from cra-

Figure 23.4 The volcano, Maat Mons, on Venus, with large lava flows in the foreground. This image was produced from radar data recorded by *Magellan* spacecraft. Simulated color is based on color images supplied by Soviet spacecraft. *(NASA/JPL)*

tered regions of the Moon and dated them radiometrically. Thus, they have established a calendar of major episodes of meteorite bombardment in the inner Solar System. The largest swarms of meteorites bombarded the terrestrial planets shortly after they formed 4.6 billion years ago and again between 4.2 and 3.9 billion years ago. With this calendar astronomers can deduce the age of the surface of a planet or moon by counting the density of craters. Few craters exist on Venus, indicating that its surface was reshaped after the major meteorite bombardments. Volcanic eruptions caused much of this resurfacing between 100 million to 1 billion years ago. Today, Venus has several large volcanic mountains, and relatively recent lava flows cover much of its surface (Fig. 23.4).

Most volcanoes on Earth occur at tectonic plate boundaries, so the discovery of volcanoes on Venus led planetary geologists to look for evidence of plate tectonic activity there. Radar images from *Magellan* and earlier *Pioneer* spacecraft show that 60 percent of Venus's surface is covered by a flat plain. Two large and several smaller mountain chains rise from the plain (Fig. 23.5). The tallest mountain is 11 kilometers high, 2 kilometers higher than Mount Everest. The images also show large crustal fractures and deep canyons (Fig. 23.6). If Earth-like horizontal motion of tectonic plates caused these features, then spreading centers and subduction zones would exist on Venus. After studying low-resolution images from *Pioneer* spacecraft, James Head and Larry Crumpler of Brown University predicted that an Africa-size highland, Aphrodite Terra, might be a spreading center, similar to the mid-oceanic ridges on Earth. However, when the higher-resolution *Magellan* data were scrutinized, no steep scarps, transform faults, or other evidence of crustal spreading was observed. Furthermore, nothing like oceanic trenches on our own planet has been seen.

Other planetary geologists have suggested that upwellings of hot rock from the planet's interior dominate tectonic activity on Venus. In some regions the rising rock has melted and erupted from volcanoes in a manner similar to processes that formed the Hawaiian Islands. In other regions, the hot Venusian mantle plumes have lifted the crust to form nonvolcanic mountain ranges.

Some geologists have suggested the term *blob tectonics* to describe the geology of Venus. Tectonics is the shaping and deformation of a planetary surface by internal processes. Clearly, Venus has experienced tectonic activity. However, its primary process seems to be rising and sinking of the surface rather than horizontal motions of lithospheric plates. (Think of blobs of rock rising and falling.)

Recall from Chapter 5 that mantle plumes on Earth may initiate rifting of the lithosphere and formation of

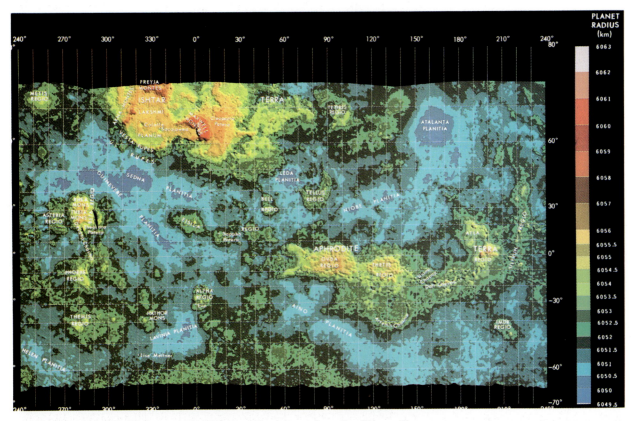

Figure 23.5   A map of Venus produced from radar images from the *Pioneer Venus Orbiter.* The lowland plains are shown in blue, and the highlands are shown in yellow and red-brown. (The colors are assigned arbitrarily.)   *(NASA/Ames Research Center)*

Figure 23.6   A portion of the surface of Venus showing a volcano, Gula Mons, in the background and a large impact crater in the foreground. Large crustal fractures appear near the crater. This image was produced from radar data recorded by *Magellan* spacecraft. Simulated color is based on color images supplied by Soviet spacecraft. *(NASA/JPL)*

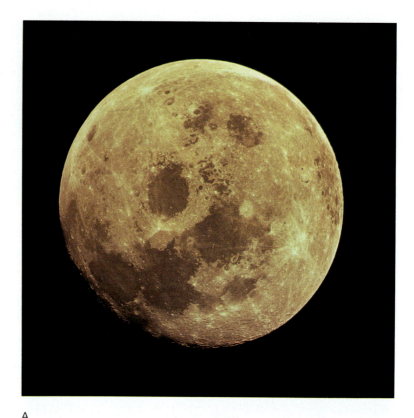

A

Figure 23.7   (A) The Moon as photographed from the *Apollo* spacecraft at a distance of 18,000 kilometers. Note the cratered regions and the flat lava flows. (B) A close-up of a portion of the heavily cratered surface of the Moon. Notice that smaller craters lie within the larger ones. The larger craters formed first.   *(NASA)*

B

a spreading center. Why has a similar process not occurred on Venus? Its surface temperature may be so high that the surface rocks are much more plastic than those on Earth. Therefore, rock stretches and lithosphere-deep cracks do not form. It is also possible that since Venus is smaller than Earth, its mantle is cooler and it has a thicker lithosphere. The internal heat is sufficient to cause vertical movement of the lithosphere but not to cause plates to move horizontally.

23.4   The Moon: Our Nearest Neighbor

Earth is the third planet from the Sun, and it has been the subject of all the foregoing chapters of this book. Most planets have small orbiting satellites called **moons**. The Earth's Moon is close enough so that we can see some of its surface features with the naked eye. In the 1600s Galileo studied the Moon with the aid of a telescope and mapped its mountain ranges, craters, and plains. Galileo thought that the plains were oceans and called them seas, or **maria**. The word *maria* is still used today, although we now know that these regions are dry, barren, flat expanses of volcanic rock (Fig. 23.7). The first close-up photographs of the Moon were taken by a Russian orbiter in 1959. A decade later the United States landed the first of six manned *Apollo* spacecraft on the Moon's surface. The Apollo program was designed to answer several questions about the Moon: How did it form? What is its geologic history? Was it once hot and molten like the Earth? If so, does it still have a molten core, and is it tectonically active?

Formation of the Moon

The question "How did the Moon form?" continues to be debated. According to the most popular theory, a giant object, perhaps a mini-planet, smashed into the Earth shortly after it formed. The collision blasted parts of this object and portions of the Earth into space. The fragments began to orbit the Earth and eventually coalesced, forming the Moon (Fig. 23.8A).

Perhaps the single most significant discovery of the Apollo program was that much of the Moon's surface consists of igneous rocks. The maria are mainly basalt flows. The highland rocks are predominantly anorthosite, a feldspar-rich igneous rock not common on Earth. Additionally, in both the maria and the highlands, the country rock was crushed by meteorite impacts and then the fragments were welded together by lava. Since igneous rocks form only from magma, it is clear that portions of the Moon were once hot and liquid.

How did the Moon become hot enough to melt? The Earth was heated initially from collisions among gases, dust, and larger particles as they collapsed under the influence of gravity. This process is called **gravitational coalescence**. Later, radioactive decay and intense meteorite bombardment heated the Earth further. But what about the Moon? Geologists have calculated that it would have taken 800 million years for radioactive decay to melt Moon rock. Radiometric age dating shows that the oldest lunar igneous rocks formed when the Moon was a mere 240 million years old. Thus, since there was insufficient time for the rock to have been melted by radioactivity, gravitational coalescence and meteorite bombardment must have been the main causes of early melting of the Moon.

History of the Moon

The Moon formed about 4.5 billion years ago, shortly after the Earth formed. Soon thereafter, meteorite bombardment melted its outermost 100 kilometers. Then the bombardment diminished and the surface cooled. The igneous rocks of the lunar highlands are about 4.4 billion years old, indicating that parts of the Moon's surface must have become solid by that time. Swarms of meteorites bombarded the Moon again between 4.2 and 3.9 billion years ago (Fig. 23.8B). Billions of meteorites, some as large as the state of Rhode Island, smashed into the surface, blasting huge craters and raising the Moon's temperature again. At the same time, radioactive decay heated the lunar interior.

By about 3.8 billion years ago, most of the Moon's interior was molten. The denser elements, such as iron and nickel, gravitated toward the center, and the less dense elements floated toward the surface. Thus, the Moon, like the Earth, has a metallic core surrounded by silicate rocks of lower density. Some of the molten rock erupted onto the Moon's surface. The lunar maria formed when lava filled circular meteorite craters (Fig. 23.8C). This volcanic activity lasted approximately 700 million years and ended about 3.1 billion years ago.

The Earth must have shared a similar history until this time, but the Moon is so much smaller that it soon cooled and has remained geologically quiet and inactive for the past 3 billion years. Seismographs left on the lunar surface by *Apollo* astronauts indicate that the energy released by moonquakes is only one billionth to one trillionth as much as that released by earthquakes on our own planet. Seismic data indicate that the core of the Moon is probably molten, or at least hot enough to be soft and plastic. However, the cool, solid upper

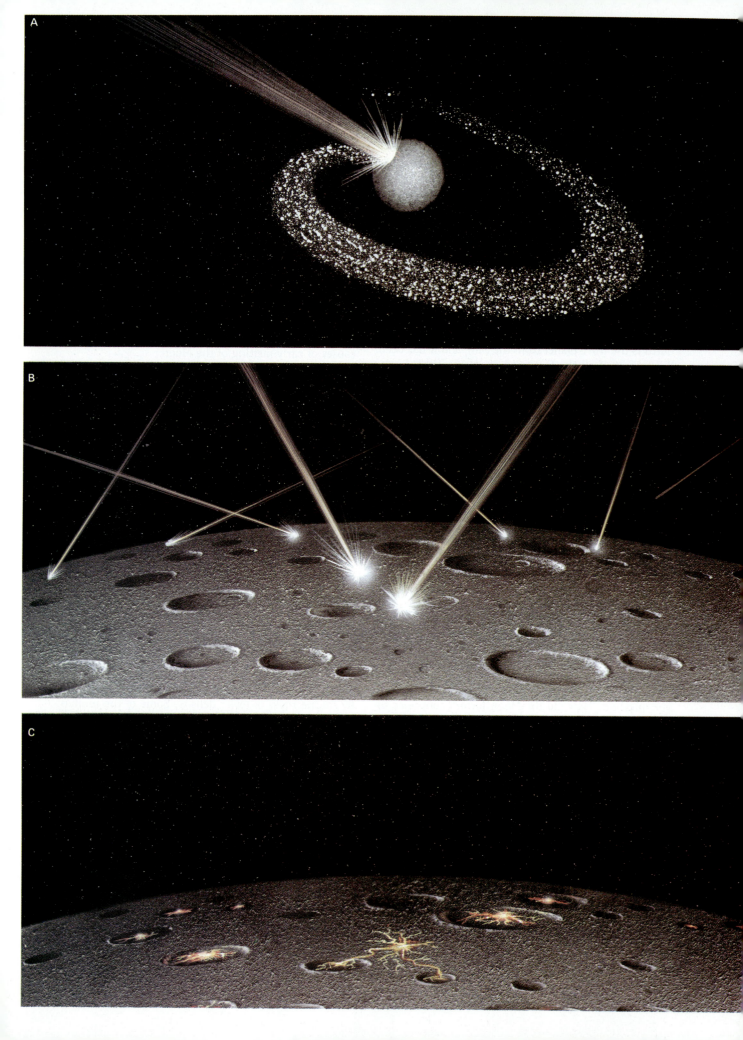

The *Apollo 11* lander returning to the command module after the first humans landed on the Moon. In the background the Earth rises over the flat lunar plain.   *(NASA)*

mantle and crust are too thick for seismic and tectonic activity. Meteorite bombardment of the Moon has continued throughout this period of tectonic dormancy, but never again have there been such intense rains of meteorites as occurred early in its history.

23.5   Mars: A Search for Extraterrestrial Life

In the 11 years between 1965 and 1976, a total of 12 United States and Russian spacecraft visited **Mars**. They included two *Viking* craft that landed on the surface of the planet and collected and analyzed samples of Martian soil.

The Geology of Mars

In many ways the geology of Mars is similar to that of Venus. It has old (heavily cratered) uplands and younger (lightly cratered) lowland plains (Fig. 23.9). Lava flows much like those on Venus and the Moon cover the plains. The 10-kilometers-high Tharsis bulge is the largest plateau, crowned by Olympus Mons, the largest volcano in the Solar System (Fig. 23.10). Olympus Mons is nearly three times as high as Mount Everest with a diameter of 500 kilometers and a height of 25 kilometers. Tremendous parallel cracks split the crust adjacent to the Tharsis bulge (Fig. 23.11). If this bulge lay near a tectonic plate boundary, we would expect to see folding or offsetting of the cracks. However, the cracks are neither folded nor offset. Therefore, geologists think that a rising mantle plume may have formed the Tharsis bulge and its volcanoes. On Earth, the lithosphere slides continuously over a mantle plume, forming a chain of volcanoes such as the Hawaiian Islands. On Mars, however, the lithosphere is stationary, allowing continuous volcanic activity in one location. The parallel cracks may be the result of stretching as the crust uplifted.

The Search for Life

One of the most exciting aspects of the exploration of Mars has been the search for life. Before the space age, photographs from earthbound telescopes showed white Martian polar regions that shrink in Martian summer

Figure 23.8   A brief history of the Moon. (A) According to one theory, the Moon was formed about 4.5 billion years ago when a Mars-size object struck the Earth, sending a cloud of vaporized rock into orbit. The vaporized rock rapidly coalesced to form the Moon. (B) During its first half-billion years, intense meteorite bombardment cratered the Moon's surface. (C) The dark, flat maria formed about 3.8 billion years ago as lava flows spread across portions of the surface. Today all volcanic activity has ceased.

Figure 23.9   A view of Mars showing mountains and lowland plains with several impact craters.   *(NASA/JPL)*

and expand in winter (Fig. 23.12). Some observers thought that the white regions were ice caps. If this hypothesis were correct, and if the ice melts in summer, water must be available. Astronomers also observed that each spring, large regions near the Martian equator darken, only to become light again in winter. Many people speculated that annual blooms of vegetation caused these color changes.

However, speculations about life on Mars have been refuted by data from orbiters and landers. High-resolution images of the Martian surface show no forests or grassy prairies. The seasonal dark patches that appear in equatorial regions each spring are not blooms of vegetation, but rather great, dry dust storms powered by seasonal winds (Fig. 23.13). When the winds subside, bright dust particles settle onto the land surface, causing it to appear light colored. The atmospheric density at Mars's surface is less than 1 percent that of Earth. Such a light atmosphere can produce high-velocity winds, but the winds cannot move much dust. Therefore, very little dust actually moves during these storms. In fact, cameras on the surface of the planet have barely detected any loss of visibility even during intense storms. The

Figure 23.10   Olympus Mons, the largest volcano on Mars and probably the largest in the Solar System.   *(NASA/JPL)*

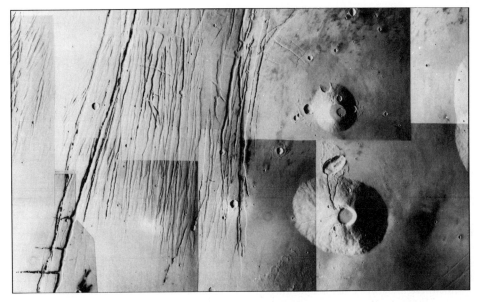

Figure 23.11 Crustal fractures in the Tharsis region of Mars. Two large volcanoes appear at right. *(NASA/JPL)*

Figure 23.12 Seasonal changes in the size of the polar ice cap and the dark equatorial markings on Mars, as seen by ground-based telescopes. *(Lowell Observatory)*

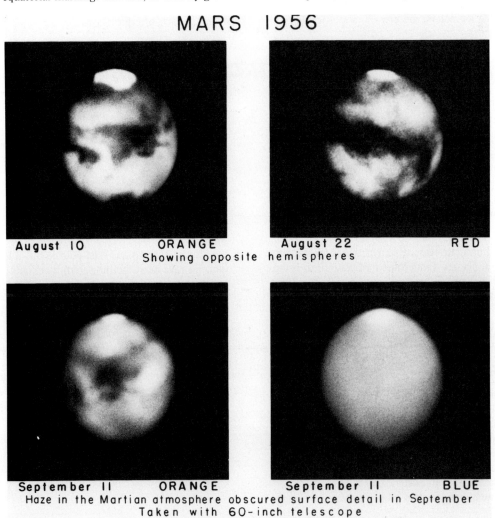

MARS 1956

August 10   ORANGE

August 22   RED

Showing opposite hemispheres

September 11   ORANGE

September 11   BLUE

Haze in the Martian atmosphere obscured surface detail in September
Taken with 60-inch telescope

For many years, people have asked whether Earth is unique in its ability to support life. Early searchers for extraterrestrial life looked to the nearby planets, Venus and Mars. But, as we have seen, Venus is too hot, and Mars, although potentially more hospitable, seems to be completely void of traces of life. Today astronomers are expanding their search for life toward more distant regions of the Solar System and other regions of our Galaxy.

Some biologists have speculated that primitive life may have evolved in the atmosphere of Jupiter. The outer surface of the Jovian atmosphere is too cold ($-140°C$) for organisms to survive, and the interior is too hot. But somewhere in between, the temperature must be favorable. The Jovian atmosphere is composed primarily of hydrogen, helium, ammonia, methane, water, and hydrogen sulfide. These compounds contain the major elements needed to build living tissue. Lightning storms occur on Jupiter, and these electrical discharges could provide the energy needed to synthesize amino acids, the basic building blocks of proteins. In short, all the ingredients are present, and it would be thrilling, but not totally surprising, to find microorganisms floating about in the clouds of the giant planet.

If living organisms are not found on any of the planets, they might occur on a moon. Perhaps biological evolution has occurred under the ice on Europa or in the methane seas of Titan.

One can only guess about life in other solar systems. An average galaxy contains roughly 100 billion stars, and millions of galaxies exist in the Universe. So many stars exist that is reasonable to believe that planetary systems have formed around some. In recent years, this statistical inference has been supported by direct evidence. Recall from Chapter 1 that astronomers have photographed a disk of particles surrounding the star *Beta Pictoris* that appears to be a solar system in the process of forming. Furthermore, the motion of several other stars indicates that they may be orbited by planets.

We live in a time when science fiction is popular, and most science fiction authors write about extraterrestrial life forms. However, in the world of science it is important to separate what is known from what is not known. The facts are summarized in the following table.

What We Do Know	*What We Have Not Found*
1. Life exists on Earth.	1. No spacecraft has found evidence of extraterrestrial life in the Solar System.
2. Some of the molecules essential to life as we know it (such as amino acids) have been formed in the laboratory by adding energy to a mixture of simple inorganic compounds. Some of these same essential substances have been found in meteorites. In addition, some organic molecules have been detected in interstellar space.	2. The transformation of simple chemical compounds essential to life to an actual living organism has never been observed.
3. We can formulate reasonable hypotheses for the formation of planets. From these formulations, it is reasonable to suppose that other stars besides the Sun have planets. Actual evidence exists in one or two cases.	3. If advanced civilizations exist outside the Solar System, we have heard nothing from them, although we are listening carefully.
4. There are very, very many stars.	

Conclusions

1. We cannot rule out the existence of extraterrestrial life. The data indicate that generation of organic molecules is an ordinary chemical process in the Universe. Many scientists deduce that, under the right conditions and if enough time is available, these molecules can combine to form living organisms. The Universe has been in existence for billions of years, and many billions of stars exist. Therefore, one argument concludes that the formation of living organisms—and even their evolution into intelligent beings—is so probable that it almost surely must have happened in many parts of the Universe.

2. On the other hand, we cannot rule out the possibility that life is unique to Earth and exists nowhere else. Remember, there is no direct proof whatsoever of living entities anywhere but on our planet.

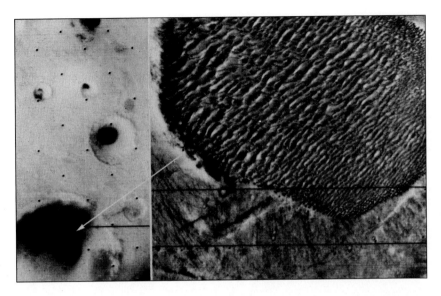

Figure 23.13   One of the mysterious "dark spots" on Mars, as seen by a ground-based telescope, is visible at the left. A close-up of the same spot taken by *Mariner 9* (right) clearly shows that the region is not vegetated, but covered with sand dunes.   *(NASA/JPL)*

darkening appears significant only when seen from above the planet.

The winter ice caps, once thought to be the source of spring floods, are mostly frozen carbon dioxide, commonly called dry ice. Water ice is present on Mars, but it lies beneath the surface and never melts. At some time in the distant past, however, rivers must have flowed across Mars's surface, eroding stream beds and canyons, for these features can be observed today (Fig. 23.14). Other photographs indicate that seas or lakes once covered parts of the Martian surface and massive floods transported sediment. But no rain has fallen for millions or hundreds of millions of years, and now Mars is a dry planet.

The Atmosphere and Climate of Mars

In trying to deduce the history of the Martian atmosphere and climate, it is helpful to compare Mars with Venus and Earth. Recall that Venus is closer to the Sun than Earth is and thus has received more solar radiation. This radiation led to greenhouse warming and increasing temperature on Venus. Most of the planet's carbon dioxide escaped into the atmosphere.

Earth receives less sunlight than Venus and therefore started out with a cooler surface. Oceans remained liquid, and additional water was frozen in glaciers. Carbon is distributed among the atmosphere, the biosphere, and rocks. This distribution has prevented excess carbon dioxide from accumulating in the atmosphere and has protected Earth from the runaway greenhouse process that has heated Venus's surface.

We see ancient volcanoes on Mars and we therefore assume that past volcanic eruptions must have released carbon dioxide and water vapor into the atmosphere,

causing greenhouse warming. Clouds formed, rain fell, and streams flowed over the surface. But then both water vapor and carbon dioxide froze into the ice caps and the planet became cooler. Today atmospheric pressure at the surface is only 0.006 that at the surface of Earth. Why did this reverse greenhouse effect occur?

Mars is half again as far from the Sun as Earth is. With less solar energy, it was initially cooler. The low temperature prevented water from evaporating, or froze water vapor that had been released during volcanic eruptions. With less water vapor, the atmosphere's insulating properties were reduced. As a result, radiation escaped into space and the planet's surface cooled. When the temperature became low enough, some of the gaseous carbon dioxide also froze into the ice caps.

In addition, Mars is smaller than Earth, so its gravitational field is weaker and some of its original water and carbon dioxide escaped into space. In short, mainly because of differences in size and distance from the Sun, Venus boiled, life evolved on Earth under moderate temperatures, and Mars froze.

Although Mars is inhospitable today, some scientists speculate that life may have evolved there during its more temperate past. If so, it is possible that fossils exist on Mars. Supporters of this hypothesis point out that Martian volcanoes must have released water, carbon dioxide, carbon monoxide, oxygen, hydrogen, nitrogen, and sulfur. On Earth these gases combined to form organic molecules—the building blocks of life. Thus, water and the other necessary compounds for the evolution of life probably once existed on Mars. This conclusion does not tell us that life did evolve, just that the components were present at one time.

The *Viking* landers searched for microorganisms in the soil and found none. However, they analyzed only

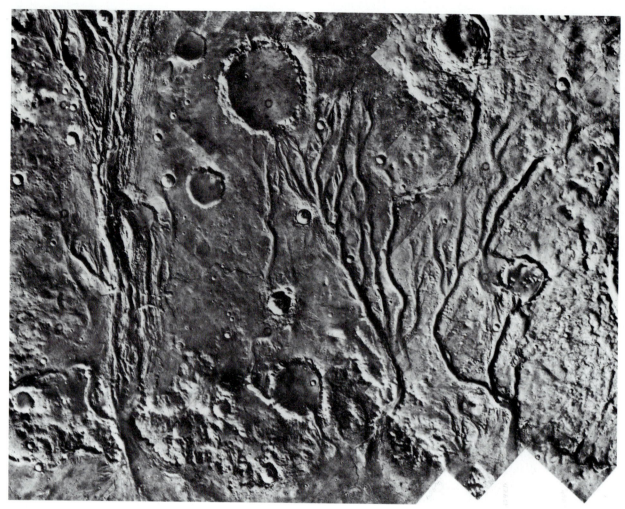

Figure 23.14   Stream beds on the surface of Mars were carved long ago by running water.   *(Viking image from NASA/JPL)*

a minuscule portion of the Martian surface. Many scientists suggest that the next spacecraft should land on an abandoned lake bed or stream channel, where fossils and organic matter are more likely to be found.

23.6   Jupiter: A Star that Failed

The Giant Planets

As explained in Section 23.1, Mercury, Venus, Earth, and Mars are the terrestrial planets. They are relatively small with rocky outer layers, and all orbit close to the Sun. Despite their differences, the four have similar chemical compositions. In contrast, the giant outer planets—Jupiter, Saturn, Uranus, and Neptune—are considerably different from the terrestrial group.

Visualize once again the primordial dust cloud that condensed to form the Solar System. The protosun and protoplanets were composed mostly of hydrogen and helium. As gravity pulled hydrogen toward the center of the Sun, the pressure became so great that hydrogen began to combine, or fuse, to form helium. Hydrogen fusion is still the source of the Sun's energy. The terrestrial planets were so close to the Sun, and their gravitational fields so weak, that most of their hydrogen, helium, and other light gases escaped and boiled off into space or were blown away by energetic particles streaming off from the Sun. In contrast, the protoplanets in the outer reaches of the Solar System were so far from the Sun that they remained cool. In addition, they were more massive and thus had stronger gravitational fields. As a result, they retained most of their gases. They may even have grown as they captured gases that escaped from the terrestrial planets. Thus, today each of the Jovian planets has a dense gaseous atmosphere, a very large liquid interior, and a much smaller solid core.

Uranus (left), Saturn, and Jupiter (right) shown to the same scale.   *(NASA/JPL)*

Structure and Composition of Jupiter

Jupiter retained most of its original hydrogen and helium. Therefore, its chemical composition is much like that of the Sun. However, Jupiter was not massive enough to generate fusion temperatures so it never became a star. Jupiter is 71,000 kilometers in radius. Its outer 12,000 kilometers are a vast sea of cold, liquid, molecular hydrogen, H_2. It has no hard, solid, rocky crust on which an astronaut could land or walk. A middle layer between the outer sea of liquid hydrogen and the core is also composed of hydrogen (Fig. 23.15). In this layer, temperatures are as high as 30,000°C and pressures are as great as 100 million times the Earth's atmospheric pressure at sea level. Under these extreme conditions, hydrogen molecules dissociate to form atoms. The pressure forces the atoms together so tightly that the electrons separate from their nuclei. These free electrons travel throughout the tightly packed nuclei much as electrons travel freely among metal atoms. As a result, the hydrogen conducts electricity and is called **liquid metallic hydrogen**. Movement within this fluid conductor generates a magnetic field ten times stronger than that of Earth.

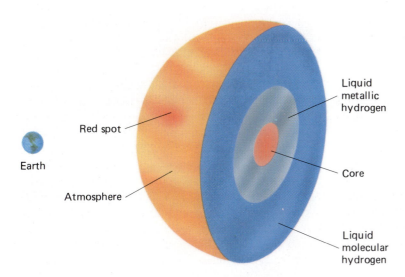

Figure 23.15   The internal structure of Jupiter, with the Earth drawn to scale on the left.

Figure 23.16 The colorful, turbulent complex cloud system of Jupiter, photographed by *Voyager*. The sphere on the left, in front of the Great Red Spot, is Io; Europa lies to the right against a white oval. *(NASA/JPL)*

Jupiter's core is a sphere about 10 to 20 times as massive as the Earth and is probably composed of metals and rock surrounded by lighter elements such as carbon, nitrogen, and oxygen.

Above the liquid layers, the Jovian atmosphere is a mixture of gases, liquid droplets, and crystals consisting mainly of hydrogen, with smaller amounts of helium, ammonia, methane, water, hydrogen sulfide, and other compounds. The atmosphere is turbulent; great storms and changing weather patterns can be seen even from earthbound telescopes (Fig. 23.16). Powerful winds blow at speeds up to 500 kilometers per hour, in alternating bands from east to west and from west to east. Most storms form, distort, and dissipate within a few hours or days, but some are surprisingly stable over long periods of time.

More than 300 years ago, two European astronomers reported seeing a **Great Red Spot** on the surface of Jupiter. Although its shape and color change from year to year, the spot remains intact to this day (Fig. 23.17). The Great Red Spot is a giant hurricane-like storm. If the entire Earth's crust were peeled off like an orange rind and laid flat, it would fit entirely within

Figure 23.17 Jupiter from a distance of 5 million kilometers, with its Great Red Spot (upper left) and the turbulent region adjacent to it. The smallest details that can be seen in this photograph are about 95 kilometers across. *(NASA)*

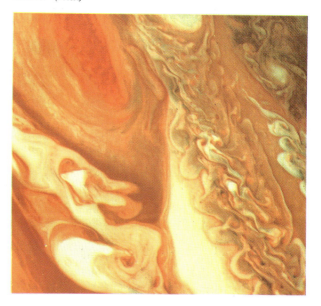

the Great Red Spot. No one knows exactly why the Great Red Spot has persisted over all these years. Hurricanes on Earth dissipate after a week or so, yet this storm on Jupiter has persisted for centuries.

The Moons of Jupiter

In 1610 Galileo discovered four tiny specks of light near Jupiter. He noted that they orbit Jupiter and reasoned that they are satellites of the giant planet. Astronomers have now found a total of 16 moons orbiting Jupiter. The four discovered by Galileo are the most widely studied. In addition to the moons, a rocky ring, similar to the rings of Saturn, lies inside the orbit of the closest moon.

Io

The innermost moon of Jupiter, **Io**, is about the size of the Earth's Moon and slightly denser. Since it is too small to have retained heat generated during its formation and by radioactive decay, many astronomers expected to see a cold, lifeless, cratered, Moonlike surface. Nothing could be further from the truth. *Voyager* photos showed huge masses of gas and rock erupting to a height of 200 kilometers above its surface (Fig. 23.18). These pictures were the first evidence of active extraterrestrial volcanism.

A few weeks before the *Voyager* photographs were transmitted to Earth, two scientists suggested on theoretical grounds that gravitational forces might heat Io's interior enough to make the planet tectonically active. Recall that the gravitational field of our own Moon

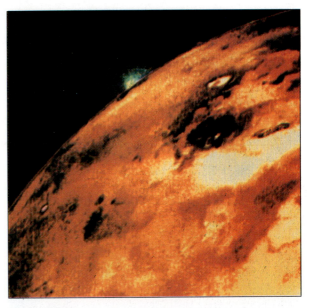

Figure 23.18   A volcanic explosion on Io can be seen on the horizon. Solid material is rising to an altitude of about 200 kilometers.   *(Voyager I photograph from NASA)*

causes the rise and fall of ocean tides on Earth. At the same time, the Earth's gravity distorts lunar rock. Thus, the Earth's gravitation may be partly responsible for moonquakes. Jupiter is 300 times more massive than Earth, so its gravitational effects on Io are correspondingly greater. In addition, the three nearby satellites, Europa, Ganymede, and Callisto, are large enough to exert significant gravitational forces on Io, but these

The *Voyager* planetary spacecraft.   *(NASA/JPL)*

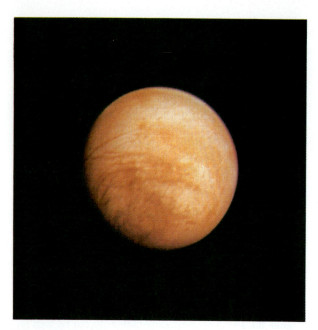

Figure 23.19   Europa, the smallest Galilean satellite, as imaged by *Voyager I*, from a distance of about 2 million kilometers. The bright areas are probably ice; the dark areas may be rocky. Note the long linear structures that crisscross the surface. Some of them are more than 1000 kilometers long and about 200 or 300 kilometers wide. They may be fractures or faults.   *(NASA)*

forces pull in different directions from that of Jupiter. Apparently, this combination of opposing and changing tidal effects causes so much frictional heating that volcanic activity is nearly continuous on Io. the frequent lava flows have obliterated all ancient landforms, giving Io a smooth and nearly crater-free surface.

Europa

Calculations show that the second closest of the Galilean moons, **Europa**, should also be subject to powerful gravitational forces, but since Europa is farther from Jupiter, the effects are less. Although no active volcanoes have been observed on Europa, its surface appears smooth and relatively crater free. Meteorites must have bombarded Europa as they did all other bodies in the Solar System. Therefore, the smooth surface indicates that old impact craters have been eliminated by tectonic or erosional processes. The interior of Europa is rocky, but its surface is covered with ice crisscrossed with a complex array of streaks, which may be fractures of faults (Fig. 23.19).

Ganymede and Callisto

Both **Ganymede** and **Callisto** are probably composed of a rocky core about the size of the Earth's Moon,

surrounded by a thick, icy crust. Their surface temperatures are so low that the ice is brittle and behaves much like rock. Two different types of terrain have been observed on Ganymede: One is densely cratered, and the other contains fewer craters and is crisscrossed by grooves and ridges (Fig. 23.20). The meteorite storms that swept through the Solar System 4 billion years ago formed the cratered regions. the smooth regions developed when the crust cracked and water from the warm interior flowed over the surface and froze, much as lava flowed over the surfaces of the terrestrial planets. The grooves and ridges may be tectonic plate boundaries and mountain ranges formed by tectonic activity.

Callisto, the outermost of the moons discovered by Galileo, is heavily cratered, indicating that its surface is very old. Its craters are shaped differently from those on either Ganymede or the Earth's Moon. Perhaps they have been modified by ice flowing slowly across its surface.

23.7   Saturn: The Ringed Giant

Saturn, the second largest planet, is similar to Jupiter. It has the lowest density of all the planets—so low, in fact, that the entire planet would float on water if there were a basin large enough to hold it. Such a low density

Figure 23.20   Ganymede is Jupiter's largest satellite, with a radius of approximately 2600 kilometers, about 1.5 times that of our Moon. Ganymede is probably composed of a mixture of rock and ice. The long white filaments radiating from the white areas may have been caused by meteorite impacts.   *(NASA/JPL)*

Figure 23.21 Saturn and its ring system. *(NASA/JPL)*

implies that it, too, must be composed primarily of hydrogen and helium with only a small core of rock and metal. Saturn's atmosphere is similar to that of Jupiter. Dense clouds and great storm systems cover the planet.

The Rings of Saturn

The most distinctive feature of Saturn is its spectacular array of rings visible from Earth even through a small telescope. Photographs from space probes show seven major rings, each containing thousands of smaller ringlets (Figs. 23.21 and 23.22). The entire ring system is only 10 to 25 meters thick, less than the length of a football field. However, the ring system is extremely wide. The innermost ring is only 7000 kilometers from Saturn's surface, whereas the outer edge of the most distant ring is 432,000 kilometers from the planet. Thus, it measures 425,000 kilometers from its inner to its outer edge. If you were to make a scale model of the entire ring system the thickness of a phonograph record, it would be 30 kilometers in diameter.

The rings are composed of dust, rock, and ice. The particles in the outer rings are only a few ten thousandths of a centimeter in diameter (about the size of a clay particle), but those in the innermost rings are chunks a few meters across. Each piece orbits the planet independently; some move faster, some more slowly in a continuous chaotic parade.

Saturn's rings may be fragments of a moon that never formed, or remnants of one that formed and was then ripped apart by Saturn's gravitational field. Imagine what happens to a small satellite orbiting a larger planet. Gravitation between the Earth and the Moon causes tides on Earth and seismic rumblings on the Moon, whereas forces between Jupiter and Io are great enough to heat and melt the rock of Io. If a moon were close enough to its planet, the tidal effects would be greater than the gravitational attraction holding the moon together. Thus, the rings of Saturn may be the debris of one or more moons that got too close to the planet.

Titan

In addition to the rings, 17 moons orbit Saturn. **Titan** is the largest. It is unique in that it is the only moon with an appreciable atmosphere. This atmosphere is primarily nitrogen with a few percent methane, CH_4. (Methane is commonly used on Earth as a fuel and

Figure 23.22 A close-up view of Saturn's A rings obtained from *Voyager II*. *(NASA/JPL)*

is the major component of natural gas.) The average temperature on the surface of Titan is −180°C, and the atmospheric pressure is 1.5 times greater than that on the Earth's surface. These conditions are close to the point where methane can exist as a solid, liquid, or vapor. Therefore, small changes in temperature or pressure on Titan could cause its atmosphere to freeze, melt, vaporize, or condense. This situation is analogous to the environment of Earth, where water can exist as liquid, gas, or solid and frequently change among those three states. Thus, methane clouds float in Titan's atmosphere, and methane seas, lakes, rivers, and ice caps may also exist.

Why is Titan the only satellite to have an atmosphere? Its large size and low temperature prevent gases from escaping into space. In addition, volcanoes may emit gases into the atmosphere, replacing those that escape.

Organic compounds do not decompose at the low temperature and in the inert atmosphere of Titan. Therefore, its surface is likely to be covered by a thick, tarry organic goo. It is possible that a similar layer also collected on the early Earth and later reacted to form life.

23.8   Uranus and Neptune: Distant Giants

Uranus and **Neptune** were unknown to the ancients. They are so distant that even today, they can be seen only poorly from Earth. In 1977, the *Voyager II* space-

craft was launched to study the Jovian planets. It flew by Jupiter in 1979 and Saturn in 1981. It encountered Uranus by 1986 and then Neptune in 1989. The journey from Earth to Neptune covered 7.1 billion kilometers and took 12 years. The craft passed within 4800 kilometers of Neptune's cloudtops, only 33 kilometers away from the planned path. The strength of the radio signals received from *Voyager* measured a ten quadrillionth of a watt ($1/10^{16}$), and it took 38 radio antennas on four continents to absorb enough radio energy to interpret the signals.

Composition of Uranus and Neptune

All four Jovian planets have dense atmospheres, liquid surfaces and interiors, and solid mineral cores, but Uranus and Neptune are denser because they have higher proportions of heavier elements. Both of these outer giants are enveloped in thick atmospheres composed primarily of hydrogen and helium with smaller amounts of compounds of carbon, nitrogen, and oxygen. Their outer layers are molecular hydrogen, but neither is massive enough to generate an interior of metallic hydrogen. Instead their interiors are composed of methane, ammonia, and water, and the cores are probably a mixture of rock and metals.

Magnetic Fields of Uranus and Neptune

Voyager recorded that the magnetic field of Uranus is tilted 58° from its axis. This was unexpected, because our explanation of the origin of the Earth's magnetic

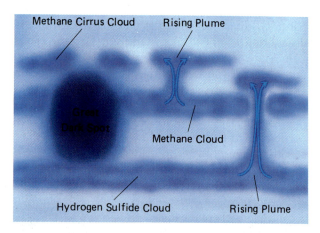

B

Figure 23.23 (A) An image of Neptune from *Voyager II*, taken through colored filters. Note the Great Dark Spot near the equator. *(NASA/JPL)* (B) A cross section of the cloud cover of Neptune.

A

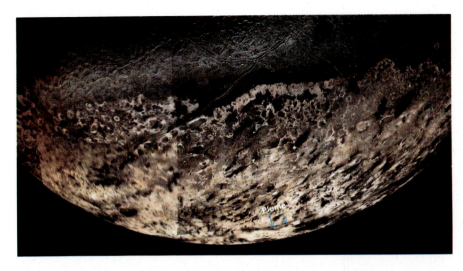

Figure 23.24   The south polar region of Triton. The small plume marked in the lower portion of the photograph is caused by a rapid vaporization of frozen nitrogen as the surface is heated by the Sun.   *(NASA/JPL)*

field suggests that the magnetic fields of all planets should be roughly aligned with the spin axis. One explanation was that *Voyager* just happened to pass Uranus during a magnetic field reversal. However, when *Voyager* reached Neptune scientists learned that its magnetic field is tilted 50° from its axis. Since the probability of catching two planets during magnetic reversals is extremely low, another explanation must exist. At present no satisfactory theory has been developed.

Neptune's Violent Weather

Whereas calm weather prevails on Uranus, Neptune is remarkably stormy. Planetary meteorologists have observed winds of at least 1100 kilometers per hour, rising and falling clouds, and a cyclonic storm system called the Great Dark Spot, similar to Jupiter's Great Red Spot (Fig. 23.23). Neptune receives only 5 percent of the solar energy that Jupiter receives, so solar heating cannot be the energy source for the violent winds. According to one fascinating but controversial theory, under the intense pressure near the core, methane decomposes into carbon and hydrogen and the carbon then crystallizes into diamond. The heat released during the formation of diamond is carried toward the surface by convection currents and then powers the winds.

Rings and Moons

A ring system and 15 moons orbit Uranus. Several of the moons are small and irregular, indicating that they may be debris from a collision with a smaller planet or moon.

Neptune has a ring system and eight moons. The largest moon is Triton, which is composed of about 75 percent rock and 25 percent ice. Like many other planets and moons, it has impact craters, mountains, and flat, crater-free plains (Fig. 23.24). Whereas the maria on the Earth's Moon are covered by lava, those on Triton are composed of ice or frozen methane.

23.9   Pluto: The Ice Dwarf

Pluto is the outermost of the known planets and has never been visited by spacecraft. Our highest-resolution photographs of Pluto are of poor quality compared with those of other planets (Fig. 23.25). Yet astronomers have deduced the properties of this planet from a variety of data.

In 1978 a satellite, Charon, was discovered orbiting Pluto. Using mathematical laws derived by Kepler and Newton in the 1600s, it is possible to calculate the

Ground based

Hubble Space Telescope

Figure 23.25   The best ground-based image of Pluto and Charon (left) and a similar view from the Hubble space telescope (right).   *(Canada-France-Hawaii Telescope and NASA/STSCI)*

GEOLOGY AND THE ENVIRONMENT

Impacts of Asteroids with the Earth

Although scientists continue to debate whether an asteroid impact led to the demise of the dinosaurs, significant impacts do occur and have been an important part of the geologic history of the Earth, the Moon, and the planets. One only has to look at the pockmarked surface of the Moon to see evidence of such impacts. Impact craters are less abundant on Earth only because most have been obliterated by erosion and tectonic activity.

Eugene Shoemaker, a geologist for the U. S. Geological Survey, estimates that about 1000 asteroids greater than 1 kilometer in diameter travel in orbits that cross the Earth's orbit. Any one of these could be sucked in by the Earth's gravity to collide with our planet. Large asteroids, with diameters greater than 10 kilometers, strike the Earth on the average of once every 40 million years; those with 1- to 10-kilometer diameters strike about once in 250,000 years. Collisions with smaller objects occur even more frequently.

On February 12, 1947, people in western Siberia saw a fireball "as bright as the Sun" when a small asteroid or swarm of meteorites crashed into the Earth. More than 100 craters formed, and 23 tons of meteorite fragments were recovered in the area. In October 1989, an asteroid with a diameter of 100 to 400 meters flew by within about 1.1 million kilometers of Earth. In January 1991, a smaller one passed within 170,000 kilometers of the Earth, half the distance from the Earth to the Moon.

The effect of a collision depends on the size of the asteroid and the location of impact. If the asteroid that missed us in 1989 had collided, it would have crashed with energy equivalent to that of a 1000-megaton nuclear bomb. If it had crashed on land, everything flammable within a radius of 100 kilometers would have been ignited instantly, and trees and structures within 250 kilometers would have been leveled by the shock waves.

How concerned should we be about the dangers of such an impact? Concern is dictated partly by your own personality, but the probability that an asteroid impact will affect you during your lifetime is greater than the chance of winning the state lottery if you only buy one ticket per drawing.

---

relative masses of a planet and its satellite if the radius of the satellite's orbit and the time required for one complete revolution of the satellite are known. When Pluto and Charon orbit each other, they periodically block one another from view. By precisely measuring these appearances and disappearances, astronomers calculate that Pluto is the smallest planet in the Solar System, smaller than the Earth's Moon. It has a mass only 1/500th that of Earth.

Once the diameter and mass are known, it is easy to calculate the density, which is 2 g/cm³. The Earth, composed of a metallic core and rocky mantle and crust, has a density of 5.5 g/cm³, and ice has a density of a little less than 1 g/cm³. Since Pluto's density is between these two values, we infer that Pluto is a mixture of rock and ice. Infrared measurements show that its surface temperature is about −220°C. Spectral analysis of Pluto's bright surface shows that it contains frozen methane. Its atmosphere is thin and composed mainly of carbon monoxide and nitrogen with some methane.

Pluto is similar in size and density to Neptune's moon, Triton. Their similarity has led some astronomers to postulate that both Pluto and Charon were once moons of Neptune and were pulled out of their orbits by a close encounter with another object. Other astronomers have suggested that many ice dwarfs exist in the outer Solar System. They are now searching for similar bodies with both optical and infrared telescopes.

23.10 Asteroids, Meteoroids, and Comets

Asteroids

Eighteenth-century astronomers noted that the dimensions of the planetary orbits increase in a regular pattern, starting with Mercury's orbit, the smallest, and continuing with the orbits of Venus, Earth, and Mars. But a gap in the pattern exists between Mars and Jupiter. The astronomers predicted that another planet might be found in the "open space" between Mars and Jupiter. Instead, they found tens of thousands of smaller bodies orbiting the Sun in a wide ring. These bodies are called **asteroids**. The largest asteroid has a diameter of 770 kilometers. Three others are about half that size, and most are far smaller. The orbit of an asteroid is not permanent, like that of a planet. If an asteroid passes too close to a nearby planet, it falls onto the planet's surface. However, if an asteroid passes near a planet without getting too close, the planet's gravity pulls the asteroid out of its current orbit and deflects it into a new orbit around the Sun. Thus, an asteroid may change its orbit frequently and erratically.

Meteoroids

Imagine tens of thousands of asteroids racing through the Solar System in changing paths. It is not surprising that many of them collide with each other. After a collision, the asteroids often break apart, forming small fragments and pieces of dust called **meteoroids**. These meteoroids leave the collision site in widely divergent directions. Some cross the orbit of the Earth and are pulled inward by our gravitational field. As the meteoroid falls to Earth, friction with the atmosphere heats the meteoroid until it glows. To our eyes it is a fiery streak in the sky, which we call a **meteor** or, colloquially, a shooting star. Most meteoroids are barely larger than a grain of sand when they enter the atmosphere and vaporize completely before they reach Earth. Larger ones, however, may reach the Earth's surface before they vaporize completely. A fallen meteoroid is called a **meteorite** (Fig. 23.26).

Meteorites may provide a window into the past by reflecting the primordial composition of the Solar System. About 90 percent of all meteorites are **stony meteorites**, composed primarily of rock similar to that of the Earth's mantle. The remaining 10 percent are metallic and consist mainly of iron and nickel, the elements that make up the Earth's core. The 90:10 ratio of stony to metallic meteorites comes quite close to the 80:20 volume ratio of the mantle to the core in the Earth.

Most stony meteorites contain small, round grains about 1 millimeter in diameter called **chondrules**,

Figure 23.26 A meteorite believed to be a fragment from the asteroid Vesta. *(NASA)*

which are composed largely of olivine and pyroxene. Many chondrules also contain fairly complex organic molecules including amino acids, the building blocks of proteins. These molecules are of the type synthesized by living organisms and form the molecular framework for life. How can we explain organic compounds coming from the cold vacuum of space? The answer must be that organic molecules form by inorganic (that is, nonliving) processes as well as by organic ones. Organic molecules have also been detected in dust clouds deep in interstellar space.

Comets

Occasionally, a glowing object appears in the sky. It travels slowly around the Sun in an elongated elliptical orbit and then disappears into space (Fig. 23.27). Such an object is called a **comet**, after the Greek word for "long haired." Despite their fiery appearance, comets are cold, and their light is reflected sunlight.

Comets originate in the outer reaches of the Solar System, and much of the time they travel through the cold void far from the Sun, even beyond the orbit of Pluto. A comet is composed of ice mixed with bits of silicate rock, metals, and frozen crystals of methane, ammonia, carbon dioxide, carbon monoxide, and other compounds.

When a comet is far out in space, millions of kilometers from the Sun, it has no tail at all but is simply a ball. As it approaches the Sun, its surface warms and some of it vaporizes. A light "wind" of radiation and ions streaming from the Sun, called the **solar wind**, blows some of the lighter particles away from the sphere to form a long tail. At this time the comet consists of a dense, solid **nucleus**, a bright outer sheath called a **coma**, and a long **tail** (Fig. 23.28). Comet tails more than 140 million kilometers long (almost as long as the distance from the Earth to the Sun) have been observed. As a comet orbits the Sun, the solar wind constantly blows the tail so that it always extends away from the Sun. There is very little matter in a comet tail. By terrestrial standards, it would represent a good, cold laboratory vacuum, yet viewed from a celestial perspective it looks like a hot, dense, fiery arrow.

Halley's comet passed through the inner Solar System between 1985 and 1987 and was studied by six spacecraft as well as by several ground-based observatories. Its nucleus is a peanut-shaped mass approximately 16 by 8 by 8 kilometers, about the same size and shape as Manhattan Island. The cold, relatively dense coma of Halley's comet had a radius of 4500 kilometers when it passed by.

Figure 23.27   Halley's Comet. *(Akira Fujii)*

Figure 23.28   A schematic view of a comet. The size of the nucleus has been enlarged several thousand times to show detail. When the comet interacts with the solar wind, magnetic field lines are generated as shown. Ions produced from gases streaming away from the nucleus are trapped within the field, creating the characteristically shaped tail.

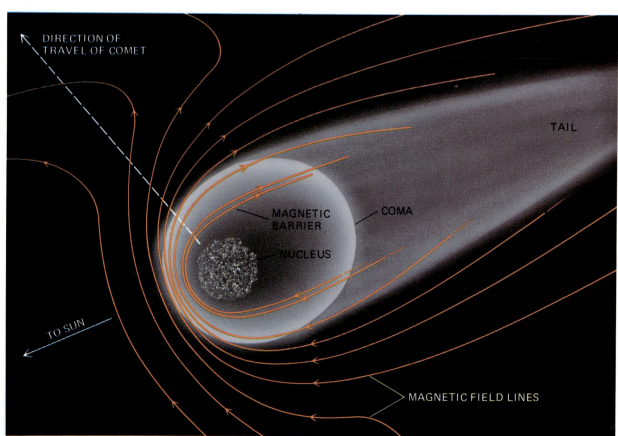

SUMMARY

Mercury is the smallest planet and the closest to the Sun. It rotates slowly on its axis and therefore experiences extremes of temperature. Its surface is heavily cratered from meteorite bombardment that occurred early in the history of the Solar System. **Venus** has a hot, dense atmosphere as a result of greenhouse warming. Its surface shows signs of recent geologic activity but no Earthlike plate tectonic activity. The Earth's Moon probably formed from the debris of a collision between an asteroid and the Earth. The Moon was heated by the energy released during condensation of the debris, by radioactive decay, and by meteorite bombardment. Evidence of ancient volcanism exists, but the Moon is cold and inactive today. **Mars** is a dry, cold planet with a thin atmosphere, but its surface bears signs of erosion and tectonic activity.

Jupiter, **Saturn**, **Uranus**, and **Neptune** are all large planets with low densities. Jupiter and Saturn have dense atmospheres, surfaces of liquid hydrogen, inner zones of liquid metallic hydrogen, and cores of rock and metal. The largest moons of Jupiter are **Io**, which is heated by gravitational forces; **Europa**, which is smooth and ice covered; and **Ganymede** and **Callisto**, which are large spheres of rock and ice. The rings of **Saturn** are made up of many small particles of dust, rock, and ice. They formed from a moon that was fragmented or from rock and ice that never coalesced to form a moon. **Titan**, the largest moon of Saturn, has an atmosphere and may be tectonically active. Uranus and Neptune have higher proportions of rock and ice than Jupiter and Saturn, and their magnetic fields are not in line with their axes of rotation. **Pluto**, the most distant planet, has a low density and is a small, icy planet.

Asteroids are small, planet-like bodies. A **meteorite** is a fallen **meteoroid**, a piece of matter from interplanetary space. Most meteorites are stony, and some contain organic molecules. About 10 percent of all meteorites are metallic. When a **comet** is in the inner part of the Solar System, it consists of a small, dense **nucleus** composed of ice, rock, and carbonaceous material; an outer sheath or **coma** composed of gases, water vapor, and dust; and a long **tail** made up of particles blown outward or left behind.

KEY TERMS

Terrestrial planets *639*	Liquid metallic	Saturn *656*	Meteor *661*
Jovian planets *639*	hydrogen *653*	Titan *657*	Meteorite *661*
Mercury *639*	Great Red Spot *654*	Uranus *658*	Chondrule *661*
Venus *640*	Io *655*	Neptune *658*	Comet *661*
Maria *645*	Europa *656*	Pluto *659*	Coma *661*
Mars *647*	Ganymede *656*	Asteroid *660*	Tail *661*
Jupiter *653*	Callisto *656*	Meteoroid *661*	

REVIEW QUESTIONS

1. List the nine planets in order of distance from the Sun, and distinguish between the terrestrial and Jovian planets.

2. Give a brief description of the planet Mercury. Include its atmosphere, surface temperature, surface features, and speed of rotation about its axis.

3. Why are very few meteorite craters visible on Venus?

4. Compare and contrast the surface topography of Mercury, Venus, the Moon, Earth, and Mars.

5. Compare and contrast the atmospheres of Mercury, Venus, the Moon, Earth, and Mars.

6. The text states that "Venus boiled, life evolved on Earth under moderate temperatures, and Mars froze." Discuss the evolution of the climates on these three planets.

7. What leads us to believe that the Moon was hot at one time in its history? How was the Moon heated?

8. Discuss the evidence that the Martian atmosphere was once considerably different than it is today.

9. Describe the composition of the planet Jupiter. How does it differ from that of Earth?

10. Explain why the mass of Jupiter was an important factor in determining its present composition and structure.

11. Compare and contrast the four Galilean moons of Jupiter.

12. Compare and contrast Saturn with Jupiter.

13. Compare and contrast Titan with the Earth.

14. Compare and contrast Pluto with Earth and Jupiter.

15. Is a comet really hot, dense, and fiery? If so, what is the energy source? If not, why do comets look like burning masses of gas?

DISCUSSION QUESTIONS

1. If Mercury rotated once every 24 hours as the Earth does, would you expect daytime temperatures on that planet to be higher or lower than they are? Defend your answer.

2. Suppose the oldest igneous rocks on the Moon formed when the Moon was 800 million years old. What conclusions would we then draw about the geologic history of the Moon? Could we answer the question "Was the Moon heated by internal radioactivity or by external bombardment?" Defend your answer.

3. Explain how we can learn about Earth's early history by studying the Moon.

4. At one time, Venus and Earth probably had similar climates, except that Venus was about 20°C warmer. If you could somehow cool the surface of Venus by 20°C, would conditions on that planet likely become similar to those on Earth? Explain.

5. Speculate on how a mantle plume could have generated faults on the surface of Venus even though it did not lead to rifting of tectonic plates.

6. Stephen Saunders, project scientist for NASA's *Magellan* mission, wrote, "Venus has been shaped by processes fundamentally similar to those that have taken place on Earth, but often with dramatically different results." Give examples to support or refute this statement.

7. Refer to Figures 23.9 and 23.14. Imagine that you know nothing about Mars except that it is a planet, and these two photographs were taken of its surface. What information can you deduce from these pictures alone? Defend your conclusions.

8. Imagine that three new planets were found between Earth and Mars. What could you tell about the geologic history of each, given the following limited data? (a) Planet X: The entire surface of this planet is covered with sedimentary rock. (b) Planet Y: This planet's atmosphere contains large quantities of water, ammonia, methane, and hydrogen sulfide. (c) Planet Z: About one third of this planet is covered with numerous impact craters. Smaller craters can be seen within the largest ones. Another third of the surface is much smoother and scattered with a few small craters. The remainder of the planet has no visible craters but is marked by great topographic relief, including mountain ranges and smooth plains but no canyons or river channels.

9. In his novel *2010: Odyssey II,* Arthur C. Clarke tells of a group of astronauts who traveled to the vicinity of Jupiter to retrieve a damaged spacecraft. At the end of the novel, Jupiter undergoes rapid changes and becomes a star. Is such a scenario plausible? Why did Mr. Clarke choose Jupiter as the planet to undergo such a change? Would any other planet have been as believable?

10. About 4 billion years from now, the Sun will probably grow significantly larger and hotter. How will this affect the composition and structure of Jupiter?

11. Would you expect to find gases in the ring system of Saturn? Why or why not?

12. Write a short science fiction story about space travel within the Solar System. You may make the plot fantastic and fictitious, but place the characters in scientifically plausible settings.

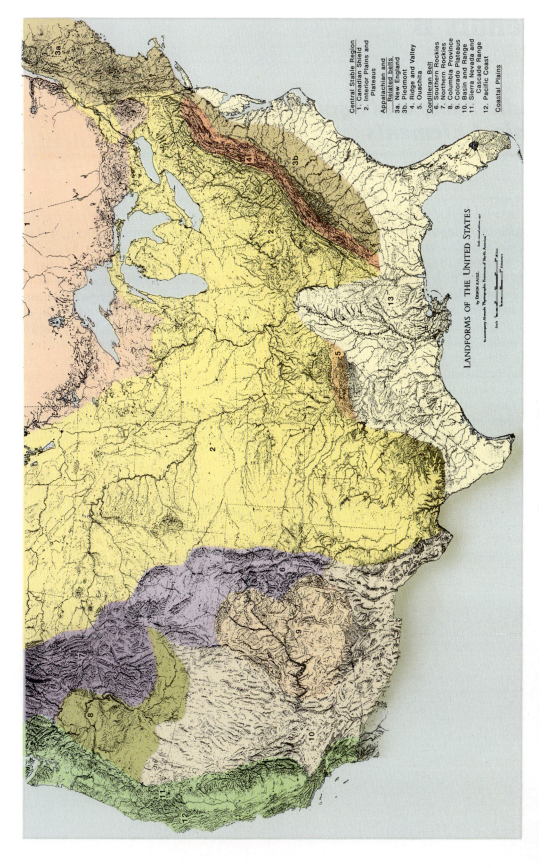

LANDFORMS OF THE UNITED STATES
by ERWIN RAISZ
to accompany Atwood's *Physiographic Provinces of North America*

Central Stable Region
1. Canadian Shield
2. Interior Plains and Plateaus

Appalachian and Related Belts
3a. New England
3b. Piedmont
4. Ridge and Valley
5. Ouachita

Cordilleran Belt
6. Southern Rockies
7. Northern Rockies
8. Columbia Province
9. Colorado Plateaus
10. Basin and Range
11. Sierra Nevada and Cascade Range
12. Pacific Coast

Coastal Plains

Appendix A   Physiographic Provinces of the United States

Appendix B

Periodic Table and Symbols for Chemical Elements

Scientists generally group or organize things that are similar so as to better understand them and determine how they relate to one another. An important step in the organization of elements was made in 1869 by Dimitri Mendeleev. Mendeleev showed that when elements are arranged in order of their increasing atomic weights, their physical and chemical properties tend to be repeated in cycles. The arrangement of elements was depicted on a chart called the periodic table of elements, a modern version of which is shown on page A-3. In the periodic table, each box contains the symbol of the element and its atomic number. Except for two long sequences set apart at the bottom, the elements appear in the increasing order of their atomic numbers. The vertical columns are called **groups** and contain elements with similar properties. For example, in column VIIA,

one finds fluorine, chlorine, bromine, and iodine. All of these elements are colored and highly reactive and share other similarities. Fluorine, however, is chemically the most active, chlorine somewhat less active, bromine still less active, and iodine the least active of the four. Except for hydrogen, all of the elements in group IA are soft, shiny metals and very reactive chemically. The horizontal rows on the table are called **periods** and contain sequences of elements having electron configurations that vary in characteristic patterns. It is thus apparent that the periodic table shows relationships among elements rather well and certainly better than an arbitrary listing.

The table on page A-4 provides the chemical names for the symbols of the elements used in the periodic table.

Periodic Table of the Elements

Key:

26
Fe
55.847

— Atomic number (Z)
— Element symbol
— Atomic mass of naturally occurring isotopic mixture; for radioactive elements, numbers in parentheses are mass numbers of most stable isotopes

IA	IIA	IIIB	IVB	VB	VIB	VIIB	VIII	VIII	VIII	IB	IIB	IIIA	IVA	VA	VIA	VIIA	O
1 **H** 1.0079																1 **H** 1.0079	2 **He** 4.00260
3 **Li** 6.941	4 **Be** 9.01218											5 **B** 10.81	6 **C** 12.011	7 **N** 14.0067	8 **O** 15.9994	9 **F** 18.998403	10 **Ne** 20.179
11 **Na** 22.98977	12 **Mg** 24.305											13 **Al** 26.98154	14 **Si** 28.0855	15 **P** 30.97376	16 **S** 32.06	17 **Cl** 35.453	18 **Ar** 39.948
19 **K** 39.0983	20 **Ca** 40.08	21 **Sc** 44.9559	22 **Ti** 47.90	23 **V** 50.9415	24 **Cr** 51.996	25 **Mn** 54.9380	26 **Fe** 55.847	27 **Co** 58.9332	28 **Ni** 58.70	29 **Cu** 63.546	30 **Zn** 65.38	31 **Ga** 69.72	32 **Ge** 72.59	33 **As** 74.9216	34 **Se** 78.96	35 **Br** 79.904	36 **Kr** 83.80
37 **Rb** 85.4678	38 **Sr** 87.62	39 **Y** 88.9059	40 **Zr** 91.22	41 **Nb** 92.9064	42 **Mo** 95.94	43 **Tc** (98)	44 **Ru** 101.07	45 **Rh** 102.9055	46 **Pd** 106.4	47 **Ag** 107.868	48 **Cd** 112.41	49 **In** 114.82	50 **Sn** 118.69	51 **Sb** 121.75	52 **Te** 127.60	53 **I** 126.9045	54 **Xe** 131.30
55 **Cs** 132.9054	56 **Ba** 137.33	57 *La 138.9055	72 **Hf** 178.49	73 **Ta** 180.9479	74 **W** 183.85	75 **Re** 186.207	76 **Os** 190.2	77 **Ir** 192.22	78 **Pt** 195.09	79 **Au** 196.9665	80 **Hg** 200.59	81 **Tl** 204.37	82 **Pb** 207.2	83 **Bi** 208.9804	84 **Po** (209)	85 **At** (210)	86 **Rn** (222)
87 **Fr** (223)	88 **Ra** 226.0254	89 †Ac 227.0278	104 **Unq** (261)	105 **Unp** (262)	106 **Unh** (263)	107 **Uns**	108	109									

*Lanthanide Series

58 **Ce** 140.12	59 **Pr** 140.9077	60 **Nd** 144.24	61 **Pm** (145)	62 **Sm** 150.4	63 **Eu** 151.96	64 **Gd** 157.25	65 **Tb** 158.9254	66 **Dy** 162.50	67 **Ho** 164.9304	68 **Er** 167.26	69 **Tm** 168.9342	70 **Yb** 173.04	71 **Lu** 174.967

†Actinide Series

90 **Th** 232.0381	91 **Pa** 231.0359	92 **U** 238.029	93 **Np** 237.0482	94 **Pu** (244)	95 **Am** (243)	96 **Cm** (247)	97 **Bk** (247)	98 **Cf** (251)	99 **Es** (252)	100 **Fm** (257)	101 **Md** (258)	102 **No** (259)	103 **Lr** (260)

Note: Atomic masses shown here are 1977 IUPAC values.

Elements and Their Chemical Symbols

Actinium	Ac	Erbium	Er	Mercury	Hg	Samarium	Sm
Aluminum	Al	Europium	Eu	Molybdenum	Mo	Scandium	Sc
Americium	Am	Fermium	Fm	Neodymium	Nd	Selenium	Se
Antimony	Sb	Fluorine	F	Neon	Ne	Silicon	Si
Argon	Ar	Francium	Fr	Neptunium	Np	Silver	Ag
Arsenic	As	Gadolinium	Gd	Nickel	Ni	Sodium	Na
Astatine	At	Gallium	Ga	Niobium	Nb	Strontium	Sr
Barium	Ba	Germanium	Ge	Nitrogen	N	Sulfur	S
Berkelium	Bk	Gold	Au	Nobelium	No	Tantalum	Ta
Beryllium	Be	Hafnium	Hf	Osmium	Os	Technetium	Tc
Bismuth	Bi	Helium	He	Oxygen	O	Tellurium	Te
Boron	B	Holmium	Ho	Palladium	Pd	Terbium	Tb
Bromine	Br	Hydrogen	H	Phosphorus	P	Thallium	Tl
Cadmium	Cd	Indium	In	Platinum	Pt	Thorium	Th
Calcium	Ca	Iodine	I	Plutonium	Pu	Thulium	Tm
Californium	Cf	Iridium	Ir	Polonium	Po	Tin	Sn
Carbon	C	Iron	Fe	Potassium	K	Titanium	Ti
Cerium	Ce	Krypton	Kr	Praseodymium	Pr	Tungsten	W
Cesium	Cs	Lanthanum	La	Promethium	Pm	Uranium	U
Chlorine	Cl	Lawrencium	Lr	Protactinium	Pa	Vanadium	V
Chromium	Cr	Lead	Pb	Radium	Ra	Xenon	Xe
Cobalt	Co	Lithium	Li	Radon	Rn	Ytterbium	Yb
Copper	Cu	Lutetium	Lu	Rhenium	Re	Yttrium	Y
Curium	Cm	Magnesium	Mg	Rhodium	Rh	Zinc	Zn
Dysprosium	Dy	Manganese	Mn	Rubidium	Rb	Zirconium	Zr
Einsteinium	Es	Mendelevium	Md	Ruthenium	Ru		

Appendix C

Identifying Common Minerals

More than 2500 minerals exist in the Earth's crust. However, of this great number, only thirty or so are common. Therefore, when you pick up rocks and want to identify the minerals, you are most likely to be looking at the same small group of minerals over and over again. The following list includes the most common and abundant minerals in the Earth's crust. A few important ore minerals and other minerals of economic value, and a few popular precious and semi-precious gems, are included because they are of special interest.

The minerals in this table fall into four categories.

1. Rock-forming minerals are shown in *red*. They are the most abundant minerals in the crust, and make up the largest portions of all common rocks. The rock-forming minerals and mineral groups are feldspar, pyroxene, amphibole, mica, clay, olivine, quartz, calcite, and dolomite. If more than one mineral of a group is common, each mineral is listed under the group name. For example, three kinds of pyroxene are abundant: augite, diopside, and orthopyroxene. All three are described under pyroxene.

2. Accessory minerals are shown in *yellow*. They are minerals that are common, but that usually occur only in small amounts.

3. Ore minerals and other minerals of economic importance are shown in *green*. They are minerals from which metals or other elements can be profitably recovered.

4. Gems are shown in *blue*. If the common gem name(s) is different from the mineral name, the gem name is given in parentheses following the mineral name. For example, emerald is the gem variety of the mineral beryl, and is listed as Beryl (emerald).

Minerals are listed alphabetically within each of the four categories for quick reference. The physical properties most commonly used for identification of each mineral, and the kind(s) of rock in which each mineral is most often found, are listed to facilitate identification of these common minerals.

Common Minerals and Their Properties

Mineral Group or Mineral	Chemical Composition	Habit, Cleavage, Fracture	Usual Color
		Rock-Forming Minerals	
Amphibole — Actinolite	$Ca_2(MgFe)_5Si_8O_{22}(OH)_2$	Slender crystals, radiating, fibrous	Blackish-green to black, dark green
Amphibole — Hornblende	$(Ca,Na)_{2-3}(Mg,Fe,Al)_5Si_6(Si,Al)_2O_{22}(OH)_2$	Elongate crystals	Blackish-green to black, dark green
Calcite	$CaCO_3$	Perfect cleavage into rhombs	Usually white, but may be variously tinted
Clay Minerals — Illite	$K_{0.8}Al_2(Si_{3.2}Al_{0.8})O_{10}(OH)_2$		White
Clay Minerals — Kaolinite	$Al_2Si_2O_5(OH)_4$		White
Clay Minerals — Smectite	$Na_{0.3}Al_2(Si_{3.7}Al_{0.3})O_{10}(OH)_2$		White, buff
Dolomite	$CaMg(CO_3)_2$	Cleaves into rhombs; granular masses	White, pink, gray, brown
Feldspar — Albite (sodium feldspar)	$NaAlSi_3O_8$ (sodic plagioclase)	Good cleavage in two directions, nearly 90°	White, gray
Feldspar — Orthoclase (potassium feldspar)	$KAlSi_3O_8$	Good cleavage in two directions at 90°	White, pink, red, yellow-green, gray
Feldspar — Plagioclase (feldspar containing both sodium and calcium)	$(Na,Ca)(Al,Si)_4O_8$	Good cleavage in two directions at 90°	White, gray
Feldspar — Biotite	$K(Mg,Fe)_3AlSi_3O_{10}(OH)_2$	Perfect cleavage into thin sheets	Black, brown, green
Feldspar — Muscovite	$KAl_2Si_3O_{10}(OH)_2$	Perfect cleavage into thin sheets	Colorless if thin
Olivine	$(MgFe)_2SiO_4$	Uneven fracture, often in granular masses	Various shades of green
Pyroxene — Augite	$Ca(Mg,Fe,Al)(Al,Si_2O_6)$	Short stubby crystals have 4 or 8 sides in cross section	Blackish-green to light green
Pyroxene — Diopside	$CaMg(Si_2O_6)$	Usually short thick prisms; may be granular	White to light green
Pyroxene — Orthopyroxene	$MgSiO_3$	Cleavage good at 87° and 93°; usually massive	Pale green, brown, gray, or yellowish
Quartz	SiO_2	No cleavage, massive and as six-sided crystals	Colorless, white, or tinted any color by impurities

Common Minerals and Their Properties *(continued)*

Hard-ness	Streak	Specific Gravity	Other Properties	Type(s) of Rock in Which the Mineral is Most Commonly Found
			Rock-Forming Minerals	
5–6	Pale green	3.2–3.6	Vitreous luster	Low- to medium-grade metamorphic rocks
5–6	Pale green	3.2	Crystals six-sided with 124° between cleavage faces	Common in many granitic to basaltic igneous rocks, and many metamorphic rocks
3	White	2.7	Transparent to opaque. Rapid effervescence with HCl	Limestone, marble, cave deposits
}			The clay minerals are so fine-grained that most physical properties cannot be identified.	Shale Shale, weathered bedrock, and soil Shale, weathered bedrock, and soil
3.5–4	White to pale gray	3.9–4.2	Effervesces slightly in cold dilute HCl.	Dolomite
6–6.5	White	2.6	Many show fine striations (twinning lines) on cleavage faces.	Granite, rhyolite, low-grade metamorphic rocks
6	White	2.6	Vitreous to pearly luster	Granite, rhyolite, metamorphic rocks
6	White	2.6–2.7	May show striations as in albite.	Basalt, andesite, medium- to high-grade metamorphic rocks
2.2–2.5	White, gray	2.7–3.1	Vitreous luster; divides readily into thin flexible sheets.	Granitic to intermediate igneous rocks, many metamorphic rocks
2–2.5	White	2.7–3	Vitreous or pearly; flexible and elastic; splits easily.	Many metamorphic rocks, granite
6.5–7	White	3.2–3.3	Vitreous, glassy luster	Basalt, peridotite
5.6	Pale green	5–6	Vitreous; distinguished from hornblende by the 87° angle between cleavage faces.	Basalt, peridotite, andesite, high-grade metamorphic rocks
5–6	White to greenish	3.2–3.6	Vitreous luster	Medium-grade metamorphic rocks
5.5	White	3.2–3.5	Vitreous luster	Peridotite, basalt, high-grade metamorphic rocks
7	White	2.6	Includes rock crystal, rose and milky quartz, amethyst, smoky quartz, etc.	Granite, rhyolite, metamorphic rocks of all grades, sandstone, siltstone

Common Minerals and Their Properties *(continued)*

Mineral Group or Mineral		Chemical Composition	Habit, Cleavage, Fracture	Usual Color
Accessory Minerals				
Apatite		$Ca_5(OH,F,Cl)(PO_4)_3$ (calcium fluorphosphate)	Massive, granular	Green, brown, red
Chlorite		$(Mg,Fe)_6(Si,Al)_4O_{10}(OH)_8$	Perfect cleavage as fine scales	Green
Corundum		Al_2O_3	Short, six-sided barrel-shaped crystals	Gray, light blue, and other colors
Epidote		$Ca_2(Al,Fe)Al_2O(SiO_4)(Si_2O_7)(OH)$	Usually granular masses; also as slender prisms	Yellow-green, olive-green, to nearly black
Fluorite		CaF_2	Octahedral and also cubic crystals	White, yellow, green, purple
Garnet	Almandine	$Fe_2Al_2(SiO_4)_3$	No cleavage, crystals 12- or 24-sided	Deep red
	Grossular	$Ca_3Al_2(SiO_4)_3$	No cleavage, crystals 12- or 24-sided	White, green, yellow, brown
Graphite		C	Foliated, scaly, or earthy masses	Steel gray to black
Hematite		Fe_2O_3	Granular, massive, or earthy	Brownish-red
Limonite		$2Fe_2O_3 \cdot 3H_2O$	Earthy fracture	Brown or yellow
Magnetite		Fe_3O_4	Uneven fracture, granular masses	Iron black
Pyrite		FeS_2	Uneven fracture cubes with striated faces, octahedrons	Pale brass yellow (lighter than chalcopyrite)
Serpentine		$Mg_3Si_2O_5(OH)_4$	Uneven, often splintery fracture	Light and dark green, yellow

Common Minerals and Their Properties *(continued)*

Hard-ness	Streak	Specific Gravity	Other Properties	Type(s) of Rock in Which the Mineral is Most Commonly Found
			Accessory Minerals	
4.5–5	Pale red-brown	3.1	Crystals may have a partly melted appearance, glassy.	Common in small amounts in many igneous, metamorphic, and sedimentary rocks
2.0–2.5	Gray, white, pale green	2.8	Pearly to vitreous luster	Common in low-grade metamorphic rocks
9	None	3.9–4.1	Hardness is distinctive.	Metamorphic rocks, some igneous rocks
6.7	Pale yellow to white	3.3	Vitreous luster	Low- to medium-grade metamorphic rocks
4	White	3.2	Cleaves easily; vitreous, transparent to translucent	Hydrothermal veins
6.5–7.5	White	4.2	Vitreous to resinous luster	The most common garnet in metamorphic rocks
6.5–7.5	White	3.6	Vitreous to resinous luster	Metamorphosed sandy limestones
1–2	Gray or black	2.2	Feels greasy; marks paper.	Metamorphic rocks
2.5	Dark red	2.5–5	Often earthy, dull appearance	Common in all types of rocks. It can form by weathering of iron minerals, and is the source of color in nearly all red rocks.
1.5–4	Brownish-yellow	3.6	Earthy masses that resemble clay	Common in all types of rocks. It can form by weathering of iron minerals, and is the source of color in most yellow-brown rocks.
5.5	Iron black	5.2	Metallic luster; strongly magnetic	Common in small amounts in most igneous rocks
6–6.5	Greenish-black	5	Metallic luster, brittle, very common	The most common sulfide mineral. Igneous, metamorphic, and sedimentary rocks; hydrothermal veins
2.5	White	2.5	Waxy luster, smooth feel, brittle	Alteration or metamorphism of basalt, peridotite, and other magnesium-rich rocks

Common Minerals and Their Properties *(continued)*

Mineral Group or Mineral	Chemical Composition	Habit, Cleavage, Fracture	Usual Color
Ore and Other Minerals of Economic Importance			
Anhydrite	$CaSO_4$	Granular masses, crystals with 2 good cleavage directions	White, gray, blue-gray
Asbestos	$Mg_3Si_2O_5(OH)_4$	Fibrous	White to pale olive-green
Azurite	$Cu_3(CO_3)_2(OH)_2$	Varied, may have fibrous crystals	Azure blue
Bauxite	$Al(OH)_3$	Earthy masses	Reddish to brown
Chalcopyrite	$CuFeS_2$	Uneven fracture	Brass yellow
Chromite	$FeCr_2O_4$	Massive, granular, compact	Black
Cinnabar	HgS	Compact, granular masses	Scarlet red to red-brown
Galena	PbS	Perfect cubic cleavage	Lead or silver gray
Gypsum	$CaSO_4 \cdot 2H_2O$	Tabular crystals, fibrous, or granular	White, pearly
Halite	$NaCl$	Granular masses, perfect cubic crystals	White, also pale colors and gray
Hematite	Fe_2O_3	Granular, massive, or earthy	Brownish-red, black
Malachite	$CuCO_3 \cdot Cu(OH)_2$	Uneven splintery fracture	Bright green, dark green
Native copper	Cu	Malleable and ductile	Copper red
Native gold	Au	Malleable and ductile	Yellow
Native silver	Ag	Malleable and ductile	Silver-white
Pyrolusite	MnO_2	Radiating or dendritic coatings on rocks	Black
Sphalerite	ZnS	Perfect cleavage in 6 directions at 120°	Shades of brown and red
Talc	$Mg_3Si_4O_{10}(OH)_2$	Perfect in one direction	Green, white, gray

Common Minerals and Their Properties *(continued)*

Hard-ness	Streak	Specific Gravity	Other Properties	Type(s) of Rock in Which the Mineral is Most Commonly Found
Ore and Other Minerals of Economic Importance				
3–3.5	White	2.9–3	Brittle; resembles marble but acid has no effect	Sedimentary evaporite deposits
1–2.5	White	2.6–2.8	Pearly to greasy luster; flexible, easily separated fibers	A variety of serpentine, found in the same rock types
4	Pale blue	3.8	Vitreous to earthy; effervesces with HCl	Weathered copper deposits
1.5–3.5	Pale reddish-brown	2.5	Dull luster, claylike masses with small round concretions	Weathering of many rock types
3.5–4.5	Greenish-black	4.2	Metallic luster, softer than pyrite	The most common copper ore mineral; hydrothermal veins, porphyry copper deposits
5.5	Dark brown	4.4	Metallic to submetallic luster	Peridotites and other ultramafic igneous rocks
2.5	Scarlet red	8	Color and streak distinctive	The most important mercury ore mineral; hydrothermal veins in young volcanic rocks
2.5	Gray	7.6	Metallic luster	The most important lead ore; commonly also contains silver; hydrothermal veins
1–2.5	White	2.2–2.4	Thin sheets (selenite), fibrous (satinspar), massive (alabaster)	Sedimentary evaporite deposits
2.5–3	White	2.2	Pearly luster, salty taste, soluble in water	Sedimentary evaporite deposits
2.5	Dark red	2.5–5	Often earthy, dull appearance, sometimes metallic luster	Huge concentrations occur as sedimentary iron ore; the most important source of iron
3.5–4	Emerald green	4	Effervesces with HCl; associated with azurite	Weathered copper deposits
2.5–3	Copper red	8.9	Metallic luster	Basaltic lavas
2.5–3	Yellow	19.3	Metallic luster	Hydrothermal quartz-gold veins, sedimentary placer deposits
2.5–3	Silver-white	10.5	Metallic luster	Hydrothermal veins, weathered silver deposits
1–2	Black	4.7	Sooty appearance	Black stains on weathered surfaces of many rocks, manganese nodules on the sea floor
3.5	Reddish-brown	4	Resinous luster; may occur with galena, pyrite	The most important ore mineral of zinc; hydrothermal veins
1–1.5	White	1–2.5	Greasy feel; occurs in foliated masses	Low-grade metamorphic rocks

Common Minerals and Their Properties *(continued)*

Mineral Group or Mineral	Chemical Composition	Habit, Cleavage, Fracture	Usual Color
Precious and Semi-Precious Gems			
Beryl (aquamarine, emerald)	$Be_3Al_2(SiO_3)_6$	Uneven fracture, hexagonal crystals	Green, yellow, blue, pink
Chrysoberyl (cat's eye, alexandrite)	$BeAl_2O_4$	Tabular crystals	Green, brown, yellow
Corundum (ruby, sapphire)	Al_2O_3	Short, six-sided barrel-shaped crystals	Gray, red (ruby), blue (sapphire)
Diamond	C	Octahedral crystals	Colorless or with pale tints
Garnet	$Fe_3Al_2(SiO_4)_3$	No cleavage, crystals 12- or 24-sided	Deep red
Jadite (jade) (a pyroxene)	$NaAl(Si_2O_6)$	Compact fibrous aggregates	Green
Olivine (peridot)	$(MgFe)_2SiO_4$	Uneven fracture, often in granular masses	Various shades of green
Opal	$SiO_2 \cdot nH_2O$	Conchoidal fracture, amorphous, massive	White and various colors
Quartz (rock crystal, amethyst, citrine, tiger eye, adventurine, carneline, chrysoprase, agate, onyx, heliotrope, bloodstone, jasper)	SiO_2	No cleavage, massive and as six-sided crystals	Colorless, white, or tinted any color by impurities
Spinel	$MgAl_2O_4$	No cleavage, rare octahedral crystals	Black, dark green, or various colors
Topaz	$Al_2SiO_4(F,OH)_2$	Cleavage good in one direction; conchoidal fracture	Colorless, white, pale tints of blue, pink
Tourmaline	$(Na,Ca)(Li,Mg,Al)(Al,Fe,Mn)_6$–$(BO_3)_3(Si_6O_{18})(OH)_4$	Poor cleavage, uneven fracture; striated crystals	Black, brown, green, pink
Turquoise	$CuAl_6(PO_4)_4(OH)_8 \cdot 4H_2O$	Massive	Blue-green
Zircon	$Zr(SiO_4)$	Cleavage poor, but often well-formed tetragonal crystals	Colorless, gray, green, pink, bluish

Common Minerals and Their Properties *(continued)*

Hard-ness	Streak	Specific Gravity	Other Properties	Type(s) of Rock in Which the Mineral is Most Commonly Found
			Precious and Semi-Precious Gems	
7.5–8.0	White	2.6–2.8	Vitreous luster	Granite, granite pegmatite, mica schist
8.5	White	3.7–3.8	Vitreous luster	Granite, granite pegmatite, mica schist
9	None	3.9–4.1	Ruby and sapphire are corundum varieties. Hardness is distinctive.	Metamorphic rocks, some igneous rocks
10	None	3.5	Adamantine luster. Hardness is distinctive.	Peridotite, kimberlite, sedimentary placer deposits
6.5–7.5	White	4.2	Vitreous to resinous luster	Metamorphic rocks, igneous rocks, placer deposits
6.7–7	White, pale green	3.3	Vitreous luster	High-pressure metamorphic rocks
6.5–7	White	3.2–3.3	Vitreous, glassy luster	Basalt, peridotite
5.5–6.5	White	2.1	Vitreous, greasy, pearly luster. May show a play of colors.	Low-temperature hot springs and weathered near-surface deposits
7	White	2.6	Colors and other features differ among the varieties.	Quartz is found in nearly all rock types, although each of the gem varieties may form in special environments.
7.5–8.0	White	3.5–4.1	Vitreous luster. Hardness is distinctive.	High-grade metamorphic rocks, dark igneous rocks
8	Colorless	3.5–3.6	Vitreous luster	Pegmatite, granite, rhyolite
7–7.5	White to gray	4.4–4.8	Vitreous, slightly resinous	Pegmatite, granite, metamorphic rocks
6	Blue-green, white	2.6–2.8	Waxy luster	Veins in weathered volcanic rocks in deserts
7.5	White	4.7	Adamantine luster	Many types of igneous rocks

Appendix D

Systems of Measurement

I The SI System

In the past, scientists from different parts of the world have used different systems of measurement. However, global cooperation and communication make it essential to adopt a standard system. The International System of Units (SI) defines various units of measurement as well as prefixes for multiplying or dividing the units by decimal factors. Some primary and derived units important to geologists are listed below.

Time

The SI unit is the **second**, s or sec, which used to be based on the rotation of the Earth but is now related to the vibration of atoms of cesium-133. SI prefixes are used for fractions of a second (such as milliseconds or microseconds), but the common words **minutes, hours,** and **days** are still used to express multiples of seconds.

Length

The SI unit is the **meter**, m, which used to be based on a standard platinum bar but is now defined in terms of wavelengths of light. The closest English equivalent is the **yard** (0.914 m). A **mile** is 1.61 kilometers (km). An **inch** is exactly 2.54 centimeters (cm).

Area

Area is length squared, as in **square meter, square foot,** and so on. The SI unit of area is the **are**, a, which is 100 sq m. More commonly used is the **hectare**, ha, which is 100 ares, or a square that is 100 m on each side. (The length of a U.S. football field plus one end zone is just about 100 m.) A hectare is 2.47 acres. An **acre** is 43,560 sq ft, which is a plot of 220 ft by 198 ft, for example.

Volume

Volume is length cubed, as in **cubic centimeter**, cm^3, **cubic foot**, ft^3, and so on. The SI unit is the **liter**, L, which is 1000 cm^3. A **quart** is 0.946 L; a U.S. liquid **gallon** (gal) is 3.785 L. A **barrel** of petroleum (U.S.) is 42 gal, or 159 L.

Mass

Mass is the amount of matter in an object. **Weight** is the force of gravity on an object. To illustrate the difference, an astronaut in space has no weight but still has mass. On Earth, the two terms are directly proportional and often used interchangeably. The SI unit of mass is the **kilogram**, kg, which is based on a standard platinum mass. A **pound** (avdp), lb, is a unit of weight. On the surface of the Earth, 1 lb is equal to 0.454 kg. A **metric ton**, also written as **tonne**, is 1000 kg, or about 2205 lb.

Temperature

The Celsius scale is used in most laboratories to measure temperature. On the Celsius scale the freezing point of water is 0° and the boiling point of water is 100°C.

The SI unit of temperature is the **Kelvin**. The coldest possible temperature, which is −273°C, is zero on the Kelvin scale. The size of 1 degree Kelvin is equal to 1 degree Celsius.

$$\text{Celsius temperature (°C)} = \text{Kelvin temperature (K)} - 273\text{ K}$$

Fahrenheit temperature (°F) is not used in scientific writing, although it is still popular in English-speaking countries. Conversion between Fahrenheit and Celsius is shown below.

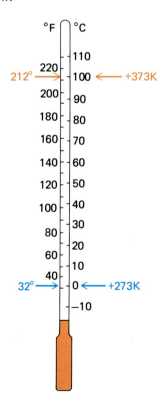

Energy

Energy is a measure of work or heat, which were once thought to be different quantities. Hence, two different sets of units were adopted and still persist, although we now know that work and heat are both forms of energy.

The SI unit of energy is the **joule**, J, the work required to exert a force of 1 newton through a distance of 1 m. In turn, a newton is the force that gives a mass of 1 kg an acceleration of 1 m/sec^2. In human terms, a joule is not much—it is about the amount of work required to lift a 100-g weight to a height of 1 m. Therefore, joule units are too small for discussions of machines, power plants, or energy policy. Larger units are

megajoule, MJ = 10^6 J (a day's work by one person)

gigajoule, GJ = 10^9 J (energy in half a tank of gasoline)

The energy unit used for heat is the **calorie**, cal, which is exactly 4.184 J. One calorie is just enough energy to warm 1 g of water 1°C. The more common unit used in measuring food energy is the **kilocalorie**, kcal, which is 1000 cal. When **Calorie** is spelled with a capital C, it means kcal. If a cookbook says that a jelly doughnut has 185 calories, that is an error—it should say 185 Calories (capital C), or 185 kcal. A value of 185 calories (small c) would be the energy in about one quarter of a thin slice of cucumber.

The unit of energy in the British system is the **British thermal unit**, Btu, which is the energy needed to warm 1 lb of water 1°F.

1 Btu = 1054 J = 1.054 kJ = 252 cal

The unit often referred to in discussions of national energy policies is the **quad**, which is 1 quadrillion Btu, or 10^{15} Btu.

Some approximate energy values are

1 barrel (42 gal) of petroleum = 5900 MJ
1 ton of coal = 29,000 MJ
1 quad = 170 million barrels of oil, or 34 million tons of coal

II Prefixes for Use with Basic Units of the Metric System

Prefix	Symbol†	Power		Equivalent
geo*		10^{20}		
tera	T	10^{12}	= 1,000,000,000,000	Trillion
giga	G	10^9	= 1,000,000,000	Billion
mega	M	10^6	= 1,000,000	Million
kilo	k	10^3	= 1,000	Thousand
hecto	h	10^2	= 100	Hundred
deca	da	10^1	= 10	Ten
——	—	10^0	= 1	One
deci	d	10^{-1}	= .1	Tenth
centi	c	10^{-2}	= .01	Hundredth
milli	m	10^{-3}	= .001	Thousandth
micro	μ	10^{-6}	= .000001	Millionth
nano	n	10^{-9}	= .000000001	Billionth
pico	p	10^{-12}	= .000000000001	Trillionth

* Not an official SI prefix but commonly used to describe very large quantities such as the mass of water in the oceans.

† The SI rules specify that its symbols are not followed by periods, nor are they changed in the plural. Thus, it is correct to write "The tree is 10 m high," not "10 m. high" or "10 ms high."

III   Handy Conversion Factors

To Convert From	To	Multiply By	
Centimeters	Feet	0.0328 ft/cm	
	Inches	0.394 in/cm	
	Meters	0.01 m/cm	(exactly)
	Micrometers (Microns)	1000 μm/cm	(")
	Miles (statute)	6.214×10^{-6} mi/cm	
	Millimeters	10 mm/cm	(exactly)
Feet	Centimeters	30.48 cm/ft	(exactly)
	Inches	12 in/ft	(")
	Meters	0.3048 m/ft	(")
	Micrometers (Microns)	304800 μm/ft	(")
	Miles (statute)	0.000189 mi/ft	
Grams	Kilograms	0.001 kg/g	(exactly)
	Micrograms	1×10^6 μg/g	(")
	Ounces (advp.)	0.03527 oz/g	
	Pounds (avdp.)	0.002205 lb/g	
Hectares	Acres	2.47 acres/ha	
Inches	Centimeters	2.54 cm/in	(exactly)
	Feet	0.0833 ft/in	
	Meters	0.0254 m/in	(exactly)
	Yards	0.0278 yd/in	
Kilograms	Ounces (avdp.)	35.27 oz/kg	
	Pounds (avdp.)	2.205 lb/kg	
Kilometers	Miles	0.6214 mi/km	
Meters	Centimeters	100 cm/m	(exactly)
	Feet	3.2808 ft/m	
	Inches	39.37 in/m	
	Kilometers	0.001 km/m	(exactly)
	Miles (statute)	0.0006214 mi/m	
	Millimeters	1000 mm/m	(exactly)
	Yards	1.0936 yd/m	
Miles (statute)	Centimeters	160934 cm/mi	
	Feet	5280 ft/mi	(exactly)
	Inches	63360 in/mi	(exactly)
	Kilometers	1.609 km/mi	
	Meters	1609 m/mi	
	Yards	1760 yd/mi	(exactly)
Ounces (avdp.)	Grams	28.35 g/oz	
	Pounds (avdp.)	0.0625 lb/oz	(exactly)
Pounds (avdp.)	Grams	453.6 g/lb	
	Kilograms	0.454 kg/lb	
	Ounces (avdp.)	16 oz/lb	(exactly)

IV Exponential or Scientific Notation

Exponential or scientific notation is used by scientists all over the world. This system is based on exponents of 10, which are shorthand notations for repeated multiplications or divisions.

A positive exponent is a symbol for a number that is to be multiplied by itself a given number of times. Thus, the number 10^2 (read "ten squared" or "ten to the second power") is exponential notation for $10 \cdot 10 = 100$. Similarly, $3^4 = 3 \cdot 3 \cdot 3 \cdot 3 = 81$. The reciprocals of these numbers are expressed by negative exponents. Thus $10^{-2} = 1/10^2 = 1/(10 \cdot 10) = 1/100 = 0.01$.

To write 10^4 in longhand form you simply start with the number 1 and move the decimal four places to the right: 10000. Similarly, to write 10^{-4} you start with the number 1 and move the decimal four places to the left: 0.0001.

It is just as easy to go the other way—that is, to convert a number written in longhand form to an exponential expression. Thus, the decimal place of the number 1,000,000 is six places to the right of 1:

$$1\,000\,000 = 10^6$$
6 places

Similarly, the decimal place of the number 0.000001 is six places to the left of 1 and

$$0.000001 = 10^{-6}$$
6 places

What about a number like 3,000,000? If you write it $3 \cdot 1,000,000$, the exponential expression is simply $3 \cdot 10^6$. Thus, the mass of the Earth, which, expressed in long numerical form is 3,120,000,000,000,000,000,000,000 kg, can be written more conveniently as $3.12 \cdot 10^{24}$ kg.

Appendix E

Rock Symbols

The symbols used in this book for types of rocks are shown below:

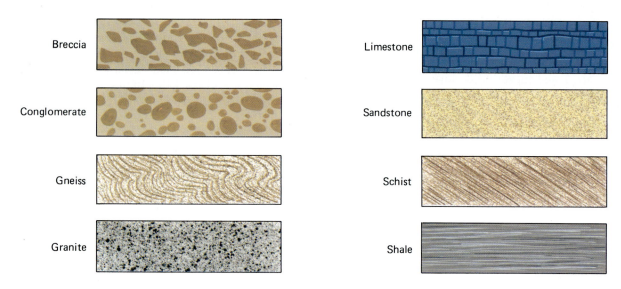

Breccia

Limestone

Conglomerate

Sandstone

Gneiss

Schist

Granite

Shale

In this book we have adopted consistent colors and style for depicting magma and layers in the upper mantle and crust.

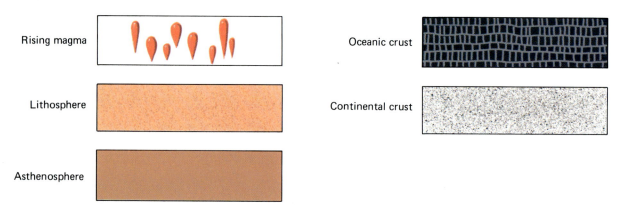

Rising magma

Oceanic crust

Lithosphere

Continental crust

Asthenosphere

Appendix F

Classification of Living Things

Early students of biology found it convenient to divide all organisms into two great realms or kingdoms, designated the Animalia and the Plantae. However, by the late nineteenth century, many biologists suggested that perhaps a third kingdom—the Protista—should be established for certain single-celled organisms that seemed to be neither plant nor animal. Even this was considered inadequate by some biologists, for as this three-kingdom classification was gaining adherents, evidence was accumulating that clearly indicated the need for still further major groupings. Studies of unicellular organisms had revealed two quite different forms of cell structures. There were, for example, prokaryotic unicellular organisms, such as bacteria and blue-green algae, that lacked a cell nucleus and possessed other traits that set them clearly apart from eukaryotic unicellular organisms, which had a true nucleus, well-defined chromosomes, and organelles. Any classification that sought to categorize organisms according to similarity of origin and fundamental differences could not ignore the contrast between eukaryotes and prokaryotes. Finally, it was recognized that the fungi deserved special taxonomic consideration also, for they are dependent upon a supply of organic molecules in their environment, as are animals, yet they absorb their food through cell membranes, as do plants. The fungi appear to have had an evolutionary radiation quite distinct from the other major groups.

In an effort to account for these differences and better represent the evolutionary relationships of organisms, R. H. Whittaker of Cornell University proposed a five-kingdom system of classification.* In this classification, plants, fungi, and animals are regarded as distinct in terms of being specialized for different modes of nutrition, photosynthesis, absorption, and ingestion. The Protista is composed of unicellular eukaryotic organisms, and the Monera include the simplest of organisms that are inferred to be similar to the primitive forms of life from which other kingdoms evolved. A modified version of Whittaker's classification, which is steadily gaining proponents, is presented here. (Geologic ranges of major fossil groups are indicated.)

---

*Whittaker, R.H. 1969. New concepts of kingdoms of organisms. *Science* 163:150–160.

KINGDOM MONERA
Prokaryotes *(Archean to Recent)*

PHYLYM CYANOPHYTA — Cyanobacteria *(Archean to Recent)*

PHYLUM MYXOBACTERIAE — Unicellular or filamentous gliding bacteria

PHYLUM SCHIZOPHYTA — True bacteria

PHYLUM ACTINOMYCOTA — Certain branching, filamentous bacteria

PHYLUM SPIROCHAETAE — Spirochetes

KINGDOM PROTOCTISTA (formerly Protista)
Solitary or colonial unicellular eukaryotic organisms that do not form tissues

PHYLUM EUGLENOPHYTA — Euglenoid organisms

PHYLUM XANTHOPHYTA — Yellow-green algae

PHYLUM CHRYSOPHYTA — Golden brown algae, diatoms, and coccolithophorids *(Paleozoic? Triassic to Recent)*

PHYLUM PYRROPHYTA — Dinoflagellates and cryptomonads *(Triassic to Recent)*

PHYLUM HYPHOCHYTRIDIOMYCOTA — Hypochytrids

PHYLUM PLASMODIOPHOROMYCOTA — Plasmodiophores

PHYLUM SPOROZOA — Sporozoans (parasitic protists)

PHYLUM CNIDOSPORIDIA — Cnidosporidians

PHYLUM ZOOMASTIGINA — Animal flagellates, protozoa that have whiplike cytoplasmic protrusions (flagellae)

PHYLUM SARCODINA — Rhizopods, protozoa with pseudopodia for locomotion *(Cambrian to Recent)*

PHYLUM CILIOPHORA — Ciliates and suctorians, movement accomplished by beating of cilia (adult suctorians are attached to objects)

KINGDOM PLANTAE

DIVISION* RHODOPHYTA — Red algae, usually marine, multicellular

DIVISION PHAEOPHYTA — Brown algae, multicellular often with large bodies, as in seaweeds and kelps

DIVISION CHLOROPHYTA — Green algae *(Proterozoic? Paleozoic to Recent)*

DIVISION CHAROPHYTA — Stoneworts

DIVISION BRYOPHYTA — Liverworts, mosses, and hornworts *(late Paleozoic to Recent)*

DIVISION PSILOPHYTA — Extinct leafless, rootless, vascular plants *(middle Paleozoic)*

DIVISION LYCOPODOPHYTA — Club mosses, with simple vascular systems and small leaves, including scale trees of Paleozoic (lycopsids) *(Silurian to Recent)*

DIVISION EQUISETOPHYTA (ARTHROPHYTA) — Horsetails, scouring rushes, including sphenopsids such as *Calamites* and *Annularia* of the late Paleozoic *(Devonian to Recent)*

DIVISION POLYPODIOPHYTA — The true ferns or pteropsids *(Devonian to Recent)*

DIVISION PINOPHYTA — The "gymnosperms" including conifers, cycads, and many evergreen plants; no true flowers *(middle Paleozoic to Recent)*

Class Lyginopteriodopsida — The seed ferns, known from fossils of the late Paleozoic and including such forms as *Neuropteris* and *Glossopteris* *(late Paleozoic to Recent)*

---

*In botany, it is conventional to use the terms *division*, *subdivision*, and so on, in place of *phylum* and *subphylum*.

Class Bennettitopsida	The extinct cycadeoids *(Triassic to Recent)*
Class Cycadopsida	Cycads *(Triassic to Recent)*
Class Ginkgoopsida	Ginkgoes *(early Miocene to Recent)*
Class Pinopsida	Conifers, as well as the extinct *Cordaites (late Paleozoic to Recent)*
Class Gnetopsida	Certain climbing shrubs and small tropical trees
DIVISION MAGNOLIOPHYTA	Flowering plants or "angiosperms"; seeds enclosed in an ovary *(Cretaceous to Recent)*
Class Magnoliopsida	Dicotyledonous plants: embryos with two cotyledons or seeds; leaves with netlike veins *(Cretaceous to Recent)*
Class Liliopsida	Monocotyledonous plants: embryos with only one seed leaf; leaves with parallel veins *(Cretaceous to Recent)*

KINGDOM FUNGI

DIVISION MYXOMYCOPHYTA	Slime molds
DIVISION EUMYCOPHYTA	True fungi

KINGDOM ANIMALIA

PHYLUM MESOZOA	Mesozoans
PHYLUM PORIFERA	Sponges; includes forms with calcareous spicules, siliceous spicules, and proteinaceous spicules; may also include the extinct stromatoporoids *(Cambrian to Recent)*
PHYLUM ARCHAEOCYATHA	Extinct spongelike organisms *(Cambrian)*
PHYLUM CNIDARIA(COELENTERATA)	Jellyfishes, corals; radially symmetric, aquatic, with body wall of two layers of cells, in the outer of which are stinging cells *(Proterozoic to Recent)*
Class Hydrozoa	Hydralike animals
Class Scyphozoa	True jellyfishes
Class Anthozoa	Corals and sea anemones
Subclass Zoantharia	Hexacorals of modern seas *(Triassic to Recent)*
Subclass Rugosa	Paleozoic tetracorals *(Cambrian to Permian)*
Subclass Tabulata	Paleozoic tabulate corals *(Ordovician to Permian)*
PHYLUM CTENOPHORA	Modern comb jellies or "sea walnuts." Not known as fossils.
PHYLUM PLATYHELMINTHES	Flatworms
PHYLUM NEMERTEA	Proboscis worms
PHYLUM NEMATODA	Roundworms
PHYLUM ACANTHOCEPHALA	Hook-headed worms
PHYLUM NEMATOMORPHA	Horsehair worms
PHYLUM ROTIFERA	Small, wormlike animals with a circle of cilia on the head
PHYLUM GASTROTRICHA	Small, wormlike animals resembling rotifers but lacking circle of cilia
PHYLUM BRYOZOA	The bryozoans, sometimes considered two phyla: *Entoprocta* and *Ectoprocta (Ordovician to Recent)*
PHYLUM BRACHIOPODA	Marine animals with two parts (valves) to their shell (dorsal and ventral) *(Cambrian to Recent)*
Class Inarticulata	Primitive brachiopods having phosphatic or chitinous valves, lacking hinge *(Cambrian to Recent)*
Class Articulata	Advanced calcareous brachiopods with valves that are hinged *(Cambrian to Recent)*

PHYLUM PHORONIDA	Wormlike marine animals that secrete and live within a leathery tube *(early Mesozoic to Recent)*
PHYLUM ANNELIDA	The segmented worms *(Proterozoic to Recent)*
PHYLUM ONYCOPHORA	Rare tropical animals considered intermediate between annelids and arthropods *(Cambrian to Recent)*
PHYLUM ARTHROPODA	Segmented animals with jointed appendages
SUBPHYLUM TRILOBITA	Tribolites, common marine arthropods of the Paleozoic Era *(Cambrian to Permian)*
SUBPHYLUM CHELICERATA	
Class Xiphosura	Horseshoe crabs *(Silurian to Recent)*
Class Eurypterida	Eurypterids *(Ordovician to Permian)*
Class Pycnogonida	Sea spiders
Class Arachnida	Scorpions, spiders, ticks, and mites
SUBPHYLUM CRUSTACEA	Lobsters, crabs, barnacles, and ostracodes *(Cambrian to Recent)*
SUBPHYLUM LABIATA	
Class Chilopoda	Centipedes
Class Diplopoda	Millipedes
Class Insecta	The 24 orders of insects *(Devonian to Recent)*
PHYLUM MOLLUSCA	Unsegmented, soft-bodied animals, usually with shells
Class Monoplacophora	Primitive forms with cap-shaped shells *(Cambrian to Recent)*
Class Amphineura	Chitons, marine forms with shells composed of eight segments
Class Scaphopoda	Tusk shells, curved tubular shells open at both ends
Class Gastropoda	Snails, abalones, asymmetric animals with single spiral conch or no shell *(Cambrian to Recent)*
Class Pelecypoda (Bivalvia)	Shells of two valves (right and left); includes clams, mussels, oysters, and scallops *(Cambrian to Recent)*
Class Cephalopoda	Marine animals with tentacles around head and well-developed eyes and nervous system
ORDER NAUTILOIDEA	Nautiloids; cephalopods with simple suture lines *(Cambrian to Recent)*
ORDER AMMONOIDEA	Ammonoids; cephalopods with complexly folded sutural lines *(Devonian to Cretaceous)*
ORDER BELEMNOIDEA	Belemnites *(late Mississippian to early Tertiary)*
ORDER SEPIOIDEA	Cuttlefishes *(Jurassic to Recent)*
ORDER TEUTHOIDEA	Squids *(Jurassic to Recent)*
ORDER OCTOPODA	Octopi
PHYLUM POGONOPHORA	Beard worms
PHYLUM CHAETOGNATHA	Arrow worms
PHYLUM ECHINODERMATA	Marine animals that are radially symmetric as adults (bilateral as larvae), have calcareous, spine-bearing plates and unique water vascular systems
Class Asteroidea	Starfishes *(Ordovician to Recent)*
Class Ophiurodea	Brittle stars and serpent stars *(Ordovician to Recent)*
Class Echinoidea	Sea urchins and sand sollars *(Ordovician to Recent)*
Class Holothuroidea	Sea cucumbers
Class Crinoidea	Sea lilies and feather stars *(Cambrian to Recent)*
Class Blastoidea	The extinct blastoids of the Paleozoic *(Silurian to Permian)*

PHYLUM HEMICHORDATA (PROTOCHORDATA)	The acorn worms; larval forms resemble echinoderm larva, adults have anterior proboscis connected by collar to wormlike body
PHYLUM CHORDATA	Bilaterally symmetric animals with notochord, dorsal hollow neural tube, and gill clefts in the pharynx
SUBPHYLUM UROCHORDATA	Sea squirts or tunicates; larval forms have notochord in tail region
SUBPHYLUM CEPHALOCHORDATA	*Branchiostoma ("Amphioxus")*; small marine animals with fishlike bodies and notochord
SUBPHYLUM VERTEBRATA	Animals with a backbone of vertebrae, definite head, ventrally located heart, and well-developed sense organs
Class Agnatha	Living lampreys and hagfish as well as extinct ostracoderms; agnatha lack jaws *(Cambrian to Recent)*
Class Acanthodii	Primitive, extinct, spiny fishes with jaws *(middle to late Paleozoic)*
Class Placodermi	Primitive, often armored, Paleozoic jawed fishes *(middle Paleozoic)*
Class Chondrichthyes	Sharks, rays, skates, and chimaeras *(middle Paleozoic to Recent)*
Class Osteichthyes	The bony fishes *(Devonian to Recent)*
Subclass Actinopterygii	Ray-finned fishes *(Devonian to Recent)*
Subclass Sarcopterygii	Lobe-finned, air-breathing fishes *(Devonian to Recent)*
ORDER CROSSOPTERYGII	Lobed-finned fishes, ancestors of amphibians
ORDER DIPNOI	Lungfishes
Class Amphibia	Amphibians, the earth's earliest land-dwelling vertebrates; include extinct Labyrinthodontia of late Paleozoic and Triassic *(Devonian to Recent)*
Class Reptilia	Reptiles, reproducing with use of amniotic eggs
Subclass Anapsida	Turtles, as well as the extinct aquatic mesosaurs and terrestrial stem reptiles called cotylosaurs *(Permian to Recent)*
Subclass Synapsida	Mammal-like reptiles, including sailback forms and therapsids *(Permian and Triassic)*
Subclass Euryapsida	Extinct, generally marine reptiles, including plesiosaurs, ichthyosaurs, and placodonts *(Permian to Cretaceous)*
Subclass Diapsida	Reptilian group that includes the extinct dinosaurs and crocodilians, lizards, snakes, and the modern tuatara *(Permian to Recent)*
Class Aves	Birds; warm-blooded, feathered, and typically winged animals; primitive forms with reptilian teeth, modern forms toothless *(Jurassic to Recent)*
Class Mammalia	Warm-blooded animals with hair covering; females with mammary glands that secrete milk for nourishing young *(Triassic to Recent)*
Subclass Eotheria	Primitive, extinct Triassic and Jurassic mammals
Subclass Prototheria	Monotremes such as the duck-billed platypus and spiny anteater. Egg-laying mammals *(Triassic to Recent)*
Subclass Allotheria	Extinct early mammals with multicusped teeth; multituberculates *(Triassic to early Cenozoic)*

Subclass Metatheria	Pouched mammals or marsupials *(Cretaceous to Recent)*
Subclass Eutheria	The placental mammals; young develop within uterus of female, obtain moisture via the placenta *(Cretaceous to Recent)*
ORDER INSECTIVORA	Primitive insect-eating mammals, including moles and shrews *(Cretaceous to Recent)*
ORDER DERMOPTERA	The colugo
ORDER CHIROPTERA	Bats *(early Tertiary to Recent)*
ORDER PRIMATES	Lemurs, tarsiers, monkeys, apes, and humans *(early Tertiary to Recent)*
ORDER EDENTATA	Living armadillos, anteaters, and tree sloths; extinct glyptodonts and ground sloths
ORDER RODENTIA	Squirrels, mice, rats, beavers, and porcupines
ORDER LAGOMORPHA	Hares, rabbits, and pikas
ORDER CETACEA	Whales and porpoises *(early Tertiary to Recent)*
ORDER CREODONTA	Extinct, ancient carnivorous placentals
ORDER CARNIVORA	Modern carnivorous placentals, including dogs, cats, bears, hyenas, seals, sea lions, and walruses
ORDER CONDYLARTHA	Extinct ancestral hoofed placentals (ancestral ungulates) *(early Tertiary)*
ORDER AMBLYPODA	Extinct primitive ungulates *(early Tertiary)*
ORDER TUBULIDENTATA	Aardvarks
ORDER PHOLIDOTA	Pangolins
ORDER PERISSODACTYLA	Odd-toed hoofed mammals, including living horses, rhinoceroses, and tapirs; and extinct titanotheres and chalicotheres *(early Tertiary to Recent)*
ORDER ARTIODACTYLA	Even-toed hoofed animals, including living antelopes, cattle, deer, giraffes, camels, llamas, hippos, and pigs; and extinct entelodonts and oreodonts *(early Tertiary to Recent)*
ORDER PROBOSCIDEA	Elephants and extinct mastodons and mammoths *(early Tertiary to Recent)*
ORDER SIRENIA	Sea cows

Glossary

aa   A lava flow that has a jagged, rubbly, broken surface.

ablation area   The lower portion of a glacier where more snow melts in summer than accumulates in winter, causing a net loss of glacial ice. (*syn:* zone of wastage)

abrasion   The mechanical wearing and grinding of rock surfaces by friction and impact.

Absaroka sequence   A sequence of Permian-Pennsylvanian sediments bounded both above and below by a regional unconformity and recording an episode of marine transgression over an eroded surface, full flood level of inundation, and regression from the craton.

absolute geologic age   The actual age, expressed in years, of a geologic material or event.

abyssal fan   A large, fan-shaped accumulation of sediment deposited at the bases of many submarine canyons adjacent to the deep-sea floor. (*syn:* submarine fan)

abyssal plain   A flat, level, largely featureless part of the ocean floor between the mid-oceanic ridge and the continental rise.

Acadian orogeny   An episode of mountain building in the northern Appalachians during the Devonian Period.

acanthodians   The earliest known vertebrates (fishes) with a movable, well-developed lower jaw, or mandible; hence, the first jawed fishes.

accreted terrain   A landmass that originated as an island arc or a microcontinent that was later added onto a continent.

accumulation area   The upper part of a glacier where accumulation of snow during the winter exceeds melting during the summer, causing a net gain of glacial ice.

active continental margin   A continental margin characterized by subduction of an oceanic lithospheric plate beneath a continental plate. (*syn:* Andean margin)

adaptation   A modification of an organism that better fits it for existence in its present environment or enables it to live in a somewhat different environment.

adaptive radiation   The diversity that develops among species as each adapts to a different set of environmental conditions.

adenosine diphosphate (ADP)   A product formed in the hydrolysis of adenosine triphosphate that is accompanied by release of energy and organic phosphate.

adenosine triphosphate (ATP)   A compound that occurs in all cells and that serves as a source of energy for physiologic reactions such as muscle contraction.

aerobic organism   An organism that uses oxygen in carrying out respiratory processes.

age   The time represented by the time-stratigraphic unit called a stage. (Informally, may indicate any time span in geologic history, as "age of cycads.")

Agnatha   The jawless vertebrates, including extinct ostracoderms and living lampreys and hagfishes.

A horizon   The uppermost layer of soil composed of a mixture of organic matter and leached and weathered minerals. (*syn:* topsoil)

algae   Any of a large group of simple plants (thallophyta) that contain chlorophyll and are capable of photosynthesis.

Allegheny orogeny   The late Paleozoic episodes of mountain building along the present trend of the Appalachian Mountains.

alluvial fan   A fanlike accumulation of sediment created where a steep stream slows down rapidly as it reaches a relatively flat valley floor.

alluvium   Unconsolidated, poorly sorted detrital sediments ranging from clay to gravel sizes and characteristically fluvial in origin.

alpha particle   A particle, equivalent to the nucleus of a helium atom, emitted from an atomic nucleus during radioactive decay.

alpine glacier   A glacier that forms in mountainous terrain.

Alpine orogeny   In general, the sequence of crustal disturbances beginning in the middle Mesozoic and continuing into the Miocene that resulted in the geologic structures of the Alps.

amino acids   Nitrogenous hydrocarbons that serve as

the building blocks of proteins and are thus essential to all living things.

ammonites Ammonoid cephalopods having more complex sutural patterns than either ceratites or goniatites.

ammonoids An extinct group of cephalopods, with coiled, chambered conch(s) and having septa with crenulated margins.

amniotic egg That type of egg produced by reptiles, birds, and monotremes. In this type, the developing embryo is maintained and protected by an elaborate arrangement of shell membranes, yolk, sac, amnion, and allantois.

amphibians "Cold-blooded" vertebrates that utilize gills for respiration in the early life stages but that have air-breathing lungs as adults.

amphibole A group of double-chain silicate minerals. Hornblende is a common amphibole.

anaerobic organism An organism that does not require oxygen for respiration, but rather makes use of processes such as fermentation to obtain its energy.

Andean margin A continental margin characterized by subduction of an oceanic lithospheric plate beneath a continental plate. (*syn:* active continental margin)

andesite A fine-grained gray or green volcanic rock intermediate in composition between basalt and granite, consisting of about equal amounts of plagioclase feldspar and mafic minerals.

angiosperms An advanced group of plants having floral reproductive structures and seeds in a closed ovary. The "flowering plants."

angle of repose The maximum slope or angle at which loose material remains stable.

angular unconformity An unconformity in which younger sediments or sedimentary rocks rest on the eroded surface of tilted or folded older rocks.

anion An ion that has a negative charge.

antecedent stream A stream that was established before local uplift started and that cut its channel at the rate at which the land was rising.

Anthropoidea The suborder of primates that includes monkeys, apes, and humans.

anticline A fold in rock that resembles an arch; the fold is convex upward and the oldest rocks are in the middle.

Antler orogeny A Late Devonian and Mississippian episode of mountain building involving folding and thrusting along a belt across Nevada to southwestern Alberta.

aquifer A porous and permeable body of rock that can yield economically significant quantities of ground water.

Archean Division of Precambrian time beginning 3.8 billion years ago and ending 2.5 billion years ago.

Archaeocyatha A group of extinct marine organisms having double, perforated, calcareous, conical-to-cylindric walls. Archaeocyathids lived during the Cambrian.

archosaurs Advanced reptiles of a group called diapsids, which includes thecodonts, "dinosaurs," pterosaurs, and crocodiles.

arcoids A group of pelecypods exemplified by species of *Arca*.

arête A sharp, narrow ridge between adjacent valleys formed by glacial erosion.

artesian aquifer An inclined aquifer that is bounded top and bottom by layers of impermeable rock so the water is under pressure.

artesian well A well drilled into an artesian aquifer in which the water rises without pumping and in some cases spurts to the surface.

artiodactyl Hoofed mammals that typically have two or four toes on each foot.

aseismic ridge A submarine mountain chain with little or no earthquake activity.

ash (volcanic) Fine pyroclastic material with particles less than 2 mm in diameter.

ash flow A mixture of volcanic ash, larger pyroclastic particles, and gas that flows rapidly along the Earth's surface as a result of an explosive volcanic eruption. (*syn:* nuée ardente)

asteroid One of numerous relatively small planetary bodies (less than 800 km in diameter) revolving around the sun in orbits lying between those of Mars and Jupiter.

asthenosphere The portion of the upper mantle beneath the lithosphere. It consists of weak, plastic rock where magma may form. It extends from a depth of about 100 kilometers to about 350 kilometers below the surface of the Earth.

atoll A circular reef that surrounds a lagoon and is bounded on the outside by deep water of the open sea.

atom The fundamental unit of elements consisting of a small, dense, positively charged center called a nucleus surrounded by a diffuse cloud of negatively charged electrons.

atomic fission A nuclear process that occurs when a heavy nucleus splits into two or more lighter nuclei, simultaneously liberating a considerable amount of energy.

atomic fusion A nuclear process that occurs when two light nuclei unite to from a heavier one. In the process, a large amount of energy is released.

atomic mass A quantity essentially equivalent to the

number of neutrons plus the number of protons in an atomic nucleus.

atomic number   The number of protons in the nuclei of atoms of a particular element. (An element is thus a substance in which all of the atoms have the same atomic number.)

australopithecines   A general term applied loosely to Pliocene and early Pleistocene primates whose skeletal characteristics place them between typically apelike individuals and those more obviously human.

autotroph   An organism that uses an external source of energy to produce organic nutrients from simple inorganic chemicals.

back arc basin   A sedimentary basin on the opposite side of a magmatic arc from the trench, either in an island arc or in an Andean continental margin.

backshore   The upper zone of a beach that is usually dry but is washed by waves during storms.

bajada   A broad depositional surface extending outward from a mountain front formed by the merging of alluvial fans.

banks   The rising slopes bordering the two sides of a stream channel.

bar   An elongate mound of sediment, usually composed of sand or gravel, in a stream channel or along a coastline.

barchan dune   A crescent-shaped dune, highest in the center, with the tips facing downwind.

barrier island   A long, narrow, low-lying island that extends parallel to the shoreline.

barrier reef   A reef separated from the coast by a deep, wide lagoon.

basal slip   Movement of the entire mass of a glacier along the bedrock.

basalt   A dark-colored, very fine-grained, mafic volcanic rock composed of about half calcium-rich plagioclase feldspar and half pyroxene.

base level   The deepest level to which a stream can erode its bed. The ultimate base level is usually sea level, but this is seldom attained.

basement rocks   The older granitic and related metamorphic rocks of the Earth's crust that make up the foundations of continents.

basin   A low area of the Earth's crust of tectonic origin.

batholith   A large plutonic mass of intrusive rock with more than 100 square kilometers of surface exposed.

baymouth bar   A bar that extends partially or completely across the entrance to a bay.

beach   Any strip of shoreline washed by waves or tides.

beach terrace   A level portion of old beach elevated above the modern beach by uplift of the shoreline or fall of sea level.

bed   The floor of a stream channel; also the smallest layer in sedimentary rocks.

bedding   Layering that develops as sediments are deposited.

bed load   That portion of a stream's load that is transported on or immediately above the stream bed.

bedrock   The solid rock that underlies soil or regolith.

belemnites   Members of the molluscan class Cephalopoda, having straight internal shells.

Benioff zone   An inclined zone of earthquake activity that traces the upper portion of a subducting plate in a subduction zone.

benthic   A bottom-dwelling organism.

bentonite   A layer of clay, presumably formed by the alteration of volcanic ash.

beta particle   A charged particle, essentially equivalent to an electron, emitted from an atomic nucleus during radioactive disintegration.

B horizon   The soil layer where ions leached from the A horizon above accumulate.

biotite   Black, rock-forming mineral of the mica group.

Bivalvia   A class of the phylum Mollusca also known as the class Pelecypoda. (The term Pelecypoda is preferred in this text so that the pelecypods are not confused with other "bivalves," such as brachiopods and ostracods.)

blastoids   Sessile (attached) Paleozoic echinoderms having a stem and an attached cup or calyx composed of relatively few plates.

blowout   A small depression created by wind erosion.

body waves   Seismic waves that travel through the interior of the Earth.

boulder   A rounded rock fragment larger than a cobble (diameter greater than 256 cm).

Bowen's reaction series   A series of minerals in which any early formed mineral crystallizing from a cooling magma reacts with the magma to form minerals lower in the series.

brachiating   Swinging from branch to branch and tree to tree by using the limbs, as among monkeys.

brachiopod   Bivalved (doubled-shelled) marine invertebrates. They were particularly common and widespread during the Paleozoic and persist in fewer numbers today.

braided stream   A stream that divides into a network of branching and reuniting shallow channels separated by mid-channel bars.

breccia   A coarse-grained sedimentary rock

composedof angular, broken fragments cemented together.

breeder reactor (nuclear) An atomic reactor that uses uranium-238, but that creates additional fuel by producing more fissionable material than it consumes.

brittle fracture A rupture that occurs when a rock breaks sharply.

Bryozoa A phylum of attached and incrusting colonial marine invertebrates.

Burgess Shale fauna A beautifully preserved fossil fauna of soft-bodied Cambrian animals discovered in 1910 by Charles Walcott in Kicking Horse Pass, Alberta.

butte A flat-topped mountain with several steep cliff faces. A butte is smaller and more tower-like than a mesa.

calcite A common rock-forming mineral, $CaCO_3$.

caldera A large circular depression caused by an explosive volcanic eruption.

Caledonian orogeny A major early Paleozoic episode of mountain building affecting Europe, which created an orogenic belt, the Caledonides, extending from Ireland and Scotland northwestward through Scandinavia.

caliche A hard soil layer formed when calcium carbonate precipitates and cements the soil.

calving A process in which large chunks of ice break off from tidewater glaciers to form icebergs.

capacity The maximum quantity of sediment that a stream can carry.

capillary action The action by which water is pulled upward through small pores by electrical attraction to the pore walls.

capillary fringe A zone above the water table in which the pores are filled with water due to capillary action.

cap rock An impermeable rock, usually shale, that prevents oil or gas from escaping upward from a reservoir.

carbon-14 A radioactive isotope of carbon with an atomic mass of 14. Carbon-14 is frequently used in determining the age of materials less than about 50,000 years old.

carbonate rocks Rocks such as limestone and dolomite made up primarily of carbonate minerals.

carbonization The concentration of carbon during fossilization.

cast (natural) A replica of an organic object, such as a fossil shell, formed when sediment fills a mold of that object.

Cataract Group A group of formations deposited during the Early Silurian and including the Whirlpool Sandstone, Manitoulin Dolomite, Cabot Head Shale, Dyer Bay Limestone, and Wingfield Shale of southern Ontario. The Cataract Group is correlative with the Albion Group of western New York.

cation A positively charged ion.

Catskill delta A buildup of Middle and Upper Devonian clastic sediments as a broad, complex clastic wedge derived from the erosion of highland areas formed largely during the Acadian orogeny.

cavern An underground cavity or series of chambers created when ground water dissolves large amounts of rock, usually limestone. (*syn:* cave)

Ceboidea The New World monkeys, characterized by prehensile tails, and including the capuchin, marmoset, and howler monkeys.

cementation The process by which clastic sediment is lithified by precipitation of a mineral cement among the grains of the sediment.

Cenozoic Era The latest of the four eras into which geologic time is subdivided; 65 million years ago to the present.

centrifugal force The apparent outward force experienced by an object moving in a circular path. Centrifugal force is a manifestation of inertia, the tendency of moving things to travel in straight lines.

ceratites One of the three larger groups of ammonoid cephalopods having sutural complexity intermediate between goniatites and ammonites.

ceratopsians The quadrupedal ornithischian dinosaurs characterized by the development of prominent horns on the head.

Cercopithecoidea The Old World monkeys (of Asia, southern Europe, and Africa), including macaques, guenons, langurs, baboons, and mandrills.

cetaceans The group of marine mammals that includes whales and porpoises.

chalicotheres Extinct perissodactyls having robust claws rather than hoofs.

chalk A very fine-grained, soft, earthy, white to gray bioclastic limestone made of the shells and skeletons of marine microorganisms.

chemical weathering The chemical decomposition of rocks and minerals by exposure to air, water, and other chemicals in the environment.

chert A hard, dense sedimentary rock composed of microcrystalline quartz. (*syn:* flint)

chlorophyll The catalyst that makes possible the reaction of water and carbon dioxide in green plants to produce carbohydrates. Photosynthesis is the reaction.

Choanichthyes That group of fishes that includes both the dipnoans (lungfishes with weak pelvic and pectoral fins) and crossopterygians (lungfishes with stout lobe-fins).

Chondrichthyes   The broad category of fishes with cartilaginous skeletons that is exemplified by sharks, skates, and rays.

chondrites   Stony meteorites that contain rounded silicate grains or chondrules. Chondrules are believed to have formed by crystallization of liquid silicate droplets.

C horizon   The lowest soil layer composed of partly weathered bedrock grading downward into unweathered parent rock.

chromosome   Threadlike microscopic bodies composed of chromatin. Chromosomes appear in the nucleus of the cell at the time of cell division. They contain the genes. The number of chromosomes is normally constant for a particular species.

chromosphere   One of the concentric shells of the sun, lying above the photosphere and telescopically visible as a thin, brilliant red rim around the edge of the Sun for a second or so at the beginning and end of a solar eclipse.

cinder cone   A small volcano, as high as 300 meters, made up of loose pyroclastic fragments blasted out of a central vent.

cinders (volcanic)   Glassy pyroclastic volcanic fragments 4 to 32 mm in size.

cirque   A steep-walled semicircular depression eroded into a mountain peak by a glacier.

clastic sediment   Sediment composed of fragments of weathered rock that have been transported some distance from their place of origin.

clastic sedimentary rocks   Rocks composed of lithified clastic sediment.

clay   Any clastic mineral particle less than 1/256 millimeter in diameter. Also a group of layer silicate minerals.

cleavage   The tendency of some minerals to break along certain crystallographic planes.

cobbles   Rounded rock fragments in the 64- to 256-mm size range; larger than pebbles and smaller than boulders.

Coccolithophorids   Marine, planktonic, biflagellate, golden brown algae that typically secrete coverings of discoidal calcareous platelets called *coccoliths*.

Colorado Mountains   Highlands uplifted in Pennsylvanian time in Colorado. Sometimes inappropriately termed the "ancestral Rockies."

column   A dripstone or speleothem formed when a stalactite and a stalagmite meet and fuse together.

columnar joints   The regularly spaced cracks that commonly develop in lava flows, forming five- or six-sided columns.

compaction   Tighter packing of sedimentary grains causing weak lithification and a decrease in porosity, usually resulting from the weight of overlying sediment.

competence   A measure of the largest particles that a stream can transport.

composite volcano   A volcano that consists of alternate layers of unconsolidated pyroclastic material and lava flows. (*syn:* stratovolcano)

compressive stress   Stress that acts to shorten an object by squeezing it.

concordant   Pertaining to an igneous intrusion that is parallel to the layering of country rock.

cone of depression   A cone-like depression in the water table formed when water is pumped out of a well more rapidly than it can be replaced by flow through the aquifer.

conformable   The condition in which sedimentary layers were deposited continuously and without interruption.

conglomerate   A coarse-grained clastic sedimentary rock, composed of rounded fragments larger than 2 mm in diameter, cemented in a fine-grained matrix of sand or silt.

conodonts   Small, toothlike fossils composed of calcium phosphate and found in rocks ranging from Cambrian to Triassic in age. Although the precise nature of the conodont-producing organism has not been determined, this uncertainty does not detract from their usefulness as guide fossils.

contact   A boundary between two different rock types or between rocks of different ages.

contact metamorphism   Metamorphism caused by heating of country rock, and/or addition of fluids, from a nearby igneous intrusion.

continental crust   The predominantly granitic portion of the crust, 20 to 70 kilometers thick, that makes up the continents.

continental drift   The theory proposed by Alfred Wegener that continents were once joined together and later split and drifted apart. The continental drift theory has been replaced by the more complete plate tectonics theory.

continental glacier   A glacier that forms a continuous cover of ice over an area of 50,000 square kilometers or more and spreads outward in all directions under the influence of its own weight. (*syn:* ice sheet)

continental margin   The region between the shoreline of a continent and the deep ocean basins, including the continental shelf, continental slope, and continental rise. Also the region where thick, granitic continental crust joins thinner, basaltic oceanic crust.

continental margin basin   A sediment-filled depression or other thick accumulation of sediment and sedimentary rocks near the margin of a continent.

continental rifting   The process by which a continent is pulled apart at a divergent plate boundary.

continental rise   An apron of sediment between the continental slope and the deep sea floor.

continental shelf   A shallow, nearly level area of continental crust covered by sediment and sedimentary rocks that is submerged below sea level at the edge of a continent between the shoreline and the continental slope.

continental slope   The relatively steep (3° to 6°) underwater slope between the continental shelf and the continental rise.

continental suture   The junction created where two continents collide and weld into a single mass of continental crust.

convection current   A circular current in a fluid or plastic material formed when heated materials rise and cooler materials sink.

convergence (in evolution)   The process by which similarity of form or structure arises among different organisms as a result of their becoming adapted to similar habitats.

convergent plate boundary   A boundary where two lithospheric plates collide head-on.

coquina   A bioclastic limestone consisting of coarse shell fragments cemented together.

Cordaites   A primitive order of treelike plants with long, bladelike leaves and clusters of naked seeds. Cordaites were in some ways intermediate in evolutionary stage between seed ferns and conifers.

core   The innermost region of the Earth, probably consisting of iron and nickel.

Coriolis effect   The deflection of winds and water currents to the right in the Northern Hemisphere and to the left in the Southern Hemisphere as a consequence of the Earth's rotation.

correlation   Demonstration of the equivalence of rocks or geologic features from different locations.

country rock   The older rock intruded by a younger igneous intrusion or mineral deposit.

crater   A bowl-like depression at the summit of a volcano. Also a similar depression formed by a meteorite impact.

craton   A segment of continental crust, usually in the interior of a continent, that has been tectonically stable for a long time, commonly a billion years or longer.

creep   The slow movement of unconsolidated material downslope under the influence of gravity.

Creodonta   Primitive, early, flesh-eating placental mammals.

crest (of a wave)   The highest part of a wave.

crevasse   A fracture or crack in the upper 40 to 50 meters of a glacier.

crinoids   Stalked echinoderms with a calyx composed of regularly arranged plates from which radiate arms for gathering food.

cross-bedding (cross-stratification)   An arrangement of laminae or thin beds transverse to the planes of stratification. The inclined laminae are usually at inclinations of less than 30° and may be either straight or concave.

Crossopterygii   That group of choanichthyan fishes ancestral to earliest amphibians and characterized by stout pectoral and pelvic fins as well as lungs.

crust   The Earth's outermost layer, about 7 to 70 kilometers thick, composed of relatively low-density silicate rocks.

crystal   A solid element or compound whose atoms are arranged in a regular, orderly, periodically repeated array.

crystal habit   The shape in which individual crystals grow and the manner in which crystals grow together in aggregates.

crystal settling   A process in which the crystals that solidify first from a cooling magma settle to the bottom of a magma chamber because the solid minerals are more dense than liquid magma.

Curie point   The temperature below which rocks can retain magnetism.

Cycadales   A group of seed plants that were especially common during the Mesozoic and were characterized by palmlike leaves and coarsely textured trunks marked by numerous leaf scars.

cyclothem   A vertical succession of sedimentary units reflecting environmental events that occurred in a constant order. Cyclothems are particularly characteristic of the Pennsylvanian System.

Cystoids   Attached echinoderms with generally irregular arrangement and number of plates in the calyx and perforated by pores or slits.

daughter isotope   An isotope formed by radioactive decay of another isotope.

debris flow   A type of mass wasting in which particles move as a fluid and more than half of the particles are larger than sand.

Deccan traps   A thick sequence (3200 meters) of Upper Cretaceous basaltic lava flows that cover about 500,000 sq km of peninsular India.

décollement   Feature of stratified rocks in which upper formations may become "unstuck" from lower formations, deform, and slide thousands of meters over underlying beds.

deflation   Erosion by wind.

deformation   Folding, faulting, and other changes in shape of rocks or minerals in response to mechanical forces, such as those that occur in tectonically active regions.

delta   The nearly flat, alluvial, fan-shaped tract of land at the mouth of a stream.

dendritic drainage pattern A pattern of stream tributaries that branches like the veins in a leaf. It often indicates uniform underlying bedrock.

deposition The laying-down of rock-forming materials by any natural agent.

depositional environment Any setting in which sediment is deposited.

depositional remanent magnetism Remanent magnetism resulting from mechanical orientation of magnetic mineral grains during sedimentation.

desert A region with less than 25 cm of rainfall a year. Also defined as a region that supports only a sparse plant cover.

desertification A process by which semiarid land is converted to desert, often by improper farming or by climate change.

desert pavement A continuous cover of closely packed rock fragments and pebbles exposed and polished as wind erodes fine sediment.

diatoms Microscopic golden brown algae (chrysophytes) that secrete a delicate siliceous frustule (shell).

differential weathering The process by which certain rocks weather more rapidly than adjacent rocks, usually resulting in an uneven surface.

differentiation The process by which a planet becomes internally zoned as when heavy materials sink toward its center and light materials accumulate near the surface.

dike A sheet-like igneous rock that cuts across the structure of country rock.

dike swarm A group of dikes that forms in a parallel or radial set.

dinoflagellates Unicellular marine algae usually having two flagella and a cellulose wall.

diorite A rock that is the medium- to coarse-grained plutonic equivalent of andesite.

dip The angle of inclination of a bedding plane, measured from the horizontal.

diploid cells Cells having two sets of chromosomes that form pairs, as in somatic cells.

Dipnoi An order of lungfishes with weak pectoral and pelvic fins; not considered ancestral to land vertebrates.

discharge The volume of water flowing downstream per unit time. It is measured in units of m^3/sec.

disconformity A variety of unconformity in which bedding planes above and below the plane of erosion or nondeposition are parallel.

discordant Pertaining to a dike that cuts across sedimentary layers or other kinds of layering in country rock, or other features that show cross-cutting relationships.

dissolved load The portion of a stream's sediment load that is carried in solution.

distributary A channel that flows outward from the main stream channel, such as is commonly found in deltas.

divergent plate boundary The boundary or zone where lithospheric plates separate from each other. (*syn:* spreading center)

DNA The nucleic acid found chiefly in the nucleus of cells that functions in the transfer of genetic characteristics and in protein synthesis.

docking The accretion of island arcs or microcontinents onto a continental margin.

Docodonts A group of small, primitive Late Jurassic mammals possibly ancestral to the living monotremes.

dolomite $CaMg(CO_3)_2$, a common rock-forming mineral.

dome A circular or elliptical anticlinal structure.

downcutting Downward erosion by a stream.

drainage basin The region that is ultimately drained by a single river.

drainage divide A ridge or other topographically high region that separates adjacent drainage basins.

drift (glacial) Any rock or sediment transported and deposited by a glacier or by glacial meltwater.

dripstone A deposit formed in a cavern when calcite precipitates from dripping water.

drumlin An elongate hill formed when a glacier flows over and reshapes a mound of till or stratified drift.

Dryopithecine In general, a group of lightly built primates that lived during the Miocene and Pliocene in mostly open savannah country and that includes *Dryopithecus,* a form considered to be in the line leading to apes.

dune A mound or ridge of wind-deposited sand.

dynamothermal (regional) metamorphism Metamorphism that has occurred over a wide region, caused by deep burial and high temperatures associated with pressures resulting from overburden and orogeny.

earthflow A flowing mass of fine-grained soil particles mixed with water. Earthflows are less fluid than mudflows.

earthquake A sudden motion or trembling of the Earth caused by the abrupt release of slowly accumulated elastic energy in rocks.

echinoderms The large group (Phylum Echinodermata) of marine invertebrates characterized by prominent pentamerous symmetry and skeleton frequently constructed of calcite elements and including spines. Cystoids, blastoids, crinoids, and echinoids are examples of echinoderms.

echo sounder An instrument that emits sound

waves and then records them after they reflect off the sea floor. The data is then used to record the topography of the sea floor.

Edentata An order of placental mammals that includes extinct ground sloths and gyptodonts, as well as living armadillos, tree sloths, and South American anteaters.

effluent stream A stream that receives water from ground water because its channel lies below the water table. (*syn:* gaining stream)

elastic deformation A type of deformation in which an object returns to its original size and shape when stress is removed.

elastic limit The maximum stress that an object can withstand without permanent deformation.

electron A fundamental particle that forms a diffuse cloud of negative charge around an atom.

element A substance that cannot be broken down into other substances by ordinary chemical means. An element is made up of only one kind of atom.

emergent coastline A coastline that was recently under water but has been exposed, either because the land has risen or sea level has fallen.

endemic population The native fauna of any particular region.

end moraine A moraine that forms at the end, or terminus, of a glacier.

entelodonts A group of extinct artiodactyls bearing a superficial resemblance to giant wild boars.

Eocambrian Name applied to the poorly fossiliferous sequence of sedimentary rocks that lie generally above Precambrian basement rocks but below readily identifiable fossiliferous Cambrian strata.

eon A major division of the geologic time scale. All of the geologic periods from the Cambrian to the Holocene comprise the Phanerozoic Eon. The term is also sometimes used to denote a span of 1 billion years.

epicenter The point on the Earth's surface directly above the focus of an earthquake.

epifaunal organisms Organisms living *on,* as distinct from *in,* a particular body of sediment or another organism.

epoch A chronologic subdivision of a geologic period. Rocks deposited or emplaced during an epoch constitute the *series* for that epoch.

era A major division of geologic time, divisible into geologic periods.

erosion The removal of weathered rocks by moving water, wind, ice, or gravity.

erratic A boulder that was transported to its present location by a glacier. It is usually different from the bedrock in its immediate vicinity.

esker A long snake-like ridge formed by deposition in a stream that flowed on, within, or beneath a glacier.

estuary A bay that formed when a broad river valley was submerged by rising sea level or a sinking coast.

eukaryote A type of living cell containing a true nucleus, enclosed within a nuclear membrane, and having well-defined chromosomes and cell organelles.

eupantotheres A group of Jurassic mammals with dentition similar to that of primitive representatives of later marsupial and placental mammals and thus thought to be ancestral to these groups.

eurypterids Aquatic arthropods of the Paleozoic, superficially resembling scorpions, and probably carnivorous.

eustatic Pertaining to worldwide simultaneous changes in sea level, such as might result from change in the volume of continental glaciers.

evaporation The transformation of a liquid to a gas.

evaporite deposit A chemically precipitated sedimentary rock that formed when dissolved ions were concentrated by evaporation of water.

evolution The continuous genetic adaptation of organisms or species to the environment.

exfoliation Fracturing in which concentric plates or shells split from the main rock mass like the layers of an onion.

extensional stress Tectonic stress in which rocks are pulled apart.

external mold A fossil cavity created in sediment by a shell or other hard body part that bears the impression of the exterior of the original.

extrusive rock An igneous rock formed from material that has erupted onto the surface of the Earth.

facies A particular aspect of sedimentary rocks that is a direct consequence of sedimentation in a particular depositional environment.

fall A type of mass wasting in which unconsolidated material falls freely or bounces down the face of a cliff.

fault A fracture in rock along which displacement has occurred.

fault creep A continuous, slow movement of solid rock along a fault, resulting from a constant stress acting over a long time.

fault zone An area of numerous closely spaced faults.

feldspar A common group of aluminum silicate rock-forming minerals that contains potassium, sodium, and calcium.

fermentation The partial breakdown of organic compounds by an organism in the absence of oxygen. The final product of fermentation is alcohol or lactic acid.

fetch The distance that the wind has travelled over the ocean without interruption.

filter-feeders (organisms) Animals that obtain their food, which usually consists of small particles or organisms, by filtering it from the water.

firn Hard, dense snow that has survived through one summer melt season. Firn is transitional between snow and glacial ice.

firn line The boundary on a glacier between permanent snow and seasonal snow. Above the firn line, winter snow does not melt completely during summer, while below the firn line it does.

fissility Fine layering along which a rock splits easily.

fjord A long, deep, narrow arm of the sea bounded by steep walls, generally formed by submergence of a glacially eroded valley.

flash flood A rapid, intense, local flood of short duration.

flood basalt Basaltic lava that erupts gently, in great volume, from cracks at the Earth's surface to cover large areas of land and form basalt plateaus.

flood plain That portion of river valley adjacent to the channel that is built by sediment deposited during floods and is covered by water during a flood.

flow Mass wasting in which individual particles move downslope as a semifluid, not as a consolidated mass.

fluvial Pertaining to sediments or other geologic features formed by streams.

flysch Term describing thick sequences of rapidly deposited, poorly sorted marine clastics.

focus The initial rupture point of an earthquake.

fold A bend in rock.

foliation Layering in rock created by metamorphism.

footwall The rock beneath an inclined fault.

foraminifers An order of mostly marine, unicellular protozoans that secrete tests (shells) that are usually composed of calcium carbonate.

foreshock Small earthquakes that precede a large quake by an interval ranging from a few seconds to a few weeks.

foreshore The zone that lies between the high and low tides; the intertidal region.

formation A lithologically distinct body of sedimentary, igneous, or metamorphic rock that can be recognized in the field and mapped.

fossil The preserved trace, imprint, or remains of a plant or animal.

fracture (1) The manner in which minerals break other than along planes of cleavage. (2) A crack, joint, or fault in bedrock.

frost wedging A process in which water freezes in a crack in rock and the expansion wedges the rock apart.

fusulinids Primarily spindle-shaped foraminifers with calcareous, coiled tests divided into a complex of numerous chambers. Fusulinids were particularly abundant during the Pennsylvanian and Permian periods.

gabbro Igneous rock that is mineralogically identical to basalt, but that has a medium- to coarse-grained texture because of its plutonic origin.

gaining stream A stream that receives water from ground water because its channel lies below the water table. (*syn:* effluent stream)

galaxy An aggregate of stars and planets, separated from other such aggregates by distances greater than those between member stars.

gamete Either of two cells (male or female) that must unite in sexual reproduction to initiate the development of a new individual.

gamma rays Very high-frequency electromagnetic waves.

gene The unit of heredity transmitted in the chromosome.

genus The major subdivision of a taxonomic family or subfamily of plants or animals, usually consisting of more than one species.

geological structure Any feature such as a fold or a fault, formed by deformation or movement of rocks. Also, the combination of all such features of an area or region.

geologic column A composite columnar diagram that shows the sequence of rocks at a given place or region, arranged to show their position in the geologic time scale.

geologic range The geologic time span between the first and last appearance of an organism.

geologic time scale A chronological arrangement of geologic events subdivided into units.

geology The study of the Earth, the materials of which it is made, the physical and chemical changes that occur on its surface and in its interior, and the history of the planet and its life forms.

geothermal energy Energy derived from the heat of the Earth.

geothermal gradient The rate at which temperature increases with depth in the Earth.

geyser A type of hot spring that intermittently erupts jets of hot water and steam. Geysers occur when ground water comes in contact with hot rock.

glacial polish   A smooth polish on bedrock created when fine particles transported at the base of a glacier abrade the bedrock.

glacial striation   Deep, parallel grooves and scratches in bedrock that form as rocks are dragged along at the base of a glacier.

glacier   A massive, long-lasting accumulation of compacted snow and ice that forms on land and moves downslope or outward under its own weight.

glauconite   A green clay mineral frequently found in marine sandstones and believed to have formed at the site of deposition.

glossopteris flora   An assemblage of fossil plants found in rocks of late Paleozoic and early Triassic age in South Africa, India, Australia, and South America. The flora takes its name from the seed fern *Glossopteris*.

gneiss   A foliated rock with banded appearance, formed by regional metamorphism.

Gondwanaland   The southern part of Wegener's Pangaea, the late Paleozoic supercontinent.

goniatites   One of the three large groups of ammonoid cephalopods with sutures forming a pattern of simple lobes and saddles and thus not as complex as either the ceratites or the ammonites.

Gowganda Conglomerate   An apparent tillite of the Canadian Shield. The Gowganda rests upon a surface of older rock that appears to have been polished by glacial action.

graben   A wedge-shaped block of rock that has dropped downward between two normal faults.

graded bedding   A type of bedding in which larger particles are at the bottom of each bed, and the particle size decreases toward the top.

graded stream   A stream with a smooth, concave profile. A graded stream is in equilibrium with its sediment supply; it transports all the sediment supplied to it without erosion or deposition in the stream bed.

gradient   The vertical drop of a stream over a specific distance.

granite   A medium- to coarse-grained, sialic, plutonic rock made predominantly of potassium feldspar and quartz.

graptolite facies   A Paleozoic sedimentary facies composed of dark shales and fine-grained clastics that contain the abundant remains of graptolites and that are associated with volcanic rocks.

graptolites   Extinct colonial marine invertebrates considered to be protochordates. Graptolites range from the late Cambrian to the Mississippian.

gravel   Unconsolidated sediment consisting mainly of rounded particles with diameters greater than 2 millimeters.

gravity anomaly   The difference between the observed value of gravity at any point on the Earth and the calculated theoretic value.

greenhouse effect   An increase in the temperature of a planet, caused when infrared-absorbing gases are introduced into the atmosphere.

greenstones   Low-grade metamorphic rocks containing abundant chlorite, epidote, and biotite and developed by metamorphism of basaltic extrusive igneous rocks. Great linear outcrops of greenstones are termed greenstone belts and are thought to mark the locations of former volcanic island arcs.

groin   A narrow wall built perpendicular to the shore to trap sand transported by currents and waves.

ground moraine   A moraine formed when a receding glacier deposits till in a relatively thin layer over a broad area.

ground water   Water contained in soil and bedrock. All subsurface water.

guide fossil   Fossil with a wide geographical distribution but narrow stratigraphic range and thus useful in correlating strata and for age determination.

Gutenberg discontinuity   The boundary separating the mantle of the Earth from the core below. The Gutenberg discontinuity lies about 2900 km below the surface.

guyot   A flat-topped seamount.

gymnosperms   An informal designation for flowerless seed plants in which seeds are not enclosed (hence "naked seeds"). Examples are conifers and cyads.

Hadean Eon   The earliest time in the Earth's history.

hadrosaurs   The ornithischian duck-billed dinosaurs of the Cretaceous.

half-life   The time it takes for half of the nuclei of a radioactive isotope in a sample to decompose.

hanging valley   A tributary glacial valley whose mouth lies high above the floor of the main valley.

hanging wall   The rock above an inclined fault.

haploid cell   A cell having a single set of chromosomes, as in gametes (*see* diploid).

hardness   The resistance of the surface of a mineral to scratching.

headland   A point of land that juts out into the sea.

headward erosion   The lengthening of a valley in an upstream direction.

Hercynian orogeny   A major late Paleozoic orogenic episode in Europe that formed the ancient Hercynian mountains. Today, only the eroded stumps of these mountains are exposed in areas where the cover of Mesozoic and Cenozoic strata has been removed by erosion.

heterotroph An organism that depends upon an external source of organic substances for its nutrition and energy.

Holocene A term sometimes used to designate the period of time since the last major episode of glaciation. The term is equivalent to Recent.

homologous organs Organs having structural and developmental similarities due to genetic relationship.

horn A sharp, pyramid-shaped rock summit eroded by glaciers where three or more cirques intersect near the summit.

hornblende A rock-forming mineral; the most common member of the amphibole group.

hornfels A fine-grained rock formed by contact metamorphism.

horst A block of rock that has moved relatively upward and is bounded by two faults.

hot spot A persistent volcanic center thought to be located directly above a rising plume of hot mantle rock.

hot spring A spring formed where hot ground water flows to the surface.

humus The dark organic component of soil composed of litter that has decomposed sufficiently so that the origins of the individual pieces cannot be determined.

hydration The chemical combination of water with another substance.

hydraulic action The mechanical loosening and removal of material by flowing water.

hydrologic cycle The constant circulation of water between the sea, the atmosphere, and the land.

hydrosphere The water and water vapor present at the surface of the Earth, including oceans, seas, lakes, and rivers.

hydrothermal metamorphism Changes in rock that are primarily caused by migrating hot water and by ions dissolved in the hot water.

Hylobatidae A group of persistently arboreal, small apes that are exemplified by the gibbons and siamangs.

hyomandibular bone (in fishes) The modified upper bone of the hyoid arch, which functions as a connecting element between the jaws and the brain case in certain fishes.

ice age A time of extensive glacial activity, when alpine glaciers descended into lowland valleys and continental glaciers spread over the higher latitudes.

iceberg A large chunk of ice that breaks from a glacier into a body of water.

ice sheet A glacier that forms a continuous cover of ice over areas of 50,000 square kilometers or more and spreads outward under the influence of its own weight. (*syn:* continental glacier, ice cap)

ichthyosaurs Highly specialized marine reptiles of the Mesozoic, recognized by their fishlike form.

Ictidosauria A group of extinct mammal-like reptiles or therapsids whose skeletal characteristics are considered to be very close to those of mammals.

igneous rock Rock that solidified from magma.

incised meander A stream meander that is cut below the level at which it originally formed, usually caused by rejuvenation.

index fossil A fossil that identifies and dates the layers in which it is found. Index fossils are abundantly preserved in rocks, are geographically widespread, and existed as species or genera for only relatively short times (*syn:* guide fossil).

infaunal organisms Organisms that live and feed within bottom sediments.

influent stream A stream that lies above the water table. Water percolates from the stream channel downward into the saturated zone. (*syn:* losing stream)

intensity (of an earthquake) A measure of the effects of an earthquake on buildings and people at a particular place.

intermediate rocks Igneous rocks with chemical and mineral compositions that are between those of granite and basalt.

internal mold A type of fossil that forms when the inside of a shell fills with sediment or precipitated minerals.

internal processes Earth processes that are initiated by movements within the Earth and those internal movements themselves—for example, formation of magma, earthquakes, mountain building, and tectonic plate movement.

intertidal zone The part of a beach that lies between the high and low tide lines.

intracratonic basin A sedimentary basin located within a craton.

intrusive rock A rock formed when magma solidifies within a pre-existing body of rock.

ion An atom with charge.

ionic substitution The replacement of one or more kinds of ions in a crystal structure by other kinds of ions of similar size and charge.

island arc A gently curving chain of volcanic islands in the ocean formed by convergence of two plates, each bearing ocean crust, and the resulting subduction of one plate beneath the other.

isopachous map   A map depicting the thickness of a sedimentary unit.

isostasy   The condition in which the lithosphere floats on the asthenosphere as an iceberg floats on water.

isostatic adjustment   The rising and settling of portions of the lithosphere to maintain equilibrium as they float on the plastic asthenosphere.

isotopes   Atoms of the same element that have the same number of protons but different numbers of neutrons.

joint   A fracture that occurs without movement of rock on either side of the break.

kame   A small mound or ridge of layered sediment deposited by a stream that flows on top of, within, or beneath a glacier.

Karoo system   A sequence of Permian to lower Jurassic rocks, primarily continental formations, which outcrop in Africa and are approximately equivalent to the Gondwana system of peninsular India.

karst topography   A type of topography formed over limestone or other soluble rock and characterized by caverns, sinkholes, and underground drainage.

Kaskaskia sequence   A sequence of Devonian-Mississippian sediments, bounded above and below by regional unconformities and recording an espisode of transgression followed by full flooding of a large part of the craton and by subsequent regression.

kettle   A depression in glacial drift created by the melting of a large chunk of ice left buried in the drift by a receding glacier. The ice prevents sediment from collecting; when the ice melts, a lake or swamp may fill the depression.

key bed   A thin, widespread, easily recognized sedimentary layer that can be used for correlation.

lacustrine   Relating to lakes, as in lacustrine sediments (lake sediments).

lagomorph   The order of small placental mammals that includes rabbits, hares, and pikas.

lagoon   A protected body of water separated from the sea by a reef or barrier island.

laminar flow   A type of flow in which water moves in straight, even paths without turbulence.

landslide   A general term for the downslope movement of rock and regolith under the influence of gravity.

Laramide orogeny   In general, those pulses of mountain building that were frequent in Late Cretaceous time and were in large part responsible for pro-

ducing many of the structures of the Rocky Mountains.

lateral moraine   A ridge-like moraine that forms on or adjacent to the sides of a mountain glacier.

laterite   A highly weathered soil rich in oxides of iron and aluminum which usually develops in warm, moist tropical or temperate regions.

Laurasia   The northern part of Wegener's Pangaea, the late Paleozoic supercontinent.

lava   Fluid magma that flows onto the Earth's surface from a volcano or fissure. Also, the rock formed by solidification of the same material.

limb   The side of a fold in rock.

limestone   A sedimentary rock consisting chiefly of calcium carbonate.

lithification   The conversion of loose sediment to solid rock.

lithofacies map   A map that shows the areal variation in lithologic attributes of a stratigraphic unit.

lithosphere   The cool, rigid, outer layer of the Earth, about 100 kilometers thick, which includes the crust and part of the upper mantle.

litopterns   South American ungulates whose evolutionary history somewhat paralleled that of horses and camels.

litter   Leaves, twigs, and other plant or animal material that has fallen to the surface of the soil but has not decomposed.

loam   Soil that contains a mixture of sand, clay, and silt and a generous amount of organic matter.

loess   A homogenous, unlayered deposit of wind-blown silt, usually of glacial origin.

Logan's line   A zone of thrust-faulting produced during the Taconic orogeny that extends from the west coast of Newfoundland along the trend of the St. Lawrence River to near Quebec and southward along Vermont's western border. (Named after the pioneer Canadian geologist Sir William Logan.)

longitudinal dune   A long symmetrical dune oriented parallel to the direction of the prevailing wind.

longshore current   A current flowing parallel and close to the coast that is generated when waves strike a shore at an angle.

longshore drift   Sediment carried by longshore currents.

lophophore   An organ, located adjacent to the mouth of brachiopods and bryozoans, which bears ciliated tentacles and has as its primary function the capture of food particles.

losing stream   A stream that lies above the water table. Water percolates from the stream channel downward into the saturated zone. (*syn:* influent stream)

low velocity layer   A region of the upper mantle

where seismic waves travel relatively slowly, approximately coincident with the asthenosphere.

Lunar maria   Low-lying, dark lunar plains filled with volcanic rocks rich in iron and magnesium.

luster   The quality and intensity of light reflected from the surface of a mineral.

lycopsids   Leafy plants with simple, closely spaced leaves bearing sporangia on their upper surfaces. They are represented by living club mosses and vast numbers of extinct late Paleozoic "scale trees."

L wave   An earthquake wave that travels along the surface of the Earth, or along a boundary between layers within the Earth. (*syn:* surface wave)

mafic rock   Dark-colored igneous rock with high magnesium and iron content, and composed chiefly of iron- and magnesium-rich minerals.

magma   Molten rock generated within the Earth.

magmatic arc   A narrow elongate band of intrusive and volcanic activity associated with subduction.

magnetic reversal   A change in the Earth's magnetic field in which the north magnetic pole becomes the south magnetic pole, and vice versa.

magnetic declination   The horizontal angle between "true," or geographic, north and magnetic north, as indicated by the compass needle. Declination is the result of the Earth's magnetic axis being inclined with respect to the Earth's rotational axis.

magnetic inclination   The angle between the magnetic lines of force for the Earth and the Earth's surface; sometimes called "dip." Magnetic inclination can be demonstrated by observing a freely suspended magnetic needle. The needle will lie parallel to the Earth's surface at the equator but is increasingly inclined toward the vertical as the needle is moved toward the magnetic poles.

magnetometer   An instrument that measures the Earth's magnetic field.

magnitude (of an earthquake)   A measure of the strength of an earthquake determined from seismic recordings. (*See also* Richter scale.)

mammoth   The name commonly applied to extinct elephants of the Pleistocene Epoch.

manganese nodule   A manganese-rich, potato-shaped mass found on the ocean floor.

mantle   A mostly solid layer of the Earth lying beneath the crust and above the core. The mantle extends from the base of the crust to a depth of about 2900 kilometers.

mantle convection   The convective flow of solid rock in the mantle.

mantle plume   A rising vertical column of mantle rock.

marble   A metamorphic rock consisting of fine- to coarse-grained recrystallized calcite and/or dolomite.

maria   Dry, barren, flat expanses of volcanic rock on the Moon that were first thought to be seas.

marsupials   Mammals of the Order Marsupialia. Female marsupials bear mammary glands, and carry their immature young in a stomach pouch.

mass spectrometer   An instrument that separates ions of different mass but equal charge and measures their relative quantities.

mass wasting   The movement of Earth material downslope, primarily under the influence of gravity.

mastodon   The group of extinct proboscideans (elephantoids), early forms of which were characterized by long jaws, tusks in both jaws, and low-crowned teeth.

meander   One of a series of sinuous curves or loops in the course of a stream.

mechanical weathering   The disintegration of rock into smaller pieces by physical processes.

medial moraine   A moraine formed in or on the middle of a glacier by the merging of lateral moraines as two glaciers flow together.

medusa   The tree-swimming, umbrella-shaped jellyfish form of the phylum Cnidaria.

meiosis   That kind of nuclear division, usually involving two successive cell divisions, that results in daughter cells having one-half the number of chromosomes that were in the original cell.

mélange   A body of intricately folded, faulted, and severely metamorphosed rocks, examples of which can be seen in the Franciscan rocks of California.

Mercalli scale   A scale that measures the intensity of an earthquake in a particular place by its effects on buildings and people. It has been replaced by the Richter scale.

mesa   A flat-topped mountain or a tableland that is smaller than a plateau and larger than a butte.

mesosphere   A zone of the Earth's mantle where pressures are sufficient to impart greater strength and rigidity to the rock.

Mesozoic Era   The part of geologic time roughly 245 to 65 million years ago. Dinosaurs rose to prominence and became extinct during this era.

metamorphic facies   A set of all metamorphic rock types that formed under similar temperature and pressure conditions.

metamorphic grade   The intensity of metamorphism that formed a rock; the maximum temperature and pressure attained during metamorphism.

metamorphic rock   A rock that forms when igneous, sedimentary, or other metamorphic rocks recrystallized in response to elevated temperature, increased pressure, chemical change, and/or deformation.

metamorphism   The process by which rocks and minerals change in response to changes in temperature, pressure, chemical conditions, and/or deformation.

Metazoa   All multicellular animals whose cells become differentiated to form tissues (all animals except Protozoa).

meteorites   Metallic or stony bodies from interplanetary space that have passed through the Earth's atmosphere and hit the Earth's surface.

meteors   Sometimes called "shooting stars," meteors are particles of solid material from interplanetary space that approach close enough to the Earth to be drawn into the Earth's atmosphere, where they are heated to incandescence. Most disintegrate, but a few land on the surface of the Earth as meteorites.

mica   A silicate sheet structure mineral with a distinctive platy crystal habit and perfect cleavage. Muscovite and biotite are common rock-forming micas.

micrite   A texture in carbonate rocks that, when viewed microscopically, appears as murky, fine-grained calcium carbonate. Micrite is believed to develop from fine carbonate mud or ooze.

mid-channel bar   An elongate lobe of sand and gravel formed in a stream channel.

mid-oceanic ridge   A continuous submarine mountain chain that forms at the boundary between divergent tectonic plates within oceanic crust.

migmatite   A rock composed of both igneous and metamorphic minerals. It forms at very high metamorphic grades when rock begins to partially melt to form magma.

Milankovitch effect   The hypothetical long-term effect on world climate caused by three known components of Earth motion. The combination of these components provides a possible explanation for repeated glacial to interglacial climatic swings.

miliolids   A group of foraminifers with smooth, imperforate test walls and chambers arranged in various planes around a vertical axis. Miliolids are common in shallow marine areas.

mineral   A naturally occurring inorganic solid with a definite chemical composition and a crystalline structure.

mineral deposit   A local enrichment of one or more minerals.

mineralization   A type of fossil in which the organic components of an organism are replaced by minerals.

mineral reserve   The known supply of ore in the ground.

mitosis   The method of cell division by means of which each of the two daughter nuclei receives exactly the same complement of chromosomes as had existed in the parent nucleus.

mobile belt   An elongate region of the Earth's crust characterized by exceptional earthquake and volcanic activity, tectonic instability, and periodic mountain building.

Mohorovičić discontinuity (Moho)   The boundary between the crust and the mantle identified by a change in the velocity of seismic waves.

Mohs hardness scale   A standard, numbered from 1 to 10, to measure and express the hardness of minerals; based on a series of ten fairly common minerals, each of which is harder than those lower on the scale.

molasse   Accumulations of primarily nonmarine, relatively light-colored, irregularly bedded conglomerates, shales, coal seams, and cross-bedded sandstones that are deposited subsequent to major orogenic events.

mold   An impression, or imprint, of an organism or part of an organism in the enclosing sediment.

mollusk   Any member of the invertebrate Phylum Mollusca, including pelecypods, cephalopods, gastropods, scaphopods, and chitons.

monocline   A fold with only one limb.

monoplacophorans   Primitive marine molluscans with simple, cap-shaped shells.

monotremes   The egg-laying mammals.

moraine   A mound or ridge of till deposited directly by glacial ice.

morganucodonts   Early mammals found in Triassic beds of Europe and Asia and characterized by their small size and their retention of certain reptilian osteologic traits.

mosasaurs   Large marine lizards of the Late Cretaceous.

mountain chain   A number of mountain ranges grouped together in an elongate zone.

mountain range   A series of mountains or mountain ridges that are closely related in position, direction, age, and mode of formation.

mud   Wet silt and clay.

mud cracks   Irregular, usually polygonal fractures that develop when mud dries. The patterns may be preserved when the mud is lithified.

mudflow   Mass wasting of fine-grained soil particles mixed with a large amount of water.

mudstone   A non-fissile rock composed of a mixture of clay and silt.

multituberculates   An early group of Mesozoic mammals with tooth cusps in longitudinal rows and other dental characteristics that suggest they may have been the earliest herbivorous mammals.

mummification A process in which the remains of an animal are preserved by dehydration.

mutation A stable and inheritable change in a gene.

mytiloids Pelecypods having rather triangular valves that are most commonly identical but in some forms unequal. The edible mussel *Mytilus* is a representative form.

nappe A large mass of rocks that have been moved a considerable distance over underlying formations by overthrusting, recumbent folding, or both.

natural levee A ridge or embankment of flood-deposited sediment along both banks of a stream channel.

nektic Pertaining to swimming organisms.

Neogene A subdivision of the Cenozoic that encompasses the Miocene, Pliocene, Pleistocene, and Holocene epochs.

neutron A subatomic particle with the mass of a proton but no electrical charge.

Nevadan orogeny In general, those pulses of mountain building, intrusion, and metamorphism that were most frequent during the Jurassic and Early Cretaceous along the western part of the Cordilleran orogenic belt.

Newark series A series of Upper Triassic, nonmarine red beds (shales, sandstones, and conglomerates), lava flows, and intrusions located within down-faulted basins from Nova Scotia to South Carolina.

New World monkeys Monkeys whose habitat is today confined to South America. They are thoroughly arboreal in habit and have prehensile tails by which they hang and swing from tree limbs.

nonconformity An unconformity developed between sedimentary rocks and older plutons or massive metamorphic rocks that had been exposed to erosion before the overlying sedimentary rocks were deposited.

nonfoliated The lack of layering in metamorphic rock.

normal fault A fault in which the hanging wall has moved downward relative to the footwall.

normal polarity A magnetic orientation that matches the Earth's modern magnetic field.

nothosaurs Relatively small early Mesozoic sauropterygians that were replaced during the Jurassic by the plesiosaurs.

notochord A rod-shaped cord of cartilage cells forming the primary axial structure of the chordate body. In vertebrates, the notochord is present in the embryo and is later supplanted by the vertebral column.

Notungulates A group of ungulates that diversified in South America and persisted until Plio-Pleistocene time.

novaculites A term applied originally to rocks suitable for whetstones and, in America, to white chert found in Arkansas. Now applied to very tough, uniformly grained cherts composed of microcrystalline quartz.

nuclear fission tracks Submicroscopic "tunnels" in minerals produced when high-energy particles from the nucleus of uranium are forcibly ejected during spontaneous fission.

nucleic acid Any of a group of organic acids that control hereditary processes within cells and make possible the manufacture of proteins from the amino acids ingested by the cells as food.

nuclides The different weight configurations of an element caused by atoms of that element having differing numbers of neutrons. Nuclides or isotopes of an element differ in number of neutrons but not in chemical properties.

nucleus The small, dense central portion of an atom, composed of protons and neutrons. Nearly all of the mass of an atom is concentrated in the nucleus.

nuée ardente A swiftly flowing, often red-hot cloud of gas, volcanic ash, and other pyroclastics formed by an explosive volcanic eruption. (*syn:* ash flow)

Nummulites Large, coined-shaped foraminifers, especially common in Tertiary limestones.

obsidian A black or dark-colored glassy volcanic rock, usually of rhyolitic composition.

oceanic crust The 7- to 10-kilometer-thick basaltic layer and the thin overlying sediment that underlie the ocean basins.

oceanic island A seamount, usually of volcanic origin, that has been built above sea level.

offlap A sequence of sediments resulting from a marine regression and characterized by an upward progression from offshore marine sediments (often limestones) to shales and finally sandstones (above which will follow an unconformity).

Ogallala aquifer The aquifer that extends for almost 1000 kilometers from the Rocky Mountains eastward beneath portions of the Great Plains.

oil shale A dark-colored shale rich in organic material that can be heated to liberate gaseous hydrocarbons.

Old World monkeys Monkeys of Asia, Africa, and southern Europe that include macaques, guenons, langurs, baboons, and mandrills.

olivine A common rock-forming mineral in mafic and ultramafic rocks with a composition that varies between Mg_2SiO_4 and Fe_2SiO_4.

onlap A sequence of sediments resulting from a marine transgression. Normally, the sequence begins with a conglomerate or sandstone deposited over an erosional unconformity and followed upward in the vertical section by progressively more offshore sediments.

ooïd A small, rounded accretionary body in sedimentary rock, generally formed of concentric layers of calcium carbonate around a nucleus such as a sand grain.

oölites Limestones composed largely of small, round or ovate calcium carbonate bodies called ooïds.

ophiolite suite An association of radiolarian cherts, pelagic muds, basaltic pillow and flow lavas, gabbros, and ultramafic rocks such as peridotite regarded as surviving masses of former oceanic crust largely destroyed in former subduction zones.

ore A natural material that is sufficiently enriched in one or more minerals to be mined profitably.

oreodonts North American artiodactyls of the middle and late Tertiary.

original horizontality (principle of) *See* principle of original horizontality.

Ornithischia An order of dinosaurs characterized by birdlike pelvic structures and including such herbivores as the ornithopods, stegosaurs, ankylosaurs, and ceratopsians.

orogenic belt Great linear tracts of deformed rocks, primarily developed near continental margins by compressional forces accompanying mountain building.

orogeny The process of mountain building; all tectonic processes associated with mountain building.

orthoclase A common rock-forming mineral; a variety of potassium feldspar ($KAlSi_3O_8$).

ostracoderms Extinct jawless fishes of the early Paleozoic.

ostracodes Small, bivalved, bean-shaped crustaceans.

outcrop An area where specific rock units are exposed at the Earth's surface or occur at the surface but are covered by surficial deposits.

outwash plain A broad, level surface formed where glacial sediment is deposited in front of or beyond a glacier.

oxbow lake A crescent-shaped lake formed where a meander is cut off from a stream and the ends of the cut-off meander become plugged with sediment.

oxidation The loss of electrons from a compound or element during a chemical reaction. In the weathering of common minerals, oxidation usually occurs when a mineral reacts with molecular oxygen.

pahoehoe A basaltic lava flow with a smooth, billowy, or "ropy" surface.

paleoclimatology The study of ancient climates.

paleoecology The study of the relationship of ancient organisms to their environment.

Paleogene A subdivision of the Cenozoic that encompasses the Paleocene, Eocene, and Oligocene epochs.

paleogeography The geography as it existed at some time in the geologic past.

paleolatitude The latitude that once existed across a particular region at a particular time in the geologic past.

paleomagnetism The earth's magnetic field and magnetic properties in the geologic past. Studies of paleomagnetism are helpful in determining positions of continents and magnetic poles.

paleontology The study of all ancient forms of life, their interactions, and their evolution.

Paleozoic Era Geologic time unit, 570 to 245 million years ago. During this era invertebrates, fishes, amphibians, reptiles, ferns, and cone-bearing trees were dominant.

Pangaea In Alfred Wegener's theory of continental drift, the supercontinent that included all present major continental masses.

Panthalassa The great universal ocean that surrounded the supercontinent Pangaea prior to its breakup.

parabolic dune A crescent-shaped dune with tips pointing into the wind.

paraconformity A rather obscure unconformity in which no erosional surface is discernible, and in which beds above and below the break are parallel.

parent rock Any original rock before it is changed by weathering, metamorphism, or other geological processes.

partial melting The process in which a silicate rock only partly melts as it is heated to form magma that is more silica-rich than the original rock.

passive continental margin A margin characterized by a firm connection between continental and oceanic crust where little tectonic activity occurs.

paternoster lake One of a series of lakes, strung out like beads and interconnected by short streams and waterfalls, created by glacial erosion.

peat A loose, unconsolidated, brownish mass of partially decayed plant matter; a precursor to coal.

pebble A sedimentary particle between 2 and 64 millimeters in diameter, and smaller than a cobble.

pectenoids Pelecypods exemplified by the scallops. They have generally subcircular shells and straight hinge lines.

pedalfer A common soil type that forms in humid environments, characterized by abundant iron and aluminum oxides and a concentration of clay in the B horizon.

pediment   A gently sloping erosional surface that forms along a mountain front uphill from a bajada, usually covered by a patchy veneer of gravel only a few meters thick.

pedocal   A soil formed in arid and semiarid climates, characterized by an accumulation of calcium carbonate.

pegmatite   An exceptionally coarse-grained igneous rock, usually with the same mineral content as granite.

pelagic sediment   Muddy ocean sediment that consists of a mixture of clay and the skeletons of microscopic marine organisms.

pelycosaurs   Early mammal-like reptiles exemplified by the sail-back animals of the Permian Period.

peneplain   A low, nearly featureless plain formed by lengthy erosion.

perched water table   The top of a localized lens of ground water that lies above the main water table, formed by a layer of impermeable rock or clay.

peridotite   A coarse-grained plutonic rock composed mainly of olivine; it may also contain pyroxene, amphibole, or mica but little or no feldspar. The mantle is thought to be made of peridotite.

period   A subdivision of an era.

Perissodactyl   Progressive, hoofed mammals, characteristically having an odd number of toes on the hind feet and usually on the front feet as well.

permafrost   A layer of permanently frozen soil or subsoil which lies from about a half meter to a few meters beneath the surface in Arctic environments.

permeability   A measure of the speed at which fluid can travel through a porous material.

permineralization   A manner of fossilization in which voids in an organic structure (such as bone) are filled with mineral matter.

petrification   The process of converting organic structures, such as bone, shell, or wood, into a stony substance, such as calcium carbonate or silica.

petroleum   A naturally occurring liquid composed of a complex mixture of hydrocarbons.

Phanerozoic   The eon of geologic time during which the Earth has been populated by abundant and diverse life. The Phanerozoic Eon followed the Cryptozoic (or Precambrian) Eon and is divided into Paleozoic, Mesozoic, and Cenozoic eras.

phenocryst   A large, early formed crystal in a fine-grained matrix in igneous rock.

phosphorite   A sediment composed largely of calcium phosphate.

photosphere   A relatively thin gaseous layer on the sun that emits nearly all the light that the sun radiates into space.

photosynthesis   The process of synthesizing carbohydrates from carbon dioxide and water, utilizing the radiant energy of light captured by the chlorophyll in plant cells.

phytoplankton   Microscopic marine planktic plants, most of which are various forms of algae.

phytosaurs   Extinct aquatic crocodilelike thecodonts of the Triassic.

pillow lava   Lava that solidified underwater, forming spheroidal lumps like a stack of pillows.

pinnipeds   Marine carnivores such as seals, sea lions, and walruses.

placer deposit   An accumulation of sediment rich in a valuable mineral or metal that has been concentrated because of its greater density.

placoderms   Extinct primitive jawed fishes of the Paleozoic Era.

placodonts   Extinct walruslike marine reptiles that fed principally upon shellfish.

plankton   Minute, free-floating aquatic organisms.

plastic   Capable of being deformed permanently without fracture.

plastic deformation   A permanent change in the original shape of a solid that occurs without fracture.

plate   A relatively rigid independent segment of the lithosphere that can move independently of other plates.

plateau   A large elevated area of comparatively flat land.

plate boundary   A boundary between two lithospheric plates.

plate tectonics theory   A theory of global tectonics in which the lithosphere is segmented into several plates that move about relative to one another by floating on and gliding over the plastic asthenosphere. Seismic and tectonic activity occur mainly at the plate boundaries.

platform sediment   The thin veneer of marine sediment found on parts of cratons overlying older igneous and metamorphic rocks of the craton.

playa   A dry desert lake bed.

playa lake   An intermittent desert lake.

Pleistocene epoch   A span of time, roughly 2 or 3 million to 8000 years ago, characterized by several advances and retreats of glaciers.

plesiosaurs   The group of extinct Mesozoic marine reptiles (sauropterygians) characterized by large, paddle-shaped limbs and broad bodies, with either very long or relatively short necks.

plucking   A process in which glacial ice erodes rock by loosening particles and then lifting and carrying them downslope.

plunging fold   A fold with a dipping or plunging axis.

pluton   An igneous intrusion.

plutonic rock   An igneous rock that forms deep (a kilometer or more) beneath the Earth's surface.

pluvial lake   A lake formed during a time of abundant precipitation. Many pluvial lakes formed as continental ice sheets melted.

point bar   A stream deposit located on the inside of a growing meander.

point source pollution   Pollution which arises from a specific site such as a septic tank or a factory.

polyp   The hydralike form of some coelenterates in which the mouth and tentacles are at the top of the body.

pore space   The open space between grains in rock, sediment, or soil.

porosity   The portion of the volume of a material that consists of open spaces.

porphyry   Any igneous rock containing relatively large crystals (phenocrysts) in a relatively fine-grained matrix.

pothole   A smooth, rounded depression in bedrock in a stream bed, caused by abrasion when circular currents circulate stones or coarse sediment.

Precambrian   All of geologic time before the Paleozoic era, encompassing approximately the first 4 billion years of Earth's history. Also, all rocks formed during that time.

precipitation   (1) A chemical reaction that produces a solid salt, or precipitate, from a solution. (2) Any form in which atmospheric moisture returns to the Earth's surface—rain, snow, hail, and sleet.

preservation   A process in which an entire organism or a part of an organism is preserved with very little chemical or physical change.

pressure relief melting   The melting of rock and the resulting formation of magma caused by a drop in pressure at constant temperature.

primary (P) wave   A seismic wave formed by alternate compression and expansion of rock. P waves travel faster than any other seismic waves.

principle of biologic succession   The principle that states that the observed sequence of life forms has changed continuously through time, so that the total aspect of life (as recognized by fossil evidence) for a particular segment of time is distinct and different from that of earlier and later times.

principle of cross-cutting relationships   The principle that states that a rock or feature must first exist before anything can happen to it or another rock cuts across it.

principle of original horizontality   The principle that states that most sediments are deposited as nearly horizontal beds, and therefore most sedimentary rocks started out with nearly horizontal layering.

principle of superposition   The principle that states that in any undisturbed sequence of sediments or sedimentary rocks, age decreases from bottom to top.

principle of temporal transgression   The principle that states that sediments of advancing (transgressing) or retreating (regressing) seas are not necessarily of the same geologic age throughout their lateral extent.

proboscidea   The elephants and their progenitors.

prokaryotes   Organisms that lack membrane-bounded nuclei and other membrane-bounded organelles.

Prosimii   The less advanced primates, such as lemurs, tarsiers, and tree shrews.

proteinoids   Extra-large organic molecules containing most of the 20 amino acids of proteins and produced in laboratory conditions simulating those found in nature.

proteins   Giant molecules containing carbon, hydrogen, oxygen, nitrogen, and usually sulfur and phosphorus; composed of chains of amino acids and present in all living cells.

Proterozoic Eon   The portion of geologic time from 2.5 billion to 570 million years ago.

proton   A dense, massive, positively charged particle found in the nucleus of an atom.

pumice   Frothy, usually rhyolitic magma solidified into a rock so full of gas bubbles that it can float on water.

pyroclastic rock   Any rock formed when lava or solid rocks erupt explosively.

pyroxene   A rock-forming silicate mineral group that consists of many similar minerals. Members of the pyroxene group are major constituents of basalt and gabbro.

quartz   A rock-forming silicate mineral, SiO_2. Quartz is a widespread and abundant component of continental rocks but it is rare in the oceanic crust and mantle.

quartzite   A metamorphic rock composed of mostly quartz formed by recrystallization of sandstone.

quartz sandstone   Sandstone containing more than 90 percent quartz.

Queenston delta   A clastic wedge of red beds shed westward from highlands elevated in the course of the Taconic orogeny.

radial pattern   A drainage pattern formed when a number of streams originate on a mountain and flow outward like the spokes on a wheel.

radioactive decay   The spontaneous emission of a particle from the atomic nucleus, thereby transforming the atom from one element to another.

radioactivity The natural spontaneous decay of unstable nuclei.

radiometric age dating The process of measuring the absolute time of geologic material by measuring the concentrations of radioactive isotopes and their decay products.

Radiolaria Protozoa that secrete a delicate, often beautifully filigreed skeleton of opaline silica.

rain shadow desert A desert formed on the lee side of a mountain range.

recessional moraine A moraine that forms at the terminus of a glacier as the glacier stabilizes temporarily during retreat.

recharge The replenishment of an aquifer by the addition of water.

recrystallization (in fossils) A process of fossil development in which the fossil retains the shape and features of the original, but the atoms rearrange to form a new mineral.

recumbent fold A fold in which the axial plane is essentially horizontal; a fold that has been turned over by compressional forces so that it lies on its side.

red beds Prevailing red, usually clastic sedimentary deposits.

reef A wave-resistant ridge or mound built by corals or other marine organisms.

reflection The return of a wave that strikes a surface.

refraction The bending of a wave that occurs when the wave changes velocity as it passes from one medium to another.

regional burial metamorphism Metamorphism of a broad area of the Earth's crust caused by elevated temperatures and pressures resulting from simple burial.

regional dynamothermal metamorphism Metamorphism accompanied by deformation, affecting an extensive region of the Earth's crust.

regional metamorphism Metamorphism that is broadly regional in extent, involving very large areas and volumes of rock. Includes both regional dynamothermal and regional burial metamorphism.

regolith The loose, unconsolidated, weathered material that overlies bedrock.

regression A general term signifying that a shoreline has moved toward the center of a marine basin. Regression may be caused by tectonic emergence of the land, eustatic lowering of sea level, or prograding of sediments, as in deltaic build-outs.

rejuvenated stream A stream that has had its gradi-

ent steepened and its erosive ability renewed by tectonic uplift or a drop of sea level.

relative geologic age The placing of an event in a time sequence without regard to the absolute age in years.

relative time Time expressed as the order in which rocks formed and geological events occurred, but not measured in years.

relief The vertical distance between a high point and a low point on the Earth's surface.

remote sensing The collection of information about a region by instruments that are not in direct contact with it.

replacement A fossilization process in which the original skeletal substance is replaced after burial by inorganically precipitated mineral matter.

reserves Known geological deposits that can be extracted profitably under current conditions.

reservoir rock Porous and permeable rock in which liquid petroleum or gas accumulates.

reversed polarity Magnetic orientations in rock which are opposite to the present orientation of the Earth's field. Also, the condition in which the Earth's magnetic field is opposite to its present orientation.

reverse fault A fault in which the hanging wall has moved up relative to the footwall.

rhynchonellids A group of brachiopods having pronounced beaks, accordionlike plications, and triangular outline.

rhyolite A fine-grained extrusive igneous rock compositionally equivalent to granite.

Richter scale A numerical scale of earthquake magnitude measured by the amplitude of the largest wave on a standardized seismograph.

rift A zone of separation of tectonic plates at a divergent plate boundary.

rift valley An elongate depression that develops at a divergent plate boundary. Examples include continental rift valleys and the rift valley along the center of the mid-oceanic ridge system.

ring of fire The belt of subduction zones and major tectonic activity, including extensive volcanism, that borders the Pacific Ocean along the continental margins of Asia and the Americas.

rip current A current created when water flows back toward the sea after a wave breaks against the shore. (*syn:* undertow)

ripple marks Small parallel ridges and troughs formed in loose sediment by wind or water currents and waves. They may then be preserved when the sediment is lithified.

rock A naturally formed solid that is an aggregate of one or more different minerals.

rock avalanche   A type of mass wasting in which a segment of bedrock slides over a tilted bedding plane or fracture. The moving mass usually breaks into fragments. (*syn:* rockslide)

rock cycle   The sequence of events in which rocks are formed, destroyed, altered, and reformed by geological processes.

rock flour   Finely ground, silt-sized rock fragments formed by glacial abrasion.

rockslide   A type of slide in which a segment of bedrock slides along a tilted bedding plane or fracture. The moving mass usually breaks into fragments. (*syn:* rock avalanche)

rounding   The sedimentary process in which sharp, angular edges and corners of grains are smoothed.

rubble   Angular particles with diameter greater than 2 millimeters.

rudists   Peculiarly specialized Mesozoic pelecypods often having one valve in the shape of a horn coral, covered by the other valve in the form of a lid.

Rugosa   The large group of solitary and colonial Paleozoic horn corals.

ruminant   A herbivorous, cud-chewing ungulate.

runoff   Water that flows back to the oceans in surface streams.

salinization   A process whereby salts accumulate in soil that is heavily irrigated.

saltation   Sediment transport in which particles bounce and hop along the surface.

salt cracking   A weathering process in which salts that are dissolved in water found in the pores of rock crystallize. This process widens cracks and pushes grains apart.

salt dome   A structural dome in sedimentary strata resulting from the upward flow of a large body of salt.

sand   Sedimentary grains that range from 1/16 to 2 millimeters in diameter.

sandstone   Clastic sedimentary rock comprised primarily of lithified sand.

saturated zone   The region below the water table where all the pores in rock or regolith are filled with water.

Sauk sequence   A sequence of upper Precambrian to Ordovician sediments bounded both above and below by a regional unconformity and recording an episode of marine transgression, followed by full flooding of a large part of the craton, and ending with a regression from the craton.

Saurischia   An order of dinosaurs with triradiate pelvic structures, including both the gigantic herbivorous sauropods and the carnivorous theropods.

scarp   A line of cliffs created by vertical displacement of land by faulting or by erosion.

schist   A strongly foliated metamorphic rock that has a well-developed parallelism of minerals such as micas.

scleractinid coral   Coral belonging to the order Scleractinia, which includes most modern and post-Paleozoic corals.

sea arch   An opening created when a cave is eroded all the way through a narrow headland.

sea-floor spreading   The hypothesis that segments of oceanic crust are separating at the mid-oceanic ridge.

seamount   A submarine mountain, usually of volcanic origin, that rises 1 kilometer or more above the surrounding sea floor.

sea stack   A pillar of rock left when a sea arch collapses or when the inshore portion of a prominence erodes faster than the tip.

secondary (S) wave   A seismic wave consisting of a shearing motion in which the oscillation is perpendicular to the direction of wave travel. S waves travel more slowly than P waves.

sediment   Solid rock or mineral fragments that are transported and deposited by wind, water, gravity, or ice, precipitated by chemical reactions, or secreted by organisms and that accumulate as layers in loose, unconsolidated form.

sedimentary rock   A rock formed when sediment is lithified.

sedimentary structure   Any structure formed in sedimentary rock during deposition or by later sedimentary processes, for example, bedding.

seismic gap   An immobile region of a fault bounded by moving segments.

seismic profiler   A device used to construct a topographic profile of the ocean floor and to reveal layering in sediment and rock beneath the sea floor.

seismic wave   Any elastic wave that travels through rock, produced by an earthquake or explosion.

seismogram   The record made by a seismograph.

seismograph   An instrument that records seismic waves.

seismology   The study of earthquake waves and the interpretation of this data to elucidate the structure of the interior of the Earth.

sessile   The bottom-dwelling habit of aquatic animals that live continuously in one place.

series   The time-rock term representing the rocks deposited or emplaced during a geologic epoch. Series are subdivisions of systems.

shale   A fine-grained clastic sedimentary rock with finely layered structure composed predominantly of clay minerals.

shear strength   The resistance of materials to being pulled apart along their cross-sections.

sheet flood   A broad, thin sheet of flowing water, typically in arid regions, that is not concentrated into channels.

shield volcano   A large, gently sloping volcanic mountain formed by successive flows of basaltic magma.

sialic rock   A rock, such as granite or rhyolite, that contains large proportions of silicon and aluminum.

silica   Silicon dioxide, SiO_2. Includes quartz, opal, chert, and many other varieties.

silicates   All minerals whose crystal structures contain silica tetrahedra. All rocks composed principally of silicate minerals.

silica tetrahedron   A pyramid-shaped structure of a silicon ion bonded to four oxygen ions, (SiO_4^{4-}).

silicoflagellates   Unicellular, tiny, flagellate marine algae that secrete an internal skeleton composed of opaline silica.

sill   A tabular or sheetlike igneous intrusion parallel to the grain or layering of country rock.

silt   All sedimentary particles from 1/256 to 1/16 millimeter in diameter.

sinkhole   A circular depression in karst topography caused by the collapse of a cavern roof or by dissolution of surface rocks.

slate   A compact, fine-grained, low-grade metamorphic rock with slaty cleavage that can be split into slabs and thin plates. Intermediate in grade between shale and phyllite.

slaty cleavage   A parallel metamorphic foliation in a plane perpendicular to the direction of tectonic compression.

slide   All types of mass wasting in which the rock or regolith initially moves as a consolidated unit over a fracture surface.

slip face   The steep lee side of a dune that is at the angle of repose for loose sand so that the sand slides or slips.

slump   A type of mass wasting in which the rock and regolith move as a consolidated unit with a backward rotation along a concave fracture.

snowline   The altitude above which there is permanent snow.

soil   Soil scientists define soil as the upper layers of regolith that support plant growth. According to another definition, soil is synonymous with regolith.

soil horizon   A layer of soil that is distinguishable from other horizons because of differences in appearance and in physical and chemical properties.

soil-moisture belt   The relatively thin, moist surface layer of soil that is wetter than the unsaturated zone beneath it.

solifluction   A type of mass wasting that occurs when water-saturated soil moves slowly over permafrost.

Sonoma orogeny   Middle Permian orogenic movements, the structural effects of which are most evident in western Nevada.

sorting   A process in which flowing water or wind separates sediment according to particle size, shape, or density.

source rock   The geologic formation in which oil or gas originates.

spar carbonate   As viewed microscopically, the clear, crystalline carbonate that has been deposited in a carbonate rock as a cement between clasts or has developed by recrystallization of clasts.

species   A unit of taxonomic classification of organisms. In another sense, a species is a population of individuals that are similar in structural and functional characteristics and that in nature breed only with one another.

specific gravity   The weight of a substance relative to the weight of an equal volume of water.

speleothem   Any mineral deposit formed in caves by the action of water.

Sphenodon (tuatara)   Large, lizardlike reptiles that have persisted from Triassic to the present and now inhabit islands off the coast of New Zealand.

Sphenopsids   A group of spore-bearing plants that were particularly common during the late Paleozoic and were characterized by articulated stems with leaves borne in whorls at nodes. Only one genus, *Equisetum*, survives.

spheroidal weathering   Weathering in which the edges and corners of a rock weather more rapidly than the flat faces, giving rise to a rounded shape.

spiracle   In cartilaginous fishes, a modified gill opening through which water enters the pharynx.

spit   A small point of sand or gravel extending from shore into a body of water.

spontaneous fission   Spontaneous fragmentation of an atom into two or more lighter atoms and nuclear particles.

spore   A usually asexual reproductive body, such as occurs in bacteria, ferns, and mosses.

spreading center   The boundary or zone where lithospheric plates rift or separate from each other. (*syn:* divergent plate boundary)

spring   A place where ground water flows out of the Earth to form a small stream or pool.

stage   The time-rock unit equivalent to an age. A stage is a subdivision of a series.

stalactite   An icicle-like dripstone, deposited from

drops of water, that hangs from the ceiling of a cavern.

stalagmite   A deposit of mineral matter that is formed on the floor of a cavern by the action of dripping water.

stapes   The innermost of the small bones in the middle ear cavity of mammals; also recognized in amphibians and reptiles.

stock   An igneous intrusion with an exposed surface area of less than 100 square kilometers.

strain   The deformation (change in size or shape) that results from stress.

stratification   The arrangement of sedimentary rocks in strata or beds.

stratified drift   Sediment that was transported by a glacier and then sorted, deposited, and layered by glacial meltwater.

stratigraphy   The study of rock strata, with emphasis on their succession, age, correlation, form, distribution, lithology, fossil content, and all other characteristics useful in interpreting their environment of origin and geologic history.

stratovolcano   A steep-sided volcano formed by an alternating series of lava flows and pyroclastic eruptions. (*syn:* composite volcano)

streak   The color of a fine powder of a mineral, usually obtained by rubbing the mineral on an unglazed porcelain streak plate.

stream   A moving body of water confined in a channel and flowing downslope.

stream piracy   The natural diversion of the headwaters of one stream into the channel of another.

stream terrace   An abandoned flood plain above the level of the existing stream.

striations   Parallel scratches in bedrock caused by rocks embedded in the base of a flowing glacier.

strike   The direction of a tilted rock surface as it intersects a horizontal plane, e.g., a bedding plane or fault. Strike is measured as a compass direction.

strike-slip fault   A fault on which the motion is parallel to its strike and is primarily horizontal.

stromatolites   Distinctly laminated accumulations of calcium carbonate having rounded, branching, or frondose shape and believed to form as a result of the metabolic activity of marine algae. They are usually found in the high intertidal to low supratidal zones.

stromatoporoids   An extinct group of reef-building organisms now believed to have affinities with the Porifera and noted for the large, often laminated masses constructed by the colonies.

subaerial   Formed or existing at or near a sediment surface significantly above sea level.

subduction   The process in which a lithospheric

plate descends beneath another plate and dives into the asthenosphere.

subduction zone (or subduction boundary)   The region or boundary where a lithospheric plate descends into the asthenosphere.

sublittoral zone   The marine bottom environment that extends from low tide seaward to the edge of the continental shelf.

submarine canyon   A deep, V-shaped, steep-walled trough eroded into a continental shelf and slope.

submarine fan   A large, fan-shaped accumulation of sediment deposited at the bases of many submarine canyons adjacent to the deep-sea floor. (*syn:* abyssal fan)

submergent coastline   A coastline that was recently above sea level but has been drowned, either because the land has sunk or because sea level has risen.

subsidence   Settling of the Earth's surface which can occur either as petroleum or ground water is removed or by natural geologic processes.

superposed stream   A stream that has downcut through several rock units and maintained its course as it encountered older geologic structures and rocks.

superposition (principle of)   *See* principle of superposition.

surf   The chaotic turbulence created when a wave breaks on the beach.

surface wave   An earthquake wave that travels along the surface of the Earth or along a boundary between layers within the Earth. (*syn:* L wave)

suspended load   That portion of a stream's load that is carried for a considerable time in suspension, free from contact with the stream bed.

suture   The junction created when two continents or other masses of crust collide and weld into a single mass of continental crust.

S wave   A seismic wave consisting of a shearing motion in which the oscillation is perpendicular to the direction of wave travel. S waves travel more slowly than P waves. (*syn:* secondary wave)

symmetrodonts   A group of primitive Mesozoic mammals characterized by a symmetric triangular arrangement of cusps on cheek teeth.

syncline   A fold that arches downward and whose core contains the youngest rocks.

system   The time–rock unit representing rock deposited or emplaced during a geologic period.

Taconic orogeny   A major episode of orogeny that affected the Appalachian region in Ordovician time. The northern and Newfoundland Appalachians were the most severely deformed during this orogeny.

talus slope An accumulation of loose angular rocks at the base of a cliff from which they have been cleared by mass wasting.

taxon (pl. taxa) Any unit in the taxonomic classification, such as a phylum, class, order, or family.

taxonomy The science of naming, describing, and classifying organisms.

tarn A small lake at the base of a cirque.

tectonics A branch of geology dealing with the broad architecture of the outer part of the Earth—specifically the relationships, origins, and histories of major structural and deformational features.

teleosts The most advanced of the bony fishes, characterized by thin, rounded scales, completely bony internal skeleton, and symmetric tail. Teleosts range from Cretaceous to Recent.

terebratulids A group of Silurian to Recent, mostly smooth-shelled brachiopods having a loop-shaped attachment for the lophomore. Terebratulids were most abundant during the Jurassic and Cenozoic.

terminal moraine An end moraine that forms when a glacier is at its greatest advance.

terminus The end, or foot, of a glacier.

terrigenous sediment Sea-floor sediment derived directly from land.

thermoremanent magnetism The permanent magnetism of rocks and minerals that results from cooling through the Curie point.

Tethys seaway A great east-west trending seaway lying between Laurasia and Gondwanaland during Paleozoic and Mesozoic time and from which arose the Alpine-Himalayan Mountain ranges.

thecodonts An order of primarily Triassic reptiles considered to be the ancestral archosaurians.

therapsids An order of advanced mammal-like reptiles.

theropods The carnivorous saurischian dinosaurs.

thrust fault A type of reverse fault with a dip of 45° or less over most of its extent.

tidewater glacier A glacier that flows directly into the sea.

till Sediment that was deposited directly by glacial ice and that has not been re-sorted by a stream.

tillite A sedimentary rock formed of lithified till.

time–stratigraphic unit (chronostratigraphic unit) The rocks formed during a particular unit of geologic time. Also called time–rock unit.

Tippecanoe sequence A sequence of Ordovician to Lower Devonian sediments bounded above and below by regional unconformities and recording an episode of marine transgression, followed by full flooding of a large region of the craton and subsequent regression.

titanotheres Large, extinct perissodactyls (odd-toed ungulates) that attained the peak of their evolutionary development during the Oligocene. *Brontotherium* is a widely known titanothere.

trace fossils Tracks, trails, burrows, and other markings made in now lithified sediments by ancient animals.

traction Sediment transport in which particles are dragged or rolled along a stream bed, beach, or desert surface.

transform fault A strike-slip fault between two offset segments of mid-oceanic ridge.

transform plate boundary A boundary between two lithospheric plates where the plates are sliding horizontally past one another.

transpiration Direct evaporation from the leaf surfaces of plants.

transport The moving of sediment by flowing water, ice, wind, or gravity.

transverse dune A relatively long, straight dune with a gently sloping windward side and a steep lee face that is oriented perpendicular to the prevailing wind.

trellis drainage pattern A drainage pattern characterized by a series of fairly straight parallel streams joined at right angles by tributaries.

trench A long, narrow depression of the sea floor formed where a subducting plate sinks into the mantle.

tributary Any stream that contributes water to another stream.

triconodonts A group of primitive Mesozoic mammals recognized primarily by the arrangement of three principal cheek tooth cusps in a longitudinal row.

trilobites Paleozoic marine arthropods of the Class Crustacea, characterized by longitudinal and transverse division of the carapace into three parts, or lobes. Trilobites were especially abundant during the early Paleozoic.

trough The lowest part of a wave.

truncated spur A triangular-shaped rock face that forms when a main valley glacier cuts off the lower portion of an arête.

tsunami A large sea wave produced by a submarine earthquake or a volcano and characterized by long wavelength and great speed.

tuff Volcanic ash that has become consolidated into rock.

turbidites Sediment deposited from a turbidity current and characterized by graded bedding and moderate to poor sorting.

turbidity current A rapidly flowing submarine current laden with suspended sediment that results from mass wasting on the continental shelf or slope.

turbulent flow A pattern in which water flows in an irregular and chaotic manner. It is typical of stream flow.

Tylopod The artiodactyl group to which camels and llamas belong.

ultimate base level The lowest possible level of downcutting of a stream, usually sea level.

ultramafic rock Rock composed mostly of minerals containing iron and magnesium, for example, peridotite.

unconformity A gap in the geological record, such as an interruption of deposition of sediments or a break between eroded igneous and overlying sedimentary strata, usually of long duration.

ungulate Four-legged mammals whose toes bear hoofs.

uniformitarianism A general principle that suggests that the past history of the Earth can be interpreted and deciphered in terms of what is known about present natural laws.

unit cell The smallest group of atoms that perfectly describes the arrangement of all atoms in a crystal, and repeats itself to form the crystal structure.

unsaturated zone A subsurface zone above the water table that may be moist but is not saturated; it lies above the zone of saturation. (*syn:* zone of aeration)

upper mantle The part of the mantle that extends from the base of the crust downward to about 670 kilometers beneath the surface.

U-shaped valley A glacially eroded valley with a characteristic U-shaped cross-section.

vagile The bottom-dwelling habit of aquatic animals capable of locomotion.

valley train A long and relatively narrow strip of outwash deposited in a mountain valley by the streams flowing from an alpine glacier.

varve A thin sedimentary layer or pair of layers that represent the depositional record of a single year.

vascular plants Plants, including all higher land plants, that have a system of vessels and ducts for distributing moisture and nutrients.

Veneroids A group of pelecypods exemplified by the common clam, *Mercenaria*.

vent A volcanic opening through which lava and rock fragments erupt.

ventifacts Cobbles and boulders found in desert environments which have one or more faces flattened and polished by windblown sand.

vesicle A bubble formed by expanding gases in volcanic rocks.

vestigial organ An organ that is useless, small, or degenerate but representing a structure that was more fully developed or functional in an ancestral organism.

volcanic bomb A small blob of molten lava hurled out of a volcanic vent that acquired a rounded shape while in flight.

volcanic neck A vertical pipe-like intrusion formed by the solidification of magma in the vent of a volcano.

volcanic rock A rock that formed when magma erupted, cooled, and solidified on the Earth's surface within a kilometer or less of the surface.

volcano A hill or mountain formed from lava and rock fragments ejected through a volcanic vent.

wash An intermittent stream channel found in a desert.

water table The upper surface of a body of ground water at the top of the zone of saturation and below the zone of aeration.

wave-cut cliff A cliff created when the lower portion of a rocky coast is eroded by waves.

wave-cut platform A flat or gently sloping platform created by erosion of a rocky shoreline.

wave height The vertical distance from the crest to the trough of a wave.

wavelength The distance between successive wave crests (or troughs).

weathering The decomposition and disintegration of rocks and minerals at the Earth's surface by mechanical and chemical processes.

welded tuff A hard, tough, glass-rich pyroclastic rock formed by the cooling of an ash flow that was hot enough to deform plastically and partly melt after it stopped moving; it often appears layered or streaky.

Williston basin A large structural basin extending from South Dakota and Montana northward into Canada; well known for the petroliferous Devonian formations deposited therein.

Zechstein sea An arm of the Atlantic that extended across part of northern Europe during the Late Permian and in which were deposited several hundred meters of evaporites, including the well-known potassium salts of Germany.

zone A bed or group of beds distinguished by a particular fossil content and frequently named after the fossil or fossils it contains. More formally known as a *biozone*.

zone of aeration   A subsurface zone above the water table that may be moist but is not saturated; it lies above the zone of saturation. (*syn:* unsaturated zone)

zone of saturation   A subsurface zone below the water table in which the soil and bedrock are completely saturated with water.

zone of wastage   The lower portion of a glacier where more snow melts in summer than accumulates in winter so that there is a net loss of glacial ice. (*syn:* ablation area)

zygote   The cell formed by the union of two gametes. Thus, a zygote is a fertilized egg.

Index

Boldface page numbers indicate where key terms are highlighted; italicized page numbers indicate a figure; t indicates a table.